KNOWING AND GUESSING

KNOWING AND GUESSING

A Quantitative Study of Inference and Information

SATOSI WATANABE

University of Hawaii

JOHN WILEY & SONS, INC.

New York · London · Sydney · Toronto

10 9 8 7 6 5 4 3 2 1

Library of Congress Catalog Card Number: 68-56165
SBN 471 92130 0
Printed in the United States of America

To my mother

Yoshiko Watanabe née Motoo

From the beginning I had promised myself that this book would be dedicated to my mother, but for various reasons inception of the idea to publication has taken nine years and as a result I should now have to say "to the memory of" instead of the simple "to" in my dedication. Were it not for my mother's constant presence in my mind, however, I should never have completed this work and for this reason I have kept the simpler version.

Preface

This book is the outgrowth of several graduate courses given intermittently from 1959 to the present in the physics, philosophy, applied science, and electrical engineering departments of universities such as Yale, Columbia, and Hawaii. Among the different topics dealt with the reader will recognize a common core: an attempt at a quantitative study of the formal aspects of the process of knowing, inferring, informing, and learning. It is quantitative because we are not satisfied with the verbal, true-or-false approach of logic and philosophy. It is formal because we are little concerned with the contents of our knowledge and information. Neither are we interested in psychology, neurophysiology, or mechanical models of the brain. We may perhaps call this work an attempt at "epistemometrics."

Both the merits and defects of this book originate mainly from the fact that one man has tried to cover problems usually considered as belonging to different disciplines or to a no-man's land in between. The result is a book written by an amateur, and amateurism is often beset by the lack of basic knowledge, absence of respect for established authority, unsystematic emphasis, pedestrian exposition, redundancy, and verbosity, In this era of experts and professionals an amateur's work is unwelcome and disdained.

Yet an amateur has a fresh sense of "amazement," a balanced bird's-eye view of tremendous scope, and a direct contact with the world of common sense which is the mother earth of all knowledge. He need not be ethereally elegant or unintelligibly succinct, but he cannot be deprived of an earthbound love of unity and soundness. He may not have an audience among experts of narrow disciplines, but he will perhaps appeal to young and untrained minds. Such unbiased minds may find this book a unified exposition of a new single field of study. This volume is addressed to them.

An amateur never knows how to evaluate his own work. What he considers his meritorious achievement may be old and trivial in the expert's eye, and what he casually remarks in passing may contain the germ of an important discovery. This book contains some definitive new ideas such as information balance in inference, nondistributive information theory, distinction between confirmation and creditation, inverse *H*-theorem, just to name a few. Even a classical topic like the Markov chain receives a rather new treatment. Yet the book must suffer without my knowledge from omission of many well-established, important, and relevant facts and also of many well-known and highly respected source documents. I must apologize in advance for my inability to put together an impartial and exhaustive bibliography.

The order of topics is intentionally unsystematic; for instance, the chapters on logic and probability should, logically speaking, be placed before the chapters on information theory, but they are pushed back until needed for an understanding of succeeding chapters. There are, however, many practical, pedagogical reasons behind this disorderly sequence. If the book is used as a text or reference, the instructor may note that it can be roughly divided into four major, somewhat overlapping parts: (a) Chapters I, II, III; (b) Chapters III, IV, V, VI; (c) Chapters VII, VIII; and (d) Chapters VII, IX. These four parts may be conveniently characterized as (a) Structure Analysis by Information Theory, (b) Mathematical Theory of Inductive and Deductive Inference, (c) Concept Formation and Classification, (d) Quantum-Mechanical Logic and Information. The reader will agree that this book is more than just a juxtaposition of these four parts.

Many of my friends, colleagues, and even some theoretical opponents have been helpful in preparing this manuscript. Also my students with their questions, complaints, and approval contributed greatly to the formation of the book. They are too many to be mentioned by name here, but my sincere thanks are due to each of them. I should, however, like to note that Professor Thomas H. Mott of Rutgers University read the entire work thoroughly and provided me with detailed comments and suggestions. Without making him responsible for any defects or lapses that may still exist in the book, I want to express my deepest gratitude for his tedious work and thoughtful advice. To my wife Dorothea goes my appreciation of her encouragement; to my son Francis my acknowledgment of the value of his criticism. To Mr. John Kinn, formerly of IBM, and to Miss Beatrice Shube of John Wiley and Sons, go my sincere thanks for their valuable help which made this publication possible. I owe my gratitude also to Professor Israel Scheffler and Professor Adolf Grünbaum for mentioning the merits of this work to various people concerned from the early stage. The major portion of this book was typed by Mrs. Catherine Dunn, Mrs. Lillian Drescher, and Miss Janet Yamamoto from an undecipherable jungle of handwritten notes. Their loyalty, patience, and accuracy are especially appreciated. It should also be acknowledged that many of my research projects which produced results included in this book were carried out at the universities mentioned earlier or at the IBM Research Center and that some of them were partly supported by U.S. Government research contracts.* The National Academy of Science of Hungary, the IBM Corporation and the Academic Press of New York have kindly given me authorization to use freely in this book the materials originally contained in my papers quoted in References as [W-20], [W-12], [W-13] and [W-18] which appeared in their respective publications.

SATOSI WATANABE

Honolulu, H.I.

* See the original papers for the exact titles of the respective contracts and the names of the funding agencies.

Contents

KNOWING AND GUESSING

1

Elements of Information Theory

1.1. PROPOSITION AND PROBABILITY

Logic, probability theory, and information theory (this last in a broader sense of the word) have one function in common, namely, to provide rational tools effective in unraveling structural relationships existing in a collection of propositions. The aspects of the structure these three disciplines deal with and the methods of attack they apply are different, yet they are related in such a way that probability theory presupposes, and is a natural extension of, logic, and information theory presupposes, and is a natural extension of, probability theory.

For this reason, in order to present a rigorous exposition of information theory, it would be necessary first to critically scrutinize and carefully define all the basic notions involved in logic and probability theory. Such an approach, however, tends to become tedious and boring and would be pedagogically unwise. Rather, we should start with the minimum necessary preparation in logic and probability theory, and envisage the new subject of information theory with fresh curiosity. After becoming familiar with manipulation of the mathematical tools peculiar to information theory, we can always come back to the more basic problems related to logic and probability. Such is more or less the course of exposition we take in this book. Thus we start here with a few brief explanatory remarks about logic and probability, which are by no means rigorous or exhaustive. The reader who wishes a fuller explanation of these matters can first read the beginning few sections of Chapter 7 and then come back to this section.

A proposition can be represented by a symbol, and is either true or false. Sentences of a certain category in ordinary language are considered to express propositions. Some propositions can be experientially determined to be true or false by a direct test, whereas the truth or falsehood of other propositions can only be inferred with some plausibility, although these propositions too have to be either true or false. We usually denote propositions in this book by capital Latin letters. If A and B are propositions, then $A \cap B$ (called conjunction of A and B or simply "A and B") represents another proposition, which is true if and only if both A and B are true. If A and B are propositions, then $A \cup B$ (called disjunction of A and B or simply "A or B") represents another proposition, which is true if and only if A or B or both is/are true. If A is a proposition, then $\neg A$ (called negation of A) is another proposition, which is true if A is false and false if A is true.

It is important to distinguish between the proposition A, which can be either true or false, and the assertion that A is true or A is the case. It is true that this last assertion itself, if judged from some external criterion, may be false, and its truth and falsehood entail respectively the truth and falsehood of A. However, within the context of our discourse, the proposition A by definition is supposed to be capable of being true or false, while the assertion that A is true, if made, is supposed to be true. This may be explained as due to a difference in the level of language. We are considering a collection of propositons, $A, B, C, \cdots$, each of which can be true or false. The operation by which a member of the collection is produced from other members of the collection belongs to the "object language." On the other hand, when we assert that A is true, this assertion does not belong to the collection of propositions. We are talking "about" the proposition. This assertion belongs to the "metalanguage."

In ordinary propositional calculus a proposition symbol like A is often used in two senses. When it is standing alone it is a proposition in the sense described above. When it is standing in a sequence of proof it is asserted to be true. A proof consists of the "premise" part and the "deduction" part. A proposition in the deduction part is asserted to be true on the ground of the premise part. In the present book, when a symbol like A is used, it always represents a proposition. When it is assumed or asserted that A is true, we write $A = \square$, where the equivalence symbol $(=)$ belongs in our terminology to the metalanguage.

In a mathematical proof in terms of propositional calculus and its extension, the premise part consists of all logical transformation rules and some specific postulates assumed in the problem. In our manner of description, depending on the context, all logical laws and some extralogical laws are assumed, and whatever follows logically from them is asserted to be true. The extralogical laws may be of semantic, pragmatic, or empirical origin.

The basic notion in the metalanguage is "entailment" or "implication." If the relation between two propositions A and B is such that if A is true B

is true, then we say that A entails B or A implies B, and write $A \rightarrow B$. The arrow represents a relation between A and B, and $A \rightarrow B$ is not a proposition in the collection of propositions we consider. In the usual symbolism, this arrow would be denoted as $\Rightarrow$. However, we use the simpler symbol $\rightarrow$ in the metalanguage and avoid the use of $\rightarrow$ and $\supset$ in the object language. If $A \rightarrow B$ and $B \rightarrow A$, then we say that A and B are equivalent, and write $A = B$. In ordinary symbolism, our equal sign would be denoted as $\Leftrightarrow$, but since we avoid using $=$ in the object language, we use $=$ in the metalanguage. It is easy to see that $A \rightarrow B$ is equivalent to both $A = A \cap B$ and $B = A \cup B$.

Propositions that are assumed or asserted to be true under all circumstances within the domain of consideration in a discussion are called constant truths. They are equivalent to one another and to a special proposition denoted by $\square$. Propositions that are assumed or asserted never to be true under any circumstances under consideration in a discussion are called constant falsehoods and are equivalent to one another and to a special proposition denoted by $\emptyset$. Thus $A = \square$ means the assertion that A is true, and $A = \emptyset$ means the assertion that A is false. From the definition of $\square$ and the definition of implication $\rightarrow$, we have $A \rightarrow \square$ for any A; hence $A \cap \square = A$ and $A \cup \square = \square$. Similarly, for any A, we have $\emptyset \rightarrow A$, $A \cap \emptyset = \emptyset$, and $A \cup \emptyset = A$.

The following well-known basic rules govern the operations of conjunction and disjunction. Idempotent laws: $A \cap A = A$, $A \cup A = A$. Commutative laws: $A \cap B = B \cap A$, $A \cup B = B \cup A$. Associative laws: $A \cap (B \cap C) = (A \cap B) \cap C$, $A \cup (B \cup C) = (A \cup B) \cup C$. Absorptive laws: $A \cap (A \cup B) = A$, $A \cup (A \cap B) = A$. Distributive laws: $A \cap (B \cup C) = (A \cap B) \cup (A \cap C)$, $A \cup (B \cap C) = (A \cup B) \cap (A \cup C)$. For the operation of negation we have the following. Law of double negation: $\neg(\neg A) = A$. Law of contradiction: $A \cap \neg A = \emptyset$. Law of the excluded middle: $A \cup \neg A = \square$. De Morgan's laws: $\neg(A \cap B) = \neg A \cup \neg B$, $\neg(A \cup B) = \neg A \cap \neg B$. Since we have $A \cap (B \cap C) = (A \cap B) \cap C = (A \cap C) \cap B$, we write $A \cap B \cap C$ to mean any one of the three; and similarly for disjunction. We have obviously also $\neg\emptyset = \square$ and $\neg\square = \emptyset$.

As explained above, the symbols $\cap$, $\cup$, and $\neg$ are used to denote operations of generating new propositions from given ones, while the symbols $=$ and $\rightarrow$ are used to denote some relations existing among propositions. The first group belongs to the object language, and the second, to the metalanguage. If one wants to be thoroughly consistent, this distinction between the two groups of logical symbols must also reflect on the proposition symbols, such as A, B, and C. In fact, what is meant by A in $A \cap B$ is not exactly the same as what is meant by A in $A \rightarrow B$. This difference becomes easy to understand if we substitute some examples for A and B. Let A and B stand respectively for "it will rain tomorrow" and "it will be cold tomorrow." In the case of $A \rightarrow B$, A stands for "*the proposition that* it will rain tomorrow," for $A \rightarrow B$

should mean "*the proposition that* it will rain tomorrow implies *the proposition that* it will be cold tomorrow," and not "it will rain tomorrow implies that it will be cold tomorrow." In the case of $A \cap B$, A must stand simply for "it will rain tomorrow," for $A \cap B$ certainly does not mean "the proposition that it will rain tomorrow and the proposition that it will be cold tomorrow," but it does mean that "it will rain tomorrow and it will be cold tomorrow." For this reason we should probably introduce quotation marks and write "A" = "B" and "A" $\rightarrow$ "B" instead of $A = B$ and $A \rightarrow B$, while we keep on writing $A \cap B$, $A \cup B$, $\neg A$, and so forth. In our discussion, however, there is very little chance of committing gross errors by failing to distinguish between A and "A." For this reason we have not adopted the symbolic use of quotation marks in this book. It suffices for our purpose to keep a clear distinction between the two groups of symbols, $(\cap, \cup, \neg)$ and $(=, \rightarrow)$. See Appendix 7.1 for more details.

If a set $\mathcal{E}$ of propositions $E_1, E_2, \cdots, E_n$,

$$\mathcal{E} = \{E_1, E_2, \cdots, E_n\}, \tag{1.1}$$

is such that any two distinct members are disjoint, that is,

$$E_i \cap E_j = \emptyset \quad \text{for} \quad i \neq j, \tag{1.2}$$

and its members are exhaustive, that is,

$$E_1 \cup E_2 \cup \cdots \cup E_n = \square, \tag{1.3}$$

then we say that $\mathcal{E}$ is a "disjoint (mutually exclusive) and exhaustive set" or, more simply, a "logical spectrum." We usually exclude $\emptyset$ from the membership of a logical spectrum. If $\square$ is a member, there could be no other proposition than $\emptyset$ that would satisfy the definition of a second member. We sometimes refer to the E_i's as "events" instead of "propositions," especially when they are experientially testable propositions. This is permissible because such a proposition can be rephrased in the form: "such and such event will happen," and a one-to-one correspondence between proposition and event can be established.

Let $\mathcal{E} = \{E_i\}$, $i = 1, 2, \cdots, n$, be a logical spectrum not containing $\emptyset$. Consider propositions that can be formed by taking disjunction of some of the E_i's. It is easy to see that a total of 2^n distinct propositions can be generated by this procedure. For there are $\binom{n}{r}$ different ways of taking r E_i's out of the given n E_i's, and $2^n = (1 + 1)^n = \sum_{r=0}^{n} \binom{n}{r}$. The one with $r = 0$ is $\emptyset$, the one with $r = n$ is $\square$, and those with $r = 1$ are the original E_i's. It can be shown that the enlarged set $\bar{\mathcal{E}}$ of 2^n members is "closed" with respect to the logical operations, in the sense that any application of logical operations on its members results in a proposition that already belongs to $\bar{\mathcal{E}}$. This extended set $\bar{\mathcal{E}}$ is called the (complemented, Boolean) lattice generated by the logical spectrum $\mathcal{E}$. The E_i's are said to be the "atoms" of the lattice $\bar{\mathcal{E}}$.

An example of logical spectrum is given by three propositions,

$$E_1:\quad \theta < 0, \qquad E_2:\quad 0 \le \theta < 32, \qquad E_3:\quad 32 \le \theta,$$

where θ is the temperature in degrees Fahrenheit at noon on January 1, 1970, at the ground level at Times Square, New York. The full lattice generated by these three will consist of $\emptyset$, $\square$, E_1, E_2, E_3, and three new propositions:

$$\begin{aligned} \neg E_1 &= E_2 \cup E_3:\quad 0 \le \theta, \\ \neg E_2 &= E_3 \cup E_1:\quad \theta < 0 \quad \text{or} \quad 32 \le \theta, \\ \neg E_3 &= E_1 \cup E_2:\quad \theta < 32, \end{aligned}$$

which makes $2^3 = 8$ propositions in total.

In the following we usually discuss the case in which n, the number of elements in a logical spectrum, is finite. Many of the results obtained will remain valid for the case in which n is countably many (enumerably infinite), although they sometimes break down when n becomes continuously many.

The concept of probability $\Pr\{E_i\}$ of a proposition E_i, in its interpretation, has two familiar aspects: on the one hand, $\Pr\{E_i\}$ is, so to speak, the degree of expectation that E_i will be the case, and, on the other, if the proposition E_i refers to the outcome of a test that can be reproduced, $\Pr\{E_i\}$ is the relative frequency of the outcome E_i in a large ensemble of similar tests. We do not discuss here the justifiability and the mutual relation of these two interpretations (see Chapter 7). Instead we resort freely to either of them as convenience calls for in each situation.

From the mathematical point of view all that is required of a probability is the following properties. For a given logical spectrum $\mathcal{E} = \{E_i\}$, $i = 1, 2, \cdots, n$, let a real number $p(E_i)$ be attached to each of the n constituent members in such a way that

$$p(E_i) \ge 0 \tag{1.4}$$

and

$$\sum_{i=1}^{n} p(E_i) = 1. \tag{1.5}$$

Let the function $p(E_i)$ thus defined in $\mathcal{E}$ now be extended to the lattice $\bar{\mathcal{E}}$ generated by $\mathcal{E}$, by requiring that the p of a member A of $\bar{\mathcal{E}}$ be given by

$$p(A) = \sum^{A} p(E_i), \tag{1.6}$$

where the summation should run over those E_i's needed to express A as a disjunction of the E_i's. Expression (1.6) tacitly implies that

$$p(\emptyset) = 0, \tag{1.7}$$

for no E_i's are needed to express $\emptyset$. Equation (1.5) means, in view of (1.3) and (1.6), that

$$p(\square) = 1. \tag{1.8}$$

A p-function thus defined on a lattice is, mathematically, called a probability.

The reader can check for himself that the basic formula of probability,

$$p(A) + p(B) = p(A \cap B) + p(A \cup B), \tag{1.9}$$

follows from the above definition, by substituting for A and B two arbitrary disjunctions of some E_i's. Then $p(A \cap B)$ will become the sum of the probabilities of those E_i's included in both A and B, and $p(A \cup B)$ will then become the sum of the probabilities of those E_i's included in either A or B or both. (1.4), (1.5), and (1.6) follow conversely from (1.7), (1.8), and (1.9).

The notion of conditional probability is defined by

$$p(A \mid B) = \frac{p(A \cap B)}{p(B)}. \tag{1.10}$$

This quantity $p(A \mid B)$, which is sometimes called "conditional probability of A given B," satisfies all the mathematical requirements of a probability of A, in accordance with (1.7), (1.8), and (1.9). It is interpreted either as the degree to which A is expected to be true when B and "nothing more specific" is known to be true or as the relative frequency of A in a collection of all systems for which a test has shown that B and "just" B is the case. The conditions in quotation marks are often forgotten, but they are critically important. Suppose that C is a proposition more specific than B and C implies B. Even then there is no guarantee that $p(A \mid B)$ and $p(A \mid C)$ should be the same. (See Section 4.2 for an explanation of the danger involved in the usual neglect of this point.)

The above definition of conditional probability gives the impression that there are two kinds of mathematical entities—conditional and unconditional probabilities—which can be easily differentiated. But in practice it is extremely seldom that we deal with a strictly unconditional probability, $p(A) = p(A \mid \square) = p(A \cap \square)/p(\square)$. Usually some condition B is tacitly understood, but not explixitly mentioned. And every time B changes, $p(A)$ has to change. For instance, in the case of the example mentioned above regarding the temperature at Times Square, the probabilities $p(E_i)$, $i = 1, 2, 3$, will change depending on what we are supposed to know. We may know only the statistics of the temperature at Times Square for the past 20 years, or we may know the weather map of the United States 24 hours before the time in question, or we may know the temperature at Times Square exactly 10 minutes before noon on January 1. To put these auxiliary conditions into the framework of (1.10) we would have to introduce a much more detailed logical spectrum than that engendered by the three E_i's. However, we usually do not construct such complicated logical spectra, and we deal with these probabilities on different conditions as if they were all unconditional probabilities. This is permissible because the conditional probability (1.10) with a fixed B, when considered as a function of A, is mathematically no different from another unconditional probability. This must be logically so, for if a proposition (B, here) is supposed to be true during a discussion or during a segment of a discussion it can be denoted as $\square$ during this period, and if we substitute $\square$ for B in (1.10), we formally obtain the unconditional probability.

This argument is related to the fact that if $B = \square$, $p(A \mid B) = p(A)$. But the converse is not true at all; namely, $p(A \mid B)$ can be equal to $p(A)$ without $p(B) = 1$. In general, when $p(A \mid B)$ and $p(A)$ are equal, we say that A and B are "probabilistically independent." This terminology is intuitively quite acceptable, since the condition it expresses is that the degree of expectation of A does not depend on whether or not B is established to be the case. It is easy to see that the concept of probabilistic independence of A and B is perfectly symmetrical with respect to A and B, for the condition can be written as

$$p(A \cap B) = p(A)p(B), \tag{1.11}$$

and neither side changes its value by interchange of A and B.

It often happens that a numerical value, say, q_i, is associated with each E_i of the n elements of the spectrum $\mathcal{E}$, and the set of n values $\{q_i\}$, $i = 1, 2, \cdots, n$, is considered as a single entity, called a stochastic variable. We say that the stochastic variable takes on value q_i when E_i occurs. The "expected (or average) value" of the stochastic variable $\{q_i\}$ is given by

$$\langle q_i \rangle_i = \sum_{i=1}^{n} q_i p(E_i). \tag{1.12}$$

In some cases the number q_i is mentioned in the definition of E_i, in some cases it is associated arbitrarily with E_i, and in other cases it is used as a convenient label for E_i (in this last case its role is no different from the index i of E_i). It often happens also that we do not use q_i at all in our discussion.

Probability can be defined on sets of objects instead of on propositions. This is because subsets of objects belonging to a set of objects form a Boolean lattice. More concretely, we can define a subset of objects as a collection of those objects that make a given proposition true. This makes a one-to-one correspondence between propositions and subsets. The propositions $\emptyset$ and $\square$

correspond respectively to the empty set and the entire set of the objects under consideration. The negation is interpreted as the complementary subset, the disjunction is interpreted as the merger of subsets, and the conjunction is interpreted as the common part of two subsets.

1.2. SURPRISE, IGNORANCE, AND INFORMATION

In this section we introduce the quantity called "information," which plays a basic role in this book. Consider a logical spectrum $\mathcal{E}$ of type (1.1), whose members E_i satisfy conditions (1.2) and (1.3). As was explained in the last section, the probability we attach to each proposition or event E_i depends on the auxiliary conditions that are supposed to be known or imposed on the ensemble. For the sake of intuitive vividness, we use the term "state of knowledge" to refer to these supposedly established auxiliary conditions. Suppose that at a certain state of knowledge we attach probability $\Pr\{E_i\} = p_i$ to each proposition E_i. These p_i must of course satisfy the relations for $p(E_i)$ in (1.4) and (1.5).

For simplicity of argument, let us assume for a moment that the E_i are experientially testable propositions. The above-mentioned probabilities p_i are assigned to the E_i without actually testing which one of the E_i's turns out to be true in the individual case presented to our experience. In other words, the state of knowledge underlying p_i does not include the result of a direct test. Now let us assume that the test is actually made. As a result of the test, the state of knowledge changes completely; namely, one of the E_i's, say, E_j, turns out to be true; that is, it receives probability unity. The probabilities p_i', $i = 1, 2, \cdots, n$, assigned to the E_i, based on the new state of information, will then be

$$p_j' = 1, \qquad p_i' = 0 \quad \text{for} \quad i \neq j. \tag{1.13}$$

Now, coming back to the probabilities p_i before the test, we can say that the higher the value of p_i, the more likely we consider the proposition E_i to turn out to be true. The probability p_i is our degree of expectation for E_i, based on a given state of knowledge. If p_i for a particular i is very small and yet E_i happens to take place in the test, we have to say that our "surprise" is very large. If, on the contrary, $p_i = 1$ and E_i happens, there is no surprise. Thus we can use a monotonically decreasing function $\phi(p_i)$ of p_i as a measure of our surprise caused by E_i, if it happens. In particular, it is convenient to take

$$\phi(p_i) = -\log p_i (\geq 0), \tag{1.14}$$

where log stands for the logarithm to any arbitrary base. The expression (1.14) is convenient not only because it becomes zero for $p_i = 1$, as is desirable, but also because it satisfies the following *additivity* condition.

Suppose that we have two spectra of propositions,

$$\mathcal{E} = \{E_1, E_2, \cdots, E_i, \cdots, E_n\} \quad \text{and} \quad \mathcal{F} = \{F_1, F_2, \cdots, F_j, \cdots, F_m\}.$$

We now make a third spectrum $\mathcal{G} = \{G_1, G_2, \cdots, G_k, \cdots, G_{nm}\}$ of "joint" propositions, of which each element G_k is a conjunction of one proposition of $\mathcal{E}$ and one proposition of $\mathcal{F}$.

$$G_k = E_i \cap F_j; \tag{1.15}$$

that is, G_k is true if and only if E_i *and* F_j are both true. We can easily see that probabilities $p(E_i)$ on $\mathcal{E}$ and the probabilities $p(F_j)$ on $\mathcal{F}$ must then be related to $p(G_k) = p(E_i \cap F_j)$ by

$$p(E_i) = \sum_{j=1}^{m} p(E_i \cap F_j); \quad p(F_j) = \sum_{i=1}^{n} p(E_i \cap F_j). \tag{1.16}$$

In fact, we have for the left side of the first equation

$$\begin{aligned} p(E_i) &= p(E_i \cap \square) = p(E_i \cap (F_1 \cup F_2 \cup \cdots \cup F_m)) \\ &= p((E_i \cap F_1) \cup (E_i \cap F_2) \cup \cdots \cup (E_i \cap F_n)) \end{aligned}$$

by the distributive law. Since $(E_i \cap F_j) \cap (E_i \cap F_k) = \phi$ for $j \neq k$, this last expression is equal, by virtue of (1.9), to the right side of the first equation of (1.16). If the propositions of $\mathcal{E}$ and the propositions of $\mathcal{F}$ are all mutually independent, we have

$$p(G_k) = p(E_i) \cdot p(F_j) \quad \text{for all} \quad i \quad \text{and} \quad j. \tag{1.17}$$

Then the surprise caused by G_k is, according to (1.14), exactly the sum of the surprise caused by E_i and the surprise caused by F_j:

$$\phi(p(G_k)) = \phi(p(E_i)) + \phi(p(F_j)). \tag{1.18}$$

This is a desirable property. There is no compelling reason to require additivity (1.18) of the quantity called "surprise," but this relation certainly makes it easy to handle the quantity ϕ, since (1.18) is an intuitively understandable relation.

Now the probability p_i implies that we shall suffer "surprise" $\phi(p_i)$ if E_i happens to occur. But at the same time it is expected that this will occur with probability p_i. Then the "expected surprise" is

$$\langle \phi(p_i) \rangle_i = \sum_{i=1}^{n} p_i \phi(p_i) = -\sum_{i=1}^{n} p_i \log p_i. \tag{1.19}$$

This may be considered a measure of our ignorance (in the "state of knowledge" before the test) with regard to the outcomes of the test performed on $\mathcal{E}$.

$$\text{ign}\,(\mathcal{E}) = -\sum_{i=1}^{n} p_i \log p_i. \tag{1.20}$$

The minimum of ign ($\mathcal{E}$) happens if and only if one of the p's becomes unity and all the other p's become zero, and

$$(\text{ign }(\mathcal{E}))_{\text{min}} = 0. \tag{1.21}$$

This corresponds to the case wherein our expectation is such that one of the E's will prove to be true with certainty. The maximum of ign ($\mathcal{E}$) happens if and only if all the p's are equal to $1/n$, and

$$(\text{ign }(\mathcal{E}))_{\text{max}} = \log n. \tag{1.22}$$

This corresponds to the case when we are unable to give any preference to one proposition over any other proposition, and thus corresponds to our complete ignorance. Between these extremities ign ($\mathcal{E}$) can be considered as a measure of the extent to which our expectation of the occurrence of propositions is widely scattered over various propositions.

The explanation of the foregoing paragraph was deliberately subjective, for we wanted to use familiar psychological or cognitive notions such as surprise, expectation, and ignorance. If we want to avoid these notions we can interpret ign ($\mathcal{E}$) of (1.20) as a measure of "uncertainty" or "indeterminacy" of the outcome in an ensemble, which is restricted by the auxiliary conditions that determine the "state of knowledge."

Suppose that we have the p_i before the test, and the corresponding ignorance (1.20). After the test the probabilities become p'_i, as given in (1.13). The ignorance after the test therefore becomes

$$\text{ign}' (\mathcal{E}) = 0. \tag{1.23}$$

The decrease in ignorance caused by the test can be considered as the *information* furnished by the test:

$$\begin{aligned} \text{information} &= \text{decrease in ignorance} \\ &= \text{ign }(\mathcal{E}) - \text{ign}' (\mathcal{E}) \\ &= -\sum_{i=1}^{n} p_i \log p_i. \end{aligned} \tag{1.24}$$

This is why this quantity is usually called "information." This, however, sometimes brings about a rather confusing situation: although the terms "ignorance" and "information" have opposite meanings, they are expressed by the same mathematical formula. If confusion happens to perturb our thinking, it is always a good policy to return to the original interpretation of the quantity as "ignorance," and then carefully apply the kind of argument expressed in (1.24) to introduce the idea of information. In the case when $\mathcal{E}$ consists of E's that are not directly testable, we cannot achieve the probability distribution (1.13) by a single test. Even in this case the interpretation of

(1.20) as the degree of ignorance or of indeterminacy is perfectly justifiable, and the consideration related to (1.23) and (1.24) is not entirely irrelevant because the probability distribution (1.13) is conceivable as an idealized limit even here.

Apart from the question of interpretation, the quantity (1.20) will in general be called the entropy function, defined by the probability distribution $\{p_i\}$ defined on $\mathcal{E}$ and denoted by $S(p)$ or $S(\mathcal{E})$, depending on whether we want to compare different sets of p's on the same spectrum or to consider the entropies of different spectra at the same time.

$$S(p) = S(\mathcal{E}) = -\sum_{i=1}^{n} p_i \log p_i. \tag{1.25}$$

When log in (1.25) stands for $\log_2$, S is said to be measured in the unit called "bit." One bit corresponds to a pair of alternatives (yes or no) with equal probability. In agreement with the fact that $\lim_{\delta\to 0} (-\delta \log \delta) \to 0$, we define the value of $0 \log 0$ as equal to 0, as was already tacitly assumed in (1.23). The use of the logarithm to a base other than 2 (but, of course, greater than 1) amounts to using another unit for the same quantity; that is, multiplying it by a constant positive factor. The use of natural logarithm (ln) often simplifies calculation involving differentiation. Physics usually uses $k \ln 2$ times the entropy expressed in bits, where k is the Boltzmann constant. This amounts to using the logarithm to the base $e^{(1/k)}$.

The reader can check for himself that if $\mathcal{E}$ and $\mathcal{F}$ are probabilistically independent in the sense of (1.17) we have the additivity of entropy function

$$S(\mathcal{G}) = S(\mathcal{E}) + S(\mathcal{F}). \tag{1.26}$$

The above explanation of the intuitive meaning of the entropy function is based on the assumption that n is finite, but it is clear that a similar explanation is possible in the case in which n is countably many (enumerably infinite), except that the maximum given by (1.22) becomes infinite. In the case wherein n is continuously many, the probability distribution p_i has to be replaced by $p(x)\,\Delta x$, where $p(x)$ is a probability density and Δx is a very small interval. This replacement in (1.25) will give $S(p) = -\Sigma\, p(x)\,\Delta x \log p(x) - \Sigma\, p(x)\,\Delta x \log \Delta x$, where the second term becomes equal to $-\log \Delta x$ if Δx does not depend on x. In the limit $\Delta x \to 0$, the first term becomes a genuine integral, but the second term explodes, implying that in the absence of a finite lower limit to the accuracy in x, the continuous spectrum conveys in general an infinite amount of information. The basic trouble here is that we cannot consider the first term alone and ignore the second term, because a transformation of variable $p(x)\,\Delta x = q(y)\,\Delta y$ keeps the sum $S(p)$ invariant, but not the two terms separately.

An objection that can be raised against taking (1.20) as the expression of ignorance is that the condition $p_i = 1/n$, leading to (1.22), may not correspond to our *complete ignorance*. One may contend that by the nature of the propositions each proposition E_i has an *a priori* weight w_i,

$$w_i \geq 0, \qquad \sum_{i=1}^{n} w_i = 1, \tag{1.27}$$

which is not all equal, so that our complete ignorance is not expressed by $p_i = 1/n$ but by

$$p_i = w_i. \tag{1.28}$$

Without concerning ourselves with what is meant by the *a priori* weight, we can define a reasonable function that resembles our original entropy function but becomes maximum when (1.28) is the case rather than when $p_i = 1/n$ is the case. One such function is

$$S^*(p) = -\sum_{i=1}^{n} p_i \log\left(\frac{p_i}{Aw_i}\right) = -\sum_{i=1}^{n} p_i(\log p_i - \log w_i) + \log A \tag{1.29}$$

and will be called the relative entropy function. We define $-p_i \log w_i = 0$ if $p_i = 0$ and $w_i = 0$. A in (1.29) is an arbitrary positive constant. A change in the value of A changes only an additive constant in $S^*(p)$. The maximum of $S^*(p)$ is realized when $p_i = w_i$, as in (1.28), and

$$(S^*(p))_{\max} = \log A. \tag{1.30}$$

The minimum of $S^*(p)$ is realized when $p_i = 1$ for a particular i for which w_i is the smallest. Thus, if

$$\min_i (w_i) = w_k, \tag{1.31}$$

which entails

$$(S^*(p))_{\min} = \log Aw_k, \tag{1.32}$$

$S^*(p)$ can be made non-negative by taking A so that $A \geq (1/w_k)$. It is reasonable to require p_i to be zero for a particular i for which $w_i = 0$. This agreement amounts to omitting from the summation in (1.29) those i's for which $w_i = 0$. This also makes it possible to choose A in such a manner that $S^*(p)$ becomes non-negative.

For a spectrum $\mathcal{G}$ of joint events defined in (1.15) satisfying the condition of independence (1.17), we can establish additivity for the relative entropy (1.29) by taking A and w for the spectrum $\mathcal{G}$ as equal respectively to the product of A's for $\mathcal{E}$ and $\mathcal{F}$ and to the product of the w's for $\mathcal{E}$ and $\mathcal{F}$; that is, $A_{\mathcal{G}} = A_{\mathcal{E}} \cdot A_{\mathcal{F}}$ and $w_{ij} = w_i \cdot w_j$.

The so-called thermodynamic entropy in physics has the form of this relative entropy function. The simple entropy function $S(p)$ of (1.25) can be considered to be a special case of the relative entropy function $S^*(p)$ of (1.29), with

$$A = n, \qquad w_i = \frac{1}{n}. \tag{1.33}$$

Conversely, a function of the type of (1.29) can be derived from (1.25) by the following procedure. Suppose that the n propositions E_i of the spectrum $\mathcal{E}$ are grouped into $\nu(< n)$ propositions forming another spectrum $\mathcal{F} = \{F_1, F_2, \cdots, F_\alpha, \cdots, F_\nu\}$, such that

$$\begin{aligned} F_1 &= E_1 \cup E_2 \cup \cdots \cup E_{n_1}. \\ F_2 &= E_{n_1+1} \cup E_{n_1+2} \cup \cdots \cup E_{n_1+n} \\ &\cdot \\ &\cdot \\ &\cdot \\ F_\nu &= E_{\sum_{\alpha=1}^{\nu-1} n_\alpha+1} \cup E_{\sum_{\alpha=1}^{\nu-1} n_\alpha+2} \cup \cdots \cup E_{\sum_{\alpha=1}^{\nu} n_\alpha}. \end{aligned} \tag{1.34}$$

In other words the proposition F_α consists of n_α propositions E's of $\mathcal{E}$, in such a way that if any one of these E's included in F_α is true, then F_α is true, and that if F_α is true, then some of the E's belonging to F_α are true. n_α is the number of E's included in F_α and satisfies

$$\sum_{\alpha=1}^{\nu} n_\alpha = n. \tag{1.35}$$

Then the probability of F_α is given by

$$p_\alpha = \sum_{i \in \alpha} p_i, \tag{1.36}$$

where the summation is extended over those E_i's that belong to the merged proposition F_α. Then the average of p_i in this merged proposition F_α is

$$\bar{p}_i = \frac{1}{n_\alpha} p_\alpha, \qquad E_i \cap F_\alpha = E_i. \tag{1.37}$$

Suppose that we have from the very beginning $\bar{p}_i$ instead of p_i; then the usual entropy function (1.25) becomes

$$\begin{aligned} S(\bar{p}) &= -\sum_{i=1}^{n} \bar{p}_i \log \bar{p}_i \\ &= -\sum_{\alpha=1}^{\nu} p_\alpha \log \left(\frac{p_\alpha}{n_\alpha}\right). \end{aligned} \tag{1.38}$$

This has the *form* of (1.29), with

$$w_\alpha = \frac{n_\alpha}{n}, \qquad A = n. \tag{1.39}$$

Thus the relative entropy function can be reinterpreted as a "coarse-grained" entropy function, in the sense that the probabilities are averaged within a "coarsely defined" group of propositions. The term "coarsely defined group" will make sense if a family of propositions of similar (or approximately equal) contents are put together as a group.

In the expression (1.38), if one of the p_α becomes 1, $S(\bar{p})$ becomes

$$S(\bar{p}) = \log n_\alpha. \tag{1.40}$$

This relation between (1.38) and (1.40) is known in physics as the relation between Gibbsian entropy, which has the form of (1.38), and Boltzmannian entropy, which has the form of the right-hand side of (1.40). Furthermore, if one puts

$$p_\alpha = \frac{n_\alpha}{n} \tag{1.41}$$

in (1.38), one obtains the maximum of $S(\bar{p})$,

$$(S(\bar{p}))_{\max} = \log n. \tag{1.42}$$

If there is among n_α ($\alpha = 1, 2, \cdots, \nu$) a particular one, say, n_μ, that is overwhelmingly larger than all the rest of the n_α, (1.42) can be approximated by

$$(S(\bar{p}))_{\max} = \log n_\mu. \tag{1.43}$$

This situation is used in physics when the entropy of the so-called micro-canonical ensemble, which corresponds to (1.42), is approximated by the entropy of the so-called Maxwell-Boltzmann cell, which corresponds to (1.43). We return to these problems of physics much later (see Section 5.3). For the moment we mention these facts only from a formal point of view.

The reader who is not very familiar with the functions introduced in this section is advised to study for himself the main geometrical features (such as zeros, maxima, and behavior of slopes) of the curves $y = -x \log x$ and $y = -x \log x - (1 - x) \log (1 - x)$ for the domain $0 \leq x \leq 1$. The second function corresponds, of course, to $S(\mathcal{E})$ for $n = 2$. It may be instructive to study also the nature of the function

$$T = 1 - \sum_{i=1}^{n} p_i^2 = 2 \sum_{i>j}^{n} \sum^{n} p_i p_j \tag{1.44}$$

in comparison with S, paying special attention to the question of additivity.

We do not want to spend time here for a bibliographical survey of various theoretical works involving the entropy function, except to mention that such a survey, if it is to be made, has to go back to the time of Maxwell and Boltzmann [B-5, M-5] and include, among others, such names as Gibbs [G-8], Szilard [S-12], Hartley [H-5], von Neumann [V-3], Wiener [W-36],

Shannon [S-6], and Brillouin [B-8, B-13] as those who contributed conspicuously to the development of various applications of the entropy function. The paper of Watanabe [W-2] may be noted for using at an early date the entropy function, not as a statistical-mechanical counterpart of thermodynamical entropy, but as a deliberately contrived measure of indeterminacy, thus marking the independence of information theory from thermodynamics. Probably the names of Fisher [F-6], MacKay [M-2], Gabor [G-1], and Carnap and Bar-Hillel [C-3] should be included in the list of those who considered the ideas of information in a broader sense than here. The reader can find good initial guidance in books by Cherry [C-4] and Pierce [P-6] if he wants to determine, in a broad context of human communication, the aspect that can be described by "quantity of information." He will find an interesting study on comparison of various concepts of information in Schutzenberger's paper [S-4].

As textbooks or reference works in English on information theory, we may mention books by Brillouin [B-13], Shannon and Weaver [S-6], Kullbach [K-16], Fano [F-2], Goldman [G-9], Yaglom and Yaglom [Y-1], Reza [R-3], and Abramson [A-2]. The books by Feinstein [F-3], Khinchin [K-9], and Wolfowitz [W-40] are highly mathematical. A concise survey on information theory is given by Elias [E-1]. A comprehensive bibliography on information theory has been compiled by Stumpers [S-11].

1.3. SOME MATHEMATICAL PROPERTIES OF ENTROPY FUNCTIONS

We discuss in this section some of the conspicuous mathematical properties of the simple entropy function (1.25), relegating their information-theoretical interpretation to later sections. As far as the mathematical gist is concerned, most of the interesting properties of the entropy function stem from three simple facts: (a) that the logarithmic function obeys the additive law, $\log xy = \log x + \log y$, (b) that z, defined by $z = -x \log x$, is non-negative for $0 \leq x \leq 1$, and (c) that z is a "concave" (toward the x-axis) function of x; that is, $d^2z/dx^2 < 0$ for $0 \leq x$. This last property gives birth to a theorem that has long been well known to physicists under the name of Gibbs' theorem and has turned out to be one of the most useful tools in information theory.

Theorem 1.1 (Gibbs' theorem). Let two sets of real numbers, $\{p_i\}$, $\{q_i\}$, $i = 1, 2, \cdots, n$, be such that

$$p_i \geq 0, \qquad q_i \geq 0 \quad \text{for all} \quad i \tag{1.45}$$

and

$$\sum_{i=1}^{n} p_i = \sum_{i=1}^{n} q_i. \tag{1.46}$$

Then we have

$$-\sum_{i=1}^{n} p_i \log p_i \leq -\sum_{i=1}^{n} p_i \log q_i, \tag{1.47}$$

where equality holds if and only if

$$p_i = q_i \quad \text{for all} \quad i. \tag{1.48}$$

It is agreed that $0 \log 0 = 0$.

Proof. Consider a function $f(x, y)$ of x and y given by

$$f(x, y) = x(\log x - \log y) - x + y \tag{1.49}$$

in the domain limited by $x \geq 0$ and $y \geq 0$. The function is finite, continuous, and differentiable as far as x and y are finite except on the x-axis; that is, at points with $y = 0$. Fixing y at a certain nonzero value, we take the x-derivative (we use the natural logarithm since the theorem does not depend on the unit),

$$\frac{\partial f(x, y)}{\partial x} = \log x - \log y. \tag{1.50}$$

This shows that the three disjoint and exhaustive cases $\partial f/\partial x \gtreqless 0$ correspond respectively to $x \gtreqless y$. This means that for a given nonzero value of y f is a convex function (toward the x-axis) of x with its minimum at $x = y$. At $x = y$ we have $f = 0$. Hence for any given i we have

$$p_i(\log p_i - \log q_i) - p_i + q_i \geq 0, \tag{1.51}$$

equality holding if and only if $p_i = q_i$. This statement is true even for $q_i = 0$, for the expression on the left side of (1.51) is $+\infty$ if $p_i \neq 0$ and $q_i = 0$, and becomes zero if $p_i = 0$ and $q_i = 0$. Summing up (1.51) for $i = 1, 2, \cdots, n$, we obtain

$$\sum_{i=1}^{n} [p_i(\log p_i - \log q_i) - p_i + q_i] \geq 0, \tag{1.52}$$

which is equivalent to (1.47) on account of (1.46). If one has equality in (1.51) for each value of i, then one will have equality in (1.52) and hence in (1.47). Conversely, if one has equality in (1.47) and hence in (1.52) as a result of (1.46), one knows, on the one hand, that each summand in (1.52) is non-negative because of (1.51) and, on the other hand, that the sum is zero. Hence one can conclude that each summand must be zero; hence $p_i = q_i$ for each i. Q.E.D.

Let us note that if p_i and q_i are two probability distributions, (1.45) and (1.46) are automatically satisfied and the left-hand side of (1.47) becomes an entropy function while the right-hand side becomes a mixed expression, not quite the entropy function unless $p_i = q_i$ for all i. The statement of the theorem for the case of equality in (1.47) is extremely powerful, since only one equation implies n relations $p_i = q_i$, $i = 1, 2, \cdots, n$, albeit only $(n - 1)$ of them are independent on account of (1.46).

The reader may ask: If Gibbs's theorem is based on the concavity of the S-function, could we not find a similar theorem also for the T-function we introduced in (1.44), which shares concavity though not additivity with the S-function? Consider a function $R(p, q)$ of two distributions p_i and q_i, defined by

$$R(p, q) = \sum_{i \neq j}^{n} \sum^{n} p_i q_j, \tag{1.53}$$

of which $T(p)$ is a special case, $R(p, p)$. The reader can check for himself that we have

$$2R(p, q) - T(p) - T(q) \geq 0, \tag{1.54}$$

where equality holds if and only if $p_i = q_i$ for all i.

In the last section we introduced in (1.15) the idea of a logical spectrum of joint events, namely, $\mathcal{G}$, which was a spectrum whose member G_k was a conjunction of a member of spectrum $\mathcal{E}$ and a member of spectrum $\mathcal{F}$. For simplicity we shall hereafter say in such a case that $\mathcal{G}$ is the product of $\mathcal{E}$ and $\mathcal{F}$ and write $\mathcal{G} = \mathcal{E} \otimes \mathcal{F}$. If E_i and F_j are independent, that is, $p(E_i \cap F_j) = p(E_i)p(F_j)$, for all i and all j as in (1.17), we say that the two spectra $\mathcal{E}$ and $\mathcal{F}$ are (probabilistically) independent.

Theorem 1.2 Let $\mathcal{E}$, $\mathcal{F}$, and $\mathcal{G}$ be logical spectra such that $\mathcal{G} = \mathcal{E} \otimes \mathcal{F}$. We have in general

$$S(\mathcal{G}) \leq S(\mathcal{E}) + S(\mathcal{F}), \tag{1.55}$$

where S is the simple entropy function defined in (1.25). The equality in (1.55) holds if and only if $\mathcal{E}$ and $\mathcal{F}$ are probabilistically independent.

Proof. This theorem uses basically Properties (a) and (c) mentioned above. The "if" part (as distinct from the "only-if" part) of the last statement is very easy to prove and was mentioned with regard to (1.26). The proof of the entire theorem is provided by Gibbs' theorem if we introduce a second probability distribution by

$$q(G_k) = q(E_i \cap F_j) = p(E_i)p(F_j) \tag{1.56}$$

besides the originally given probability distribution $p(G_k) = p(E_i \cap F_j)$. The $p(E_i)$ and the $p(F_j)$ of (1.56) are derived from $p(E_i \cap F_j)$ by the procedure given in (1.16). Gibbs' theorem applied to $\mathcal{G}$ will then state

$$\begin{aligned} S(\mathcal{G}) &= -\sum_{i}^{n} \sum_{j}^{m} p(E_i \cap F_j) \log p(E_i \cap F_j) \leq -\sum_{i}^{n} \sum_{j}^{m} p(E_i \cap F_j) \log q(E_i \cap F_j) \\ &= S(\mathcal{E}) + S(\mathcal{F}), \end{aligned} \tag{1.57}$$

for a summation with respect to k over $k = 1, 2, \cdots, mn$ is equivalent to a double summation with respect to i and to j over $i = 1, 2, \cdots, n$ and $j = 1, 2, \cdots, m$.

The equality symbol in (1.57) holds if and only if $p(E_i \cap F_j) = q(E_i \cap F_j)$ for all i and j. This ends the proof. The fact that the right-hand side of (1.57) becomes a sum of two S-functions is due to the additivity of the logarithm function.

A generalization of the foregoing result is immediate. Take N logical spectra $\mathcal{E}^{(a)} = \{E_{i_a}^{(a)}\}$, $a = 1, 2, \cdots, N$, and assume that the ath spectrum $\mathcal{E}^{(a)}$ consists of n_a elements; that is, $i_a = 1, 2, \cdots, n_a$. Consider a logical spectrum $\mathcal{G} = \{G_k\}$, each member of which is a conjunction of one member of $\mathcal{E}^{(1)}$, one member of $\mathcal{E}^{(2)}$, $\cdots$, and one member of $\mathcal{E}^{(N)}$.

$$G_k = \bigcap_{a=1}^{N} E_{i_a}^{(a)} = E_{i_1}^{(1)} \cap E_{i_2}^{(2)} \cap \cdots \cap E_{i_N}^{(N)}. \tag{1.58}$$

There will be a total of $m = n_1 \times n_2 \times \cdots \times n_N$ disjoint elements of this kind, and the disjunction of all these m elements will be equivalent to $\square$. In other words, the index k used for $\mathcal{G}$ will run from 1 to $m = n_1 \times n_2 \times \cdots \times n_N$. Under these conditions we say that $\mathcal{G}$ is the product of $\mathcal{E}^{(1)}, \mathcal{E}^{(2)}, \cdots, \mathcal{E}^{(N)}$ and write

$$\mathcal{G} = \bigotimes_{a=1}^{N} \mathcal{E}^{(a)} = \mathcal{E}^{(1)} \otimes \mathcal{E}^{(2)} \otimes \cdots \otimes \mathcal{E}^{(N)}. \tag{1.59}$$

The probability distribution $p(G_k)$ on $\mathcal{G}$ must be such that, for instance, the probability distribution $p(E_{i_1}^{(1)})$ on $\mathcal{E}^{(1)}$ can be derived from it by

$$p(E_{i_1}^{(1)}) = \sum_{i_2=1}^{n_2} \cdots \sum_{i_N=1}^{n_N} p(E_{i_1}^{(1)} \cap E_{i_2}^{(2)} \cap \cdots \cap E_{i_N}^{(N)}). \tag{1.60}$$

If it so happens that

$$p(E_{i_1}^{(1)} \cap E_{i_2}^{(2)} \cap \cdots \cap E_{i_N}^{(N)}) = \prod_{a=1}^{N} p(E_{i_a}^{(a)}) \tag{1.61}$$

for all i_1, all i_2, $\cdots$, and all i_N, then we say that $\mathcal{E}^{(1)}, \mathcal{E}^{(2)}, \cdots, \mathcal{E}^{(N)}$ are all independent. Now, by imitating the proof of Theorem 1.2, the reader will be able to prove Theorem 1.3 with ease.

Theorem 1.3 If a logical spectrum $\mathcal{G}$ is the product of N logical spectra $\mathcal{E}^{(a)}$, $a = 1, 2, \cdots, N$, as in (1.59), we have

$$S(\mathcal{G}) \leq \sum_{a=1}^{N} S(\mathcal{E}^{(a)}), \tag{1.62}$$

where S is the simple entropy function defined in (1.25). The equality in (1.62) holds if and only if all the $\mathcal{E}^{(a)}$'s are independent in the sense of (1.61).

It may be noted that this theorem can be proven, as suggested above, either by identifying the p-function and the q-function of Gibbs' theorem with the left-hand side and the right-hand side of (1.61) or by applying Theorem 1.2 step by step, first to $\mathcal{G} = \mathcal{E}^{(1)} \otimes \mathcal{F}^{(1)}$ [with $\mathcal{F}^{(1)} = \mathcal{E}^{(2)} \otimes \cdots \otimes \mathcal{E}^{(N)}$] and then to $\mathcal{F}^{(1)} = \mathcal{E}^{(2)} \otimes \mathcal{F}^{(2)}$ [with $\mathcal{F}^{(2)} = \mathcal{E}^{(3)} \otimes \cdots \otimes \mathcal{E}^{(N)}$], and so forth.

We can obtain further insight into the question of dependence and independence by introducing what is known as conditional entropy. Take two logical spectra $\mathcal{E}$ and $\mathcal{F}$ and consider the conditional probabilities

$$p(F_j \mid E_i) = \frac{p(E_i \cap F_j)}{p(E_i)}, \tag{1.63}$$

which becomes equal to the unconditional probability of F_j if and only if E_i and F_j are independent and $p(E_i) \neq 0$. Now if we fix our attention on an E_i, the quantity $p(F_j \mid E_i)$ becomes a probability distribution defined on $\mathcal{F}$, satisfying all the requirements of the probability concept. Hence we can define the entropy function with this probability distribution by the usual formula (1.25). The quantity thus obtained is called conditional entropy of $\mathcal{F}$ given E_i, and is given by

$$\begin{aligned} S(\mathcal{F} \mid E_i) &= -\sum_{j=1}^{m} p(F_j \mid E_i) \log p(F_j \mid E_i) \\ &= -\sum_{j=1}^{m} \frac{p(E_i \cap F_j)}{p(E_i)} \log \frac{p(E_i \cap F_j)}{p(E_i)} . \end{aligned} \tag{1.64}$$

For later use we note the obvious fact that

$$S(\mathcal{F} \mid E_i) \geq 0, \tag{1.65}$$

which is nothing but Property (b) mentioned at the beginning of this section.

Considering $S(\mathcal{F} \mid E_i)$ as a stochastic variable for E_i, we can obtain its average by

$$S(\mathcal{F} \mid \mathcal{E}) = \langle S(\mathcal{F} \mid E_i) \rangle_i = \sum_{i=1}^{n} S(\mathcal{F} \mid E_i) p(E_i), \tag{1.66}$$

which we call (average) conditional entropy of $\mathcal{F}$ given $\mathcal{E}$. Writing out the right-hand side of (1.66), we obtain

$$\begin{aligned} S(\mathcal{F} \mid \mathcal{E}) &= \sum_{i=1}^{n} \sum_{j=1}^{m} - p(E_i \cap F_j)[\log p(E_i \cap F_j) - \log p(E_i)] \\ &= S(\mathcal{G}) - S(\mathcal{E}). \end{aligned} \tag{1.67}$$

If there is an E_i for which $p(E_i) = 0$, $S(\mathcal{F} \mid E_i)$ becomes indeterminate for this E_i; hence, strictly speaking, the term in (1.66) corresponding to such an E_i has no sense. However, the explicit expression on the first line of (1.67) makes sense even for E_i for which $p(E_i) = 0$; namely, the contribution to (1.67) from the term involving E_i for which $p(E_i) = 0$ is zero. Note that if $p(E_i) = 0$ then $p(E_i \cap F_j) = 0$ and that we agreed on $0 \log 0 = 0$. We employ this property of (1.67) as a supplement to complete the definition of $S(\mathcal{F} \mid \mathcal{E})$. This can be easily "read into" original definition (1.66) by interpreting $S(\mathcal{F} \mid E_i)$ for E_i whose $p(E_i) = 0$ as standing for some unspecified but *finite* value. From (1.65) we get

$$S(\mathcal{F} \mid \mathcal{E}) \geq 0. \tag{1.68}$$

Now, considering two tests determining the truth or falsehood of E_i and F_j, we may say that F_j is completely dependent on E_i if the outcome of F_j is deterministically decided by fact that E_i is found to be the case, that is, if

$$p(F_j \mid E_i) = 0 \quad \text{or} \quad 1. \tag{1.69}$$

If all the F_j's of $\mathcal{F}$ are completely dependent on an E_i whose probability $p(E_i) \neq 0$, then we have $S(\mathcal{F} \mid E_i) = 0$ according to (1.64). Furthermore, if all the F_j's of $\mathcal{F}$ are completely dependent on each one of those E_i's of $\mathcal{E}$ whose probability is not zero, then we have $S(\mathcal{F} \mid \mathcal{E}) = 0$. Conversely, suppose that we are first given $S(\mathcal{F} \mid \mathcal{E}) = 0$. Since, the summand on the right-hand side of the first line of (1.67) is non-negative, $S(\mathcal{F} \mid \mathcal{E}) = 0$ implies that for each pair (i, j) we should have $p(E_i) = p(E_i \cap F_j) = 0$, or else either $p(E_i) \neq 0$, $p(E_i \cap F_j) = 0$ or $p(E_i) \neq 0$, $p(E_i) = p(E_i \cap F_j)$. In other words, $S(\mathcal{F} \mid \mathcal{E}) = 0$ implies (1.69) for all pairs (i, j) provided that $p(E_i) \neq 0$. Thus we have proven Theorem 1.4.

Theorem 1.4 The average conditional entropy $S(\mathcal{F} \mid \mathcal{E})$ is always non-negative, and is equal to zero if and only if $\mathcal{F}$ is completely dependent on $\mathcal{E}$; that is, all F_j's of $\mathcal{F}$ are completely dependent on each one of those E_i of $\mathcal{E}$ for which $p(E_i) \neq 0$ in the sense of (1.69).

The result expressed in the second line of (1.67), in combination with Theorem 1.2, leads to the following corollary.

Corollary 1.1 The average conditional entropy $S(\mathcal{F} \mid \mathcal{E})$ is never larger than the unconditional entropy $S(\mathcal{F})$, that is,

$$S(\mathcal{F} \mid \mathcal{E}) \leq S(\mathcal{F}), \tag{1.70}$$

and they are equal if and only if $\mathcal{E}$ and $\mathcal{F}$ are independent.

This result is in agreement with the remark made just below (1.63), because independence of $\mathcal{E}$ and $\mathcal{F}$ means $p(F_j) = p(F_j \mid E_i)$ for all i and all j, which entails $S(\mathcal{F}) = S(\mathcal{F} \mid E_i)$, hence $S(\mathcal{F}) = S(\mathcal{F} \mid \mathcal{E})$.

Since $\mathcal{E}$ and $\mathcal{F}$ are symmetrical throughout our discussion, we also have

$$S(\mathcal{E} \mid \mathcal{F}) = S(\mathcal{G}) - S(\mathcal{F}) \tag{1.71}$$

in parallel with (1.67) and

$$S(\mathcal{E} \mid \mathcal{F}) \leq S(\mathcal{E}) \tag{1.72}$$

in parallel with (1.70). Vanishing of $S(\mathcal{E} \mid \mathcal{F})$ means a complete dependence of $\mathcal{E}$ on $\mathcal{F}$. We should be a little careful here, for the concept of independence of $\mathcal{E}$ and $\mathcal{F}$ is symmetrical with respect to $\mathcal{E}$ and $\mathcal{F}$, but the concept of dependence of $\mathcal{F}$ on $\mathcal{E}$ is not. The equality sign in (1.70) and the equality sign in (1.72) are equivalent, but vanishing of $S(\mathcal{F} \mid \mathcal{E})$ and $S(\mathcal{E} \mid \mathcal{F})$ are not.

It is instructive to note that the average conditional entropy is a special case of the relative entropy introduced in (1.29). In fact, we can rewrite (1.67) as

$$S(\mathcal{F} \mid \mathcal{E}) = -\sum_{k=1}^{nm} p(G_k) \log \frac{p(G_k)}{Aw_k}, \tag{1.73}$$

with

$$Aw_k = p(E_i), \tag{1.74}$$

where E_i is uniquely determined by G_k because of the condition that $G_k \cap E_i = G_k$.

The quantity claimed to be non-negative in Theorem 1.2 and Corollary 1.1, namely,

$$J(\mathcal{E}, \mathcal{F}) = S(\mathcal{E}) + S(\mathcal{F}) - S(\mathcal{E} \otimes \mathcal{F}) \tag{1.75}$$

$$= S(\mathcal{F}) - S(\mathcal{F} \mid \mathcal{E}) \tag{1.76}$$

$$= S(\mathcal{E}) - S(\mathcal{E} \mid \mathcal{F}), \tag{1.77}$$

is called "interdependence between $\mathcal{E}$ and $\mathcal{F}$" and plays a very important role in Chapter 2, where we also discuss the information-theoretical meanings of various mathematical formulae derived in the present section. This quantity $J(\mathcal{E}, \mathcal{F})$, in contrast to entropy and conditional entropy, has a special property of allowing a converging counterpart in the continuous case when the accuracy of the variables increases indefinitely. Let us substitute $p(xy)\,\Delta x\,\Delta y$ for $p(E_i \cap F_j)$ in our formulas. Then, corresponding to our $J(\mathcal{E}, \mathcal{F}) = S(\mathcal{E}) + S(\mathcal{F}) - S(\mathcal{E} \otimes \mathcal{F})$, we have

$$-\sum p_x(x)\,\Delta x \log p_x(x)\,\Delta x - \sum p_y(y)\,\Delta y \log p_y(y)\,\Delta y$$
$$+ \sum\sum p(x, y)\,\Delta x\,\Delta y \log p(x, y)\,\Delta x\,\Delta y, \tag{1.78}$$

with

$$p_x(x) = \sum_y p(x, y)\,\Delta y \quad \text{and} \quad p_y(y) = \sum_x p(x, y)\,\Delta x. \tag{1.79}$$

In the limit $\Delta x \to 0$, $\Delta y \to 0$, the quantity (1.78) will converge to an integral

$$-\iint p(x, y)\,dx\,dy \log \frac{p_x(x)p_y(y)}{p(x, y)}, \tag{1.80}$$

with

$$p_x(x) = \int p(x, y)\, dy \quad \text{and} \quad p_y(y) = \int p(x, y)\, dx. \tag{1.81}$$

The quantity (1.80) is invariant for a transformation $(x, y) \rightarrow (\xi = \xi(x), \eta = \eta(y))$, such that $p(x, y)\, dx\, dy = q(\xi, \eta)\, d\xi\, d\eta$.

The formula involving the conditional entropy introduced in (1.67)

$$S(\mathcal{E} \otimes \mathcal{F}) = S(\mathcal{E}) + S(\mathcal{F} \mid \mathcal{E}), \tag{1.82}$$

suggests various generalizations. A few elementary ones are mentioned here. We can consider (1.82) as having been derived from the formula

$$p(E_i \cap F_j) = p(E_i) \times \frac{p(E_i \cap F_j)}{p(E_i)} \tag{1.83}$$

by taking the logarithm of each side of (1.83) and averaging with probability $p(E_i \cap F_j)$. If we are given three spectra,

$$\mathcal{E} = \{E_i\}, \qquad \mathcal{F} = \{F_j\}, \qquad \mathcal{H} = \{H_k\},$$

then a natural generalization of (1.83) is

$$p(E_i \cap F_j \cap H_k) = p(E_i) \times \frac{p(E_i \cap F_j)}{p(E_i)} \times \frac{p(E_i \cap F_i \cap H_k)}{p(E_i \cap F_j)}. \tag{1.84}$$

If we take the logarithm of each side of (1.84) and average with the probability $p(E_i \cap F_j \cap H_k)$, we obtain

$$S(\mathcal{E} \otimes \mathcal{F} \otimes \mathcal{H}) = S(\mathcal{E}) + S(\mathcal{F} \mid \mathcal{E}) + S(\mathcal{H} \mid \mathcal{E} \otimes \mathcal{F}). \tag{1.85}$$

Since we have, because of (1.82),

$$S(\mathcal{E} \otimes \mathcal{F} \otimes \mathcal{H}) - S(\mathcal{E}) = S(\mathcal{F} \otimes \mathcal{H} \mid \mathcal{E}), \tag{1.86}$$

(1.85) can be rewritten as

$$S(\mathcal{F} \otimes \mathcal{H} \mid \mathcal{E}) = S(\mathcal{F} \mid \mathcal{E}) + S(\mathcal{H} \mid \mathcal{E} \otimes \mathcal{F}). \tag{1.87}$$

For the three spectra $\mathcal{E}$, $\mathcal{F}$, and $\mathcal{H}$ there are three formulas of this type. Extension of (1.85) and (1.87) for the case with more than three spectra is immediate. Obviously we have

$$S(\mathcal{E}^{(1)} \otimes \mathcal{E}^{(2)} \otimes \cdots \mathcal{E}^{(N)}) = S(\mathcal{E}^{(1)}) + S(\mathcal{E}^{(2)} \mid \mathcal{E}^{(1)})$$
$$+ S(\mathcal{E}^{(3)} \mid \mathcal{E}^{(1)} \otimes \mathcal{E}^{(2)}) + \cdots + S(\mathcal{E}^{(N)} \mid \mathcal{E}^{(1)} \otimes \mathcal{E}^{(2)} \otimes \cdots \otimes \mathcal{E}^{(N-1)}) \tag{1.88}$$

and

$$S(\mathcal{E}^{(2)} \otimes \cdots \otimes \mathcal{E}^{(N)} \mid \mathcal{E}^{(1)}) = S(\mathcal{E}^{(2)} \mid \mathcal{E}^{(1)})$$
$$+ (\mathcal{E}^{(3)} \mid \mathcal{E}^{(1)} \otimes \mathcal{E}^{(2)}) + \cdots + S(\mathcal{E}^{(N)} \mid \mathcal{E}^{(1)} \otimes \mathcal{E}^{(2)} \otimes \cdots \otimes \mathcal{E}^{(N-1)}). \tag{1.89}$$

In the last and present sections we have followed more or less the line of exposition in which the definition of the S-function was first accepted and the additivity and other important properties of the S-function were derived therefrom. An inverse line of exposition is also enlightening, which consists of defining first the properties to be required of the yet-to-be-introduced function and showing that the function that satisfies these requirements is bound to be the S-function. One may call such an attempt an axiomatic approach, if the requirements correspond to intuitive properties that people agree to regard as desirable. There are several ways of carrying out such an axiomatic plan, but the following one seems attractive because the notions involved are all very simple. It has also another merit. In the last section we started from the general form of the S function given in (1.20) or (1.25) and derived there from the form $\log n$ in (1.22) as a special case. The second part of the following argument allows us, so to speak, to go backward, starting from a special case and ending with the general case. This delicate argument, as far as is known to me, was first proposed by Brillouin [B-10]. In the first reading the reader may omit the remainder of this section, for it digresses from the main topic of this chapter.

Our requirements are as follows:

1. $S(\mathcal{E})$ is a real symmetrical function of the probabilities $p(E_i)$ assigned to the members E_i, $i = 1, 2, \cdots, n$ of a logical spectrum $\mathcal{E}$.
2. If three logical spectra $\mathcal{E}$, $\mathcal{F}$, and $\mathcal{G}$ are such that $\mathcal{G} = \mathcal{E} \otimes \mathcal{F}$ and $\mathcal{E}$ and $\mathcal{F}$ are probabilistically independent, $S(\mathcal{G}) = S(\mathcal{E}) + S(\mathcal{F})$.
3. Let $F(r)$ be the value of $S(\mathcal{E})$ for the special case in which $p(E_i) = 1/r$ for r out of the n E_i's of $\mathcal{E}$ and $p(E_i) = 0$ for the remaining $(n - r)E_i$'s $(r \leq n)$. This $F(r)$ is a universal function of r, independent of $\mathcal{E}$, and monotonically increasing with r in the strict sense.

Our proof that $S(\mathcal{E})$ satisfying these conditions is bound to be the one given in (1.25) consists of two steps. In the first step we show that $F(r)$ must be $\log r$. Consider $\mathcal{G} = \otimes_{a=1}^{N} \mathcal{E}^{(a)}$, where $\mathcal{G} = \{G_k\}$, $k = 1, 2, \cdots, \prod_{a=1}^{N} n_a$ and $\mathcal{E}^{(a)} = \{E_{i_a}^{(a)}\}$, $i_a = 1, 2, \cdots, n_a$. Let the probability for $G_k = \bigcap_{a=1}^{N} E^{(a)}$ be constant and given by $p(G_k) = 1/\prod_{a=1}^{N} n_a$, so that $p(E_{i_a}^{(a)}) = 1/n_a$. This is obviously a case of probabilistic independence. By a repeated application of Item 2, we obtain, with the help of the F-function introduced in Item 3,

$$F\left(\prod_{a=1}^{N} n_a\right) = \sum_{a=1}^{N} F(n_a). \tag{1.90}$$

By substituting special values $n_2 = n_3 = \cdots = n_N = 1$, we obtain

$$F(n_1) = F(n_1) + (N - 1)F(1). \tag{1.91}$$

Hence

$$F(1) = 0 \tag{1.92}$$

and, since $F(r)$ is monotonically increasing with r in the strict sense, $F(r)$ must be positive for $r > 1$. If all the n 's are equal to n, (1.90) becomes

$$F(n^N) = NF(n). \tag{1.93}$$

Let us take two integers n and m and assume that $n \neq 1$ and $m \neq 1$. For three given integers n, m, and N there will always be an integer M such that

$$m^M \leq n^N < m^{M+1}; \tag{1.94}$$

from which follows

$$\frac{M}{N} \leq \frac{\log n}{\log m} < \frac{M + 1}{N}, \tag{1.95}$$

where the base of the logarithm is arbitrary. Since $F(n)$ is a strictly monotonic function of n according to Item 3, we get from (1.94)

$$F(m^M) \leq F(n^N) < F(m^{M+1}), \tag{1.96}$$

which entails, on account of (1.93),

$$MF(m) \leq NF(n) < (M + 1)F(m). \tag{1.97}$$

Since $F(m) > 0$, we can divide (1.97) by $F(m)$ and obtain

$$\frac{M}{N} \leq \frac{F(n)}{F(m)} < \frac{M + 1}{N}. \tag{1.98}$$

Combining (1.95) and (1.98) we obtain

$$\left| \frac{F(n)}{F(m)} - \frac{\log n}{\log m} \right| < \frac{1}{N}. \tag{1.99}$$

Since N can be arbitrarily large, we conclude that $F(n) = \text{constant} \times \log n$. But this constant must be positive in order for $F(n)$ to be monotonically increasing. Hence we can include this arbitrary constant in the arbitrariness of the base of the logarithm and obtain

$$F(n) = \log n. \tag{1.100}$$

This ends the first step.

Now we go back to $\mathcal{G} = \bigotimes_{a=1}^{N} \mathcal{E}^{(a)}$ introduced above, where we assumed that all n_a's are equal to n. But this time the probability distribution is different from before. We assume that the probability $p(E_{i_a}^{(a)}) = p_i$ depends on i, though not on a. $p(G_k)$ is to be determined by the condition of probabilistic independence. To make the explanation shorter we give a special type to the proposition $E_{i_a}^{(a)}$. When i_a is equal to a number i, $E_{i_a}^{(a)}$ should stand for "the

ath object satisfies the ith predicate Q_i." There are n disjoint and exhaustive predicates and they are applicable to each of N objects. Then a statement G_k concerns the state of the collection of N objects, specifying the predicate each object satisfies, whereby the probability of each object satisfying a predicate is independent. Now we assume that N is extremely large. Then, in the collection of N objects,

$$\text{“}Np_i\text{”} = N_i \tag{1.101}$$

objects will satisfy predicate Q_i, where the quotation marks mean "the integer nearest to $\cdots$." By increasing N indefinitely, we can make all the "Np_i" as close to some integers as we want. By increasing N sufficiently, we can also make the relative fluctuation of the fraction of objects satisfying Q_i negligible. As far as $\mathcal{G}$ is concerned, there are n^N possible G_k in total, and each possibility G_k specifies a number of objects satisfying each predicate Q_i. Under the present circumstances only those G_k's that satisfy the condition (1.101) occur. All the other G_k's have probability 0 of occurring.

Now how many G_k's satisfy the condition (1.101)? The answer is: the number ν of ways in which one can put N_i objects out of N in the ith box, $i = 1, 2, \cdots, n$; that is,

$$\nu = \frac{N!}{N_1!\, N_2! \cdots N_n!}. \tag{1.102}$$

We know, on one hand, that the sum of the probabilities of these ν possibilities is 1, and, on the other hand, that all of these ν possibilities must have the same probability because all the N objects are on the same footing. Consequently, by Item 3, we have to conclude that

$$S(\mathcal{G}) = F(\nu) = \log \frac{N!}{N_1!\, N_2! \cdots N_n!}. \tag{1.103}$$

Remembering the Sterling formula $\ln N_i! = N_i(\ln N_i - 1)$ and noting that $\log_b = \log_b e \ln$, we can rewrite (1.103) as

$$S(\mathcal{G}) = -\sum N_i \log \frac{N_i}{N}. \tag{1.104}$$

The special case we dealt with before, with $N_i/N = p_i = 1/n$, leads to the entropy value $S(\mathcal{G}) = \log(1/n^N)$ because there are n^N equally probable cases. Equation (1.104) is in agreement with this. In the present case all the $\mathcal{E}$'s are independent; hence we should have, according to item (2),

$$\begin{aligned} S(\mathcal{E}^{(a)}) &= \frac{1}{N} S(\mathcal{G}) = -\sum_{i=1}^{n} \frac{N_i}{N} \log \frac{N_i}{N} \\ &= -\sum_{i=1}^{n} p_i \log p_i, \end{aligned} \tag{1.105}$$

where p_i is the probability that the ith of the n members of $\mathcal{E}^{(a)}$ is true. Thus we have derived the desired formula from the three requirements.

Before passing on to the next section, a few words will be devoted here to what may be characterized as a natural extension of the idea suggested by Gibbs' theorem. If the quantity

$$F(p, q) = -\sum_{i=1}^{n} p_i(\log q_i - \log p_i) \tag{1.106}$$

is non-negative and vanishes if and only if two probability distributions $\{p_i\}$ and $\{q_i\}$, $i = 1, 2, \cdots, n$, coincide perfectly, as (1.47) states, then this quantity may be used as a measure of discrepancy between the two probability distributions. If we use ε_i defined by

$$q_i = p_i + \varepsilon_i, \qquad \left(\sum_{i=1}^{n} \varepsilon_i = 0, \qquad -p_i \le \varepsilon_i \le 1 - p_i\right) \tag{1.107}$$

to characterize the second distribution $\{q_i\}$, we can express (1.106) by the use of Taylor's expansion (assuming the logarithm to be the natural logarithm) as

$$F(p, q) = -\sum_i{}' p_i \log\left(1 + \frac{\varepsilon_i}{p_i}\right) = \tfrac{1}{2}\sum_i{}' p_i \frac{\varepsilon_i^{\,2}}{(p_i + \theta_i\varepsilon_i)^2}, \qquad 0 \le \theta_i \le 1, \tag{1.108}$$

where the summation $\sum_i'$ does not include those i for which $p_i = 0$. For small ε_i, the denominator, which is between p_i^2 and q_i^2, may be replaced by p_i^2 or q_i^2 or $p_i q_i$.

If we interpret $\{q_i\}$ as the probability distribution derived from a theory or a hypothesis and p_i as the relative frequency of event E_i in a collection of N samples, we may consider (1.108) as the measure of discrepancy between the hypothesis and the collection of N empirical data. In fact, if we assume the ε_i to be small, (1.108) becomes $(1/2N)$ times the quantity well known as χ^2:

$$F(p, q) \doteqdot \tfrac{1}{2}\sum_i \frac{(p_i - q_i)^2}{q_i} = \frac{1}{2N}\chi^2. \tag{1.109}$$

It is historically interesting that as early as 1935 Wilks [W-38] pointed out that there is no theoretical reason why χ^2 should be used in preference to $-2\log\lambda$, where λ is the likelihood ratio based on the multinomial distribution. The quantity $-2\log\lambda$ in this case turns out to be exactly $2N$ times our $F(p, q)$ in its exact form (1.106). (See Perez [P-4] and Kullbach [K-16] for related matters.)

Independently of the traditional development of the theory of hypothesis testing, I myself was led to introduce the quantity σ, defined below as a measure of the degree of confirmation of hypothesis that gives $\{q_i\}$ by a population of experimental data characterized by $\{p_i\}$ (see [W-13] and Chapter 6 of the present book).

$$\frac{1}{\sigma} = 1 + \frac{F(p, q)}{S(p)}. \qquad (1.110)$$

We return to the problem of hypotheses testing in Chapter 6.

1.4. ACQUISITION OF INFORMATION THROUGH SUCCESSIVE POLYCHOTOMIC OBSERVATIONS

This section is meant to serve at least five purposes. (a) It gives the reader more insight into the intuitive meaning of the quantity "information"; (b) It prepares a mathematical solution to the problem of "noiseless" coding, discussed in the next section; (c) it provides a useful model in a general discussion of the process of observation in Chapter 5; (d) it furnishes the groundwork for a simple illustration to the inverse *H*-theorem in Chapter 6; and (e) it gives an introductory glimpse of the group of problems known as "search problems."

In the last section we learned that the state of knowledge about a logical spectrum $\mathcal{E} = \{E_i\}$ expressed by the probability distribution p_i over the cases E_i can be characterized by the initial ignorance,

$$S_I = -\sum_{i=1}^{n} p_i \log p_i, \qquad (1.111)$$

and that after the "test" one of the p's becomes unity and the other p's become zero, so that the final state of knowledge is characterized by final ignorance

$$S_F = 0. \qquad (1.112)$$

The decrease in ignorance was interpreted as the information obtained,

$$\text{information} = S_I - S_F = S_I. \qquad (1.113)$$

However, it is often the case that a test or observation does not determine at one stroke whether the searched object or the "true" case falls on an E_i, but only whether it belongs to a certain group of E's. After we have located the object in a certain group of E's we can further locate the object in a subgroup of this group. Repeating this process, one may finally locate the object in an individual case E_i. Methods of measurement in the physical sciences are often of this type. We consider in this section the information gained in such a step-by-step process of testing.

The following old mathematical quiz has a close relation to the general problem we discuss in this section. There is a certain number of, say, eight, coins of the same appearance. It is known, however, that one of them is counterfeit and is lighter than a genuine one. We are allowed to use a balance without standard weights in our attempt to identify the falsified one. What is the minimum number of weighing operations necessary to reach a definitive answer? The method of observation we are allowed to use here consists first of dividing a collection of coins into three groups, say, A, B, and C, in such a way that A and B have the same number of coins. C could be empty. Then we put A and B respectively on the left and right pans of the balance. If the left pan goes down, it means that the falsified coin is in group B; if the right pan goes down, it means that it is in group A; finally, an equal poise means that it is in group C. We can apply this trichotomy again in the group in which the falsified one is located. If the number of coins is eight (or nine), it is obvious that we can get a definitive answer in two steps. (For an elegant pre-information-theoretical solution of this problem, see Dyson [D-11].) The classical counterfeit coin problem differs from our general problem in three ways: (a) its threefold partition into A, B, and C has the restriction that two out of the three must have the same number of coins (alternatives); (b) in its usual version, at least, the probability of each coin being fake is assumed to be the same; and (c) it is sometimes assumed that the counterfeit coin is known to be of a different weight, but not known to be lighter or heavier than the rest. With regard to (a) we can change the old quiz somewhat to conform with our general problem type by allowing each arm of the balance to be variable and to take any integral multiple of a certain unit length. In this way we can detect the false coin even though the left and right pans hold different numbers of coins. With regard to (b) our problem is more general than the classical quiz, and thus we do not need to be worried. With regard to (c) we can limit ourselves, for simplicity, to the version in which we know whether the fake coin is lighter or heavier than the genuine ones.

Let the set $\mathcal{B}^{(0)} \equiv \mathcal{E} = \{E_1, E_2, \cdots, E_n\}$ be a logical spectrum with n non-$\emptyset$ members. We divide this set $\mathcal{B}^{(0)}$ of n objects into m ($\leq n$) nonempty subsets $\mathcal{B}_\mu^{(1)}$, $\mu = 1, 2, \cdots, m$, in such a way that each member of $\mathcal{B}^{(0)}$ belongs to one and only one of the $\mathcal{B}_\mu^{(1)}$'s:

$$\begin{gathered} \mathcal{B}^{(0)} = \mathcal{B}_1^{(1)} \vee \mathcal{B}_2^{(1)} \vee \cdots \vee \mathcal{B}_m^{(1)}, \\ \mathcal{B}_\mu^{(1)} \wedge \mathcal{B}_\nu^{(1)} = \Phi \quad \text{for} \quad \mu \neq \nu, \\ \mathcal{B}_\mu^{(1)} \neq \Phi, \end{gathered} \tag{1.114}$$

where $\wedge$ and $\vee$ are set-theoretical conjunction and disjunction applied to sets consisting of objects E's. $\mathcal{B}_1 \wedge \mathcal{B}_2$ is the collection of those members common to $\mathcal{B}_1$ and $\mathcal{B}_2$. $\mathcal{B}_1 \vee \mathcal{B}_2$ is the collection of those members that belong to either $\mathcal{B}_1$ or $\mathcal{B}_2$ or both. Φ is an empty set. If (1.114) is the case, we speak of an m-fold partition, or a polychotomy of degree m, or an m-chotomy, of the set $\mathcal{B}^{(0)}$. A polychotomy of degree 1 is called monochotomy and amounts to doing no polychotomy. A polychotomy of degree 2 is a dichotomy. Each of the subsets $\mathcal{B}_{\mu}^{(1)}$ can further be subjected to another polychotomy. For instance, the subset $\mathcal{B}_{\mu}^{(1)}$ can be subdivided into k nonempty sub-subsets $\mathcal{B}_{\mu\nu}^{(2)}$ by a k-chotomy in such a way that

$$\begin{gathered}\mathcal{B}_{\mu}^{(1)} = \mathcal{B}_{\mu 1}^{(2)} \vee \mathcal{B}_{\mu 2}^{(2)} \vee \cdots \vee \mathcal{B}_{\mu k}^{(2)},\\ \mathcal{B}_{\mu\nu}^{(2)} \wedge \mathcal{B}_{\mu\kappa}^{(2)} = \Phi \qquad (\nu \neq \kappa), \qquad \mathcal{B}_{\mu\nu}^{(2)} \neq \Phi.\end{gathered} \tag{1.115}$$

We can continue this process until each individual case E_i is isolated; that is, until the resulting subsets of a polychotomy is no longer divisible into more than one nonempty set. We can thus construct a complete tree of polychotomies, such that its trunk is $\mathcal{B}^{(0)} = \mathcal{E}$ and its peripheral branches are sets with single members E_i's. The intermediate branches are some subsets in $\mathcal{B}^{(0)}$ consisting of more than one E_i. The $\mathcal{B}$'s are very much like logical spectra because they are sets of disjoint propositions, but except for $\mathcal{B}^{(0)}$ they are not logical spectra because their constituent propositions are not exhaustive. The polychotomic tree considered here is sometimes called an "additive" polychotomic tree, in contrast to a "multiplicative" polychotomic tree, which will be considered in the next chapter.

Our interpretation of (1.114) and (1.115) is a *set-theoretical* one, but we can give them also a logical reinterpretation by letting a subset (such as $\mathcal{B}_{\mu\nu}^{(\alpha)}$) stand for a proposition that is the logical disjunction of the constituent E_i's included in the subset. In this reinterpretation the set consisting of a single E_i is understood as E_i itself; $\vee$ and $\wedge$ are understood as $\cup$ and $\cap$, and Φ as $\emptyset$. The entire set $\mathcal{B}^{(0)}$ corresponds to $\square$. This reinterpretation also allows us to attach probabilities to sets $\mathcal{B}_{\mu\nu}^{(\alpha)}$ of E_i's, for in the logical reinterpretation $\mathcal{B}_{\mu\nu}^{(\alpha)}$ is a disjunction of disjoint E_i's and hence its probability must be the sum of the probabilities of these E_i's.

In the illustration of coins and balance, each E_i stands for a proposition that "the falsified coin is the ith coin." The disjunction $E_1 \vee E_2 \vee E_3$, for instance, means set-theoretically just a collection of three propositions, E_1, E_2, and E_3. In the logical reinterpretation it means that "the falsified coin is either coin 1, or coin 2, or coin 3." In the following we no longer distinguish between $\wedge$, $\vee$, Φ and $\cap$, $\cup$, $\emptyset$.

A complete polychotomic tree consists of "branches," corresponding to the $\mathcal{B}$-sets, connected at "branching points" (see Figure 1.1). To each branch $\mathcal{B}$ are attached four (not independent) quantities: (a) probability $p(\mathcal{B})$, (b) polychotomic fraction $q(\mathcal{B})$, (c) rank (in the tree) $r(\mathcal{B})$, and (d) branching cost $c(\mathcal{B})$. The probability $p(B)$ of a $\mathcal{B}$ is defined, as explained above, as the sum of the probabilities of those E_i's included in the $\mathcal{B}$. The polychotomic

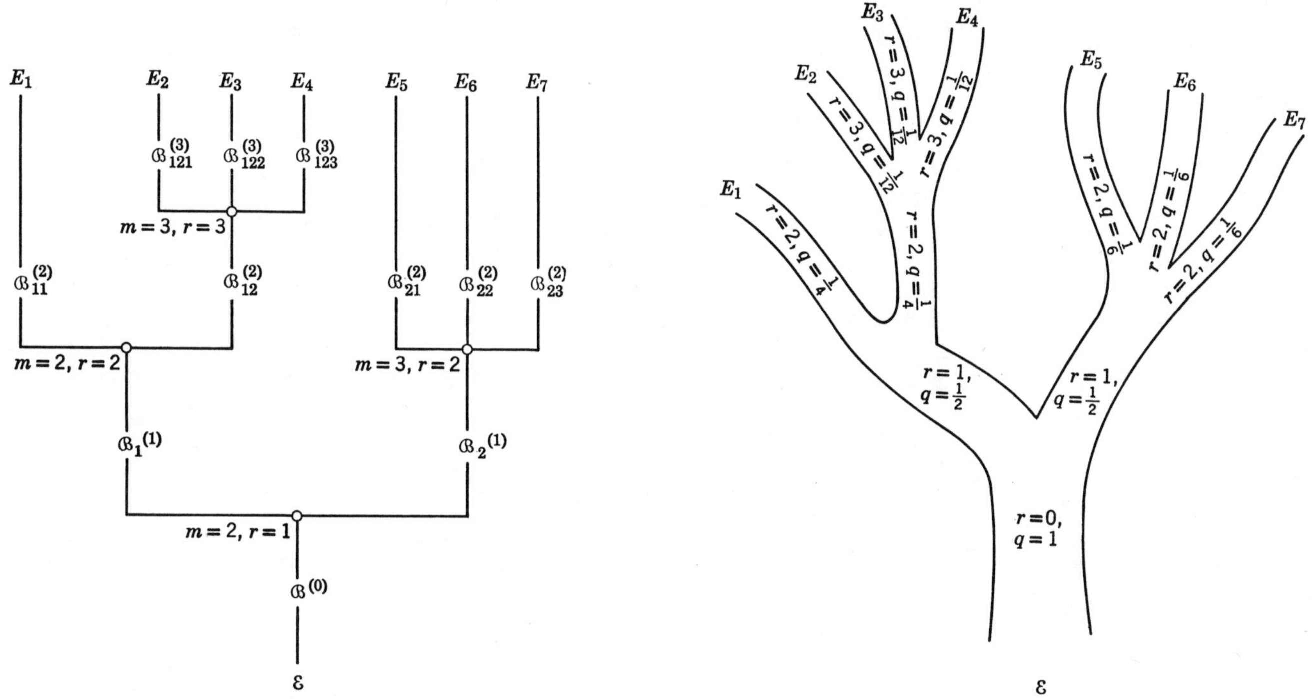

Figure 1.1 The drawing at the left shows a polychotomic tree scheme defined on a set with seven E's. Each branch is given a name according to (1.114) and (1.115) and the degree and rank of each branching point are indicated as defined in the text. The drawing at the right, which represents the same polychotomic tree, shows the polychotomic fraction and the rank of each branch, defined in relation to (1.116). This illustration clarifies the intuitive meaning of Theorem 1.5.

fraction $q(\mathcal{B})$ of a $\mathcal{B}$ is defined as the reciprocal of the product of the degrees of the polychotomies that are found between the branch and the trunk $\mathcal{B}^{(0)}$. Let $m^{(1)}(\mathcal{B})$ be the degree of the first polychotomy applied to $\mathcal{B}^{(0)}$, $m^{(2)}(\mathcal{B})$ be the degree of the second polychotomy applied to that branch of the first polychotomy to which the branch $\mathcal{B}$ is connected, $\cdots$, and $m^{(r)}(\mathcal{B})$ be the degree of the last polychotomy of which $\mathcal{B}$ is one of the direct offshoots. Then

$$q(\mathcal{B}) = \prod_{s=1}^{r} \frac{1}{m^{(s)}(\mathcal{B})}. \tag{1.116}$$

The integer r is the number of branching points one has to go through starting from the trunk $\mathcal{B}^{(0)}$ to reach $\mathcal{B}$. This number is uniquely determined by $\mathcal{B}$ provided that one excludes monochotomies from consideration, although $q(\mathcal{B})$ is not affected by inclusion of monochotomies in (1.116). This number $r(\mathcal{B})$ is called the rank of the branch $\mathcal{B}$ in the tree. The branching cost $c(\mathcal{B})$ of $\mathcal{B}$ is defined by

$$c(\mathcal{B}) = -\log q(\mathcal{B}) = \sum_{s=1}^{r} \log m^{(s)}(\mathcal{B}). \tag{1.117}$$

To each branching point in the tree are attached three numbers: degree, cost, and rank. The degree of a branching point is nothing but the degree of the corresponding polychotomy. If m is the degree of a branching point, $\log m$ is called its cost. Equation (1.117) shows that the branching cost of a branch is the sum of the costs of the branching points that lie between $\mathcal{B}^{(0)}$ and $\mathcal{B}$. The rank r of a branching point is equal to the rank of the branches that are its direct offshoots. Hence, counting from $\mathcal{B}^{(0)}$ toward the periphery, the rth branching point has rank number r.

It should be noted that the quantities $q(\mathcal{B})$, $r(\mathcal{B})$, and $c(\mathcal{B})$ characterizing a branch $\mathcal{B}$ are all determined, once the scheme of a polychotomic tree is given, and are independent of the probabilities attached to the E_i. A simple consequence of the definition of polychotomic fraction is the following theorem.

Theorem 1.5 Let $q(E_i)$ be the polychotomic fraction, defined in (1.116), of the peripheral branch consisting of E_i in a complete polychotomic tree constructed in a set $\mathcal{E} = \{E_i\}$, $i = 1, 2, \cdots, n$. Then we have

$$\sum_{i=1}^{n} q(E_i) = 1. \tag{1.118}$$

Let us picture the complete polychotomic tree as a real tree, as in Figure 1.1, and assign a thickness to each branch in such a way that, at each branching point representing an m-chotomy, a branch of thickness θ is divided into m branches of thickness θ/m. If the thickness of the original trunk is used as the unit, the thickness θ of a branch $\mathcal{B}$ is nothing but its polychotomic fraction. Equation (1.118) asserts the evident fact that the total sum of the thickness of the peripheral branch is exactly the same as the thickness of the trunk. This explanation makes it clear that we have in general

$$\sum q(\mathcal{B}) = 1, \tag{1.119}$$

where the summation extends over a collection of branches that would simultaneously become peripheral branches if we "pruned" the tree in any arbitrary fashion.

Let us introduce at this stage a usage of words (justified later in applications) that consists of saying that we spend cost $\log m$ everytime we perform a polychotomic observation of degree m in our effort to identify the particular E_i that is true. The content of (1.117) can be paraphrased as the total cost we spend in successive polychotomies that end up with E_i being equal to the negative of the logarithm of the polychotomic fraction of E_i. Now we are prepared to introduce the next theorem.

Theorem 1.6 If an observation, aiming at identification of the "true" case E_i in a set of possibilities $\mathcal{E} = \{E_i\}$ associated with a certain probability distribution, is carried out by a tree of successive polychotomic tests, then the information obtained by the observation can never be larger than the average cost of the E_i's in this polychotomic tree.

Proof. By Gibbs' theorem, we have, writing q_i for $q(E_i)$,

$$\langle c(E_i)\rangle_i = -\sum_{i=1}^{n} p_i \log q_i \geq -\sum_{i=1}^{n} p_i \log p_i = S_I, \tag{1.120}$$

where the left side is the average of the cost we spend in identifying the true case as an E_i and the right side is the original ignorance, which is destroyed by identification of the true case. The application of Gibbs' theorem to the q's is justified by Theorem 1.5.

Corollary 1.2 In an m-bounded polychotomic tree defined on a set $\mathcal{E} = \{E_i\}$ of E_i's with a certain probability distribution, the average rank of the E_i's is larger than or equal to $S_I/\log m$, where S_I is the simple entropy defined by the probabilities attached to the E_i's.

Proof. By an "m-bounded" polychotomic tree is meant one in which none of the polychotomies has degree larger than m. From (1.117) we get

$$c(E_i) \leq r(E_i) \log m, \tag{1.121}$$

which yields, in combination with (1.120),

$$\langle r(E_i)\rangle_i \geq \frac{\langle c(E_i)\rangle_i}{\log m} \geq \frac{S_I}{\log m}. \tag{1.122}$$

When $\mathcal{E} = \{E_i\}$ with some probability distribution is given, the ideal polychotomic tree would mean one with the smallest average cost $\langle c(E_i)\rangle_i$. If the values of the p_i's and the value of n are such that we can choose the q_i's so as to satisfy $q_i = p_i$ for all i, then we know, by applying Gibbs' criterion to (1.120), that a polychotomic tree giving these values of q_i would be an ideal polychotomic tree. The question of finding an ideal polychotomic tree in the case when it is impossible to satisfy $q_i = p_i$ for all i is a topic of the next section. For the present we can vaguely say that it is desirable to match the q's with the given p's as closely as possible.

It is important to note that the statement of Theorem 1.6 refers to the *average* cost. In an individual case the actual cost may be smaller than the information obtained. The reader can check this in the following example. Suppose that all the probabilities $p(E_i)$ are equal to $1/n$ with $n > 2$, that the first polychotomy is a dichotomy dividing the $\mathcal{E}$ into E_1 and all the rest, and that a second polychotomy of degree $(n - 1)$ is applied to this latter group. If E_1 is the "true" one (which happens seldom), the cost (log 2) is smaller than the information (log n). If one of the remaining E_i's is true (which happens often), the opposite is the case.

It is suggested that the reader apply conclusions of Theorem 1.6 and Corollary 1.1 to our illustration of falsified coins and balance, assuming that all the probabilities for eight (or nine) eventualities are all equal. This will convince him that a two-stage test is the shortest. Note: $S_I = \log 8$ and $\log m = \log 3$ for the case of eight coins, and $S_I = \log 9$ and $\log m = \log 3$ for the case of nine coins. He can compare two polychotomic tree schemes, in one of which two stages are necessary and sufficient to reach a conclusion in every eventuality, and in the other of which sometimes one stage is enough but on the average more than two stages are required.

In connection with (1.120) we compared only the initial ignorance and the final ignorance, although we were discussing the case in which the test is not a one-stroke affair but a step-by-step polychotomic determination. Let us now consider the information gain at each stage of polychotomy. Starting from the probability distribution p_i referring to $\mathcal{B}^{(0)}$ (i.e., $\mathcal{E}$), let us assume that we have first located the object in a subset $\mathcal{B}_\mu^{(1)}$, where μ is one particular value out of m possibilities. Then the probability distribution must be revised so that the probabilities for those E's outside $\mathcal{B}_\mu^{(1)}$ become zero. But within $\mathcal{B}_\mu^{(1)}$ the *relative* probabilities of the E's must remain the same as before. This means that the revised probabilities $p_i^{(1)}$ must be given by

$$\begin{aligned} p_i^{(1)} &= 0 \qquad \text{for} \quad E_i \notin \mathcal{B}_\mu^{(1)} \\ &= \frac{p_i^{(0)}}{p_\mu} \qquad \text{for} \quad E_i \in \mathcal{B}_\mu^{(1)}, \end{aligned} \tag{1.123}$$

where

$$p_\mu = \sum_{i\in\mu} p_i^{(0)}, \qquad \sum_{\mu=1}^{m} p_\mu = 1. \tag{1.124}$$

Here $p_i^{(0)}$ is the original $p_i = p(E_i)$. p_μ is the probability (in the state of knowledge represented by $p_i^{(0)}$) of finding the object system in $\mathcal{B}_\mu^{(1)}$. The ignorance after having located the object in $\mathcal{B}_\mu^{(1)}$ then is

$$\begin{aligned} S_\mu^{(1)} &= -\sum_{i=1}^{n} p_i^{(1)} \log p_i^{(1)} \\ &= -\sum_{i\in\mu} \frac{p^{(0)}}{p_\mu} \log \frac{p_i^{(0)}}{p_\mu}. \end{aligned} \tag{1.125}$$

The information gain is

$$I_\mu^{(1)} = S^{(0)} - S_\mu^{(1)}, \tag{1.126}$$

where $S^{(0)}$ is the same as S_I in (1.111), with $p_i^{(0)} = p_i$.

It may be noted that this $I_\mu^{(1)}$ is not necessarily non-negative. The reader can check the case in which $p_1^{(0)}$ is very close to unity and the small difference $(1 - p_1^{(0)})$ is evenly distributed among the remaining $(n - 1)$ eventualities, and in which a first dichotomic observation determines that the true one is not E_1; that is, it is among these $(n - 1)$ cases. There is a certain way of avoiding negative information (other than taking an average) and yet arriving at the same conclusion (see Appendix 1.1).

On the other hand, the average of $I_\mu^{(1)}$ is non-negative. Since there is probability p_μ of reaching the value of ignorances $S^{(1)}$, the average ignorance after the first polychotomic test is

$$\langle S_\mu^{(1)} \rangle_\mu = \sum_{\mu=1}^{m} p_\mu S_\mu^{(1)} = S^{(0)} + \sum_{\mu=1}^{m} p_\mu \log p_\mu. \tag{1.127}$$

Thus the average information obtained in the first stage is

$$\langle I_\mu^{(1)} \rangle_\mu = S^{(0)} - \langle S_\mu^{(1)} \rangle_\mu = -\sum_{\mu=1}^{m} p_\mu \log p_\mu, \tag{1.128}$$

which is non-negative.

The relation (1.127) can be considered as a special case of (1.67) or (1.71) that reads

$$S(\mathcal{E} \mid \mathcal{F}) = S(\mathcal{E} \otimes \mathcal{F}) - S(\mathcal{F}), \tag{1.129}$$

where $\mathcal{E}$ is the present $\mathcal{E}$, which is $\{E_i\}$ with probability p_i, and $\mathcal{F}$ is to be interpreted as $\mathcal{F} = \{\mathcal{B}_1^{(1)}, \mathcal{B}_2^{(1)}, \cdots, \mathcal{B}_m^{(1)}\}$ with probability p_μ. The spectrum $\mathcal{E} \otimes \mathcal{F}$ consists formally of nm elements, but all but n elements are $\emptyset$, having probability 0. In other words, $\mathcal{E} \otimes \mathcal{F}$ is no different from $\mathcal{E}$ itself. The probability $p(E_i \mid F_j)$ can be written $p(E_i \mid \mathcal{B}_\mu^{(1)})$, which is given by (1.123).

Thus we have $S(\mathcal{E} \mid \mathcal{F}) = \langle S^{(1)} \rangle$, $S(\mathcal{E} \otimes \mathcal{F}) = S^{(0)}$, and $S(\mathcal{F}) = -\sum_{\mu=1}^{m} p_\mu \log p_\mu$, which, when substituted in (1.129), yield (1.127).

Now, noting that the maximum value of $-\sum_{\mu=1}^{m} p_\mu \log p_\mu$ is $\log m$, we obtain

$$\langle I^{(1)} \rangle \leq \log m^{(1)}. \tag{1.130}$$

The superscript (1) is attached to m to indicate that it refers to the first-stage polychotomy. In words, (1.130) means Theorem 1.7 (at least for the first polychotomy).

Theorem 1.7 The maximum average information that can be obtained in an m-chotomy is its cost, $\log m$.

The restriction to the first polychotomy will be removed presently. This is a more general theorem than the theorem according to which the maximum of $-\sum_{i=1}^{n} p_i \log p_i$ is $\log n$ [see (1.22)]. The present theorem reduces to the latter if each $\mathcal{B}_\mu^{(1)}$ consists of only one E and $S^{(1)} = 0$. Note that in this special case the information obtained is always the same, as a result of which the qualification "average" is not necessary.

After we have located the object in $\mathcal{B}_\mu^{(1)}$, we further apply another polychotomy of degree, say, $m_\mu^{(2)}$, so that $\mathcal{B}_\mu^{(1)}$ is again subdivided into subsets $\mathcal{B}_{\mu 1}^{(2)}, \mathcal{B}_{\mu 2}^{(2)}, \cdots, \mathcal{B}_{\mu k}^{(2)}$, where k stands for $m_\mu^{(2)}$. The probability distribution will be further revised according to the outcome of this $m_\mu^{(2)}$-chotomy. For instance, if the object is located in $\mathcal{B}_{\mu\nu}^{(1)}$, the new probability distribution $p_i^{(2)}$ will be

$$\begin{aligned} p_i^{(2)} &= 0 \qquad \text{for} \quad E_i \notin \mathcal{B}_{\mu\nu}^{(2)} \\ &= \frac{p_i^{(1)}}{p_{\mu\nu}} \quad \text{for} \quad E_i \in \mathcal{B}_{\mu\nu}^{(1)}, \end{aligned} \tag{1.131}$$

with

$$p_{\mu\nu} = \sum_{i \in \mu\nu} p_i^{(1)}, \qquad \sum_{\nu=1}^{k} p_{\mu\nu} = 1. \tag{1.132}$$

The mathematical structure is exactly the same as in the first-stage polychotomy, whence all the results referring to the first-stage polychotomy apply here too, including (1.130), where $m^{(1)}$ should be replaced by $m_\mu^{(2)}$. The ignorance after the object has been located in $\mathcal{B}_{\mu\nu}^{(2)}$ is

$$S_{\mu\nu}^{(2)} = -\sum_{i=1}^{n} p_i^{(2)} \log p_i^{(2)}, \tag{1.133}$$

with $p_i^{(2)}$ given in (1.131), and the information obtained by the second-stage polychotomy is, corresponding to (1.126),

$$I_{\mu\nu}^{(2)} = S_\mu^{(1)} - S_{\mu\nu}^{(2)}. \tag{1.134}$$

If we average over ν we get

$$\langle I_{\mu\nu}^{(2)}\rangle_\nu = -\sum_{\nu=1}^{m_\mu^{(2)}} p_{\mu\nu} \log p_{\mu\nu} \leq \log m_\mu^{(2)}, \tag{1.135}$$

confirming Theorem 1.7 for the second-stage polychotomy. It is a matter of simple rephrasing to make the proof of Theorem 1.7 valid for any stage number.

Consider now a particular case of successive testing. We first locate the object in a particular group μ and then in a particular group $\mu\nu$, and so on, until finally we locate it in a particular E_i. Thus the path we follow in this search can be characterized by a sequence of the type $\{\mu, \mu\nu, \mu\nu\kappa, \cdots, \mu\nu\kappa\lambda\cdots\omega\}$, where the last label $\mu\nu\kappa\lambda\cdots\omega$ specifies the particular E_i. The amount of ignorance goes through a sequence $\{S^{(0)}, S_\mu^{(1)}, S_{\mu\nu}^{(2)}, \cdots, S_{\mu\nu\kappa\lambda\cdots\omega}^{(r)}\}$, where the last one $S_{\nu\kappa\cdots\omega}^{(r)}$, must be zero because we have located the object in a single E_i. The superscript (r) stands for $r(E_i)$. In terms of the information obtained at each stage, this means that

$$\begin{aligned} I_\mu^{(1)} &+ I_{\mu\nu}^{(2)} + \cdots + I_{\mu\nu\kappa\cdots\omega}^{(r)} \\ &= (S^{(0)} - S_\mu^{(1)}) + (S_\mu^{(1)} - S_{\mu\nu}^{(1)}) + \cdots + (S_{\mu\nu\kappa\cdots\rho}^{(r-1)} - S_{\mu\nu\kappa\cdots\rho\omega}^{(r)}) \\ &= S^{(0)}. \end{aligned} \tag{1.136}$$

Although some of the I's involved here may be negative, the total sum of information is non-negative and is equal to the origin ignorance, as it should be. This shows also that each $I_{\mu\nu\kappa\cdots}^{(k)}$ may differ from one case to another depending on the final destination E_i, but the sum of the I's is always the same. Now let us take the average of $I_{\mu\nu\kappa\cdots}^{(k)}$ on different paths indicated by $(\mu\nu\kappa\cdots)$ leading to different E's. The $r(E_i)$ differs from one E_i to another, but we can make it equal for all i by artificially including monochotomies that do not contribute anything to the information or to the cost. First, the average of $I^{(1)}$ is given by $\langle I^{(1)}\rangle$ of (1.128) and was found in (1.130) not to be larger than $c^{(1)} = \log m^{(1)}$. At the second stage, we found the result (1.135). A further averaging over μ would give

$$\langle I_{\mu\nu}^{(2)}\rangle_{\mu\nu} = \sum_{\mu=1}^{m} p_\mu \langle I_{\mu\nu}^{(2)}\rangle_\nu \leq \sum_{\mu=1}^{m} p_\mu \log m_\mu^{(2)} \tag{1.137}$$

or simply

$$\langle I^{(2)}\rangle \leq \langle c^{(2)}\rangle, \tag{1.138}$$

where $\langle I^{(2)}\rangle$ is the overall average information obtained at the second stage and $\langle c^{(2)}\rangle$ is the average cost at the second stage. We can obtain a similar result for the third, fourth, $\cdots$ stages. Adding them together gives

$$\langle I^{(1)}\rangle + \langle I^{(2)}\rangle + \cdots + \langle I^{(r)}\rangle \leq c^{(1)} + \langle c^{(2)}\rangle + \cdots + \langle c^{(r)}\rangle. \tag{1.139}$$

But according to (1.136) the left-hand side of (1.139) should be equal to $S^{(0)}$. Therefore

$$S^{(0)} \leq \langle c \rangle, \tag{1.140}$$

where $\langle c \rangle$ is the total average cost in the successive polychotomic search. Thus we come back to the same conclusion as Theorem 1.6.

1.5. OPTIMAL *m*-BOUNDED POLYCHOTOMIC TREE AND "NOISELESS" CODING

In the last section we introduced the concept of an additive polychotomic tree, which has a large number of applications. What is done in this section may be summarized as introducing a measure of "goodness" of a polychotomic tree and giving an algorithm for finding the optimal polychotomic tree that maximizes this measure of "goodness."

We are given a logical spectrum $\mathcal{E} = \{E_i\}$, $i = 1, 2, \cdots, n$, with a probability distribution $\{p_i\}$. Consider a complete m-bounded polychotomic tree that gives rank r_i to proposition E_i, excluding monochotomies. The term "m-bounded" means that none of the branching points (polychotomies) has degree larger than m. For a given probability distribution $\{p_i\}$ and a given value of m, a complete m-bounded polychotomic tree that minimizes the average rank

$$\langle r_i \rangle_i = \sum_{i=1}^{n} p_i r_i \tag{1.141}$$

is called an optimal m-bounded polychotomic tree. We see examples later in which the average rank is the quantity it is desirable to reduce as much as possible. In a search process the average rank is the average number of steps required to locate the desired object. The prescription for constructing an optimal m-bounded polychotomic tree was discovered first by Huffman [H-15] and later, but independently, by Zimmerman [Z-1], the former treating the problem as one of "noiseless" coding and the latter treating it as one of search procedure. Before Huffman, Shannon [S-6] and Fano [F-1] discussed allied problems in communication and prepared the ground for Huffman.

To describe a polychotomic tree scheme we can adopt either one of two opposite views: namely, either we look at the tree, as we did in the last section, as successive divisions, or, inversely, we look at it as successive mergers. The latter is more convenient for the purpose of this section. In fact, any branching point can be considered as a place where one branch is divided up into several offshoots, or as a place where several branches merge to form a single stalk. Suppose that at one of the last-stage branching points (i.e., those branching points whose direct offshoots are E_i's) a branch $\mathcal{B}$ is divided into k E's, say, $E_1, E_2, \cdots$, and E_k. Consider a new logical spectrum $\mathcal{G}$, which consists of $(n - k + 1)$ members, $E_1 \cup E_2 \cup \cdots \cup E_k$ and $E_{k+1}, E_{k+2}, \cdots, E_n$, for which the probability distribution is given by $\sum_{i=1}^{k} p(E_i)$ and $p(E_{k+1}), \cdots, p(E_n)$. If we "prune" the tree for $\mathcal{E} = \{E_i\}$ in such a way that the above-mentioned branching point is eliminated, then the remaining part of the tree is a polychotomic tree for $\mathcal{G}$. The passage from the tree for $\mathcal{G}$ to the tree for $\mathcal{E}$ corresponds to a division of a branch, whereas the passage from $\mathcal{E}$ to $\mathcal{G}$ corresponds to a "partial merger" of some members of $\mathcal{E}$.

We can successively apply partial mergers to a tree for $\mathcal{E}$ until finally the resultant merger becomes $\square = E_1 \cup E_2 \cup \cdots \cup E_n$. If we specify a complete sequence of such partial mergers, then a polychotomic tree is uniquely specified. On the other hand, we have to note that for a given polychotomic tree there are usually many complete sequences of partial mergers, for in any tree there are in general more than one last-stage branching point, thus providing arbitrariness as to which one of them should be taken as the first partial merger. Furthermore, there may also be more than one tree that minimizes the average rank $\langle r_i \rangle_i$. The strategy we are looking for is one that allows us to draw up one of the optimal sequences, where an optimal sequence means one that determines a polychotomic tree such that no other tree has a smaller average rank.

We mention first some of the simple facts about an optimal polychotomic tree and an optimal sequence of partial mergers.

Lemma 1.1 In an m-bounded optimal polychotomic tree, if $p_i > p_j$, $r_i \leq r_j$.

Proof. If $r_i > r_j$, we could decrease $\langle r_i \rangle_i$ by interchanging the positions of E_i and E_j in the tree.

Lemma 1.2 In any polychotomic tree there are at least two E_i's whose ranks r_i are equal to $R = \max_i r_i$, provided that $n \geq 2$.

Proof. The case in which there is none is excluded by the definition of $\max_i r_i$. The case in which there is only one is excluded because it would mean that the last branching for this E_i is a monochotomy, which is excluded from our consideration.

Lemma 1.3 In an optimal m-bounded polychotomic tree the rank of an unsaturated branching point, if there is any, is equal to $R = \max_i r_i$. An unsaturated branching point is one whose degree is less than its upper bound m.

Proof. The rank of a branching point is by definition equal to the rank of its direct offshoots. The lemma says that no saturated (degree m) branching points can be removed further than unsaturated branching points from the stem of the tree. This is so because, if an unsaturated branching point had a rank less than some saturated branching point, we could reduce $\langle r_i \rangle_i$ by taking one of the offshoots from the latter branching part and attach it to the former.

The reader can conclude from Lemma 1.3 that a tree of the type shown in Figure 1.1, considered as an m-bounded tree with $m = 3$, can never be an optimal tree.

Lemma 1.4 For a given $\mathcal{E}$ with $\{p_i\}$ there is an optimal m-bounded polychotomic tree in which there is not more than one unsaturated branching point.

Proof. By Lemma 1.3, all the unsaturated branching points have the same rank R. Hence we can move an E_i from one unsaturated branching point to another without changing the value of $\langle r_i \rangle_i$. Thus we can fill up unsaturated branching points one by one, leaving at most one partially filled branching point and possibly some monochotomies.

Lemma 1.5 An optimal m-bound polychotomic tree of the type described in Lemma 1.4 has no unsaturated branching point if $(n - 1)$ is divisible by $(m - 1)$. If not, it will have an unsaturated branching point of degree μ, which is determined by

$$\mu = n - (a - 1)(m - 1) \tag{1.142}$$

and

$$2 \leq \mu \leq m - 1, \tag{1.143}$$

where a is some positive integer.

Proof. Suppose first that the tree has no unsaturated branching points. Then each partial merger will reduce the number of peripheral branches by $m - 1$. We should be able to repeat this process until only one branch (i.e., the stem) is left. Hence, if a is number of branching points, $1 = n - a(m - 1)$. Suppose next that the tree has one unsaturated branching point of degree μ. Because it has rank R, we can start our merging process with it. This reduces the number of peripheral branches by $\mu - 1$. After this first partial merger the situation becomes the same as in the first case. Hence $1 = n - (\mu - 1) - (a - 1)(m - 1)$, which is (1.142). The rest of the proof consists of showing that the divisible case ($\mu = 1$) and $(m - 2)$ other cases ($\mu = 2, 3, \cdots, m - 1$) involved in (1.142) and (1.143) constitute a set of mutually exclusive and exhaustive cases. The easiest way to see this is to rewrite the above criterion in the form

$$\mu - 1 = n - 1 \mod m - 1 \tag{1.144}$$

and to interpret the case $\mu = 1$ as meaning the absence of any unsaturated branching point; see Lemma 1.2.

It may be noted that if $m = 2$ there can be no unsaturated branching point. Lemma 1.3, combined with the result of Lemma 1.5, tells us that the number of E_i's that have the largest rank (R) in an optimal tree is $bm + \mu$ with some non-negative integer b.

Lemma 1.6 Let one of the outermost branching points in an optimal polychotomic tree for $\mathcal{E}$ be characterized by a partial merger $\mathcal{E} \to \mathcal{G}$. Then the new tree obtained by eliminating this branching point from the first tree is an optimal tree for $\mathcal{G}$.

Proof. It has already been established that the new tree is a possible polychotomic tree for the logical spectrum $\mathcal{G}$. If $\mathcal{G}$ is obtained from $\mathcal{E}$ by merging k E_i's, say, $E_1, E_2, \cdots, E_k$, then the average rank in $\mathcal{G}$, $\langle r \rangle_\mathcal{G}$, and that in $\mathcal{E}$, $\langle r \rangle_\mathcal{E}$, are connected by

$$\langle r \rangle_\mathcal{E} = \langle r \rangle_\mathcal{G} + \sum_{i=1}^{k} p_i. \tag{1.145}$$

This is because the difference between these two lies in the fact that $\mathcal{G}$ has a peripheral branch $E_1 \cup E_2 \cup \cdots \cup E_k$ whose rank is less than the rank of each of $E_1, E_2, \cdots, E_k$ in $\mathcal{E}$ by 1, and the probability for $E_1 \cup E_2 \cup \cdots \cup E_k$ is equal to the sum of the probabilities of $E_1, E_2, \cdots, E_k$. When $\mathcal{E} = \{E_i\}$ with $\{p_i\}$ is given, and if it is decided which E_i's are to be merged to form $\mathcal{G}$, then $\sum_{i=1}^{k} p_i$ is determined. Hence a tree that minimizes $\langle r \rangle_\mathcal{E}$ will also minimize $\langle r \rangle_\mathcal{G}$.

We are now prepared to introduce the algorithm for building an optimal polychotomic tree. Because of Lemma 1.6 it will suffice for us to find a general algorithm for one single partial merger, since if the algorithm is so formulated as to apply to any arbitrary optimal tree we can reiterate the same algorithm for the result of a previous partial merger. Therefore our task will be accomplished if we succeed in characterizing in concrete terms a type of last-stage branching point that has to exist in an optimal tree. In fact, if we know the way to find such a branching point, we can take this for the first partial merger $\mathcal{E} \to \mathcal{G}$.

Lemma 1.7 In the case when $(n - 1)$ is not divisible by $(m - 1)$ (see Lemma 1.5) there is an optimal m-bounded polychotomic tree that has an unsaturated branching point of rank $R = \max_i r_i$, at which μ least probable E_i's are merged.

Proof. We have already seen that the number of unsaturated branching points can be reduced to one, which has rank R, and that the number of E_i's merged at this point is μ given by (1.142) and (1.143).

An optimal tree may or may not have branching points of rank R outside the unsaturated branching point. Consider first the case in which there is only the unsaturated branching point with rank R and all the remaining branching points have rank less than R. Lemma 1.1 then tells us that any of the E_i's merged in the unsaturated branching point must have a probability no larger than those E_i's merged at saturated branching points. This practically determines the first merger; that is, μ least probable E_i's should be merged at the unsaturated branching point. The only possible ambiguity stems from the E_i's with equal probabilities, but an arbitrary choice in this regard does not affect the value of $\langle r_i \rangle_i$ in any way. Next suppose that there are b saturated branching points of rank R besides the unsaturated branching points. Then Lemma 1.1 tells us that $bm + \mu$ least probable E_i's should have rank R. Furthermore, we know that since they have all the same rank, any interchange of their positions among themselves does not affect the value of $\langle r_i \rangle_i$. Hence one possibility is certainly that μ least probable E_i's are merged at the unsaturated branching point.

Lemma 1.8 In the case in which $(n - 1)$ is divisible by $(m - 1)$ there is an optimal m-bounded polychotomic tree such that at one of the branching points of the maximum range R, m least probable E_i's are merged.

Proof. The proof is the same as the second half of the proof of Lemma 1.7.

This leads to the following final result.

Theorem 1.8 Let us be given a logical spectrum $\mathcal{E} = \{E_i\}, i = 1, 2, \cdots, n$, with a probability distribution $\{p_i\}$. An optimal m-bounded polychotomic tree can be built by reiterated application of the following algorithm of partial merger. Step 1 is needed only for the first partial merger.

Step 1. Determine whether by (1.142) and (1.143) there is an unsaturated branching point and, if so, find the values of μ.

Step 2. If there is an unsaturated branching point of degree μ, take μ least probable E_i's and merge them. If there is not, take m least probable E_i's and merge them. A group of $\mu(m)$ least probable E_i's is so defined that each member of the group is not more probable than any nonmember of the group.

The branch which results from this partial merger is to be treated as if it were an E at the next stage, and its probability is to be equated to the sum of the probabilities of the merged branches.

The reader is advised to build an optimal tree for an $\mathcal{E}$ with $n = 8$, assuming one time that the probabilities of the eight alternatives are all equal to $\frac{1}{8}$, and another time that they are 0.4, 0.3, 0.17, 0.031, 0.03, 0.029, 0.021, and 0.019. Also calculate the average rank in each case. Note that the optimum tree thus obtained in the second case happens to violate the auxiliary constraints that exist in the original coin problem (see Figure 1.2).

The algorithm given in Theorem 1.8 makes it possible for us to build an m-bounded polychotomic tree that minimizes $\langle r_i \rangle_i$ when $\mathcal{E} = \{E_i\}$ and $\{p_i\}$, $i = 1, 2, \cdots, n$, are given. We know from Corollary 1.2 that

$$\langle r_i \rangle_i \geq \frac{S_I}{\log m}. \tag{1.146}$$

But there is no simple formula that expresses $\min_{\text{tree}} \langle r_i \rangle_i$ in terms of the given p_i. Instead we can give an upper bound to $\min_{\text{tree}} \langle r_i \rangle_i$, which does not help much as the estimate of $\min_{\text{tree}} \langle r_i \rangle_i$, but has a simple and useful expression.

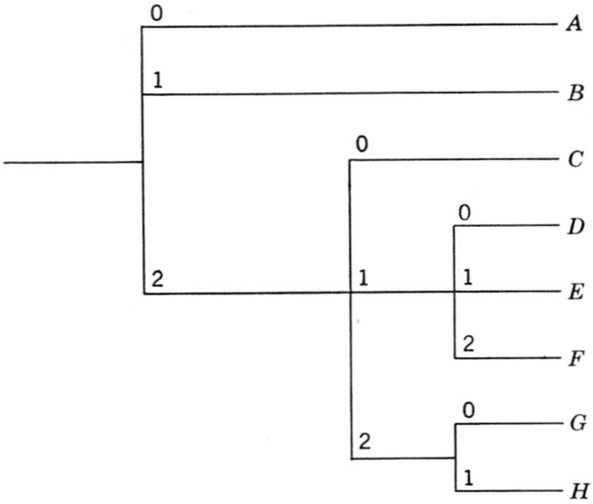

Figure 1.2 The problem concerns the optimal ternary-number representation of eight letters, A, B, C, D, E, F, G, and H, whose probabilities are assumed to be $P_A = 0.4$, $p_B = 0.3$, $p_C = 0.17$, $p_D = 0.031$, $p_E = 0.03$, $p_F = 0.029$, $p_G = 0.021$, and $p_H = 0.019$. Thus $n = 8$, $m = 3$. Equation (1.144) gives $\mu = 2$. Hence we first have to bunch together the two least probable letters, G and H to produce a new branch $p_{G \cup H} = 0.04$. Among the seven available branches the three least probable are D, E, and F, each of which is less probable then $G \cup H$. This provides a new branch $D \cup E \cup F$, with probability $p_{D \cup E \cup F} = 0.09$. The next-stage merger will unite the the three least probable branches C, $D \cup E \cup F$, and $G \cup H$, leaving three branches, A, B, and $C \cup D \cup E \cup F \cup G \cup H$, to be united in the last-stage merger. The resulting code is $A = 0$, $B = 1$, $C = 20$, $D = 210$, $E = 211$, $F = 212$, $G = 220$, and $H = 221$. The average length is $\langle r_i \rangle_i = 1.43$ and $S_I/\log m = 1.37$.

Theorem 1.9 For a given logical spectrum $\mathcal{E} = \{E_i\}$ with $\{p_i\}$, $i = 1, 2, \cdots, n$, there exists an m-bounded polychotomic tree such that its average rank is not larger than $(S_I/\log m) + 1$.

$$\frac{S_I}{\log m} + 1 \geq \langle r_i \rangle_i. \tag{1.147}$$

Proof. Let n_ρ, $\rho = 1, 2, \cdots, \nu$, be the number of those E_i's whose r_i are equal to ρ. $\nu = \max r_i$. Obviously we have $\sum_{\rho=1}^{\nu} n_\rho = n$. Let us consider the condition about n_ρ, under which we can certainly construct an m-bounded polychotomic tree. If we have

$$m \geq n_1 \tag{1.148}$$

we can connect the n_1 E_i's directly to the first polychotomy whose degree is m (or less). After we have used n_1 branches to accommodate those E_i's whose $r_i = 1$, we have $m - n_1$ remaining branches available, each of which can be divided into m or less branches by a second polychotomy. This makes at most $(m - n_1)m$ branches available for those E_i's whose $r_i = 2$. Hence if

$$m^2 - n_1 m \geq n_2, \tag{1.149}$$

we can accommodate $n_2 E_i$'s whose $r_i = 2$. Note that the condition $n_2 \geq 0$, which is always understood, combined with (1.149), incorporates (1.148). We repeat this process and obtain

$$m^\nu - n_1 m^{\nu-1} - n_2 m^{\nu-2} - \cdots - n_{\nu-1} m \geq n_\nu \tag{1.150}$$

as the sufficient condition for existence of a polychotomic tree, with the understanding that all the n_ρ are non-negative. The condition (1.150) can be rewritten as

$$1 \geq \sum_{\rho=1}^{\nu} n_\rho \left(\frac{1}{m}\right)^\rho, \tag{1.151}$$

or, in terms of ranks,

$$1 \geq \sum_{i=1}^{n} \left(\frac{1}{m}\right)^{r_i}. \tag{1.152}$$

It should be noted that if the equality sign in (1.150) does not hold, a certain number of the last-stage branches will remain empty. This may in some cases reduce the actual rank of certain E_i's, unless we count monochotomy as a polychotomy. This shows only the feasibility of an m-bounded tree for which some r_i are smaller than determined by the n. Hence, for the purpose of the proof of Theorem 1.9, we can proceed with (1.152).

The condition (1.152), as a necessary condition for existence of a polychotomic tree, is obvious in the light of Theorem 1.5, since

$$\left(\frac{1}{m}\right)^{r_i} \leq q_i \tag{1.153}$$

because of the m-boundedness and (1.116). What is new here is that (1.152) is also a sufficient condition. The criterion (1.152) considered in communication problems is attributed to L. K. Kraft (see [K-14, F-1]). The r_i are quantities we should like to make as small as possible, while (1.152) tells us that if we make the r_i sufficiently large, we can build a polychotomic tree. In the absence of unsaturated branching points, equality will hold in (1.153), since $r_i \log m = -\log q_i$, and minimization of the average range $\langle r_i \rangle_i$ and minimization of the average cost $\langle c(E_i) \rangle_i$ will become equivalent. This latter, according to (1.120), would be realized if we could take $p_i = q_i$ for all i; that is, if we could take

$$r_i = -\frac{\log p_i}{\log m}. \tag{1.154}$$

But this is generally impossible because the right-hand side would not give an integer. Hence one of the possible strategies will consist of taking r_i as an integer defined by

$$-\frac{\log p_i}{\log m} + 1 > r_i \geq -\frac{\log p_i}{\log m} \tag{1.155}$$

or equivalently

$$\frac{p_i}{m} < \left(\frac{1}{m}\right)^{r_i} \leq p_i. \tag{1.156}$$

This choice of r_i does not necessarily lead to the optimal polychotomic tree, but it has the advantage of being defined by a simple formula. The reader can try for exercise the case in which $n = 10$, $p_i = \frac{1}{10}$, and $m = 2$. The strategy (1.156) gives $n_4 = 10$, $\langle r_i \rangle_i = 4$. The tree built according to the algorithm explained in connection with (1.150) results in $n_2 = 2$, $n_4 = 8$, $\langle r_i \rangle_i = 3.6$ because of the empty branches that occur at $\nu = 4$. The true optimal tree has $n_3 = 6$, $n_4 = 4$, and $\langle r_i \rangle_i = 3.4$.

The values of r_i given by (1.156) satisfy the condition (1.152); hence there exists a corresponding polychotomy. Taking the average of (1.155), we obtain

$$S_I + \log m > \langle r_i \rangle_i \log m \geq S_I, \tag{1.157}$$

which completes the proof of Theorem 1.9. The second half of (1.157) is a necessary condition, as we have already seen in (1.122), and the left half of (1.157) means that within this necessary condition there exists actually a tree satisfying the condition (1.147). Hence the optimal tree will lie between these two limits.

The problem of m-bounded polychotomic trees is known among engineers as that of "noiseless" coding, and we now show briefly how this latter can be considered as a special application of the former. A message written in a language is a string of elementary building blocks, which may be called "elementary symbols." It is important to note that there is a large degree of ambiguity as to what are to be considered elementary symbols. An English sentence can be considered as consisting of alphabetic letters, spaces, punctuation, and numerals or as consisting of words, spaces, punctuation, and numerals. It can also be considered as consisting of artificial building blocks, each of which consists of a certain fixed number of elementary symbols in the usual sense.

Consider two languages, $\mathcal{A}$, consisting of n elementary symbols, a_i, $i = 1, 2, \cdots, n$, and $\mathcal{B}$, consisting of m elementary symbols, b_j, $j = 1, 2, \cdots, m$, under a restriction $n \geq m$. (This restriction will be lifted later.) Encoding of messages in $\mathcal{A}$ by $\mathcal{B}$ consists of establishing a rule determining correspondence between each elementary a-symbol and a sequence of b-symbols (of not necessarily fixed length), and of writing a b-string corresponding to a given a-string according to this rule. Decoding consists of retrieving, in the presence of a b-string, the a-string from which this b-string was derived by the process of encoding. In the following, a b-sequence corresponding to an a_i is called a b-letter code.

The essential requirement here is that the result of a consecutive application of encoding and decoding on an a-message be unique and equivalent to an identity transformation. For this to be the case it is first necessary that two different a-symbols do not correspond to a single b-sequence. Conversely, however, two different b-sequences may correspond to a single a-symbol. This fact can be exploited to diminish the harmful effect of "noise," which here means an uncontrollable alteration of b-strings. In other words, sometimes it is possible to make the rule of correspondence in such a way that the error may alter the b-sequence into which an a_i was encoded, but the altered b-sequence will remain with high probability within the same group of sequences that correspond to the original a-symbol. This is a basic idea underlying the technique of self-correcting codes in the presence of noise. In "noiseless" coding, however, we assume that only one b-sequence corresponds to an a-symbol.

These conditions are sufficient to make the encoding-decoding cycle an identity transformation (in the absence of noise) as far as one a-symbol is concerned. But in order also to realize the identity condition for a-strings of arbitrary length, we have to require one more condition. In fact, the decoder, in the presence of a long b-string obtained from an a-string by the encoding procedure described above, has to know where one b-letter code ends and where another b-letter code starts, so that he can carry out the decoding procedure. A simple method for meeting this requirement is to adopt the following prescription (hereafter referred to as the rule of stringwise uniqueness): "addition of any number of b-symbols to the end of a b-letter code (corresponding to an a-symbol) gives rise to no other b-letter code (corresponding to an a-symbol)." If this rule is followed, the decoder, provided that he can recognize the beginning of a b-string corresponding to an a-string, can determine where the first b-letter code ends, and then start to tackle the next b-letter code.

This guarantees that for a given a-string there will be a unique b-string, and for a given b-string, which is the translation of an a-string, there will be a unique a-string, provided that the starting point of the b-string is clearly indicated. However, this should not be misconstrued as a one-to-one correspondence between all possible a strings and all possible b strings, because there can be many b-strings that do not correspond to any a-strings.

The reader is now prepared to see for himself that a given rule of correspondence between language $\mathcal{A}$ with n symbols and language $\mathcal{B}$ with $m(\leq n)$ symbols satisfying the rule of stringwise uniqueness corresponds to an m-bounded polychotomic tree built in a logical spectrum with n members, and vice versa. A choice of a branch at a branching point corresponds to a selection of one b symbol. The rule of stringwise uniqueness corresponds to the obvious agreement that each peripheral branch in a tree ends up with only one E_i and cannot be continued to correspond to another E_j. Furthermore, if we make correspondence between the relative frequency of an a symbol a_i and the probability p_i attached to proposition E_i, the average length of b-strings is given by the average rank $\langle r_i \rangle_i$ in the polychotomic tree. This gives pragmatic significance to the foregoing discussion about the optimal tree.

In fact, one of the most important considerations regarding coding pertains to the length of the b-string. For economy in time or money it is usually desired to make the average of the length as short as possible. It is quite obvious that to achieve this end, more frequent a-symbols should get shorter b-letter codes. We see that this idea has long been actually put in practice in the Morse code, which assigns the shortest code (a single dot) to the most frequent letter, *e*. Recent research has made it clear that this very significant idea is to be attributed to Alfred Vail, who helped to develop telegraphy in the first half of the nineteenth century as a collaborator of Samuel Morse. The entire theory of noiseless coding is nothing but a quantitative elaboration of Vail's idea. See Figure 1.2 for a numerical example of optimal code.

Inequalities (1.157) show that there exists a coding rule such that the average length per a symbol is between $(S^{(1)}/\log m)$ and $(S^{(1)}/\log m) + 1$, where $S^{(1)}$ is the usual entropy function defined in terms of the probabilities of the a symbols. In the case of coding there is, as noted at the beginning, ambiguity as to what is to be regarded as elementary building blocks a_i. We shall show that by considering a group of elementary building blocks as a new single building block and by applying the coding method to these new a-symbols, we can usually reduce the average length of the encoded message even below the lower bound calculated for the encoding based on the old a-symbols. To make the discussion concrete, let us compare two methods of encoding, one based on alphabetic letters and the other based on blocks of g letters (g-grams), g being a fixed integer. The spaces, numerals, and punctuation are also counted as alphabet letters here, and their total number will be denoted as n_0.

Let us cut a long a-string arbitrarily in blocks (sequences) of length g. Each of g positions in this a-sequence can be occupied by any one of the n_0 different letters. Hence there are n_0^g different a-sequences of length g; that is, n_0^g is the n of the language $\mathcal{A}$ in the new point of view. We may assume that language $\mathcal{A}$ has a certain probability for each of these n_0^g sequences. Let us denote this probability by $p(i_1, i_2, \cdots, i_g)$, where each i_k can be any one of n_0 possible values $1, 2, \cdots, n_0$. Our assumption is that this probability is uniquely determined by the set of values $i_1, i_2, \cdots, i_g$, no matter where this sequence of length g is located in an actual a-string. The same kind of assumption was actually understood when we used p_i in the foregoing consideration, which dealt with a-sequences of unit length. Now if we sum up $p(i_1, i_2, \cdots, i_k, \cdots, i_g)$ with respect to all i's except a particular one, say, i_k, we should obtain $p(i_k)$, which must be the same distribution as p_i.

$$\sum_{i_1=1}^{n_0} \sum_{i_2=1}^{n_0} \cdots \sum_{i_{k-1}=1}^{n_0} \sum_{i_{k+1}=1}^{n_0} \cdots \sum_{i_g=1}^{n_0} p(i_1, i_2, \cdots, i_k, \cdots, i_g) = p(i_k). \quad (1.158)$$

However, unless all the i's are independent, this $p(i_1, i_2, \cdots, i_g)$ is different from

$$q(i_1, i_2, \cdots, i_g) = p(i_1)p(i_2), \cdots, p(i_g). \quad (1.159)$$

Now by the use of Gibbs' theorem we obtain

$$\begin{aligned} S^{(g)} &= -\sum_{i_1=1}^{n_0} \sum_{i_2=1}^{n_0} \cdots \sum_{i_g=1}^{n_0} p(i_1, i_2, \cdots, i_g) \log p(i_1, i_2, \cdots, i_g) \\ &\leq -\sum_{i_1=1}^{n_0} \sum_{i_2=1}^{n_0} \cdots \sum_{i_g=1}^{n_0} p(i_1, i_2, \cdots, i_g) \log q(i_1, i_2, \cdots, i_g) \\ &= -\sum_{k=1}^{g} \sum_{i_k=1}^{n_0} p(i_k) \log p(i_k) = gS^{(1)}, \end{aligned} \quad (1.160)$$

equality holding if and only if $p = q$ for all i's; that is, if and only if the letters in a block of length g are all independent.

Now let us make a b-sequence correspond to each of n_0^g a-sequences of length g. The problem here is no different in nature from the one we discussed before; the only variation is that the number of code sequences is n_0^g instead of n, and the probability for each code sequence is $p(i_1, i_2, \cdots, i_g)$ instead of p_i. Let us denote the whole of the variables $(i_1, i_2, \cdots, i_g)$ by a single variable $\lambda = 1, 2, \cdots, n_0^g$, and the length of the b-sequence assigned to sequence λ by L_λ. We also use the strategy used in (1.155); then we have

$$-\frac{\log p_\lambda}{\log m} + 1 > L_\lambda \geq -\frac{\log p_\lambda}{\log m}. \quad (1.161)$$

Then the average length $\langle L \rangle = \sum_{\lambda=1}^{n_0^g} p_\lambda L_\lambda$ will be

$$\frac{S^{(g)}}{\log m} + 1 > \langle L \rangle \geq \frac{S^{(g)}}{\log m}. \tag{1.162}$$

Now $\langle L \rangle$ is the average length of b-sequences, each corresponding to an a-sequence of g letters. Hence the average length $\langle l \rangle$ of b-sequence per a-letter is $\langle L \rangle / g$, and will satisfy

$$\frac{S^{(g)}}{g \log m} + \frac{1}{g} > \langle l \rangle \geq \frac{S^{(g)}}{g \log m}. \tag{1.163}$$

Making $g \to \infty$, we obtain

$$\lim_{g \to \infty} \langle l \rangle = \frac{S^{(g)}}{g \log m}. \tag{1.164}$$

Using the result (1.160), we can derive

$$\lim_{g \to \infty} \langle l \rangle \leq \frac{S^{(1)}}{\log m}, \tag{1.165}$$

equality holding in the independent case $p_\lambda = q_\lambda$ for all λ.

Thus, using sequence-to-sequence correspondence instead of letter-to-sequence correspondence, and making the length of a-sequences arbitrarily long, we can always achieve the average length $S^{(1)}/\log m$ even in the most unfavorable case of independent a_i's. If $p_\lambda \neq q_\lambda$ for some of the λ, the average length can be made even smaller than $S^{(1)}/\log m$. A problem of coding from an n-letter language $\mathcal{A}$ into an m-letter language $\mathcal{B}$ with $n < m$ makes sense when a sequence-to-sequence or sequence-to-symbol translation, instead of a symbol-to-sequence translation, is considered.

2

Structure Analysis

2.1. ORGANIZATION OF INDIVIDUALS IN AN ASSEMBLY

The aim of this chapter is to show that the mathematical tools developed in Chapter 1 are peculiarly powerful in analyzing a certain aspect of the organizational structure existing in an assembly of stochastically behaving individuals. It would be erroneous to claim that these mathematical tools are capable of giving ultimate solutions to all the old problems connected with the whole and its parts, yet they certainly give a deep insight into the problem of "organization" particularly when the word "organization" is understood in the sense that the more organized the system is, the better defined the behavior of the whole becomes in spite of the behavioral indeterminacy of individuals.

As stated at the beginning of Chapter 1, information theory deals with relationships existing among *propositions*, but when these propositions correspond to the states of *systems* we can simplify our statement by saying that information theory deals with the relationships existing among the systems. In this manner of speech we state here that our mathematical formulas are useful in investigating the organizational relation between the whole and the parts. Even at the logical level we can often eliminate the term "proposition" from our speech without causing confusion. When, for instance, a testable proposition A implies a testable proposition B, we know that if we observe a physical state making A true then we have to expect with certitude a physical state making B true. Hence we can talk about a necessary relation between facts instead of talking about a logical relation between propositions.

Historically speaking, it may be an exaggeration but not a misrepresentation to contend that information theory was born out of the need for a mathematical tool to investigate the problem of organization in an assembly of individuals. In the mid-1930s Niels Bohr, in order to explain peculiarities of nuclear reactions, came to the idea that an assembly of particles under the mutual influence of nuclear forces behaves entirely differently from an assembly of particles, say, electrons, under electromagnetic interaction, in the sense that the state of a nucleon in the former is critically dependent on the state of the neighboring particles with which it strongly interacts, while an electron in the latter is moving freely and independently under the smeared-out average influence of all the remaining particles (see [B-4]). A research topic that Werner Heisenberg suggested to me as a young student in the late 1930s concerned the consequences of this peculiarity of nuclear forces in the stable state of atomic nuclei, as distinct from the excited state, which Bohr had mainly discussed. Having previously become familiar, while writing my dissertation [W-1], with von Neumann's ideas expressed in his textbook [V-3], I decided after discussions with Heisenberg and Hans Euler to use von Neumann's concept of microscopic entropy to characterize the peculiar nature of an assembly of nucleons (see [W-2]). To von Neumann this quantity was nothing much beyond a simplified version of thermodynamic entropy and was used by him as a useful tool to demonstrate the irreversibility of the process of observation. In my paper of 1939 [W-2] this quantity was shown to be a good measure of two properties. First, of course, it measures the degree of indeterminacy of the state of a single particle taken apart; second, it measures the degree of interaction and interdependence of particles constituting an organized system. The above-mentioned properties of an excited state of an atomic nucleus pointed out by Bohr would indeed imply large values of these two measures. According to the modern point of view of information theory, the first corresponds to "ignorance" and the second corresponds to "organization," which in communication theory is called "redundancy."

If Szilard's 1929 paper [S-12] was the first to report that a decrease of thermodynamic entropy is accompanied by acquisition of information, my 1939 paper [W-2] was the first to introduce a quantity expressing lack of information, entirely divorced from thermodynamics, and to show that this quantity is also a measure of interdependence and organization. In the field of communication, Nyquist in 1924 [N-3] and Hartley in 1928 [H-5] had the idea that the amount of communication could be expressed by the logarithm of the number of alternative symbols [corresponding to our (1.22) or (1.42)], but the idea of redundancy came much later, after communications engineers started to use the entropy function in the late 1940s (see Wiener and Shannon, among others [W-36, S-6]). The reader can find a typical discussion of redundancy in Shannon's 1951 paper [S-5]. The fact that redundancy meant organization was emphasized with enthusiasm by Jerome Rothstein around

1951 (see [R-7, R-8]). The works of McGill [M-6] in 1954 and Garner in 1956 [G-6] and 1962 [G-5] must be mentioned as having applied from an early date the entropic expression of interdependence to psychological data; see also Mitra's paper [M-7] of 1955. I myself came back to the problem of interdependence in a paper published in 1954 [W-6], in which I applied the idea to time sequences. In papers of 1958 [W-11] and 1960 [W-12] I continued to perfect a theory of interdependence based on entropic functions. Sections 2 and 3 of this chapter are based on these last two papers and discuss "simultaneous" interdependence. Sections 4 and 5 of this chapter discuss "temporal" interdependence in time sequences. Applications of entropic measures of interdependence to the problem of concept formation, initiated by papers published in 1961 [W-15, W-17] and 1962 [W-18], are discussed in Chapter 8.

It should be mentioned that the 1939 paper on information theory [W-2] had a theoretical generality that none of the subsequent papers (except [W-18]) shared; that is, it was conceived in a framework in which our probabilistic knowledge of a system is expressed as a density matrix (a positive-definite Hermitian matrix), which is a more general notion than a simple probability distribution. This is a necessary generality in order to build an information theory applicable also to atomic systems. We discuss a generalized quantum mechanical information theory in Chapter 9.

When one hears of a mathematical expression of interdependence, the first idea that occurs may be that of correlation coefficient. The entropic measure is quite different in nature from the correlation coefficient. Consider the product $\mathcal{G} = \mathcal{E} \otimes \mathcal{F}$ of two logical spectra $\mathcal{E}$ and $\mathcal{F}$ with a certain probability distribution $p(G_k) = p(E_i \cap E_j)$. Consider further two stochastic variables, $\mathfrak{Q} = \{Q_i\}$ defined on $\mathcal{E}$ and $\mathfrak{R} = \{R_j\}$ defined on $\mathcal{F}$. The correlation coefficient between $\mathfrak{Q}$ and $\mathfrak{R}$ is given by

$$\frac{\langle (Q_i - \langle Q_i \rangle_i)(R_j - \langle R_j \rangle_j) \rangle_{ij}}{[(\langle Q_i^2 \rangle_i - \langle Q_i \rangle_i^2)(\langle R_j^2 \rangle_j - \langle R_j \rangle_j^2)]^{\frac{1}{2}}}, \tag{2.1}$$

which measures, so to speak, the degree of agreement of the values of $\mathfrak{Q}$ and $\mathfrak{R}$. On the other hand, the entropic measure of interdependence for the case of two logical spectra $\mathcal{E}$ and $\mathcal{F}$ indicates the degree of departure from the probabilistic independence, that is, from the condition that $p(E_i \cap E_j) = p(E_i) \cdot p(F_j)$ for all i and all j. It is true that if $\mathcal{E}$ and $\mathcal{F}$ are probabilistically independent, the correlation between $\mathfrak{Q}$ and $\mathfrak{R}$ vanishes. But the converse is not true, which means that the entropic measure can uncover a relation between $\mathcal{E}$ and $\mathcal{F}$ that the correlation coefficient may not uncover. See Chapter 7 for more about probabilistic independence. It should be further noticed that (2.1) is invariant only for a (nonhomogeneous) linear transformation of Q_i and R_j and is strongly dependent on what stochastic variables are used and how they are measured. The method of factor analysis (see

Appendix 8.5), which depends on correlation coefficient, shares these shortcomings. The entropic measure depends only on probabilities and has nothing to do with the stochastic variables defined on $\mathcal{E}$ and $\mathcal{F}$.

The exact mathematical expression of the entropic measure of interdependence will be introduced later, but we can already anticipate the kind of formula we shall obtain. We previously stated that the degree of "organization" becomes larger if the degree of indeterminacy of the system as a whole decreases in spite of a large degree of individual indeterminacy. If there is no organization, a large amount of indeterminacy of individuals will result in a large amount of indeterminacy of the whole. Thus the strength of organization will be measured by the balance between the indeterminacy of the parts and the indeterminacy of the whole. If we remember that entropy is a measure of indeterminacy, then it will not be too far-fetched to assume that the degree of organization is given by

$$\text{organization} = (\text{sum of entropies of parts}) - (\text{entropy of whole}). \quad (2.2)$$

In fact, the very kernel of the entire discussion of this chapter is summarized by (2.2).

In usual information theory the entropy of the whole can never be smaller than the entropy of any one of its parts. But in the quantum mechanical information theory, which is explained in Chapter 9, the entropy of the whole can become smaller than that of one of its parts. In fact, if the entire system is in a pure quantum state, the entropy of the whole becomes zero. As a result, we can see from (2.2) that in such a case the entropy of individual particles becomes a measure of organization. This justifies the interpretation used in the 1939 paper [W-2] (see Chapter 9 for more details).

In the past the method of analyzing organizational structure with entropic functions was sometimes called ITCA, standing for information-theoretical correlation analysis [W-15], but in this book it is called IDA, standing for interdependence analysis. This change has been made in order to make it clear that we are dealing not with probabilistic correlation, but with probabilistic dependence.

2.2. IDA (INTERDEPENDENCE ANALYSIS) AND MULTIPLICATIVE POLYCHOTOMIC TREES

Let us start with the case of a system consisting of two constituent parts or, rather, the case in which we regard any given system as consisting of two parts and we want to investigate the interdependence of these two parts. Let $\mathcal{E}$ stand for the set of propositions, each of which describes the state of one of the two constituent parts to a desired accuracy. Similarly, let $\mathcal{F}$ stand for the set of state descriptions of the other part. Then $\mathcal{G} = \mathcal{E} \otimes \mathcal{F}$ will be the set of descriptions of the entire system. To each member $G_k = E_i \cap F_j$ of $\mathcal{G}$ is assigned a probability $p(G_k) = p(E_i \cap F_j)$, from which the probability $p(E_i)$ for a member E_i of $\mathcal{E}$ and the probability $p(F_j)$ for a member F_j of $\mathcal{F}$ can be derived by (1.16).

In Section 1.3 we discussed various relations governing three entropy functions, $S(\mathcal{E})$, $S(\mathcal{F})$, and $S(\mathcal{E} \otimes \mathcal{F})$, and two (average) conditional entropy functions, $S(\mathcal{E} \mid \mathcal{F})$ and $S(\mathcal{F} \mid \mathcal{E})$. Those relations are all derivable from three basic facts. (a) An entropy function is non-negative, that is,

$$S \geq 0, \tag{2.3}$$

where S can be any one of the five S functions mentioned above. (b) The definition of conditional entropy entails

$$S(\mathcal{F}) + S(\mathcal{E} \mid \mathcal{F}) = S(\mathcal{E} \otimes \mathcal{F}) = S(\mathcal{E}) + S(\mathcal{F} \mid \mathcal{E}). \tag{2.4}$$

(c) Theorem 1.2 implies that

$$S(\mathcal{E} \otimes \mathcal{F}) \leq S(\mathcal{E}) + S(\mathcal{F}). \tag{2.5}$$

Combining (2.4) and (2.5), we obtain

$$S(\mathcal{F} \mid \mathcal{E}) \leq S(\mathcal{F}) \quad \text{and} \quad S(\mathcal{E} \mid \mathcal{F}) \leq S(\mathcal{E}). \tag{2.6}$$

Combining (2.3) and (2.4), we obtain, among others,

$$S(\mathcal{E}) \leq S(\mathcal{E} \otimes \mathcal{F}) \quad \text{and} \quad S(\mathcal{F}) \leq S(\mathcal{E} \otimes \mathcal{F}). \tag{2.7}$$

It is fairly easy to attach intuitive meanings to these formulas, with the help of either the notion of ignorance (indeterminacy) or the notion of information. For instance, the left half of (2.4) can be interpreted as follows. The total ignorance about the entire system $\mathcal{E} \otimes \mathcal{F}$ is the sum of the initial ignorance about the partial system $\mathcal{F}$ and the average remaining ignorance about the other partial system $\mathcal{E}$ after we have determined the state of $\mathcal{F}$. In other words, the total information about $\mathcal{E} \otimes \mathcal{F}$ is the information obtained by observing $\mathcal{F}$ plus the additional information obtained by observing $\mathcal{E}$ after having already obtained knowledge about $\mathcal{F}$. Formula (2.5) says that the information obtained by observing the total system $\mathcal{G} = \mathcal{E} \otimes \mathcal{F}$ is less than the sum of the information obtained from $\mathcal{E}$ and the information obtained from $\mathcal{F}$ separately. This is because, owing to interdependence, there is overlapping between the information carried by $\mathcal{E}$ and the information carried by $\mathcal{F}$. The first relation of (2.6) means that the total ignorance about (hence the total information carried by) $\mathcal{F}$ is larger than, or at least equal to, the average remaining ignorance about (hence the additional information provided by) $\mathcal{F}$ after the state of $\mathcal{E}$ has been made known. This is because the interdependence between $\mathcal{E}$ and $\mathcal{F}$ makes it possible to derive some information about $\mathcal{F}$ from observation of $\mathcal{E}$. Relations of (2.7) express what was already stated in Section 2.1, namely, that the ignorance about (the information carried by) a combined system $\mathcal{E} \otimes \mathcal{F}$ is larger than (or at least equal to) the ignorance about (the information carried by) its part $\mathcal{E}$, or $\mathcal{F}$. (This last is a special case of the relation, which does not hold in quantum mechanical information theory.)

Taking advantage of the inequality in (2.5), we can introduce a non-negative quantity $J(\mathcal{G}; \mathcal{E}, \mathcal{F})$, or simply $J(\mathcal{E}, \mathcal{F})$, by

$$J(\mathcal{G}; \mathcal{E}, \mathcal{F}) = S(\mathcal{E}) + S(\mathcal{F}) - S(\mathcal{E} \otimes \mathcal{F}) \tag{2.8}$$

$$= S(\mathcal{F}) - S(\mathcal{F} \mid \mathcal{E}) \tag{2.9}$$

$$= S(\mathcal{E}) - S(\mathcal{E} \mid \mathcal{F}), \tag{2.10}$$

where the second and third lines are obtained from the first with the help of (2.4). Theorem 1.2 states that

$$J(\mathcal{G}; \mathcal{E}, \mathcal{F}) \geq 0, \tag{2.11}$$

where equality holds if and only if $\mathcal{E}$ and $\mathcal{F}$ are probabilistically independent. Formulas (2.9,) and (2.10), on the other hand, tell us that

$$J(\mathcal{G}; \mathcal{E}, \mathcal{F}) \leq \min\,(S(\mathcal{E}), S(\mathcal{F})). \tag{2.12}$$

If $\mathcal{E}$ is entirely dependent on $\mathcal{F}$, then $S(\mathcal{E} \mid \mathcal{F}) = 0$, and, in view of (2.4), $S(\mathcal{F}) = S(\mathcal{E}) + S(\mathcal{F} \mid \mathcal{E}) \geq S(\mathcal{E})$. Hence $\min\,(S(\mathcal{E}), S(\mathcal{F})) = S(\mathcal{E})$ in this case, and $J(\mathcal{G}; \mathcal{E}, \mathcal{F})$, in fact, takes its formal maximum set by (2.12), for substitution of $S(\mathcal{E} \mid \mathcal{F}) = 0$ in (2.10) yields

$$J(\mathcal{G}; \mathcal{E}, \mathcal{F}) = S(\mathcal{E}). \tag{2.13}$$

If $\mathcal{E}$ is entirely dependent on $\mathcal{F}$ and $\mathcal{F}$ is entirely dependent on $\mathcal{E}$, then

$$S(\mathcal{E}) = S(\mathcal{F}) = J(\mathcal{G}; \mathcal{E}, \mathcal{F}). \tag{2.14}$$

Thus the lower bound set by (2.11) and the upper bound set by (2.12) correspond respectively to the minimum and maximum interdependence between $\mathcal{E}$ and $\mathcal{F}$. This makes it natural to call the quantity $J(\mathcal{G}; \mathcal{E}, \mathcal{F})$ "interdependence between $\mathcal{E}$ and $\mathcal{F}$," or "interdependence existing in $\mathcal{G}$ with respect to $\mathcal{E}$ and $\mathcal{F}$." Another argument for giving this name to J is provided by (2.9) and (2.10). In fact, $S(\mathcal{F})$ in (2.9) is the information obtained by observing $\mathcal{F}$ only, while $S(\mathcal{F} \mid \mathcal{E})$ is the average additional information obtained by observing $\mathcal{F}$ when the outcome of $\mathcal{E}$ is already known. These two are not the same, for the knowledge about $\mathcal{E}$ can give some information about the outcome of $\mathcal{F}$ if $\mathcal{F}$ is somehow correlated with $\mathcal{E}$. Therefore the difference is the part that is no longer considered as new information when $\mathcal{F}$ is observed after $\mathcal{E}$ has been observed; hence it corresponds to the strength of interdependence. In other words, only the part $S(\mathcal{F} \mid \mathcal{E})$ of $S(\mathcal{F})$ can be considered as the information exclusively carried by $\mathcal{F}$ itself unobtainable through $\mathcal{E}$; hence $S(\mathcal{F}) - S(\mathcal{F} \mid \mathcal{E})$ is the part of information about $\mathcal{F}$ that is doubly carried by both $\mathcal{E}$ and $\mathcal{F}$. This is redundancy and can be attributed to interdependence.

We now turn to the case of a system consisting of N parts, and for that purpose we first introduce a quantity that is a generalization of $J(\mathcal{G}; \mathcal{E}, \mathcal{F})$. Consider a product $\mathcal{G} = \bigotimes_{a=1}^{N} \mathcal{E}^{(a)}$ of N logical spectra $\mathcal{E}^{(a)}, a = 1, 2, \cdots, N$, as defined by (1.58)–(1.60). If we define the "interdependence existing in $\mathcal{G}$ with respect to the $\mathcal{E}^{(a)}$, $a = 1, 2, \cdots, N$," or "interdependence among the $\mathcal{E}^{(a)}$, $a = 1, 2, \cdots, N$," by

$$J(\mathcal{G}; \mathcal{E}^{(1)}, \mathcal{E}^{(2)}, \cdots, \mathcal{E}^{(N)}) = \sum_{a=1}^{N} S(\mathcal{E}^{(a)}) - S(\mathcal{G}), \tag{2.15}$$

we know from Theorem 1.3 that

$$J(\mathcal{G}; \mathcal{E}^{(1)}, \mathcal{E}^{(2)}, \cdots, \mathcal{E}^{(N)}) \geq 0, \tag{2.16}$$

where equality holds if and only if the $\mathcal{E}^{(a)}$ are all probabilistically independent as in (1.61). In later parts of this book we sometimes neglect mentioning $\mathcal{G}$ as an argument of J for simplicity.

Writing $\mathcal{G} = \mathcal{E}^{(1)} \otimes \mathcal{F}^{(1)}$ with

$$\mathcal{F}^{(1)} = \mathcal{E}^{(2)} \otimes \mathcal{E}^{(3)} \otimes \cdots \otimes \mathcal{E}^{(N)},$$

we have, from (2.7),

$$S(\mathcal{E}^{(1)}) \leq S(\mathcal{G}). \tag{2.17}$$

Since this relation holds for any $\mathcal{E}^{(a)}$. we also have

$$\max_{a} S(\mathcal{E}^{(a)}) \leq S(\mathcal{G}). \tag{2.18}$$

Hence we have

$$J(\mathcal{G}; \mathcal{E}^{(1)}, \mathcal{E}^{(2)}, \cdots, \mathcal{E}^{(N)}) \leq \sum_{a=1}^{N} S(\mathcal{E}^{(a)}) - \max_{a} S(\mathcal{E}^{(a)}). \tag{2.19}$$

If $\mathcal{F}^{(1)}$ depends entirely on $\mathcal{E}^{(1)}$, that is, if

$$S(\mathcal{E}^{(2)} \otimes \mathcal{E}^{(3)} \otimes \cdots \otimes \mathcal{E}^{(N)} \mid \mathcal{E}^{(1)}) = 0, \tag{2.20}$$

then

$$S(\mathcal{E}^{(1)}) = S(\mathcal{G}) = \max_{a} S(\mathcal{E}^{(a)})$$

because of (2.4) and (2.18), and J of (2.15) becomes equal to the formal upper bound given in (2.19). We may intuitively interpret $J(\mathcal{G}; \mathcal{E}^{(1)}, \mathcal{E}^{(2)}, \cdots, \mathcal{E}^{(N)})$ as the intensity of cohesion, or the strength of interdependence, with which the parts represented by $\mathcal{E}^{(1)}, \mathcal{E}^{(2)}, \cdots, \mathcal{E}^{(N)}$ are organized to form the entire system represented by $\mathcal{G}$. To be precise, we should probably say "the entire system" as depicted by the probabilities assigned to $\mathcal{G}$.

Coming back to the general problem of a whole and its parts, we should realize that complication, hence also richness, of arguments concerning this problem stems at least partly from the fact that the same system can be divided into parts in many different ways. Even if the system is composed of a finite number of "indivisible" building blocks, people are not necessarily interested in dividing up the system in one stroke into these ultimate building blocks. They may be, for instance, first interested in the interdependence between the upper and lower halves of the system, and then in the interdependence between the upper left and the upper right quarters of the system. Some other people may be interested in the interdependence among the regions cut out from the system by concentric spheres with the center located somewhere. Other persons may be interested in knowing the way to divide the system into two parts so as to minimize the interdependence between them, as if they were looking for a natural division or a hidden crack. A question arising most naturally, in view of the existence of different partitions and corresponding interdependences in the same system, is the relationship that exists among these different partitions and interdependences. Since they are defined within the same system, they cannot be entirely unrelated after all.

We had better start with introducing a systematic way of describing different ways of partitioning the system into subsystems and of partitioning the subsystems into sub-subsystems, and so on. Let the entire system be denoted by $O^{(0)}$ and let us assume that it consists of N "indivisible" atomic building blocks $A_1, A_2, \cdots, A_N$. By an "indivisible" system is meant one such that we are not interested in knowing the state of its internal structure. We denote an intermediate system, consisting of some of these A's also, by O, but with a suitable superscript. Starting from $O^{(0)}$ we can first divide it into $M^{(0)}$ subsystems $O_\mu^{(1)}$, $\mu = 1, 2, \cdots, M^{(0)}$, in such a way that each A of $O^{(0)}$ belongs to one and only one of the $O^{(1)}$. Borrowing the set-theoretical notation, we may write

$$O^{(0)} = O_1^{(1)} \vee O_2^{(1)} \vee \cdots \vee O_{M^{(0)}}^{(1)}$$

and

$$O_\mu^{(1)} \wedge O_\nu^{(1)} = \Phi, \qquad \mu \neq \nu, \tag{2.21}$$

although we have a concrete physical system rather than an abstract set in mind. We can then divide each $O_\mu^{(1)}$ into $M_\mu^{(1)}$ portions $O_{\mu\nu}^{(2)}$'s, $\nu = 1, 2, \cdots, M_\mu^{(1)}$, in such a way that

$$O_\mu^{(1)} = O_{\mu 1}^{(2)} \vee O_{\mu 2}^{(2)} \vee \cdots \vee O_{\mu M_\mu^{(1)}}^{(2)}$$

and

$$O_{\mu\nu}^{(2)} \wedge O_{\mu\kappa}^{(2)} = \Phi, \qquad \nu \neq \kappa. \tag{2.22}$$

We can continue this process until finally each resulting system consists only of one atom A. This completes a polychotomic tree of a real system $O^{(0)}$. For any given partition of a subsystem of $O^{(0)}$ there is always at least one polychotomic tree of this kind, such that one of the branching points coincides with it.

Now let logical spectrum $\mathcal{E}^{(a)}$ be the set of descriptions of the atomic subsystem A_a, and let $\mathcal{G}^{(0)} = \bigotimes_{a=1}^{N} \mathcal{E}^{(a)}$ be the logical spectrum describing the entire system $O^{(0)}$. Then, corresponding to the first branching point (2.21), we have

$$\mathcal{G}^{(0)} = \mathcal{G}_1^{(1)} \otimes \mathcal{G}_2^{(1)} \otimes \cdots \otimes \mathcal{G}_{M^{(0)}}^{(1)}, \tag{2.23}$$

where $\mathcal{G}_\mu^{(1)}$ describes the subsystem $O_\mu^{(1)}$, that is, is the product of all those $\mathcal{E}^{(a)}$'s that correspond to A_a's constituting $O_\mu^{(1)}$. Generally, to a branching point that divides a system $O^{(\lambda)}$ into systems $O_1^{(\lambda+1)}, O_2^{(\lambda+1)}, \cdots, O_{M(\lambda)}^{(\lambda+1)}$, that is, to a partition (with κ standing for a sequence of subscripts)

$$O_\kappa^{(\lambda)} = \bigvee_{\nu=1}^{M_\kappa^{(\lambda)}} O_{\kappa\nu}^{(\lambda+1)}, \tag{2.24}$$

will correspond a factorization

$$\mathcal{G}_\kappa^{(\lambda)} = \bigotimes_{\nu=1}^{M_\kappa^{(\lambda)}} \mathcal{G}_{\kappa\nu}^{(\lambda+1)}, \tag{2.25}$$

where each $\mathcal{G}$ is a product of those $\mathcal{E}$'s corresponding to the A's of the corresponding O. In the polychotomic tree of Chapter 1, at each branching point (or rather a merging point), sets of propositions $\mathcal{B}$'s were "added." But in the present polychotomic tree, at each branching point (or rather a merging point) sets of propositions, $\mathcal{G}$'s are "multiplied." For this reason we can call this latter a multiplicative polychotomic tree. (Figure 2.1). It should not be misunderstood that the "relations" between subsystems are lost in this kind of description. If for instance one subsystem is located "to the left of" another subsystem, this relation is imbedded in the respective locations of the two subsystems.

At each branching point of the type (2.25) we can define an interdependence by

$$J(\mathcal{G}_\kappa^{(\lambda)}; \mathcal{G}_{\kappa1}^{(\lambda+1)}, \mathcal{G}_{\kappa2}^{(\lambda+1)}, \cdots, \mathcal{G}_{\kappa M_\kappa^{(\lambda)}}^{(\lambda+1)}) = \sum_{\nu=1}^{M_\kappa^{(\lambda)}} S(\mathcal{G}_{\kappa\nu}^{(\lambda+1)}) - S(\mathcal{G}_\kappa^{(\lambda)}), \tag{2.26}$$

where $\mathcal{G}_\kappa^{(\lambda)}$ is the "incoming" branch and $\mathcal{G}_{\kappa\nu}^{(\lambda+1)}$ are the "outgoing" branches. Now suppose that we add all the J's taken at all the branching points of any

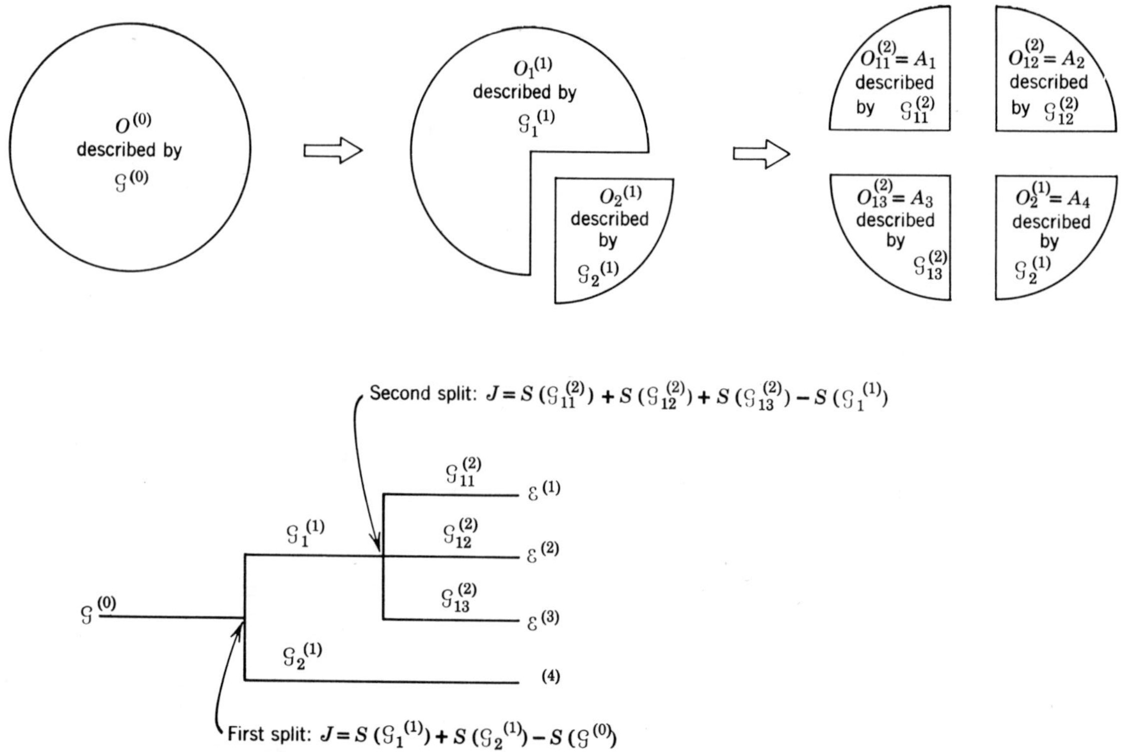

Figure 2.1 Multiplicative polychotomic tree, The total system $O^{(0)}$ consisting of four elementary building blocks A_1, A_2, A_3, and A_4, is first split into $O_1^{(1)}$ consisting of A_1, A_2, and A_3, and $O_2^{(1)}$ consisting of A_4. In the second split $O_1^{(1)}$ is decomposed into its components.

single tree. We shall certainly obtain

$$\sum_{\text{all branching points}} J(\text{branching point}) = \sum_{a=1}^{N} S(\mathcal{E}^{(a)}) - S(\mathcal{G}^{(0)}) \tag{2.27}$$

because, except for the trunk ($\mathcal{G}^{(0)}$) and the peripheral branches ($\mathcal{E}^{(a)}$), all the other branches will appear in (2.27) once as an incoming branch and another time as an outgoing branch.

Theorem 2.1 The total sum of the interdependences taken at all branching points in a complete multiplicative polychotomic tree whose trunk is $\mathcal{G}^{(0)}$ and whose peripheral branches are $\mathcal{E}^{(a)}$, $a = 1, 2, \cdots, N$, is independent of the way the tree is chosen and is equal to $J(\mathcal{G}^{(0)}; \mathcal{E}^{(1)}, \mathcal{E}^{(2)}, \cdots, \mathcal{E}^{(N)})$.

Hereafter we write

$$J_{\text{tot}}(\mathcal{G}^{(0)}) = J(\mathcal{G}^{(0)}; \mathcal{E}^{(1)}, \mathcal{E}^{(2)}, \cdots, \mathcal{E}^{(N)}) = \sum_{a=1}^{N} S(\mathcal{E}^{(a)}) - S(\mathcal{G}^{(0)}) \tag{2.28}$$

and call it the total interdependence in $\mathcal{G}^{(0)}$. This is the interdependence at the only branching point of the particular tree that divides $\mathcal{G}^{(0)}$ directly into N peripheral branches $\mathcal{E}^{(a)}$, $a = 1, 2, \cdots, N$. The following are some of the simple corollaries that ensue from Theorem 2.1.

Corollary 2.1 $J_{\text{tot}}(\mathcal{G}^{(0)})$ is the largest interdependence that can be defined within the system described by $\mathcal{G}^{(0)}$.

Corollary 2.2 If

$$\mathcal{G}^{(0)} = \bigotimes_{\mu=1}^{M} \mathcal{G}_{\mu}^{(1)} \tag{2.29}$$

then

$$J_{\text{tot}}(\mathcal{G}^{(0)}) = J(\mathcal{G}^{(0)}; \mathcal{G}_1^{(1)}, \cdots, \mathcal{G}_M^{(1)}) + \sum_{\mu=1}^{M} J_{\text{tot}}(\mathcal{G}_{\mu}^{(1)}) \geq \sum_{\mu=1}^{M} J_{\text{tot}}(\mathcal{G}_{\mu}^{(1)}). \tag{2.30}$$

Corollary 2.3 If the system described by $\mathcal{H}$ is a part of the system described by $\mathcal{G}$, then

$$J_{\text{tot}}(\mathcal{G}) \geq J_{\text{tot}}(\mathcal{H}). \tag{2.31}$$

Corollary 2.4 Suppose that the system described by $\mathcal{H}$ is part of the system describing by $\mathcal{G}$. If $J_{\text{tot}}(\mathcal{G}) = 0$, then $J_{\text{tot}}(\mathcal{H}) = 0$; that is, if $J_{\text{tot}}(\mathcal{H}) > 0$ then $J_{\text{tot}}(\mathcal{G}) > 0$. The converse, as a general statement, is false.

The reader can make up a counterexample for himself or refer to one of the examples given in the next section. If fact, it is easy to make an example with $N = 3$, where all two-member interdependences vanish, that is,

$$J_{\text{tot}}(\mathcal{E}^{(1)} \otimes \mathcal{E}^{(2)}) = J_{\text{tot}}(\mathcal{E}^{(2)} \otimes \mathcal{E}^{(3)}) = J_{\text{tot}}(\mathcal{E}^{(3)} \otimes \mathcal{E}^{(1)}) = 0,$$

but the three-member interdependence does not, that is,

$$J_{\text{tot}}(\mathcal{E}^{(1)} \otimes \mathcal{E}^{(2)} \otimes \mathcal{E}^{(3)}) \neq 0.$$

In a case like this interdependence is a purely more-than-bilateral property. See Section 8.3.

It is sometimes useful to rewrite the content of Theorem 2.1 as

$$J_{\text{tot}}(\mathcal{G}^{(0)}) = \sum_{\text{all branching points in a tree}} J(\text{branching point}) \tag{2.32}$$

and consider this as an expansion of $J_{\text{tot}}(\mathcal{G}^{(0)})$ into components. Each tree will offer a new expansion. We can imagine ourselves engaged in a building-demolishing business. There are all kinds of ways of reducing the building into individual bricks, and some part of the work may be very hard, some easy. But the total amount of labor involved is always the same, irrespective of where we start and how we proceed. That is what our theorem says. The strong point of this expansion (2.32) is that each term is non-negative and expresses adequately the strength of interdependence corresponding to a well-defined partition. Compare with this the following expansion proposed by some authors (see page 57 of [F-1]). Let $\mathcal{G}^{(0)}$ be the product of N $\mathcal{E}^{(a)}$'s. Take a set of $L(\leq N)\mathcal{E}^{(a)}$'s out of these N, and call the product of these $\mathcal{E}^{(a)}$'s $\mathcal{F}$. For a given integer L there are $\binom{N}{L}$ different $\mathcal{F}$'s, but for the moment let us fix our attention on one particular $\mathcal{F}$. The entropy of $\mathcal{F}$ will be denoted by $X^{(L)}(\mathcal{F})$. Within this $\mathcal{F}$ we can take $\binom{L}{K}$ different sets of K $\mathcal{E}^{(a)}$'s, and each set defines the entropy of the corresponding product of $\mathcal{E}^{(a)}$'s. Let $X^{(K)}(\mathcal{F})$ denote the sum of the entropies of all these $\binom{L}{K}$ different products of K $\mathcal{E}^{(a)}$'s. Define $Y(\mathcal{F})$ by

$$Y(\mathcal{F}) = \sum_{K=1}^{L} (-1)^{L-K+1} X^{(K)}(\mathcal{F}). \tag{2.33}$$

Then we can prove an expansion theorem

$$J_{\text{tot}}(\mathcal{G}^{(0)}) = \sum_{\mathcal{F}} Y(\mathcal{F}), \tag{2.34}$$

where the summation extends over all the possible $2^N - (N + 1)$ subproducts $(L \geq 2)\mathcal{F}$ that can be taken within $\mathcal{G}^{(0)}$. Unfortunately I cannot see much beauty or utility in this expansion, mainly because the terms on the right side are not necessarily non-negative; neither does there seem to be any clear intuitive meaning to their signs.

The following expansion [W-11], which is a special case of (2.32), enjoys a particular simple interpretation:

$$J_{\text{tot}}(\mathcal{G}^{(0)}) = \sum_{r=2}^{N} J\left(\bigotimes_{a=1}^{r} \mathcal{E}^{(a)}; \quad \bigotimes_{b=1}^{r-1} \mathcal{E}^{(b)}, \quad \mathcal{E}^{(r)}\right). \tag{2.35}$$

Suppose that we construct the whole system $O^{(0)}$ by juxtaposing its constituent parts, one by one, by joining first A_2 to A_1, then A_3, then A_4, and so on, until finally all N atomic constituents are put together. Then each term on the right side of (2.35) represents the interdependence between the existing part and the new building block being added.

2.3. SIMPLE ILLUSTRATIONS OF IDA—STRUCTURE OF ABSTRACT PATTERNS

Structure means the absence of certain possible configurations, and chaos means the presence, factual or potential, of all possible configurations. Whether or not the popular etymology of "gas" as a derivative from "chaos" is correct, each molecule in a gas is flying almost independently of the other molecules, and all possible relative positions and all possible relative motions of molecules (within the conservation laws) are to be expected to occur sooner or later. On the other hand, in a crystal, which shows a strong structural organization, neighboring molecules interfere with each other so strongly that they can do nothing but align themselves in a certain regular way, and any other configurations (except for small thermal agitations) are all but excluded. In a case like this we need locate only a few molecules; the positions of all other molecules will be then automatically determined. If we compare the position of molecules to signals or symbols in communication, this last situation will correspond to a redundant message, for we need read only the beginning few words, from which we can infer all the rest. This simple explanation will be enough to convince ourselves that the three notions—restriction in possible configurations, organizational structure, and redundancy—are intimately related. In this section we take extremely simple examples and show that the method of IDA developed in the last section is useful in quantitatively analyzing problems regarding organizations and redundancy. A more sophisticated use of IDA will be added in Section 8.3. The actual examples we use in this section are "spatial" patterns. However, since the method of IDA does not take into account any geometrical properties, such as contiguity, distance, and angle, we speak of "abstract" patterns here.

Suppose that we have a square divided into four equal square cells, as in Figure 2.2. Each cell can be black or white. Then there are $2^4 = 16$ different figures. In order to identify each one of them, let us introduce four sets of propositions, $\mathcal{E}^{(1)}$, $\mathcal{E}^{(2)}$, $\mathcal{E}^{(3)}$, and $\mathcal{E}^{(4)}$, which correspond respectively to the

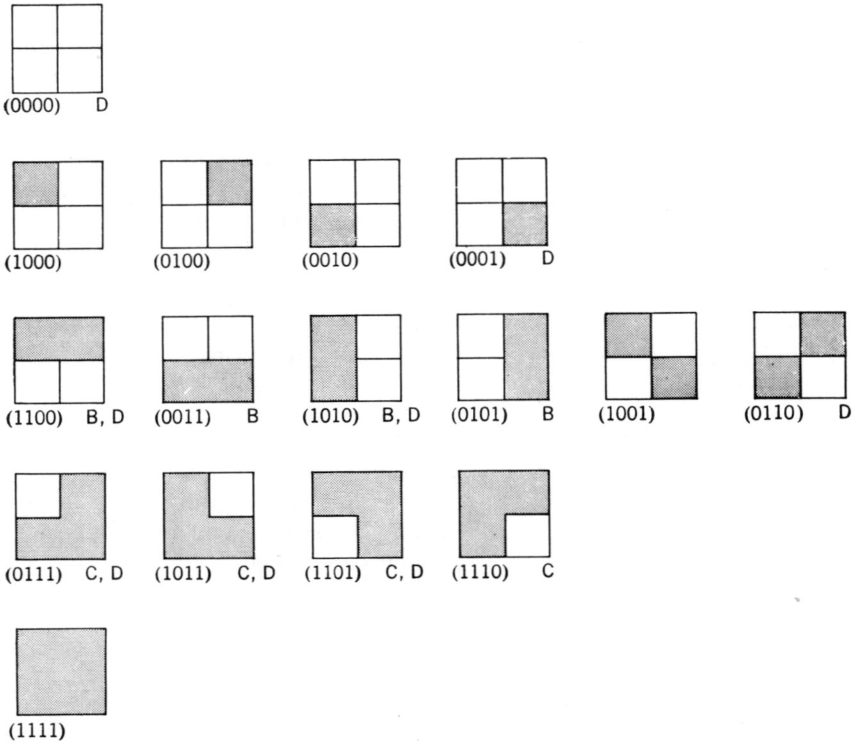

Figure 2.2 Structure of abstract patterns. Language A uses all 16 figures with equal probabilities. Languages B, C, and D use the figures shown above with equal probabilities.

upper left, upper right, lower left, and lower right cells. Each set consists of two propositions: "the cell is white" and "the cell is black." To make this simpler, let us introduce four stochastic variables, y_1, y_2, y_3, and y_1, corresponding to $\mathcal{E}^{(1)}$, $\mathcal{E}^{(2)}$, $\mathcal{E}^{(3)}$, and $\mathcal{E}^{(4)}$. Each variable can be 0 or 1, 0 meaning white and 1 meaning black. Thus, for instance, a joint expression ($y_1 = 1$, $y_2 = 1$, $y_3 = 0$, $y_4 = 0$), or simply (1100), means that the upper row is black and the lower row is white.

If all 16 figures are allowed there is no redundancy, for one has to determine the states of all four cells in order to identify the figure. But if the set of allowed figures is limited to a fraction of 16 figures, redundancy will appear. If one compares "cell," "figure," and "set of figures," respectively, to word, message, and language, what we are going to do is analyze the interdependence of words.

To fix our idea, let us first assume that the language consists of four figures (1100), (1010), (0011), and (0101), that is, upper horizontal line, left vertical line, lower horizontal line, and right vertical line, and that they are used with equal probability (see explanation of Language B in Figure 2.2).

$$p(1100) = p(1010) = p(0011) = (0101) = \tfrac{1}{4}. \tag{2.36}$$

Then, obviously, we have the total entropy $S(\mathcal{G}^{(0)})$, equal to

$$S(y_1y_2y_3y_1) = 2 \text{ bits}, \tag{2.37}$$

corresponding to the fact that each figure can carry 2 bits of information since it verifies one out of four equally probable cases. Similarly, the entropy functions of three variables are all equal to 2 bits, since four different configurations can appear with equal probability.

$$S(y_2y_3y_4) = S(y_3y_4y_1) = S(y_4y_1y_2) = S(y_1y_2y_3) = 2 \text{ bits}. \tag{2.38}$$

But the entropy functions of two variables divide themselves into two categories. If the two variables are taken horizontally or vertically, their entropies are 2 bits.

$$S(y_1y_2) = S(y_3y_4) = S(y_1y_3) = S(y_2y_4) = 2 \text{ bits}. \tag{2.39}$$

This is because there are four possible cases, black-white, white-black, black-black, and white-white, with an equal probability. But if the two variables are taken diagonally their entropies are 1 bit.

$$S(y_1y_4) = S(y_2y_3) = 1 \text{ bit}. \tag{2.40}$$

This is because there are only two possible cases, black-white and white-black. The entropy function of one variable is obviously 1 bit, since black and white appears with equal probability in each cell.

$$S(y_1) = S(y_2) = S(y_3) = S(y_4) = 1 \text{ bit}. \tag{2.41}$$

Observation of one cell gives only 1 bit of information; hence it is sufficient to select two figures out of four possibilities but not sufficient to identify one figure out of four. Observation of two cells placed diagonally also gives 1 bit, which is not sufficient to identify the entire figure. But observation of two cells placed horizontally or vertically gives 2 bits and is sufficient to identify the figure. This is obvious by inspection of the picture, but it is interesting to see the mathematical expression of the situation.

Now the total interdependence is

$$\begin{aligned} J_{\text{tot}} &= S(y_1) + S(y_2) + S(y_3) + S(y_1) - S(y_1y_2y_3y_4) \\ &= 4 - 2 = 2 \text{ bits}. \end{aligned} \tag{2.42}$$

This corresponds to the fact that with $p(0) = p(1) = \frac{1}{2}$ one could at maximum convey 4 bits of information (using all 16 figures with equal probability), but by limiting oneself to the four figures one can send only 2 bits. The difference $4 - 2 = 2$ is the loss of information, or redundancy.

Now if we divide the set of four variables (y_1, y_2, y_3, y_4) into a group of three variables and one remaining variable, the correlation corresponding to this branching is independent of the way the division is made and is equal to

$$J[(y_1y_2y_3y_4); (y_1y_2y_3), y_4] = 2 + 1 - 2 = 1 \text{ bit.} \tag{2.43}$$

This corresponds to the fact that a group of three variables conveys 2 bits of information, while one variable conveys 1 bit of information, but altogether they can convey only 2 bits of information. This is because variable y_4 is completely determined by the other three variables. This case should be compared with (2.13) with

$$\mathcal{E} = y_4, \qquad \mathcal{F} = (y_1y_2y_3), \qquad \mathcal{G} = (y_1y_2y_3y_4). \tag{2.44}$$

Here $\mathcal{E}$ is completely determined by $\mathcal{F}$, but $\mathcal{F}$ is not determined by $\mathcal{E}$.

More interesting is the case of dichotomy of the four variables into two groups of two variables. There are two cases, typified by

$$J[(y_1y_2y_3y_4); (y_1y_2), (y_3y_4)] = 2 + 2 - 2 = 2 \text{ bits} \tag{2.45}$$

and

$$J[(y_1y_2y_3y_4); (y_1y_4), (y_2y_3)] = 1 + 1 - 2 = 0 \text{ bit.} \tag{2.46}$$

The first case (2.45) means, in the light of (2.14), that (y_1y_2) and (y_3y_4) are mutually completely dependent. If (y_1y_2) is black-black, then (y_3y_4) is white-white, and vice versa. If (y_1y_2) is black-white, then (y_3y_4) is also black-white and vice versa. In the second case (2.46) each group (y_1y_4) and (y_2y_3) means a diagonal and can be black-white or white-black. Even if we know that (y_1y_4) is one of the two, say, black-white, there is still probability $\frac{1}{2}$ that (y_2y_3) will be, black-white and probability $\frac{1}{2}$ that (y_2y_3) will be white-black. Therefore this is a case of complete independence, corresponding to the equals sign in (2.11); (2.45) shows the maximum interdependence and (2.46) shows the minimum interdependence.

As far as the interdependence between two variables is concerned, there are two cases. For a pair of variables that are horizontal or vertical neighbors there is no interaction, as exemplified by

$$J[(y_1y_2); y_1, y_2] = 1 + 1 - 2 = 0. \tag{2.47}$$

For a pair of variables that are diagonal neighbors there is an interaction of the type

$$J[(y_1y_1); y_1, y_4] = 1 + 1 - 1 = 1. \tag{2.48}$$

This is because if one of them is black, the other must be white.

The expansion (2.32) can be done in different ways, depending on how the polychotomy is made; for instance, if $(y_1y_2y_3y_4)$ is first divided into (y_1y_2) and (y_3y_1) by a dichotomy, and then both (y_1y_2) and (y_3y_4) are subjected to a further dichotomy, one obtains

$$\begin{aligned} 2 &= J[(y_1y_2y_3y_4);\, y_1,\, y_2,\, y_3,\, y_4] \\ &= J[(y_1y_2y_3y_4);\, (y_1y_2),\, (y_3y_4)] \\ &\quad + J[(y_1y_2);\, y_1,\, y_2] + J[(y_3y_4);\, y_3,\, y_4] \\ &= (2 + 2 - 2) + (1 + 1 - 2) + (1 + 1 - 2) = 2 + 0 + 0, \end{aligned} \tag{2.49}$$

which shows that the entire interdependence can be attributed to the relation between the upper and lower halves of the figure. But we have to realize that two horizontal (or vertical) neighbor cells are entirely unrelated to each other. On the other hand, if we first divide the four cells into two pair of diagonals and then separate each diagonal into its constituent parts, we obtain $2 = 0 + 1 + 1$, where the first 0 corresponds to the first dichotomy, and each of the next two 1's corresponds to a pair of diagonal neighbors that are strongly linked. The reader can try the expansions (2.33) and (2.35) as exercises.

By way of digression, we add here a remark of general nature, which can easily be understood with the help of the example we have just been discussing. The total interdependence has the general expression, as in (2.28),

$$J_{\text{tot}}(\mathcal{G}^{(0)}) = \sum_{a=1}^{N} S(\mathcal{E}^{(a)}) - S(\mathcal{G}^{(0)}). \tag{2.50}$$

Hence it can become larger if individual $S(\mathcal{E}^{(a)})$ become larger and/or the total $S(\mathcal{G}^{(0)})$ becomes smaller. For instance, compare two cases: in Case A all 16 figures are used with equal probability; in Case B only four of them are used with equal probability, namely, the ones we have used, (1100), (0011), (1010), and (0101). In both cases $S(\mathcal{E}^{(a)})$ is 1 bit, but $S(\mathcal{G}^{(0)})$ is 4 bits in Case A and 2 bits in Case B. This means that the redundancy in Case B was brought about by reducing $S(\mathcal{G}^{(0)})$. On the other hand, compare Case B with a third case, C, in which also only four of the 16 figures are used with equal probability, but this time we use (0111), (1011), (1101), and (1110). It can be easily seen that each $S(\mathcal{E}^{(a)})$ in Case C is 0.8113 bit, instead of 1 bit as in Case B, whereas $S(\mathcal{G}^{(0)})$ is still 2 bits. This means that Case B has higher redundancy than C because $S(\mathcal{E}^{(a)})$ are larger. This remark supplements the discussion following (2.2).

We now introduce a more intriguing case than those discussed above. Suppose that the language consists of eight figures, (0000), (0001), (1100), (0110), (1010), (1101), (0111), and (1011) (see the explanation of Language D in Figure 2.2). At first glance this collection seems to be arbitrary, but it has very good symmetry. We assume again that these eight figures appear with equal probability. We can easily confirm that all one-cell entropies are 1 bit.

$$S(y_i) = 1, \qquad i = 1, 2, 3, 4 \tag{2.51}$$

and all two-cell entropies are 2 bits

$$S(y_i y_j) = 2, \qquad i \neq j, \qquad i, j = 1, 2, 3, 4. \tag{2.52}$$

But there are two different values for three-cell entropies; namely, for the particular triplet $(y_1 y_2 y_3)$ we have

$$S(y_1 y_2 y_3) = 2, \tag{2.53}$$

but for the rest we have

$$S(y_i y_j y_k) = 3, \qquad (i, j, k) \neq (1, 2, 3). \tag{2.54}$$

The total entropy is also 3 bits. Since we discuss this case in detail in an illustration of Section 8.3, we limit ourselves here to just a few remarks. First, we note that all two-cell interdependence

$$J_{\text{tot}}(y_i y_j) = J[(y_i y_j);\, y_i y_j] = 0, \qquad i \neq j, \;\; i, j = 1, 2, 3, 4, \tag{2.55}$$

whereas three-cell interdependence

$$J_{\text{tot}}(y_i y_j y_k) = J[(y_i y_j y_k);\, y_i, y_j, y_k] \tag{2.56}$$

is 1 for $(y_1 y_2 y_3)$ and is 0 for all others. Since the total interdependence is also 1 (= 4 − 3) bit, the 1-bit interdependence in the group $(y_1 y_2 y_3)$ represents the entire interdependence. We can see, therefore, that y_4 has no linkage with the rest. Within the group of $(y_1 y_2 y_3)$, however, there exists no two-cell interdependence. Hence the interdependence in the group $(y_1 y_2 y_3)$ is purely three-body interdependence. The reader who wants to pursue the topics treated in the first three sections of the present chapter may take up Section 8.3 at this point before going to the next section.

2.4. PROBABILISTIC DEPENDENCE IN STATIONARY STOCHASTIC CHAINS

Any phenomenon, as presented to our cognizance, is in general a sequence of "situations" or "states" arranged on a time axis. These successive states usually are partly correlated and partly haphazard. The stochastic chain, which will presently be defined more precisely, is a mathematical model for such a sequence of observed data. For this reason the study of a stochastic chain particularly its structure, is extremely important in a theory of cognition. Although the technical problem of communication does not belong to the main topics of this book, it may be noted that the theory of communication also starts with regarding messages of communication as stochastic chains. This section is devoted to explaining the basic notions pertaining to stochastic chains, such as stationarity, range of interdependence, Markovianity, and ergodicity. The next section shows how to adapt the method of IDA to deal with stochastic chains. The reader who is not particularly interested in the rigorous build-up of concepts may omit this section in his first reading and proceed to Section 2.5.

We introduce an integer parameter t to designate an "instant" in time or a "position" in a sequence. Usually we let t take any integer value, positive or negative or zero, but occasionally we limit it to a non-negative domain. Besides this parameter, we consider a set of n possible states, $\{z_i\}$, $i = 0, 1, \cdots, n-1$. If we attach one of the n possible states z_i to each value of t—that is, consider the index i of z_i as a function of t, $i = i(t)$—we obtain an infinitely long sequence of states arranged on a discrete "time axis." Each of such infinite sequences (considered as sequences either of z_i or of i) is denoted by x. The symbol x here is not a quantity, but merely an abstract symbol standing for an object (which in this case is an infinitely long sequence), in just the same way as we speak of a proposition A or a point P. We consider the set of all possible x's and call it X. It is easy to see that there are continuously many x's in X, although n is finite. In fact, suppose that we allow ourselves to use a half-infinity of positions ($t = 0, 1, 2, \cdots$), each of which is occupied by 0 or 1 (i.e., $n = 2$), and imagine that a binary point is placed to the left of the starting position, $t = 0$. Then each sequence considered as a binary number will express a real number in the interval $[0, 1]$ and, conversely, any real number in this domain can be expressed as a sequence of this type. Hence X contains at least as many x's as the number of real numbers in the interval $[0, 1]$.

We have to pay attention to one of the consequences of this situation. If the probabilities are given to sets of real numbers ξ between 0 and 1 in some smooth manner, for instance, in such a way that $\Pr\{0 \le \xi < a\} = a$, then a set of enumerably many numbers, such as the set of all rational numbers, will receive probability zero. This implies that in a population of continuously many samples events of a certain class may happen infinitely many (countably many) times and yet have zero probability.

In this set X we can consider various subsets, taking different collections of x's. But the most natural and useful kind of subset is characterized by a condition of the following type: the tth position is in state i_0, the $(t+1)$st position is in state $i_1, \cdots$, and the $(t+r-1)$st position is in state i_{r-1}. Letting $x(t)$ stand for the state of the tth position of the sequence x, we can define such a set of x's (called "cylinder") by

$$\{x \mid x \in X, x(t+a) = i_a, \qquad a = 0, 1, 2, \cdots, r-1\}, \tag{2.57}$$

which we denote by $C[t, \{i_0, i_1, \cdots, i_{r-1}\}]$ to bring out that this is a set of x's on which a given ordered set of r numbers $\{i_0, i_1, \cdots, i_{r-1}\}$ occupies r positions starting at position t. When there are two cylinders with the same set of numbers $\{i_0, i_1, \cdots, i_{r-1}\}$ but with different values of t, we shall speak as if they were the same object located at different places.

For a given r there are n^r different sequences $\{i_0, i_1, \cdots, i_{r-1}\}$, hence n^r different cylinders of length r at a given position t. Each $x \in X$ must belong to one and only one of these n^r cylinder sets. If we let $\mathcal{C}^{(r)}(t)$ denote this set of n^r disjoint and exhaustive cylinder sets, we can interpret $\mathcal{C}^{(r)}(t)$ also as a logical spectrum, by making a one-to-one correspondence between a cylinder and a proposition stating that x belongs to that cylinder. A Boolean lattice $\bar{\mathcal{C}}^{(r)}(t)$ generated by $\mathcal{C}^{(r)}(t)$ is a set of all those sets of x's that can be expressed as a disjunction (merger) of some cylinders, including the empty set Φ and the total set X. The number of members of $\bar{\mathcal{C}}^{(r)}(t)$ is 2^{n^r} because there are n^r atoms of the lattice. The logical spectrum $\mathcal{C}^{(r)}(t)$ of length r can obviously be regarded as a product [in the sense of (1.59)] of r logical spectra of length 1:

$$\mathcal{C}^{(r)}(t) = \mathcal{C}^{(1)}(t) \otimes \mathcal{C}^{(1)}(t+1) \otimes, \cdots, \otimes \mathcal{C}^{(1)}(t+r-1). \qquad (2.58)$$

If two logical spectra $\mathcal{C}^{(r)}(t)$ and $\mathcal{C}^{(s)}(u)$ are such that the range of positions covered by the second includes the range of positions covered by the first, then each member of the lattice $\bar{\mathcal{C}}^{(r)}(t)$ is a member of the lattice $\bar{\mathcal{C}}^{(s)}(u)$. For instance, we have

$$C[t, \{i_0, i_1\}] = \bigcup_{i2=0}^{n-1} C[t, \{i_0, i_1, i_2\}], \qquad (2.59)$$

showing that a cylinder defining $\bar{\mathcal{C}}^{(2)}(t)$ can be expressed as a disjunction of cylinders defining $\bar{\mathcal{C}}^{(3)}(t)$. We can write this situation as

$$\bar{\mathcal{C}}^{(r)}(t) \subset \bar{\mathcal{C}}^{(s)}(u) \quad \text{if} \quad u \leq t \quad \text{and} \quad t + r \leq u + s, \qquad (2.60)$$

where the set-theoretical inclusion symbol $\subset$ is used in the sense that $\bar{\mathcal{C}}^{(r)}(t)$ and $\bar{\mathcal{C}}^{(s)}(u)$ are considered *sets* of x's.

If a probability is given to each cylinder $C[t, \{i_0, i_1, \cdots, i_{r-1}\}]$ of the logical spectrum $\mathcal{C}^{(r)}(t)$, in accordance with (1.4) and (1.5), then all members of the lattice $\bar{\mathcal{C}}^{(r)}(t)$ will be given probabilities according to (1.6), satisfying (1.7), (1.8), and (1.9). If $\bar{\mathcal{C}}^{(r)}(t) \subset \bar{\mathcal{C}}^{(s)}(u)$, and if the probabilities are given to members of $\bar{\mathcal{C}}^{(s)}(u)$, then the probabilities of the members of $\bar{\mathcal{C}}^{(r)}(t)$ are automatically determined. If the probabilities are determined on $\bar{\mathcal{C}}^{(r)}(t)$, then any set of x's definable in terms of $x(t), x(t+1), \cdots, x(t+r-1)$, is given a probability. In order that any set whatsoever of x's in X may be given a probability, it is necessary that the probabilities be determined on a σ-algebra

$\bar{\mathfrak{C}}^{(\infty)}(-\infty)$ generated by all possible cylinders that can be taken at any place on the domain of t from $-\infty$ to $+\infty$. In an unrigorous manner of speech $\bar{\mathfrak{C}}^{(\infty)}(-\infty)$ may be considered as a sort of lattice corresponding to $\bigotimes_{t=-\infty}^{t=+\infty} \mathfrak{C}^{(1)}(t)$. A σ-algebra is very much the same as a lattice, but the difference is that in a σ-algebra conjunction and disjunction of countably many elements are allowed. It must be admitted that the mathematical definitions and theorems given in Chapter 1 are not powerful enough to deal with a σ-algebra. Fortunately, however, most of its properties can be guessed from knowledge about the lattice. See Section 7.2 for more about σ-algebra.

For actual empirical data obtained by observations made along real time there are no infinitely long sequences, nor is the number of available sequences infinite. In principle, however, we can obtain a very large collection of very long experimental sequences resulting from repeated similar observations, and we can actually determine the relative frequency of the group of those sequences that satisfy the condition defining a cylinder. This may be regarded as an empirical analogue of the probability of a cylinder. For theoretical purposes we have to simplify the matter by using the idealized mathematical entities described in this section.

We may note here that it is sometimes convenient to use the notion of stochastic variables in order to interpret the meaning of probabilities. To each value of t we attach a stochastic variable whose value $x(t)$ can be any one of n possibilities. In other words, each sequence of numbers is considered here as produced by a sequence of stochastic variables. If we further attach an intuitive sense of time to the parameter t, we obtain a picture of a sequence of numbers being produced, digit after digit, by some stochastic rule. In this sense each sequence of numbers, inasmuch as it is considered a member of the set of sequences governed by such a stochastic rule, is called a "stochastic chain."

In terms of the $x(t)$, which may now be regarded either as the tth digit of an infinite sequence x or as the value of the stochastic variable at t, we can write the probability of a cylinder of the type (2.57) as

$$\begin{aligned}\Pr\{x(t) = i_0, x(t+1) = i_1, \cdots, x(t+r-1) = i_{r-1}\}\\ = \Pr\{C[(t), \{i_0, i_1, \cdots, i_{r-1}\}]\}.\end{aligned} \tag{2.61}$$

When the probabilities of this type depend only on the set of numbers $\{i_0, i_1, \cdots, i_{r-1}\}$ and do not depend on t, for any arbitrary r, we say that the stochastic chain is stationary. Our discussion in this book is restricted to stationary chains. This assumption allows us to use the notation

$$p^{(r)}(x_1, x_2, \cdots, x_r) \tag{2.62}$$

to denote $\Pr\{x(t+1) = x_1, x(t+2) = x_2, \cdots, x(t+r) = x_r\}$, which does not here depend on t. The superscript (r) is often useful though admittedly redundant. (Hereafter we count integers from 1 instead of 0 for convenience.)

Evidently the quantity (2.62) has to obey

$$\sum_{x_r=1}^{n} p^{(r)}(x_1, x_2, \cdots, x_r) = p^{(r-1)}(x_1, x_2, \cdots, x_{r-1}),$$
$$\sum_{x_1=1}^{n} p^{(r)}(x_1, x_2, \cdots, x_r) = p^{(r-1)}(x_2, x_3, \cdots, x_r), \tag{2.63}$$

and

$$\sum_{x_1=1}^{n} \sum_{x_2=1}^{n} \cdots \sum_{x_r=1}^{n} p^{(r)}(x_1, x_2, \cdots, x_r) = 1. \tag{2.64}$$

The conditional probability of x_r when the $(r - 1)$ preceding positions are known to be $(x_1, x_2, \cdots, x_{r-1})$ is given by

$$p^{(r)}(x_r \mid x_1, \cdots, x_{r-1}) = \frac{p^{(r)}(x_1, x_2, \cdots, x_r)}{p^{(r-1)}(x_1, x_2, \cdots, x_{r-1})}. \tag{2.65}$$

If the stochastic chain under consideration is such that

$$p^{(r)}(x_r \mid x_1, \cdots, x_{r-1}) = p^{(r-1)}(x_r \mid x_2, \cdots, x_{r-1}) \tag{2.66}$$

for all values of $(x_1, x_2, \cdots, x_r)$, then the knowledge of the state of the $(r - 1)$st neighbor to the left of a certain position, on top of the knowledge of the states of the positions in between, does not affect the probability of the state at the position in question. If the condition (2.66) holds for all $(x_1, x_2, \cdots, x_r)$ and for all r such that $r > \nu$, and breaks down for at least one choice of $(x_1, x_2, \cdots, x_r)$ for $r = \nu$, we say that the "range of dependence" of the stochastic chain is ν. A chain with $\nu = 2$ is called a (simple) Markov chain. A chain with $\nu = 1$ is nothing but a purely random sequence. We shall occasionally deal with chains with an infinite range of dependence, which means that there is no finite integer ν such that for all $r > \nu$ (2.66) holds for all $(x_1, x_2, \cdots, x_r)$.

This definition of range of dependence is not complete unless we have an agreement as to how to understand the condition (2.66) when one or both of the conditional probabilities involved there become indeterminate due to the expression (2.65) having zero as numerator and denominator. Our agreement is that we consider the condition (2.66) as upheld when one (actually the left one) or both of the conditional probabilities become indeterminate. Justification for this apparently arbitrary covenant will be offered presently.

It is of some theoretical interest to note that a stochastic chain with a finite range $(\nu < \infty)$ of dependence can be formally reduced to a (simple) Markov chain ($\nu = 2$) by considering the state of a block of several consecutive positions as the state at one position. If ν is the range of dependence we then consider $\mathcal{F}(t) = \mathcal{C}^{(\nu-1)}(t) = \mathcal{C}^{(1)}(t) \otimes \mathcal{C}^{(1)}(t+1) \otimes, \cdots, \otimes \mathcal{C}^{(1)}(t+\nu-2)$ as the logical spectrum characterizing the state of position t. If we correspondingly let y_r stand collectively for $(x_r, x_{r+1}, \cdots, x_{r+\nu-2})$, then we have indeed

$$\begin{aligned} p^{(3)}(y_3 \mid y_1 y_2) &= p^{(\nu+1)}(x_{\nu+1} \mid x_1, x_2, \ldots, x_\nu) \\ &= p^{(\nu)}(x_{\nu+1} \mid x_2, x_3, \cdots, x_\nu) \\ &= p^{(2)}(y_3 \mid y_2). \end{aligned} \tag{2.67}$$

In the foregoing we derived the conditional probability $p^{(r)}(x_r \mid x_1, \cdots, x_{r-1})$ from the "unconditional" probability $p^{(r)}(x_1, x_2, \cdots, x_r)$, which corresponds to the stationary relative frequency of a cylinder $(x_1, x_2, \cdots, x_r)$ in an ensemble of infinite sequences. We can also proceed the other way around; that is, starting with the conditional probability $p^{(\nu)}(x_\nu \mid x_1, \cdots, x_{\nu-1})$, we can build an ensemble of infinite sequences and derive therefrom the "unconditional" probabilities $p^{(r)}(x_1, x_2, \cdots, x_r)$ (see [W-6]). Suppose that (2.66) is satisfied for all $r > \nu$ and that we are given the values of $p^{(\nu)}(x_\nu \mid x_1, \cdots, x_{\nu-1})$ for all $(x_1, x_2, \cdots, x_\nu)$. There are $n^{\nu-1}$ different sequences of $\nu - 1$ numbers $(x_1, x_2, \cdots, x_{\nu-1})$, but for the moment we consider only those infinite sequences that start with one particular $(\nu - 1)$-position sequence $(x_1, x_2, \cdots, x_{r-1})$. Then the probability $p^{(\nu)}(x_\nu \mid x_1, \cdots, x_{\nu-1})$ will give the fraction of infinite sequences whose ν first positions are $(x_1, x_2, \cdots, x_\nu)$. Similarly, from (2.66), the probability $p^{(\nu+1)}(x_\nu, x_{\nu+1} \mid x_1, x_2, \cdots, x_{\nu-1}) = p^{(\nu)}(x_{\nu+1} \mid x_2, x_3, \cdots, x_\nu) p^{(\nu)}(x_\nu \mid x_1, x_2, \cdots, x_{\nu-1})$ will give the fraction of sequences whose first $(\nu + 1)$ positions are $(x_1, x_2, \cdots, x_\nu, x_{\nu+1})$ among the sequences whose first $(\nu - 1)$ positions are $(x_1, x_2, \cdots, x_{\nu-1})$. Repeating this process, we can build an ensemble of half-infinite sequences that start with $(x_1, x_2, \cdots, x_{\nu-1})$ and are governed by a conditional probability $p^{(\nu)}(x_\nu \mid x_1, \cdots, x_{\nu-1})$. If a stochastic chain is produced by a conditional probability of length ν, which cannot be reduced to a conditional probability of length less than ν, then, of course, the range of interdependence in the chain becomes ν.

More generally, we can give the probability $w(x_1, x_2, \cdots, x_{\nu-1})$ to each of the $n^{\nu-1}$ possible initial sequences, and build an ensemble in which $w(x_1, x_2, \cdots, x_{\nu-1})$ represents the fraction of those sequences that start with $(x_1, x_2, \cdots, x_{\nu-1})$. The sequences thus obtained start at a certain finite position t, but we can shift this starting position gradually to the left at the same time as (but more slowly than) we prolong the sequence to the right by the method explained above, and obtain an ensemble of sequences that extend indefinitely to the left and to the right as discussed before. Of course we can determine the relative frequency and hence the "unconditional" probability of any cylinder of any length at any place from this ensemble. This amounts to deriving the probability to any member of $\overline{\mathcal{C}}^{(\infty)}(-\infty)$ from the given conditional probabilities. We must know, however, that this simplification was possible by virtue of two major assumptions: stationarity and finiteness of range of dependence. Besides this basic fact there are a few more remarks to make regarding this process of producing stochastic chains from conditional probabilities.

First, it should be noted that even if the conditional probability $p^{(\nu)}(x_\nu \mid x_1, \cdots, x_{\nu-1})$ is independent of the position t, as this notation already implies, it is not at all guaranteed that the stochastic chain produced by it will be stationary and that the time-independent unconditional probability $p^{(\nu)}(x_1, x_2, \cdots, x_\nu)$ will be definable. For instance, take a simple case of a Markov chain ($\nu = 2$) with only two states ($n = 2$), defined by

$$\begin{pmatrix} p(1 \mid 1) & p(1 \mid 2) \\ p(2 \mid 1) & p(2 \mid 2) \end{pmatrix} = \begin{pmatrix} 0 & 1 \\ 1 & 0 \end{pmatrix}. \tag{2.68}$$

This conditional probability (transition probability) will produce a nonstationary chain except in the case in which the initial sequence (one position) has probability distribution $w(1) = w(2) = \frac{1}{2}$. In fact, if we have, for instance, $w(1) = 1$ for starting position $t = 0$, then any odd-numbered position will be in State 2 with certainty and any even-numbered position will be in State 1 with certainty; hence the probabilities $\Pr\{x(t) = i\}$, $i = 1, 2$, will depend on t. It goes without saying that nonstationary chains can occur in a variety of ways other than the one explained here.

Furthermore, even if the time-independent unconditional probability is definable, it can happen that the conditional probability becomes irretrievable. For instance, take the case

$$\begin{pmatrix} p(1 \mid 1) & p(1 \mid 2) \\ p(2 \mid 1) & p(2 \mid 2) \end{pmatrix} = \begin{pmatrix} a & 0 \\ 1 - a & 1 \end{pmatrix}, \tag{2.69}$$

with $0 < a < 1$. If a position is in State 1, there will be probability $(1 - a) \neq 0$ of the next position being in State 2. Once a position has become State 2, all the subsequent positions will remain in State 2. This implies that every sequence sooner or later enters State 2 and remains there no matter what the initial distribution $w(1)$, $w(2)$ may be. If we start the sequence at a finite position, say, $t = 0$, and if we put $w(1) \neq 0$, then $\Pr\{x(t) = 1\}$ will not be zero for a finite value of t; in fact, it will equal $w(1)a^t$; but if we shift the starting point to $-\infty$, the $\Pr\{x(t) = 1\}$ will become zero for any finite value of t, provided that $a < 1$. This gives a stationary chain, and we shall have $p^{(1)}(1) = p^{(2)}(1, 2) = p^{(2)}(2, 1) = p^{(2)}(1, 1) = 0$ and $p^{(1)}(2) = p^{(2)}(2, 2) = 1$. It is noteworthy that the constant a of (2.69) disappears completely from these unconditional probabilities. If we apply (2.65) to these probabilities in an attempt to obtain the conditional probabilities, we encounter the difficulties mentioned before; namely, $p^{(2)}(1 \mid 1) = p^{(2)}(1, 1)/p^{(1)}(1)$ and $p^{(2)}(2 \mid 1) = p^{(2)}(1, 2)/p^{(1)}(1)$ become indefinite. This example suggests that the indefinite conditional probabilities can be interpreted as having unspecified values, since any arbitrary value of a, provided that $a \neq 1$, leads to the indefinite forms. This is the basis of the convention we have made regarding the condition (2.66) for the case of indeterminate conditional probabilities.

In the rest of this section we briefly explain the concept of ergodicity in a stationary stochastic chain with finite range of interdependence, leading up to a theorem that connects ergodicity with a certain kind of probabilistic independence. Most of the theorems will be made plausible but are not rigorously demonstrated. Since any stochastic chain with a finite range of interdependence can be reduced theoretically to a Markov chain, we discuss mainly the latter case. Some of the topics dealt with in the following will recur in Section 5.2, but will be discussed there from a different point of view.

Ergodicity, roughly speaking, means that no matter where a chain starts or what state a chain passes through at a certain time, say, $t = 0$, it will enter "with probability 1" sooner or later any given state, say, z_i, and return to it infinitely many times thereafter. This "everywhere-migrating" property is not at all guaranteed by the stationarity of the stochastic chain; it is a more restrictive notion than stationarity. What is true of any stationary chain is the following "recurrence" property. If a chain x passes through a state z_i at a certain time, say, $t = 0$, then it comes back to z_i after any specified time $t > 0$, provided that z_i has a nonzero stationary probability. This theorem becomes quite plausible if we notice that, were this not true, the probability of z_i at t in the ensemble would gradually decrease with time t, which means either that the stationary probability of z_i is zero or else that it is not stationary. This theorem means that a chain passing through z_i at a certain time may perhaps migrate to some other states but will keep on coming back to z_i from time to time. Let A_i be the set of all those states visited by chains that have once passed through z_i, and let it be called "migration domain of z_i." This is a time-independent notion. Then if z_k is a member of A_i the migration domain A_k of z_k must be included in A_i by virtue of the definition. On the other hand, because of the recurrence theorem, A_i must be included in A_k. This is so because if an x passes first z_i and then later z_k, it has to go back to z_i after it has visited z_k, provided that z_i has a nonzero probability. Hence $A_i = A_k$ for all $z_k \in A_i$ except for states of zero probability. Such a subset is an "ergodic subset." On the other hand, a subset of z's such that the set of sequences that crosses its border has probability 0 is called an "invariant subset." An ergodic subset is an invariant subset that does not contain any other invariant subset within it except the empty set. The entire set of z_i and the empty set are obviously also invariant sets. Now, excluding the z's with probability zero, the entire set of z_i is thus divided up into ergodic subsets. The Markov chain is called ergodic if there is only one ergodic subset that has nonzero stationary probability.

An ergodic subset A can consist of more than one cyclic class B_μ, $\mu = 0, 1, \cdots, \lambda - 1$, of z_i's, $A = \bigvee_{\mu=0}^{\lambda-1} B_\mu$, and $B_\mu \wedge B_\kappa = \Phi$ for $\mu \neq \kappa$, such that if $x(t)$ is in B_μ, the probability is 1 that $x(t+1)$ will be in $B_{\mu+1}$, with the understanding that addition in the suffix is done modulo λ. In the example of (2.68), if $w(1) = w(2) = \frac{1}{2}$, we have a stationary chain, and the chain is ergodic because $\{z_1, z_2\}$ is an ergodic subset. But this ergodic subset splits into two cyclic classes $B_1 = \{z_1\}$ and $B_2 = \{z_2\}$, because any chain is a successive alternation of z_1 and z_2. In general each B can consist of more than one z.

We defined an invariant set and an ergodic set in terms of the z's, but we can similarly define the corresponding sets in terms of the x's. Consider the set of all those x's whose tth position $x(t)$ is in an invariant (ergodic) set of states; then their $(t+1)$st position $x(t+1)$ will be also in the same set. In other words, a set of x's is an invariant set if it is identical with the set of x^*'s, where x^* is obtained from x by shifting t by one unit. The stationary probability of z_i is of course derived from the probability originally defined on the set of x's.

It is now very important to note that in the nonergodic case, in which there are more than one ergodic subset with nonzero probability, there exists a strong interdependence between two positions infinitely apart. If z_j does not belong to the migration domain A_i of z_i, the joint probability of z_i and z_j will not be equal to the product of the probabilities of z_i and z_j.

$$\Pr\{x(t) = i, x(t_1) = j\} = 0, \tag{2.70}$$

even though

$$\Pr\{x(t) = i\} \neq 0 \quad \text{and} \quad \Pr\{x(t_1) = j\} \neq 0. \tag{2.71}$$

This probabilistic dependence remains no matter how far t and t_1 are separated. On the other hand, if it is an ergodic case, each chain behaves as if it has "forgotten" its own remote past.

One aspect of "forgetfulness" is manifest in the following theory, which can be considered as a consequence of Birkoff's ergodic theorem, but which can also be proven independently. The so-called strong law of large numbers in Markov chains states that if the Markov chain is ergodic, the relative frequence of appearance of a state z_i on each individual sequence x is equal, with probability 1, to the stationary probability of z_i. This means that no matter what state a chain may be in at a certain time, it will thereafter not only visit any other state z_i but also visit it with a predetermined relative frequency common to all sequences, except those sequences with probability

0. (The term ergodicity is overworked and also sometimes means the strong law of large numbers.)

Consider, for example, the case of a Markov chain governed by the conditional probability

$$\begin{pmatrix} p(1 \mid 1) & p(1 \mid 2) \\ p(2 \mid 1) & p(2 \mid 2) \end{pmatrix} = \begin{pmatrix} 1 - \varepsilon & \varepsilon \\ \varepsilon & 1 - \varepsilon \end{pmatrix}. \tag{2.72}$$

If we start at $t = 0$ with $w(1)$ and $w(2)$, then $\lim_{t\to\infty} \Pr\{x(t) = 1\} = \lim_{t\to\infty} \Pr\{x(t) = 2\} = \frac{1}{2}$ no matter what values $w(1)$ and $w(2)$ may have, provided that $\varepsilon \neq 0$ and $1 - \varepsilon \neq 0$. It is easy to see that a single sequence that starts with State 1 at $t = 0$ has probability 0 of remaining forever in State 1, for such a probability is given by $\lim_{k\to\infty} (1 - \varepsilon)^k = 0$, provided that $\varepsilon \neq 0$. Hence State 2 will appear at a finite t with probability 1. Once State 2 appears, the rest is the same as for a sequence that started with State 2. Since the transition between States 1 and 2 is entirely symmetrical, there is no reason why State 1 or State 2 should appear more frequently in a single sequence in the long run. Thus the relative frequencies of State 1 and State 2 will be $\frac{1}{2}$ in each individual sequence. Insofar as $\varepsilon \neq 0$, the Markov chain defined by (2.72) is ergodic. But if $\varepsilon = 0$, then a sequence that starts with State 1 will remain in State 1, and similarly for State 2. Thus the set of states $\{z_1, z_2\}$ is split into two ergodic subsets $\{z_1\}$ and $\{z_2\}$, and the Markov chain is no longer ergodic.

"Forgetfulness' is more explicitly brought out in the following theorem. The sufficiency of the condition mentioned in the theorem may be considered to be a very simple special case of a theorem that Abbott proved with great generality [A-1].

Theorem 2.2 A stationary Markov chain is ergodic if

$$\lim_{t_1\to\infty} [\Pr\{x(t_1) = j \mid x(t) = i\} - \Pr\{z_j\}] = 0 \tag{2.73}$$

for all i and j. If a stationary Markov chain is ergodic and has only one cyclic class, then the above condition is true.

Explanation. The first half is immediately proven because, if the chain is not ergodic, it will have at least two ergodic subsets, and if z_i and z_j belong to two different ergodic subsets, (2.73) will certainly be violated. The second half is the content of usual versions of ergodic theorem in Markov chains. Less restrictive than the strong law of large numbers of Markov chains is the formula (Yoshida-Kakutani law) that can be obtained by taking the average of the expression under $\lim_{t_1\to\infty}$ in (2.73) for the time interval from t to t_1

and bringing t_1 to infinity. [Y-13]. If there is more than one cyclic class, the actual value of Pr $\{x(t_1) = j \mid x(t) = i\}$ may oscillate with t_1, but its average may converge to Pr $\{z_i\}$. Limitation of Theorem 2.2 to the case of one cyclic class is intended to avoid this eventuality. See Theorems 5.6 and 5.12.

Now for a stochastic chain with a finite range ν of dependence, a cylinder of length $(\nu - 1)$ must be substituted for a cylinder of length 1 in the above discussion, as was seen in (2.67). [In (2.67), the letter y stood for a $(\nu - 1)$-position state, but in (2.74) it stands for a one-position state.] By introducing the notation

$$p(x_1, x_2, \cdots, x_\lambda \,\|[\tau]\|\, y_1, y_2, \cdots, y_\kappa) = \frac{\sum_{z_1=1}^{n} \sum_{z_2=1}^{n} \cdots \sum_{z_\tau=1}^{n} p(y_1, y_2, \cdots, y_\kappa, z_1, z_2, \cdots, z_\tau, x_1, x_2, \cdots, x_\lambda)}{p(y_1, y_2, \cdots, y_\kappa)} \tag{2.74}$$

we could generalize (2.73) and write the characterization of erodicity as

$$\lim_{\tau \to \infty} [p(x_1, x_2, \cdots, x_{\nu-1} \,\|[\tau]\|\, y_1, y_2, \cdots, y_{\nu-1}) - p(x_1, x_2, \cdots, x_{\nu-1})] = 0 \tag{2.75}$$

for all $(x_1, x_2, \cdots, x_{\nu-1})$ and all $(y_1, y_2, \cdots, y_{\nu-1})$. In the case of an infinite range of interdependence we require (2.75) for all possible lengths ν.

The classical reference text on Markov chains is M. Fréchet's book [F-9]. In English, two textbooks by Kemeny and Snell [K-6] and by Feller [F-4], among others, may be mentioned. For more advanced descriptions, see Khinchin [K-8], Hopf [H-12], Halmos [H-2], Kakutani [K-1], and Chung [C-5].

2.5. IDA IN STOCHASTIC CHAINS

The essence of the method of interdependence analysis lies in the possibility of expansion of the total interdependence existing in an assembly of stochastic variables into various terms, each of which has a clear meaning. Each position in a stochastic chain can be considered, as was explained in the last section, as a stochastic variable, and the expansions explained in Section 2.2 are all applicable to these stochastic variables. However, because of the invariance with respect to a shift in positions—that is, because of the stationarity of the chains—a very special expansion becomes very useful (see [W-6, W-11, W-12]).

To make the derivation easier, let us first introduce a simple mathematical formula. Consider a function $F(r)$ of an integer variable r, and define the "derivatives" by

$$[F(r)]_1 = F(r) - F(r-1)$$
$$[F(r)]_2 = [F(r)]_1 - [F(r-1)]_1 = F(r) - 2F(r-1) + F(r-2)$$
$$[F(r)]_3 = [F(r)]_2 - [F(r-1)]_2 = F(r) - 3F(r-1) + 3F(r-2) - F(r-3)$$
$$\vdots$$
$$[F(r)]_t = [F(r)]_{t-1} - [F(r-1)]_{t-1} = \sum_{s=0}^{t} (-1)^s \binom{t}{s} F(r-s). \tag{2.76}$$

If we have the condition

$$F(0) = 0 \tag{2.77}$$

we can obviously write

$$F(r) = \sum_{s=1}^{r} [F(s)]_1, \qquad r \geq 1. \tag{2.78}$$

Applying this formula again to each $[F(s)]_1$ in (2.78) and assuming that

$$[F(0)]_1 = F(0) - F(-1) = 0, \tag{2.79}$$

or, by virtue of (2.77),

$$F(-1) = 0, \tag{2.80}$$

we obtain

$$F(r) = \sum_{s=1}^{r} \sum_{t=1}^{s} [F(t)]_2, \qquad r \geq 1. \tag{2.81}$$

In (2.81) $[F(1)]_2$ appears r times, $[F(2)]_2$ appears $(r-1)$ times, and so on, and $[F(r)]_2$ appears once (see Figure 2.3). Therefore (2.81) can be rewritten as

$$F(r) = \sum_{s=1}^{r} (r - s + 1)[F(s)]_2, \qquad r \geq 1. \tag{2.82}$$

This is the formula we need for the expansion we use in this section, but we also mention here some formulas that lead to other expansions. Assuming

$$F(r) = 0 \quad \text{for} \quad r \leq 0, \tag{2.83}$$

we obtain

$$F(r) = \sum_{s=1}^{r} \binom{r - s + t - 1}{t - 1} [F(r)]_t \quad \text{for} \quad r \geq 1, \quad t \geq 1, \tag{2.84}$$

where $[F(r)]_t$ is the tth derivative in the sense of (2.76). We also have

$$F(r) = \sum_{s=1}^{r} \binom{r}{s} [F(s)]_s. \tag{2.85}$$

This last formula, which we prove later and use for another purpose, can also be related to the expansion (2.32).

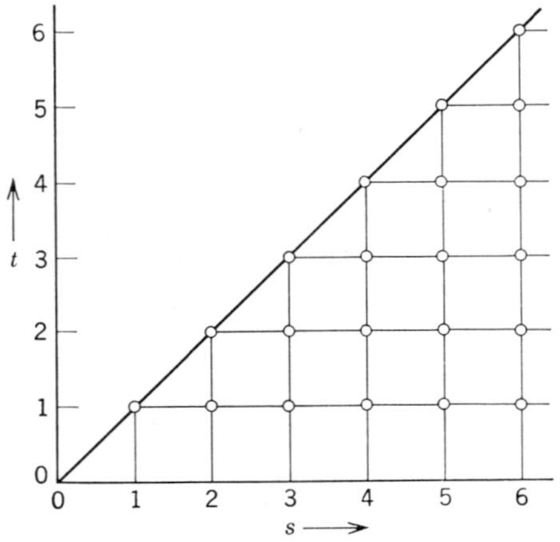

Figure 2.3 Domain of summation in (2.81). The summation $\sum_{s=1}^{r}\sum_{t=1}^{s}$ extends over the terms corresponding to the dots in the figure.

Now, assuming that the stochastic chain is stationary, we can use the probability distribution (2.62) obeying (2.63) and (2.64) to define entropy functions. Thus the r-position entropy is defined by

$$S^{(r)} = -\sum_{x_1}\sum_{x_2}\cdots\sum_{x_r} p^{(r)}(x_1, x_2, \cdots, x_r)\log p^{(r)}(x_1, x_2, \cdots, x_r). \tag{2.86}$$

Obviously

$$S^{(0)} = 0. \tag{2.87}$$

Since $S^{(r)}$ is not defined for $r < 0$, we can formally add the definitions

$$S^{(r)} = 0 \quad \text{for} \quad r < 0. \tag{2.88}$$

Then we can consider $F(r)$ in the foregoing formulas to be $S^{(r)}$, and obtain various formulas. Particularly interesting is (2.82) which becomes

$$S^{(r)} = -\sum_{s=1}^{r}(r - s + 1)W^{(s)} = rS^{(1)} - \sum_{s=2}^{r}(r - s + 1)W^{(s)}, \tag{2.89}$$

with

$$W^{(r)} = -S^{(r)} + 2S^{(r-1)} - S^{(r-2)}. \tag{2.90}$$

For $r = 1$ and 2 we have, with the help of (2.87), (2.88), and (2.90),

$$W^{(1)} = -S^{(1)} \tag{2.91}$$

and

$$W^{(2)} = -S^{(2)} + 2S^{(1)}. \tag{2.92}$$

For $r \geq 3$ we always have three S's involved in a $W^{(r)}$.

The total correlation existing in a sequence of r positions is, according to (2.28),

$$J_{\text{tot}}^{(r)} = rS^{(1)} - S^{(r)}. \tag{2.93}$$

We can now rewrite (2.89) as

$$J_{\text{tot}}^{(r)} = \sum_{s=2}^{r} (r - s + 1)W^{(s)}. \tag{2.94}$$

The expansions (2.89) and (2.94) were first introduced in [W-6], and the interesting meaning of the quantity $W^{(r)}$ was explained there; see also [W-11, W-12]. Some authors rediscovered the interesting properties of $W^{(r)}$ in later papers; see [H-6], for example. The quantity $W^{(r)}$ with $r \geq 2$ was called in my older papers "correlation index of range r" but has been renamed "interdependence index of range r" in this book. As we shall presently see, we always have

$$W^{(r)} \geq 0 \quad \text{for} \quad r \geq 2. \tag{2.95}$$

[$W^{(1)}$ (2.91) is the only renegade.] For this reason (2.94) can be considered as another expansion of the total interdependence into a certain kind of partial interdependence. Note that the coefficients of the W's in (2.94) are also all non-negative.

The meaning of $W^{(r)}$ is rather interesting. Consider a probability distribution for r variables given by

$$q^{(r)}(x_1, x_2, \cdots, x_r) = \frac{p^{(r-1)}(x_1, x_2, \cdots, x_{r-1})p^{(r-1)}(x_2, x_3, \cdots, x_r)}{p^{(r-2)}(x_2, x_3, \cdots, x_{r-1})}. \tag{2.96}$$

This $q^{(r)}$ can be considered a probability distribution, since it is non-negative and satisfies the normalization condition because of (2.63) and (2.64). According to the Gibbs theorem we have

$$-(\textstyle\sum)^r p^{(r)} \log p^{(r)} \leq -(\textstyle\sum)^r p^{(r)} \log q^{(r)}, \tag{2.97}$$

where equality holds if and only if $p^{(r)}(x_1, x_2, \cdots, x^r)$ and $q^{(r)}(x_1, x_2, \cdots, x_r)$ are equal for all values of $(x_1, x_2, \cdots, x_r)$. This condition can be rewritten as

$$\frac{p^{(r)}(x_1, x_2, \cdots, x_r)}{p^{(r-1)}(x_1, x_2, \cdots, x_{r-1})} = \frac{p^{(r-1)}(x_2, x_3, \cdots, x_r)}{p^{(r-2)}(x_2, x_3, \cdots, x_{r-1})} \tag{2.98}$$

or, in terms of the conditional probabilities,

$$p^{(r)}(x_r \mid x_1, x_2, \cdots, x_{r-1}) = p^{(r-1)}(x_r \mid x_2, x_3, \cdots, x_{r-1}). \tag{2.99}$$

This is identical with (2.66) and means that knowledge of x_1 over and above the knowledge of $(x_2, x_3, \cdots, x_{r-1})$ does not affect the probability of x_r. Hence this relation, if it holds for all $(x_1, x_2, \cdots, x_r)$, can be interpreted as the absence of "interdependence of range r." The largest value of r for which this relation breaks down is the "range of dependence" of the stochastic chain, as defined in the last section. It should be clearly understood that condition (2.99) is different from the probabilistic independence of x_1 and x_r, which can be expressed as $p(x_r \mid x_1) = p(x_r)$. The type of correlation, whose absence is characterized by $p(x_r \mid x_1) = p(x_r)$ instead of (2.99), is discussed briefly at the end of this section. Now it should be further noted that if one or both sides of (2.99) becomes indeterminate, it follows necessarily that $p^{(r)}(x_1, x_2, \cdots, x_r)$ is zero. This entails that the set of numbers $(x_1, x_2, \cdots, x_r)$ that makes this happen contribute an equal amount (zero in this case) to the left and right sides of (2.97), in just the same way as a set of numbers $(x_1, x_2, \cdots, x_r)$ for which (2.98) holds. This agrees perfectly with the convention we made in the last section regarding the indeterminate conditional probabilities.

Now the left-hand side of (2.97) is of course $S^{(r)}$, and the right-hand side becomes, by virtue of (2.96), equal to $2S^{(r-1)} - S^{(r-2)}$. Therefore (2.97) becomes

$$W^{(r)} \equiv -\Sigma^{(r)} p^{(r)}(\log q^{(r)} - \log p^{(r)}) \geq 0. \tag{2.100}$$

Equality in (2.100) holds if and only if (2.99) is the case for all $(x_1, x_2, \cdots, x_r)$; that is, if and only if the interdependence of range r is absent. On the other hand, all the W's are non-negative and contribute to the total interdependence. As a consequence $W^{(r)}$ can be considered as expressing the strength of interdependence of range r. It is, incidentally, also interesting to know that the prefix "inter" of interdependence is appropriate here in the following very special sense. For the operation of "time reversal," which may be realized by changing the sign of the parameter t, S functions remain invariant. The condition of absence of interdependence of range r as expressed by (2.99) does not have a symmetrical form with respect to the future and the past, but, expressed in the form of $W^{(r)} = 0$, it is certainly invariant for time reversal.

If there is a number ν such that

$$W^{(r)} = 0, \qquad r \geq \nu + 1 \tag{2.101}$$

and

$$W^{(\nu)} \neq 0, \tag{2.102}$$

then ν is what was called the range of interdependence of the stochastic chain. It should be noted that the fact that $W^{(r)}$ vanishes for a certain value of r, say, r_0, does not necessarily imply that $W^{(r)}$ for $r > r_0$ vanishes. The Markov chain has $\nu = 2$; that is,

$$W^{(r)} = 0 \quad \text{for} \quad r \geq 3 \tag{2.103}$$

and

$$W^{(2)} \neq 0. \tag{2.104}$$

Formally the case $W^{(2)} = 0$ may be included in the Markov chain, but actually this means that each position is independent; that is, it is the case of Bernouilli trials. Expansion (2.94) has no more than $\nu - 1$ nonvanishing terms. The expansion is valid whether r in it is $\geq \nu$ or $< \nu$.

In (1.67) take as $\mathcal{G}$ a sequence of r positions, as $\mathcal{E}$ a partial sequence of $(r - 1)$ positions, and as $\mathcal{F}$ a sequence of one position. Then, noting that $S(\mathcal{F} \mid \mathcal{E}) \geq 0$, we obtain, as we also did in (2.7),

$$S^{(r)} \geq S^{(r-1)}. \tag{2.105}$$

Next, in (1.55), take as $\mathcal{G}$ a sequence of $(r - 1)$ positions, as $\mathcal{E}$ a partial sequence of $(r - 2)$ positions, and as $\mathcal{F}$ a sequence of one position. Then (1.55) or, equivalently, (2.5) becomes

$$S^{(1)} + S^{(r-2)} - S^{(r-1)} \geq 0. \tag{2.106}$$

Adding (2.105) and (2.106), we obtain an upper bound for $W^{(r)}$,

$$S^{(1)} \geq W^{(r)}, \qquad r \geq 2. \tag{2.107}$$

From (2.101) ensue the following useful formulas: Adding $W^{(r)}$ for $r = \nu + 1, \nu + 2, \cdots, \nu + k$, we obtain

$$\begin{aligned} 0 = \sum_{s=1}^{k} W^{(\nu+s)} &= -\sum_{s=1}^{k} S^{(\nu+s)} + 2\sum_{s=1}^{k} S^{(\nu+s-1)} - \sum_{s=1}^{k} S^{(\nu+s-2)} \\ &= -S^{(\nu+k)} + S^{(\nu)} + S^{(\nu+k-1)} - S^{(\nu-1)} \end{aligned}$$

or

$$S^{(\nu+k)} - S^{(\nu+k-1)} = S^{(\nu)} - S^{(\nu-1)}, \qquad k \geq 0. \tag{2.108}$$

This must be so because (2.101) means that

$$[S^{(\nu+k)}]_2 = 0, \tag{2.109}$$

which yields, by "integration,"

$$[S^{(\nu+k)}]_1 = [S^{(\nu)}]_1, \qquad k \geq 0, \tag{2.110}$$

which is equivalent to (2.108). "Integrating" once more (2.110) with respect to k from 1 to k, we obtain

$$S^{(\nu+k)} - S^{(\nu)} = k(S^{(\nu)} - S^{(\nu-1)}) \tag{2.111}$$

or

$$S^{(\nu+k)} = (k + 1)S^{(\nu)} - kS^{(\nu-1)} \quad \text{for} \quad k \geq 0. \tag{2.112}$$

Relation (2.111) or (2.112) can, of course, be obtained by summing (2.108) from $k = 1$ to $k = k$. These relations hold actually also for $k = -1$.

It is often important to handle the information per position, that is,

$$G^{(r)} \equiv \frac{S^{(r)}}{r} = S^{(1)} - \left(\frac{J_{\text{tot}}^{(r)}}{r}\right)$$
$$= S^{(1)} - \sum_{s=2}^{r} \frac{r-s+1}{r} W^{(s)}, \qquad r \geq 1, \tag{2.113}$$

which follows from (2.89) and (2.94). For $r \to \infty$ we can write (if the limits exist)

$$G^{(\infty)} = S^{(1)} - \lim_{r\to\infty} \sum_{s=2}^{r} W^{(s)} + \lim_{r\to\infty} \sum_{s=2}^{r} \frac{s-1}{r} W^{(s)}. \tag{2.114}$$

We presently demonstrate convergence of $G^{(\infty)}$, but at this stage we limit ourselves to noting that if $\sum_{s=2}^{r} W^{(s)}$ converges, $\sum_{s=2}^{r} [(s-1)/r)]W^{(s)}$ has to become zero. This is because for a sufficiently large r_0 we can make $\sum_{s=r_0}^{\infty} [W^{(s)}]$ smaller than any arbitrary positive number ε, and the corresponding summation $\lim_{R\to\infty} \sum_{s=r_0}^{R} [(s-1)/R]W^{(s)}$ will become smaller than ε. Thus we write

$$G^{(\infty)} = S^{(1)} - \sum_{s=2}^{\infty} W^{(s)}, \tag{2.115}$$

where the summation will actually break off at some finite place if ν is finite. The last series can also be interpreted as interdependence per position:

$$\lim_{r\to\infty} \frac{J_{\text{tot}}^{(r)}}{r} = \sum_{r=2}^{\infty} W^{(r)}. \tag{2.116}$$

In the case of a finite range we can obtain from (2.112), for $r > \nu$,

$$G^{(\nu+k)} = \left(\frac{k+1}{k+\nu}\right) S^{(\nu)} - \frac{k}{k+\nu} S^{(\nu-1)}, \qquad k \geq 0, \tag{2.117}$$

and therefrom

$$G^{(\infty)} = S^{(\nu)} - S^{(\nu-1)}. \tag{2.118}$$

It may be noted that $S^{(\nu)} - S^{(\nu-1)}$ is by definition

$$-(\Sigma)^{\nu} p^{(\nu)}(x_1, x_2, \cdots, x_\nu) \log [p^{(\nu)}(x_1, x_2, \cdots, x_\nu)/p^{(\nu-1)}(x_1, x_2, \cdots, x_{\nu-1})],$$

hence is nothing but the average conditional entropy computed with the conditional probability $p^{(\nu)}(x_\nu \mid x_1, \cdots, x_{\nu-1})$.

Writing $\mathfrak{X}_r$ for the logical spectrum of stochastic variable whose value is x_r, one may write (2.118) in the form

$$G^{(\infty)} = S(\mathfrak{X}_\nu \mid \mathfrak{X}_1, \mathfrak{X}_2, \cdots, \mathfrak{X}_{\nu-1}). \tag{2.119}$$

If the chain is Markovian, we have from (2.112)

$$S^{(r)} = (r-1)S^{(2)} - (r-2)S^{(1)}, \qquad r \geq 2 \tag{2.120}$$

and

$$G^{(\infty)} = S^{(2)} - S^{(1)} = S(\mathfrak{X}_2 \mid \mathfrak{X}_1). \tag{2.121}$$

It may be helpful to visualize the situation by a geometrical representation. The first "derivative," that is, the slope of the $S^{(r)}$ curve, is

$$[S^{(r)}]_1 = S^{(r)} - S^{(r-1)} \geq 0, \tag{2.122}$$

where the last double symbol originates from (2.105). The second derivative (curvature) is

$$[S^{(r)}]_2 = -W^{(r)} \leq 0, \qquad r \geq 2. \tag{2.123}$$

This means that $S^{(r)}$ as a function of r is a monotonically increasing curve with a gradually decreasing slope (concave toward the r axis). Since there is a lower bound set by (2.122) for the slope, there has to be a limiting slope. Because the curvature vanishes for $r > \nu$, the slope must remain constant thereafter. The slope, meaning the information increase per position, becomes $G^{(\infty)}$ after it has reached its final value. This will make it easy to understand (2.118). Equations like (2.112) also acquire an easy geometrical meaning (see Figure 2.4).

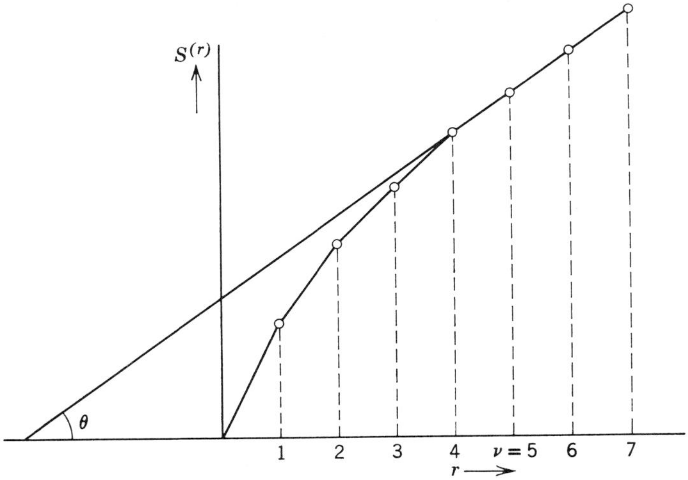

Figure 2.4 Entropy of a segment in a stochastic chain with finite range of interdependence. $S^{(r)}$ as a function of r is a monotonically increasing concave curve whose slope reaches a limiting value at $r = \nu$. The first derivative (slope) is the increase of information obtained by an additional position. The second derivative is the negative of the interdependence index $W^{(r)}$; $\tan\theta$ is the information per position $G^{(\infty)}$ in an infinitely long chain. Since the curve becomes straight after $r = \nu - 1$, $S^{(\nu+k)}$ must be given by $S^{(\nu)} + k\tan\theta$, for $k = -1, 0, 1, 2, \cdots$, which is (2.112). The graph assumes $\nu = 5$.

Let us now examine the limiting behavior of the $G^{(r)}$ and $W^{(r)}$ for the case when ν is not necessarily finite. It is important to note that the expansion (2.113) is valid for any $r \geq 0$ whether or not the maximum range of interdependence ν is finite. From (2.113) it follows easily that

$$G^{(r)} - G^{(r+1)} = \sum_{s=2}^{r+1} \frac{(s-1)}{r(r+1)} W^{(s)} (\geq 0), \tag{2.124}$$

which shows that $G^{(r)} > G^{(r+1)}$ if there is any interdependence of range less than $r + 1$. Expression (2.124) shows that the sequence $\{G^{(r)}\}$; $r = 1, 2, \cdots$, is a monotonically decreasing sequence. Since $G^{(r)}$ is $S^{(r)}/r$, which is non-negative, each term in the sequence has a lower bound 0. Consequently we have the following theorems.

Theorem 2.3 If a stochastic chain is stationary, whether or not the range of interdependence is finite, the information per position, $G^{(r)}$, converges to a limit for $r \to \infty$.

Another important consequence that can be drawn from (2.124) is that, since $\{G^{(r)}\}$ converges, the quantity on the right side of (2.124) must tend to zero for $r \to \infty$. Similarly, since $\{G^{(r)}\}$ converges, the sum in (2.113) must converge to some finite value. Thus we can state Theorem 2.4.

Theorem 2.4 In a stationary stochastic chain the correlation index $W^{(r)}$ becomes small for large r in such a fashion that

$$\lim_{r\to\infty} \sum_{s=2}^{r} \frac{r-s+1}{r} W^{(s)} \tag{2.125}$$

converges to a finite (i.e., $<\infty$) value and

$$\lim_{r\to\infty} \sum_{s=2}^{r+1} \frac{(s-1)}{r(r+1)} W^{(s)} \tag{2.126}$$

converges to zero.

It is instructive to reinterpret the formulas obtained above from the point of view of the degree of our ignorance regarding the state of a position. We take any one position in the infinite sequence, say, the 0th position ($t = 0$), and propose to guess the state x_0 of this position. Then the ignorance (or degree of uncertainty) is given by

$$\text{ign}^{(1)} = S^{(1)} = -\sum_{x_0} p^{(1)}(x_0) \log p^{(1)}(x_0). \tag{2.127}$$

Next suppose that we already know that the preceding position, that is, the (-1)st position ($t = -1$), was in a certain state, say, x_{-1}. Then our ignorance regarding the state of the 0th position becomes

$$-\sum_{x_0} p(x_0 \mid x_{-1}) \log p(x_0 \mid x_{-1})$$
$$= -\sum_{x_0} \frac{p^{(2)}(x_{-1}, x_0)}{p^{(1)}(x_{-1})} \log \frac{p^{(2)}(x_{-1}, x_0)}{p^{(1)}(x_{-1})} . \tag{2.128}$$

However, the (—1)st position will turn out to be in the state x_{-1} with probability $p^{(1)}(x_{-1})$. Therefore the ignorance about the state of the 0th position, on the basis of the knowledge of the state of the (—1)st position, becomes on the average

$$\text{ign}^{(2)} = -\sum_{x_{-1}} p^{(1)}(x_{-1}) \sum_{x_0} \frac{p^{(2)}(x_{-1}, x_0)}{p^{(1)}(x_{-1})} \log \frac{p^{(2)}(x_{-1}, x_0)}{p^{(1)}(x_{-1})}$$
$$= S^{(2)} - S^{(1)}. \tag{2.129}$$

Similarly, ignorance about the state of the 0th position on the basis of the knowledge about the states of r preceding positions, that is, the (—1)st, (—2)th, · · · , ($-r$)th positions, will be

$$\text{ign}^{(r+1)} = S^{(r+1)} - S^{(r)}. \tag{2.130}$$

The decrease in ignorance, or increase in knowledge, about the state of the 0th position as a result of knowing one more position in the past is given by

$$\text{ign}^{(r)} - \text{ign}^{(r+1)} = -S^{(r+1)} + 2S^{(r)} - S^{(r-1)} = W^{(r+1)}. \tag{2.131}$$

This consideration further clarifies the meaning of the interdependence index $W^{(r)}$. The condition $W^{(r+1)} = 0$ means that additional knowledge about the ($-r$)th position does not change our prediction about the 0th position. The more we learn about the past, the less our ignorance about the present becomes. The original ignorance is $S^{(1)}$, the second-stage ignorance is $S^{(1)} - W^{(2)}$, the third-stage ignorance is $S^{(1)} - W^{(2)} - W^{(3)}$, and so on. Thus the minimum ignorance is

$$(\text{ign})_{\min} = S^{(1)} - W^{(2)} - W^{(3)} - \cdots. \tag{2.132}$$

This ignorance represents the uncertainty in guessing the state of the 0th position with the best knowledge of the past. The moment observation determines this state this uncertainty disappears; that is, the ignorance becomes zero. This decrease in ignorance is the so-called "information." Thus it is not surprising that $G^{(\infty)}$ of (2.115) and $(\text{ign})_{\min}$ of (2.132) coincide. $W^{(r)}$ represents, on the one hand, the decrease of information per position due to dependence of range r and, on the other hand, the average increase of our knowledge about the state of a position from knowing the state $(r - 1)$ positions prior to it over and above our knowledge about the states of the positions in between.

As an illustration to show how interdependence indices can be used in analyzing the structure in a time sequence, we adduce an experiment made on the IBM 704 computer, simulating the process of shuffling of playing cards. (For details see [W-12].) We know that by shuffling, the "orderliness" existing in a stack of cards gradually disappears. "Orderliness" can mean dependence, and such a dependence will decay in the process of shuffling. To simplify the problem, we assume that there are only two kinds of cards, one marked with 0 and the other marked with 1. The stack of cards is now replaced by a chain of 0's and 1's. Incidentally, the method used here provides an interesting approach to the problem of random numbers in general. The concept of "randomness" usually refers to an infinite sequence of numbers, while the concept of randomness introduced here can be applied to a finite sequence of numbers.

The original sequence of binary numbers was produced according to certain transition probabilities, but these probabilities were so chosen that the probability was very high that the chain would take the pattern

$$\cdots 00001111000011110000 1111 \cdots. \tag{2.133}$$

Let us assume for the moment that the sequence has a rigid rule (2.133); that is, a "run" of four 1's and a "run" of four 0's appear alternately. Suppose that we give ourselves the task of guessing the value x_0 of the variable $\mathfrak{X}_0$. Without any preliminary knowledge of other digits, there is an equal probability that $x_0 = 0$ and $x_0 = 1$. Now suppose that we know that $\mathfrak{X}_{-1}$ is 1; that is, $x_{-1} = 1$. Then this position $\mathfrak{X}_{-1}$ may be, with equal probability, any one of the four possible positions in a run of four 1's. If $\mathfrak{X}_{-1}$ is the first, second, or third position in this run, x_0 will be 1. If $\mathfrak{X}_{-1}$ is the last position of the four, x_0 will be 0. Hence the probability of x_0 being 1 is now $\frac{3}{4}$ and the probability of its being 0 is $\frac{1}{4}$. If we know x_{-1} and x_{-2}, our prediction of x_0 will become more accurate. Finally, if we know x_{-1}, x_{-2}, x_{-3}, and x_{-4}, the prediction of x_0 is no longer probabilistic, but deterministic. Further knowledge of x_{-5} will no longer change our prediction about x_0. Therefore $W^{(5)} \neq 0$, and $W^{(r)} = 0$, $r \geq 6$.

Of course a sequence that strictly obeys the rule (2.133) is not a *stationary* stochastic sequence in the sense of (2.62), except when a particular probability distribution is given to initial sequences. Therefore the sequence we used in experiment was produced by the conditional probabilities of range 5,

$$p^{(5)}(x_r \mid x_{r-4}, x_{r-3}, x_{r-2}, x_{r-1}), \tag{2.134}$$

which obeys the following rule. The probability of a run ending at a position does not depend on what happened before the run started. If a run has lasted for a length r less than 4 ($r = 1, 2, 3$), the probability of its continuing one more is $1 - \varepsilon$. If a run has lasted four or more places, the probability of its terminating at the next place is $1 - \varepsilon$. If $\varepsilon = 0$, (2.134) will produce (2.133) rigorously. If $\varepsilon = \frac{1}{2}$, the sequence will become purely random. Since the stochastic chain is produced by the five-position conditional probabilities (2.134), we have $W^{(r)} = 0$ for $r \geq 6$.

The shuffling of a long sequence of numbers has been done in the following way. We divided this sequence into smaller segments in a random fashion so that the average length of segments became l. We inverted the order of these segments (but not the order of numbers inside each segment) and put them together to obtain a "shuffled" sequence. This operation was considered as one shuffle, and many shuffles were made successively on a sequence.

We produced, in our experiment on the IBM 704, a binary number of 105,000 digits obeying the probability rule (2.134) with $\varepsilon = \frac{1}{16}$. Then we shuffled this number with average segment length $l = 17$. We should have (barring inevitable errors due to computational inaccuracies and due to the finiteness of the sequence) $W^{(r)} = 0$ and $r \geq 6$ before shuffling and also after many number of shuffles, because a purely random sequence will result after an infinite number of shuffles. At intervening stages $W^{(r)}$ for $r \geq 6$ may depart slightly from zero, but this is not of much interest.

We are interested rather in $W^{(r)}$ for $2 \leq r \leq 5$, which measures the strength of the structure of the type (2.133), which will gradually be destroyed in the measure as we continue to shuffle. Since shuffling has the tendency to destroy a structure of larger length, we should expect that $W^{(r)}$ with larger r will decrease faster with shuffling. Figure 2.5 gives the $W^{(r)}$'s for different values of the parameter σ, which is the number of shuffles we applied to the sequence.

The evaluation of $W^{(r)}$ was carried out by taking the relative frequencies of various segments of length 1 through 6 that were taken in the long sequence of length 105,000 produced after each shuffle. We see in Figure 2.5 a beautiful graphical representation in terms of the $W^{(r)}$'s of all the expected tendencies regarding the decay of "orderliness" in a stack of cards by the process of shuffling.

So much for the illustration, and now for a few more remarks of general nature. As soon as the quantity $W^{(r)}$ is theoretically defined, it is quite natural to evaluate it from the relative frequency of various segments in a population of empirical data. In the above explanation we deliberately avoided making a distinction between $W^{(r)}$ and its evaluation. Chacko Abraham was the first to give an interpretation of this procedure of evaluation in the context of the traditional theory of hypothesis testing (see the footnote by Abraham at the end of [W-33]). A few years later Kullbach et al. worked out this idea in detail [K-17]. In fact, when the $p^{(r-1)}(x_1, x_2, \cdots, x_{r-1})$ are known, $q^{(r)}(x_1, x_2, \cdots, x_r)$, given in (2.96), can be considered as a theoretical value of $p^{(r)}(x_1, x_2, \cdots, x_r)$, based on the hypothesis that the interdependence of range r vanishes. Hence

$$W^{(r)} = -\sum p^{(r)}(\log q^{(r)} - \log p^{(r)}) \tag{2.135}$$

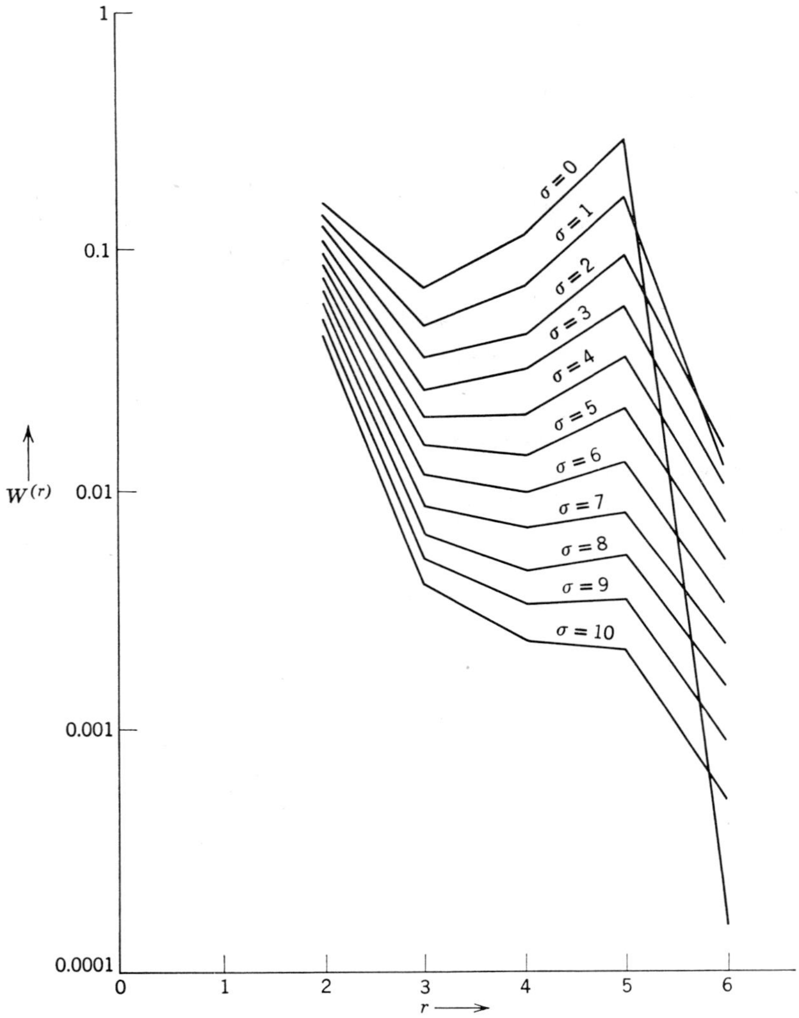

Figure 2.5 Decay of orderliness by shuffling. The indices of interdependence $W^{(r)}$, $2 \leq r \leq 6$, decrease with an increasing number σ of shuffles.

may be considered as Wilks' $-2 \log \lambda$ (divided by $2N$), which becomes λ^2 when $p^{(r)}$ and $q^{(r)}$ are not very different [see (1.106) and (1.109)].

The interdependence between two positions in a chain considered in this section is characterized by the fact that we assume knowledge of intermediate positions that separate these two positions. For this reason this concept of interdependence may be called "beading" interdependence. In contrast to this, the following concept of interdependence may be called "bridging" interdependence. Let us denote by $\mathcal{F}$ a sequence of κ positions and by $\mathcal{E}$ a later sequence of λ positions such that there are τ positions in between. The interdependence between $\mathcal{E}$ and $\mathcal{F}$ is given by

$$V^{(\kappa,\tau,\lambda)} = S(\mathcal{E}) + S(\mathcal{F}) - S(\mathcal{E} \otimes \mathcal{F}) = S(\mathcal{E}) - S(\mathcal{E} \mid \mathcal{F}). \quad (2.136)$$

It is easy to see that this $V^{(\kappa,\tau,\lambda)}$ can be expressed in the following way. Take the probability defined in (2.74), define the conditional entropy of $(x_1, x_2, \cdots, x_\lambda)$ for given $(y_1, y_2, \cdots, y_\kappa)$, and then average this with the help of the probability $p(y_1, y_2, \cdots, y_\kappa)$. This gives $S(\mathcal{E} \mid \mathcal{F})$. Substract this from the entropy $S(\mathcal{E})$ of $(y_1, y_2, \cdots, y_\kappa)$; the difference is (2.136), which can be written

$$V^{(\kappa,\tau,\lambda)} = \Big(\sum_x\Big)^\lambda \Big(\sum_y\Big)^\kappa \Big[\Big(\sum_z\Big)^\tau p(y_1, \cdots, y_\kappa, z_1, \cdots, z_\tau, x_1, \cdots, x_\lambda)\Big]$$
$$\times \log \frac{\Big[\Big(\sum_z\Big)^\tau p(y_1, \cdots, y_\kappa, z_1, \cdots, z_\tau, x_1, \cdots, x_\lambda)\Big]}{p(y_1, \cdots, y_\kappa) p(x_1, \cdots, x_\lambda)}. \quad (2.137)$$

Compare this with (1.100), in which the probability distributions $\{p_i\}$ and $\{q_i\}$ are to be replaced respectively by $(\sum_z)^\tau\ p(y_1, \cdots, y_\kappa, z_1, \cdots, z_\tau, x_1, \cdots, x_\lambda)$ and $p(y_1, \cdots, y_\kappa)p(x_1, \cdots, x_\lambda)$. Then by the use of Theorem 1.1 we can claim that the condition

$$\lim_{\tau \to \infty} V^{(\kappa,\tau,\lambda)} \to 0 \quad (2.138)$$

and the condition

$$\lim_{\tau \to \infty} \Big| \Big(\sum_z\Big)^\tau p(x_1, \cdots, x_\kappa, z_1, \cdots, z_\tau, y_1, \cdots, y_\lambda) - p(x_1, \cdots, x_\kappa) p(y_1, \cdots, y_\lambda) \Big| \to 0 \quad (2.139)$$

for all $(x_1, \cdots, x_\kappa)$ and all $(y_1, \cdots, y_\lambda)$ imply each other. If $p(y_1, \cdots, y_\lambda) = 0$, then $p(x_1, \cdots, x_\kappa, z_1 \cdots, z_\tau, y_1, \cdots, y_\lambda) = 0$. Hence, excluding this trivial case, we can divide (2.139) by $p(y_1, \cdots, y_\lambda)$. Put $\kappa = \lambda = \nu - 1$ in the resulting expression, and we obtain (2.75). In particular, combining this result with Theorem 2.2, we obtain the following.

Theorem 2.5 A stationary Markov chain is ergodic if

$$\lim_{\tau \to \infty} V^{(1,\tau,1)} \to 0.$$

This is also a necessary condition for ergodicity if the Markov chain has only one cyclic class.

2.6. EFFECT OF COARSE OBSERVATION ON INTERDEPENDENCE

Accuracy of observation is not determined by the observed object but depends on various factors, such as instruments of measurement, which are intentionally adopted or accidentally made available, and on how such instruments are hooked up and operated. On the other hand, there is little doubt that our concepts and theories depend strongly on the correlation and probabilistic dependences we discover in a great number of observed data. Now if the structure of probabilistic dependence depends, as we shall prove, on the accuracy of observation, we shall have to admit that our theoretical picture of the "world" is influenced in a very essential fashion by factors that do not belong to the "world," and it will become a task of the next step to find out the transformation rule governing the passage from a picture based on one degree of accuracy to a picture based on another degree of accuracy.

The contents of the present section should be understood as a modest first step in this general direction. We discuss the change in dependence properties of a stochastic chain when the observation becomes more "coarse," so that several states appear to be the same state. For brevity we often refer to the original stochastic chain and the coarsely redefined stochastic chain, respectively, as a microscopic chain and a macroscopic chain. Some examples in which such macroscopization plays a role may be helpful.

In communication the "noise" can be such that a certain number of emitted symbols are received as an identical symbol at the receiving end. Then the relation between the emitted message and the received message is one between microscopic and macroscopic chains, as explained above. The amount of information that can be sent through a communication channel with this kind of "noise" may be discussed from the present point of view.

In the so-called irreversible thermodynamics, Onsager's reciprocal relation is one of the most fundamental laws. The usual derivation of this law is based on two crucial facts: (a) under certain conditions the macroscopic observed data taken at certain intervals can be considered as a Markov chain, and (b) the fundamental laws of nature are invariant for time reversal. I have shown elsewhere that Markovianity depends critically on the length of intervals between observations [W-10]. But as we shall presently see Markovianity is an elusive property that is very easily lost as a result of a slight change in the assumed accuracy of observation. To determine the condition of validity of Onsager's law, we should therefore investigate among other things the mutual conditioning of the time interval, the geometrical size of the physical system, and the accuracy of observation that are required to uphold Markovianity.† In the following we do not discuss these topics, which involve

† Even if Onsager's relation can be proven without Markovianity, these conditions on observation may remain necessary.

physical laws, but we point out that they are closely related to the mathematical problem discussed in this section.

There are innumerable other examples in which grouping of states plays a crucial role in our understanding of our surroundings. But there has not been much theoretical investigation of this subject except for a few scattered papers of a rather preliminary nature. The papers of Blackwell [B-3*a*] and Harris [H-4] are among the earliest reliable studies on this subject. The present section, which is concerned with comparison of the range of dependence and the information contents in a microscopic and a macroscopic chains, is essentially an improved version of the paper by Watanabe and Abraham [W-33]. Birch's more recent paper [B-4] is concerned with a somewhat similar topic, but the method and approach are entirely different. Onicescu discusses the subject matter in a chapter in his recent book under the title "theory of Watanabe and Abraham" [O-2].

It can be expected without calculation that grouping of states will cause the information content of a chain to decrease. However, it is an interesting fact that if the original microscopic chain has redundancy, the loss of information can sometimes be made to vanish. One important task is then to obtain a formula by which to estimate the decrease of information due to macroscopization in the presence of redundancy. Another noteworthy fact is that the range of correlation changes by grouping of states. For instance, if a "macroscopic" chain is Markovian, the corresponding "macroscopic" chain can have a range smaller or larger than 2. The latter case means that a macroscopic chain has a longer "memory" than the original microscopic chain. (The term "memory" should not be understood too literally.) An interesting question then is: How far does this "memory" go back, and in what manner does this "memory" attenuate with time? This question, among others, is discussed below.

Each position in an infinite sequence is supposed to take any one of n states, $1, 2, \cdots, n$. It is further assumed that there exists a unique probability $p^{(r)}(x_1, x_2, \cdots, x_r)$ that a sequence of r consecutive positions will take a set of states $x_1, x_2, \cdots, x_r$, and that this probability is independent of the location of the segment in the infinite sequence; that is, the stochastic chain is stationary. States $1, 2, \cdots, n$ are now grouped into n' classes, $1, 2, \cdots, n'$, with $n' \leq n$, so that no class is empty and each state belongs to one and only one of the classes. We introduce variables ξ_i to label the class to which the state of the ith position belongs, where $\xi_i = 1, 2, \cdots,$ or n'. We sometimes refer to "states" and "classes," respectively, as "microstates" and "macrostates." The chain described in terms of ξ_i is sometimes called a macroscopic chain or a functional chain because ξ_i is a function of x_i. The probability that a sequence of length r is in macrostates $(\xi_1, \xi_2, \cdots, \xi_r)$ is given by

$$\pi^{(r)}(\xi_1, \xi_2, \cdots, \xi_r) = \sum_{x_1}^{\in\xi_1} \sum_{x_2}^{\in\xi_2} \cdots \sum_{x_r}^{\in\xi_r} p^{(r)}(x_1, \cdots, x_2, \cdots, x_r), \quad (2.140)$$

where the summation symbol $\sum_{x_1}^{\in\xi_1}$ means that the summation is made with respect to x_1 over those microstates included in macrostate ξ_1. The existence of a stationary probability $\pi^{(r)}$ is guaranteed by the existence of a stationary probability $p^{(r)}$. The macroprobabilities, π's, satisfy the basic relations similar to (2.63) and (2.64), in which the x's should be here replaced by ξ's, and n by n'. In the following, with some exceptions, Latin and Greek symbols refer respectively to the microscopic and macroscopic aspects, while German symbols refer to some relations between these two aspects.

The macroinformation content of a sequence of length r then is

$$\Sigma^{(r)} = -\sum_{\xi_1=1}^{n'}\sum_{\xi_2=1}^{n'}\cdots\sum_{\xi_r=1}^{n'}\pi^{(r)}(\xi_1, \xi_2, \cdots, \xi_r)\log\pi^{(r)}(\xi_1, \xi_2, \cdots, \xi_r), \quad (2.141)$$

corresponding to (2.86), and the macroinformation per position is

$$\Gamma^{(r)} = \frac{\Sigma^{(r)}}{r}. \quad (2.142)$$

We have, corresponding to (2.113), the expansion

$$\Gamma^{(r)} = \Sigma^{(1)} - \frac{1}{r}\sum_{s=2}^{r}(r - s + 1)\Omega^{(s)}, \quad (2.143)$$

with the help of the macrodependence indices $\Omega^{(r)}$ defined, by analogy with (2.90), by

$$\Omega^{(s)} = -\Sigma^{(s)} + 2\Sigma^{(s-1)} - \Sigma^{(s-2)}. \quad (2.144)$$

It may be noted that $0 \leq \Gamma^{(r)} \leq \Sigma^{(1)}$ and $\Sigma^{(1)} \leq \log n'$, but these upper and lower bounds are not so interesting to our problem.

If there is an integer ν' such that

$$\Omega^{(r)} = 0 \quad \text{for} \quad r > \nu', \qquad \Omega^{(\nu')} \neq 0, \quad (2.145)$$

ν' is the range of dependence in the macroscopic chain. If ν' is finite, it is obvious that we can write, as in (2.115),

$$\Gamma^{(\infty)} = \Sigma^{(1)} - \sum_{r=2}^{\infty}\Omega^{(r)}. \quad (2.146)$$

However, as was shown in the paragraph just below (2.114), even if $\Omega^{(r)}$ does not vanish completely for larger r, (2.146) still follows from (2.143) if $\sum_{r=2}^{\infty}\Omega^{(r)}$ converges. If ν' is finite, then we have, in accordance with (2.112) and (2.118) (putting $\nu' + k = r$),

$$\Sigma^{(r)} = (r - \nu' + 1)\Sigma^{(\nu')} - (r - \nu')\Sigma^{(\nu'-1)} \quad (2.147)$$

and

$$\Gamma^{(\infty)} = \Sigma^{(\nu')} - \Sigma^{(\nu'-1)}. \quad (2.148)$$

So much for the macroscopic quantities whose microscopic counterparts are already familiar to us. We now introduce quantities that connect the macroscopic and microscopic aspects of a stochastic chain. The first quantity we introduce is a measure of the degree of coarseness of the macroscopic observation relative to the microscopic observation. "Coarseness," defined by

$$\mathfrak{K} = S^{(1)} - \Sigma^{(1)}, \tag{2.149}$$

measures the loss of information due to the macroscopization of measurement when we observe only one position separately from the rest of the chain. If $2m$ equally probable microstates are pairwise merged to make m macrostates, then the coarseness becomes 1 bit. The information loss $\mathfrak{L}^{(r)}$ per position in a segment of length r is given by

$$\mathfrak{L}^{(r)} = G^{(r)} - \Gamma^{(r)}, \tag{2.150}$$

of which $\mathfrak{K}$ is a special case for $r = 1$, $\mathfrak{K} = \mathfrak{L}^{(1)}$. Sometimes the condition $\mathfrak{L}^{(\infty)} = 0$ does not necessarily mean absence of loss, because the original information $G^{(\infty)}$ can be also zero. To avoid this inconvenience the percentual information loss per position

$$\mathfrak{H}^{(r)} = \frac{\mathfrak{L}^{(r)}}{G^{(r)}} \tag{2.151}$$

is sometimes useful.

The conditional probability of a microscopic cylinder $(x_1, x_2, \cdots, x_r)$ given a macroscopic cylinder $(\xi_1, \xi_2, \cdots, \xi_r)$ is given by

$$\begin{aligned} p(x_1, x_2, \cdots, x_r \mid \xi_1, \xi_2, \cdots, \xi_r) &= \frac{p(x_1, x_2, \cdots, x_r)}{\pi(\xi_1, \xi_2, \quad, \xi_r)} \quad \text{if} \quad x_i \in \xi_i \quad \text{for all} \quad i = 1, 2, \cdots, r. \\ &= 0 \quad \text{otherwise.} \end{aligned} \tag{2.152}$$

The inverse conditional probability satisfies

$$\begin{aligned} p(\xi_1, \xi_2, \cdots, \xi_r \mid x_1, x_2, \cdots, x_r) &= 1 \quad \text{if} \quad x_i \in \xi_i \quad \text{for all} \quad i = 1, 2, \cdots, r \\ &= 0 \quad \text{otherwise.} \end{aligned} \tag{2.153}$$

If we designate by $\mathcal{E}$ and $\mathcal{F}$ the stochastic variables corresponding respectively to $(\xi_1, \xi_2, \cdots, \xi_r)$ and $(x_1, x_2, \cdots, x_r)$, (2.153) shows that $\mathcal{E}$ is entirely dependent on $\mathcal{F}$. Hence $S(\mathcal{E} \mid \mathcal{F}) = 0$ and we have

$$S(\mathcal{F}) - S(\mathcal{E}) = S(\mathcal{F} \mid \mathcal{E}) \geq 0, \tag{2.154}$$

as follows from (2.4). The average conditional entropy calculated with the help of (2.152) is $S(\mathcal{F} \mid \mathcal{E})$, whereas $S^{(r)}$ and $\Sigma^{(r)}$ correspond to $S(\mathcal{F})$ and $S(\mathcal{E})$.

Hence we have

$$\mathfrak{L}^{(r)} = \frac{1}{r}(S^{(r)} - \Sigma^{(r)}) \geq 0, \tag{2.155}$$

which can be formulated as Theorem 2.6.

Theorem 2.6 The loss of information by coarse observation of a stochastic chain is non-negative. This implies also that the measure $\mathfrak{R}$ of coarseness of observation defined in (2.149) is non-negative.

From (2.154) follows also the following theorem, which can be skipped by those readers who are not particularly interested in the analogy of the present problem with a communication channel. The concept of channel capacity used in this theorem will be discussed more fully in the next chapter, but will be explained in the proof of the theorem to the extent needed to understand the meaning of the theorem.

Theorem 2.7 If the macroscopization of a stochastic chain is regarded as an information channel, its channel capacity per position is $\log n'$, where n' is the number of macrostates available.

Proof. Let $\mathcal{F}$ and $\mathcal{E}$ represent the microscopic and macroscopic segments of length r; then $\mathcal{F}$ and $\mathcal{E}$ are compared here to the input and output of a communication channel. Before observation of the output $\mathcal{E}$ we have ignorance about the input $\mathcal{F}$ in the amount $S(\mathcal{F})$. After the observation of the output $\mathcal{E}$ this ignorance becomes $S(\mathcal{F} \mid \mathcal{E})$. Hence the information (which is by definition the decrease in ignorance) about the input obtained through the output is $S(\mathcal{F}) - S(\mathcal{F} \mid \mathcal{E})$, which, according to (2.8), (2.9), and (2.10), is equal to $S(\mathcal{E}) - S(\mathcal{E} \mid \mathcal{F})$ and to $J(\mathcal{E}, \mathcal{F})$. Now this quantity $J(\mathcal{E}, \mathcal{F})$ is a function of the probability distribution $\{p(F_j)\}$ and of the conditional probability distributions $\{p(E_i \mid F_j)\}$. The maximum of $J(\mathcal{E}, \mathcal{F})$ obtained by varying $\{p(F_j)\}$ with a fixed set of $\{p(E_i \mid E_j)\}$ corresponds to the concept of channel capacity. The channel capacity per position C is defined as the limit for $r \to \infty$ of the maximum in the above sense of $J(\mathcal{E} \mid \mathcal{F})$ devided by r. In our special case $S(\mathcal{E} \mid \mathcal{F})$ is zero, as a result of which we have $J(\mathcal{E}, \mathcal{F}) = S(\mathcal{E}) = \Sigma^{(r)}$, hence $J(\mathcal{E}, \mathcal{F})/r = \Gamma^{(r)}$, and $C = \lim_{r \to \infty} \max \Gamma^{(r)}$. Now in the expression (2.143) of $\Gamma^{(r)}$ we can make all the nonpositive terms involving Ω's vanish by choosing $\{p(F_j)\}$ suitably, namely, by choosing the microscopic probability $p^{(r)}(x_1, x_2, \cdots, x^r)$ in such a way that all the W's vanish. We prove this point in Corollary 2.7. Then $\Gamma^{(r)}$ becomes $\Sigma^{(1)}$, whose maximum is $\log n'$. This can be attained by choosing $p^{(1)}(x_1)$ in such a way that each macrostate receives an equal probability $1/n'$. Q.E.D.

About the range of dependence in the macroscopic chain we have the following theorem.

Theorem 2.8 If ν and ν' are respectively the range of interdependence in a microscopic chain and that of the corresponding macroscopic chain, there exist in fact cases belonging to each of three possibilities: (a) $\nu > \nu'$, (b) $\nu = \nu'$, and (c) $\nu < \nu'$.

Proof. We need only show concrete cases illustrating each of the three categories. As an example of Category (a), take a (microscopic) Markov chain $\nu = 2$ defined by the transition probabilities given by

$$\|p(x_2 \mid x_1)\| = \begin{array}{c} \\ x_2 \\ \| \\ 1 \\ 2 \\ 3 \\ 4 \end{array} \overset{x_1 = 1 \quad 2 \quad 3 \quad 4}{\begin{pmatrix} 0 & \frac{1}{2} & \frac{1}{4} & \frac{1}{4} \\ \frac{1}{2} & 0 & \frac{1}{4} & \frac{1}{4} \\ \frac{1}{4} & \frac{1}{4} & 0 & \frac{1}{2} \\ \frac{1}{4} & \frac{1}{4} & \frac{1}{2} & 0 \end{pmatrix}}, \tag{2.156}$$

which gives $W^{(2)} = 0.5$ and $W^{(r)} = 0$ for $r > 2$. If we now group Microstates 1 and 2 together to form a Macrostate 1 and group Microstates 3 and 4 together to form Macrostate 2, we obtain $\Omega^{(r)} = 0$ for all $r \geq 2$. $\nu' = 1$. This is because no matter what microscopic state may stand at position t, the probability of the next position $t + 1$ being in either Macrostate 1 or Macrostate 2 is equal to $\frac{1}{2}$. The range has thus been reduced from 2 to 1.

For Cases (b) and (c), consider a Markov chain defined by

$$\|p(x_2 \mid x_1)\| = \begin{array}{c} \\ x_2 \\ \| \\ 1 \\ 2 \\ 3 \\ 4 \end{array} \overset{x_1 = 1 \qquad 2 \qquad 3 \qquad 4}{\begin{pmatrix} \alpha & \alpha & \alpha & 1 - 3\alpha \\ 1 - 3\alpha & \alpha & \alpha & \alpha \\ \alpha & 1 - 3\alpha & \alpha & \alpha \\ \alpha & \alpha & 1 - 3\alpha & \alpha \end{pmatrix}}. \tag{2.157}$$

For the present purpose α can be put equal to zero, although that would require a special initial probability distribution in order to secure a stationary chain. Equation (2.157) yields $W^{(2)} = 2$ and $W^{(r)} = 0$ for $r > 2$.

Now if we group the four microstates as $\{1, 2\}$ and $\{3, 4\}$, we obtain $\Omega^{(2)} = 0$, $\Omega^{(3)} = 1$, and $\Omega^{(r)} = 0$ for $r > 3$. Thus $\nu' = 3$. The range has increased from 2 to 3. If we group the microstates as $\{1, 3\}$ and $\{2, 4\}$, we obtain $\Omega^{(2)} = 1$ and $\Omega^{(r)} = 0$ for $r > 2$, meaning $\nu' = 2$. Thus the range remains unchanged. This completes the proof.

The reader is strongly advised to derive for himself the values of Ω's in these last two cases and, in particular, to attempt to obtain intuitive insight into the mechanism by which the range increases. To do this he should note that the microscopic chain has only one type,

$$\cdots 123412341234 \cdots, \tag{2.158}$$

which will appear in the first macroscopic observation, $\{1, 2\}$, $\{3, 4\}$, as

$$\cdots 112211221122 \cdots \tag{2.159}$$

and in the second macroscopic observation, $\{1, 3\}$, $\{2, 4\}$, as

$$\cdots 121212121212 \cdots. \tag{2.160}$$

The W's and Ω's can be calculated by considering the probabilities with which different segments appear in each of the sequences (2.158), (2.159), and (2.160).

In the macroscopic sequence (2.159) suppose that we know only that $\xi = 1$ at t; then there is 50% probability that the next position $t + 1$ will be 1 and 2. But if we know that $\xi = 1$ also at $t - 1$, the probability that the next position $t + 1$ will be 1 becomes 0. This is because the additional knowledge that $\xi = 1$ at $t - 1$ reveals the fact that the $\xi = 1$ at t originated from the microstate $x = 2$. This is the reason why $\Omega^{(3)}$ is not zero here.

We now give a few lemmas leading up to a theorem that will permit us to evaluate the information loss, hence also the information content of a macroscopic chain, for the case in which the microscopic chain has a finite range of interdependence. This theorem will be particularly useful when the macroscopic range ν' becomes infinite, so that it becomes important to know how far we should go in an expansion like (2.143) or (2.146) to obtain the desired accuracy.

Lemma 2.1 The information loss per position $\mathfrak{L}^{(r)}$, defined by (2.150), converges for $r \to \infty$.

Proof. This is because $G^{(r)}$ and $\Gamma^{(r)}$ in (2.150) converge according to Theorem 2.3.

Lemma 2.2 The information loss per position $\mathfrak{L}^{(r)}$, defined by (2.150), satisfies

$$\mathfrak{L}^{(r+s)} \leq \frac{1}{r+s}\,(r\mathfrak{L}^{(r)} + s\mathfrak{L}^{(s)}). \tag{2.161}$$

Proof. Consider two probability distributions p and q defined on $(x_1, \cdots, x_r, x_{r+1}, \cdots, x_{r+s})$ by

$$\begin{aligned} p(x_1, \cdots, x_{r+s} \mid \xi_1, \cdots, \xi_{r+s}) &= \frac{p(x_1, \cdots, x_{r+s})}{\pi(\xi_1, \cdots, \xi_{r+s})} \quad \text{if} \quad x_i \in \xi_i, \qquad i = 1, \cdots, r+s \\ &= 0 \quad \text{otherwise} \end{aligned} \tag{2.162}$$

and

$$q(x_1, \cdots, x_{r+s} \mid \xi_1, \cdots, \xi_{r+s})$$

$$= \frac{p(x_1, \cdots, x_r)}{\pi(\xi_1, \cdots, \xi_r)} \cdot \frac{p(x_{r+1}, \cdots, x_{r+s})}{\pi(\xi_{r+1}, \cdots, \xi_{r+s})} \quad \text{if} \quad x_i \in \xi_i, \qquad i = 1, \cdots, r+s$$

$$= 0 \quad \text{otherwise.} \tag{2.163}$$

Let $\mathfrak{D} = \mathfrak{D}(\xi_1, \cdots, \xi_{r+s})$ be defined by

$$\mathfrak{D} = -\sum_{x_1=1}^{n} \cdots \sum_{x_{r+s}=1}^{n} p(\log p - \log q). \tag{2.164}$$

Then Gibbs' theorem says that $\mathfrak{D} \leq 0$. Now take the average of $\mathfrak{D}(\xi_1, \cdots, \xi_{r+s})$ with the help of the probability $\pi(\xi_1, \cdots, \xi_{r+s})$. Then we get, from (2.162), (2.163), (2.86), and (2.141),

$$\langle \mathfrak{D} \rangle = (S^{(r+s)} - \Sigma^{(r+s)}) - (S^{(r)} - \Sigma^{(r)}) - (S^{(s)} - \Sigma^{(s)}), \tag{2.165}$$

which should be nonpositive according to Gibbs' theorem. Hence, by the definitions of $G^{(r)}$, $\Gamma^{(r)}$, and $\mathfrak{L}^{(r)}$, we obtain (2.161).

An immediate consequence of this lemma is the following.

Lemma 2.3 The information loss per position $\mathfrak{L}^{(r)}$ obeys

$$\mathfrak{L}^{(2r)} \leq \mathfrak{L}^{(r)}. \tag{2.166}$$

Lemma 2.4 For any integer $r \geq 1$ we have

$$\mathfrak{L}^{(r)} \geq \mathfrak{L}^{(\infty)}, \tag{2.167}$$

where $\mathfrak{L}^{(\infty)}$ is the limit alluded to in Lemma 2.1.

Proof. Note that $\lim_{\rho\to\infty} \mathfrak{L}^{(2\rho r)} = \mathfrak{L}^{(\infty)}$, which would contradict $\mathfrak{L}^{(r)} < \mathfrak{L}^{(\infty)}$ in view of (2.166).

Lemma 2.5 Define a sequence $\{\mathfrak{M}^{(r)}\}$, $r = 1, 2, \cdots$, by

$$\mathfrak{M}^{(r)} = \mathfrak{R} - \sum_{s=2}^{r} (W^{(s)} - \Omega^{(s)}). \tag{2.168}$$

Then, if the microscopic chain has a finite range ν of interdependence and if $\sum_{s=2}^{r} \Omega^{(r)}$ converges for $r \to \infty$, $\{\mathfrak{M}^{(r)}\}$ is a monotonically increasing sequence sequence for $r \geq \nu$, converging to $\mathfrak{L}^{(\infty)}$.

Proof. We can write for $r \geq \nu$

$$\mathfrak{M}^{(r)} = \mathfrak{R} - \sum_{s=2}^{\nu} W^{(s)} + \sum_{s=2}^{r} \Omega^{(s)}, \tag{2.169}$$

where $\Omega^{(s)}$ is non-negative, and we have

$$\mathfrak{M}^{(\infty)} = G^{(\infty)} - \Gamma^{(\infty)} = \mathfrak{L}^{(\infty)} \tag{2.170}$$

under the assumptions of the theorem.

Theorem 2.9 The sequence $\{\mathfrak{L}^{(r)}\}$ of information loss per position for $r \geq r_0$ is a monotonically decreasing sequence converging to $\mathfrak{L}^{(\infty)}$, provided that the sequence $\{\mathfrak{M}^{(r)}\}$ defined in (2.168) for $r > r_0$ is a monotonically increasing sequence converging to $\mathfrak{L}^{(\infty)}$.

Proof. From the definition of $\mathfrak{L}^{(r)}$,

$$\mathfrak{L}^{(r)} = \mathfrak{K} - \sum_{s=2}^{r} \frac{r-s+1}{r} (W^{(s)} - \Omega^{(s)}), \tag{2.171}$$

it follows that

$$r\mathfrak{L}^{(r)} - (r-1)\mathfrak{L}^{(r-1)} = \mathfrak{M}^{(r)} \tag{2.172}$$

or

$$\mathfrak{L}^{(r)} - \mathfrak{L}^{(r-1)} = \frac{1}{r-1} (\mathfrak{M}^{(r)} - \mathfrak{L}^{(r-1)}) \leq \frac{1}{r-1} (\mathfrak{L}^{(\infty)} - \mathfrak{L}^{(r-1)}), \qquad r > r_0 \tag{2.173}$$

because $\mathfrak{M}^{(r)} \leq \mathfrak{M}^{(\infty)} = \mathfrak{L}^{(\infty)}$ for $r > r_0$ by the premise. Now according to Lemma 2.4 the right-hand side of (2.173) is nonpositive; hence

$$\mathfrak{L}^{(r)} - \mathfrak{L}^{(r-1)} \leq 0, \qquad r > r_0, \tag{2.174}$$

implying $\mathfrak{L}^{(r_0)} \geq \mathfrak{L}^{(r_0+1)} \geq \mathfrak{L}^{(r_0+2)} \geq \cdots \geq \mathfrak{L}^{(\infty)}$. Q.E.D.

Corresponding to the case $r_0 = 1$, $\nu = 2$, we have Corollary 2.5.

Corollary 2.5 If the microscopic chain is Markovian, then

$$\mathfrak{K} = \mathfrak{L}^{(1)} \geq \mathfrak{L}^{(2)} \geq \cdots \geq \mathfrak{L}^{(\infty)}. \tag{2.175}$$

Corollary 2.6 If ν is finite and if $\Omega^{(s)}$ is known up to $s = r \geq \nu$, then $\mathfrak{L}^{(\infty)}$ can be estimated by

$$\mathfrak{M}^{(r)} \leq \mathfrak{L}^{(\infty)} \leq \mathfrak{L}^{(r)}. \tag{2.176}$$

Theorem 2.10 Whether or not ν is finite, we have in general

$$\mathfrak{L}^{(1)} \geq \mathfrak{L}^{(r)}, \qquad r \geq 1. \tag{2.177}$$

Proof. The proof runs parallel to the proof of Lemma 2.2 if one takes p and q as defined by

$$p(x_1, \cdots, x_r \mid \xi_1, \cdots, \xi_r) = \frac{p(x_1, \cdots, x_r)}{\pi(\xi_1, \cdots, \xi_r)}$$

and

$$q(x_1, \cdots, x_r \mid \xi_1, \cdots, \xi_r) = \frac{p(x_1)}{\pi(\xi_1)} \cdots\cdots \frac{p(x_r)}{\pi(\xi_r)} \tag{2.178}$$

if $x_i \in \xi_i$, $i = 1, 2, \cdots, r$, and $p = q = 0$ otherwise.

Theorem 2.11 The total interdependence in a segment of length r in the microscopic chain is never smaller than the total interdependence of a segment of length r in the corresponding macroscopic chain.

Proof. Theorem 2.10, in view of (2.171), implies that

$$\sum_{s=2}^{r} (r - s + 1)(W^{(s)} - \Omega^{(s)}) \geq 0, \tag{2.179}$$

which allows the interpretation given in Theorem 2.11 because of (2.94).

Corollary 2.7 If the microscopic chain is free from interdependence, the macroscopic chain is also free from interdependence.

This fact was used previously in the proof of Theorem 2.7. It must be noted that (2.179) does not guarantee that for each s, $W^{(s)} \geq \Omega^{(s)}$. Hence $W^{(s)} = 0$ for a particular s does not imply $\Omega^{(s)} = 0$ for this s. Furthermore, the converse of Corollary 2.7 is not true. In fact, we have seen with reference to (2.156) that it can happen that $\sum_{s=2}^{\infty} W^{(s)} \neq 0$ while $\sum_{s=2}^{\infty} \Omega^{(s)} = 0$.

Consider now the quantity

$$\mathfrak{r}^{(r)} = \frac{\sum_{s=2}^{r} (r - s + 1)\Omega^{(s)}}{\sum_{s=2}^{r} (r - s + 1)W^{(s)}}, \tag{2.180}$$

which is finite insofar as the microscopic chain has some interdependence. Theorem 2.11 states that we have in general

$$\mathfrak{r}^{(r)} \leq 1, \tag{2.181}$$

insofar as there is some microinterdependence. The statement (2.181) is equivalent to $\mathfrak{L}^{(r)} \leq \mathfrak{L}^{(1)} = \mathfrak{K}$, meaning that the information loss per position in a segment of length larger than 1 is never larger than the "coarseness" of the macroscopization. But a small value of $\mathfrak{L}^{(r)}$ does not necessarily imply a small fractional loss, because $G^{(r)}$ itself may be small too. For this reason the quantity $\mathfrak{H}^{(r)}$ defined in (2.151) becomes important and the condition $\mathfrak{H}^{(r)} \leq \mathfrak{H}^{(1)}$ may be considered as an appropriate expression for the fractional recovery of information due to the interdependence. In this respect we have the following theorem.

Theorem 2.12 The three mutually exclusive cases $\mathfrak{H}^{(r)} \lesseqgtr \mathfrak{H}^{(1)}$ are equivalent respectively to three mutually exclusive cases,

$$\mathfrak{r}^{(r)} \lesseqgtr 1 - \frac{\mathfrak{R}}{S^{(1)}}, \tag{2.182}$$

unless there is no interdependence in the microscopic chain. In the latter case $\mathfrak{H}^{(r)} = \mathfrak{H}^{(1)}$.

Proof. By the use of definition (2.151) we can rewrite the three conditions $\mathfrak{H}^{(r)} \lesseqgtr \mathfrak{H}^{(1)}$ as

$$\frac{r\Sigma^{(r)} - \sum_{s=2}^{r} (r - s + 1)\Omega^{(s)}}{rS^{(r)} - \sum_{s=2}^{r} (r - s + 1)W^{(s)}} \lesseqgtr \frac{\Sigma^{(1)}}{S^{(1)}}, \tag{2.183}$$

which is equivalent to

$$\frac{\sum_{s=2}^{r} (r - s + 1)\Omega^{(s)}}{\sum_{s=2}^{r} (r - s + 1)W^{(s)}} \lesseqgtr \frac{\Sigma^{(1)}}{S^{(1)}} \tag{2.184}$$

provided that the denominator is nonzero. Equation (2.184) is another way of writing (2.182). Q.E.D.

We should note that the upper two signs ($\leq$) impose a more restrictive condition than (2.181), as they should, because $\mathfrak{H}^{(r)} \leq \mathfrak{H}^{(1)}$ is a stronger condition than $\mathfrak{L}^{(r)} \leq \mathfrak{L}^{(1)}$, and the latter is always true.

Consider as an illustration the case of (2.157), where the value of α can be between 0 and $\frac{1}{3}$. At the point $\alpha = \frac{1}{4}$, all the entries in the matrix (2.157) become equal, and the chain becomes a pure Bernouilli sequence; as a result $W^{(r)} = 0$ for all $r \geq 2$. Except at this point, we have a proper Markov chain $W^{(r)} = 0$, $r \geq 0$, and $W^{(2)} \neq 0$. As α deviates from $\frac{1}{4}$, the chain acquires stronger interdependence, resulting in a larger value of $W^{(2)}$. We have seen that $W^{(2)}$ reaches the value 2 at $\alpha = 0$. Macroscopization is defined by merger of States 1 and 2 and States 3 and 4. As regards the macroscopic interdependence $\Omega^{(r)}$, we can easily prove that $\Omega^{(2)} = 0$ for any value of α, but the other $\Omega^{(r)}$, with $r > 2$, depend considerably on the value of α. In the purely random case ($\alpha = \frac{1}{4}$) we have of course $\Omega^{(r)} = 0$ for all $r \geq 2$. On the other hand, we have noted that at $\alpha = 0$ we have $\Omega^{(r)} = 0$ for $r \geq 4$ and $\Omega^{(3)} = 1$. At in-between values of α, however, the sequence of $\{\Omega^{(r)}\}$ does not break off at a finite value of r, although $\Omega^{(r)}$ decreases very rapidly with r. In the region of values of α not far from the purely random case, $\alpha = \frac{1}{4}$, we can show that $\{\Omega^{(r)}\}$ obeys approximately the "law of exponential

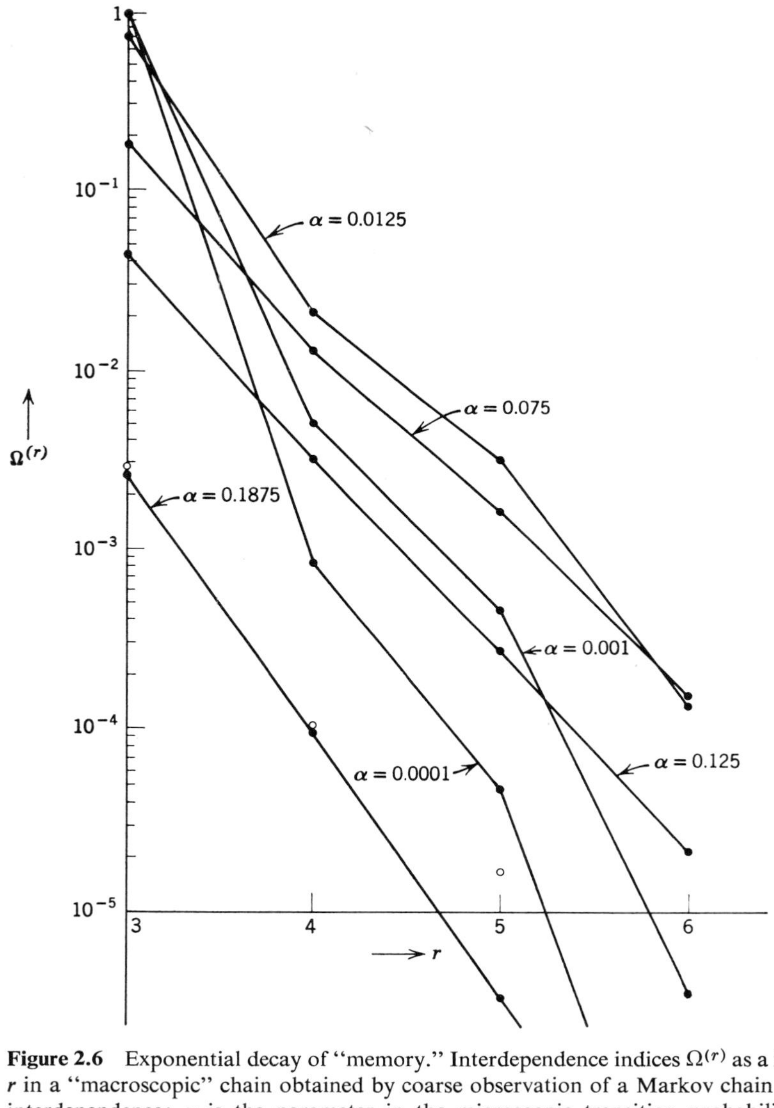

Figure 2.6 Exponential decay of "memory." Interdependence indices $\Omega^{(r)}$ as a function of r in a "macroscopic" chain obtained by coarse observation of a Markov chain with weak interdependence; α is the parameter in the microscopic transition probability (2.157). Macrostate 1 is the merger of Microstates 1 and 2, and Macrostate 2 is the merger of Microstates 3 and 4. The black dots are theoretical values and the unfilled circles are Monte Carlo values (for $\alpha = 0.1875$).

decay of memory":

$$\Omega^{(r)} \propto e^{-\gamma r} \tag{2.185}$$

for $r \geq 3$, with $\gamma > 0$. In a case like this; in which $\Omega^{(r)}$ decays fast with r, the sequence $\{\mathfrak{M}^{(r)}\}$ gives a closer approximation to $\mathfrak{L}^{(\infty)}$ than the sequence $\{\mathfrak{L}^{(r)}\}$ at a finite value of r. This is because $\mathfrak{M}^{(r)} = \mathfrak{K} - W^{(2)} + \sum_{s=2}^{r} \Omega^{(s)}$ does not change very much with r whereas $\mathfrak{L}^{(r)} = \mathfrak{K} - [1 - (1/r)]W^{(2)} + \sum_{s=2}^{\infty} [1 - (s-1)/r]\Omega^{(s)}$ varies considerably for smaller values of r. In the present example we have $S^{(1)} = 2$, $\Sigma^{(1)} = 1$, and if $|\alpha - (\frac{1}{4})|$ is of the order of ε ($\ll 1$), then $W^{(2)}$ will be of the order of ε^2, and $\Omega^{(3)}$ as well as $\sum_{s=2}^{\infty} \Omega^{(s)}$ will be of the order of ε^4. Hence the coarseness $\mathfrak{K}$ is 1, and the loss per position $\mathfrak{L}^{(r)}$ is less than $\mathfrak{K}$ by a difference of the order of ε^2. As regards the fractional loss $\mathfrak{H}^{(r)}$, we see that the left-hand side of (2.184) is of the order of $\varepsilon^{(2)}$ and hence is smaller than the right side, which is $\frac{1}{2}$. This means that the fractional loss $\mathfrak{H}^{(r)}$ for $r \geq 2$ is smaller than $\mathfrak{H}^{(1)}$.

Figure 2.6 shows for various values of α between 0 and $\frac{1}{4}$ the calculated values of $\Omega^{(r)}$ for (2.157) with merger scheme $\{1, 2\}$, $\{3, 4\}$, as discussed in the foregoing paragraph. The calculation was carried out by Miss Carol Shannesy (now Mrs. Wade) with the use of the double-precision method on an IBM 704 computer. We see that the law of exponential decay of "memory" is obeyed fairly well for values $\alpha \geq 0.075$. Michael Greene produced an actual stochastic sequence of 208,800 numbers obeying (2.157) with $\alpha = 0.1875$, carried out the macroscopization $\{1, 2\}$, $\{3, 4\}$, and computed $\Omega^{(r)}$ from the actual frequencies of various macroscopic segments. The unfilled circles in Figure 2.6 are the Monte Carlo values obtained by Mr. Greene.

3

Prediction and Retrodiction

3.1. FUNDAMENTAL ASYMMETRY IN PROBABILISTIC INFERENCE—SOME BASIC THEOREMS

In this chapter and for the rest of the book we are concerned mainly with application of the information theory (Chapter 1) and the method of IDA (Chapter 2) to problems connected with inferential processes. In this section I should like to point out what I consider to be one of the most fundamental features of man's inferential processes with reference to the experimental data about his environment—a feature that will be reflected in the necessity of an asymmetrical interpretation of certain mathematically symmetrical formulas. Recognition of this asymmetry as a principle governing inference not only helps to clarify the nature of problems in communication but also sheds a new sidelight on the problem of one-wayness of the "flow" of time (Sections, 3.2, 5.3, and 6.2) and on the nature of inductive inference (Sections 3.2 and 4.4). It is characteristic of this discussion that no new mathematical formulas are involved. What is new is a viewpoint in interpretation, or rather the establishment of such an interpretational viewpoint as one of the basic principles concerning human inference. The concept of retrodiction, which plays an important role in this discussion, was introduced in 1952 [W-5] and further developed in papers of 1955 [W-9] and 1965 [W-19].

We start again with a logical spectrum $\mathcal{G} = \{G_k\}$, which is a product of two others, $\mathcal{E} = \{E_i\}$ and $\mathcal{F} = \{F_j\}$, so that $G_k = E_i \cap F_j$ and $\mathcal{G} = \mathcal{E} \otimes \mathcal{F}$, as in (1.15). The probabilities $p(E_i)$ and $p(F_j)$ for E_i and F_j can be derived from the probability $p(E_i \cap F_j)$ for G_k by (1.16). The definition of conditional probability (1.10) permits us to write

$$p(E_i)p(F_j \mid E_i) = p(E_i \cap F_j) = p(F_j)p(E_i \mid F_j), \tag{3.1}$$

which shows perfect symmetry between $\mathcal{E}$ and $\mathcal{F}$.

However, in most (but not all) of the practical cases that require inference, there appears a basic asymmetry between the role of $\mathcal{E}$ and that of $\mathcal{F}$. In the physical sciences, for instance, we are often interested in the initial state and the final state of a phenomenon. We can make $\mathcal{E}$ correspond to the initial state and $\mathcal{F}$ to the final state. E_i may, for instance, be a proposition such as "the physical system as observed at $t = 0$ is in state i," and F_j may be a proposition such as "the physical system as observed at $t = t$ is in state j." Now a physical law usually gives a probabilistic prediction such as "if E_i is true, then there is such-and-such probability of F_j being true." This means that the physical law gives only $p(F_j \mid E_i)$, which is not sufficient, in the general case, to determine the other four probabilities involved in (3.1): $p(E_i)$, $p(E_i \mid F_j)$, $p(F_j)$, and $p(E_i \cap F_j)$. These depend on how the collection of samples is prepared for the experiment, or with what probability each initial state is presented.

In a similar way, in the problem of communication, E_i may be "the symbol emitted to a communication device is a_i" and F_j may be "the symbol received from a communication device is a_j." Then the nature of the communication device (information transducer), which has a certain probability of making errors, will determine $p(F_j \mid E_i)$. However, the probability $p(E_i)$ is determined by the frequency with which the user of the communication device uses the symbol a_i.

In these interpretations $\mathcal{E}$ and $\mathcal{F}$ may be said to represent "cause" and "effect" in a probabilistic sense, and the matrix $\|p(F_j \mid E_i)\|$ may be called a "causal matrix." To avoid connotational contamination due to various semantic nuances associated with the overworked terms "cause" and "effect," we should probably specify our usages of these terms by stating that we often encounter problems involving two sets $\mathcal{E}$ and $\mathcal{F}$ of propositions for which we are given only $p(F_j \mid E_i)$ by the nature of $\mathcal{E}$ and $\mathcal{F}$, and that in such cases $\mathcal{E}$ and $\mathcal{F}$ are called "cause" and "effect," respectively. They are "causally related." To borrow terminology from communication technology, we might also call $\mathcal{E}$ and $\mathcal{F}$ "input" and "output," respectively.

Prediction is an act of guessing which consists of attaching a weight to the occurrence of each event F_j when a certain cause event E_i is known to have taken place. This weight, or probability, is $p(F_j \mid E_i)$. Now the converse act of guessing the E_i when F_j is known is called "retrodiction" (see [W-5, W-9]). The retrodictive probability is $p(E_i \mid F_j)$, which cannot be derived in the general case from $p(F_j \mid E_i)$. In other words, retrodiction is impossible in general. (Exceptions will be studied later.)

The case of a Markov chain may tend to give a wrong impression in this respect. If $\mathcal{E}$ and $\mathcal{F}$ respectively represent the state at time t and the state at time $t + 1$, then the transition matrix $p(F_j \mid E_i)$ alone will determine in a stationary case the unconditional probability distribution over the states, which may be also written $p(E_i)$. But this is an entirely different problem, because we consider here an infinite chain of causative links $\mathcal{E} \rightarrow \mathcal{F}$ in such a way that $\mathcal{F}$ of one link becomes $\mathcal{E}$ of the next link. The unconditional probability $p(E_i)$ is obtained in the limit of infinitely long chains. This fact does not contradict the general impossibility of pure retrodiction.

In order to see precisely what we need know to determine $p(E_i \mid F_j)$ when $p(F_j \mid E_i)$ is given, we first rewrite (3.1) as

$$p(E_i \mid F_j) = \frac{p(E_i \cap F_j)}{p(F_j)} = \frac{p(E_i)p(F_j \mid E_i)}{p(F_j)}. \tag{3.2}$$

If the division by $p(F_j)$ is not justified there will result an indeterminate value of these fractions. Besides (3.2) we have two relations,

$$p(F_j) = \sum_i p(E_i)p(F_j \mid E_i) = \sum_i p(E_i \cap F_j) \tag{3.3}$$

and

$$p(E_i) = \sum_j p(F_j)p(E_i \mid F_j) = \sum_j p(E_i \cap F_j), \tag{3.4}$$

which can be obtained from (3.1) by summation with respect to i or j and using the normalization condition. Let us first notice that when $p(F_j \mid E_i)$ is prescribed, knowledge of $p(E_i)$ and knowledge of $p(E_i \cap F_j)$ are interchangeable, in the sense that they are mutually derivable from each other. The only nontrivial remark to be made in this respect is that when $p(F_j \mid E_i)$ is predetermined, $p(E_i)$ can be chosen arbitrarily, but $p(E_i \cap F_j)$ cannot be chosen entirely arbitrarily because it has to satisfy

$$\frac{p(E_i \cap F_j)}{\sum_k p(E_i \cap F_k)} = p(F_j \mid E_i). \tag{3.5}$$

Now, in order to obtain the retrodictive probability $p(E_i \mid F_j)$ when the predictive probability $p(F_j \mid E_i)$ is given, we have, according to (3.2), to know $p(E_i)$ and $p(F_j)$, except in a special case in which the quotient $p(E_i)/p(F_j)$ can be known while $p(E_i)$ and $p(F_j)$ separately cannot be known. According to (3.3) it is necessary and sufficient to know the $p(E_i)$'s to derive $p(F_j)$. Consequently all we need is knowledge of the $p(E_i)$'s except for the special case mentioned above. This exception happens in an extraordinary case when the predictive probability $p(F_j \mid E_i)$ is 0 or 1 in such a way that for a given F_j there is only one E_i that makes $p(F_j \mid E_i)$ unity. In other words, this is the case when $\|p(F_j \mid E_i)\|$ is a permutation matrix, in which by definition each column as well as each row has all zero entries except one, which is unity. In this case, for a pair (i^*, j^*) that makes $p(F_{j*} \mid E_{i*})$ unity, we have $p(E_{i*})/p(F_{j*}) = 1$ [see (3.3)] and $p(E_{i*} \mid F_{j*}) = 1$ [see (3.2)]. Since we also have $\sum_{i=1}^{n} p(E_i \mid F_{j*}) = 1$, we have $p(E_i \mid F_{j*}) = 0$ for $i \neq i^*$. Thus we get in this special case $p(E_i \mid F_j) = p(F_j \mid E_i)$ for all i, and j. To borrow terminology from the consideration of causality, this is a bilaterally deterministic case. Except for this special case, pure retrodiction, or derivation of the retrodictive probabilities from the predictive probabilities only, is impossible.

If, on the other hand, we know $p(E_i)$ as well as $p(F_j \mid E_i)$, we can calculate $p(F_j)$ by (3.3) and then obtain $p(E_i \mid F_j)$ by (3.2). Writing explicitly, we have

$$p(E_i \mid F_j) = \frac{p(E_i)p(F_j \mid E_i)}{\sum_k p(E_k)p(F_j \mid E_k)}. \tag{3.6}$$

This is known as Bayes' formula. It means that if we do not know the outcome of $\mathcal{F}$, then we associate probability $p(E_i)$ with proposition E_i, but after knowing that F_j has taken place, we associate a revised probability $p(E_i \mid F_j)$, given by (3.6), with proposition E_i. $p(E_i)$ is often called prior probability or *a priori* probability, while $p(E_i \mid F_j)$ is called *a posteriori* probability.

However, it is often the case that the man who tries to make a retrodiction does not know $p(E_i)$ although he knows $p(F_j \mid E_i)$ from physical laws or from some other source. The last resort for him may be to invoke the principle of ignorance and give the same value to all $p(E_i)$, although the validity of this principle is quite questionable. Under this assumption he gets

$$p(E_i \mid F_j) = \frac{p(F_j \mid E_i)}{\sum_k p(F_j \mid E_k)}. \tag{3.7}$$

This kind of retrodiction is called "blind retrodiction" (see [W-9]). The right-hand side of (3.7) is what is known as the "likelihood" of E_i.

Slightly better is the case in which the retrodictor knows, besides $p(F_j \mid E_i)$, the probability distribution $p(F_j)$. This can very well happen, since he may have observed the "effects" $\mathcal{F}$ (received symbols in communication) quite often so that he may have a rough estimate of $p(F_j)$. Assuming that $p(E_i)$ remains unknown, can he then successfully retrodict from his knowledge of $p(F_j \mid E_i)$ and $p(F_j)$? This amounts to asking if one can solve m equations

$$p(F_j) = \sum_{i=1}^{n} p(E_i)p(F_j \mid E_i) \tag{3.8}$$

for n unknowns $p(E_i)$. If n is larger than m, of course, it is generally impossible to determine $p(E_i)$, and therefore $p(E_i \mid F_j)$. Even when $n = m$, this is not possible if

$$\det [p(F_j \mid E_i)] = 0. \tag{3.9}$$

See Appendix A3.1 for a summary of the above discussed situation expressed from a slightly different viewpoint.

Consider the following game between two persons, X and Y. There are two buttons, B_1 and B_2, and two lamps, L_1 and L_2. The probability that lamp L_j will be lighted when button B_i is pushed is denoted by

$$p(L_j \mid B_i). \tag{3.10}$$

Player X is supposed to push one or the other of the two buttons with fixed independent probabilities; that is, $p(B_i)$ is fixed. Player Y is supposed to observe the lamps, and is thus capable of determining $p(L_j)$ in the long run. Both X and Y know the probabilities $p(L_j \mid B_i)$. Player X tries to "predict" the observed data of Player Y. Player Y tries to "retrodict" which button Player X has pushed. Which player has a better chance? If a player guesses right at each trial, he is said to be perfectly successful. If he can give the right probability, he is said to be statistically successful, although the rate of success at individual trials may be low.

In the *bilaterally deterministic* case, which can be expressed as either

$$\begin{array}{cc} & B_1 \quad B_2 \\ \begin{array}{c} L_1 \\ L_2 \end{array} & \begin{pmatrix} 1 & 0 \\ 0 & 1 \end{pmatrix} \end{array} \quad \text{or} \quad \begin{array}{cc} & B_1 \quad B_2 \\ \begin{array}{c} L_1 \\ L_2 \end{array} & \begin{pmatrix} 0 & 1 \\ 1 & 0 \end{pmatrix} \end{array}, \tag{3.11}$$

both players will be perfectly successful, not only statistically but also in each individual try. In the *unilaterally* (*causally*) *deterministic* case, in which one has

$$\begin{array}{cc} & B_1 \quad B_2 \\ \begin{array}{c} L_1 \\ L_2 \end{array} & \begin{pmatrix} 1 & 1 \\ 0 & 0 \end{pmatrix} \end{array} \quad \text{or} \quad \begin{array}{cc} & B_1 \quad B_2 \\ \begin{array}{c} L_1 \\ L_2 \end{array} & \begin{pmatrix} 0 & 0 \\ 1 & 1 \end{pmatrix} \end{array}, \tag{3.12}$$

Player X will be perfectly successful in each individual trial, while Player Y will be completely helpless, since he will observe the same result all the time. This last case belongs to the case of (3.9).

In a case like

$$\begin{array}{cc} & B_1 \quad B_2 \\ \begin{array}{c} L_1 \\ L_2 \end{array} & \begin{pmatrix} \frac{2}{3} & \frac{1}{3} \\ \frac{1}{3} & \frac{2}{3} \end{pmatrix} \end{array} \tag{3.13}$$

Players X and Y will both be statistically successful because (3.9) is not the case. If X presses button B_1, for instance, then he will predict successfully that L_1 will be lighted with probability $\frac{2}{3}$ and L_2 will be lighted with probability $\frac{1}{3}$. Conversely, Player Y will also be statistically successful if X abides with the rule that he should not change $p(B_1)$. Y will obtain empirically $p(L_1)$ and $p(L_2)$, and can solve (3.8) for $p(B_i)$. His retrodictive probabilities $p(B_i \mid L_j)$

will then be

$$\begin{array}{c} \\ B_1 \\ \\ B_2 \end{array} \begin{array}{cc} L_1 & L_2 \\ \left(\begin{array}{cc} \dfrac{\frac{4}{3}p(L_1) - \frac{2}{3}p(L_2)}{p(L_1)} & \dfrac{\frac{2}{3}p(L_1) - \frac{1}{3}p(L_2)}{p(L_2)} \\ \dfrac{-\frac{1}{3}p(L_1) + \frac{2}{3}p(L_2)}{p(L_1)} & \dfrac{-\frac{2}{3}p(L_1) + \frac{4}{3}p(L_2)}{p(L_2)} \end{array}\right) \end{array}. \tag{3.14}$$

The denominators, $p(L_1)$ and $p(L_2)$, will never vanish in this case.

Finally, in the perfectly random case in which the predictive probability is given by

$$\begin{array}{c} \\ L_1 \\ L_2 \end{array} \begin{array}{c} \begin{array}{cc} B_1 & B_2 \end{array} \\ \begin{pmatrix} \frac{1}{2} & \frac{1}{2} \\ \frac{1}{2} & \frac{1}{2} \end{pmatrix} \end{array}, \tag{3.15}$$

Player X will be statistically successful if he simply predicts that half the time lamp L_1 will be lighted and half the time lamp L_2 will be lighted. On the other hand, Y will always observe $p(L_1) = p(L_2) = \frac{1}{2}$, but this gives him no clue whatsoever as to what Player X is doing. This is because Player Y cannot solve the equations (3.8) because of (3.9). If Y is very simple-minded he may infer that Player X must have pushed B_1 and B_2 equally frequently on the ground that everything known to him [i.e., $p(L_j \mid B_i)$ and $p(L_j)$] is completely symmetrical between B_1 and B_2 and between L_1 and L_2. But he may be completely wrong, since Player X may be, for instance, pressing B_1 all the time. This is a case of blind retrodiction.

In summary, we noted that there is a large and important class of cases of inference that can be described in terms of two logical spectra, $\mathcal{E}$ and $\mathcal{F}$, such that the (predictive) conditional probabilities $p(F_j \mid E_i)$ are determined by the essential nature of the problem whereas the (retrodictive) conditional probabilities $p(E_i \mid F_j)$ depend on contingent factors that are not determined by the essential nature of the problem. As we shall see in later sections, we sometimes have to deal with a product of more than two factors (logical spectra) to describe a complex of factors that enter our consideration in an inferential process. Even in such a case, if we pick a pair of factors and keep the other factors constant, asymmetry of the kind explained above appears again. Under the assumed asymmetry between $\mathcal{E}$ and $\mathcal{F}$ we asked the question of obtaining the retrodictive probabilities from the predictive probabilities. We noted first that derivation of the $p(E_i \mid F_j)$ from only the $p(F_j \mid E_i)$ is impossible except in the bilaterally deterministic case (impossibility of pure retrodiction), and next that derivation of the $p(E_i \mid F_j)$ from the $p(F_j \mid E_i)$ and $p(F_j)$ is sometimes possible and sometimes impossible. Finally, if the $p(E_i)$ are known in addition to the $p(F_j \mid E_i)$, the $p(E_i \mid F_j)$ are determined automatically. In this last case the complete symmetry exhibited in (3.1) can be exploited to a full extent, and any asymmetry is expelled from the mathematical description of the problem. But this does not imply that the assymmetry on the interpretational level disappears completely. On the contrary, we must become that much more cautious not to be deceived by the apparent symmetry. In the following we repeat (2.3)–(2.7), which have a formal symmetry between $\mathcal{E}$ and $\mathcal{F}$, but this time we give interpretations to these formulas assuming that $\mathcal{E}$ and $\mathcal{F}$ have an interpretational asymmetry as discussed in this section.

Since $p(E_i)$ is assumed to be known in the following, observation of E_i reduces the ignorance about the cause from $S(\mathcal{E})$ to zero, implying that this observation gives information in the amount of $S(\mathcal{E})$. Thus, in prediction of $\mathcal{F}$ based on $\mathcal{E}$, $S(\mathcal{E})$ may be called "observed cause information." Before observation of E_i, the observer has ignorance in the amount of $S(\mathcal{F})$ about the effect. But after observation of E_i, ignorance of the effect reduces on the average to $S(\mathcal{F} \mid \mathcal{E})$. Therefore the average amount of information about the effect as inferred from the cause is

$$K(\mathcal{F} \mid \mathcal{E}) = S(\mathcal{F}) - S(\mathcal{F} \mid \mathcal{E}). \qquad (3.16$$

This quantity may be called "inferred effect information." In prediction we do not actually observe F_j directly, but $S(\mathcal{F})$ in (3.16) represents the amount of information that would be obtained by direct observation of F_j. In an exact symmetry, in retroduction, the amount of effect information as obtained by direct observation of the effect itself is $S(\mathcal{F})$, and the average amount of information about the cause as inferred from the observation of the effect is

$$K(\mathcal{E} \mid \mathcal{F}) = S(\mathcal{E}) - S(\mathcal{E} \mid \mathcal{F}), \qquad (3.17)$$

which may be called "inferred cause information."† We have the following theorem.

Theorem 3.1 The inferred effect information $K(\mathcal{F} \mid \mathcal{E})$, the inferred cause information $K(\mathcal{E} \mid \mathcal{F})$, and the interdependence $J(\mathcal{E} \otimes \mathcal{F}; \mathcal{E}, \mathcal{F})$ existing in $\mathcal{E} \otimes \mathcal{F}$ with respect to $\mathcal{E}$ and $\mathcal{F}$ are equal to one another.

$$K(\mathcal{F} \mid \mathcal{E}) = K(\mathcal{E} \mid \mathcal{F}) = J(\mathcal{E} \otimes \mathcal{F}; \mathcal{E}, \mathcal{F}). \qquad (3.18)$$

This result was mentioned under (2.8), (2.9), and (2.10). We also have, from (2.12), Theorems 3.2 and 3.3.

† $K(\mathcal{F} \mid \mathcal{E})$, which is non-negative, is the average of $S(\mathcal{F}) - S(\mathcal{F} \mid E_i)$, which is the information about $\mathcal{F}$ provided by E_i. This latter is not necessarily non-negative. There is, however, a way of giving a similar meaning to $K(\mathcal{F} \mid \mathcal{E})$ without using such possibly negative quantities (see Appendix A1.1). The same remark applies also to $K(\mathcal{E} \mid \mathcal{F})$.

Theorem 3.2. In prediction the inferred effect information is never larger than the information that would have been obtained by a direct observation of the effect.

$$K(\mathcal{F} \mid \mathcal{E}) \leq S(\mathcal{F}). \tag{3.19}$$

Theorem 3.3 In retrodiction the inferred cause information is never larger than the information that would have been obtained by a direct observation of the cause.

$$K(\mathcal{E} \mid \mathcal{F}) \leq S(\mathcal{E}). \tag{3.20}$$

Theorems 3.2 and 3.3 may be expressed somewhat loosely as follows.

Theorems 3.2 and 3.3 (Paraphrase). Information obtained through direct observation of an event is never less than the information obtained by inference from other observed events dependent on it.

The Chinese proverb, "One seeing is better than hundred hearings," is more pertinent in this context than the familiar saying, "Seeing is believing." Implication of Theorem 3.2 in a deductive inference is discussed in Chapter 5. Theorem 3.3 allows for the following interpretation. $\mathcal{E}$ corresponds to the original information, which we cannot directly obtain, and by a medium of transmission of information of some kind, we receive only $\mathcal{F}$, from which we try to reconstruct $\mathcal{E}$. The amount of "retrieved information" is $K(\mathcal{E} \mid \mathcal{F})$, which is, according to Theorem 3.3, less than or, at very best, equal to the original information. The cause E_i can be, for instance, an input message of a certain length that is put into a communication channel, and the effect F_j then will be the output or received message. This channel may or may not have a so-called memory, but this does not affect the general conclusion obtained above. $K(\mathcal{E} \mid \mathcal{F})$ will sometimes be called "transduced information," when used in connection with communication.

Theorem 3.3 (Paraphrase). An information transducer can never increase the amount of information transduced.

Theorem 3.4 In prediction the inferred effect information is never larger than the observed cause information.

$$K(\mathcal{F} \mid \mathcal{E}) \leq S(\mathcal{E}). \tag{3.21}$$

Theorem 3.5 In retrodiction the inferred cause information is never larger than the observed effect information.

$$K(\mathcal{E} \mid \mathcal{F}) \leq S(\mathcal{F}). \tag{3.22}$$

Because of Theorem 3.1, Theorems 3.4 and 3.5 are respectively equivalent to Theorems 3.3 and 3.2. An obvious paraphrase of these two theorems is the following.

Theorems 3.4 and 3.5 (Paraphrase). The amount of information contained in an inferred consequence is never larger than the amount of information contained in the source data on the basis of which the inference is made. (The consequence could be cause or effect in the foregoing paragraphs.)

These theorems were previously put forward in a slightly different context (see [W-9]).

Theorem 3.6 In prediction the inferred effect information is non-negative.

$$K(\mathcal{F} \mid \mathcal{E}) \geq 0. \tag{3.23}$$

Theorem 3.7 In retrodiction the inferred cause information is non-negative.

$$K(\mathcal{E} \mid \mathcal{F}) \geq 0. \tag{3.24}$$

This is, as mentioned before, a consequence of the combination of (2.4) and (2.5). As we saw in Chapter 2, insofar as there is a slightest dependence between $\mathcal{E}$ and $\mathcal{F}$, $J = K$ never vanishes. Hence the following paraphrase.

Theorems 3.6 and 3.7 (Paraphrase). Insofar as effect and cause are not entirely independent, we can obtain some positive information about one of them from observation of the other. [The equals sign in (3.23) or (3.24) corresponds to the independent case.]

The preceding six theorems are concerned with inference based on cause or effect, and comparison was made between inferred information and observed information. The theorems are formally symmetrical with respect to cause and effect. The next category of problems concerns a comparison of the cause entropy $S(\mathcal{E})$ and the effect entropy $S(\mathcal{F})$. Without making reference to the possibility of inferring $\mathcal{F}$ from $\mathcal{E}$ or $\mathcal{E}$ from $\mathcal{F}$, we merely ask what is the amount of information obtainable directly by observation of $\mathcal{E}$ and what is the amount of information obtainable directly by observation of $\mathcal{F}$, and then we compare the two answers. In terms of the concept of ensemble, we may say that we have N objects, of which $N_i = Np(E_i)$ are in the "cause state" E_i and $N_j = Np(F_j)$ are in the "effect state" F_j. $S(\mathcal{E})$ represents the degree of uncertainty of state at the cause, and $S(\mathcal{F})$ represents the degree of uncertainty of state at the effect. Obviously we first have Theorem 3.8.

Theorem 3.8 If the causal matrix $p(F_i \mid E_j)$ is deterministic, the cause uncertainty is never less than the effect uncertainty,

$$S(\mathcal{E}) \geq S(\mathcal{F}). \tag{3.25}$$

Proof. Since $p(F_j \mid E_i) = 0$ or 1, we have $S(\mathcal{F} \mid \mathcal{E}) = 0$, which implies (3.25), with the help of (2.4), $S(\mathcal{E}) + S(\mathcal{F} \mid \mathcal{E}) = S(\mathcal{F}) + S(\mathcal{E} \mid \mathcal{F})$, and $S(\mathcal{E} \mid \mathcal{F}) \geq 0$. Equality in (3.25) holds if and only if the $p(F_j \mid E_i)$ are bilaterally deterministic. In other words, the inequality in (3.25) corresponds to the case of unilateral (cause-to-effect) determinism.

On the other hand, if $p(F_j \mid E_i)$ is not deterministic, the situation, so to speak, reverses itself. In particular, when the number n of states in $\mathcal{E}$ and the number m of states in $\mathcal{F}$ are the same, we have the following H-theorem.

Theorem 3.9 If $n = m$ and $\sum_{i=1}^{n} p(F_j \mid E_i) = 1$, then

$$S(\mathcal{E}) \leq S(\mathcal{F}). \tag{3.26}$$

The proof of this theorem is given in Section 5.2. The case in which $\sum_{i=1}^{n} p(F_j \mid E_i) \neq 1$ is treated in the same section, where it is shown that (3.26) still holds if a "relative" entropy of the type (1.29) is used instead of the simple entropy function. An immediate corollary to (3.26) is

$$S(\mathcal{F} \mid \mathcal{E}) \geq S(\mathcal{E} \mid \mathcal{F}).$$

We have seen that Bayes' formula (3.6) means that the probability of "cause" E_i changes from its *a priori* value $p(E_i)$ to its *a posteriori* value $p(E_i \mid F_j)$ as a result of observation of an "effect" F_j. Assume that one E_{i0} of the n possible causes is actually at work, and that this cause E_{i0} (and none other) results in a particular probability distribution $\gamma_j = p(F_j \mid E_{i0})$. We do not know, however, which E_i is the true cause E_{i0}, and we determine only the *a posteriori* probabilities $p^{(\nu)}(E_i)$ of various E_i's on the basis of ν repeated independent experiments. If we denote by j_ν the result (effect) of the νth experiment, then, by the repeated use of (3.6), we obtain

$$p^{(\nu)}(E_i) = \frac{p^{(\nu-1)}(E_i)p(j_\nu \mid E_i)}{\sum_k p^{(\nu-1)}(E_k)p(j_\nu \mid E_k)}. \tag{3.27}$$

Now define the inductive entropy

$$U^{(\nu)} = -\sum_{i=1}^{n} p^{(\nu)}(E_i) \log p^{(\nu)}(E_i). \tag{3.28}$$

If the j_ν are distributed according to the multinomial distribution corresponding to $\gamma_j = p(F_j \mid E_{i0})$, we can show the inverse H-theorem,

$$U^{(\nu+1)} \leq U^{(\nu)}, \tag{3.29}$$

for a sufficiently large ν. This is another asymmetrical result, but its tendency is opposite to the H-theorem (3.26). A more detailed explanation of this matter is given in Section 6.2.

In connection with the materials presented in this section, we should probably introduce some elementary terminologies and notions commonly used in the theory of communication, since the basic theorems that underlie them have been mentioned above. The inferred cause (input) information is $K(\mathcal{E} \mid \mathcal{F}) = S(\mathcal{E}) - S(\mathcal{E} \mid \mathcal{F})$, where $S(\mathcal{E} \mid \mathcal{F})$ represents the average uncertainty about the input when the output is received. For this reason $S(\mathcal{E} \mid \mathcal{F})$ is sometimes called "equivocation" (see [S-6]). When each input as well as output is a sequence of N symbols, $(1/N)K(\mathcal{E} \mid \mathcal{F})$ may be considered the transduced information per symbol through the transducer, or "channel." The limit, $R = \lim_{N\to\infty} (1/N)K(\mathcal{E} \mid \mathcal{F})$, is usually called "rate of information transmission" or "rate of information transfer." This quantity depends not only on the physical characteristics [which are described by the causal matrix $p(F_j \mid E_i)$], but also on the input probability distribution $p(E_i)$. The maximum (or, more precisely, supremum) of R obtained by varying $p(E_i)$ while keeping $p(F_j \mid E_i)$ fixed is called the "channel capacity" C. Sometimes this maximization of R is done by varying $p(E_i)$ under a certain restrictive condition imposed on $p(E_i)$. An example of such a restricted channel capacity is the so-called "ergodic channel capacity," which is obtained by varying $p(E_i)$ within the limits such that the input and output are ergodic. If the equivocation is zero, that is, if $S(\mathcal{E} \mid \mathcal{F}) = 0$, we say that the channel is loss-free, for then $K(\mathcal{E} \mid \mathcal{F}) = S(\mathcal{E})$. The condition $S(\mathcal{E} \mid \mathcal{F}) = 0$ implies $p(E_i \mid F_j) = 0$ or 1. This does not imply $p(F_j \mid E_i) = 0$ or 1. This last condition, which is equivalent to $S(\mathcal{F} \mid \mathcal{E}) = 0$, means that the output is uniquely determined by the input; hence we say that the channel is noiseless. But it may happen that $p(F_j \mid E_i) = p(F_j \mid E_{i'}) = 1$ for $i \neq i'$. Hence noiselessness does not imply freedom from loss. Sometimes the word "noiseless" is used to imply both lossless and noiseless. The "nonconstructive" definition of C as the maximum of R is due to Shannon [S-6]. For the case when N is finite, Muroga [M-9] discovered a method that gives an explicit expression of max $(1/N)K(\mathcal{E} \mid \mathcal{F})$ in terms of the causal matrix elements $p(F_j \mid E_i)$.

Before leaving this section we should mention a kind of erroneous argument often used in explaining the past. For instance, if we find a deck of playing cards in numerical order and separated into the four suits, we naturally infer that there must have been a human intervention in the past. Explaining this natural inference, people often say that the special order of cards has only probability $1/(52!)$, and hence there must have been some special cause to bring about such an improbable event. But this argument is incorrect, because any particular order of cards in which the deck can be found has, combinatorially speaking, the same probability $1/(52!)$. The only difference is that the deductive probability $p(F \mid E)$ is high and the prior probability $p(E)$ is quite appreciable, if we take as F the well-ordered stack and as E a human intervention. If we take as F any other particular order of cards, then we cannot formulate an E for which $p(F \mid E)$ and $p(E)$ are large. A similar erroneous argument is constantly used even by critical thinkers. If we find a footprint on a beach, we naturally infer that somebody must have walked on that place. This is not because the footprint has a small probability, as some philosophers have said, but because the acceptable hypothesis that a man has walked there makes the probability of the footprint large. Arguing along this erroneous line, they sometimes conclude that the retrodiction is certain whereas the prediction is uncertain. This "certain" retrodiction, however, depends on the prior probability $p(E)$, which is derivable neither from the deductive law [which determines $p(F \mid E)$] nor from the evidence (which is F). This retrodiction is therefore based on an extra-evidential, non-nomological factor (see Chapter 4) and is different in nature and in reliability from nomological deduction.

In connection with the explanation of the past, we have to mention another childishly erroneous argument, which maintains that prediction and explanation are perfectly symmetrical. When we make a prediction we cannot and do not include all the details of the present state (see [W-23] for a special-relativistic reason for the impossibility of learning the present state, and see Chapter 7 for the argument for a deliberate neglect of certain present facts in prediction). But a neglected factor can sometimes have a serious consequence in the future. When we explain a past event from hindsight, we can of course bring in a factor that has been neglected in prediction on a legitimate ground.

3.2. ORIGIN OF ASYMMETRY IN PREDICTION AND RETRODICTION IN PHYSICS—CAUSALITY AND FREEDOM

In this section we discuss in more detail the nature and origin of the fundamental asymmetry we talked about in the last section, limiting ourselves this time to prediction and retrodiction in the narrow sense of the terms; that is, to the probabilistic inference relating the results of a prior and an ulterior observation of the same physical system. It seems that physical science has a tacit understanding, which may be formulated as in the following postulate. This understanding is inescapably bound up with our inherent causal pattern of thinking, which in turn may be a result of adaptation to nature by generations and generations of conscious beings that have preceded us in evolution.

Postulate 3.1 (Postulate of probabilistic causality). Let $\mathcal{E} = \{E_i\}$ and $\mathcal{F} = \{F_j\}$ be state descriptions of an isolated physical system at two different times, t_1 and t_2, respectively. If $t_1 < t_2$, the conditional probabilities $p(F_j \mid E_i)$ for all i and j are determined by the nature of the physical system alone.

The qualification "isolated" is attached to the physical system to avoid a complicated discussion about influences from outside the system, although one could treat these influences also as uncontrolled stochastic events. To the "nature" of the physical system belong the physical laws governing the system as well as those properties that average samples of the system would usually share, but not the specific past history, which varies from one sample to another. In the following we usually understand that $\mathcal{E}$ and $\mathcal{F}$ are the same type of observation performed at two different times, and use the same value of the indices, i and j, to indicate the same observational result.

In the last section we discussed the impossibility of obtaining retrodictive probabilities $p(E_i \mid F_j)$ from the predictive probabilities $p(F_j \mid E_i)$ that were supposed to be given. At the present juncture we are interested in whether the $p(E_i \mid F_j)$ and the $p(F_j \mid E_i)$ are determined by the nature of the system. It would be interesting if we could state that the condition $t_1 < t_2$ is equivalent to the condition that the $p(F_j \mid E_i)$, but not the $p(E_i \mid F_j)$, are determined by the nature of the system. Unfortunately this is an oversimplification, but exceptions are few and well definable. Let us first not specify whether $t_1 < t_2$ or $t_1 > t_2$; we exclude the case $t_1 = t_2$. Formally there can be four cases, according to whether or not the $p(F_j \mid E_i)$ and the $p(E_i \mid F_j)$ are determined. First, it is impossible that neither of them is determined, because, according to Postulate 3.1, at least one of them must be given insofar as the description is adequate. In the case when the $p(F_j \mid E_i)$ are given and the $p(E_i \mid F_j)$ are not given, we have to conclude that $t_1 < t_2$, because if $t_1 > t_2$ then $p(E_i \mid F_j)$ must be given according to Postulate 3.1. Similarly, if the $p(E_i \mid F_j)$ are given and the $p(F_j \mid E_i)$ are not given, we have to conclude that $t_1 > t_2$. Last, the case in which both the $p(E_i \mid F_j)$ and the $p(F_j \mid E_i)$ are determined by the system divides itself into two subcases. In one of them the $p(E_i)$ and $p(F_j)$ remain undetermined. This is the case of bilateral determinism, which we discussed in the last section. In this case the matrices $\|p(F_j \mid E_i)\|$ and $\|p(E_i \mid F_j)\|$ are permutation matrices, and we should have $p(F_j \mid E_i) = p(E_i \mid F_j)$. The remaining case is an extremely rare one, in which the $p(E_i)$, $p(F_j)$, $p(E_i \mid F_j)$, and $p(F_j \mid E_i)$ are all determined by the nature of the system. In this case we shall speak of an "uncontrollable" system; the reason for this name will become clear presently. If we exclude these last two subcases, the remaining ones are two in which one and only one of the two sets of probabilities $p(F_j \mid E_i)$ and $p(E_i \mid F_j)$ is given. Thus we have the following theorem.

Theorem 3.10 (Theorem of impossibility of lawlike retrodiction). Under the same notation used in Postulate 3.1, the condition $t_1 < t_2$ is equivalent, except in the uncontrollable case and in the bilaterally deterministic case, to the condition that the $p(F_j \mid E_i)$ but not the $p(E_i \mid F_j)$ are determined by the nature of the physical system.

Now, referring to (3.1), we have noted that when only the $p(F_j \mid E_i)$ are given we still have complete freedom in the choice of the $p(E_i)$. This includes our freedom to pick a particular E_k as the initial condition; that is, to make

$p(E_k) = 1$. On the other hand, we cannot choose $p(F_j)$ freely when the $p(F_j \mid E_i)$ are given (see Appendix A3.1). The reader can see this impossibility by trying to make $p(F_k) = 1$ for a particular F_k. This is because we have to satisfy not only the n equations of (3.8), but also $p(E_i) \geq 0$ and $\sum_{i=1}^{n} p(E_i) = 1$. The same difficulty does not occur in the former case because the predictive probabilities $p(F_j \mid E_i)$ by definition satisfy $p(F_j \mid E_i) \geq 0$ and

$$\sum_{i=1}^{n} p(F_j \mid E_i) = 1,$$

as a consequence of which the conditions $p(E_i) \geq 0$ and $\sum_{j=1}^{n} p(E_i) = 1$ guarantee the conditions $p(F_j) \geq 0$ and $\sum_{j=1}^{n} p(F_j) = 1$. The freedom of choice we are talking about here means only "not imposed by the physical nature of the system," and does not necessarily imply a human capability of creating any arbitrary initial state, for human action is also limited by various other factors. Conversely, however, in order that a human agent can prepare an initial state freely, this kind of underlying indetermination is necessary.

Theorem 3.11 (Theorem of free initial choice). In the same notations as in Postulate 3.1, the condition $t_1 < t_2$ is equivalent, except in the bilaterally deterministic and uncontrollable cases, to the condition that the physical nature of the system imposes no restriction on the choice of $p(E_i)$ but does impose certain constraints on the choice of $p(F_j)$.

In the bilaterally deterministic case either $p(E_i)$ or $p(F_j)$ can be chosen freely, but in the more general case only a particular one, $p(E_i)$, can be chosen freely. This temporal asymmetry in basic indetermination (freedom of choice) is extremely significant in connection with the role played by science in human activities. Man conceives a goal ("end") that he desires to realize, and tries to know the right "means" leading to achievement of this end. The advice science gives to man can usually be formulated in the form "if you do A, then there is such and such probability of your getting B." If B is his "end" and if this probability is very high, then A is an appropriate "means" for him to take. A is a "cause" leading to "effect" B. If he is capable of creating the situation designated by A, then he will probably get B, which he wanted. Thus, in order that man's freedom to pursue his objectives may not be in vain, it is necessary that he be able to choose the means. Theorem 3.11 may not be sufficient, but is necessary for this pattern of human thinking and action to be possible. Thus we see that the temporal asymmetry described in Theorem 3.11 is quite natural and even harmonious with the very nature of life. The reader is advised to read the concluding section of [W-22] for the philosophical background of this paragraph.

In theoretical physics there are many laws describing temporally one-way phenomena, such as Ohm's law and the law of heat conduction. However, these laws are usable only after one knows in which sense time is flowing. The only law with any semblance of deriving one-wayness from a more basic principle is the H-theorem, or a group of H-theorems, of which a simple version was mentioned as Theorem 3.9. As far as classical physics is concerned, it must be admitted that nothing essentially new has been added to the interpretation of the H-theorems since the time of the famous critical review written in 1910 by the Ehrenfests [E-3]. If my paper of 1952 [W-5] suggested a new angle to the problem, it is because it not only emphasized that the H-theorem as a mathematical theorem shows no privileged sense in the time direction but also claimed that the asymmetrical conclusion one draws from it actually stems from the asymmetry between prediction and retrodiction (see also [W-9] and [W-19]). We relegate the H-theorems and allied problems to Chapter 5, and limit out discussion here to the problem of this basic asymmetry. The discussion takes an entirely different appearance according to whether one assumes classical physics or quantum physics as the underlying theoretical framework. We discuss mainly the case of classical physics in the following. A most conspicuous feature of this theoretical scheme may be stated as the law of bilateral determinism of classical physics.

LAW 3.1 (Law of bilateral determinism of classical physics). According to classical physics, if the state description is maximal, then the state at t_1 and the state at t_2 of the same isolated physical system correspond to each other one to one.

A "maximal" description of a system is the most accurate description of the system that is considered to be possible according to the underlying theoretical scheme. The so-called microscopic state usually means the maximally described state. A consequence of this law is that, as far as the maximal description is concerned, there is no asymmetry (of the type we are considering) in physical phenomena. Physicists should note here that the term "symmetrical" as used here is a considerably less restrictive concept than the adjective "reversible."

It should also be noted that in the maximal description the state of a physical system in classical physics is represented by a point in the so-called phase space. Hence a description by discrete sets such as $\mathcal{E}$ and $\mathcal{F}$ cannot correspond to the classical description. However, this shortcoming is not so great as it appears at first glance. If the system is spatially limited and if the available energy has an upper bound, the volume of the phase space that is needed becomes finite. Furthermore, if we take a semi-quantum-mechanical point of view, this volume can be considered as consisting of elementary volumes of size h^N, each of which corresponds to a quantum state (h is Planck's constant; $2N$ is the number of dimensions of the phase space). Then the number of possible states in the maximal description becomes finite, and the use of discrete sets $\mathcal{E}$ and $\mathcal{F}$ becomes permissible. Classical physics should be understood as the limiting case of this description for $h \to 0$. Our explanation in terms of descrete $\mathcal{E}$ and $\mathcal{F}$ is a model theory, but it is adequate to convey the gist of the problem.

Since microscopic description implies in classical physics bilateral determinism, which in turn implies temporal symmetry, we can conclude the following law from Law 3.1.

LAW 3.2 (Law of macroscopic origin of asymmetry). If physical phenomena in a certain mode of description show a temporal asymmetry of the kind dealt with in Theorem 3.10, it is, according to classical physics, due to the fact that the description is macroscopic.

This shows that the one-wayness of temporal sequences of physical phenomena reckoned as a basic attribute of nature is due to a veil of Maya, as far as classical physics is concerned, since whether or not we observe it, the microscopic state is supposed to exist in reality and its behavior is symmetrical. Law 3.2 establishes that if there is one-wayness in a description, then the description is macroscopic. It will be then instructive to check the converse of the theorem. Such a converse consideration will reveal whether and when macroscopic description entails one-wayness, and the "when" part of this consideration can be hoped to give some clue as to in which direction the past-to-future direction should lie. Before entering into a detailed calculation, we can expect in principle three different cases in macroscopic description: (a) bilateral determinism, (b) uncontrollable system, and (c) determination of only one of the two sets of conditional probabilities.

According to Law 3.1 and section 3.1, we have here a microscopic transition matrix $\|p(F_j \mid E_i)\|$, $i, j = 1, 2, \cdots, n$, which is a permutation; that is, in each column (E_i fixed) all entries are 0 except one, which is 1, and in each row (F_j fixed) all entries are 0 except one, which is 1. In order to satisfy (3.1) we should have, for all i and all j, $p(F_j \mid E_i) = p(E_i \mid F_j)$ and, for pairs (i, j), such that $p(F_j \mid E_i) = 1$, relations $p(E_i) = p(F_j)$. It is understood that the same measurement is made at t_1 and at t_2, and the same value of indices i and j refers to the same result. There are n different possible results specified in different states, and now these n states are divided into n' nonempty, nonoverlapping groups, $n' < n$. As a result the n states E_i at t_1 are grouped into n' groups of states Ψ_μ, $\mu = 1, 2, \cdots, n'$ and similarly the n states F_j at t_2 are grouped into n' groups of states, Φ_ν, $\nu = 1, 2, \cdots, n'$. The "states" and "groups of states" may also be called respectively microstates and macrostates. The macroscopic joint probability of Ψ_μ and Φ_ν is given by

$$\pi(\Psi_\mu \cap \Phi_\nu) = \sum_{j \in \nu} \sum_{i \in \mu} p(E_i) p(F_j \mid E_i) = \sum_{i \in \mu} \sum_{j \in \nu} p(F_j) p(E_i \mid F_j), \quad (3.30)$$

where the summation $\sum_{i\in\mu}$ means a summation over those E_i's included in Ψ_μ. Equation (3.30) is analogous to (2.140) for $r = 2$. From (3.30) follow the expressions of conditional probabilities

$$\pi(\Phi_\nu \mid \Psi_\mu) = \frac{\sum_{j\in\nu}\sum_{i\in\mu} p(E_i)p(F_j \mid E_i)}{\sum_{i\in\mu} p(E_i)} = \frac{\sum_{i\in\mu}\sum_{j\in\nu} p(F_j)p(E_i \mid F_j)}{\sum_{i\in\mu}\sum_{j}^{\text{all}} p(F_j)p(E_i \mid F_j)} \tag{3.31}$$

and

$$\pi(\Psi_\mu \mid \Phi_\nu) = \frac{\sum_{j\in\nu}\sum_{i\in\mu} p(E_i)p(F_j \mid E_i)}{\sum_{i}^{\text{all}}\sum_{j\in\nu} p(E_i)p(F_j \mid E_i)} = \frac{\sum_{i\in\mu}\sum_{j\in\nu} p(F_j)p(E_i \mid F_j)}{\sum_{j\in\nu} p(F_j)}, \tag{3.32}$$

satisfying

$$\pi(\Psi_\mu)\pi(\Phi_\nu \mid \Psi_\mu) = \pi(\Psi_\mu \cap \Phi_\nu) = \pi(\Phi_\nu)\pi(\Psi_\mu \mid \Phi_\nu), \tag{3.33}$$

where the unconditional probabilities $\pi(\Psi_\mu)$ and $\pi(\Phi_\nu)$ are respectively the denominator in (3.31) and the denominator in (3.32).

The important point here is that the macroscopic conditional probabilities $\pi(\Phi_\nu \mid \Psi_\mu)$ and $\pi(\Psi_\mu \mid \Phi_\nu)$ depend not only on the physically fixed microscopic conditional probabilities $p(F_j \mid E_i) = p(E_i \mid F_j)$, but also on the arbitrary unconditional probabilities $p(E_i)$ or $p(F_j)$. This would in general contradict directly Postulate 3.1 applied to a macroscopic description, which would require either $\pi(\Phi_\nu \mid \Psi_\mu)$ or $\pi(\Psi_\mu \mid \Phi_\nu)$ to be determined by nature or by the nature of the problem. Postulate 3.1 however, would be satisfied if some special circumstances made $p(E_i)$ or $p(F_j)$ disappear from (3.31) or (3.32). One very special situation happens if each macrostate consists of the same number of microstates and for each pair (μ, ν) all the $p(F_j \mid E_i) = 0$ for all $i \in \mu$ and $j \in \nu$ or else $p(F_j \mid E_i) = 1$ for each $i \in \mu$ and some $j \in \nu$. This results in a bilaterally deterministic macroscopic causal matrix, irrespective of the $p(E_i)$ and $p(F_j)$: $\pi(\Phi_\nu \mid \Psi_\mu) = \pi(\Psi_\mu \mid \Phi_\nu) = 0$ or 1. Not only can this special grouping hardly be expected to happen, but it is also an uninteresting case, for it would not exhibit any temporal asymmetry. Another extreme is the macroscopically uncontrollable case, in which both the $\pi(\Phi_\nu \mid \Psi_\mu)$ and the

$\pi(\Psi_\mu \mid \Phi_\nu)$ are determined by the system but are not related by the relation $\pi(\Phi_\nu \mid \Psi_\mu) = \pi(\Psi_\mu \mid \Phi_\nu) = 0$ or 1. We show later how this can happen, but this is again uninteresting for the present consideration because neither $\pi(\Psi_\mu)$ nor $\pi(\Phi_\nu)$ can be freely chosen.

Physicists have long noticed that the only natural way to replace (3.31) by an expression that does not depend on $p(E_i)$ is to put

$$\pi^{(0)}(\Phi_\nu \mid \Psi_\mu) = \frac{1}{n_\mu} \sum_{i\in\mu} \sum_{j\in\nu} p(F_j \mid E_i). \tag{3.34}$$

This will uphold Postulate 3.1 applied to the macroscopic description with $t_1 < t_2$. n_μ is the number of microscopic states E_i in the macroscopic state Ψ_μ. Equation (3.34) can be considered a special case of the general expression (3.31), for which either (a) the $p(E_i)$ have a constant value within each macroscopic state Ψ_μ or (b) the $p(F_j \mid E_i)$ have a constant value within each macroscopic state Ψ_μ. Formally the case of macroscopic bilateral determinism and the macroscopic uncontrollable case also belong to (3.34), but they require further restrictions in addition to (3.34). Now when (3.34) is the case, we generally still have freedom of choice of the macroscopic unconditional probability $\pi(\Psi_\mu) = \sum_{i\in\mu} p(E_i)$. If $\pi(\Psi_\mu)$ is such an arbitrarily chosen unconditional probability, then all the other macroscopic probabilities can be derived from $\pi(\Phi_\nu \mid \Psi_\mu)$ and $\pi(\Psi_\mu)$. In particular, we have a macroscopic Bayes theorem:

$$\pi^{(0)}(\Psi_\mu \mid \Phi_\nu) = \frac{\pi^{(0)}(\Psi_\mu)\pi^{(0)}(\Phi_\nu \mid \Psi_\mu)}{\sum_\kappa \pi^{(0)}(\Psi_\kappa)\pi^{(0)}(\Phi_\nu \mid \Psi_\kappa)}. \tag{3.35}$$

Now a natural question is: What happens if we interchange the roles of $\mathcal{E}$ and $\mathcal{F}$ in the above arguments? Insofar as we do not specify $t_1 < t_2$ or $t_1 > t_2$, there should be no difference between the two points of view. Corresponding to (3.34), we have

$$\pi^*(\Psi_\mu \mid \Phi_\nu) = \frac{1}{n_\nu} \sum_{i\in\mu} \sum_{j\in\nu} p(E_i \mid F_j), \tag{3.36}$$

which we can obtain from the second line of (3.32) by assuming either that the $p(F_j)$ are constant within each macrostate Φ_ν or that for each given i the $p(E_i \mid F_j)$ are constant within each macrostate $F_j \in \Phi_\nu$. The assumption (3.36) entails, in symmetry with (3.35),

$$\pi^*(\Phi_\nu \mid \Psi_\mu) = \frac{\pi^*(\Phi_\nu)\pi^*(\Psi_\mu \mid \Phi_\nu)}{\sum_\kappa \pi^*(\Phi_\kappa)\pi^*(\Psi_\mu \mid \Phi_\kappa)}. \tag{3.37}$$

Because of the microscopic bilateral determination $p(F_j \mid E_i) = p(E_i \mid F_j)$

we can rewrite (3.36) as

$$\pi^*(\Psi_\mu \mid \Phi_\nu) = \frac{1}{n_\nu} \sum_{i \in \mu} \sum_{j \in \nu} p(F_j \mid E_i), \tag{3.38}$$

which entails an interesting relation between the π^* and $\pi^{(0)}$:

$$n_\nu \pi^*(\Psi_\mu \mid \Phi_\nu) = n_\mu \pi^{(0)}(\Phi_\nu \mid \Psi_\mu). \tag{3.39}$$

Although this equation has a little similarity to the expression of the so-called detailed balance, which we discuss later, the two are different.

Now are these two viewpoints reconcilable? In general, no, because (3.34) contains nothing but quantities determined by the system whereas(3.37) contains arbitrary quantities $\pi^*(\Phi_\nu)$, which are not determined by the system, similarly for (3.35) and (3.36). This general rule of irreconcilability breaks down, however, in two special cases: (a) when $\pi^{(0)}(\Phi_\nu \mid \Psi_\mu) = \pi^{(0)}(\Psi_\mu \mid \Phi_\nu) = \pi^*(\Phi_\nu \mid \Psi_\mu) = \pi^*(\Psi_\mu \mid \Phi_\nu)$, $\pi^{(0)}$ and π^* are permutation matrices, and (b) when $\pi^*(\Phi_\nu) = n_\nu / \sum_{\kappa=1}^{n'} n_\kappa$, $\pi^{(0)}(\Psi_\mu) = n_\nu / \sum_{\kappa=1}^{n'} n_\kappa$. Case (a) is the case of macroscopic bilateral determinism, and Case (b) is a macroscopically uncontrollable case. Theorem 3.10 tells us then that except for these two cases the condition $t_1 < t_2$ corresponds to the condition (3.34) and the condition $t_1 > t_2$ corresponds to (3.36).

To fix our idea, let us assume $t_1 < t_2$. In order that the condition (3.34) may be true and the condition (3.36) may be false, we have to admit the "observation" at the initial instant t_1, and the "observation" at the final instant t_2 must be different in nature. In fact, at least in laboratory experiments, the first one is actually the preparation of a macroscopic state, while the second one is just a passive measurement. We prepare the initial state in such a way that (3.34) may be true at least to a very good approximation, and this is in order that a macroscopic causal description of the system may become possible. In fact, we seem to give a time larger than the "relaxation" time to the system before we start the experiment, sufficient for the system to reach equilibrium within the fixed constraints that characterize the macroscopic definition of the initial state. For instance, the initial macroscopic state may be specified by the condition that a vessel contains a gas at a given temperature, but the left half has twice the density of the right half. Then in order to prepare this initial state we have to insert a separation wall between the two parts (of which one contains twice as much gas as the other) and bring the whole vessel, whose external wall is diathermal, in contact with a large thermostat (heat bath) of the given temperature and let the system settle down under these constraints. The experiment itself may be one of diffusion; then the separation wall must be removed at the moment the experiment starts, but the preparation of the state is done by waiting for equilibrium under constraints. Now such a time of preparation is usually much shorter than the so-called Poincaré cycle, during which the system behaves ergodically by going through all the microscopic states possible under the constraints in such a way that the time it spends in a certain region of the phase space will be proportional to the volume of the region. If the time of preparation is comparable to the Poincaré cycle, the probability (3.34) will become the time average of the exact conditional probability for the time period of preparation, and its use in prediction will be warranted by the fact that the system must be in one of the microscopic states with the assumed probability, although the conjecture may be true that there exists a large region of the phase space of initial states from which the microscopic conditional probability $p(F_j \mid E_i)$ remains constant. Now the time of preparation is usually very short compared to the Poincaré cycle of the object system of usual size, but there seems to be good reason to believe that during the time of preparation the system goes randomly through not all but a large number of representative points, so that the time average during this time can be very well approximated by the "ergodic" average (3.34). Summarizing the above argument, we may state the following law.

LAW 3.3 (Law of random initial state). In a predictive experiment in terms of macroscopic observations in physics, the experiment is so arranged that the probability that the system will initially be located in a subset of the set of the microscopic states corresponding to the prescribed macroscopic initial state is approximately proportional to the natural measure given to this particular subset. This is necessary and sufficient to make possible a causal description in terms of macroscopic states.

The "natural measure" means the one proportional to the volume in the phase space, which, in the semiquantum picture, is proportional to the number of quantum states contained in the volume. We shall see in Section 4.2 that a prototype of Law 3.3 is already at work at the common-sense level of deductive inference. Law 3.3 in conjunction with Law 3.2 implies that the experimenter projects his one-way direction of time on physical phenomena, whereas there is no such direction in the physical world in microscopic description. This statement, taken alone apart from the other facts about the empirical world, is extremely misleading in the sense that the one-wayness of the physical world has a subjective origin. In order to re-establish the objectivity of time direction, we need at least two more laws. One of them must state in essence that all humans, as well as all living matter share the same direction of time. The second must state that the past-to-future direction of all living matter coincides with the particular direction of time discernible in the present span of time in this portion of the universe, such as the direction in which the universe is expanding and the motion of the moon is slowing down. The last direction of time is a matter of definition. If the system is not perfectly in equilibrium, the entropy of this portion of the universe is increasing in one direction of time and decreasing in the other direction of time. We take one of them and call it, say, the positive direction. The truly interesting fact is that all living matter seems to have a particular sense of past-to-future direction, which it undoubtedly derived from this positive direction of the surrounding inanimate universe. A confusing fact is that human beings carry out their laboratory experiments in such a way that the entropy increase of their experimental systems agrees with the entropy increase of the surrounding portion of the universe. (See [W-22] for a detailed discussion of the "direction" of time.)

The above seems to be an inevitable consequence of classical physics. Quantum physics presents an opportunity for an entirely different interpretation. The reason is that the law of bilateral determinism (Law 3.1) is no longer true in quantum mechanics. In the accepted usage of quantum mechanics, we have to know from the beginning the past-to-future direction of time, because the natural law is supposed here to give only the past-to-future microscopic conditional probability $p(F_j \mid E_i)$ with $t_1 < t_2$. For an attempt to reformulate quantum mechanics so that it gives not only the microscopic predictive probability but also the microscopic retrodictive probability, see [W-9]. But these two aspects do not agree unless one accepts the unrealistic principle underlying "blind" retrodiction. The general conclusion that as far as laboratory experiments are concerned the experimenter is projecting his direction of time on physical phenomena remains unchanged even in quantum mechanics. So much for the general theme of this section; in the remainder we explain some technical details that may interest only students of theoretical physics.

We stated that we wait for a period comparable to the "relaxation time" before we start out experiment, so that the system comes to an equilibrium under the given constraints. But the relaxation time very rapidly becomes small as the size of the system decreases, and it may indeed become negligible compared not only with the intervals we usually place between two successive observations but also with the duration of each operation of observation. Under these circumstances the system will go through many representative microscopic states in a macroscopic state with approximately right probability during each operation of observation, whether the macrostate refers to the initial condition or to the final condition. In other words, the averaging implied by (3.34) is no longer the result of our own free choice in preparation of the initial state, but is here physically inevitable in a passive observation, and for the same reason the averaging implied by (3.36) cannot be avoided either. Thus, in a passive observation of small systems, the two views (3.35) and (3.37) have to become compatible. In fact, this case becomes what we called an uncontrollable case and corresponds to a temporally stationary microcanonical distribution. We can thus assume in irreversible thermodynamics that an equilibrium of this kind is approximately taking place in each local system that is a spatially small portion of the larger system in nonequilibrium (see, for instance, [L-3]). This local equilibrium changes slowly (in the time scale of the relaxation time of the local system) so that the total system as a whole is in nonequilibrium (in the time scale of the relaxation time of the total system).

Let us first introduce, by way of definition, a stationary macroscopic unconditional probability characterized by the condition that the initial distribution $\pi(\Psi_\mu)$ and the final distribution $\pi(\Phi_\nu)$ are the same; that is,

$$\pi(\Psi_\mu) = \pi(\Phi_\mu), \tag{3.40}$$

where $\pi(\Psi_\mu)$ and $\pi(\Phi_\nu)$ are related by

$$\pi(\Phi_\nu) = \sum_\mu \pi(\Psi_\mu)\pi(\Phi_\nu \mid \Psi_\mu). \tag{3.41}$$

In other words, the distribution (3.40) is the eigenvector of the matrix $\pi(\Phi_\nu \mid \Psi_\mu)$ corresponding to the eigenvalue 1. Since the matrix elements are non-negative, we know, from one of Frobenius' theorems that for any given $\pi(\Phi_\nu \mid \Psi_\mu)(\geq 0)$ we can make all components of $\pi(\Psi_\mu)$ non-negative. See Section 5.2. In general we have to know all the matrix elements to solve (3.41) under (3.40), but in some cases a less precise knowledge about the matrix is sufficient to obtain the solution. In physics we usually use the so-called "theorem of detailed balance,"

$$n_\mu \pi(\Phi_\nu \mid \Psi_\mu) = n_\nu \pi(\Phi_\mu \mid \Psi_\nu), \tag{3.42}$$

to derive the values of $\pi(\Psi_\mu) = \pi(\Phi_\mu)$. Note well that, unlike (3.39), (3.42) has predictive probabilities on both sides. The condition (3.42) results in a solution of (3.41) given by the "microcanonical distribution,"

$$\pi(\Psi_\mu) = \pi(\Phi_\mu) = \frac{n_\mu}{\sum_\kappa n_\kappa}, \tag{3.43}$$

which is unique if the matrix is irreducible. We should note, however, that the solution (3.43) can satisfy (3.40) and (3.41) under a condition different from (3.42). For instance, if we have an equation of the type (3.39),

$$n_\nu \pi(\Psi_\mu \mid \Phi_\nu) = n_\mu \pi(\Phi_\nu \mid \Psi_\mu), \tag{3.44}$$

instead of (3.42), we can still satisfy (3.40) and (3.41) by (3.43). We exploit this last fact later. The condition (3.44) will be called "condition of retrodictive balance." At the end of this section we consider a further liberalization of this condition.

The condition of detailed balance (3.42) is usually derived from reversibility (invariance from time reversal) or from inversibility (invariance for the product of time reveral and space inversion.) Mathematically either of these invariances results in the condition that for each microscopic state E_i there exists its "conjugate" E_i^* such that $(E_i^*)^* = E_i$, and that if E_i belongs to Ψ_μ then E_i^* also belongs to Ψ_μ, and $p(F_j \mid E_i) = p(F_k \mid E_l)$ if $F_k = E_i^*$ and $E_l = F_j^*$. Then we have

$$\pi(\Phi_\nu \mid \Psi_\mu) = \frac{1}{n_\mu} \sum_{i \in \mu} \sum_{j \in \nu} p(F_j \mid E_i)$$

$$= \frac{1}{n_\mu} \sum_{k \in \mu} \sum_{l \in \nu} p(F_k \mid E_l) = \frac{n_\nu}{n_\mu} \pi(\Phi_\mu \mid \Psi_\nu). \tag{3.45}$$

Now coming back to the problem of compatibility of (3.34) and (3.36), we note that if the unconditional probability $\pi^{(0)}(\Psi_\mu)$ is given by (3.43)

$$\pi^{(0)}(\Psi_\mu) = \frac{n_\mu}{\sum_\kappa n_\kappa}, \tag{3.46}$$

then (3.35), which derives from (3.34), will coincide with (3.36) provided that Law 3.1 holds. In fact, from (3.39), which is true as far as Law 3.1 is true, it follows that

$$\sum_\mu \pi^{(0)}(\Phi_\nu \mid \Psi_\mu) n_\mu = n_\nu. \tag{3.47}$$

Substitution of (3.34) and (3.47) in (3.35) yields

$$\pi^{(0)}(\Psi_\mu \mid \Phi_\nu) = \frac{1}{n_\nu} \sum_{i \in \mu} \sum_{j \in \nu} p(F_j \mid E_i), \tag{3.48}$$

which becomes equal to $\pi^*(\Psi_\mu \mid \Phi_\nu)$ of (3.36) because of the relation

$$\sum_{i \in \mu} \sum_{j \in \nu} p(F_j \mid E_i) = \sum_{i \in \mu} \sum_{j \in \nu} p(E_i \mid F_j), \tag{3.49}$$

which follows directly from Law 3.1 and is equivalent to (3.39). Thus we have shown that if we have Law 3.1 and the unconditional probability (3.46),

$$\pi^{(0)}(\Psi_\mu \mid \Phi_\nu) = \pi^*(\Psi_\mu \mid \Phi_\nu). \tag{3.50}$$

Similarly, starting from

$$\pi^*(\Phi_\nu) = \frac{n_\nu}{\sum_\kappa n_\kappa}, \tag{3.51}$$

we can show with the help of Law 3.1 that

$$\pi^*(\Phi_\nu \mid \Psi_\mu) = \pi^{(0)}(\Phi_\nu \mid \Psi_\mu). \tag{3.52}$$

Finally, as we already stated, (3.46) and (3.39) result in

$$\pi^{(0)}(\Phi_\nu) = \frac{n_\nu}{\sum_\kappa n_\kappa}, \tag{3.53}$$

and similarly (3.51) and (3.39) result in

$$\pi^*(\Psi_\mu) = \frac{n_\mu}{\sum_\kappa n_\kappa}, \tag{3.54}$$

permitting

$$\pi^{(0)}(\Psi_\mu) = \pi^*(\Psi_\mu) \quad \text{and} \quad \pi^*(\Phi_\nu) = \pi^{(0)}(\Phi_\nu). \tag{3.55}$$

Thus two points of view agree. However, as we stated earlier, this is achieved at the cost of our freedom in the sense of Theorem 3.11.

A by-product of this argument is the theorem that the condition of detailed balance (hence also microscopic reversibility or inversibility, which is usually used to derive this condition) is not necessary to prove that the microcanonical distribution is stationary and the condition of retrodictive balance (which can be derived from the perfect microscopic retrodictability, Law 3.1) is sufficient for this. It may be noted further that all we used (3.42) or (3.44) for was the assurance of the condition

$$\sum_\mu n_\mu \pi(\Phi_\nu \mid \Psi_\mu) = n_\nu, \tag{3.56}$$

and for this we actually do not even need (3.44) or Theorem 3.10. Suppose that we have the double stochasticity of the microscopic transition matrix, which means, besides the usual normalization, that

$$\sum_i^{\text{all}} p(F_j \mid E_i) = 1. \tag{3.57}$$

Combined with an equation of the type (3.34), namely,

$$\pi(\Phi_\nu \mid \Psi_\mu) = \frac{1}{n_\mu} \sum_{i\in\mu} \sum_{j\in\nu} p(F_j \mid E_i), \tag{3.58}$$

the double stochasticity (3.57) guarantees (3.56); hence also the fact that (3.43) is a solution of (3.41). The permutation matrix considered in Law 3.1 is of course a special case of the double-stochastic matrix. The invariance for "conjugation" used in (3.45) is also sufficient to guarantee double stochasticity, but not necessary. See [W-19] for a discussion of the case of quantum mechanics.

3.3. CONCEPT OF INFORMATION BALANCE

The concept of "information balance" plays a very important role in the analysis of inferential processes described in the following chapters. It should be pointed out from the beginning that the notion of "balance" in this case is quite different in nature from the notion of energy balance or monetary balance. In the latter cases balance is associated with a law of conservation or temporal *constance* of the total amount of the variable attached to an isolated system. Thus in general the difference between the influx and the efflux of a quantity of this kind for a subsystem in equal to the simultaneous increase in the content of the quantity in the subsystem. We consider in Chapter 5 the question whether or not a certain reformulation of theory allows a conservation law of this type to hold for the quantity "information," but it is safe to understand that there is no conservation law of information in general, at least in the current formulation of the theory. What we mean by an "information balance" in the present section and elsewhere in this book has nothing to do with such a conservation balance. On the other hand, it must be acknowledged that in order to talk justifiably about a balance, there must be some kind of equality underlying the phenomenon. For instance, one can talk about a balance of force, because the vectorial sum of forces must equal zero in equilibrium. In the case of information balance a typical equality is the equation (3.18)

$$S(\mathcal{E}) - S(\mathcal{E} \mid \mathcal{F}) = S(\mathcal{F}) - S(\mathcal{F} \mid \mathcal{E}), \tag{3.59}$$

whose meaning is now re-examined from a new angle.

Suppose that we have two logical spectra $\mathcal{X} = \{X_i\}$ and $\mathcal{Y} = \{Y_j\}$, which are not independent. They can be respectively $\mathcal{E}$ and $\mathcal{F}$ or $\mathcal{F}$ and $\mathcal{E}$ of Sections 3.1 and 3.2, but in order not to commit ourselves to a causal sequence at this stage we use two different symbols, $\mathcal{X}$ and $\mathcal{Y}$. We consider our inference of $\mathcal{Y}$ on the ground of $\mathcal{X}$. We assume knowledge of the probabilities attached to $\mathcal{X} \otimes \mathcal{Y}$, but we first suppose that we make no observation on the outcomes X_i or Y_j. Then we have an amount of ignorance $S(\mathcal{X})$ about the outcome of $\mathcal{X}$ and an amount of ignorance $S(\mathcal{Y})$ about the outcome of $\mathcal{Y}$. Now suppose that we make an observation of $\mathcal{X}$ and get a particular result X_i, canceling the ignorance $S(\mathcal{X})$; that is, we acquire an amount of information $S(\mathcal{X})$. This acquisition of information also results in reduction of ignorance about $\mathcal{Y}$ from $S(\mathcal{Y})$ to $S(\mathcal{Y} \mid X_i)$; that is, the information obtained about $\mathcal{Y}$ on the basis of X_i is

$$\begin{aligned} \inf(\mathcal{Y} \mid X_i) &= K(\mathcal{Y} \mid X_i) = S(\mathcal{Y}) - S(\mathcal{Y} \mid X_i) \\ &= -\sum_j p(Y_j) \log p(Y_j) - \sum_j p(Y_j \mid X_i) \log p(Y_j \mid X_i). \end{aligned} \tag{3.60}$$

Since X_i occurs with probability $p(X_i)$, the information about $\mathcal{Y}$ obtained through inference on the ground of $\mathcal{X}$ is on the average

$$\inf(\mathcal{Y} \mid \mathcal{X}) = \sum_i p(X_i) \inf(\mathcal{Y} \mid X_i) = S(\mathcal{Y}) - S(\mathcal{Y} \mid \mathcal{X}). \tag{3.61}$$

This inferred information is drawn from the source information of the type "$\mathcal{X}$ is X_i," $i = 1, 2, \cdots, n$, each of which contains an amount of information $S(\mathcal{X})$. Hence it would make good sense to compare $\inf(\mathcal{Y} \mid \mathcal{X})$ with $S(\mathcal{X})$ and to ask what part of the source information $S(\mathcal{X})$ is used in the inference and what part is unused. This way of looking at the problem is justified because of the inequality discussed in Theorems 3.4 and 3.5, which can be written

$$\inf(\mathcal{Y} \mid \mathcal{X}) = S(\mathcal{X}) - S(\mathcal{X} \mid \mathcal{Y}) \leq S(\mathcal{X}). \tag{3.62}$$

From (3.59) or the first half of (3.62) we can now set up a balance sheet of the following type:

$$\begin{aligned}(\text{source information}) = {} & (\text{information useful for inference}) \\ & + (\text{irrelevant information}),\end{aligned} \tag{3.63}$$

where

$$\begin{aligned}(\text{source information}) &= S(\mathcal{X}), \\ (\text{information useful for inference}) &= \inf(\mathcal{Y} \mid \mathcal{X}) = S(\mathcal{Y}) - S(\mathcal{Y} \mid \mathcal{X}), \\ \text{and} \qquad & \\ (\text{irrelevant information}) &= W(\mathcal{Y} \mid \mathcal{X}) = S(\mathcal{X} \mid \mathcal{Y}).\end{aligned} \tag{3.64}$$

One might object to this interpretation on the ground that the quantity $K(\mathcal{Y} \mid X_i)$ of (3.60) can become negative whereas negative information does not make much sense. The reader is referred to Appendix A3.1, where a non-negative quantity instead of $K(\mathcal{Y} \mid X_i)$ of (3.60) is used, but leads to the same expression (3.61). The "balance sheet" (3.63) is not affected by this modification.

The above is the mere core of the idea of information balance, and in its actual applications it takes many variations. One of the very common types of complications is that we have to consider more than two interrelated logical spectra. In Chapters 4 and 5 we see important applications involving three logical spectra. Relegating detailed discussions to these chapters, we limit ourselves here to pointing out that the prototype of tripartite interrelations already exists in the logical schemata. For instance, in the classical syllogism there are three propositions: major premise, minor premise, and consequence. Similarly, in classical initial-value problems of differential equations, there are also three mutually related statements: a differential equation, initial condition, and the required final state.

It may not be a superficial remark that the human mind has an easy grasp of two-party relations but encounters considerable difficulty in intuitively grasping and theoretically handling more-than-two-party relations. Our mathematical formalism is usually capable of encompassing such more-than-two-party relations, but we cannot easily interpret and manipulate them.

The case of three interrelated logical spectra does not seem to be an exception. Thus in this case we start with projecting the three-party relation on a two-party relation, in particular to a two-party information balance.

Suppose that we have three logical spectra $\mathfrak{X} = \{X_i\}$, $\mathfrak{Y} = \{Y_j\}$, $\mathfrak{Z} = \{Z_k\}$ that are not entirely mutually independent; that is, there is at least one set of three numbers (i, j, k) such that $p(X_i \cap Y_j \cap Z_k)$ is not equal to $p(X_i)p(Y_j)p(Z_k)$. The simplest way to reduce this to a two-party problem is to fix one of the three variables. For instance, in a problem whose aim is to guess the outcome of $\mathfrak{Z}$, we may temporarily fix either X_i or Y_j. For a fixed outcome X_i the quantity $p(X_i \cap Y_j \cap Z_k)/p(X_i) \equiv p_i(Y_j \cap Z_k)$ defined in $\mathfrak{X} \otimes \mathfrak{Y} \otimes \mathfrak{Z}$ behaves as a probability defined on $\mathfrak{Y} \otimes \mathfrak{Z}$. Hence we can repeat all the foregoing two-party arguments using $p_i(Y_j \cap Z_k)$, provided that we do not forget that a third factor, X_i, is fixed for convenience during the discussion.

A natural extension of this point of view consists of considering $\mathfrak{X} \otimes \mathfrak{Y} \otimes \mathfrak{Z}$ in two steps, namely, first $\mathfrak{X}$ and $\mathfrak{Z}$ and then $\mathfrak{Y}$ and $\mathfrak{X} \otimes \mathfrak{Z}$. In other words, in guessing the outcome of $\mathfrak{Z}$, we first ask what information (inf_1) about $\mathfrak{Z}$ the outcome X_i brings, and then ask what information (inf_2) about $\mathfrak{Z}$ Y_j brings, when the outcome of $\mathfrak{X}$ is already known to be X_i. The answer is obviously

$$\text{inf}_1 = S(\mathfrak{Z}) - S(\mathfrak{Z} \mid X_i) \tag{3.65}$$

and

$$\text{inf}_2 = S(\mathfrak{Z} \mid X_i) - S(\mathfrak{Z} \mid X_i \cap Y_j). \tag{3.66}$$

If we take the average we have

$$\langle \text{inf}_1 \rangle = S(\mathfrak{Z}) - S(\mathfrak{Z} \mid \mathfrak{X}) = S(\mathfrak{X}) - S(\mathfrak{X} \mid \mathfrak{Z}) \leq S(\mathfrak{X}) \tag{3.67}$$

and

$$\langle \text{inf}_2 \rangle = S(\mathfrak{Z} \mid \mathfrak{X}) - S(\mathfrak{Z} \mid \mathfrak{X} \otimes \mathfrak{Y}) = S(\mathfrak{Y} \mid \mathfrak{X}) - S(\mathfrak{Y} \mid \mathfrak{X} \otimes \mathfrak{Z}) \leq S(\mathfrak{Y} \mid \mathfrak{X}). \tag{3.68}$$

The first formula (3.67) is no different from (3.61) and (3.62), except that $\sum_j p(X_i \cap Y_j \cap Z_k)$ is used as a distribution over $\mathfrak{X} \otimes \mathfrak{Z}$. The second formula (3.68) can be verified by writing out each term involved in full, or else by the help of (1.87), which reads

$$S(\mathfrak{Y} \otimes \mathfrak{Z} \mid \mathfrak{X}) = S(\mathfrak{Y} \mid \mathfrak{X}) + S(\mathfrak{Z} \mid \mathfrak{X} \otimes \mathfrak{Y}), \tag{3.69}$$

or, by interchange of $\mathfrak{Y}$ and $\mathfrak{Z}$,

$$S(\mathfrak{Z} \otimes \mathfrak{Y} \mid \mathfrak{X}) = S(\mathfrak{Z} \mid \mathfrak{X}) + S(\mathfrak{Y} \mid \mathfrak{X} \otimes \mathfrak{Z}). \tag{3.70}$$

Since we have $S(\mathfrak{Y} \otimes \mathfrak{Z} \mid \mathfrak{X}) = S(\mathfrak{Z} \otimes \mathfrak{Y} \mid \mathfrak{X})$, (3.68) follows from (3.69) and

(3.70). We can fit each of the two steps (3.67) and (3.68), as well as their combination, into the framework of information balance expressed by (3.63). Step (3.67) is obvious, since one need only replace $\mathfrak{Y}$ in (3.64) by $\mathfrak{Z}$. The second step (3.68) can be interpreted in accordance with the point of view of (3.63) by putting

$$\begin{aligned}&\text{(source information)} = S(\mathfrak{Y} \mid \mathfrak{X}),\\ &\text{(information useful for inference)} = \langle \mathrm{inf}_2 \rangle = S(\mathfrak{Z} \mid \mathfrak{X}) - S(\mathfrak{Z} \mid \mathfrak{X} \otimes \mathfrak{Y}),\\ &\text{(irrelevant information)} = S(\mathfrak{Y} \mid \mathfrak{X} \otimes \mathfrak{Z}).\end{aligned} \tag{3.71}$$

For the combination of the two steps we have the sum of (3.67) and (3.68), which can be interpreted in the framework of (3.63) by putting

$$\begin{aligned}&\text{(source information)} = S(\mathfrak{X}) + S(\mathfrak{Y} \mid \mathfrak{X}) = S(\mathfrak{X} \otimes \mathfrak{Y}),\\ &\text{(information useful for inference)} = \langle \mathrm{inf}_1 \rangle + \langle \mathrm{inf}_2 \rangle = S(\mathfrak{Z}) - S(\mathfrak{Z} \mid \mathfrak{X} \otimes \mathfrak{Y}),\\ &\text{(irrelevant information)} = S(\mathfrak{X} \mid \mathfrak{Z}) + S(\mathfrak{Y} \mid \mathfrak{X} \otimes \mathfrak{Z}) = S(\mathfrak{X} \otimes \mathfrak{Y} \mid \mathfrak{Z}).\end{aligned} \tag{3.72}$$

The simplification in the first line is allowed on account of (1.67), and that in the third lıne, on account of (1.87). The intuitive meaning of the balance sheet (3.72) is quiteclear. Observation of $\mathfrak{X}$ and then of $\mathfrak{Y}$ reduces the ignorance about $\mathfrak{Z}$ from $S(\mathfrak{Z})$ to $S(\mathfrak{Z} \mid \mathfrak{X} \otimes \mathfrak{Y})$. By observation of $\mathfrak{X}$ one obtains information in the amount $S(\mathfrak{X})$, and by observation of $\mathfrak{Y}$ after the determination of $\mathfrak{X}$ one obtains information in the amount $S(\mathfrak{Y} \mid \mathfrak{X})$ on the average. The sum of these two is the total source information. However, the indeterminacy in $\mathfrak{X} \otimes \mathfrak{Y}$ in the amount $S(\mathfrak{X} \otimes \mathfrak{Y} \mid \mathfrak{Z})$ is solely due to the fact that the factor $\mathfrak{Z}$ cannot determine entirely the outcome of $\mathfrak{X} \otimes \mathfrak{Y}$. Hence the information obtained by complete determination of $\mathfrak{X} \otimes \mathfrak{Y}$ includes a part corresponding to $S(\mathfrak{X} \otimes \mathfrak{Y} \mid \mathfrak{Z})$, which has no relevance to $\mathfrak{Z}$.

The point of view of (3.67) and (3.68) is based on averaging of (3.65) and (3.66) with respect to X_i and Y_j by the use of $p(X_i \cap Y_j)$. Often in practice we want to fix either X_i and Y_j and take the average of one with respect to the other. Suppose that we fix X_i and take the average with respect to Y_j. Then the first step remains the same, but the second step changes.

$$\mathrm{inf}_1 = S(\mathfrak{Z}) - S_i(\mathfrak{Z}) \tag{3.73}$$

and

$$\begin{aligned}\langle \mathrm{inf}_2^* \rangle &= S_i(\mathfrak{Z}) - S_i(\mathfrak{Z} \mid \mathfrak{Y})\\ &= S_i(\mathfrak{Y}) - S_i(\mathfrak{Y} \mid \mathfrak{Z}) \leq S_i(\mathfrak{Y}).\end{aligned} \tag{3.74}$$

The entropy functions with suffix i means that they are defined in terms of the conditional probabilities given X_i, that is, those probabilities derivable from $p_i(Y_j \cap Z_k) = p(X_i \cap Y_j \cap Z_k)/p(X_i)$. Equation (3.73) is the same as (3.65). Equation (3.74) can be interpreted in terms of an information balance (3.63) by putting

$$\begin{aligned}(\text{source information}) &= S_i(\mathfrak{Y}),\\ (\text{useful information}) &= S_i(\mathfrak{Z}) - S_i(\mathfrak{Z} \mid \mathfrak{Y}),\\ (\text{irrelevant information}) &= S_i(\mathfrak{Y} \mid \mathfrak{Z}).\end{aligned} \tag{3.75}$$

This form proves very useful in an analysis of information balance in deductive inference in Chapter 5.

To help widen the applicability of the idea of information balance it is quite helpful to recall the ensemble view of the concepts of probability and of information. Ignorance in this view means the degree of indeterminacy of the state of a system in the ensemble. Information means reduction of ignorance which amounts to a renewal of the ensemble such that the indeterminacy decreases. Loosely, information means more stringent restriction on the membership of the ensemble. In this terminology, to infer $\mathfrak{Z}$ means to restrict the corresponding ensemble. $S(\mathfrak{Z})$ corresponds to the unrestricted ensemble with respect to $\mathfrak{Z}$. $S(\mathfrak{Z} \mid X_i)$ means its more restricted ensemble, such that $\mathfrak{X}$ gives X_j. In this sense "information useful for inference" may be also called "information resulting in restricting the ensemble." To illustrate this point as well as the basic ideas of this section in general, let us consider a very simple example.

Suppose that there are two pushbuttons, B_1 and B_2 and four lamps, L_1, L_2, L_3, and L_4, which are connected in such a way that if a button is pressed only one lamp is lighted; namely, if B_1 is pressed there is probability $\frac{1}{2}$ each that L_1 or L_2 will be lighted, and if B_2 is pressed there is probability $\frac{1}{2}$ each that L_3 or L_4 will be lighted. Let X_i, $i = 1, 2$, stand for "B_i is pressed," and let Y_j, $j = 1, 2, 3, 4$, stand for "L_j is lighted." We assume that $p(X_1) = p(X_2) = \frac{1}{2}$, or that there is an equal probability that either of two buttons will be pushed. Then $S(\mathfrak{X}) = 1$ and $S(\mathfrak{Y}) = 2$. This means that in guessing the outcome of $\mathfrak{Y}$ there is an indeterminacy of 2 bits, because four lamps have equal probabilities. If we know which button is pushed, however, this becomes 1 bit, because the possibilities are now reduced to two lamps with equal probability. In fact, $S(\mathfrak{Y}) = 2$ and $S(\mathfrak{Y} \mid \mathfrak{X}) = 1$, meaning that the "information useful for inference" is 1 bit. But to achieve this we used the "source information" of 1 bit, because we identified one of the two equally probable buttons $S(\mathfrak{X}) = 1$. This means that to reduce ignorance by 1 bit we used the source information of 1 bit. In other words, there was no "wasted or irrelevant information." In fact, in the present case $S(\mathfrak{X} \mid \mathfrak{Y}) = 0$. Now we can see in the foregoing that we can replace "guessing the outcome of $\mathfrak{Y}$" by "determining (or restricting) the ensemble of $\mathfrak{Y}$." In fact, by determining that one or the other of the buttons has been pressed, we restrict $\mathfrak{Y}$ from the four equally probable cases to two equally probable cases, because one out of only two lamps instead of four lamps can now be lighted.

The reader may try the inverse process of guessing or determining the buttons from the data about the lamps. In this case the wasted or irrelevant information is not zero, corresponding to the fact that the source information regarding which of the four lamps is lighted is unnecessarily detailed insofar as the determination of the pressed button is concerned, in the sense that whether L_1 or L_2 is lighted makes no difference, and similarly whether L_3 or L_4 is lighted makes no difference.

4

Deduction and Induction

4.1. VERIFIABLE PROPOSITIONS, HYPOTHESES, AND EXPERIENTIAL PROPOSITIONS

Figuratively speaking, scientific activity describes a circular motion, one half of whose circumference represents an inductive inference leading from "particular experimental propositions" (or "experiential propositions") to a "general theoretical proposition" (or "hypothesis"), and the other half represents a deductive inference leading from the general theoretical proposition to a particular experimental proposition. The formal rules governing the procedure of successively producing propositions in deductive inference constitute what is usually called "logic" or, more precisely, "deductive logic." It is well known that more than two centuries ago David Hume made it clear that inductive inference does not obey the logical rules [H-16, H-17]. However, one cannot deny that there are some "rational" elements in inductive inference, and it is an intellectually challenging task to formulate those elements in a theoretical system, whether or not such a system may be called "inductive logic." (Hereafter the word "logic" without an adjective usually refers to deductive logic.)

The importance of science in life stems primarily from its usage in prediction of the future. Thus the alleged circular motion may better be depicted as a helical motion, in the sense that we "extract" a general rule from "past" experimental facts and apply it to a future (not yet tested) case in order to foretell the experimental results that will be obtained. Thus the starting and ending points of the circle usually have temporalistically different locations, and a new experimental result frequently causes a reformulation of the general rule.

It is noteworthy that what is considered in scientific inference as an experimental fact is often a general statement induced from more directly experimental facts. This implies that there exists a hierarchy of general statements, in which the upward direction points to more generality and the downward direction points to more concrete experience. This hierarchy is noticeable in any theoretical system. Another point, which may be adduced to illustrate the intricacy of the actual inductive-deductive cycle, is that any inductive process involves a deductive process. This is because in each induction the general statement must be so formulated that the past experimental facts are deductively derivable from it.

Furthermore, it is instructive that many practical problems of a deductive nature can be solved more efficiently by a type of thinking that may be called inductive. In solving a problem in plane geometry, for instance, we are given the axioms A, the premise P, and the conclusion C, and we are asked to show that C follows deductively from the conjunction of A and P. It is obvious that if we examine systematically all the logical consequences of A and P, C must be found to be one of them if C is "provable" at all. However, a human solver, as well as its mechanical analogue (see [G-7]), often goes in an inverse direction. Facing C, he looks for a general statement G from which C would logically follow, and he then determines whether this G is a logical consequence of A and P. This mental production of G from C is in some sense similar to an inductive inference.†

We do not intend to give answers to all the intricate problems connected with inductive-deductive inference in this book. We shall be satisfied with a schematic model of this inferential procedure insofar as it reproduces what may be considered as the basic features of the procedure. By the nature of our approach the model is probabilistic, but we first assume a nonprobabilistic position and then examine where such a position leads us to.

The first question pertains to the definition of what is to be called a general proposition (or hypothesis) in the inductive-deductive inference. In tackling this problem we assume existence of a nonempty set $\mathcal{E}$ of experiential propositions such that (a) any member E of $\mathcal{E}$ (symbolically $E \in \mathcal{E}$) can be directly determined by an "admitted empirical method" to be true or not, (b) E is not

† It may be mentioned that in Euclidean geometry as a branch of mathematics, the system of axioms A is to be considered as an arbitrary choice and not an empirical general rule. However, Euclidean geometry as a branch of empirical science can be considered as a set of general statements extracted from the experimental results obtained by an operationally defined method of measurements of length and angle applied to solid bodies. Historically, geometry started as a branch of empirical science, but it is significant that it gradually became a branch of mathematics, in which adequacy of a deductive inference is no longer judged by its connection with experimental facts, but only by logical correctness of the deductive inference. This fact does not change the validity of the argument that inductive methods are sometimes effective in a deductive task.

identically true ($\square$), and (c) E is not identically false ($\emptyset$). It is clearly understood that the fact that a proposition is experiential has nothing to do with its being experimentally proven to be true. In an exact science the "admitted empirical method" may be rephrased as "operationally defined experimental method" or sometimes "measurement method." An example of an identically true proposition is "the color of this (colored) object is red or nonred." An example of an identically false proposition is "the color of this (colored) object is red and nonred." What are called here "experiential propositions" will be renamed "observational propositions" in the next section, and propositions of a wider class are called "experiential propositions." But our discussion in this section is limited to the case in which these two classes coincide; hence no real discrepancy in the usage of the words exists. We understand in the following that Condition (a) also implies that if $E_1 \in \mathcal{E}$ and $E_2 \in \mathcal{E}$, $E_1 \cap E_2 \in \mathcal{E}$ unless $E_1 \cap E_2 = \emptyset$; that if $E_1 \in \mathcal{E}$ and $E_2 \in \mathcal{E}$, $E_1 \cup E_2 \in \mathcal{E}$ unless $E_1 \cup E_2 = \square$; and that if $E \in \mathcal{E}$, $\neg E \in \mathcal{E}$. This may be justifiable because these simple logical operations can be mechanically realized by a suitable hook-up inside the observational apparatus in such a way that observation directly gives the corresponding logical functions.

Now a hypothesis or general statement H must be such that an experiential proposition E can be logically derived from it without H itself being an experiential statement. However, it must be noted that in such a logical derivation one often uses, along with the hypothesis H, some other auxiliary knowledge A, which is an experiential proposition or another hypothesis. Take a simple example. From the hypothesis H, "on no August day (of any year) does it snow (in New York)" we logically derive an experiential proposition E, "it will not snow today," with the help of an auxiliary proposition A, "today is an August day." In this case A is another experiential proposition. Taking Newton's second law as H and Newton's law of gravitation as A, one logically derives an experiential proposition E, "the planet Earth describes an elliptical orbit." In this case A is another hypothesis. It may be noted that A alone cannot imply E, for, if so, H would be entirely superfluous in deriving E. This consideration leads to a formulation essentially identical with the famous (but unfortunately unsuccessful) definition of "verifiable propositions" considered by some logical positivists, such as Ayer [A-8, A-9, C-6]).

Let us briefly summarize what these philosophers intended to do with some modification in the details. A proposition V_1 is said to be verifiable if it is experiential or if, in conjunction with a tautology $\square$ or another verifiable proposition V_2, it leads logically to an experiential proposition E, whereas V_2 alone does not logically lead to E. Let $\mathcal{V}$ be the set of all verifiable propositions, and let $\mathcal{E}$ be the set of all experiential propositions. Then $\mathcal{V} \supset \mathcal{E}$, that is, $\mathcal{V}$ contains $\mathcal{E}$, meaning that any proposition belongs to $\mathcal{E}$ necessarily belongs to $\mathcal{V}$. A proposition that is verifiable but not experiential then may be called a hypothesis. Let $\mathcal{H}$ be the set of hypotheses; then $\mathcal{V} = \mathcal{E} \vee \mathcal{H}$, meaning that a verifiable proposition is either a hypothesis or an experiential proposition, and $\mathcal{E} \wedge \mathcal{H} = \Phi$, meaning that a proposition cannot be a hypothesis and at the same time an experiential proposition. (The symbol Φ denotes an empty set, and the symbols $\supset$, $\wedge$, and $\vee$ have set-theoretical meanings.)†

Now $\mathcal{V}$ is the set of proposition V_1, such that at least one of the following three conditions is satisfied: (a) V_1 belongs to $\mathcal{E}$ ($V_1 \in \mathcal{E}$); (b) there exists $E \in \mathcal{E}$, such that V_1 logically implies E, that is, $V_1 \rightarrow E$; or (c) there exist $V_2 \in \mathcal{V}$ and $E \in \mathcal{E}$, such that

$$V_1 \cap V_2 \rightarrow E \tag{4.1}$$

and

$$V_2 \not\rightarrow E, \tag{4.2}$$

where $\rightarrow$ and $\not\rightarrow$ mean respectively "implies" and "does not imply." One may note that the first two rather trivial cases (a) and (b) can be brought to the form (4.1) and (4.2) if we allow V_2 to be a constant truth $\square$, since (4.1) would then become $V_1 \cap \square = V_1 \rightarrow E$, which also allows for $V_1 = E$ as a possibility, while (4.2) becomes $V_2 = \square \not\rightarrow E$, which is always true because $\square$ is not an experiential proposition. Hence we can take (4.1) and (4.2) as the essence of the definition under examination. One may object to this definition and say that it is logically circular because in order to define $V_1 \in \mathcal{V}$, we assumed $V_2 \in \mathcal{V}$. But this is not so. Since $\mathcal{E}$ is a part of $\mathcal{V}$, one can first take V_2 as a member of $\mathcal{E}$, which is supposed to be well defined; this will allow us to define some V_1 ($\notin \mathcal{E}$), and then next time we can use one of these V_1 as a V_2. We can repeat this process and enlarge the set $\mathcal{V}$ from $\mathcal{E}$ until we cannot enlarge it any longer. Such a final set is $\mathcal{V}$. It must be clearly understood that a proposition's being verifiable has nothing to do with its being "verified" or being "true." It should be noted that the derivation of E from V_1 in the sense of (4.1) is a primitive model of deductive process. Introduction of V_1 in the presence of E_1 is a model of induction.

A trivial remark that can be immediately made is that the foregoing definition of $\mathcal{V}$ allows it to contain absurdity $\emptyset$ as its member. In fact, $\emptyset$ implies any statement, hence any experiential proposition E, whether alone or in conjunction with any other V_2. But to say that an absurd statement is a "verifiable" proposition is against common sense. Therefore we add to the definition that V_1 should not be absurdity $\emptyset$. By the same token, if $V_1 \cap V_2 = \emptyset$, then (4.1) is automatically satisfied, which means that any proposition that contradicts a verifiable proposition is a verifiable. This is a dangerously

† This is contrary to the usage of some authors who employ $\cap$ and $\cup$ for set-theoretical operations and $\vee$ and $\wedge$ for logical operations. In the present book we sometimes use $\cap$ and $\cup$ for set-theoretical purposes, too, when no confusion is foreseeable.

generous admission policy. Thus we should supplement (4.1) with an additional condition,

$$V_1 \cap V_2 \neq \emptyset. \tag{4.3}$$

The meaning of this condition will become easier to understand if we replace (4.1) and (4.3) by their equivalents,

$$V_1 \cap V_2 \rightarrow E \quad \text{and} \quad V_1 \cap V_2 \not\rightarrow \neg E. \tag{4.4}$$

In the following we do not mention (4.3) or the second relation of (4.4) each time we use (4.1). The reason for this omission is that inclusion of (4.3) does not affect the main point we intend to make in the following. In passing it may be noted that (4.4) is a good illustration of the fact that the implication relation $\alpha \rightarrow \beta$ in general has to be divided into two cases that are quite different in interpretation, (a) $\alpha \neq \emptyset$ and (b) $\alpha = \emptyset$, which may also be characterized as (a) $\alpha \not\rightarrow \neg\beta$ and (b) $\alpha \rightarrow \neg\beta$. A deep probabilistic reason for this distinction is discussed at the end of Section 7.3.

Ayer's original intention in proposing (4.1) and (4.2) was to expel from philosophical consideration a class of metaphysical propositions that have no bearing on experiential fact, by identifying "meaninglessness" with "non-verifiability." He hoped that the meaningful language and the meaningless language would thereby be completely divorced, so that, for instance, the negation of a meaningless proposition would be also meaningless. For if a proposition whose negation is meaningless were meaningful, then the language consisting of meaningful statements would not become "closed" for the operation of negation. Church showed [C-6], however, that Ayer's plan does not work thereby demonstrating that any arbitrary proposition *or* its negation is always verifiable according to Ayer's definition. This obviously excludes the possibility of a proposition and its negation both being meaningless. We are not particularly interested in the problem of meaninglessness, but Church's proof has a serious consequence in our definition of hypothesis, for it implies that any arbitrary proposition or its negation becomes an experiential proposition or a hypothesis. This would be obviously too broad a definition of hypotheses to be practically useful. Church's proof is given in Appendix A4.1.

The whole trouble obviously lies in the fact that if V_1 is a verifiable proposition, then any proposition that implies V_1 becomes verifiable; that is, if $V_1 \in \mathcal{V}$ and $U \rightarrow V_1$, then $U \in \mathcal{V}$ according to Ayer's definition. In fact, if the ground on which V_1 is considered as a verifiable proposition is the relations (4.1) and (4.2), then $U \cap V_2 = U \cap V_1 \cap V_2 \rightarrow V_1 \cap V_2 \rightarrow E$, and $V_2 \not\rightarrow E$. Therefore U is equally qualified as a verifiable proposition. The extreme case $U = \emptyset$ has already been eliminated by (4.3), but without going to that extreme we can imagine a ridiculous case. Let V_1 stand for the second law of Newton, and let N stands for a statement "all unicorns have five legs."

Then $U = V_1 \cap N$ ($\rightarrow V_1$), which states that "Newton's second law is true and all unicorns have five legs" becomes a verifiable proposition with the same merit as V_1 itself. This fact has a close connection with Hempel's first paradox in the theory of confirmation, which we discuss in Section 4.5. In view of this situation we are tempted to impose a further constraint on the definition (4.1) and (4.2) by requiring that there does not exist a proposition V_1^* such that

$$V_1 \rightarrow V_1^*, \qquad V_1 \neq V_1^*$$

and

$$V_1^* \cap V_2 \rightarrow E. \tag{4.5}$$

If we use a loose expression, "α is larger than β," to mean $\beta \rightarrow \alpha$ and $\beta \neq \alpha$, then the V_1 satisfying (4.5) on top of (4.1) and (4.2) is the largest among those V_1 that satisfy (4.1) and (4.2) for a given pair E and V_2. Actually this largest V_1 can be written explicitly as

$$V_1 = E \cup \neg V_2. \tag{4.6}$$

We can see this easily, by noticing that if any V satisfies $V \cap V_2 \rightarrow E$, we can write $V = (V \cap V_2) \cup (V \cap \neg V_2) \rightarrow E \cup \neg V_2$, where the first step (equivalence) is a tautology (distributive law) and the second step (inplication) is an application of the obvious relation: if $\alpha \rightarrow \gamma$ and $\beta \rightarrow \delta$, then $\alpha \cup \beta \rightarrow \gamma \cup \delta$. This shows that conditions (4.1), (4.2), and (4.5) can be rephrased as the following provision: V_1 is a verifiable proposition if there exist $V_2 \in \mathcal{V}$ and $E \in \mathcal{E}$ such that $V_2 \nrightarrow E$ and (4.6) is satisfied. To see the relation of (4.6) to (4.1), we may note that (4.6) entails

$$V_1 \cap V_2 = E \cap V_2, \tag{4.7}$$

which satisfies (4.1). But we must be careful because (4.6) does not conversely follow from (4.7). We should not lose sight of the fact that the basic relations (4.1) and (4.2) are not violated by (4.6); only an additional restriction is added. It may be noted that $V_2 \rightarrow E$ is equivalent to $V_1 = \square$.

This condition (4.6) may seem to be unduly strict, whereas Ayer's condition was unduly liberal. In fact, any proposition V, such that $\emptyset \neq V \rightarrow V_1$, where V_1 is given by (4.6), satisfies Ayer's condition $V \cap V_2 \rightarrow E$. One might think that we have gone to the opposite extreme. But this situation is considerably liberalized if we add the stipulation that any proposition that is a conjunction of more than one verifiable proposition is again a verifiable proposition. If V_1 is a verifiable proposition in the sense of (4.6), then any V that implies V_1 satisfies the relation $V \cap V_2 \rightarrow E$ with the same E and V_2. But such a V ($\neq V_1$) cannot be claimed to be a verifiable proposition on the ground of $V \cap V_2 \rightarrow E$. However, we should note that such a V may yet qualify as a verifiable proposition because there may be another pair, E^* and V_2^*, for which $V = E^* \cup \neg V_2^*$. Besides this trivial possibility, there arises a new possibility for such a V to qualify as a verifiable proposition, owing to the above proposed new stipulation. Suppose that a proposition X is established as a verifiable proposition; then $V = X \cap V_1$ becomes a verifiable proposition as a result of this new agreement, and unless $V_1 \rightarrow X$, this new V is "smaller" than V_1; that is, V_1 is larger than V. (It is necessary that V_1 and X be not disjoint in order to avoid $V = \emptyset$, which is excluded from $\mathcal{V}$.)

To summarize, our tentative definition of the set $\mathcal{V}$ of verifiable propositions is that $V_1 \in \mathcal{V}$ if $V_1 \neq \emptyset$ and (i) if there exists $E \in \mathcal{E}$ such that $V_1 = E$; (ii) if there exists a pair (V_2, E), $V_2 \in \mathcal{V}$, $E \in \mathcal{E}$, such that $V_1 = E \cup \neg V_2$; or (iii) if there exists a set of propositions $W_1, W_2, \cdots$, with $W_1 \in \mathcal{V}$, $W_2 \in \mathcal{V}$, $W_3 \in \mathcal{V}, \cdots$, such that $V_1 = W_1 \cap W_2 \cap W_3 \cap \cdots = \bigcap_i W_i$.

This definition may still be a little narrow, but we are on safe ground. In fact, it is shown in Appendix A4.1 that Church's argument against Ayer's definition no longer works if we take $\mathcal{V}$ defined here as the set of meaningful propositions. The condition (4.2), $V_2 \not\rightarrow E$, may be added to (ii), but it will serve only to avoid $V_1 = \square$ here.

A further safe liberalization of $\mathcal{V}$ will be effected if we take into account the fact, already mentioned at the beginning of this section, that a hypothesis of more concrete nature is often considered as an experimental fact for a hypothesis of more abstract nature. This leads us to replace (ii) above by (ii′) "if there exists a pair (V_2, V_3), $V_2 \in \mathcal{V}$, $V_3 \in \mathcal{V}$, such that $V_1 = V_3 \cup \neg V_2$.

Now it must be noted that neither $\mathcal{E}$ nor $\mathcal{V}$ is closed with respect to the basic logical operations. This should not disturb us too much, however, because, for instance, the negation of a hypothesis often has absolutely no power of predicting any observable result. On the other hand, it cannot be denied that in scientific language we constantly negate a hypothesis. Thus we can reach a kind of simplified model of scientific language by identifying this with a set of propositions $\mathcal{V}^*$ "generated" by $\mathcal{V}$ and closed with respect to the operations of disjunction, conjunction, and negation. This set $\mathcal{V}^*$ of propositions "generated" by $\mathcal{V}$ is intuitively visualized as being obtained from $\mathcal{V}$ by gradually augmenting the set with new members, which are obtained by logical operation applied to old members, until finally no new members can be produced. Such a "scientific language" may be close to what Ayer had in mind when he tried to define meaningful propositions.

A more important question pertains to the kind of conjunction allowed in the conditions (iii), denoted there as $\bigcap_i$. Must the index i be limited to a finite number? Can we include a conjunction of countably many (i.e., an enumerably infinite number of) propositions? How about the case of continuously many propositions?†

† As an example of continuously many propositions, consider the collection of all sentences, each of which says "We say it is mild when the temperature is θ degrees," where θ is any real number between 60 and 75.

First, it is mathematically the safest to limit the conjunction to a finite number of terms, but this would hardly be satisfactory as a definition of verifiable propositions, because the simplest type of hypothesis has the form "all that is X is P," which entails a conjunction of propositions: "Object 1, which is X, is P," and "Object 2, which is X, is P," and "Object 3, which is X, is P," and so on. In the case of $\mathcal{E}$ we can legitimately object to allowing a countable conjunction $\bigcap_i^\infty E_i$ as a member of $\mathcal{E}$, where $E_i \in \mathcal{E}$, $i = 1, 2, \cdots$, because there is no observational method that determines the truth or falsehood of $\bigcap_i^\infty E_i$ in finite time. But in the case of $\mathcal{V}$ it is hardly justifiable to exclude a countable conjunction, because a hypothesis is a general statement that usually encompasses at least countably many objects.

This raises two delicate problems. One stems from the fact that the usual formulation of mathematical logic avoids an expression like $\bigcap_i^\infty W_i$. Another is connected with the fact that in the predicate calculus the proposition $\forall i\, P(i)$" [which allows an interpretation: for all i, $P(i)$, where $i = 1, 2, \cdots$] and the conjunction proposition "$P(1)$ and $P(2)$ and $P(3) \cdots$" are not necessarily mutually derivable from each other. The latter is derivable from the former, but not vice versa, in the so-called ω-inconsistent case. As far as the first question about $\bigcap_{i=1}^\infty W_i$ is concerned, it should be mentioned that in set theory a countable conjunction is readily allowed. In fact, σ algebra (or the Borel field), which is a kind of set on which the concept of probability can be defined, is supposed to allow for countable conjunctions and countable disjunctions. The De Morgan laws for countable conjunction and disjunction are also upheld. The reason why $\bigcap_{i=1}^\infty W_i$ (understood set-theoretically) converges to a set is intuitively understandable if one remembers that a monotonically decreasing numerical sequence with a lower bound converges. In the present case we similarly have $\bigcap_{i=1}^n W_i \supset \bigcap_{i=1}^{n+1} W_i$ and $\bigcap_{i=1}^n W_i \supset \Phi$. If the set-proposition parallelism is accepted on a very broad basis, there should be no reason why we should shy away from a countable conjunction of the type $\bigcap_{i=1}^\infty W_i$. In fact, the proposition mentioned above, "$P(1)$ and $P(2)$ and $P(3) \cdots$," may be understood as being equivalent to $\bigcap_{i=1}^\infty P(i)$. As far as the second question, pertaining to the difference between $\forall i\, P(i)$ and $\bigcap_{i=1}^\infty P(i)$, is concerned, we can be perfectly satisfied by assuming that a general statement in the induction-deduction cycle means only an infinite conjunction $\bigcap_{i=1}^\infty$ and not an $\forall$-proposition. The only caution we must take is that the labeling system in terms of i must be exhaustive in the sense that any given object under consideration has a finite integer (even very large) as its label. Some authors, for instance, Davis [D-2], write $\bigwedge_{i=1}^\infty$ to mean an $\forall$-proposition. We deviate from such usage.

Let us explain as an illustration how a proposition of the type "all X's are P" can be shown to be a verifiable proposition in our scheme, where "being X" and "being P" are experimentally determinable, and predicate P is meaningfully applicable to an object X. Take, as an example, "all ravens are black" as V_1. Denote the proposition that "bird i ($i = 1, 2, 3, \cdots$) is a raven" by Y_i. It is assumed that there is an experimental procedure by which a bird can be determined to be a raven or a non-raven. Then Y_i is an experiential proposition and therefore a verifiable proposition. The conjunction $V_1 \cap Y_i$ implies E_i, that "bird i is black." According to Ayer's criterion (4.1), (4.2), this fact with a particular i is a sufficient ground to declare V_1 to be a verifiable proposition. From our present point of view this is not sufficient. We first have to form, according to Condition (ii), $W_i = E_i \cup \neg Y_i$, which states that "bird i is either black or non raven." (The word "bird" can be replaced by the word "object," if so preferred.)

This W_i is certainly not equivalent to V_1, which says that all ravens are black, although $V_1 \rightarrow W_i$. Now make the countable conjunction of W_i with respect to all i's. We obtain a statement $W = \bigcap_{i=1}^{\infty} W_i$, which reads "all birds are either black or non-ravens." This last statement W is equivalent to V_1. Indeed, if W is true, then if any bird is a raven it has to be black, which shows that $W \rightarrow V_1$. Conversely, if V_1 is true, any bird must be either a raven, which is black, or a non-raven, showing that $V_1 \rightarrow W$. This establishes that V_1 is a verifiable proposition, for it is a (countable) conjunction of verifiable propositions $W_i = E_i \cup \neg Y_i$. It must be clear that this argument has nothing to do with the truth or falsehood of the propositions involved. All we assumed is that P is experimentally testable and that membership in X is experimentally decidable. It is also important to note that for the purpose of this definition no question is raised as to technical feasibility of the test $\bigcap_{i=1}^{\infty} E_i$, for the process of countable conjunction was carried out in the framework of $\mathcal{V}$ and not in that of $\mathcal{E}$.

Many hypotheses concern more than one object, but they can be reduced to hypotheses concerning one composite object. For instance, consider the general proposition that it never rains on two consecutive days. Let i or j label a day. Let $C(i,j)$ stand for the proposition that j is the next day of i. Let $R(i)$ stand for the proposition that i is a rainy day. Then, by introducing $F(i,j) = \neg R(i) \cup \neg R(j)$, which means that i and j are not both rainy, one can write the hypothesis as $\bigcap_{ij}[\neg C(i,j) \cup F(i,j)]$. The point is that in this expression each object is a pair of days (i, j) and $\bigcap$ extend over all the pairs.

In many practical cases the limitation to "countable" conjunction in our definition is not broad enough. However, we do not have any mathematical scheme that allows for a conjunction of continuously many elements, and for this reason we cannot generalize our definition *in toto* to the continuous case. However, the following generalization, which applies only to a special class of continuous cases, seems to be broad enough to include most of the practically interesting cases. The special class of cases I have in mind may be called the class of parameterizable cases.

Take first a countable case, such as "all ravens are black." A statement equivalent to this statement is "for any natural number i it is true that raven i is black," whereby the label i is assumed to belong to a certain exhaustive way of numbering ravens. In this case we can see that for each particular i the statement "raven i is black" is an experiential proposition, and when parameter i is allowed to take any value in the specified range (all natural numbers) then the statement is no longer an experiential proposition, but a verifiable proposition. In this way the countable conjunction is replaced by a statement containing a parameter, whereby the parameter is allowed to take any value in a specified domain. Alternatively, a statement including a parameter whose domain is specified is a verifiable proposition if the statement becomes an experiential proposition for each particular value of the parameter, taken from this domain. There is no problem in extending this to the case in which a finite number of parameters is involved. We can generalize this easily to the continuous case simply by allowing the domain to be continuous. For instance, the statement, "a mass point thrown vertically upwards from the earth's surface with velocity v centimeters per second returns to the earth's surface after $2v/g$ second, provided that $0 \leq v \leq v_0$," is not an experiential proposition, but it is a verifiable proposition, because for any particular value v satisfying $0 \leq v \leq v_0$ the statement becomes an experiential proposition. In order for the statement to be true, within a certain allowable error, v_0 must be set so that the motion is limited within the region where g can be considered to be constant (resistance is ignored), but verifiability has nothing to do with truth of a statement. We could generalize the definition a step further and replace (iii) by (iii′), "a proposition that contains a finite number of parameters whose values have specified continuous or discrete domains is verifiable if it reduces to a verifiable proposition for each set of particular values of these parameters." It is understood, of course, that the determination of the values of the parameters is operationally feasible.

This is about all we are in a position to do in improving Ayer's original idea insofar as we remain within the nonprobabilistic framework. It is our task in the next section to reconsider the whole matter from a probabilistic point of view. Such an effort will hopefully lead us to a more realistic model of deduction.

4.2. PROBABILISTIC DEDUCTION—PROBABILISTIC EXPERIENTIAL PROPOSITIONS

So far we have improved Ayer's definition more or less only along his line of thought, but now we have to raise more basic objections. In the first place, we agreed that an experiential proposition E must be defined by an operationally defined method of observation or measurement. But the result of such a method of observation usually yields more than just two propositions, E and its negation $\neg E$. Sometimes the result even spreads over a continuous range of values. For instance, an ammeter shows a continuous range of amperes within certain limits. Obviously, because of inevitable inaccuracy and errors, we can divide the continuous range into a finite number of possible discrete cases. Even then the number of cases is usually a large integer. It is true that if the number of cases is finite, we can reduce it to the conjunction of a finite number of dichotomical classifications. For instance, if there are four cases, $q = 1, 2, 3, 4$, we can take the propositions E_1, "q is even" (i.e., 2 or 4), and E_2, "q is equal to or larger than 3" (i.e., $q = 3$ or 4). Then $E_1 \cap E_2$ is equivalent to $q = 4$, $\neg E_1 \cap E_2$ is equivalent to $q = 3$, $E_1 \cap \neg E_2$ is equivalent to $q = 2$, and $\neg E_1 \cap \neg E_2$ is equivalent to $q = 1$. However, such a decomposition into dichotomies is artificial and cannot be very efficient in analyzing the problems. Therefore it is much more advisable to characterize an observation by a set $\mathfrak{D}$ of possible outcomes (data) D_i of observation: $\mathfrak{D} = \{D_1, D_2, \cdots, D_i, \cdots, D_n\}$. We assume that the D's are mutually exclusive and exhaustive, that is, it is impossible that two D's can become true by a single observation ($D_i \cap D_j = \emptyset$, $i \neq j$), and that one of the D's has to turn out to be true by an observation ($D_1 \cup D_2 \cup \cdots \cup D_n = \Box$). For instance, D_1, D_2, D_3, and D_4 may respectively mean "the weather at a certain location at a certain time is fair, cloudy, rainy, and snowing or hailing." Or, in another example, D_i may mean "the ammeter shows that the current I is in the range $(i - 1)\varepsilon \leq I < i\varepsilon$ $(i = 1, 2, 3, \cdots, n)$ amperes," where ε may be considered as the order of magnitude of an inevitable error. We say that each D_i is an "observational proposition." Each D_i is by definition a proposition, but in order to save notations we also sometimes use D_i to denote the value of the variable measured. The integer n is the number of possible outcomes, and we usually assume it to be finite. The previous analysis of verifiability referred to the binary or dichotomical case, $n = 2$; E and $\neg E$ may be regarded as D_1 and D_2. The most important agreement here is that the observational operation is well defined. The symbol $\mathfrak{D}$ is primarily defined as the set of possible outcomes in a certain well-defined experiment; hence it can also be considered as a set of integers, each of which denotes one of the possible outcomes. But in order to economize symbols, we sometimes use $\mathfrak{D}$ to mean the observational operation itself.

A second objection to the original definition of an experiential proposition is directed against the tacit assumption, made in the previous analysis, that a conclusion from a hypothesis is of the type: Such and such observational result will be obtained or not obtained. This is certainly not a realistic picture of human inference even in the dichotomic case. A hypothesis (plus some auxiliary knowledge) gives rise in the general case to a proposition that indicates only the probability of an observational result being obtained. The probabilistic nature of an experiential proposition is true not only of almost all scientific prediction and scientific knowledge, but also of some observational results. For instance, the purpose of temperature measurement in a gas is not to measure a physical quantity, but to obtain the probabilistic distribution of kinetic energy of the gas (see Mandelbrot [M-4]). In any event, we have to set up a class of propositions S (for stochastic) of the type "the probability of D_i being true is p_i for $i = 1, 2, \cdots, n$":

$$S: \bigcap_{i=1}^{n} (\text{probability of } D_i \text{ being true, equals } \pi_i). \tag{4.8}$$

In other words, S is a proposition that is the conjunction of n propositions, each of which states that the probability of D_i is π_i. We call a proposition of this type a pure probabilistic experiential proposition, in contrast to each D_i, which is an observational proposition. If S is such that $p_j = 1$ for some j and $p_j = 0$ for $i \neq j$, then S is effectively equivalent to D_j (insofar as the number of trials is finite). In addition to this, if $n = 2$, S becomes a categorical, dichotomic experiential proposition E, which we discussed previously.

One may contend that a probabilistic experiential proposition S thus defined is nothing but a special type of the categorical (i.e., true-or-false) experiential proposition, since if we repeat the same observation N times, the number N_i of times we obtain D_i will become Np_i as $N \to \infty$. It then would suffice to consider a single categorical proposition,

$$E: \quad N_1 = Np_1,\ N_2 = Np_2, \cdots, \quad \text{and} \quad N_n = Np_n.$$

Then this proposition E is either true or false. This contention is formally supported by the law of large numbers, but such a proposition E can hardly be considered a categorical "experiential" proposition because the condition $N \to \infty$ precludes any feasibility in human experience. Therefore it is more natural not to reduce every probabilistic proposition S to a categorical proposition E.

As we emphasized in Section 1.1, there is in practice no such thing as an unconditional probability. Even when we do not mention any condition, some conditions are tacitly understood or agreed upon. This means that there is a background reasoning leading to a statement of the type S of (4.8). If we want to specify the conditions on the basis of which a proposition of the type S is asserted, we have to resort to another type of proposition,

$$T: \bigcap_{i=1}^{n} [p(D_i \mid C) = \pi_i], \tag{4.9}$$

where C represents the conditions. We may call a proposition of the type T a conditional probabilistic experiential proposition. The set of conditions C

often consists of a theory or hypothesis H and some auxiliary factual conditions A about the system. The condition A must include also the description of the experimental set-up as well as the initial and boundary conditions. In such a case the conditional probability in (4.9) is written as $p(D_i \mid H \cap A)$. In this expression the eventuality that H is false is not entirely excluded from consideration. If H is assumed to be true all through a discussion, we can take H from the condition and mention only A.

The true meaning of C in T is often misunderstood and therefore requires a careful explanation. What T states can be formulated verbally as follows:

> T: if C is established to be true and if no proposition G is established to be true such that $G \rightarrow C$ and $G \neq C$, then S (with the given π_i).†

The reason why we have to put a second condition that no G is true is that $p(D_i \mid C)$ and $p(D_i \mid G)$ are not necessarily the same even if G implies C. In other words, if not only C but some other condition K, which is not implied by C, is true, then $C \cap K$ implies C but is different from C. Hence if any further information K in addition to C is available, the probability of D_i can change. The most common mistake is committed by using the following type of inference: C and T; therefore S. Obviously this does not follow if T is correctly interpreted as in the above. But this type of inference is a careless carryover from the familiar nonprobabilistic inference, in which the counterpart of T takes the form "if C, then S," where the probabilities involved in S become 1 or 0. In this nonprobabilistic case the type of inference becomes perfectly legitimate (this is a modus ponens). We now show how this special case of nonprobabilistic inference is actually embedded in the more general probabilistic consideration. The gist of the relation between the two cases can be summarized in the following.

Theorem 4.1 The formula $p(F \mid C \cap K) = p(F \mid C)$ holds for any arbitrary K which satisfies $p(C \cap K) \neq 0$, if and only if $p(F \mid C) = 0$ or 1.

Proof. Write $p(F \mid C \cap K) = x$ and $p(F \mid C \cap \urcorner K) = y$. This yields $p(F \cap C \cap K) = xp(C \cap K)$ and $p(F \cap C \cap \urcorner K) = yp(C \cap \urcorner K)$, hence $p(F \cap C) = xp(C \cap K) + yp(C \cap \urcorner K)$. On the other hand, putting $p(F \mid C) = a$, we obtain $p(F \cap C) = ap(C) = ap(C \cap K) + ap(C \cap \urcorner K)$. Thus

$$xp(C \cap K) + yp(C \cap \urcorner K) = a[p(C \cap K) + p(C \cap \urcorner K)]. \quad (4.10)$$

Now there are three possible cases satisfying this equation: (a) if $a = 1$,

† C itself is a nonprobabilistic proposition, but we can consider the probability of C being true.

then $x = 1$; (b) if $a = 0$, then $x = 0$; and (c) if $a \neq 0$ and $a \neq 1$, then x is not necessarily equal to a. [The condition $p(C \cap K) \neq 0$ is used in deriving this consequence.] Thus, if $a = 0$ or 1, we have $a = x$. This means that if $p(F \mid C) = 0$ or 1, $p(F \mid C \cap K) = p(F \mid C)$ for any K. This proves one half of the theorem. Next suppose that $a \neq 0$ and $a \neq 1$ and take $K = F \cap C$; then $x = p(F \cap C \cap K)/p(C \cap K) = 1 \neq a$. [The denominator $p(C \cap K) = p(F \cap C)$ is not zero because $a \neq 0$.] This means that if $a \neq 0$ and $a \neq 1$, there exists some K such that $p(F \mid C \cap K) \neq p(F \mid C)$. By contraposition, if $p(F \mid C \cap K) = p(F \mid C)$ for all K, then a must be either 0 or 1. This proves the other half of the theorem.

A direct paraphrase of the theorem is the following corollary.

Corollary 4.1 When $G \to C$, $G \neq C$, and T(4.9) is true, we can legitimately claim S on the ground of G (4.8) if and only if the π_i in S are either 1 or 0.

Obviously the corollary justifies the modus ponens and the syllogism for categorical cases, but warns against an indiscriminate use of these modes of inference in probabilistic cases. We must be very clear, however, that this consequence need not be interpreted as a deviation from the usual logic, if we treat T as a proposition and if we do not force the condition C involved in T in the role of a premise in syllogism or in modus ponens. In fact, if there is a proposition, say, U, that implies T and is not a constant absurdity $\emptyset$, then having U and C and "not having G," we can conclude S just as we can conclude S from T, C, and "not having G," where the phrase "not having G" means that we do not have any G established such that $G \to C$ and $G \neq C$.

In fact, as we already mentioned, C often consists of two ingredients, hypothesis H and contingent experimental data A. Often, if not always, if there is H^*, which implies H and differs from H, we still have

$$p(D_i \mid H \cap A) = p(D_i \mid H^* \cap A), \qquad i = 1, 2, \cdots, n. \qquad (4.11)$$

On the other hand, if there is A^*, which implies A but differs from A, we often do not have

$$p(D_i \mid H \cap A) = p(D_i \mid H \cap A^*), \qquad i = 1, 2, \cdots, n. \qquad (4.12)$$

For this reason we can often treat H as if it were a logical premise of T, whereas we have to treat A as a probabilistic condition in the sense of Theorem 4.1. Similarly, if we decide to stick to one particular H during a discussion and not change it, we can also treat H as if it were a logical premise of T because we are not interested in the functional dependence of the probability p on H. In this mode of description we can replace (4.9) by

$$T: \bigcap_{i=1}^{n} [p(D_i \mid A) = \pi_i] \qquad (4.13)$$

and complement it with

$$H \rightarrow T. \tag{4.14}$$

This suggests that the formalism may be interpreted as meaning that T is, in a traditional sense, a deductive consequence with regard to an experiment $\mathfrak{D}$ derived from theory H. But we must be careful, because it is not always guaranteed that T of (4.9) with $C = H \cap A$ can be transformed into the form (4.13) and (4.14). At the other extreme, if we include all hypotheses, tacit and explicit, in C of (4.9), T has to become a tautology, because T itself does not say anything about the truth of condition C, and asserts the derivation of S. In (4.14) the truth of T depends on the truth of H, although the implication ($\rightarrow$) itself is meant to be tautological.

Apart from this theoretical consideration, it is advisable in practice to exclude from C those hypotheses whose truth we do not want to argue about in a discussion, and include only the remaining hypotheses in C. This practical rule also applies to A. The portion of contingent data that we want to assume to be true throughout a discussion can be excluded from C. With this understanding we use the standard form of T:

$$T:\ \bigcap_{i=1}^{n} [p(D_i \mid H \cap A) = \pi_i]. \tag{4.15}$$

We have gone much too far into a formal argument; we should go back to a more concrete level to become familiar with the kinds of things we are actually talking about. For instance, $\mathfrak{D}$ may be an observation of weather at New York City at a certain date, described in terms of four classes: fair, cloudy, rainy, and "snowy or hailing." The condition A may be a description in these four terms of the weather on the previous day in New York. The general law H may be a Markovian rule with a set of transition probabilities (4×4 matrix). The values of $p(D_i \mid H \cap A)$ with the same $\mathfrak{D}$ can change every time we take a different $H \cap A$. For instance, A could be the collection of all meteorological data in the world for the last 24 hours, and H could be an elaborate probabilistic theory that partially takes thermodynamic and hydrodynamic equations into consideration and can be applied to the particular type of initial condition A. Empirically the latter may give a better prediction on the average, but we are not interested in that fact. Nor are we interested in the credibility of H. We are interested, however, in the fact that the set of probabilities expressed in S is "deducted" on the ground of some general rule H with the help of some facts A, and that its values vary depending on H and A, even though $\mathfrak{D}$ remains the same.

An important consequence of this situation is that the conclusion with regard to an individual case derived from a statistical general rule is critically dependent on the ensemble of which the individual is considered, in this derivation, to be a randomly picked member (see Theorem 4.1). This means that the condition characterizing the ensemble to which the individual belongs is to be taken as the sole available information about the individual; all other information is to be ignored. In the foregoing example, for instance, the first derivation classifies a particular day by the weather of the preceding day at the same place described in terms of the four categories, and all other information is to be ignored. In the second derivation the particular day is characterized by the condition imposed by the worldwide meteorological data of the last 24 hours but is not further restricted by anything else. The conclusions may very well be different in the two derivations, but this does not involve anything paradoxical if one reflects a little on the nature of the concept of conditional probability.

Consider, to fix the idea, the case in which H_1: fraction r of all G's are B, and fraction $(1 - r)$ of them are non-B, and A_1: an individual object a is a G, and $\mathfrak{D}$ is a test of B-ness applied to object a.† Then we have

$$T_1: \quad p(B \mid H_1 \cap A_1) = r, \tag{4.16}$$

stating that the probability that object a will turn out to be B is r, "provided that $H_1 \cap A_1$." The expression "provided that X" here is an abbreviation for "provided that an individual case is taken randomly from the population characterized by just X, not more, not less," or "provided that the information about the individual case taken into account is just X, not more, not less." It can be said more simply that the individual must be a "fair example" of X. If the systems under consideration are small physical objects, we should imagine that we put *all the objects* satisfying X in a bag, shake it well, and blindfolded pull one piece out of the bag at random. This will give a "fair sample" of X. Of course, because we cannot do this operation in reality, the experiment must realize a situation as close as possible to this kind of random sampling. From the above T_1 we may expel H_1 from the condition as in (4.13) and (4.14) and apply the method of random sampling to a collection of all G's.

An easy way to avoid confusion attributable to the particular meaning of "condition" is to introduce the following rule of interpretation for a probabilistic statement.

RULE 4.1. Let there be given a statement of the type, "the probability of an object a, which is A, being B, is p." In this statement it should be understood that a is a randomly picked sample from a collection of all A's or, in other words, that any knowledge or information beyond a being A is either unobtainable or deliberately ignored.

The reader will remember that we encountered a similar interpretation rule in Law 3.3, p. 122. Even an everyday life sentence like "an A is usually B" requires this kind of interpretation, for the first predicate, A, and the

† This G has nothing to do with the G considered previously in this section.

second predicate, B, have quite different roles. The first one (subject-attributive use) requires a "fair" representative of A, whereas the second one (object-predicative use) requires that the object be B no matter how exceptional or unusual it may be with respect to other properties.

Those who did not understand the true meaning of these elementary consequences of the concept of conditional probability were disturbed by the fact that $p(D_i \mid C)$ varies with C and wanted to have a unique value by requiring the so-called principle of total evidence. This principle states that all the available facts pertaining to an individual must be taken as the basis of evaluation of the probability; that is, they must be introduced into C (see [C-2]). This advice was motivated by fear due to ignorance and leads to no practical advantage. The usefulness of probabilistic inference originates precisely from the fact that it allows us to ignore some of the pertinent information and yet to arrive faster at practical conclusions that serve as guidance for action. It is usually too tedious to exhaust all available pertinent information. But if all possible information were taken into account it would lead in many cases to a deterministic consequence, superseding, according to the principle, all other probabilistic consequences. But this would obviate the entire purpose of our studying probabilistic inference.

Incidentally, most of the philosophers ([C-2, H-9, etc.]) usually refer to the kind of inference we have been discussing in the above as inductive inference, simply because they involve probabilities. But this is a flagrant, though traditionally consecrated, misnomer because in such an inference a statement about an individual case is derived from a general statement. The distinction between induction and deduction has nothing to do with use of probability.

An apparently paradoxical situation (called "inductive inconsistency" by Hempel) can be generated by juxtaposing next to (4.16) another probabilistic deduction,

$$T_2\colon\ p(B \mid H_2 \cap A_2) = 1 - r, \tag{4.17}$$

with H_2: fraction r of all F's are non-B and fraction $(1 - r)$ of all F's are B, and A_2: individual object a is an F and the B-ness test is made on a.

Now object a may be both G and F, and H_1, H_2, A_1, and A_2 can be all true, but if r is very large then T_1 would say that a is very probably B but T_2 would say that a is very probably non-B (Cooley-Hempel paradox) [C-7, H-9]. From our point of view the trick is transparent, because T_1 does not say that a is probably B but only that if object a is randomly taken from the collection of G's then its probability of being B is very high, which does not at all contradict T_2, which only says that if object a is randomly taken from the collection of F then its probability of being B is very low. The point is that the conditions to be met with in T_1 and the conditions to be met with in

T_2 cannot be simultaneously satisfied. The object can very well be simultaneously G and F, but it cannot in general be a random sample of G and a random sample of F at the same time. In other words, we cannot ignore all other information outside its being G and yet take into account that it is also F. If we take into account that a is both G and F, there will be another probability of its being B, but this value cannot be derived from H_1 and H_2. This is because $p(B \mid G \cap F)$ cannot be calculated from $p(B \mid G)$ and $p(B \mid F)$, in spite of the fact that $G \cap F \rightarrow G$ and $G \cap F \rightarrow F$.

Let us now briefly examine whether we can translate our previous discussion regarding verifiable propositions into the present probabilistic language. First, we should note that (4.9) or (4.10) involves three basic types of propositions. One is D_i, on which the probability is defined. The second is S, which is not explicitly written in (4.9) or (4.10) but is the "consequence" of T in the sense explained in the verbal interpretation of T given below (4.9). The third one is C, which is the "condition" of T as explained in the verbal interpretation. One may notice that S defined on D_i roughly corresponds to E_i of the last section, and $C = H \cap A$ roughly corresponds to $V_1 \cap V_2$ of formulas such as (4.1). The newcomer T is between these two, taking the place of logical implication. We shall see that the probabilistic consideration sheds a new light on the difficulties inherent in the nonprobabilistic case studied in the last section.

Let us start with noting that what we have said regarding the logical operations allowed for the experiential propositions E in the categorical (nonprobabilistic) case can be carried over to observational propositions of the type D_i in the present case, for the definitions of these two kinds of propositions are essentially the same. That is, there may be infinitely many observational propositions, and a combination of observational propositions built with a finite number of negations, a finite number of conjunctions, and a finite number of disjunctions is another observational proposition. Let us write $\mathcal{E}$ for the merger of all $\mathfrak{D}$'s, that is, for the class of all D's belonging to all possible $\mathfrak{D}$'s. Next, note that the probabilistic experiential propositions S with respect to a certain operation $\mathfrak{D}$ give some probability distributions for the possible outcomes of $\mathfrak{D}$. Let us write $\mathcal{S}(\mathfrak{D})$ for the set of all such probabilistic experiential propositions S with respect to a $\mathfrak{D}$, and write simply $\mathcal{S}$ for the merger of such $\mathcal{S}(\mathfrak{D})$'s for all $\mathfrak{D}$'s. Some members of $\mathcal{S}(\mathfrak{D})$ will be categorical and give probability 1 to some members of $\mathfrak{D}$. These are equivalent to some observational propositions and create no new situation. But those members of $\mathcal{S}(\mathfrak{D})$ that are not categorical create a new situation. In fact, the logical negation of such a probabilistic experiential proposition S is not a probability distribution and hence does not belong to $\mathcal{S}(\mathfrak{D})$. In some considerations the logical negation of S can be replaced by its complement in $\mathcal{S}(\mathfrak{D})$, that is, by the disjunction of the other members of $\mathcal{S}(\mathfrak{D})$. But this disjunction is not a probabilistic experiential proposition.

Starting with the hard-core verifiable propositions consisting of $\mathcal{E}$ and $\mathcal{S}$, we can now augment the class of verifiable propositions in a way similar to (4.1) and (4.2); namely, as a condition for a proposition V_1 to be verifiable, we may tentatively require that there exist an experiment $\mathfrak{D}$ and a verifiable proposition V_2 such that a conditional experiential proposition

$$T: \quad \{p(D_i \mid V_1 \cap V_2) = \pi_i\} \tag{4.18}$$

follows from the contents of V_1 and V_2, and that $p(D_i \mid V_2)$ be either indeterminate or different from π_i.

The trouble in the categorical case is that if V_1 were a verifiable proposition, any proposition that implies V_1 would automatically become a verifiable proposition according to Ayer's scheme (4.1) and (4.2). But according to (4.18) this is not always so. An extreme example is a constant absurdity $\emptyset$, which, if substituted for V_1, would satisfy (4.1) but not (4.18). However, we cannot rely entirely on (4.18) to avoid a conjunction of an honest verifiable proposition with purely fictitious nonsense that has no bearing on experience. To guard ourselves against such eventualities, we may add another condition that there exist no proposition V_1^* ($\neq V_1$) such that we can write

$$V_1 = V_1^* \cap V_1^{**} \quad \text{(i.e., } V_1 \to V_1^*\text{)}, \tag{4.19}$$

where V_1^* satisfies the condition (4.18) with the same values of π_i. This condition has the effect of stripping off all metaphysical ornaments (V_1^{**}) from hypotheses. The combination of the two conditions referring to (4.18) and (4.19) may serve as a sufficient condition for a verifiable proposition. The class $\mathcal{V}$ of verifiable propositions now consists of $\mathcal{E}$, $\mathcal{S}$, and those propositions that satisfy (4.18) and (4.19). Those satisfying (4.18) and (4.19) but not belonging to $\mathcal{E}$ and $\mathcal{S}$ belong to the class $\mathcal{H}$ of hypotheses. Since $\mathcal{S}$ is placed between $\mathcal{E}$ and $\mathcal{H}$, we sometimes speak of $\mathcal{S}$ as hypotheses. It may also be noted that the availability of a direct experimental observation can affect the distinction between a hypothesis and an observational proposition. But this ambiguity does not affect the definition of $\mathcal{V}$. We may further augment this class of verifiable propositions by a requirement (iii) at the end of Section 4.1. After this we still have a narrow language. For instance, the negation of a hypothesis is not a hypothesis. Further requiring closure with respect to all logical operations, we may finally build a "scientific language." We may add here that we cannot deny some danger that the condition referring to (4.19) is too stringent, and by stripping off all propositions of the type V_1^{**} we may deprive a theory of a unifying principle, which by itself has no bearing on experimental results yet serves the purpose of building a simple and elegant rational theoretical structure.

The status of propositions of the type T is unique. If we mention all conditions, tacit and explicit, considered as conditions of the probabilities.

T itself should become a tautology. T does not assert S, but only states that if C and not more than C, then S, and this does not contain an empirical truth. If we exclude part of hypotheses from the condition as we did in (4.13) and (4.14) the proposition T asserts empirical truth only to the extent that H in (4.14) asserts empirical truth. However, within a discussion in which the truth of H is not disputed, T can be considered as a logical truth. The most important point is that a probability distribution is derived from a theory or hypothesis, either in a logical fashion, as H in (4.14), or in a conditional sense, as H in (4.10). This necessary derivation is the kernel of deduction and the probability we are dealing with here is a *deductive probability* (see Section 4.4).

Let me insert here a concrete example of the extension implied by (iii) in the present probabilistic case. A proposition that contains unspecified parameters and/or unspecified probability distributions is verifiable if substitution of definite values of parameters and/or definite probability distributions for them makes it an experiential proposition. For instance, consider the following proposition of the Markov chain type. Let $w(1)$ and $w(0)$ be the probabilities that the weather of a given day will be fair and non-fair, and let $u(1)$ and $u(0)$ be the corresponding probabilities for the preceding day. The proposition V in question is "the $w(j)$, $j = 0, 1$, are given by $w(j) = \sum_{i=0}^{i=1} p_{ji}u(i)$, where the p_{ji} are given constants." As it stands, V may not look like an experiential proposition, for it does not give any definite probability distribution over the possible weather on the given day. But the moment we insert some values of $u(0)$ and $u(1)$ V becomes a genuine experiential proposition of type S. Hence V is a verifiable proposition. Another way to look at the same case is to consider V as a hypothesis H in (4.15) and the data $u(0)$ and $u(1)$ as A in (4.15). This makes V a verifiable proposition provided that the p_{ji} have definite values. If the p_{ji} are left as unspecified parameters, we have to resort to (iii′) to call V with unspecified p's a verifiable proposition. According to the classification of propositions we used before, those verifiable propositions that are not experiential should be called hypotheses. But since we have subdivided experiential propositions into those that are observational propositions and those that are not, we occasionally use the term hypothesis somewhat loosely to designate any verifiable proposition that is not observational. In this usage of the word "hypothesis" what we denoted by S of (4.8) is also a hypothesis. This is in a sense justifiable, because S itself cannot be determined to be true or false by a finite number of experiments. In most applications, however, H in $p(D_i \mid H \cap A)$ is a general proposition and A is some observational result. H and A are then called respectively a hypothesis and an auxiliary condition. We consider later a set $\mathcal{H}$ of H's and a set $\mathcal{A}$ of A's, besides the set $\mathfrak{D}$ of D's already mentioned. The probabilistic experiential proposition T gives a probability distribution over $\mathfrak{D}$ for a given member of $\mathcal{H}$ and a given member of $\mathcal{A}$. Sometimes it is more convenient to consider $H \cap A$ as a single entity and call it a compound hypothesis.

In connection with the negation of a hypothesis, let us note the following facts. Going back to the categorical dichotomic case, take as a hypothesis the statement "on no August day does it snow." The negation of this statement is "on some August day(s) it snows." The conjunction of this last statement with "today is an August day" yields only a self-evident statement "today it may or may not snow." Similarly, the negation of a probabilistic hypothesis in general does not lead to a probabilistic experiential proposition with definite values of probabilities. For this reason the negation $\neg H$ of a hypothesis H is not in general a hypothesis in the usage of the term adopted here. However, as we see in Section 4.4, we understand in usual language the statement $\neg H$ as meaning that there exists a probability distribution of D_i that is different from $p(D_i \mid H)$ for at least one i (hence at least for two i's on account of the normalization of probabilities). In other words, we make a conjunction $\neg H \cap B$ of $\neg H$ and B, where B stands for "there exists a unique, definite probability distribution for D_i." Then $\neg H \cap B$ entails that $p(D_i \mid \neg H \cap B) \neq p(D_i \mid H)$ for at least one i, and therefore at least for two i's. In this reinterpretation of $\neg H$, by substitution of $\neg H \cap B$ for $\neg H$, we can say something more about $\neg H$, although obviously $\neg H$ and $\neg H \cap B$ are not equivalent. We shall say more about the negation of a hypothesis later.

It often happens that each observational proposition D_i, a member of $\mathfrak{D}$, can be decomposed as $D_j^{(1)} \cap D_k^{(2)}$, where $D_j^{(1)}$ is a member of another set $\mathfrak{D}^{(1)} = \{D_1^{(1)}, D_2^{(1)}, \cdots, D_j^{(2)}, \cdots, D_{n(1)}^{(1)}\}$ of observational propositions, and $D_k^{(2)}$ is a member of a third set $\mathfrak{D}^{(2)} = \{D_1^{(2)}, D_2^{(2)}, \cdots, D_k^{(2)}, \cdots, D_{n(2)}^{(2)}\}$ of observational propositions. $\mathfrak{D} = \mathfrak{D}^{(1)} \otimes \mathfrak{D}^{(2)}$. Suppose further that the experiential proposition $S(\mathfrak{D}) \in \mathcal{S}(\mathfrak{D})$, on condition C, gives a probability distribution satisfying the condition of probabilistic independence in the sense that $p(D_i \mid C)$ is equal to $p(D_j^{(1)} \mid C)p(D_k^{(2)} \mid C)$ for all $D_i = D_j \cap D_k$. Then we can decompose S on C into two entirely independent deductive processes, $S^{(1)}$ on C and $S^{(2)}$ on C, where $S^{(1)}$ gives the probability distribution $p(D_j^{(1)} \mid C)$ and $S^{(2)}$ gives the probability distribution $p(D_k^{(2)} \mid C)$. It is further possible that C is decomposable as $C^{(1)} \cap C^{(2)}$ in such a way that $p(D_j^{(1)} \mid C) = p(D_i^{(1)} \mid C^{(1)})$ and $p(D_k^{(2)} \mid C) = p(D_k^{(2)} \mid C^{(2)})$ for all j and k. One may remark that the possibility of scientific theory depends greatly on decomposability of this kind. The world need not always be studied as a single whole, as some metaphysicians have suggested.

Before leaving this section we should mention a special kind of proposition similar to S, but not quite the same. Let us interpret $\mathfrak{D} = \{D_i\}$ as a set of possible types that each of the objects under consideration can take. Let $\mathcal{G}$ be a subset of $\mathfrak{D}$ corresponding to a certain genus of objects in the sense that each item of the genus has a type within $\mathcal{G}$ and each type in $\mathcal{G}$ has a representative in the genus. Let G be a proposition that an object a belongs to this genus, and let the observation $\mathfrak{D}$ be operated on this object a. Then we have from G a quasiprobabilistic proposition:

$$T^*: \quad p(D_i \mid G) \neq 0 \quad \text{for} \quad D_i \in \mathcal{G}; \qquad p(D_i \mid G) = 0 \quad \text{for} \quad D_i \notin \mathcal{G}.$$

Such a proposition T^* is called a generic proposition and has some application later (for instance, 5.1). It is sometimes convenient to make a probabilistic proposition out of T by assigning suitable nonzero values to those $p(D_i \mid G)$ that correspond to $D_i \in \mathcal{G}$. It may be noted for later use that the number of possible genera $\mathcal{G}$ that can exist in $\mathfrak{D}$ that consists of n elements is $2^n - 1$, excluding an empty set. This is clear because there are $\binom{n}{s}$ different genera possible having s members taken from n possibilities, and $\sum_{s=0}^{n} \binom{n}{s} = 2^n$.

4.3. PROBABILISTIC INDUCTION—CONFIRMATION AND CREDITATION

Inductive inference must include at least two distinct processes. One is creation or production of new hypotheses in the presence of experiential facts that cannot be explained adequately by existing hypotheses, and the other is evaluation of available hypotheses with reference to a body of pertinent evidence. C. S. Pierce called the former "abduction" and the latter "induction" proper [P-3]. However, characterization of induction by these two processes only—generation of hypotheses and "evidential" evaluation of hypotheses—tends to let us overlook an equally important aspect of inductive inference, namely, "extra-evidential" evaluation of hypotheses. This latter is necessary for induction because there are usually more than one, and often very many, hypotheses that agree equally well with the experimental facts at hand, and a choice of one, or a few, from these therefore has to be based on something else than "evidence." In fact, precisely because of this extra-evidential judgment, induction becomes a subtle topic of study. The term "evidence" is used here in the following limited sense. A hypothesis, alone or in conjunction with a known fact, gives rise deductively to a prediction (experiential proposition) about the outcome of a certain well-defined experiment. The outcome of this experiment offers a basis for "evidential" valuation. Any other considerations outside of, or prior to, the direct experiential data obtained in such experiments are called "extra-evidential" here. It can easily be recognized from the beginning that a hypothesis as a general statement usually contains an infinite number of individual cases, whereas experiential evidence contains a finite number of cases, and hence there can be an infinite number of possible hypotheses from which the entire body of existing evidence can be derived.

The domain of all possible and potential hypotheses pertinent to an experience is never clearly known; indeed, the history of science shows that a new theoretical breakthrough strikes people as a hypothesis that has never been dreamt of; that is, it is beyond what has been thought of as the domain of possible hypotheses. This leads to rather a mystical view of abduction: that the creation of a new hypothesis is an act of genius, which is essentially unpredictable and can never be put in a framework of rational theorization. Another school of thought about abduction, however, is that a hypothesis is not created as a new idea or out of a boundless unknown sea of possible hypotheses, but has been selected out of a limited domain of possible hypotheses imposed by the capacity of the individual brain, by past experience, and by the given evidence. This school usually adds to this thesis another ingredient, the dictum that there is no new idea other than a new "combination" of old ideas. Even if this "selection" view is accepted, the mystery still remains as to how a "good" hypothesis is selected out of the finite number of embryo hypotheses. It is evident that this selection is not of the same kind as the one at work in the evaluation of available hypotheses in induction proper, for the former is in most actual cases a matter of "hunch" and is not as much concerned with detailed rational comparison with the evidence as the latter. As is well known, H. Poincaré had an interesting explanation for the mechanism of this process of selection in abduction [P-7]. Without literally following his description, which is related to a discovery in mathematics, let us try to adapt and translate freely his idea to our present problem. This point of view may be interpreted as meaning that the savant's unconscious mind is constantly grinding out new candidate hypotheses by combining ingredients of past experience and knowledge, and that there is a guardian sitting at the threshold between the conscious and the unconscious, who, according to his "esthetic" taste, admits only "elegant" hypotheses to the conscious. After such a hypothesis has emerged to the conscious, it will be submitted to rational scrutiny in connection with the evidence. This viewpoint has to admit that there is an important personal, even subjective, factor at work in abductive "selection." Furthermore, this factor is a matter of "feeling" rather than of "thinking." It is quite questionable whether the content of feeling can ever be subjected to representation in terms of symbols to which logical and mathematical operations can be applied. I should rather leave this question unanswered, except for a simple remark that what man performs with the help of "feeling" in a rapid and penetrating fashion can sometimes (though probably not always) be achieved by the tedious labor of rational thinking. If the spirit of the selection theory of abduction were to be upheld, one would have to argue that abductive selection must rely on feeling only, because the rational procedure needed to achieve the same goal would become prohibitively laborious because of the immense number of possibilities to choose from.

In any event, if abduction is simply a matter of selection, whether based on personal feeling or not, then we can envision a certain parallelism between abduction and induction, since both are concerned with choosing from a given collection of hypotheses. It is often understood that selection in induction proper is based solely on comparison with the evidence—evidence in the sense of a collection of past experiential data about which the hypothesis could give a deductive prediction. But, as indicated at the beginning of this section, this view certainly does not correspond to reality. We can mention at least five different kinds of extra-evidential considerations at work in the evaluation of hypotheses in induction.

Competitive Hypotheses. The evaluation of a particular hypothesis depends on other available hypotheses. Suppose that a hypothesis, which at a certain time is the only available hypothesis, agrees to a good extent with the body of evidence. This hypothesis receives a high evaluation. Now suppose that many other hypotheses suddenly become available and agree equally as well with the evidence as the first hypothesis. The privileged position of the first hypothesis is then lost and its evaluation has to go down. Furthermore if a new hypothesis that agrees with the evidence much better than the first hypothesis appears, the evaluation of the first one has to suffer considerably. This shows that evaluation of a hypothesis is not determined solely by comparison with the evidence.†

Position in Theory Making. A careful observation of actual inductive processes in a scientist reveals that he is constantly concerned not only with evidence but also with the relation of the hypothesis in question with other hypotheses or laws, that govern other domains. This is because he has a vision of a theoretical system in which this particular hypothesis and other laws should be coherently coordinated. His judgment regarding such a broad concern is usually bound to be highly personal and subjective. And this kind of concern is extraneous to the evidence in question, or, to be more precise, is far more comprehensive than his concern about the limited evidence in question.

Preference of Predicates. Induction is generalization by the use of one of the many predicates satisfied by individual examples. Too bold a generalization by the use of a very broad predicate risks quick disproval by a counter-example. Too timid a generalization by the use of a very narrow predicate

† It is therefore not surprising to see in the history of science that a theory is not immediately discarded by a case of discord with experiment, but it is discarded by the appearance of a theory which agrees better with the experiment.

does not greatly serve the purpose of induction. Consider three hypotheses—H_1: "all ravens are black"; H_2: "all ravens of the Western Hemisphere and all seagulls of the Eastern Hemisphere are black"; and H_3: "all ravens are black in even-numbered centuries and white in odd-numbered centuries." Suppose that you observe ravens in New York on December 1, 1969, and discover that they are black. This evidence certainly agrees with all three hypotheses mentioned above. However, somehow we would not take two of the hypotheses too seriously even if the amount of similar evidence increases. We are not inclined to believe in H_2 because we feel that a class of objects consisting of ravens of the western hemisphere and seagulls of the eastern hemisphere is not a "natural" class that obeys a single natural law. (See Section 7.6 for further remarks on the notion of "natural class.") We do not have much confidence in H_3 because we believe that the color description of a species does not depend on time. (See Section 4.5 for more about "time-dependent" predicates.) This example suggests that we have a preference for certain predicates (or classes defined by them) used in hypotheses apart from confirmation by the evidence. It is universally recognized that a hypothesis using "simpler" predicates is preferred, provided that it is adequately supported by the evidence. It is true that "undesirable" predicates are usually so deemed by us because we anticipate that the hypotheses in terms of them will be disproved by future evidence. However, anticipated evidence is not evidence in the present context, and its effect on the evaluation of a hypothesis must be distinguished from the effect of the evidence at hand.

Circumstantial Appraisal. Suppose that the hypothesis H, "this coin is double-headed," will be tested by tossing the coin. The number of consecutive heads necessary for an average man to conclude that the hypothesis H is highly plausible will depend considerably on the knowledge K of circumstantial facts about the coin, such as the fact that the coin came from the pocket of a psychology professor or from the cash register of a grocery store. In this case K is not (as it is with an evidential fact) a deductive consequence of H; rather, it deductively influences the plausibility of H. Here is an example of another kind of extra-evidential fact that influences the credibility of a hypothesis. One may speak of "circumstantial" factors in such a case.

Parallel Experiments. This may be counted as another circumstantial appraisal, but because of its frequent occurrence it may be considered separately. There is a large urn containing a large number of red and white balls in an unknown ratio. We take 10 balls at random and put them in a smaller urn A, and similarly we prepare another smaller urn B with 10 balls. Hypotheses under consideration are of the type "urn A contains n red balls," where n can be any integer between 0 and 10, inclusive. An evidential experiment consists of taking ball randomly out of urn A, examining its color, and returning it to urn A. Suppose that we do the same experiment on urn B.

The result of this second experiment is not an evidential test for the hypotheses about urn A, because the probability that a ball from B will be red cannot be deductively derived from the hypothesis. (See Section 6.4 for a mathematical treatment of the urn problem.) Yet the experimental result about B will affect our evaluation of the hypotheses about A. For instance, if every ball coming from B turns out to be red, the hypotheses that assign a large number of red balls to A will become more plausible. Here is an example in which our evaluation of a hypothesis about a system is affected by a test made on another system. The relation between the two depends on another hypothesis on a different level. It should also be noted that in many cases a single hypothesis has two domains of application and its evaluation depends on two separate sets of evidence. This means that a hypothesis regarded as governing one of the two domains receives an evaluation dependent on something apart from the evidence in the domain under consideration. (See the discussion of conjunctive experimentation, Section 6.3.)

So much for the existence of two kinds of evaluation, evidential and extra-evidential; let us pass on to the procedure of evaluation. The most primitive idea about the connection between a hypothesis and the evidence is concerned with the dichotomic judgment as to whether the hypothesis is "confirmed" or "disconfirmed" by the evidence. In this context a hypothesis is supposed to be capable of stating only that such-and-such an observational proposition is bound to be true or bound to be false. If an experimental fact which is predicted by the hypothesis is presented, the hypothesis is said to be confirmed by the evidence. This is the essence of what is usually known as Nicod's rule, for which a more rigorous formulation is given later. If an experimental fact that contradicts the hypothesis presents itself, the hypothesis is said to be disconfirmed. For instance, if the hypothesis says that all ravens are black, an observed black raven confirms it and a white raven disconfirms it. Disconfirmation implies that the hypothesis is logically refuted and therefore disqualified as a hypothesis. Nobody can disagree with this procedure, and any sophisticated theory of induction should incorporate the process of logical refutation. The theory of induction must be such that a hypothesis is automatically discarded if it precludes the occurrence of a certain experimental event and such an event actually presents itself, insofar as the information about this occurrence is perfectly reliable. On the other hand, what is called confirmation here is a misleading concept. Of course, existence of confirmation is, in some sense, "better" for a hypothesis than its absence. But it is also obvious that confirmation does not imply that the hypothesis is true. Indeed, even if the first 100 ravens we observe are all black, the 101st raven might be white. How dangerous this concept of confirmation can become in application is vividly illustrated by two famous paradoxes contrived by Hempel. We discuss them from our point of view in Section 4.5.

A slightly more advanced idea about the relation of hypothesis to evidence is the concept of "degree of confirmation," or "confirmability." This concept is obviously inevitable, because everybody will agree that a body of evidence consisting of 1000 black ravens (more precisely, 1000 randomly taken ravens turning out to be black) confirms the said hypothesis more strongly than a body of evidence consisting of 10 black ravens, other factors remaining the same. This means that we agree that the more often a hypothesis is "confirmed" in the previous sense—that is, the more events we observe in agreement with the conclusion of the hypothesis—the more likely the hypothesis seems to be true. The idea of degree of confirmation, at least in its primitive version, is at work even at the common-sense level of human recognition. But the 1945 paper of Hempel and Oppenheim's [H-10] may have been the first to deal with it in a modern context of methodological studies. It is to Carnap's merit that he consistently used the continuous degree of confirmation $C(H, \mathcal{B})$ in his discussion of inductive logic, where C is supposed to be a function of the hypothesis H and the evidence $\mathcal{B}$ (see [C-2]). However, his attempt was doomed to fail, because he wanted the same quantity to play a double role: as a degree of confirmation as a function of the hypothesis and the evidence, and also as our over-all evaluation of the hypothesis. (For a more detailed criticism of Carnap's concept, see [W-20].) The over-all evaluation must reflect, as we have emphasized, not only the evidential confirmation but also the extra-evidential evaluation, which a function of the type $C(H, \mathcal{B})$ cannot represent. In the following we write $\Gamma(H, \mathcal{B})$ for an unspecified function of $(H, \mathcal{B})$ that measures the degree of evidential confirmation.

Thus we have to introduce a quantity that indicates our over-all evaluation of a hypothesis or over-all confidence, which reflects both evidential and extra-evidential evaluation. We call such a quantity "credibility" to express the idea that it is supposed to measure the degree of confidence we place in the hypothesis H_I, and denote it by $q(H_I)$ or $q(I)$. We require this quantity to satisfy the probability postulates. This last requirement amounts essentially to the condition that the numerical expression of confidence we place in the event that either hypothesis H_1 or hypothesis H_2 is correct is the sum of the numerical expressions of confidence we place in the event that each of them is correct, provided that H_1 and H_2 cannot be correct simultaneously. Hence we can safely say that this requirement does not introduce any artificial restriction on our concept of over-all evaluation. (See Section 7.4 for the notion of probability as a measure of confidence.)

In order that such a "probability" may be defined, we have to be given a set of hypotheses $\mathcal{H} = \{H_1, H_2, \cdots, H_N\}$. This fact is important, because all too often people talk about probability or the "goodness" of a hypothesis without mentioning competitive hypotheses. In such a case one might say that the competition is tacitly understood to be the negation of the hypothesis. But the logical negation of a hypothesis is not a hypothesis, and has to be interpreted as the disjunction of all competitive hypotheses. Hence, if the competitive hypotheses are not given, we cannot talk about the probability of a hypotheses. No probability measure can be introduced in a subset if the total set is not given first. This criticism also applies to Carnap's approach (see [*W*-20]).

Now this credibility $q(H_1)$ is a probability of H_I in the set $\mathcal{H}$, and must depend on the evidence $\mathcal{B}$ at hand. Hence we may also denote it as $q(H_I \mid \mathcal{B})$. However, the crucial point is that, unlike $\Gamma(H_I, \mathcal{B})$, this $q(H_I \mid \mathcal{B})$ depends not only on H_I and $\mathcal{B}$, but also on "extraneous" factors, or extra-evidential factors, which we denote by X. X includes at least the five different extra-evidential elements we mentioned above. Furthermore, it should not be forgotten that various auxiliary conditions A under which the experiment is performed must also be brought into consideration. Thus the credibility of a hypothesis is actually a function of at least H_I, $\mathcal{B}$, X, and A. We can expect at this stage that it will be practically impossible to give an explicit mathematical form to the dependence of q on X. We have to be satisfied with a mathematical formalism that distinguishes credibility from confirmability and places them in a right relationship. In some fortunate cases we can discuss certain quantitative effects of X on q, but in general we cannot.

In the actual process of induction in everyday life, assignment of a numerical value to credibility may be difficult, and the theory we develop is a mathematical model that limits itself to cases in which numerical assignment is possible. The results obtained with such a model give a general orientation about the matter even when numerical assignment is difficult or impossible, and clarify the nature of the main factors at work in every process of induction. As we shall see in Section 7.4, the degree of confidence that a person places on an alternative can often be measured through his behavior in spite of the fact that he himself may not be able to express it quantitatively.

Let us try to anticipate what kind of properties the quantity $q(I \mid \mathcal{B})$ will have if a theory about it is to be formulated properly. (a) As already stated, if any event prohibited by a hypothesis actually occurs (i.e., if a counter-example is presented), the hypothesis must be discarded.† This process of logical refutation (disconfirmation, infirmation) must for formulated in terms of $q(I \mid \mathcal{B})$ as follows. If H_I gives a zero probability to an event and if this event is included in $\mathcal{B}$, $q(I \mid \mathcal{B})$ must become 0. Some logicians argue that zero probability does not imply nonoccurrence (which is true), and hence occurrence of an event with zero probability should not invalidate the hypothesis. We show later that this last conclusion is wrong. (b) The essence of scientific method, in general, resides not in discovering an absolute truth

† To K. Popper, this process is the only source of knowledge (which is wrong), [P-8].

but in successive "improvement" of our knowledge. Reflecting this general nature of scientific method, the theory of induction must be based on some basic rule governing the successive change of $q(I \mid \mathcal{B})$ with an increasing body of evidence $\mathcal{B}$. (c) The relation between confirmability $\Gamma(H_I, \mathcal{B})$ and credibility $q(I \mid \mathcal{B})$ must be such that the larger the value of Γ, the larger the value of $q(I \mid \mathcal{B})$, provided that other factors remain the same. (d) As stated before, the quantity $q(I \mid \mathcal{B})$ must be affected by a judgment based on factors other than the test under consideration; that is, it must depend on X. However, the effect of X on $q(I \mid \mathcal{B})$ must gradually fade with increasing $\mathcal{B}$, because the experiential evidence must be the final judge of the validity of a hypothesis. (e) Since a hypothesis refers to an infinite number of individual cases, we should not theoretically (i.e., as long as we ignore certain aspects of human behavior) declare any hypothesis to be a law on the basis of a finite body of evidence. This may be interpreted as meaning that any $\mathcal{B}$ of finite size should not lead to both $\Gamma(H_I, \mathcal{B}) = 1$ (maximum value) and $q(I \mid \mathcal{B}) = 1$ for any H_I. (f) However, if we entirely exclude the possibility $\Gamma = 1$ and $q = 1$ from the theory, it would almost amount to denying any scientific truth. Hence the possibility must be left for $\Gamma = 1$ and $q = 1$ at least in the limiting case in which the size of $\mathcal{B}$ becomes infinite. (g) Roughly speaking, induction is a process in which the weights on various hypotheses become more and more concentrated on fewer and fewer possibilities (hypotheses). This is contrary to the H-theorem, according to which probability distribution gradually spreads out over more and more possibilities. Therefore there must be a theorem, which may be called inverse H-theorem, regarding the probability distribution $q(I \mid \mathcal{B})$ as a function of the increasing body of evidence $\mathcal{B}$. All these seven properties are satisfied by the theory developed in the next section, although more detailed explication will be needed for some of the words used in the above explanation.

There is a subtle aspect of inductive process that we have not discussed so far; that is, a phenomenon that may be called "sudden certainty." This means that one becomes suddenly convinced that a certain hypothesis is true, which may mean that $q(I \mid \mathcal{B})$ jumps up to a value equal to, or close to, unity. This phenomenon is discussed separately in Section 6.1.

Before leaving this section, we should mention the so-called Duhemian thesis (or the Duhem-Quine thesis) of unfalsifiability of hypotheses, which one might (wrongly) construe as contradicting what I said with regard to logical refutation of a hypothesis (see [D-10, G-12, G-13, Q-2]). Since this thesis is conceived in the framework of nonprobabilistic inference, it may be more appropriate to explain it in the corresponding nonprobabilistic language. Suppose that a hypothesis H implies an observational proposition E; that is, $H \rightarrow E$. If E is observed, one feels that H has become more probable, but this by no means entails that H is true. But on the other hand, if $\neg E$ is observed, one has to repudiate H. This is because if $H \to E$ then $\neg E \to \neg H$. In general, however, H alone seldom implies an E. We take an additional verifiable proposition A, and we derive E from $H \cap A$ as $H \cap A \to E$. If $\neg E$ is now observed, we can conclude only that $\neg(H \cap A) = \neg H \cup \neg A$; that is, H or A or both must be false. If A can be false, then H need not necessarily be false here. What we claimed in the foregoing paragraphs is that if A is known to be true then $\neg E$ implies $\neg H$ in this case. This argument is perfectly legitimate, and also in practice we as scientists constantly use this type of argument with success. However, there is a pitfall in the practical application of this inferential procedure. This is because, as we mentioned in Chapter 1 in connection with the concept of conditional probability, we usually assume so many things as facts and take so many conditions for granted. To be rigorous, all these must be enumerated in A in conjunction. Failure of any one of these factors results in $\neg A$. Therefore $\neg E$ may not necessarily mean $\neg H$, the fault originating in some overlooked factor in A. Under these circumstances one is tempted to suspect that any arbitrary hypothesis can be saved in the presence of apparently adverse evidence simply by changing A in some way. Modification of A may be quite possible, because A is usually a conjunction of many factors, not all of which are experientially imposed on us. Some are just hypotheses belonging to our theoretical view. This Duhemian idea is very good practical advice to a working scientist, because it urges scientists to search for and scrutinize all tacitly assumed assumptions or conditions before repudiating a hypothesis under evaluation.

From a formal point of view, the Dunhemian thesis may be formulated as follows. Given arbitrary propositions H and E, there always exists an A such that

$$H \cap A \to E. \tag{4.20}$$

This is obviously always true because we can take an A such that $H \cap A = \emptyset$, which automatically satisfies (4.20). We can avoid this extremely trivial case by requiring, besides (4.20),

$$H \cap A \nrightarrow \neg E. \tag{4.21}$$

Generally speaking, it is not surprising that we can satisfy a relation of the type (4.20) by requiring a lot of conditions on A. It would become less trivial if (4.20) were satisfied by a less restrictive proposition A. For this reason, if we want the least trivial case, we should look for an A that is as "large" as possible yet satisfies (4.20) (where the word "large" is used in the sense that if $\alpha \to \beta$ and $\alpha \neq \beta$ then β is "larger" than α). The answer to this search for the largest A is readily given by the little theorem we proved in Section 4.1: A^* defined by $A^* = E \cup \neg H$ is such that any A satisfying (4.20) will imply

A^*; that is, $A \rightarrow A^*$. A^* is the least restrictive of all A's satisfying (4.20) and the relation (4.20) with A^* replacing A may be considered as the least trivial case. Unless H and E directly contradict each other, $H \cap A^*$ is not $\emptyset$, since $H \cap A^* = H \cap E$. A reputed philosopher of science [G-12, G-13] disclaimed the Duhem-Quine thesis, arguing first that the reply in terms of A^* is trivial and second that the existence of a nontrivial A is not only a non sequitur, but actually false in a special case. But this line of attack is a wasted effort from the beginning, because if A^* is trivial, any other A, which necessarily implies A^*, is more trivial. (People do not seem to have realized that A^* is the largest A.) The anti-Duhemianist might contend that this sense of triviality was not agreeable to him, but he himself had no definition for it, although his entire argument rests crucially on this undefined notion. There is always a possibility, as Quine pointed out, that a hypothesis can be saved by a modification of logic. We discuss a non-Boolean logic in Chapter 9.

It should be mentioned in passing that the problem of induction is intimately bound up with the problem of class formation. This can easily be seen by noting that the prototype of induction consists in deriving the proposition of the type "all X's are P" from observation of a few X's. These few objects may be considered as samples of the class X, but they may be also considered as samples of some other class. Two yellow roses and three red roses may be samples of roses as well as samples of the class of nonwhite objects. Thus we can see that ambiguity (i.e., the necessity of extra-evidential factors) in induction is inseparable from ambiguity in class formation from a few sample objects. The problem of class formation is discussed separately in Chapter 8. The nonexistence of "natural" classes is formulated there as a theorem of the ugly duckling (Section 7.6).

4.4. DEDUCTION AND INDUCTION AS PREDICTION AND RETRODICTION—MATHEMATICAL MODEL OF INDUCTION

In Sections 4.2 and 4.3 we saw that both deduction and induction have to invoke probabilities involving hypothesis H, auxiliary condition A, and experimental result D. In Section 4.3 we used the symbol $\mathcal{B}$ to designate a body of evidence that is a sequence of experimental results in most cases. If the sequence of experiments giving these results is considered as a single compound experiment, $\mathcal{B}$ can be considered a single D also. Thus the most general approach would be to consider probabilities defined on a product logical spectrum, $\mathcal{H} \otimes \mathcal{A} \otimes \mathcal{D}$, where $\mathcal{H} = \{H_I\}$ is the set of available hypotheses, $\mathcal{A} = \{A_k\}$ is the set of possible auxiliary conditions, and $\mathcal{D} = \{D_i\}$ is the set of possible experimental outcomes.

In the case of $\mathcal{A}$ and $\mathcal{D}$, the condition of mutual exclusion and the condition of exhaustion of their members can be easily satisfied by adjustment of definitions, and do not raise any difficult problem. In the case of $\mathcal{H}$, however, we must be a little cautious because $\mathcal{H}$ is not the set of all possible hypotheses but that of all available hypotheses. This entails at least two consequences. First, any probability distribution defined over $\mathcal{H}$, by its nature, adds up to unity. This implies that one or another member of $\mathcal{H}$ must be true. It can happen that one of them receives probability unity in an actual calculation. But in reality it is possible that none of the available hypotheses is the true one in the absolute sense. This dilemma can be solved if we assume that all the probabilities defined on $\mathcal{H}$ are conditional probabilities, with the condition that $\mathcal{H}$ contains a true hypothesis. Thus, if the probability of H_I is 1, for example, it means that if $\mathcal{H}$ contains a true hypothesis it must be H_I.

Another necessary remark is that within the framework of $\mathcal{H}$ the negation of a hypothesis H_I corresponds to the disjunction of all the members of $\mathcal{H}$ except H_I, that is, the complement of H_I in $\mathcal{H}$. In fact, the probability of H_I and the probability of its complement in $\mathcal{H}$ will always add up to unity. This complement has usually very little to do with the formal negation of H_I in the broader language, except that the former implies the latter. If, for instance, $\mathcal{H}$ consists of only two hypotheses, H_1, "all ravens are black," and H_2, "all ravens are yellow," the negation of H_1 within $\mathcal{H}$, that is, the complement of H_1, is H_2, whereas the formal negation of H_1 states that some ravens are not black. Here again, the discrepancy can be eliminated if we interpret the probabilities on $\mathcal{H}$ as conditional probabilities. In fact, in the above example, if it is the case that either H_1 or H_2 is assumed to be true, the negation of H_1 becomes equivalent to H_2. But for simplicity we do not mention the "condition" each time.

Apart from this, we should note that people very often talk about the negation of a hypothesis H without mentioning $\mathcal{H}$. Although $\mathcal{H}$ is not specified, they do not necessarily mean the unconditional negation $\neg H$ of H in the broad language either, because they usually proceed as if the negation $\neg H$ were another hypothesis, which is usually not the case with the unconditional negation $\neg H$. We have already stated that $\neg H$ often has to be interpreted as conjunction of $\neg H$ with some other proposition, which amounts to defining a tacit set $\mathcal{H}$.

In any event, for a discussion of the product $\mathcal{H} \otimes \mathcal{A} \otimes \mathcal{D}$, the basic quality must be the triple unconditional probability

$$p(H_I \cap A_k \cap D_i), \tag{4.22}$$

from which all other conditional and unconditional probabilities can be derived. (We no longer mention the tacit condition that $\mathcal{H}$ contains a true hypothesis.) We assume that $\mathcal{H}$, $\mathcal{A}$, and $\mathcal{D}$ are all finite, and, in particular, we use N and n as the numbers of members respectively of $\mathcal{H}$ and $\mathcal{D}$: $I = 1, 2, \cdots, N$, and $i = 1, 2, \cdots, n$. Of particular importance are two types of conditional probabilities,

$$p(D_i \mid H_I \cap A_k) \tag{4.23}$$

and

$$p(H_I \mid D_i \cap A_k). \tag{4.24}$$

The first (4.23) is the probability that an experimental datum D_i will be obtained if hypothesis H_I is true and when the auxiliary condition A_k has been found to be the case. This is nothing but the deductive probability, which appeared in (4.9) or (4.15), and its value is determined solely by the definition of H_I for given D_i and A_k. On the other hand, the second conditional probability (4.24) is the probability (credibility) of hypothesis H_I when the auxiliary condition A_k has been found to be the case and if the experimental datum D_i has been obtained. This is what was denoted by $q(H_I \mid \mathcal{B})$ in the last section. Since this is the probability of a general rule based on a particular experimental result, it can be called *inductive probability*. The probability is not determined by the nature of H_I, D_i, and A_k alone and can be given by Bayes' formula,

$$p(H_I \mid D_i \cap A_k) = \frac{p(H_I \mid A_k)p(D_i \mid H_I \cap A_k)}{\sum_{J=1}^{N} p(H_J \mid A_k)p(D_i \mid H_J \cap A_k)}, \tag{4.25}$$

which is a consequence of the obvious definition:

$$\begin{aligned} p(H_I \mid D_i \cap A_k)p(D_i \cap A_k) &= p(H_I \cap D_i \cap A_k) \\ &= p(D_i \mid H_I \cap A_k)p(H_I \cap A_k). \end{aligned} \tag{4.26}$$

To obtain (4.25) one should note that

$$p(D_i \cap A_k) = \sum_I p(D_i \mid H_I \cap A_k)p(H_I \cap A_k)$$

and

$$p(H_I \cap A_k) = p(H_I \mid A_k)q(A_k). \tag{4.27}$$

Since we usually consider the problem under a fixed A_k, it is more appropriate to write $p(H_I \mid A_k) = p_k(H_I)$, $p(D_i \mid H_I \cap A_k) = p_k(D_i \mid H_I)$, and $p(H_I \mid D_i \cap A_k) = p_k(H_I \mid D_i)$. Then (4.25) assumes a familiar appearance of the Bayes formula,

$$p_k(H_I \mid D_i) = \frac{q_k(H_I)p_k(D_i \mid H_I)}{\sum_{J=1}^{N} p_k(H_J)p_k(D_i \mid H_J)}, \tag{4.28}$$

where $p_k(D_i \mid H_I)$ is the deductive probability determined by D_i and H_I for a given A_k, and $p_k(H_I \mid D_i)$ is the inductive probability, which is not determined by D_i, H_I, and A_k only, but depends on the *a priori* probability $p_k(H_I)$. Thus we see that in the terminology of Section 3.1, *deduction and induction are respectively prediction and retrodiction.*

In the rest of this section we often assume A_k to be fixed during a discussion. Hence we usually use the abbreviated form of (4.28),

$$q(H_I \mid D_i) = \frac{q(H_I)p(D_i \mid H_I)}{\sum_{J=1}^{N} q(H_J)p(D_i \mid H_J)}, \tag{4.29}$$

where we use q instead of p for the probabilities of hypotheses. One might be tempted to argue that because the truth or plausibility of a hypothesis H_I per se cannot be affected by the contingent condition A_k, the probability $p_k(H_I)$ in (4.28) will not depend on A_k and hence should be written $p(H_I)$ anyway, whether or not A_k is fixed during a discussion. But this argument is wrong, for the condition A_k can limit the domain of applicability of a hypothesis and the hypothesis can be correct only within this domain. We come back to this point at the end of this section, and for the time being we proceed with a fixed A_k.

Regarding comparison of the formulas thus obtained with actual experience, the most important caution to take is to comply strictly with the conditions of the conditional probabilities. We noted in Section 4.2 that neglect of this elementary rule can lead to disastrous confusion in deductive inference. In both deductive and inductive probabilities the procedures and conditions for the experimental test have to be clearly defined, and comparison of the theory with experience must be done in accordance with these procedures and conditions. In particular, the actual conditions under which the experiment is performed should be neither more nor less stringent than those implied by the conditional probabilities. In other words, the individual case that can be used in induction must be a randomly picked sample from the entire collection of cases obeying the prescribed conditions. We call this admonition *rule of conditioned randomness* for convenience. Rule 4.1 expresses the same idea in a different context. We see a good example of the danger involved in violation of this rule in the next section.

In the last sections we mentioned a few important requirements of a probabilistic theory of inductive inference. For the rest of this section we derive immediate consequences of the probabilistic formula of induction (4.29), and see whether the earlier requirements are met by them.

The first thing to note in (4.29) is that its denominator is common to all hypotheses H_I; hence it serves the purpose of normalization so that the conditional probabilities for different hypotheses add up to unity. The numerator consists of two factors, the deductive probability $p(D_i \mid H_I)$ and the unconditional probability $q(H_I)$. The former is determined by the nature of the hypothesis and the experimental result D_i. The latter, on the contrary, is not determined by the hypothesis or the experiment, and has to be determined by a consideration outside the evidence D_i. This is precisely the place where an extra-evidential consideration can be reflected. It is a very satisfactory feature of the Bayes theorem that it provides room for extra-evidential influence. The two probabilities $q(H_I)$ and $q(H_I \mid D_i)$ are respectively called "prior" and "posterior" probabilities of H_I, which convention we also use at times, but they should perhaps be called "extra-evidential" and "comprehensive" probabilities. Sometimes we call them "*a priori* credibility" and "*a posteriori* credibility." The influence of competitive hypotheses on the credibility of a hypothesis is felt through the denominator. In the same way, the value of $q(H_I)$ is affected by other $q(H_J), J \neq I$, for the normalization condition must be satisfied. What I called blind retrodiction (Section 3.1) corresponds to a standpoint from which all the extra-evidential factors are equal; that is, $q(H_I) = 1/N$. We explain later that the so-called classical theory of hypothesis testing deals mainly with such "blind induction" (see Section 6.3, the subsection on maximum likelihood method).

Let us take two hypotheses H_1 and H_2 from $\mathcal{H}$ and compare their credibilities after an experimental result D_i has been obtained:

$$\frac{q(H_1 \mid D_i)}{q(H_2 \mid D_i)} = \frac{q(H_i)}{q(H_2)} \cdot \frac{p(D_i \mid H_1)}{p(D_i \mid H_2)}. \tag{4.30}$$

This shows that after the experimental result D_i, the ratio of credibilities increases in favor of H_1 if H_1 gives a larger deductive probability than H_2 to D_i. Suppose a hypothesis H_1 of the type "all R is B." Let us assume that $\mathcal{H}$ contains such other hypotheses as "an arbitrary R is B with probability a (< 1)." Let us take the disjunction of all these hypotheses other than H_1 and call it H_2 for convenience. The experiment consists of testing the B-ness of an arbitrary R. Then $p(B \mid H_1) = 1$ and $p(B \mid H_2) < 1$. Therefore, if we observe that an arbitrary R is B,

$$\frac{q(H_1 \mid B)}{q(H_2 \mid B)} > \frac{q(H_1)}{q(H_2)}, \tag{4.31}$$

meaning that the credibility ratio increases in favor of the hypothesis stating that all R is B in comparison with the hypothesis that does not precisely state this. If H_1 and H_2 are two alternatives so that $q(H_1) + q(H_2) = 1$ and $q(H_1 \mid B) + q(H_2 \mid B) = 1$, then (4.31) implies that

$$q(H_1 \mid B) > q(H_1) \quad \text{and} \quad q(H_2 \mid B) < p(H_2). \tag{4.32}$$

This fact may be considered a more rigorous version of the so-called Nicod rule, which says, "If we observe an object a that satisfies $R(a) \supset B(a)$, the general rule $\forall x(R(x) \supset B(x))$ is 'confirmed.' "† A very important thing (which is often ignored by other theorists) is that the rule of conditioned randomness must be strictly observed here, as in any other case; that is, the object must be randomly taken from a collection of R's and then tested as to its B-ness.

After having stated these basic remarks, let us examine the more specific conditions required of a theory in the last section. First is the property of "logical refutation" or falsification, which was mentioned under (a). The general rule is as follows: if H_J gives a zero probability, $p(D_j \mid H_J) = 0$, to an event D_j, if there is at least one H_I in $\mathcal{H}$ that gives a nonzero probability to this D_j, $p(D_j \mid H_I) \neq 0$, and if this event D_j happens, then the credibility of H_J becomes zero, $q(H_J \mid D_j) = 0$, or H_j is logically refuted or falsified. This is a direct consequence of (4.29).

Insofar as $\{D_i\}$ has a finite number of members, if a special event D_i happens, making both numerator and denominator in (4.29) vanish, then we should falsify the tacit basic hypothesis that $\mathcal{H}$ contains a true hypothesis. Even if the set $\{D_i\}$ has an infinite number of members, the general rule of falsification mentioned above remains valid provided that the conditions enumerated in the foregoing paragraph are satisfied. But in general, when $\{D_i\}$ has an infinite number of members, the occurrence of an indefinite form 0/0 in (4.29) does not necessarily mean that $\mathcal{H}$ does not contain a correct hypothesis. This is because, when $\{D_i\}$ has an infinite number of members, each member may receive, according to any hypothesis in $\mathcal{H}$, a vanishing probability. A wrong argument in this connection, however, is the following one. Suppose that $\{D_i\}$ has a finite number of members, that $p(D_j \mid H_J) = 0$ for a particular H_J and a particular D_j, and that there exists another H_I such that $p(D_j \mid H_I) \neq 0$. The argument in question is based on the fact that the relative frequency of occurrence of an event is the number of trials times probability. Hence by repeating an infinite number of times even an event with zero probability may happen. Therefore, the argument continues, the occurrence of D_i cannot logically falsify H_J even if $p(D_j \mid H_J) = 0$, and $p(D_j \mid H_I) \neq 0$. This reasoning is wrong, intuitively because we can never perform infinitely many trials, and formally because (4.29) does not allow such an argument. In fact, $q(H_J \mid D_j)$ becomes zero whereas $q(H_I \mid D_j)$ does not.

To compare our results with the other requirements of the last section, we have to rewrite (4.29) for the case of repeated trials. If the body of evidence $\mathcal{B}$ consists of a sequence of ν outcomes $\{D_{i_1}, D_{i_2}, D_{i_3}, \cdots, D_{i\nu}\}$, the D_i of

† $\alpha \supset \beta$ is equivalent to $\neg\alpha \cup \beta$, and the assertion that $\neg\alpha \cup \beta$ is true is equivalent to $\alpha \rightarrow \beta$. See Appendix A7.1.

(4.29) should be understood as standing for such a sequence. If the hypothesis concerns the occurrence of each of these events, and if, under the experimental condition, the hypothesis predicts no probabilistic dependence among successive outcomes, the probability $p_k(D_i \mid H_I)$ can be written

$$\prod_{\mu=1}^{\nu} p(D_{i\mu} \mid H_I), \tag{4.33}$$

which we write hereafter simply $\prod_{\mu=1}^{\nu} p(i_\mu \mid I)$, omitting D and H. Writing further $q^{(\nu)}(I)$ for $p(H_I \mid D_i)$ because it now refers to ν consecutive experiments, we have from (4.29)

$$q^{(\nu)}(I) = \frac{q^{(0)}(I) \prod_{\mu=1}^{\nu} p(i_\mu \mid I)}{\sum_{J=1}^{N} q^{(0)}(J) \prod_{\mu=1}^{\nu} p(i_\mu \mid J)}, \tag{4.34}$$

where $q^{(0)}(I)$ is the *a priori* or extraevidential credibility. We have to assume that $q^{(0)}(I) \neq 0$ for all $I = 1, 2, \cdots, N$, since if $q^{(0)}(I) = 0$, $q^{(\nu)}(I) = 0$ for all ν, and this amounts to ignoring H_I entirely among the other members of $\mathcal{H}$. It may be noted that insofar as ν is finite, the order of the factors $p(i_\mu \mid I)$ under $\prod_{\mu=1}^{\nu}$ is immaterial. An interesting and satisfactory feature of the Bayes formula of repeated independent trials is that we have, as can be proved by (4.34),

$$q^{(\nu)}(I) = \frac{q^{(\nu-1)}(I) p(i_\nu \mid I)}{\sum_{J=1}^{N} q^{(\nu-1)}(J) p(i_\nu \mid J)}, \tag{4.35}$$

which means that $q^{(\nu-1)}(I)$ can be considered as the *a priori* credibility before the νth experiment and we can use (4.29) for the νth experiment to obtain $q^{(\nu)}(I)$. This fact also suggests a certain analogy with a Markov chain in the sense that the entire past history before $q^{(\nu)}(I)$ is summarized in a nutshell by $q^{(\nu-1)}(J)$, $J = 1, 2, \cdots, N$, and $q^{(\nu)}(I)$ does not depend explicitly on $q^{(\mu)}(J)$ with $\mu < \nu - 1$. However, because the dependence of $q^{(\nu)}(I)$ on $q^{(\nu-1)}(J)$ is not linear, the process is not a Markov chain. In this connection it may be noted that a Markov chain without a sink is connected with the H-theorem, whereas the successive Bayesian sequence (4.35) is connected, as we shall see later, with an inverse H-theorem (see Sections 5.2 and 6.1).

It may be useful to interject a remark here that is not a direct consequence of the Bayes formula itself, but follows from the interpretation given to $q^{(\nu)}(I)$. What is the predictive probability, that is, the deductive probability of outcome D_i, given by an observer in the $(\nu + 1)$st experiment when he has made ν observations $(i_1, i_2, \cdots, i_\nu)$? The answer is obviously

$$p^{(\nu+1)}(i) = \sum_{I=1}^{N} p(i \mid I) q^{(\nu)}(I), \tag{4.36}$$

for $q^{(\nu)}(I)$ is the probability of the "condition" I, and $p(i \mid I)$ is the probability of i, given "condition" I.

We can give (4.34) another form, which is in some respect more accessible to intuitive understanding. Let ν_i be the number of times that D_i appears in the sequence $\mathcal{B}$ of ν outcomes $\{D_{i_1}, D_{i_2}, \cdots, D_{i_\nu}\}$ and $\alpha_i^{(\nu)}$ be its relative frequency.

$$\alpha_i^{(\nu)} = \frac{\nu_i}{\nu}, \qquad \sum_{i=1}^{n} \alpha_i^{(\nu)} = 1, \qquad \sum_{i=1}^{n} \nu_i = \nu. \tag{4.37}$$

Then (4.34) becomes

$$q^{(\nu)}(I) = \frac{q^{(0)}(I) G^{(\nu)}(I)}{\sum_{J=1}^{N} q^{(0)}(J) G^{(\nu)}(J)}, \tag{4.38}$$

with

$$G^{(\nu)}(I) = \left\{ \prod_{i=1}^{n} [p(i \mid I)]^{\alpha_i^{(\nu)}} \right\}^{\nu}. \tag{4.39}$$

Inside the braces 0^0, if it happens, should be equated to 1, since it corresponds to the absence of the factor in (4.34). Apart from the denominator, which serves the purpose of normalization, $q^{(\nu)}(I)$ is proportional to two factors, the extra-evidential factor $q^{(0)}(I)$ and the purely evidential factor $G^{(\nu)}(I)$.

This $G^{(\nu)}(I)$ may be considered for a given ν as a measure of the degree of matching between the theoretical prediction and the experimental fact. However, an inconvenience here is that $G^{(\nu)}(I)$ diminishes with increasing ν, whereas the concept of matching should essentially be based on the comparison of $p(i \mid I)$ and $\alpha_i^{(\nu)}$. To obtain a related quantity that does not have this inconvenience and becomes unity when the perfect matching

$$\alpha_i^{(\nu)} = p(i \mid I), \qquad i = 1, 2, \cdots, n, \tag{4.40}$$

takes place, let us take the logarithm of $G^{(\nu)}(I)$,

$$\log G^{(\nu)}(I) = \nu \sum_{i=1}^{n} \alpha_i^{(\nu)} \log p(i \mid I). \tag{4.41}$$

This quantity with a negative sign (the logarithm here is nonpositive) is, according to the Gibbs theorem, always larger than $-\nu \sum_{i=1}^{n} \alpha_i^{(\nu)} \log \alpha_i^{(\nu)}$ and becomes equal to this if and only if (4.40) happens. Hence it is quite natural to introduce the ratio of these two as the desired measure,

$$\Gamma^{(\nu)}(I) = \frac{-\sum_{i=1}^{n} \alpha_i^{(\nu)} \log \alpha_i^{(\nu)}}{-\sum_{i=1}^{n} \alpha_i^{(\nu)} \log p(i \mid I)}, \tag{4.42}$$

which we call degree of confirmation or *confirmability* of H_I in the presence of the body of evidence $\mathcal{B}$.† In terms of $\Gamma^{(\nu)}(I)$, $G^{(\nu)}(I)$ of (4.39) becomes

$$G^{(\nu)}(I) = \exp\left[-\frac{-\nu \sum a_i^{(\nu)} \log \alpha_i^{(\nu)}}{\Gamma^{(\nu)}(I)}\right]. \tag{4.43}$$

It is advisable to agree that if the right-hand side of (4.42) becomes 0/0, then we put $\Gamma^{(\nu)}(I) = 1$. This is because the simultaneous vanishing of the numerator and denominator here implies $\alpha_i = p(i \mid I) = 0$ or 1 for all i. Expression (4.43) assumes that the logarithms used in it are all natural logarithms.

It is important to know when the credibility $q^{(\nu)}(I)$ becomes zero and when it becomes unity. First we consider the problem under the assumption that ν is finite. Since none of the $q^{(0)}(I)$ is supposed to be zero, $q^{(\nu)}(I) = 0$ happens, as we have seen, if and only if $G^{(\nu)}(I) = 0$ and there is at least another $G^{(\nu)}(J)$ that is not zero. The second condition means that there is at least one hypothesis in $\mathcal{H}$ that has not been logically refuted by the ν outcomes. Next, $q^{(\nu)}(I) = 1$ for a finite ν happens if and only if the factor $G^{(\nu)}$ vanishes for all hypotheses but one. This means that all hypotheses except one have been logically refuted by the end of ν observations.

As far as the confirmability $\Gamma^{(\nu)}(I)$ is concerned, it becomes unity if and only if a perfect match (4.40) takes place between the prediction and the observed sequence $\mathcal{B}$. $\Gamma^{(\nu)}(I) = 0$ happens either when the numerator in (4.42) is zero and the denominator is not zero, or when the denominator is infinite. The latter case corresponds to logical refutation. The former case happens when only one particular outcome has repeated itself in the experience, whereas the hypothesis does not give a probability of unity to this outcome.

For the limiting case $\nu \to \infty$, we have to proceed more carefully than in the foregoing paragraph, since it is likely that the $G^{(\nu)}(I)$ for all I will tend to zero for $\nu \to \infty$, but in such a fashion that $q^{(\nu)}(I)$ in (4.38) will tend to a finite value. To see this, it is convenient to rewrite (4.38) as

$$q^{(\nu)}(I) = \frac{q^{(0)}(I)F^{(\nu)}(I)}{\sum_{J=1}^{N} q^{(0)}(J)F^{(\nu)}(J)}, \tag{4.44}$$

with

$$F^{(\nu)}(I) = \frac{[A^{(\nu)}(I)]^\nu}{\sum_{J=1}^{N} [A^{(\nu)}(J)]^\nu} = \frac{G^{(\nu)}(I)}{\sum_{J=1}^{N} G^{(\nu)}(J)} \tag{4.45}$$

† Logically speaking, "confirmability" should rather be called "confirmedness," but for euphony we shall occasionally use the terminology confirmability.

and

$$A^{(\nu)}(I) = \prod_{i=1}^{n} [p(i \mid I)]^{\alpha_i^{(\nu)}}, \qquad G^{(\nu)}(I) = [A^{(\nu)}(I)]^{\nu}. \tag{4.46}$$

If the outcomes of $\mathfrak{D}$ occur according to a certain set of probabilities and if each experiment is independent, we have the fixed limiting values of $\alpha_i^{(\nu)}$ for $\nu \to \infty$:

$$\lim_{\nu\to\infty} \alpha_i^{(\nu)} = \gamma_i. \tag{4.47}$$

Correspondingly, the $A^{(\nu)}(I)$ will have the limiting values

$$A(I) = \lim_{\nu\to\infty} A^{(\nu)}(I) = \prod_{i=1}^{n} [p(i \mid I)]^{\gamma_i}. \tag{4.48}$$

(Note that the superscript (ν) is dropped to indicate the limiting value.) The limiting value of $F^{(\nu)}(I)$ will be equal to the limiting value of the middle expression of (4.45), in which $A^{(\nu)}(I)$ is replaced by $A(I)$:

$$\lim_{\nu\to\infty} F^{(\nu)}(I) = \lim_{\nu\to\infty} \frac{[A(I)]^{\nu}}{\sum_{J=1}^{N} [A(J)]^{\nu}}. \tag{4.49}$$

Since $A(I)$ is between 0 and 1, each $[A(I)]^{\nu}$ tends to zero for $\nu \to \infty$ [except when there is i such that $p(i \mid I) = \gamma_i = 1$], and (4.49) can have only two possible values, 0 or $1/m$,

$$\lim_{\nu\to\infty} F^{(\nu)}(I) = \frac{1}{m} \quad \text{for} \quad A(I) = \max_J A(J), \tag{4.50a}$$

where m is the number of I's that satisfy the last condition, and

$$\lim_{\nu\to\infty} F^{(\nu)}(I) = 0 \quad \text{for} \quad A(I) \neq \max_J A(J). \tag{4.50b}$$

Since we have from (4.42), (4.47), and (4.48)

$$\Gamma(I) = \lim_{\nu\to\infty} \Gamma^{(\nu)}(I) = \frac{\gamma_i \log \gamma_i}{-\log A(I)}, \tag{4.51}$$

the condition in (4.50a) can also be written as

$$\Gamma(I) = \max_J \Gamma(J). \tag{4.52}$$

In terms of credibilities, this conclusion means that

$$\lim_{\nu\to\infty} q^{(\nu)}(I) = \frac{q^{(0)}(I)}{\sum_{J\in\mathscr{K}} q^{(0)}(J)} \quad \text{for} \quad H_I \in \mathfrak{K}, \tag{4.53}$$

where $\mathcal{K}$ is the set of hypotheses that satisfy (4.52), and

$$\lim_{\nu\to\infty} q^{(\nu)}(I) = 0 \quad \text{for} \quad H_I \notin \mathcal{K}. \tag{4.54}$$

The condition (4.52) does not necessarily imply a perfect match (4.40) for $\nu \to \infty$. Conversely, however, a perfect match for $\nu \to \infty$ implies (4.52). Hence, if only one hypothesis shows a perfect match in the long run, that hypothesis will win out with $q^{(\nu)}(I) \to 1$. It is important that this limiting value be independent of $q^{(0)}(I)$ insofar as this is not zero. It may be noted that, provided the condition of statistical independence is satisfied, $\Gamma(I) = 1$ means a complete satisfaction of the empirical test, whereas $\Gamma^{(\nu)}(I) = 1$ with a finite ν can happen by accident for any hypothesis that is not logically refuted. On the other hand, $q^{(\nu)}(I) \to 1$ for $\nu \to \infty$ can happen without a perfect match for $\nu \to \infty$, whereas $q^{(\nu)}(I) = 1$ with a finite ν implies that all hypotheses except this particular one have been logically refuted.

It is instructive to rewrite (4.30) in terms of the F's:

$$\frac{q^{(\nu)}(I)}{q^{(\nu)}(J)} = \frac{q^{(0)}(I)}{q^{(0)}(J)} \cdot \frac{F^{(\nu)}(I)}{F^{(\nu)}(J)}, \tag{4.55}$$

where $q^{(0)}(I) \neq 0$, $q^{(0)}(J) \neq 0$ by agreement, and $F^{(\nu)}(I) \neq 0$ and $F^{(\nu)}(J) \neq 0$ for a finite ν provided that neither H_I nor H_J is refuted logically. Let us assume that H_J shows a better matching with experiments than H_I; that is, $F^{(\nu)}(I)/F^{(\nu)}(J) \to 0$ as $\nu \to \infty$. We then have the following dual consequence. (a) No matter how H_I is favored by experiment as compared with H_J, we can always make $q^{(\nu)}(J) > q^{(\nu)}(I)$ for a finite ν by giving a sufficiently large *a priori* credibility to H_J, unless H_J is logically refuted until the νth experiment. (b) No matter how strongly H_J is favored over H_I on extra-evidential grounds, there will always be a number ν_0 such that $q^{(\nu)}(I) > q^{(\nu)}(J)$ for $\nu > \nu_0$ if the prediction of H_I agrees with experiment better than that of H_J. This represents, so to speak, a perpetual seesaw between the empirical and the *a priori*, preserving, however, a final victory for the empirical in the long run. On the other hand, the finiteness of our experience forces us to introduce the *a priori* and, collaterally, the lack of necessity of an inductive conclusion.

In this connection, however, we have to note that in the actual situation of daily life and scientific research one is fairly restricted in choosing *a priori* credibilities if one wants to be "reasonable." The reason is that the *a priori* credibilities for hypotheses of different areas are interrelated in a theoretical scheme and cannot be changed arbitrarily just for one hypothesis for one type of experiment. It is indeed true that one can envision the *a priori* credibility for one hypothesis as being derived from a previous induction of higher level or of other domain, and in that induction there must also have been arbitrary *a priori* elements, and so on. Thus we can never eliminate *a priori* elements completely. However, a definite tendency seems to lie in the normal human mind to adjust *a priori* evaluations in such a way that the over-all picture of the world contains no flagrant contradiction between the *a priori* and the empirical. A "flagrant contradiction" may be interpreted as meaning a very high (low) credibility given to a hypothesis with a very low (high) degree of confirmation after a long series of experiments. Referring to this property of the human mind, we may speak of *the rule of consistent docility*, in which "consistent" emphasizes our effort to construct a contradiction-free over-all world picture and "docility" expresses our readiness to be taught by nature and our environment.

Let us look back and check if the requirements enumerated in the last section (pages 160–161) are actually satisfied by our theory. We have seen earlier that Item (a) is satisfied. Item (b) concerns the successive nature of acquisition of scientific knowledge, and this is certainly very explicitly embodied in our basic formula (4.35). Item (c) requires that the larger the $\Gamma^{(\nu)}(I)$, the larger the $q^{(\nu)}(I)$, and this can be clearly seen in (4.38) and (4.43). Item (d) says that the effect of the value of $q^{(0)}(I)$ must become less important as the experiment goes on. This can be best seen by taking the logarithm of the numerator of (4.38),

$$\log q^{(0)}(I) + \nu \sum_{i=1}^{n} \alpha_i^{(\nu)} \log p(i \mid I), \tag{4.56}$$

where the factor represented by $\sum_{i=1}^{n}$ tends to a constant value independent of ν. Hence the second term is essentially proportional to ν and will overwhelm the first term, which is a constant independent of the value of ν. Item (e), which states that for a finite ν we should not have $\Gamma^{(\nu)}(I) = q^{(\nu)}(I) = 1$, was confirmed by our analysis except in the very extraordinary case in which all hypotheses other than H_I have been logically refuted by the end of the νth experiment and this H_I shows perfect matching exactly at the end of νth experiment. Item (f) demands the existence of the possibility of $\Gamma^{(\infty)}(I) = q^{(\infty)}(I) = 1$. This, of course, happens if H_I is the only hypothesis that shows perfect matching for $\nu \to \infty$. As far as Item (g) is concerned, we shall give a rigorous proof in Section 6.2, but we can already see why this must be true. Suppose that we define the inductive entropy $U^{(\nu)}$ by

$$U^{(\nu)} = -\sum_{i=1}^{N} q^{(\nu)}(I) \log q^{(\nu)}(I). \tag{4.57}$$

If we use the right-hand side of (4.49) as the approximate expression of $F^{(\nu)}(I)$, we see that the $F^{(\nu)}(I)$ of a few previleged hypotheses become gradually large while the $F^{(\nu)}(I)$ of all other hypotheses will gradually diminish. This means that the $q^{(\nu)}(I)$ will be gradually concentrated on fewer hypotheses, resulting in a gradual decrease of $U^{(\nu)}$ with $\nu \to \infty$.

Formulation of an inductive process as a "retrodiction" has, as we saw at the beginning of this section, a compelling reason for its adoption from a very basic viewpoint about the probabilistic nature of inference. In addition to this, we have just seen that this formulation leads to consequences that faithfully describe some of the most salient features of the process of inductive inference.

Now that we have established our probabilistic model of induction on fairly solid ground, we may ask ourselves whether (and to what extent) the inherent ambiguity regarding legitimate generalization in inductive process has been eliminated in this model. The answer to this question is negative. We have to say that although compliance with the rule of conditioned randomness justifies generalization within the class conditioned by A_k, any wider generalization depends on an extra-evidential, non-necessary inference. Suppose that there are only two possible outcomes, D_1 and D_2, and two hypotheses, H_1 and H_2, and that $p(D_1 \mid A_k \cap H_1) = 1$ and $p(D_1 \mid A_k \cap H_2) \neq 1$. An individual case obeying A_k gives outcome D_1. Can this be considered as a support of H_1 *under* A_k? The answer is yes, if the individual case can be considered as a random sample from the collection of individual cases obeying A_k. This increases the credibility of H_1 as applicable to any other random sample within the class A_k. This is already a step forward, because a traditional argument against inductive generalization would be that each individual case has its specifications B_k beyond $A_k (B_k \to A_k, B_k \neq A_k)$, and this individual case supports H_1 only under the condition B_k, which could be so narrow that only one item in the universe can satisfy, prohibiting any generalization.

However, even with the help of the rule of conditioned randomness, we can not go any further than A_k. This is because, as we already cautioned, what we denoted by $q(H_I)$ in our foregoing argument is in reality what should have been denoted as $q_k(H_I)$ or $q(H_I \mid A_k)$. The unconditional degree of creditability should be given by

$$q(H_I) = \sum_k q(H_I \mid A_k) p(A_k), \tag{4.58}$$

which is a sort of average of $q_k(H_I)$ under various A_k, but we do not and cannot know $q_k(H_I)$ for all possible k.

There are two mutually complementing rules of action in connection with (4.58). First, the set $\mathcal{A} = \{A_k\}$ should be truly exhaustive so that it includes all possible circumstances under which H_I can engender any kind of testable prediction. It is obvious that we cannot test all the cases in $\mathcal{A}$, but we have to think of all possible testable consequences implied by H_I, and test H_I from as many different angles and in as many representative cases as possible. We call this *the rule of all-round test*. To make this rule workable we have to know which cases to skip. The rule here, if any, is of course extra-evidential in nature (with regard to the tests involved), and is bound to be very vague, sometimes bordering on the intuitive. In default of a good name, we call the guiding rule at work here *the rule of intentional omission*, because it encourages us to omit, consciously, many of the tests on extra-evidential grounds, and allows us, right or wrong, to arrive quickly at a conclusion regarding the veracity of a hypothesis as a whole. The adjective "intentional" is added to remind us that an inadvertent omission of tests because of ignorance of the existence of certain testable cases is dangerous.

Referring to (4.58), we may characterize the effect of these two rules somewhat as follows. The rule of all-round test alone urges us to carry out experimentation for every possible A_k to the highest possible number ν of observation. The rule of intentional omission, on the other hand, allows us to omit experimentation for many A_k's, that is, to remain with $\nu = 0$ for these A_k's. In other words, it allows us to use the *a priori* or extra-evidential values of credibilities $q(H_I \mid A_k)$ for these omitted A_k's. These extra-evidential or *a priori* credibilities can of course be affected by, or even estimated with the help of, the evidential or *a posteriori* ($\nu \neq 0$) credibilities for other A_k's that have not been omitted.

But which A_k can be omitted and which A_k cannot be omitted depends on a very subtle balance of various considerations, all external to the test under A_k in question. Some factors in these considerations may be based on experience in a similar but not identical area, and other factors may be very close to our esthetic sense. The old principle of simplicity, principle of elegance, principle of "*natura non facit saltum*," principle of interpolation, and so on, also come into play.

An example in which the rule of intentional omission can be used safely, is as follows. Suppose that A_k specifies the value of a parameter involved in the hypothesis H_I, that the value of predictive probability $p(D_i \mid H_I \cap A_k)$ varies very slowly with changes in A_k, and that this last point is also confirmed by experiment in a certain domain. Then people will probably take fairly large intervals between sampling points in the domain of A_k. On the other hand, the next example presents a case in which our extra-evidential consideration does not allow us any omission. There are two urns, each containing 10 balls, of which some are red and some are white. H_I states that Urn 1 contains n_1 white and $(10 - n_1)$ red balls, and Urn 2 contains n_2 white and $(10 - n_2)$ red. A_1 states that the color of one ball randomly taken from Urn 1 is tested, and A_2 states that the color of one ball randomly taken from Urn 2 is tested. Then there is no extra-evidential reason to believe that the value of $q(H_I \mid A_1)$ can have any relation to the value of $q(H_I \mid A_2)$. Hence we have no extra-evidential grounds for skipping the test of either of the urns. If one knows, however, from some other source of information, that each urn is filled with 10 balls taken randomly from a single, large urn containing many red and white balls, then the *a priori* value of $q(H_I \mid A_2)$ for $\nu = 0$ will be influenced by the *a posteriori* value of $q(H_I \mid A_1)$ for $\nu \neq 0$.

The rule of all-round test is a warning against too broad a generalization, whereas the rule of omission is a warning against too timid a generalization. To infer that everything in the world is black on the evidence of few objects that happen to be black is too broad a generalization. To claim that Chinese ravens may be nonblack because we have found ravens black only outside China is too timid an attitude toward generalization.

If

$$p(D_i \mid H_I \cap A_k) = p(D_i \mid H_I^* \cap A_k) \tag{4.59}$$

for all i, and if $H_I \to H_I^*$ and $H_I^* \neq H_I$, then H_I could be suspected to be too general. (Note that "X is more general than Y" means in this connotation "X is smaller than Y" in the previous usage of the terms.) In fact, in this case, there may be a superfluous (at least with respect to A_k) element C such that

$$H_I = H_I^* \cap C. \tag{4.60}$$

It may be noted that this process of eliminating a superfluous element is the same as in (4.19), where we wanted to eliminate nonverifiable elements.

On the other hand, replacement of H_I by H_I^* risks the loss of useful generality, which is after all the desired fruit of inductive inference. Indeed, even if H_I and H_I^* may stand in relation (4.59), H_I^* may not act as a hypothesis for another condition, say, A_j, that is, H_I^* may not engender any probability distribution $p(D_i \mid H_I^* \cap A_j)$, whereas H_I acts as a hypothesis and $p(D_i \mid H_I \cap A_j)$ is confirmed by experiment. What is needed is a "sound" extra-evidential judgement between the two extremes.

The problem can also be considered as a problem of composite hypothesis, since we can write

$$H_I = \bigcap_k H_{Ik}^*. \tag{4.61}$$

To be on the safe side, one should perhaps make conjunction of only those H_{Ik}^* that have been tested. But such a procedure is in general not fruitful. We usually make a conjunction of many more untested factors, but how many of them, and which ones, depend on an extra-evidential judgment. In some very fortunate cases an extra-evidential evaluation of hypotheses can be linked with some evidential evaluation of higher level, or wider range, or in some related cases. In the majority of practical cases, however, no quantitative model seems to be powerful and complex enough to simulate the human process of extra-evidential evaluation.

This admitted difficulty should not be confused with a less well-founded objection of a sweeping nature, to the effect that this kind of mathematical model in nonsense because no real numerical values of probabilities can ever be known in practice. The answer to this is threefold. First, personal probabilities can become observable quantities by cleverly devised psychological and behavioral experiments (see Section 7.4). Second, the logical structure of $\mathcal{H} \otimes \mathcal{D} \otimes \mathcal{A}$ is such that the probabilities *can be defined on it*, and more specific results, such as the Bayes formula, follow directly from the definition of probabilities without any further assumption. Thus the impossibility or difficulty of knowing their precise numerical values cannot be used as an argument denying their existence. Third, the fact that all the major features of inductive inference can be rediscovered in this mathematical model may be considered as a good evidential support for it.

The use of the Bayesian formula in connection with hypothesis testing has a long history (see, for instance, [S-1]). But the context (in connection with the epistemological problems of inductive inference) and the manner in which the formula and allied results are introduced and interpreted, as in this textbook, may have something fresh and suggestive. The contents of this section derive mainly from [W-13] and [W-20].

4.5. CLASSICAL PARADOXES

This section is devoted to a discussion of some of the "paradoxes" about induction that philosophers have ingeniously contrived in the past. Although our way of resolving the paradoxes will often be at variance with the ones offered by the respective authors, we should appreciate the educational value and entertaining quality of those paradoxes, as well as their past contribution to the development of various theories about induction. Our purpose in presenting these paradoxes lies not only in testing our point of view in the presence of these apparent difficulties, but also in having opportunities for elaborating some of the points that have been made only vaguely on an abstract level in the foregoing sections. The paradoxes to which I have attached the name of Hempel may not be solely his creation, but he should be credited for having presented them clearly and having called the attention of a wider audience to them than achieved by the original authors.

All these paradoxes are conceived in the nonprobabilistic framework. The crux of some of them is directly related to the limitations inherent in the nonprobabilistic view, but the gist of others has nothing to do with the nonprobabilistic view, and we can discuss them using the nonprobabilistic rule of Nicod, which, as was shown, can be given a certain justifiable interpretation from the probabilistic theory. For classical literature on the paradoxes, see Hempel [H-7, H-8, H-10], Carnap [C-2], Goodman [G-10], and Scheffler [S-2, S-3].

Hempel's Paradox Regarding Conjunction of Conflicting Hypotheses

This paradox has been used as a counterexample against an apparently innocuous rule of induction for a conjunctive hypothesis that states that if D_1 confirms H_1 and if D_2 confirms H_2, $D_1 \cap D_2$ confirms $H_1 \cap H_2$. This rule of conjunctive hypothesis is a direct consequence of the more basic Nicod rule for a nonconjuctive hypothesis: if $H \to D$, then D confirms H. Indeed, if $H_1 \to D_1$ and $H_2 \to D_2$, obviously $H_1 \cap H_2 \to D_1 \cap D_2$. [Note this last relation is equivalent to $(H_1 \cap H_2) \cap (D_1 \cap D_2) = H_1 \cap H_2$, which follows directly from $H_1 \cap D_1 = H_1$ and $H_2 \cap D_2 = H_2$.] We shall see that we can derive two lessons, one concerning the definition of a hypothesis and another concerning a composition of two hypotheses.

Let H_1 now stand for "all objects are black," and D_1, for "an object is found to be black." Then D_1 confirms H_1 in Nicod's sense. Next let H_2 stand for "all objects are yellow," and D_2, for "an object is found to be yellow." Then D_2 confirms H_2. If we use the rule mentioned above, we have to say that $D_1 \cap D_2$, which means that "one object is found to be black and another object is found to be yellow" should confirm $H_1 \cap H_2$, which means that "all objects are black and yellow (at one time)." This is ridiculous.

Such an application of the rule of conjunctive hypothesis can be rejected by the rule we made in Section 4.2, that a constant absurdity is not a verifiable proposition, and therefore is not a hypothesis. Here $H_1 \cap H_2$ is a "constant absurdity," and therefore cannot be inferred as a hypothesis. One should remember that the justification I gave to this rule (of exclusion of absurdity as a hypothesis) was that an absurdity can have two conflicting consequences. If one wants to have a more systematic way of defining the limitations of the rule of conjunctive hypothesis, one should resort to Theorem 4.1, p. 145. By this theorem we can conclude that

$$p(D_1 \mid H_1 \cap H_2) = 1 \tag{4.62}$$

from

$$p(D_1 \mid H_1) = 1, \tag{4.63}$$

provided that

$$p(H_1 \cap H_2) \neq 0. \tag{4.64}$$

Similarly, we can conclude that

$$p(D_2 \mid H_1 \cap H_2) = 1 \tag{4.65}$$

from

$$p(D_2 \mid H_2) = 1, \tag{4.66}$$

provided (4.64). Combining (4.62) and (4.65), we obtain

$$p(D_1 \cap D_2 \mid H_1 \cap H_2) = 1. \tag{4.67}$$

The reason why (4.67) follows from (4.62) and (4.65) is that in the axiom of probability, $p(A) + p(B) = p(A \cap B) + p(A \cup B)$, if $p(A) = p(B) = 1$, each of $p(A \cap B)$ and $p(A \cup B)$ has to be unity too. In the case of the above-mentioned paradox, the condition (4.64) is violated, and hence we cannot derive (4.67). As a matter of fact, we have in this case $p(H_1 \cap H_2) = 0$, because $H_1 \cap H_2 = \emptyset$. Here is a case in which $A \to B$ but not $p(B \mid A) = 1$.

The Nicod rule of the simple case itself should be amended: if $H \to D$ and $p(H) \neq 0$, then D "confirms" H, because if $p(H) = 0$ for sure, nothing can increase its credibility. From this and the above analysis, we derive the following Nicodian rule of a compound hypothesis: if $H_1 \to D_1$ and $H_2 \to D_2$ and if $p(H_1 \cap H_2) \neq 0$ [which also implies $p(H_1) \neq 0$ and $p(H_2) \neq 0$], D_1 and D_2 confirm $H_1 \cap H_2$. This amendment becomes particularly important in the case of a compound hypothesis because we are apt to forget to test $p(H_1 \cap H_2) \neq 0$ when we know $p(H_1) \neq 0$ and $p(H_2) \neq 0$. If one does not want to use the concept of probability in this context, he may loosely replace $p(A) \neq 0$ by $A \neq \emptyset$ in the foregoing sentences.

The above argument is limited to cases in which the probabilities involved are either 0 or 1. In more general cases there appears an additional complication, mainly because we cannot apply Theorem 4.1 any longer. First, we should note that exclusion of $\emptyset$ as a hypothesis becomes a special case of the rule that a proposition H that cannot engender the deductive probabilities $p(D_i \mid H)$, $i = 1, 2, \cdots, n$, is not a hypothesis (see Section 4.2). Second, the value of $p(D_1 \cap D_2 \mid H_1 \cap H_2)$ cannot, in general, be related to the values of $p(D_1 \mid H_1)$ and $p(D_2 \mid H_2)$, as can be seen from the consideration with regard to Theorem 4.1. (See our remark on conjunctive experimentation in Section 6.3 for more details.) Only in a special case can the former become the product of the latter two probabilities:

$$p(D_1 \cap D_2 \mid H_1 \cap H_2) = p(D_1 \mid H_1)p(D_2 \mid H_2). \tag{4.68}$$

[Equations (4.63), (4.66), and (4.67) represent a very special case of (4.68).] This can happen if we have a condition of statistical independence

$$p(D_1 \cap D_2 \mid H_1 \cap H_2) = p(D_1 \mid H_1 \cap H_2)p(D_2 \mid H_1 \cap H_1)$$

and conditions of "irrelevance" in the sense that $p(D_1 \mid H_1 \cap H_2) = p(D_1 \mid H_1)$ and $p(D_2 \mid H_1 \cap H_2) = p(D_2 \mid H_2)$. More generally, consider two sets of hypotheses $\mathcal{H}^{(1)} = \{H_I^{(1)}\}$ and $\mathcal{H}^{(2)} = \{H_J^{(2)}\}$, which can respectively give rise to conditional probabilities $p(D_i^{(1)} \mid H_I^{(1)})$ and $p(D_j^{(2)} \mid H_J^{(2)})$, where $D_j^{(1)}$ and $D_j^{(2)}$ are outcomes of two separate experiments, $\mathcal{D}^{(1)} = \{D_i^{(1)}\}$ and $\mathcal{D}^{(2)} = \{D_j^{(2)}\}$. If we have

$$p(D_i^{(1)} \cap D_j^{(2)} \mid H_i^{(1)} \cap H_J^{(2)}) = p(D_i^{(1)} \mid H_I)p(D_j^{(2)} \mid H_J) \tag{4.69}$$

for all i, j, I, and J, we speak of a multiplicative case. Conversely, if a set of hypotheses $\mathcal{H} = \{H_k\}$ is such that we can write $H_k = H_I^{(1)} \cap H_J^{(2)}$ so that (4.69) is true, then we speak of a *separable* case. Of course, even if the hypotheses are separable, actual observed data $D_j^{(1)} \cap D_j^{(2)}$ may be such that the condition of statistical independence

$$\frac{\nu_{ij}}{\nu} = \frac{\nu_i^{(1)}}{\nu} \cdot \frac{\nu_j^{(2)}}{\nu} \quad \text{(for all} \quad i, j) \tag{4.70}$$

is not upheld. Obviously, then, the confirmability of separable hypotheses can not be 100%. In (4.70) it is assumed that ν observations of combined experiment $\mathfrak{D}^{(1)} \otimes \mathfrak{D}^{(2)}$ are made, that ν_{ij} of them gave the result $D_i^{(1)} \cap D_j^{(2)}$, and that

$$\nu_i^{(1)} = \sum_{\text{all } j} \nu_{ij}^{(2)} \quad \text{and} \quad \nu_j^{(2)} = \sum_{\text{all } i} \nu_{ij}.$$

Hempel's Paradox Regarding Conjunction of Heterogeneous Hypotheses

This paradox or, rather, our resolution of this paradox is a good example to show that the process of induction cannot be freed from extra-evidential factors and be reduced to formal rules. Suppose a hypothesis H that states, "all ravens are black and all August days in New York are snowy." Let us adopt, for the moment, the following two themes about induction and deduction: (a) If H implies the occurrence of an observational result D, and if D is actually observed, then the credibility of H is increased; (b) If the credibility of H is increased, the reliability (i.e., confidence we place on) of any consequence of H (i.e., observational proposition deductively derivable from H) increases. Now let us suppose that you open a window of your hotel room on Fifth Avenue, New York, on an August morning and see one black raven. The fact that you have seen the black raven increases your confidence in hypothesis H and causes you to put on your heavy overcoat before leaving the hotel. This, of course, is ridiculous. It is true that (a) and (b) are not quite in accord with our general approach. But we can easily rephrase (a) and (b) in conformity with our language without destroying the main point of the paradox. Theme (a) is Nicod's rule, and (b) can be justified on the same ground as our formula (4.36). The main catch of the paradox lies obviously not in these details but in the fact that the hypothesis H consists of two components, of which the observed fact D is "irrelevant" to the second of these two components, and that the predictive consequence is drawn from the second component.

The explanation usually given for this paradox runs somewhat as follows (see, for example, [G-10]). Any hypothesis H, at least of a certain class, being a general statement, can be rephrased as "all X's satisfy the predicate P." From this follows deductively proposition D_i of the type "sample i of X satisfies predicate P" with a particular i. The usual criterion states that observational verification of D_i confirms only H, which stands in the above stated relation with D_i. (D_i is a strict instantiation of H.) Thus it might be claimed that according to this criterion the observed black raven confirms only that "all ravens are black" and not that "all ravens are black and all August days are snowy in New York." But we can easily show that this argument is wrong. The conjunctive hypothesis mentioned above can be brought to the form of a single hypothesis, "all X's are P," in such a way that either a black raven (a raven that turns out to be black) or a snowy August day in New York (an August day in New York that turns out to be snowy) can be legitimately considered as its strict instantiation.

The apparently compound hypothesis H is: "all ravens are black and all August days in New York are snowy." Let R be the predicate such that any object satisfying R is a raven and any object not satisfying R is not a raven. Similarly, let A be the defining predicate for an object being an August day in New York. Let B and S stand respectively for black and snowfall. We may admit $R \cap A = \emptyset$, since a bird cannot be a day and a day cannot be a bird. Then the hypothesis H is, as will be shown in the next paragraph, equivalent to the propositions that all objects that satisfy the predicate $X = R \cup A$ also satisfy the predicate $P = (R \cap B) \cup (A \cap S)$. A particular raven that turns out to be black is a genuine instance of this general statement, which has the type of all X's satisfy P. Hence, according to the criterion introduced previously, the black raven has to be considered as confirming the hypotheses H *in toto*.

The original H states that R implies B, that is,

$$R = R \cap B, \tag{4.71}$$

and A implies S, that is,

$$A = A \cap S. \tag{4.72}$$

The second hypothesis K, which is claimed here to be equivalent to H under the condition C,

$$R \cap A = \emptyset,$$

states that X implies P,

$$X = X \cap P, \tag{4.73}$$

where $X = R \cup A$ and $P = (R \cap B) \cup (A \cap S)$. First, let us show that H implies K (actually without C). Make the disjunction of (4.71) and (4.72), side by side,

$$R \cup A = (R \cap B) \cup (A \cap S), \tag{4.74}$$

which means $X = P$ and hence $X \rightarrow P$, which is equivalent to K given in (4.73).

Next let us show that K implies H under C. Write (4.73) in the form

$$R \cup A = (R \cup A) \cap P = (R \cap P) \cup (A \cap P). \tag{4.75}$$

Make the conjunction of the left and right sides of (4.75) with R, and we get, from the left side of (4.75), $R \cap (R \cup A)$, which becomes R because of the absorptive law. From the right of (4.75) we get $[R \cap (R \cap P)] \cup [R \cap (A \cap P)] = R \cap P = R \cap B$ because of C. Hence, $R = R \cap B$. Similarly $A = A \cap S$. Thus we have derived (4.71) and (4.72) from (4.73) with the help of C. This completes the proof that H and K are equivalent.

This shows, contrary to the earlier view of some logicians, that as far as logical structure is concerned, a black raven is a strict instance of H and that this paradox cannot be eliminated by a purely formal prescription. In fact, the above proof makes it clear that this is another case of a question of degree of generalization. In the foregoing argument, for instance, we can interpret R as ravens in Manhattan and A as ravens outside Manhattan and put $S = B$. Then the original intention of Hempel and Goodman would amount to a prohibition from considering a black raven in Manhattan as evidence of a general rule that covers any territory broader than Manhattan. On the other hand, if we interpret R as a particular raven seen one morning in Manhattan and give any crazy interpretation to A and S, Hempel and Goodman's prescription about strict instantiation cannot prevent us from legitimately inferring any general rule about anything outside that one bird. When we feel that a black raven in Manhattan is evidence of the general rule regarding the blackness of all ravens, and consider the inference about snow in August, from a black raven, as a wrong inference, we are tacitly using an extralogical, extraevidential consideration. We may be using various hypotheses of a higher order, such as one stating that coloring of birds of the same species is very often the same all over the world. We may also point out that we know that ravens in Manhattan may fly to New Jersey and Connecticut, and so on, so that the rule of conditioned randomness is satisfied to some extent if we limit ourselves to ravens of the Eastern region. On the other hand, we do not have the slightest guarantee that we have been presented with a fair sample out of the mixed ensemble of all ravens and all August days in New York.

For a more detailed discussion of this problem, particularly a proof of our contention by the use of predicate calculus instead of propositional calculus, see [W-20].

Hempel's Paradox of Indoor Ornithology

This example usefully illustrates three important points made in the last section: (a) that a nonquantitative (i.e., nonprobabilistic) explication of inductive process is inadequate, (b) that an experiential proposition must be accompanied by a specification of the experimental procedure, and (c) that a high degree of confirmation does not necessarily mean a high degree of creditation.

The paradox is conceived in the framework of a black and white (nonquantitative) confirmation theory whose starting postulate is Nicod's rule that if hypothesis H implies D and if D is observed, then H is "confirmed." The "paradox" consists of the following situation: let H be the proposition that all ravens are black, which is, of course, equivalent to its contraposition, the proposition that all nonblack objects are nonravens. Then H must be confirmed when we happen to observe a black raven as well as when we happen to observe a nonblack nonraven in our living room. To quote Goodman [G-10, page 71], "The prospect of being able to investigate ornithological theories without going out in the rain is so attractive that we know there must be a catch in it," whence the name "indoor ornithology."

To make sense out of this situation, we have to make various statements involved above more precise. Of course, the most objectionable concept is the black-and-white theory of confirmation, but let us return to it later and try to stay as long as possible in the framework of thought assumed here. (The first person to note that resolution of the paradox requires a quantitative consideration seems to have been Janina Hosiasson-Lindenbaum [H-13].) I stated, "you happen to observe a black raven," but this may mean that "when you open your eyes, the first thing you observe happens to be a black raven" or that "you have a collection of black objects, and the one you arbitrarily pick happens to be a raven" or that "you have a collection of ravens and the one you arbitrarily pick happens to be black." Each of these statements corresponds to a certain experimental procedure. The hypothesis H, either in its original version or in its contraposition, implies only the last of these three observational propositions. Continuing on the same line of consideration, we note that the hypothesis H, in either form, implies that "an arbitrarily picked nonblack object is a nonraven" but does not imply that "an arbitrarily picked nonraven is nonblack." In either form of H there are two experiments about which H has a definite assertion. One is the test of blackness of an arbitrary raven, and the other is the test of raven-ness of an arbitrary nonblack object. Let us take the first kind of experiment and denote by $D_1^{(1)}$ the observational proposition that an arbitrarily picked raven is black. The alternative result of the same experiment is denoted by $D_2^{(1)}$, which states that an arbitrarily picked raven is nonblack. Similarly, regarding the second kind of experiment, let $D_1^{(2)}$ denote the proposition that an arbitrarily picked nonblack object is a nonraven, and let $D_2^{(2)}$ be its negation. The paradox consists in the fact that although H implies $D_1^{(1)}$ as well as $D_1^{(2)}$, a man of good common sense would not be inclined to consider $D_1^{(2)}$ as a confirmation of H but would so accept $D_1^{(1)}$. It is a common defect of the explanations we find in the literature that they do not rigorously state what kind of observation is envisioned (see, for instance, page 72 of [G-10]).

Why does anybody of good sense not consider $D_1^{(2)}$ above as good confirmation of the hypothesis that all ravens are black or, equivalently, that all nonblack objects are nonravens? The answer is obvious, because even if all ravens are nonblack, the chance is very slim that an arbitrary nonblack object will be a raven. Indeed, we know that there are in the actual world a tremendous number of nonblack objects that are not ravens even if all ravens are nonblack. This means that the chance is very high that $D_1^{(2)}$ will turn out to be true independently of the color of ravens. The chance of $D_1^{(2)}$ being true is essentially determined by the ratio of the number of nonblack nonravens to the total number of all nonblack objects, which is almost 1 no matter what color the ravens may have. Expressed the other way around, this means that any hypothesis regarding the color of ravens has to admit a very high probability to $D_1^{(2)}$. Indeed, in the limiting case in which the number of ravens in the world (or living room) is negligible, this probability will become unity, provided that there is a nonblack object in the world at all. In this limiting case $D_1^{(2)}$ "confirms" any hypothesis regarding the color of ravens. On the other hand, the chance of $D_1^{(1)}$ being true is essentially determined by the ratio of black ravens to the total number of ravens. The hypothesis that gives a perfect certainty to $D_1^{(1)}$ is only the hypothesis H, which says that all ravens are black or, equivalently, that all nonblack objects are nonravens. Thus, as far as the hypotheses regarding the color of ravens are concerned, $D_1^{(1)}$ confirms H more strongly than any other hypothesis. This is true no matter how small the number of ravens is compared with the number of all nonblack objects.

A lesson to be drawn from this consideration is, first, that the evidence must be formulated operationally in the form of a proposition. The substantive, "a black raven," is not evidence, but a sentence, "The object I see under such-and-such a condition is a black raven," or a sentence, "A raven is found to be black," could constitute evidence. Once the evidence is incorporated in a well-formed observational sentence, we have to consider the type of hypothesis that can say something about the evidence. In the above example of $D_1^{(2)}$, the hypotheses that come into consideration must say something about the nature (i.e., whether raven or nonraven) of things that are nonblack. Therefore, rigorously speaking, an arbitrary hypothesis regarding the color of ravens, such as "half of all the ravens are black and the other half are yellow," is not relevant to $D_1^{(2)}$. But a hypothesis regarding the color of ravens in conjunction with another hypothesis regarding the ratio of the number of all nonblack objects to the number of all ravens can become relevant to $D_1^{(2)}$. For this conjunction can say something about the chance of a nonblack object being a raven or not. What I said before is that if the second component of such a compound hypothesis says that the number of all ravens is negligible compared with the number of all nonblack objects, then a compound hypothesis of which the first component says any arbitrary thing about the color of ravens "implies" $D_1^{(2)}$; hence $D_1^{(2)}$ "confirms" any such compound hypothesis. Now a hypothesis about the ratio of all ravens to all nonblack objects can be tested separately, and I am sure the hypothesis that this ratio is negligible can be considered as a fact in the actual world. Of course, negligibly small is not zero, and the term "implies" here actually means "gives a very high probability to," and "confirms" here means "is given a very high probability by." If many hypotheses give a high probability to a certain observational result, and if such an observational result is obtained, their confirmability may be increased, but not their credibility, for credibility is a competitive idea. This is why a man of good common sense does not think that the hypothesis H would be made more creditable by an observation that a nonblack object was found to be a nonraven.

We took for granted in the above argument that the number of ravens is negligibly small in comparison with the total number of nonblack objects, and hence in comparison with the total number of all objects. Suppose now that ravens start suddenly to reproduce themselves superabundantly, keeping the same ratio of black to nonblack variations. Also suppose that the world thus becomes such that the overwhelming majority of existing things are ravens. Then consider a set of all nonblack objects. This set will be composed of nonblack ravens and of nonblack nonravens. But unless all ravens are black, that is, unless the ratio of black to nonblack ravens is infinite, the set of nonblack objects will also be overwhelmingly populated by ravens. In this circumstance suppose that $D_1^{(2)}$ happens to take place. This strikes out practically all possibilities that a finite (nonzero) fraction of ravens will be nonblack. For, if a finite fraction of ravens were nonblack, these nonblack ravens would occupy practically the entire set of nonblack objects and $D_1^{(2)}$ would have practically no chance of taking place. This means that a man of good common sense has to conclude that the fraction of nonblack ravens is practically nil, and that we can no longer say that $D_1^{(2)}$ is a bad argument for concluding that almost all ravens are black, when the world consists almost entirely of ravens. The original paradox suggested that $D_1^{(1)}$ and $D_1^{(2)}$ are equally good support for H. This can very well happen between two extremities: the world with a negligible population of ravens and the world that consists exclusively of ravens. We discuss this matter presently in a more quantitative fashion.

So far I have tried to stay as much as possible in the framework of black-and-white confirmation theory. But our argument gradually has become more and more quantitative. Actually the entire argument must be based on the concept of credibility. We show how the theory of credibility can be applied to this problem. Suppose two propositions H_1 and H_2, where H_1 means that all ravens are black and H_2 means that not all ravens are black or, equivalently, that some ravens are nonblack. As already stated, H_2 as it stands is not a hypothesis. However, if we make a conjunction of H_2 with a proposition A that there is a definite probability of an arbitrary raven's being black, then $H_2 \cap A$ implies that the probability of $D_1^{(1)}$ is α and that $\alpha \neq 1$. Thus we obtain

$$\begin{aligned} p(D_1^{(1)} \mid H_1) &= 1, \qquad p(D_2^{(1)} \mid H_1) = 0, \\ p(D_1^{(1)} \mid H_2 \cap A) &= \alpha \neq 1, \qquad p(D_2^{(1)} \mid H_2 \cap A) = 1 - \alpha \neq 0. \end{aligned} \tag{4.75}$$

Similarly, if we make a conjunction of H_2 with a proposition B that there is a definite probability of an arbitrary nonblack object's being a nonraven, then $H_2 \cap B$ implies that the probability of $D_1^{(2)}$ is β and that $\beta \neq 1$. Thus we obtain

$$\begin{aligned} p(D_1^{(2)} \mid H_1) &= 1, \qquad p(D_2^{(2)} \mid H_1) = 0, \\ p(D_1^{(2)} \mid H_2 \cap B) &= \beta \neq 1, \qquad p(D_2^{(2)} \mid H_2 \cap B) = 1 - \beta \neq 0. \end{aligned} \tag{4.77}$$

Since the propositions A and B seem to be quite justifiable, let us simply write H_2 for $H_2 \cap A$ and for $H_2 \cap B$.

Now suppose that we made an observation $\mathfrak{D}^{(1)} = \{D_1^{(1)}, D_2^{(1)}\}$ and obtained $D_1^{(1)}$. This means that the color of a raven was found to be black. Supposing that H_1 and H_2 have credibilities $q^{(0)}(H_1)$ and $q^{(0)}(H_2)$ before the observation, we obtain from (4.21) the ratio of credibilities after the observation:

$$\frac{q^{(1)}(H_1)}{q^{(1)}(H_2)} = \frac{q^{(0)}(H_1)\, p(D_1^{(1)} \mid H_1)}{q^{(0)}(H_2)\, p(D_1^{(1)} \mid H_2)} = \frac{q^{(0)}(H_1)}{q^{(0)}(H_2)} \frac{1}{\alpha}. \tag{4.78}$$

Now let us assume that we make the other observation, $\mathfrak{D}^{(2)} = \{D_1^{(2)}, D_2^{(2)}\}$, and obtained $D_1^{(2)}$. This means that a nonblack object was found to be a nonraven. Then by this observation the ratio of credibilities becomes

$$\frac{q^{(1)}(H_1)}{q^{(1)}(H_2)} = \frac{q^{(0)}(H_1)\, p(D_1^{(2)} \mid H_1)}{q^{(0)}(H_2)\, p(D_1^{(2)} \mid H_2)} = \frac{q^{(0)}(H_1)}{q^{(0)}(H_2)} \frac{1}{\beta}. \tag{4.79}$$

Let us apply the frequency point of view of probability to α and β, assuming that the numbers of objects involved are extremely large. Then we get

$$1 - \alpha = \frac{\text{number of nonblack ravens}}{\text{number of all ravens}} \tag{4.80}$$

and

$$1 - \beta = \frac{\text{number of nonblack ravens}}{\text{number of all nonblack objects}}. \tag{4.81}$$

Hence we have

$$\frac{1-\beta}{1-\alpha} = \frac{\text{number of all ravens}}{\text{number of all nonblack objects}}. \tag{4.82}$$

It should be clearly understood that the numbers of various objects mentioned in these formulas are those indirectly implied by the point of view associated with H_2. Of course, in order for this point of view to be tenable, all these numbers must agree with the factual values—which, however, we do not know. But in order to see the structure of the relations involved here, it is convenient to discuss the ratio of numbers on the right side of (4.82) as if it were an experimentally determinable parameter. We can then consider (4.82) as a constraint imposed on α and β, within which they can vary. Suppose that the number of all ravens is extremely small compared with the number of all nonblack objects; then, from (4.82), β is practically unity. This entails, according to (4.79), that the ratio $[q^{(1)}(H_1)]/[q^{(1)}(H_2)]$ be practically the same as $[q^{(0)}(H_1)]/[q^{(0)}(H_2)]$. This means that the observation of $D_1^{(2)}$ does not favor H_1 at all. On the other hand, if the number of ravens is so large that practically all objects in the world are ravens, then the number of nonblack objects will become practically equal to the number of nonblack ravens, insofar as α is not exactly 1. This means that the ratio of (4.82) becomes the reciprocal of the ratio of (4.80). This in turn means that β must be 0. From (4.79) it follows that $q^{(1)}(H_2)$ can be made 0 provided that $\alpha \neq 1$, no matter what the ratio $[q^{(0)}(H_1)]/[q^{(0)}(H_2)]$ may be. This means that $D_1^{(2)}$ kills any H_2 in favor of H_1 insofar as H_2 (or, more rigorously, $H_2 \cap A$) states that a finite (nonzero) fraction of ravens is nonblack.

Now the original paradox was that $D_1^{(2)}$ was as good (or as bad) as $D^{(1)}$ as evidence for H_1. When does this happen according to our formalism? This, of course, requires $\alpha = \beta$. Substituting $\alpha = \beta$ in (4.82), we obtain

$$\frac{\text{number of all ravens}}{\text{number of all nonblack objects}} = 1. \tag{4.83}$$

This means that in order to make living room ornithology possible it is necessary to have just as many ravens in the room as all nonblack objects (assuming that the ratios α and β are constant everywhere, hence also in the living room). This would mean that if there are 1000 nonblack objects in the room, there will also be 1000 ravens in the room. This is not precisely the kind of world we live in.

Hempel [H-7, H-8, H-10] was right when he boldly argued that $D_1^{(2)}$ is a confirmation of H, but the extent to which he was right may be compared to the extent to which the statement that 10^{-6} cm is a nonzero length is correct in everyday life, where the minimum measurable length may be 10^{-2} cm.

Goodman's Paradox of the Grue Emerald

This paradox, attributable to N. Goodman, will keep its fame for the ingenuity of its creator and its pedagogic and entertaining value, even if it can be stripped of its magical appeal rather easily. (See N. Goodman [G-10], page 73; also the excellent exposition on this and allied topics by I. Scheffler [S-2].) In the context of the present book this example is particularly valuable in showing (a) that our preference for certain predicates is based on an extrasyntactical consideration and (b) that, at least in this case, we can fairly well determine the ground on which we prefer certain predicates to others.

Consider a hypothesis H that all emeralds are green. Then the body of evidence $\mathcal{B}$ that all individual emeralds so far examined have been found to be green confirms hypothesis H. (Any other hypothesis one may have, if it does not give deductive probability 1 to $\mathcal{B}$, will be given less confirmability than H.) Now consider a hypothesis K that all emeralds examined before the end of the year 1999 will be green and all emeralds examined after the beginning of the year 2000 will be blue. For short, Goodman proposed to introduce a new predicate "is grue," which stands for "is found to be green if examined before t_0 or is found to be blue if examined after t_0," where t_0 can be any fixed time, but in the above case it stands for midnight between the year 1999 and the year 2000. Then hypothesis K can be expressed as "all emeralds are grue." This K gives deductive probability 1 to the evidence $\mathcal{B}$ mentioned above, and hence K must be confirmed by $\mathcal{B}$ to the same degree as H.

However, the prediction about an experiment made after t_0 based on K and that based on H are contradictory, whereas both have exactly the same degree of confirmation from the experience before t_0. If we want to make any difference between K and H in our scheme, it must be attributed to the *a priori* credibility. But on what ground are we entitled to give a higher *a priori* credibility to H then to K? If the reader says it is because hypothesis K contains dependence on a particular time t_0 and that the expression of a natural law must not refer to a particular time, then I have to join Goodman in rejecting his argument. Let A stand for "(emerald) is green" and B for "(emerald) is blue." Similarly, following Goodman, let C stand for "(emerald) is grue" and D for "(emerald) is bleen," where "bleen" means "is blue before t_0 or is green after t_0." This means that C stands for "A before t_0 or B after t_0," while D stands for "B before t_0 or A after t_0." This definition seems to give time dependence to C and D, whereas A and B seem to be time-independent. This is only because we started from A and B and defined C and D in terms of A and B. It is clear that we can go the other way around and define A and B in terms of C and D. In this case A obviously stands for "C before t_0 or D after t_0" and B stands for "D before t_0 or C after t_0." In other words, hypothesis H states that "all emeralds will be grue before t_0 and bleen after t_0," and hypothesis K states simply that "all emeralds are grue." Expressed in this way, H is a time-dependent hypothesis and K is a time-independent hypothesis. Thus there is complete symmetry between K and H, at least insofar as the syntactical construction of the hypothesis is concerned.

Goodman's answer to this difficulty seems, in brief, to be that there is such a thing as a degree of "entrenchment" (in the historical record of the usage of a language) for each predicate: the past biography of the word "green" shows a higher degree of entrenchment than that of the word "grue." When he says that "green" is better entrenched than "grue," he is not concerned with whether or not hypotheses using "green" have been more successful in the past than hypotheses using "grue," but he is pointing to the fact that "green" was more frequently used in the past than "grue" in formulating hypotheses. Well-entrenched predicates and hypotheses using only such predicates are "projectible," whereas poorly entrenched predicates or those hypotheses using such predicates are not projectible. There are, according to Goodman, several rules to cope with marginal or borderline cases such as the rule that a hypothesis that contradicts a projectible hypothesis is unprojectible, and so on. From all this I seem to be entitled to conclude that if Goodman were placed in a primitive tribe speaking a language in which the word "evil spirit" was entrenched, all hypotheses formulated in terms of "evil spirit" would become projectible for him.

To understand the import and limitations of Goodman's assertion, we have to know the kind of task he is imposing on himself. I think it is not far from the truth to say that he is trying to give an "explication" of what is regarded as projectible, that is, what is usually thought to be a good inductive generalization. He is not just arbitrarily giving a "definition" to the term projectible, nor is he directly recommending his criterion to practicing scientists. An explication is often considered to be successful if the concept is expressed in clear and distinct words, and the extension of the newly formulated precise concept (explicatum) overlaps more or less with the major part of the extension of the vaguely conceived primary concept (explicandum). However, there can be more than one explicatum for a given explicandum, and there arises a question as to the degree of desirability of each explicatum. Such comparison of the desirability of each explicatum is possible, not in terms of its extension, but only in terms of the relation of its intension to the other explicated concepts of the surrounding area. From this point of view we have to say that Goodman's explication in terms of "entrenchment" is not a particularly happy one, because the phenomenon of entrenchment is so heterogeneous with respect to the other explicated features of inductive inference. The following is an attempt to give a different and better explication to the categorical distinction that exists between "green" and "grue."

However, we have to agree with Goodman at the outset on one major point, namely, that the apparent symmetry between "green" and "grue" can be broken only if reference is made to an extrasyntactical empirical fact (entrenchment is an empirical fact) that provides, so to speak, a fixed anchoring point. My main complaint is that the linguistic fact of entrenchment is only remotely related to other features of inductive process. Let us go back to the comparison of K and H, and try again to discover any possible asymmetry between them. Suppose that we have an instrument called a "green-meter" (a black box that may contain some sort of spectrometer), which tells the experimenter whether or not an object under examination is green. By the same token, let us assume that we also have a "blue-meter." Similarly, we can invent a "grue-meter" and "bleen-meter," which tell the experimenter "yes" if the object is grue or bleen, respectively. Now the point boils down to an ostensibly insignificant distinction of the structure of these four meters. The grue-meter and bleen-meter have to have inside their black boxes some kind of timepiece, other than a spectrometer, whereas the green-meter and blue-meter need not have any timepiece. In this way "green" and "grue" can be characterized respectively as "operationally time-independent" and "operationally time-dependent."

Actually the instrumental embodiment of an operationally time-dependent predicate need not contain a clock, properly called. It can be characterized as an instrument whose crucial physical property changes in time in relation to other "fixed" properties of many physical objects. By rephrasing my criterion in this way, I am obviously inviting a naive formal objection to the effect that any change in time is definable and testable only in relation to some supposedly fixed reference; hence, by changing the reference, what seems to have been changing in time would become fixed in time, and what seems to have been fixed would become changing in time.

The answer to this is that it is indeed everybody's freedom to call any quantity fixed, but the common-sense view (in agreement with the more formalized science of physics) has its own well-defined choice, and this choice is such that any deviation from it makes a coherent understanding of the physical world well-nigh impossible without violating the basic form of our reasoning.

Interchangeability of a fixed frame and a uniformly moving frame in the theory of special relativity is a very exceptional case. In all other cases, interchanging a fixed quantity and a temporally changing quantity (for instance, weight) causes our description of nature to become chaotic. Our agreement to use a rigid rod as a meterstick is not just a matter of convention; it is imposed by the necessity of making an intelligible quantitative picture of the physical world. The fact that we assign a thermal expansion coefficient to each metal is a good indication that nature is telling us what to take as a fixed quality. For this reason we agree that the meterstick must be kept at 0°C and normal pressure. It is important that by nature's insistence and our theoretical thinking, the concept of a fixed standard length can be "anchored" in a material object, such as the meter standard at Sèvres, or in some spectral line.

Once we agree with the basic viewpoint of physics (and also with the common-sense view) on this point, the symmetry between the predicate "green" and the predicate "grue" disappears and the former distinguishes itself by being time-independent. It must be clearly understood that the logical symmetry between these two predicates, taken apart from other predicates, is indisputable, and that the asymmetry arises only when they are considered in relation to other predicates in the context of physical theory. It is important to note that fixed quantities in physical theory are not conceptually so defined, but are anchored in material objects, such as the standard meterstick or a spectral line. Therefore the criterion based on the theory can be reduced to an operational criterion. My suggestion that one examine the green-meter and the grue-meter for the presence or absence of a timepiece (or some time-dependent element) is exactly such an operational method. It is obvious that even this operational criterion presupposes a certain theoretical backing. But the role of physical theory is reduced to a bare minimum; that is, the theory is used only for two major purposes: the length of a solid body such as the metal scale on the spectrometer is to be considered approximately constant, and a certain mechanism is recognized as a timepiece (or a time-dependent element). This level of physical theory may be characterized as common-sense physics. Hence, in order to emphasize the point that the physical theory has an observational anchoring point, the proposed criterion may be called an "operational" one, provided that we do not lose sight of the fact that we cannot recognize or understand any result of an operationally defined observation without the help of some sort (even in the most primitive form) of "theory."

It would have been very stimulating if somebody had argued that "green" and "grue" are basically indistinguishable. But Goodman himself never doubted the human capability of telling which is which. What I did with the grue-meter just made explicit what had been tacitly understood even by Goodman. My point is that this basis of distinction is more important than the outward sign of entrenchment.†

Thus we may conclude (a) that the predicates "green" and "grue" are not symmetrical in the light of the basic structure of physical theory as imposed by nature and by our basic form of cognition; (b) that this asymmetry is characterized by a contrast between "time independence" and "time

† We should guard ourselves against confusion of the present consideration with the often-made speculation that if all rigid bodies change their sizes in time at the same rate, and if all other physical quantities involving the "dimension" of length change correspondingly, then there will be no change in the physical world as we observe it. If such a change is in principle unobservable, I doubt very much that we are entitled to talk about it. However, if we want to leave room for such a speculation, we should say that the lengths of all rigid bodies are relatively fixed and that we use them as our standard in formulation of physical theory. This rephrasing does not affect the previous argument.

dependence"; and (c) that this asymmetry can be detected by an operationally defined method. The rest of our argument is that a hypothesis formulated in terms of time-independent predicates is preferable to one formulated in terms of time-dependent predicates. My contention is that preference for a predicate because it is time-independent is much better founded than preference for a predicate because it is entrenched in the history of a language. Indeed, the idea of time independence has an intimate relation with the very nature of inductive inference in natural science, while the idea of entrenchment has not.

One of the main functions of theory formation resides essentially in identifying some constancy or unity in spatial and temporal varieties. A hypothesis or law represents a unity in the variety of different individual cases. A dynamic law represents an atemporal unity in temporal varieties, for the law itself is fixed in time. My criterion is in harmony with this general feature of induction, for it rests on the concept of constancy in time. I do not discuss here whether there exists unity or constancy in the physical world itself, or whether we look instead at the physical world in such a way that we can discover unity or constancy. This problem is irrelevant to our present discussion. It may be noted that, for the same reason empirically time-independent predicates are preferred, empirically space-independent predicates are preferred for use in induction. Operationally, this means that a preferred predicate corresponds to measurement at different locations with-the same instrument or an instrument of the same construction. As far as time dependence is concerned, I can now state that a hypothesis that is syntactically time-independent and uses operationally (physically) time-independent predicates is more projectible than a hypothesis that is syntactically time-dependent and uses operationally time-independent predicates or a hypothesis that is syntactically time-independent and uses operationally time-dependent predicates. A fourth possibility—a hypothesis that is syntactically time-dependent using time-dependent predicates—must be judged after being transformed into one of the first three forms.

5

Deduction and Observation—*H*-Theorem and Negentropy Principle

5.1. INFORMATION BALANCE IN DEDUCTION

As described at the beginning of Chapter 4, scientific activity consists of two parts, upward and downward, connected at both upper and lower ends. The two parts are induction and deduction and the two connecting endpoints are theory (hypothesis) and observation (experiment). The purpose of Chapters 5 and 6 is a "dynamic" study of the amount of information involved in four processes: induction, deduction, hypotheses evaluation, and observation. The term "dynamic" is loosely used here to indicate that emphasis is laid on the temporal changes of knowledge (acquisition or loss of information) in these processes. Analogy with thermo-"dynamics" may perhaps justify a neologism, "information dynamics."

In the present section and in the first section of Chapter 6 the idea of information balance introduced in Section 3.3 is applied to the processes of deduction and induction, in which three basic elements—hypotheses ($\mathcal{H}$), experimental data ($\mathfrak{D}$), and auxiliary conditions ($\mathcal{A}$)—play the major roles (see Chapter 4). In deduction information about $\mathfrak{D}$ is derived from $\mathcal{H}$ and $\mathcal{A}$, whereas in induction information about $\mathcal{H}$ is derived from $\mathcal{A}$ and $\mathfrak{D}$. In a special case of deduction, in which we fix a hypothesis (H) and consider the past experimental result as the auxiliary condition (A) and the future experimental result of the same observation as the data (D), our consideration of information balance leads naturally to the H-theorem. Section 5.2 provides mathematical proof of different versions of the H-theorem. The H-theorem concerns deductive prediction of "future" experiments on the basis of "past" experiments. An important question, which naturally arises next, pertains to the information obtained at the moment of the observation about which the prediction has been made. This consideration leads to the problem of the relationship between the information acquired (negentropy) by observation and the change of thermodynamic entropy of the physical system observed—a problem first intelligently treated by Brillouin. In the last section of this chapter Brillouin's principle of negentropy is critically reviewed and a generalized and revised version of the principle is suggested. Implications of this principle seem to be more far-reaching than commonly suspected.

In the next chapter, after information balance in the sense of Chapter 3 is considered with regard to the inductive process (Section 6.1), we concentrate on changes in the credibility of hypotheses as a result of an increasing body of evidence. As anticipated in chapter 4, this leads to the inverse H-theorem of induction. Whereas the H-theorem discusses the change of the probability measure on $\mathfrak{D}$ as a result of deduction, the inverse H-theorem discusses the change of the probability measure on $\mathcal{H}$ as a result of induction. Section 6.2 gives a rigorous mathematical proof of the theorem. In the remainder of Chapter 6 we shall show that the inverse H-theoremic trends are not limited to inductive learning but are a general feature of all kinds of learning processes.

For the remainder of the present section we are concerned with the information balance in deductive processes. We start out with three sets of propositions: a set of hypotheses, $\mathcal{H} = \{H_I\}$, $I = 1, 2, \cdots, N$; a set of auxiliary conditions, $\mathcal{A} = \{A_k\}$, $k = 1, 2, \cdots, m$; and a set of experimental outcomes, $\mathfrak{D} = \{D_i\}$, $i = 1, 2, \cdots, n$. Deduction consists of deriving a probabilistic prediction $p(D_i \mid H_I \cap A_k)$ over experimental outcomes on the basis of a hypothesis and an auxiliary condition, whereas induction consists of deriving a credibility distribution $p(H_I \mid D_i \cap A_k)$ over competing hypotheses on the basis of available experimental evidence and known auxiliary conditions. In the case of deduction, therefore, our predictive information has two sources, hypothesis and auxiliary fact. It would be an excitingly new achievement in epistemology if we succeed in stating something quantitative about the amount of information derivable from these two sources. The purpose of the following discussion is precisely an attempt at that. The essence of the results reported here was first published in my paper of 1960 [W-13].

The reader should compare $\mathcal{H}$, $\mathcal{A}$, and $\mathfrak{D}$ of this section with the $\mathfrak{X}$, $\mathfrak{Y}$, and $\mathfrak{Z}$ in Section 3.3 in the following discussion of information balance in deductive inference. We assume an underlying triple-joint probability distribution $p(H_I \cap A_k \cap D_i)$ over $\mathcal{H} \otimes \mathcal{A} \otimes \mathfrak{D}$, from which we can derive three double-joint probability distributions, three single probability distributions, and various conditional probabilities. When we know nothing more than

$p(H_I \cap A_k \cap D_i)$, our ignorance of the outcome of the experiment is given by $S(\mathfrak{D})$, which is defined by the probability $p(D_i) = \sum_I \sum_k p(H_I \cap A_k \cap D_i)$. If we know which of the hypotheses is appropriate, this ignorance changes from $S(\mathfrak{D})$ to $S(\mathfrak{D} \mid H_I)$, which is defined by the probability $p(D_i \mid H_I) = \sum_k p(H_I \cap A_k \cap D_i)/\sum_k \sum_i p(H_I \cap A_k \cap D_i)$; H_I is the hypothesis assumed to be the right one. The difference

$$\inf(\mathfrak{D} \mid H_I) = S(\mathfrak{D}) - S(\mathfrak{D} \mid H_I) \tag{5.1}$$

is the information about the outcome furnished by H_I [see (3.73)].

In addition to this, if we know that A_k among $\{A_k\}$ is the case, the ignorance about the outcome changes to $S(\mathfrak{D} \mid H_I \cap A_k)$, defined by the probability $p(D_i \mid H_I \cap A_k) = p(H_I \cap A_k \cap D_k)/\sum_i p(H_I \cap A_k \cap D_i)$. Hence the information provided by H_I and A_k with regard to the experimental outcome is

$$\inf(\mathfrak{D} \mid H_I \cap A_k) = S(\mathfrak{D}) - S(\mathfrak{D} \mid H_I \cap A_k). \tag{5.2}$$

The quantity

$$\begin{aligned}\inf_I(\mathfrak{D} \mid A_k) &= \inf(\mathfrak{D} \mid H_I \cap A_k) - \inf(\mathfrak{D} \mid H_I)\\ &= S(\mathfrak{D} \mid H_I) - S(\mathfrak{D} \mid H_I \cap A_k)\end{aligned} \tag{5.3}$$

can be interpreted as the additional information provided by A_k when it is already established that H_I is the right hypothesis.

As often discussed in earlier chapters, these quantities (5.1)–(5.3) can become negative. If we want to deal only with positive quantities as information, we can either modify the definition of entropy, as in Appendix A1.1 (see Example 5.2) or take the average values of (5.1), (5.2), and (5.3) with regard to the "conditions" of the conditional probabilities involved. The average of the first of the three quantities is [corresponding to (3.67)]

$$\begin{aligned}\inf(\mathfrak{D} \mid \mathfrak{H}) &= \sum_I p(H_I) \inf(\mathfrak{D} \mid H_I) = S(\mathfrak{D}) - S(\mathfrak{D} \mid \mathfrak{H})\\ &= S(\mathfrak{H}) - S(\mathfrak{H} \mid \mathfrak{D}) \le \min\{S(\mathfrak{D}), S(\mathfrak{H})\} \le S(\mathfrak{H}).\end{aligned} \tag{5.4}$$

This can be interpreted as follows. The source of information is the knowledge that H_I is the right one. The probability of H_I changes from the original value $p(H_I) = \sum_k \sum_i p(H_I \cap A_k \cap D_i)$ to 1, releasing information in the amount $S(\mathfrak{H})$. Of this source information, the amount $S(\mathfrak{H} \mid \mathfrak{D})$ is irrelevant to the prediction of $\mathfrak{D}$ and is not used. The difference $S(\mathfrak{H}) - S(\mathfrak{H} \mid \mathfrak{D})$ is precisely the information $\inf(\mathfrak{D} \mid \mathfrak{H})$, which is used on the average for the prediction of $\mathfrak{D}$ on the basis of $\mathfrak{H}$. $S(\mathfrak{H})$ sets the upper bound to $\inf(\mathfrak{D} \mid \mathfrak{H})$. Of course, if $S(\mathfrak{D})$ is smaller than $S(\mathfrak{H})$, $S(\mathfrak{D})$ supersedes $S(\mathfrak{H})$ as the upper bound. $\inf(\mathfrak{D} \mid \mathfrak{H})$ is the interdependence between $\mathfrak{D}$ and $\mathfrak{H}$ and is non-negative according to (2.11).

In exactly the same fashion, we can decompose the average of $\inf(\mathfrak{D} \mid H_I \cap A_k)$ as follows:

$$\inf(\mathfrak{D} \mid \mathcal{H} \otimes \mathcal{A})$$
$$= \sum_I \sum_k p(H_I \cap A_k) \inf(\mathfrak{D} \mid H_I \cap A_k) = S(\mathfrak{D}) - S(\mathfrak{D} \mid \mathcal{H} \otimes \mathcal{A})$$
$$= S(\mathcal{H} \otimes \mathcal{A}) - S(\mathcal{H} \otimes \mathcal{A} \mid \mathfrak{D}) \leq \min\{S(\mathfrak{D}), S(\mathcal{H} \otimes \mathcal{A})\} \leq S(\mathcal{H} \otimes \mathcal{A}), \tag{5.5}$$

where $S(\mathcal{H} \otimes \mathcal{A})$ is the source information provided by the hypothesis and the auxiliary condition of which the irrelevant part $S(\mathcal{H} \otimes \mathcal{A} \mid \mathfrak{D})$ is wasted, leaving useful information $\inf(\mathfrak{D} \mid \mathcal{H} \otimes \mathcal{A})$ for the prediction of $\mathfrak{D}$. Equality in both occurrences of $\leq$ in (5.5) holds if and only if $S(\mathcal{H} \otimes \mathcal{A} \mid \mathfrak{D}) = 0$; that is, if and only if the $p(H_I \cap A_k \mid D_i)$ are 0 or 1 for all I, all k, and all i. $\inf(\mathfrak{D} \mid \mathcal{H} \otimes \mathcal{A})$, being the interdependence between $\mathfrak{D}$ and $\mathcal{H} \otimes \mathcal{A}$, is non-negative.

As regards the third quantity (5.3), we first note that it can be rewritten as

$$\inf_I(\mathfrak{D} \mid A_k) = S_I(\mathfrak{D}) - S_I(\mathfrak{D} \mid A_k). \tag{5.6}$$

The subscript I indicates the probability measure on $\mathcal{A} \otimes \mathfrak{D}$ defined by $p_I(A_k \cap D_i) = p(H_I \cap A_k \cap D_i)/p(H_I)$. In other words, $S_I(\mathfrak{D})$ is defined by $p_I(D_i) = p(H_I \cap D_i)/p(H_I)$, and $S_I(\mathfrak{D} \mid A_k)$ is defined by $p_I(A_k \cap D_i)/p_I(A_k) = p(H_I \cap A_k \cap D_i)/p(H_I \cap A_k)$. Fixing I, but averaging over A_k with the help of $p_I(A_k) = p(H_I \cap A_k)/p(H_I)$, we obtain

$$\inf_I(\mathfrak{D} \mid \mathcal{A}) = S_I(\mathfrak{D}) - S_I(\mathfrak{D} \mid \mathcal{A}) = S_I(\mathcal{A}) - S_I(\mathcal{A} \mid \mathfrak{D}) \leq S_I(\mathcal{A}). \tag{5.7}$$

This shows that when we fix the hypothesis, the information provided by $\mathcal{A}$ is on the average $S_I(\mathcal{A})$, of which $S_I(\mathcal{A} \mid \mathfrak{D})$ is unused, leaving the relevant information $\inf_I(\mathfrak{D} \mid \mathcal{A})$. The last equality in (5.7) holds if and only if $S_I(\mathcal{A} \mid \mathfrak{D}) = 0$; that is, if and only if the $p_I(A_k \mid D_i)$ are 0 or 1 for all k and i for a given I.

If we further average (5.7) with respect to H_I with the help of $p(H_I)$, we obtain [corresponding to (3.68)]

$$\begin{aligned} \langle \inf_I(\mathfrak{D} \mid \mathcal{A}) \rangle_I &= S(\mathfrak{D} \mid \mathcal{H}) - S(\mathfrak{D} \mid \mathcal{H} \otimes \mathcal{A}) \\ &= S(\mathcal{A} \mid \mathcal{H}) - S(\mathcal{A} \mid \mathcal{H} \otimes \mathfrak{D}) \leq S(\mathcal{A} \mid \mathcal{H}). \end{aligned} \tag{5.8}$$

The equals sign in $\leq$ holds if and only if $S(\mathcal{A} \mid \mathcal{H} \otimes \mathfrak{D}) = 0$; that is, if and only if $p(A_k \mid H_I \cap D_i)$ is 0 or 1 [note that $p(A_k \mid H_I \cap D_i) = p_I(A_k \mid D_i)$]. If we add (5.4) and (5.8), we come back to (5.5), with

$$\inf(\mathfrak{D} \mid \mathcal{H}) + \langle \inf_I(\mathfrak{D} \mid \mathcal{A}) \rangle_I = \inf(\mathfrak{D} \mid \mathcal{H} \otimes \mathcal{A}). \tag{5.9}$$

This last equation is the average of the first half of the definition (5.3), namely, $\inf(\mathcal{D} \mid H_I \cap A_k) = \inf(\mathcal{D} \mid H_I) + \inf_I(\mathcal{D} \mid A_k)$.

In some practical applications, depending on the meanings assigned to $\mathcal{H}$, $\mathcal{A}$, and $\mathcal{D}$, $\inf(\mathcal{D} \mid H_I)$ becomes independent of I. In such a case $\inf(\mathcal{D} \mid H_I)$ becomes equal to $\inf(\mathcal{D} \mid \mathcal{H})$, whose upper bound is $S(\mathcal{H})$. In some problems $\inf_I(\mathcal{D} \mid A_k)$ does not depend on A_k; in such cases $\inf_I(\mathcal{D} \mid A_k)$ becomes equal to $\inf_I(\mathcal{D} \mid \mathcal{A})$, whose upper bound is $S_I(\mathcal{A})$. This $S_I(\mathcal{A})$ is often independent of I, and then is equal to $S(\mathcal{A} \mid \mathcal{H})$. In a particular class of "deterministic" cases, the hypothesis and the auxiliary condition provide sufficient information to select one and only one outcome. This happens when $\inf(\mathcal{D} \mid H_I \cap A_k)$ is just sufficient to cancel the original ignorance $S(\mathcal{D})$. This happens, of course, if and only if $S(\mathcal{D} \mid H_I \cap A_k) = 0$; then we have $S(\mathcal{D}) = \inf(\mathcal{D} \mid H_I) + \inf_I(\mathcal{D} \mid A_k)$. Logicians' discussions about deduction correspond to this particular case (see Section 4.1).

It may be worthwhile to insert at this point a general remark regarding the roles of "laws" and "auxiliary conditions" in determining the development of a natural phenomenon. Very often in various applications of information theory, particularly in the nonphysical sciences, people seem to forget to incorporate the role of $\inf(\mathcal{D} \mid H)$ in their consideration and are surprised to find that $S_H(\mathcal{A})$ is much smaller than $S(\mathcal{D})$ (which is of the order of $\log n$). To their eyes, this appears to imply that a choice out of a tremendous number (n) of possibilities is determined by a very small amount of input information $S_H(\mathcal{A})$. Actually, $\inf(\mathcal{D} \mid H)$, which is sometimes difficult to evaluate, must also be included in the input information. If a living cell, for instance, grows into a large and complex biological entity, the entire specification of this complex entity need not be encoded in the original cell because a large portion of information can come from $\inf(\mathcal{D} \mid H)$, which lies in the environment or in nature itself. This remark, although rather elementary, may shed some new light on many similar problems (see Example 5.1). (See also a paper entitled "Where does the Information Come from?" [W-26].)

At the end of Section 4.2 we spoke of "generic propositions," G. It is clear that a generic proposition can be regarded as a special type of hypothesis. A generic hypothesis G does not specify exact values of $p(D_i \mid G)$, but states whether $p(D_i \mid G) = 0$ or $\neq 0$. If G is true, one of those outcomes belonging to a certain subclass, say, $\mathcal{G}$, of $\mathcal{D}$ may happen, but any of those outcomes belonging to its complement $\mathcal{D} - \mathcal{G}$ $(= \mathcal{D} \cap \neg\mathcal{G})$ is certain not to happen. The conditional probability $p(D_i \mid G)$ in this discussion may be interpreted as $p(D_i \mid G) = \sum_k p(G \cap A_k \cap D_i)/\sum_{k,i} p(G \cap A_k \cap D_i)$. To make a quantitative discussion possible, it is sometimes convenient to assume

$$p(D_i) = \frac{1}{n}, \qquad D_i \in \mathcal{D} \tag{5.10}$$

and

$$p(D_i \mid G) = \frac{1}{m} \quad \text{for} \quad D_i \in \mathcal{G}$$
$$= 0 \quad \text{for} \quad D_i \in D - \mathcal{G}, \tag{5.11}$$

where n is the number of D's in $\mathcal{D}$ and m is the number of D's in $\mathcal{G}$.

The information provided by G regarding the outcome is

$$\inf(\mathcal{D} \mid G) = S(\mathcal{D}) - S(\mathcal{D} \mid G), \tag{5.12}$$

which becomes

$$\inf(\mathcal{D} \mid G) = \log \frac{n}{m} \tag{5.13}$$

under the assumption of (5.10) and (5.11). If we apply (5.4) to this problem, we can state that the average of $\inf(\mathcal{D} \mid G)$ is not larger than n, because there can be 2^n different G's. But this upper bound $S(\mathcal{H})$ is unimportant, because the other upper bound $S(\mathcal{D}) = \log n < n$. An easy intuitive meaning of this formula (5.13) is that the more restrictive the generic hypothesis (i.e., the smaller the m), the larger will be the information it provides. In this sense. $\inf(\mathcal{D} \mid G)$ may be called the "predictive power" of the generic hypothesis. By extension, even if the hypothesis is not generic, its predictive power may be defined by $\inf(\mathcal{D} \mid H)$ of (5.1).

Coming back to the case of generic hypotheses, we may note that $S(\mathcal{D} \mid G)$, which is $\log m$, is a measure of the extension of cases (D's) comprehended in G. For this reason $S(\mathcal{D} \mid H)$ may be called "comprehension" of H whether or not H is generic. Thus (5.12) or (5.1) can be interpreted as meaning

$$\text{predictive power} + \text{comprehension} = \text{constant},$$

where

$$\text{comprehension} = S(\mathcal{D} \mid H) \quad \text{and} \quad \text{predictive power} = \inf(\mathcal{D} \mid H). \tag{5.14}$$

In (5.14) "constant" [which is $S(\mathcal{D})$] means "independent of H."

If we interpret G as a statement that an object belongs to the class ($\mathcal{G}$) corresponding to a concept, the predictive power and comprehension can be respectively interpreted as the intension and extension of the concept:

$$\text{intension} + \text{extension} = \text{constant}. \tag{5.15}$$

For this matter, see Sections 7.5 and 8.3.

To avoid a possible misunderstanding, it must be made clear that the "predictive power" of a hypothesis, or the "information provided" by a hypothesis, does not mean the "information content" of a hypothesis. The latter is to be considered infinite if the hypothesis is, as it is usually, a general statement. This may be understood in the following simple example. Suppose that the hypothesis says that all ravens are black. This corresponds to an observation determining that the color of a raven is black (D_1) or is nonblack (D_2). If we can assume that D_1 and D_2 have equal probability before our knowledge of the hypothesis, then the hypothesis gives us information in the amount of 1 bit. For we have in this case $n = 2$ and $p(D_1 \mid H) = 1$, $p(D_2 \mid H) = 0$ in (5.1). Thus the predictive power is 1 bit. However, the hypothesis tells us more than just this. Suppose that we observe the colors of two ravens; then there are *a priori* four possibilities, each having *a priori* probability $\frac{1}{4}$. But the hypothesis chooses only one of them with equal probability. Thus the hypothesis provides information in the amount of 2 bits. In a similar way, for the observation of ν ravens, the hypothesis provides ν bits of information. By the nature of a general statement, however, this ν can be infinite. Hence the hypothesis can provide an infinite amount of information. This may be rephrased as a general assertion that the hypothesis "contains" an infinite amount of information. The argument we used was a categorical dichotomic case, but it is easy to extend the argument to a general case. (See Section 5.4 for further discussion of related problems.)

Now let us give two examples of information balance in deductive inference.

Example 5.1. Difference Equation and Initial Condition. Suppose that we are given a differential equation $d^r y/dx^r = 0$ and are asked to solve it by the initial conditions. We then have to know r constants, $(d^s y/dx^s)_{x=0} = \alpha^s$, $s = 0, 1, 2, \cdots, r - 1$. The solution picks one out of a continuous number of possible curves in the xy plane. This selection is made possible by the information provided by the hypothesis $d^r y/dr^r = 0$ combined with the information provided by the auxiliary conditions $(d^s y/dx^s)_{x=0} = \alpha^s$. It will be very interesting if we can quantitatively separate the portion of information provided by the hypothesis and the portion provided by the auxiliary conditions. We shall see that the former is overwhelmingly larger than the latter, as anticipated in the earlier part of this section. The conclusion that can be derived from the hypothesis alone is bound to be a generic one. If the equation has $r = 2$, for instance, the allowed solutions form a set of straight lines. With the help of the initial condition, the conclusion becomes categorical in the sense that one and only one straight line is specified.

The problem as such is outside the scope of our theory because it involves a continuous number of possibilities. Now, we set up an exact analogue of the problem within the domain of finite possibilities and calculate the information balance with respect to this simplified problem. This will allow us to infer the situation in a continuous case by a limiting process.

Instead of unbounded continuous variables x and y let us take a finite domain $0 \leq x \leq P - 1$ and $0 \leq y \leq P - 1$ of integer variables x and y, where P is a prime number. (This last assumption is introduced for convenience of calculation.) The objects considered are single-valued functions y of x; that is, we assume that there is one and only one value of y for each value of x. There are $n = P^P$ such objects (curves), since the value of y for each x can be any one out of P possibilities. Hence the number of possible hypotheses (concepts) is $2^{P^P} - 1$, excluding the hypothesis with zero comprehension. From any given curve $y = f(x)$, we can calculate "derivatives" by

$$f'(x) = f(x + 1) - f(x),$$

$$f''(x) = f'(x + 1) - f'(x) = f(x + 2) - 2f(x + 1) + f(x),$$

and

$$f^{(r)}(x) = \sum_{s=0}^{r} (-1)^{r-s} \binom{r}{s} f(x + s). \tag{5.16}$$

Since $f(x)$ is defined only for $0 \leq x \leq P - 1$, the rth derivative is defined only for x such that

$$0 \leq x \leq (P - 1) - r. \tag{5.17}$$

In other words, at x, derivatives of orders higher than $(P - 1) - x$ do not exist. Although y is defined as non-negative, the derivatives as defined in (5.16) can be positive, zero, or negative. With the help of these derivatives we can obtain a "Taylor expansion."

Consider the expression

$$Y \equiv \sum_{r=0}^{u} f^{(r)}(x_0) \binom{u}{r} = \sum_{r=0}^{u} \sum_{s=0}^{r} (-1)^{r-s} \binom{u}{r} \binom{r}{s} f(x_0 + s), \tag{5.18}$$

where the expression of $f^{(r)}(x)$ given in (5.16) is used. Noting that

$$\sum_{r=0}^{u} \sum_{s=0}^{r} = \sum_{s=0}^{u} \sum_{r=s}^{u} \quad \text{and} \quad \binom{u}{r}\binom{r}{s} = \binom{u - s}{r - s}\binom{u}{s},$$

which becomes 1 when $u = s$, we can rewrite (5.18) as

$$Y = \sum_{s=0}^{u} \sum_{r=s}^{u} (-1)^{r-s} \binom{u - s}{r - s} \binom{u}{s} f(x_0 + s). \tag{5.19}$$

Changing the variable r to t, which is defined by $t = r - s$, we can further rewrite

$$Y = \sum_{s=0}^{u} \binom{u}{s} f(x_0 + s) \sum_{t=0}^{u-s} (-1)^t \binom{u - s}{t}. \tag{5.20}$$

But, by the binomial expansion, we have

$$0 = (1 - 1)^{u-s} = \sum_{t=0}^{u-s} (-1)^t \binom{u - s}{t}$$

insofar as $u > s$. Hence only the term for $u = s$ remains in (5.20); that is,

$$Y = f(x_0 + u). \tag{5.21}$$

This gives the Taylor expansion for $x = x_0 + u \geq x_0$:

$$\begin{aligned} f(x) &= \sum_{r=0}^{x-x_0} f^{(r)}(x_0) \binom{x - x_0}{r} \\ &= f(x_0) + \frac{f'(x_0)}{1!}(x - x_0) + \frac{f''(x_0)}{2!}(x - x_0)(x - x_0 - 1) + \cdots \\ &\quad + \frac{f^{(r)}(x_0)}{r!}(x - x_0)(x - x_0 - 1) \cdots (x - x_0 - r + 1) + \cdots \\ &\quad + \frac{f^{(x-x_0)}(x_0)}{(x - x_0)!}(x - x_0)(x - x_0 - 1) \cdots (1). \end{aligned} \tag{5.22}$$

At x_0 there exists only up to the $[(P - 1) - x_0]$th derivative, inclusive; hence, even for $x = P - 1$, the expansion ends with the highest existing derivative. If x is less than $(P - 1)$, the expression terminates before coming to the term involving the highest derivative. The value of $f(x)$ in (5.22) must lie in the domain $[0, P - 1]$. Because the binomial coefficient $\binom{x - x_0}{r}$, with $x - x_0 \geq r \geq 0$, is an integer, we see from (5.18) that we can add any multiple of P to each derivative $f^{(r)}(x_0)$ without changing the value of $f(x)$, if we understand the addition in (5.22) as an addition modulo P and require $0 \leq f(x) \leq P - 1$. Hereafter we use mainly the expansion for $x_0 = 0$:

$$\begin{aligned} f(x) = f(0) + \frac{f'(0)}{1}x + \frac{f''(0)}{2!}x(x - 1) + \cdots \\ + \frac{f^{(x)}(0)}{x!}x(x - 1) \cdots 2 \cdot 1, \qquad (\text{mod } P), \end{aligned} \tag{5.23}$$

where each derivative is considered to be determined up to an arbitrary additive constant that is a multiple of P. The curve in Figure 5.1 can be expressed in the form of (5.23), with $P = 7$ and $f(0) = 2$, $f^{(1)}(0) = 1$, $f^{(2)}(0) = 0$, $f^{(3)}(0) = 5$, $f^{(4)}(0) = 6$, and $f^{(r)}(0) = 0$ for $r \geq 5$. Figure 5.2 shows how the derivatives can be computed from the curve.

We now need the following theorem, whose proof is given in Appendix A5.1.

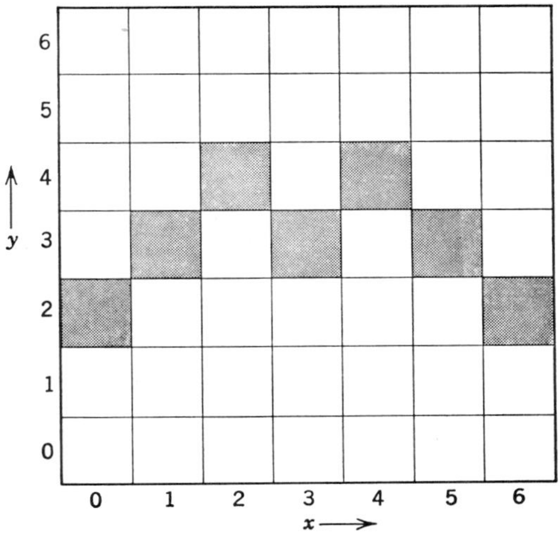

Figure 5.1 This M-shaped curve can be expressed as $y(x) = 2 + x + 2x(x - 1)(x - 2) + 2x(x - 1)(x - 2)(x - 3) \pmod 7$.

x	y	$y^{(1)}$	$y^{(2)}$	$y^{(3)}$	$y^{(4)}$	$y^{(5)}$	$y^{(6)}$
0	2	1	0	5	6	0	0
1	3	1	5	4	6	0	
2	4	6	2	3	6		
3	3	1	5	2			
4	4	6	0				
5	3	6					
6	2						

Figure 5.2 The values of the derivatives are calculated from the values of $y(x)$ taken from Figure 5.1. The numbers below the stair-shaped heavy line are not necessary if the law $y^{(5)}(x) = 0$ is known.

Theorem 5.1 Each single-valued function y of x with $0 \leq x \leq P - 1$, $0 \leq y \leq P - 1$, corresponds one to one to an expression

$$f(x) = b_0 + b_1 x + b_2 x(x-1) + \cdots + b_{P-1}x(x-1)\cdots(x-P+2)$$
$$= \sum_{r=0}^{P-1} b_r x(x-1)\cdots(x-r+1), \qquad (\text{mod } P), \tag{5.24}$$

where the constant coefficients b_r are in the domain $0 \leq b_r \leq P - 1$.

We stated already that there are P^P possible curves. This fact is now reflected in the fact that there are P coefficients b_r $(r = 0, 1, 2, \cdots, P-1)$, each of which can take any of P possible values, $0, 1, 2, \cdots, P-1$. We now simulate the equation $d^s y/dx^s = 0$ by $f^{(s)}(x) = 0$ and the initial condition $(d^r y/dx^r)_{x=0} = \alpha_r$ by $f^{(r)}(0) = \alpha_r$, $r = 0, 1, \cdots, s-1$. The equation limits the number of possible curves, and the initial condition selects one out of these allowed curves. First, we note that if we assume the "law" $f^{(s)}(x) = 0$ $[0 \leq x \leq (P-1) - s]$, we shall have $f^{(t)}(x) = 0$ $[0 \leq x \leq (P-1) - t]$ for $s \leq t \leq P - 1$. Hence, by substituting $f^{(t)}(0) = 0$ $[s \leq t \leq P-1]$ in (5.23), we obtain $b_s = b_{s+1} = \cdots = b_{P-1} = 0$ in (5.24). This leaves s coefficient $b_0, b_1, \cdots, b_{s-1}$ free, corresponding to P^s possible curves. These are exactly the coefficients determined by the initial conditions, $f^{(r)}(0) = \alpha_r$, $r = 0, 1, \cdots, s-1$. One might wonder whether we used the equation to the full extent, because $f^{(s)}(x) = 0$ refers to different x's whereas we seem to have used it only at $x = 0$. This is not actually so, since the x in $f^{(s)}(x) = 0$ are allowed to take $P - s$ values $[0 \leq x \leq (P-1) - s]$, and the $P - s$ relations $f^{(t)}(x) = 0$ for $s \leq t \leq P - 1$, which we used, are derived from them. The curve in Figure 5.1 can be considered as one of the 7^5 possible solutions of the equation $f^{(5)}(x) = 0$.

Let us now turn to the problem of information balance. The set $\mathcal{H}$ of hypotheses consists of $P + 1$ members H_s: $f^{(s)}(x) = 0$, $s = 0, 1, 2, \cdots, P$. The set $\mathcal{A}$ of auxiliary conditions consists of P^P possible initial conditions A_k: $(b_0, b_1, \cdots, b_P)$, $k = 1, 2, \cdots, P^P$. The set $\mathcal{D}$ of observed outcomes consists of P^P different curves D_i, $i = 1, 2, \cdots, P^P$. Let us designate by $\mathcal{G}_s$ the subset of curves allowed by H_s. $\mathcal{G}_s$ consists of P^s curves. As we have seen, initial conditions and curves correspond one to one. Let us write $A_k \sim D_i$ to denote this correspondence. Hence some initial conditions are compatible with a given hypothesis, but some are not. Let us designate by $\mathcal{B}_s$ the subset of $\mathcal{A}$ consisting of those A's that are compatible with hypothesis H_s. If A_k: $(b_0, b_1, \cdots, b_P)$ is such that $b_{t-1} \neq 0$ and $b_t = b_{t+1} = \cdots = b_P = 0$, $A_k \in \mathcal{B}_s$ for all $s \geq t$ and $A_k \notin \mathcal{B}_s$ for all $s < t$. The relations $D_i \in \mathcal{G}_s$ and $A_k \sim D_i$ imply $A_k \in \mathcal{B}_s$, but $D_i \in \mathcal{G}_s$ and $A_k \in \mathcal{B}_s$ do not necessarily imply $D_i \sim A_k$. For a given D_i let t be such that $D_i \in \mathcal{G}_s$ for $t \leq s$ and $D_i \notin \mathcal{G}_s$ for $s < t$. Then there are $(P - t + 1)$ hypotheses that are satisfied by D_i. There are N_t curves D_i that have the same t, where $N_t = P^t - P^{t-1}$ for $t = 1, 2, \cdots$, P and $N_0 = 1$.

The most natural probability measure on $\mathcal{H} \otimes \mathcal{A} \otimes \mathcal{D}$ would be such that $p(H_s)$ is proportional to the number of curves P^s (2 to the exponent "comprehension" of H_s) covered by H_s, and that $p(A_k \mid H_s)$ is $1/P^s$ insofar as $A_k \in \mathcal{B}_s$ and 0 if $A_k \notin \mathcal{B}_s$. Such a probability measure is defined by

$$p(H_s \cap A_k \cap D_i) = \frac{P-1}{P^{P+1}-1} \quad \text{if} \quad A_k \in \mathcal{B}_s \quad \text{and} \quad A_k \sim D_i$$
$$= 0 \quad \text{otherwise.} \tag{5.25}$$

All the other probabilities can be derived from (5.25).

The information regarding $\mathcal{D}$ provided by the fact that hypothesis H_s is the applicable one is $\inf(\mathcal{D} \mid H_s) = S(\mathcal{D}) - S(\mathcal{D} \mid H_s)$, and the information provided by A_k when H_s is already chosen is given by $\inf_s(\mathcal{D} \mid A_k) = S_s(\mathcal{D}) - S_s(\mathcal{D} \mid A_k)$. Since $S(\mathcal{D} \mid H_s) = S_s(\mathcal{D})$ in general and since (5.25) is such that $S_s(\mathcal{D} \mid A_k) = 0$, we have

$$\inf(\mathcal{D} \mid H_s) + \inf_s(\mathcal{D} \mid A_k) = \inf(\mathcal{D} \mid H_s \cap A_k) = S(\mathcal{D}), \tag{5.26}$$

which expresses the fact, which we already knew, that a unique curve is determined by H_s and A_k. The original ignorance $S(\mathcal{D})$, which is canceled by $\inf(\mathcal{D} \mid H_s \cap A_k)$, is $S(\mathcal{D}) = \log(P^{P+1} - 1) - \log(P-1) - [(P-1)/(P^{P+1}-1)] \sum_{t=0}^{P} (P - t + 1)N_t \log(P - t + 1)$. $\inf_s(\mathcal{D} \mid A_k)$, which is the same as $S(\mathcal{D} \mid H_s)$ here, is $S_s(\mathcal{D}) = s \log P$.

We should note that for large P, $\inf(\mathcal{D} \mid H_s)$ will be of the order of $P \log P$ whereas $\inf_s(\mathcal{D} \mid A_k)$ will be of the order of $s \log P$. This result is not sensitive to the choice of the probability measure adopted, because for large P, $\inf(\mathcal{D} \mid H_s)$ is essentially the logarithm of the number of possible curves minus the logarithm of the number of curves allowed by the hypothesis, whereas $\inf_s(\mathcal{D} \mid A_k)$ is essentially equal to the latter logarithm, which is much smaller than the former. This indicates that in the continuous case the information from the "law" is overwhelmingly larger than the information from the initial condition.

The average of the information provided by the hypotheses is $\inf(\mathcal{D} \mid \mathcal{H}) = S(\mathcal{D}) - S(\mathcal{D} \mid \mathcal{H}) = S(\mathcal{H}) - S(\mathcal{H} \mid \mathcal{D}) \leq S(\mathcal{H})$. In the present case we have the wasted information

$$S(\mathcal{H} \mid \mathcal{D}) = [(P-1)/(P^{P+1}-1)] \sum_{t=0}^{P} (P - t + 1)N_t \log(P - t + 1),$$

which is not zero; hence the source information $S(\mathcal{H})$ is not used with perfect efficiency, where

$$S(\mathcal{H}) = \log(P^{P+1} - 1) - \log(P-1)$$
$$- \{P(P^{P+2} - P^{P+1} - P^P + 1)/[(P^{P-1} - 1)(P-1)]\} \log P.$$

As far as the second stage, regarding the information provided by A_k, is concerned, we have $\inf_s(\mathcal{D} \mid A_k) = S_s(\mathcal{D}) - S_s(\mathcal{D} \mid A_k) = s \log P - 0 = s \log P$, which does not depend on A_k. Hence it is the same as its average with respect to A_k, which is $\inf_s(\mathcal{D} \mid \mathcal{A}) = S_s(\mathcal{D}) - S_s(\mathcal{D} \mid \mathcal{A}) = S_s(\mathcal{A}) - S_s(\mathcal{A} \mid \mathcal{D}) \leq S_s(\mathcal{A})$. Since $S_s(\mathcal{A} \mid \mathcal{D})$ is obviously zero (A_k is uniquely determined by D_i), there is no waste of information. The source information $S_s(\mathcal{A}) = s \log P$ is exhaustively exploited to supply $\inf_s(\mathcal{D} \mid \mathcal{A}_k) = s \log P$.

The waste of information in this case stems from the fact that a single curve belongs in general to more than one hypothesis. The reader can check this by assuming a probability measure such that $p(A_k \mid H_s) \neq 0$ only for those A_k that satisfy H_s but not H_u, with $u < s$. There are N_s such A_k's (hence curves), where $N_s = P^s - P^{s-1}$ for $s > 0$ and $N_0 = 1$. If $p(H_s)$ is proportional to N_s and the nonvanishing $p(A_k \mid H_s)$ are all equal to $1/N_s$, then $\inf(\mathcal{D} \mid \mathcal{H}) = S(\mathcal{H}) = -\sum_s (N_s/P^P) \log (N_s/P^P)$, and there is no loss in the first stage. The second stage is described by $\inf_s(\mathcal{D} \mid A_k) = S_s(\mathcal{A}) = \log N_s$, which shows no loss either. Equation (5.26) holds here, too, and the prediction is deterministic because the original ignorance $S(\mathcal{D}) = P \log P$ is precisely canceled by the sum of $\inf(\mathcal{D} \mid H_s) = \log (P^P/N_s)$ and $\inf_s(\mathcal{D} \mid A_k) = \log N_s$. For large P we again have $\inf(\mathcal{D} \mid H_s) = P \log P \gg s \log P = \inf_s(\mathcal{D} \mid A_k)$.

Example 5.2. Aged Markov Chain. The weather on each day at a certain location is described as either fair or nonfair, and the hypothesis H is that the succession of these two cases is governed by a Markov chain that gives transition probability p from fair to nonfair and q from nonfair to fair, where p and q are not both 0. The hypothesis further assumes that the Markov chain started a very long time ago so that it can be considered at present as an "aged" Markov chain (see below for a quantitative definition of this). We are interested in predicting tomorrow's weather; that is, the set of observational results $\mathcal{D} = \{D_1, D_2\}$ represents tomorrow's weather, where D_1 is fair and D_2 is nonfair. Our auxiliary knowledge is today's weather; thus there are two A's, A_1 meaning that today is fair and A_2 meaning that today is nonfair. What is the amount of information $\inf_H(\mathcal{D} \mid A_i)$ provided by A_i ($i = 1, 2$) in predicting tomorrow's weather? Is it equal to the average information $S_H(\mathcal{A})$ provided by today's weather as such?

That the Markov chain is aged means that the probability a that an arbitrary day will be fair and the probability $b = 1 - a$ that an arbitrary day will be nonfair are given as the solution of the equation

$$\begin{pmatrix} 1-p & q \\ p & 1-q \end{pmatrix} \begin{pmatrix} a \\ b \end{pmatrix} = \begin{pmatrix} a \\ b \end{pmatrix}, \tag{5.27}$$

which gives

$$a = \frac{q}{(p+q)} \quad \text{and} \quad b = \frac{p}{(p+q)}. \tag{5.28}$$

The probability that today is fair and tomorrow will also be fair is given by $p_H(D_1 \cap A_1) = a(1-p)$. Similarly, $p_H(D_2 \cap A_1) = ap$, $p_H(D_1 \cap A_2) = bq$, and $p_H(D_2 \cap A_2) = b(1-q)$. From these relations follow

$$p_H(D_1) = \sum_{i=1}^{2} p_H(D_1 \cap A_i) = a(1-p) + bq = a = p_H(A_1)$$

and

$$p_H(D_2) = \sum_{i=1}^{2} p_H(D_2 \cap A_i) = ap + b(1-q) = b = p_H(A_2).$$

This is what should be the case according to the definition (5.27). The p and q are the conditional probabilities $p_H(D_i \mid A_k)$, namely, $p_H(D_2 \mid A_1) = p$ and $p_H(D_1 \mid A_2) = q$. The probabilities so far introduced do not alone determine $p(D_i) = \sum_H p_H(D_i)p(H)$; indeed, we have not said anything about competing hypotheses. For the sake of concreteness, we may, however, assume $p(D_1) = p(D_2) = \frac{1}{2}$. Then the original ignorance about $\mathfrak{D}$ without knowledge that a particular hypothesis is correct will be $S(\mathfrak{D}) = 1$ bit. Hence the information about $\mathfrak{D}$ brought about by H is

$$\begin{aligned}\inf(\mathfrak{D} \mid H) &= S(\mathfrak{D}) - S_H(\mathfrak{D}) = 1 + a \log a + b \log b \\ &= 1 + \frac{q}{p+q} \log \frac{q}{p+q} + \frac{p}{p+q} \log \frac{p}{p+q}. \end{aligned} \tag{5.29}$$

Under the assumption of H, the information about tomorrow's weather supplied by the fact that today is fair is now

$$\begin{aligned}\inf_H(\mathfrak{D} \mid A_1) &= S_H(\mathfrak{D}) - S_H(\mathfrak{D} \mid A_1) \\ &= -\frac{q}{p+q} \log \frac{q}{p+q} - \frac{p}{p+q} \log \frac{p}{p+q} \\ &\quad + p \log p + (1-p) \log (1-p) \end{aligned} \tag{5.30}$$

and the information about tomorrow's weather supplied by the fact that today is nonfair is

$$\begin{aligned}\inf_H(\mathfrak{D} \mid A_2) &= S_H(\mathfrak{D}) - S_H(\mathfrak{D} \mid A_2) \\ &= -\frac{q}{p+q} \log \frac{q}{p+q} - \frac{p}{p+q} \log \frac{p}{p+q} \\ &\quad + q \log q + (1-q) \log (1-q). \end{aligned} \tag{5.31}$$

The peculiarity of the aged Markov chain is that

$$S_H(\mathfrak{D}) = S_H(\mathcal{A}). \tag{5.32}$$

Hence we have in this case

$$\inf_H (\mathfrak{D} \mid A_i) = S_H(\mathfrak{D}) - S_H(\mathfrak{D} \mid A_i) \leq S_H(\mathfrak{D}) = S_H(\mathcal{A}) \qquad (5.33)$$

in addition to the general rule given in (5.7),

$$\inf_H (\mathfrak{D} \mid \mathcal{A}) = \langle \inf_H (\mathfrak{D} \mid A_i) \rangle_i \leq S_H(\mathcal{A}), \qquad (5.34)$$

where $S_H(\mathcal{A})$ is the source information carried by observation of today's weather. So much for the upper bound of $\inf_H (\mathfrak{D} \mid A_i)$.

As far as the lower bound of $\inf_H (\mathfrak{D} \mid A_i)$ is concerned, we may note that $\inf_H (\mathfrak{D} \mid A_i)$ can become negative. For instance, if $p = \frac{1}{2}$ and $q \neq \frac{1}{2}$, $\inf_H (\mathfrak{D} \mid A_1)$ is obviously negative. If we impose double stochasticity on the matrix (5.27), that is, if $\sum_j p_H(D_i \mid A_j) = 1$, then $p = q$ and $\inf_H (\mathfrak{D} \mid A_1)$ as well as $\inf_H (\mathfrak{D} \mid A_2)$ becomes non-negative, since $S_H(\mathfrak{D}) = 1$ and $S_H(\mathfrak{D} \mid A_i) \leq 1$. If $p + q = 1$, then $\inf_H (\mathfrak{D} \mid A_1) = \inf_H (\mathfrak{D} \mid A_2) = 0$. This is because $p_H(D_i) = p_H(D_i \mid A_j)$ if $p + q = 1$. In any event, we know that the average of $\inf_H (\mathfrak{D} \mid A_i)$ is non-negative:

$$\inf_H (\mathfrak{D} \mid \mathcal{A}) = S_H(\mathfrak{D}) - S_H(\mathfrak{D} \mid \mathcal{A}) = S_H(\mathcal{A}) - S_H(\mathcal{A} \mid \mathfrak{D}) \geq 0. \qquad (5.35)$$

This means that on the average the knowledge of today's weather sharpens the prediction of tomorrow's weather. It should be noted, however, that even if $\inf (\mathfrak{D} \mid A_1) < 0$, that is, even if the prediction about tomorrow's weather becomes less sharp because of A_1, it does not mean at all that a prediction that takes today's weather into account is worse in quality than one that does not. On the contrary, it is obviously better to take today's weather into account. For if we take a large number of pairs of consecutive days starting with a fine day, the probability distribution of the second day will be $1 - p$ and p instead of a and b.

From the point of view of relative entropy (see Appendix A1.1) the above argument is based on the *a priori* probability $\frac{1}{2}$ and $\frac{1}{2}$ for fair and nonfair days. If we use a and b as the *a priori* probabilities for fair and nonfair days, then the relative entropy function for any probability distribution π_1 and $\pi_2 = 1 - \pi_1$ for fair and nonfair days, with reference to the probability distribution a and b, must be defined as

$$S^* = -\pi_1 \log \frac{\pi_1}{2a} - \pi_2 \log \frac{\pi_2}{2b}. \qquad (5.36)$$

According to (1.29) or (A1.1.2) as well as this standardization, $S^*_H(\mathfrak{D})$ becomes 1 instead of $-a \log a - b \log b$. Similarly,

$$S^*_H(\mathfrak{D} \mid A_1) = -(1 - p) \log \frac{1 - p}{2a} - p \log \frac{p}{2b}. \qquad (5.37)$$

As a result, we obtain

$$\inf^* (\mathfrak{D} \mid A_1) = 1 + (1 - p) \log \frac{1 - p}{2a} + p \log \frac{p}{2b}, \qquad (5.38)$$

which is always non-negative and becomes zero if and only if $(1 - p) = a$ and $p = b$, that is, $p + q = 1$. The same is true for $\inf (\mathfrak{D} \mid A_2)$; hence it is also true for $\langle \inf^* (\mathfrak{D} \mid A_i) \rangle_i$.†

5.2. MATHEMATICAL PROOFS OF *H*-THEOREMS AND THE MARKOV CHAIN

In Example 5.2 we regarded $\mathcal{A}$ and $\mathfrak{D}$ as experimental results of the same kind of observation, one referring to an earlier instant and the other to a later instant. The probability distributions on $\mathcal{A}$ and $\mathfrak{D}$ were connected by an equation of the Markovian type,

$$p_H(D_i) = \sum_k p_H(D_i \mid A_k) p_H(A_k). \qquad (5.39)$$

We assumed that the Markov chain was an aged one, so that $p_H(D_i) = p_H(A_i)$, where the same index i on D_i and A_i means the same value of the experimental measurement. This entails $S_H(\mathfrak{D}) = S_H(\mathcal{A})$ [see (5.32)]. But the assumption of "agedness" need not hold in general, and it becomes meaningful to ask whether $S_H(\mathfrak{D})$ is larger or smaller than $S_H(\mathcal{A})$. If $S_H(\mathfrak{D})$ is larger than $S_H(\mathcal{A})$ our predictive knowledge about the outcome of an experiment becomes less accurate through the Markov process, and if $S_H(\mathfrak{D})$ is smaller than $S_H(\mathcal{A})$ it becomes sharper. This is the central question answered by *H*-theorems, which express, in general, a degeneration of predictive knowledge by the Markov process. The comparison of $S_H(\mathfrak{D})$ and $S_H(\mathcal{A})$, which is the subject of the *H*-theorems, should not be confused with the comparison of $S_H(\mathfrak{D} \mid A_k)$ or $S_H(\mathfrak{D} \mid \mathcal{A})$ with $S_H(\mathcal{A})$ [or with $S_H(\mathfrak{D})$ in the aged chain], which was discussed in Example 5.2. Relegating discussions of the epistemomentrical implications of *H*-theorems to the next section, we limit ourselves here to the mathematical side of *H*-theorems.

We consider an ensemble of strings, each string consisting of an infinite sequence of states $x(t)$. Each position on the infinite string is labeled by an integer index t. Each position can be in any one of a finite number n of "states" $\mathfrak{Z} = \{z_i\}$, $i = 1, 2, \ldots, n$. The conditional probability (in the ensemble) that position t will be in state z_j, given the state z_i of the preceding

† What is discussed in this section is a mathematical model. Apart from the model, one can evaluate $p(D_i \mid A_i)$ and $p(D_i) = p(A_i)$ from the frequencies in actual meteorological statistics. Many years ago S. Fujiwhara showed that the persistence coefficients $\gamma_i = p(D_i \mid A_i)/p(D_i)$ are larger than unity [F-11]. γ_1 and γ_2 correspond respectively to $(1 - p)/a$ and $(1 - q)/b$ in the present notation. See also Besson [B-1] and Borel [B-6].

position $(t - 1)$, can be denoted in various ways:

$$p_{ji}(t) = p_t(i \to j) = p_t(j \mid i) = \Pr\{x(t) = z_j \mid x(t-1) = z_i\}. \quad (5.40)$$

The basic assumption of the Markov chain is that the conditional probability of $x(t) = z_j$ does not change when the states of positions earlier than $t - 1$ are also taken into the condition. In the terminology of Section 2.4, this means $\nu = 2$. In that section we assumed that $p_{ji}(t)$ does not depend on t. We liberalize that point slightly in this section and assume that $p_{ji}(t)$ can depend on t insofar as the following is the case: if the relation $p_{ji}(t) \neq 0$ is true for one value of t, then it is true for all values of t and for $t \to \infty$ [$\lim_{t\to\infty} p_{ji}(t) \neq 0$]. This assumption, of course, entails that if the relation $p_{ji}(t) = 0$ is true for one value of t, it is true for all values of t, and hence also for $t \to \infty$. In other words, we assume for any given pair (i, j) either that $p_{ji}(t) = 0$ for all t or that $p_{ji}(t) > p_0$ for all t, where p_0 does not depend on (i, j) or t. We call this condition assumption of invariant connectivity. In the so-called doubly stochastic case we assume only the condition of invariant connectivity. In the non-doubly-stochastic case we assume a more restrictive version of invariant connectivity, which will be explained later.

We have to distinguish between two kinds of averaging processes based on two different kinds of probabilities of a state z_i in the ensemble: time average and ensemble average. If, for instance, we take a single infinite string (identified by index ξ) and seek the relative frequency $\pi_i(\xi)$ of state z_i in it, we are interested in the time average. If, on the other hand, we take a particular time and seek the relative frequency $w_i(t)$ of state z_i at t in the ensemble of strings, we are interested in the ensemble average. By definition of $p_{ji}(t)$, we have, of course,

$$w_j(t) = \sum_{i=1}^{n} p_{ji}(t) w_i(t-1). \quad (5.41)$$

Equation (5.39) can be regarded as a case of (5.41) for a particular value of t. The probability $w_i(t)$ is sometimes called "occupancy probability."

The obvious condition

$$p_{ji}(t) \geq 0, \qquad \sum_{j=1}^{n} p_{ji}(t) = 1 \quad (5.42)$$

guarantees that if $w_i(t-1) \geq 0$ and $\sum_{i=1}^{n} w_i(t-1) = 1$ then

$$w_i(t) \geq 0, \qquad \sum_{i=1}^{n} w_i(t) = 1. \quad (5.43)$$

In some cases we have, in addition to (5.42), the inverse normalization,

$$\sum_{i=1}^{n} p_{ji}(t) = 1. \quad (5.44)$$

In such cases we speak of a doubly stochastic Markov chain.

Ignoring the mathematician's requirement of rigor, let us describe roughly the different concepts of ergodicity. First, we note that insofar as the ensemble average and the time average exist, we can interchange the order of averaging processes and obtain

ensemble average of time average = time average of ensemble average;

that is,

$$\langle \pi_i(\xi) \rangle_\xi = \langle w_i(t) \rangle_t, \tag{5.45}$$

where the symbol $\langle \mathcal{Q}(\alpha) \rangle_\alpha$ means the average with respect to α of the quantity $\mathcal{Q}(\alpha)$. The strictest concept of ergodicity requires both

$$\pi_i(\xi) = \pi_i \tag{5.46}$$

and

$$\lim_{t\to\infty} w_i(t) = w_i, \tag{5.47}$$

where π_i is independent of ξ and w_i is independent of t. Because of (5.45) we must then have, under the conditions (5.46) and (5.47),

$$\pi_i(\xi) = \pi_i = \langle \pi_i(\xi) \rangle_\xi = \langle w_i(t) \rangle_t = w_i = \lim_{t\to\infty} w_i(t), \tag{5.48}$$

which may be read,

$$\text{time average} = \text{ensemble average}. \tag{5.49}$$

Note that if (5.47) is the case, then, for the overwhelming majority of values of t, we can put $w_i(t) \doteqdot w_i$. Equation (5.48) is the gist of the strong law of large numbers for the Markov chain. The rigorous formulation of this law must not assume outright the existence of $\pi_i(\xi)$ in an infinite string but has to start with a finite portion of the string. Let $\pi_i(\xi, \tau)$ be the relative frequency of state z_i in the segment of length τ on a particular string ξ. Then the theorem states that for a given $\varepsilon > 0$,

$$\lim_{\tau\to\infty} \Pr\{|\pi_i(\xi, \tau) - w_i| \geq \varepsilon\} = 0, \tag{5.50}$$

independently of ξ. The reader should note that this is essentially equivalent to Theorem 2.2.

A less restrictive concept of ergodicity is expressed by the convergence (5.47), without discussing the individual difference indicated by ξ. This may be called ensemble-average ergodicity. A further liberalization of the concept will require only

$$\lim_{t\to\infty} \frac{1}{t} \sum_{t=1}^{t} w_i(t) = w_i, \tag{5.51}$$

which is implied by, but does not necessarily imply, (5.47). This may be called time-and-ensemble-average ergodicity. In this section we are interested mainly in the ensemble-average ergodicity (5.47) and time-and-ensemble-average ergodicity (5.51), but not in individual ergodicity (5.50). This is because the first two kinds of ergodicity have a close relation to the H-theorem.

The entropy function $S(t)$ that satisfies the H-theorem (see proof later),

$$S(t) \geq S(t-1), \tag{5.52}$$

in the doubly stochastic case is defined by

$$S(t) = -\sum_{i=1}^{n} w_i(t) \log w_i(t). \tag{5.53}$$

Since there is an upper limit $S(t) \leq \log n$, we can immediately conclude from (5.52) the convergence $\lim_{t\to\infty} S(t) = S(\infty)$. This convergence can also be concluded, of course, from the ensemble average convergence (5.47). On the other hand, the convergence of $S(t)$ for $t \to \infty$ alone does not necessarily imply the convergence (5.47). However, if for instance we know $S(\infty) = \log n$, we can conclude $w_i(t) \to 1/n$ for $t \to \infty$. This suggests that even in the cases in which $S(\infty) \neq \log n$, a careful study of the limiting behavior of $S(t)$ may lead to some useful conclusions regarding the limiting behavior of $w_i(t)$. This is the line of approach we adopt in this section.

Before we start the main part of our discussion, it is convenient to introduce a certain classification of states z_i. In Section 2.4 we introduced a classification of z_i, but now we introduce a more detailed consideration of the matter from a slightly different point of view. The classification utilizes basically only the conditions $p_{ji}(t) = 0$ or $\neq 0$; hence it is not affected whether the $p_{ji}(t)$ remain constant or vary in compliance with the assumption of invariant connectivity. The total number of states n is assumed to be finite; hence we are dealing with a "finite" Markov chain.

Let the "influx" and the "efflux" of a subset $\mathfrak{C}$ of the set of states $\mathfrak{Z}$ be defined at t by

$$\begin{aligned} \text{influx}\,(\mathfrak{C}) &= \text{efflux}\,(\mathfrak{Z} - \mathfrak{C}) = \sum_{j}^{\mathfrak{C}} \sum_{i}^{\mathfrak{Z}-\mathfrak{C}} p_{ji}, \\ \text{efflux}\,(\mathfrak{C}) &= \text{influx}\,(\mathfrak{Z} - \mathfrak{C}) = \sum_{j}^{\mathfrak{Z}-\mathfrak{C}} \sum_{i}^{\mathfrak{C}} p_{ji}. \end{aligned} \tag{5.54}$$

Because we are interested only in the properties of these quantities that do not depend on t, we need not mention t as an argument. Of course,

$$\text{influx}\,(\mathfrak{Z}) = \text{efflux}\,(\mathfrak{Z}) = 0. \tag{5.55}$$

Furthermore, we define the "intraflux" of $\mathfrak{C}$ by

$$\text{intraflux}\,(\mathfrak{C}) = \sum_{j}^{\mathfrak{C}} \sum_{i}^{\mathfrak{C}} p_{ji}. \tag{5.56}$$

If we denote by $N(\mathfrak{C})$ the number of states in $\mathfrak{C}$, we obtain

$$\text{efflux}(\mathfrak{C}) + \text{intraflux}(\mathfrak{C}) = \sum_j^{\mathfrak{Z}} \sum_i^{\mathfrak{C}} p_{ji} = N(\mathfrak{C}) \tag{5.57}$$

because of (5.42). If the Markov chain is doubly stochastic then

$$\text{influx}(\mathfrak{C}) + \text{intraflux}(\mathfrak{C}) = \sum_j^{\mathfrak{C}} \sum_i^{\mathfrak{Z}} p_{ji} = N(\mathfrak{C}) \qquad \text{(doubly stochastic case)} \tag{5.58}$$

because of (5.44). Combining (5.57) and (5.58), we obtain

$$\text{influx}(\mathfrak{C}) = \text{efflux}(\mathfrak{C}) \qquad \text{(doubly stochastic case).} \tag{5.59}$$

Coming back to the general case, it is sometimes convenient to define the "flow" from a subset $\mathfrak{C}$ to another subset $\mathfrak{D}$ by

$$\text{flow}(\mathfrak{C} \to \mathfrak{D}) = \sum_j^{\mathfrak{D}} \sum_i^{\mathfrak{C}} p_{ji}. \tag{5.60}$$

we have evidently

$$\text{influx}(\mathfrak{C}) = \text{flow}(\mathfrak{Z} - \mathfrak{C} \to \mathfrak{C}), \qquad \text{efflux}(\mathfrak{C}) = \text{flow}(\mathfrak{C} \to \mathfrak{Z} - \mathfrak{C}),$$

$$\text{intraflux}(\mathfrak{C}) = \text{flow}(\mathfrak{C} \to \mathfrak{C}), \qquad \text{influx}(\mathfrak{C}) + \text{intraflux}(\mathfrak{C}) = \text{flow}(\mathfrak{Z} \to \mathfrak{C}),$$

$$\text{efflux}(\mathfrak{C}) + \text{intraflux}(\mathfrak{C}) = \text{flow}(\mathfrak{C} \to \mathfrak{Z}). \tag{5.61}$$

If the subset $\mathfrak{C}$ in $\mathfrak{Z}$ is a proper subset of $\mathfrak{Z}$ such that

$$\text{influx}(\mathfrak{C}) = \text{efflux}(\mathfrak{C}) = 0, \tag{5.62}$$

$\mathfrak{C}$ and $\mathfrak{Z} - \mathfrak{C}$ are completely isolated in the sense that the $w_i(t)$ in $\mathfrak{C}$ and the $w_i(t)$ in $\mathfrak{Z} - \mathfrak{C}$ have nothing to do with each other. In such a case we can consider the Markov chains in $\mathfrak{C}$ and $\mathfrak{Z} - \mathfrak{C}$ separately. Hence we can assume without loss of generality that there is no such isolated proper subset $\mathfrak{C}$ in $\mathfrak{Z}$.

Suppose next that there is a subset $\mathfrak{C}$ in $\mathfrak{Z}$ such that

$$\text{efflux}(\mathfrak{C}) = 0, \qquad \text{influx}(\mathfrak{C}) \neq 0;$$

that is, (5.63)

$$\text{influx}(\mathfrak{Z} - \mathfrak{C}) = 0, \qquad \text{efflux}(\mathfrak{Z} - \mathfrak{C}) \neq 0.$$

Thus $\mathfrak{C}$ is an efflux-free (nonisolated) subset and $\mathfrak{Z} - \mathfrak{C}$ is an influx-free (nonisolated) subset. If $\mathfrak{D}$ is an influx-free (nonisolated) subset such that there is no smaller subset in it that is efflux-free (nonisolated), we say that $\mathfrak{D}$ is an elementary influx-free (nonisolated) subset. It is obvious that the sum of w's within such an elementary influx-free (nonisolated) subset $\mathfrak{D}$ gradually vanishes with time. For this reason such a $\mathfrak{D}$ is called a vanishing subset. Since the sum of the w's in $\mathfrak{Z}$ is a constant, we can visualize w as a kind of fluid. In this picture the probability will have flowed out of $\mathfrak{D}$ practically

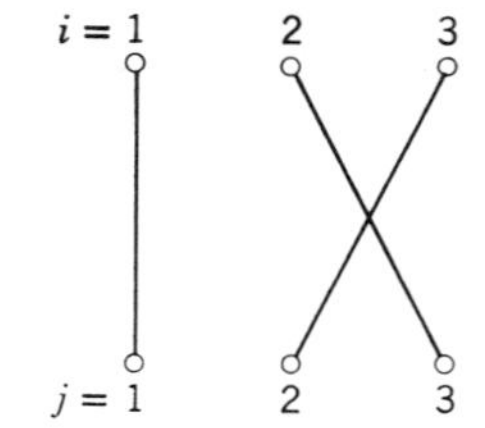

Figure 5.3 This graph, which can happen in a doubly stochastic chain as well as in a non-doubly-stochastic chain in the absence of the irreducibility assumption, is forbidden by the assumption; z_i is the initial state and z_j is the final state; a straight line connecting z_i and z_j means that $p_{ji} \neq 0$.

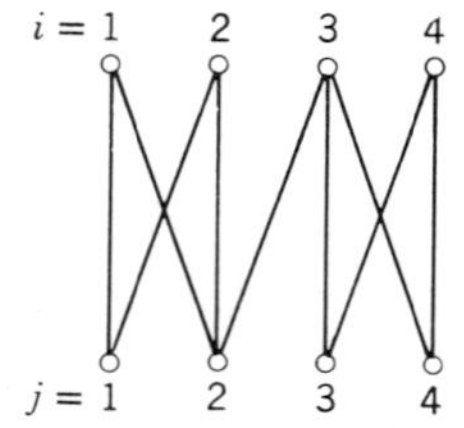

Figure 5.4 This graph, which is forbidden in a doubly stochastic chain but not in a non-doubly-stochastic chain in the absence of the irreducibility assumption, is forbidden by the assumption. influx $(z_3 \cup z_4) =$ efflux $(z_1 \cup z_2) = 0$. influx $(z_1 \cup z_2) =$ efflux $(z_3 \cup z_4) \neq 0$.

completely after a sufficient length of time. Hence, if we are not interested in the initial transient period, we might as well ignore completely those states z_i that belong to vanishing subsets, such as $\mathfrak{D}$. If we eliminate those z_i from $\mathfrak{Z}$ and delete those rows and columns in the matrix $\|p_{ji}\|$ corresponding to the eliminated states, the remaining efflux-free (nonisolated) subsets become "isolated." It can happen that a new vanishing subset appears in the newly separated isolated subset. But the probability in those new vanishing subsets will also vanish with time; thus we can again expel them from our consideration, and are left with separate isolated subsets without any efflux-free or influx-free sub-subsets within them. Each such isolated subset can be studied separately.

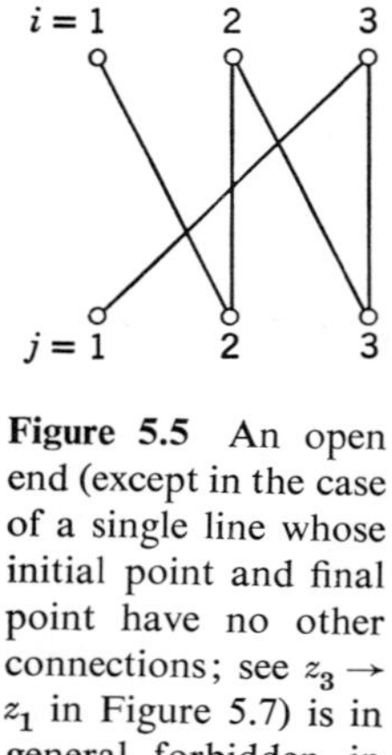

Figure 5.5 An open end (except in the case of a single line whose initial point and final point have no other connections; see $z_3 \to z_1$ in Figure 5.7) is in general forbidden in a doubly stochastic chain. influx (z_2) + intraflux $(z_2) > 1$ and influx (z_1) + intraflux $(z_1) < 1$ both violate (5.58).

For this reason we assume from the beginning that there are no isolated proper subsets and no efflux-free or influx-free (nonisolated) subsets within $\mathfrak{Z}$. [In the doubly stochastic case (5.63) cannot happen because of (5.59) anyway.] We call this assumption one of irreducibility because it means that the matrix $\|p_{ji}\|$ is an irreducible matrix. See Figures 5.3–5.8 for various forbidden and allowed graphs.

We introduce the concept of "terminally connected" states by the following definition, in which $z_i \overset{T}{\sim} z_j$ denotes that z_i and z_j are terminally connected.

DEFINITION 5.1

(a) If $z_i \overset{T}{\sim} z_j$ and $z_j \overset{T}{\sim} z_k$ then $z_i \overset{T}{\sim} z_k$.

(b) For given z_i and z_j in $\mathfrak{Z}$, if there exists in $\mathfrak{Z}$ a state z_k such that $p_{ki} \neq 0$ and $p_{kj} \neq 0$, then $z_i \overset{T}{\sim} z_j$.

From (b) follows the reflexivity ($z_i \overset{T}{\sim} z_i$), but not the transitivity. Hence the transitivity (a) is an additional condition, which has the effect of enlarging the extension of the concept. If we use arrows in the sense that $z_i \to z_j$ as well as $z_j \leftarrow z_i$ means that $p_{ji} \neq 0$, $z_i \overset{T}{\sim} z_j$ means that there exists a sequence of states starting with z_i and ending with z_j such that we can place right arrows ($\to$) and left arrows ($\leftarrow$) alternately between consecutive z's, starting with a right arrow and ending with a left arrow. Starting from any one state, if we collect states that are terminally connected with it, using Definition 5.1, until we can no longer enlarge the collection, we obtain a family of terminally connected states. It may be noted that each z_i has at least one partner z_k such that $z_i \to z_k$, since $\sum_j p_{ji} = 1$, where z_k could be z_i itself or some other state. Hence each state belongs to one and only one family of terminally connected states. It can very well happen that the family consists of a single member. Now if we invert the role of initial and final states, we obtain the concept of initially connected states $z_i \overset{I}{\sim} z_j$ and the concept of a family of "initially connected" states.

DEFINITION 5.2

(a) If $z_i \overset{I}{\sim} z_j$ and $z_j \overset{I}{\sim} z_k$ then $z_i \overset{I}{\sim} z_k$.

(b) For given z_i and z_j, if there exists z_k such that $z_k \to z_i$ and $z_k \to z_j$, then $z_i \overset{I}{\sim} z_j$. ($z_k \to z_i$ means $p_{ik} \neq 0$.)

It is true that in a non-doubly-stochastic, nonirreducible Markov chain a given state z_i may have no initial state z_k such that $z_k \to z_i$. But this is excluded by the irreducibility assumption. Thus each state belongs to one and only one family of initially connected states. Take a family $\mathcal{A}$ of terminally connected states, and let z_i be any one member of $\mathcal{A}$ and let z_j be such that $z_i \to z_j$. Let $\mathcal{B}$ be the family of initially connected states to which z_j belongs. It is clear by Definitions 5.1 and 5.2 that this $\mathcal{B}$ is uniquely determined by $\mathcal{A}$, independent of the choice of z_i in $\mathcal{A}$ and z_j in $z_i \to z_j$. We call $\mathcal{B}$ the terminal counterpart of $\mathcal{A}$ and write $\mathfrak{T}(\mathcal{A}) = \mathcal{B}$. Similarly, inverting the roles of initial and final states, we obtain a unique family $\mathcal{A}$ of terminally connected states starting from a family $\mathcal{B}$ of initially connected states. We call $\mathcal{A}$ the initial counterpart of $\mathcal{B}$ and write $\mathfrak{I}(\mathcal{B}) = \mathcal{A}$. Furthermore, it is obvious that $\mathfrak{T}(\mathfrak{I}(\mathcal{A})) = \mathcal{A}$ and $\mathfrak{I}(\mathfrak{T}(\mathcal{B})) = \mathcal{B}$. All states in $\mathcal{A}$ pass to some states in $\mathcal{B}$ and all states in $\mathcal{B}$ originate from some states in $\mathcal{A}$. Thus we obtain the following lemma.

Lemma 5.1 In an irreducible Markov chain the entire set $\mathfrak{Z}$ of states is decomposed into an exhaustive and disjoint set of families of terminally connected states, $\mathcal{A}_\alpha$, $\alpha = 1, 2, \cdots, \rho$, and also into an exhaustive and disjoint set of families of initially connected states, $\mathcal{B}_\alpha$, $\alpha = 1, 2, \cdots, \rho$, such that $\mathfrak{T}(\mathcal{A}_\alpha) = \mathcal{B}_\alpha$ and $\mathfrak{I}(\mathcal{B}_\alpha) = \mathcal{A}_\alpha$. Thus $\mathfrak{Z} = \bigcup_{\alpha=1}^{\rho} \mathcal{A}_\alpha = \bigcup_{\alpha=1}^{\rho} \mathcal{B}_\alpha$.

It may be noted that in a doubly stochastic chain the lemma still holds even if we remove the qualification "irreducible" from its statement. This is so because a state z_i in a doubly stochastic chain, irreducible or reducible, always has at least one initial partner z_j such that $z_j \rightarrow z_i$ as well as one terminal partner z_k such that $z_i \rightarrow z_k$, whereas a state in a non-doubly-stochastic, reducible Markov chain is guaranteed to have only a terminal partner. In the case of a doubly stochastic chain we have another conspicuous fact:

$$N(\mathcal{A}_\alpha) = N(\mathcal{B}_\alpha) \qquad \text{(doubly stochastic case)}, \tag{5.64}$$

where $\mathcal{A}_\alpha$ and $\mathcal{B}_\alpha$ are related as in Lemma 5.1. The proof is immediate: $N(\mathcal{A}_\alpha) = \text{flow}\ (\mathcal{A}_\alpha \rightarrow \mathfrak{Z}) = \text{flow}\ (\mathcal{A}_\alpha \rightarrow \mathcal{B}_\alpha) = \text{flow}\ (\mathfrak{Z} \rightarrow \mathcal{B}_\alpha) = N(\mathcal{B}_\alpha)$. The first and the last equals signs here come from (5.57) and (5.58), while the second and the third equals signs come from the definition of initial and terminal partners.

We now introduce the notion of associated states, by which we write $z_i \overset{A}{\sim} z_j$ to mean that z_i and z_j are associated.

DEFINITION 5.3

(a) If $z_i \overset{A}{\sim} z_j$ and $z_j \overset{A}{\sim} z_k$ then $z_i \overset{A}{\sim} z_k$.

(b) If $z_i \overset{T}{\sim} z_j$ then $z_i \overset{A}{\sim} z_j$.

(c) If $z_i \overset{I}{\sim} z_j$ then $z_i \overset{A}{\sim} z_j$.

(d) If $z_i \rightarrow z_k$, $z_j \rightarrow z_l$, and $z_k \overset{A}{\sim} z_l$ then $z_i \overset{A}{\sim} z_j$.

(e) If $z_k \rightarrow z_i$, $z_l \rightarrow z_j$, and $z_k \overset{A}{\sim} z_l$ then $z_i \overset{A}{\sim} z_j$.

We call a set of mutually associated states a family of associated states. It is important to know that in (d) and (e) z_i and z_j belong to one family and z_k and z_l belong to one family, but these two families may or may not be the same. It may be easier to understand this definition by considering a process by which we can build up a family of associated states by merging $\mathcal{A}$'s and $\mathcal{B}$'s step by step. There are three allowed ways of merger. An $\mathcal{A}$ and a $\mathcal{B}$ can be merged if they have a common element. Starting with an $\mathcal{A}$, we can add a $\mathcal{B}$, and then another $\mathcal{A}$, and so forth. Next, two $\mathcal{A}$'s, $\mathcal{A}_\alpha$ and $\mathcal{A}_\beta$, can be merged if $\mathfrak{T}(\mathcal{A}_\alpha)$ and $\mathfrak{T}(\mathcal{A}_\beta)$ are known to belong to a family (this last family may or may not be the family into which $\mathcal{A}_\alpha$ and $\mathcal{A}_\beta$ are to be incorporated). Similarly, two $\mathcal{B}$'s, $\mathcal{B}_\alpha$ and $\mathcal{B}_\beta$, can be merged if $\mathfrak{I}(\mathcal{B}_\alpha)$ and $\mathfrak{I}(\mathcal{B}_\beta)$ are known to belong to a family. It may be pointed out that there is a certain degree of redundancy in Definition 5.3. In fact, (c) ensues from (b) and (e); similarly, (b) ensues from (c) and (d). From Definition 5.3 it is clear that any state belongs to one and only one family of associated states. Hence the following lemma.

Lemma 5.2 In an irreducible Markov chain the entire set $\mathfrak{Z}$ of states is decomposed into an exhaustive and disjoint set of families of associated states $\mathcal{C}_\lambda$, $\lambda = 1, 2, \cdots, \tau$, where each family $\mathcal{C}_\lambda$ can be considered as a merger of families $\mathcal{A}_\alpha$ of terminally connected states as well as a merger of families $\mathcal{B}_\alpha$ of initially connected states. Thus

$$Z = \bigcup_{\lambda=1}^{\tau} C_\lambda, \qquad C_\lambda = \bigcup_{\alpha}^{\text{some}} A_\alpha, \qquad C_\lambda = \bigcup_{\beta}^{\text{some}} B_\beta.$$

The last part of the lemma is obvious because a family $\mathcal{A}_\alpha$ of terminally connected states has either to belong or not to belong to a $\mathcal{C}_\lambda$ with all its members and cannot belong to $\mathcal{C}_\lambda$ partially; similarly for the $\mathcal{B}$'s. It should be noted that nothing is said as to whether or not the union $\bigcup_\alpha^{\text{some}}$ and the union $\bigcup_\beta^{\text{some}}$ in the last two equations of the lemma may cover the same values of the indices α and β.

Let us now generalize slightly the usuage of the symbols $\mathcal{T}$ and $\mathcal{I}$ so that we can apply them to $\mathcal{C}$'s. The obvious meaning is that if $\mathcal{C}_\lambda = \bigcup_\alpha \mathcal{A}_\alpha$ then $\mathcal{T}(\mathcal{C}_\lambda) = \bigcup_\alpha \mathcal{T}(\mathcal{A}_\alpha) = \bigcup_\alpha \mathcal{B}_\alpha$, where the unions cover the same values of the α. Similarly, if $\mathcal{C}_\lambda = \bigcup_\alpha \mathcal{B}_\alpha$, $\mathcal{I}(\mathcal{C}_\lambda) = \bigcup_\alpha \mathcal{I}(\mathcal{B}_\alpha) = \bigcup_\alpha \mathcal{A}_\alpha$ and $\mathcal{T}(\mathcal{I}(\mathcal{C}_\lambda)) = \mathcal{I}(\mathcal{T}(\mathcal{C}_\lambda)) = \mathcal{C}_\lambda$. It is easy to see that $\mathcal{T}(\mathcal{C}_\lambda)$ is a $\mathcal{C}_\mu$, where μ may or may not be the same as λ. This is so because, thanks to Item (e) of Definition 5.3, any two states included in $\mathcal{T}(\mathcal{C}_\lambda)$ are associated, and thanks to Item (d) of Definition 5.3, if $z_k \overset{A}{\sim} z_l$ and $z_k \in \mathcal{T}(\mathcal{C}_\lambda)$ then $z_l \in \mathcal{T}(\mathcal{C}_\lambda)$ too. We are now prepared to introduce the following interesting theorem.

Theorem 5.2 In an irreducible Markov chain the entire set $\mathcal{Z}$ of states is decomposable into families of associated states as $\mathcal{Z} = \bigcup_{\lambda=1}^{\tau} \mathcal{C}_\lambda$, where we can suitably arrange the labeling of the index λ so that

$$\mathcal{T}(\mathcal{C}_\lambda) = \mathcal{C}_{\lambda+1} \quad \text{and} \quad \mathcal{I}(\mathcal{C}_{\lambda+1}) = C_\lambda, \tag{5.65}$$

where λ is $1, 2, \cdots, \tau$, and $\lambda = \tau + 1$ means $\lambda = 1$.

Proof. We consider a sequence $\mathcal{C}_\lambda$, $\mathcal{T}(\mathcal{C}_\lambda)$, $\mathcal{T}(\mathcal{T}(\mathcal{C}_\lambda))$, $\mathcal{T}(\mathcal{T}(\mathcal{T}(\mathcal{C}_\lambda)))$, $\cdots$, $\mathcal{T}^{(r)}(\mathcal{C}_\lambda)$, $\cdots$, where each term is a $\mathcal{C}$. We first show that at a certain stage r (r is the number of repetitions of $\mathcal{T}$) the term must come back to $\mathcal{C}_\lambda$. Since there is a finite number of $\mathcal{C}$'s in $\mathcal{Z}$, the sequence must repeat the same $\mathcal{C}$'s. Hence, for some s and r ($s < r$), we should have $\mathcal{T}^{(s)}(\mathcal{C}_\lambda) = \mathcal{T}^{(r)}(\mathcal{C}_\lambda)$. Then $\mathcal{T}^{(r-s)}(\mathcal{C}_\lambda) = C_\lambda$. Let r be the smallest number such that $C_\lambda = \mathcal{T}^{(r)}(\mathcal{C}_\lambda)$. Then the sequence must be a repetition of the r-term sequence, $\mathcal{C}_\lambda$, $\mathcal{T}(\mathcal{C}_\lambda)$, $\cdots$, $\mathcal{T}^{(r-1)}(\mathcal{C}_\lambda)$. This sequence, however, has to include all the $\mathcal{C}$'s in $\mathcal{Z}$, because it it did not it would mean that $\mathcal{Z}$ is divided into more than one isolated set, a case we excluded by the irreducibility assumption. As a result r must be τ of the theorem, which may be called the "period" of the cycle. This shows that what we call a family $\mathcal{C}_\lambda$ of associated states in this section is precisely what we called a cyclic class B_μ in Section 2.4. We shall soon see in the present section that the $\mathfrak{Z}$ under the irreducible assumption corresponds to what we called an ergodic subset.

It may be noted that as a consequence of (5.64), which states that $N(\mathcal{A}_\alpha) = N(\mathcal{C}(\mathcal{A}_\alpha))$, we have in the doubly stochastic case $N(\mathcal{C}_\lambda) = N(\mathcal{C}(\mathcal{C}_\lambda)) = N(\mathcal{C}_{\lambda+1})$. Hence

$$N(\mathcal{C}_\lambda) = \frac{N(\mathfrak{Z})}{\tau} = \frac{n}{\tau}$$

(doubly stochastic case). (5.66)

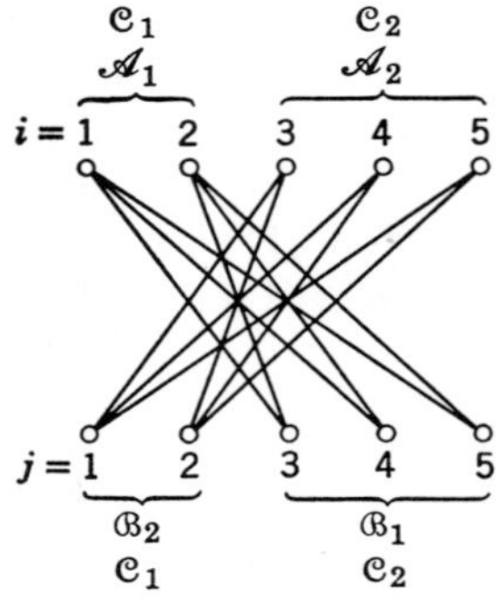

Figure 5.6 There are two $\mathcal{A}$'s, two $\mathcal{B}$'s and two $\mathcal{C}$'s. $\mathcal{C}(\mathcal{C}_1) = \mathcal{C}_2$ and $\mathcal{C}(\mathcal{C}_2) = \mathcal{C}_1$. This graph, however, cannot happen in a doubly stochastic chain, because 2 = efflux ($\mathcal{C}_1$) $\neq$ influx ($\mathcal{C}_1$) = 3, violating (5.59).

(See Figures 5.3–5.8 for various cases of $\mathcal{C}$'s.) So much for the classification of states in $\mathfrak{Z}$; we now go to the H-theorem and the associated ensemble-average ergodic theorem.

We consider two probability distributions p_i and q_i over n cases, $i = 1, 2, \cdots, n$, with $p_i \geq 0$, $\sum_{i=1}^n p_i = 1$, and $q_i \geq 0$, $\sum_{i=1}^n q_i = 1$. We interpret this as meaning that the stochastic variable q takes the value q_i with probability p_i. Let $f(q)$ be a single-valued function defined in the domain $0 \leq q \leq 1$, and let its average be denoted by

$$\langle f(q) \rangle = \sum_{i=1}^{n} p_i f(q_i). \tag{5.67}$$

In the following lemmas the symbol $\psi(q)$ means the function defined by

$$\psi(q) \equiv -q \ln q, \qquad \psi(0) = 0 \tag{5.68}$$

in the domain $0 \leq q \leq 1$. Actually the property of ψ we need in the lemmas is essentially its convexity upward,

$$\frac{d^2\psi(q)}{dq^2} < 0, \tag{5.69}$$

but we use the form (5.68) for concreteness.

Lemma 5.3 We always have

$$\langle \psi(q) \rangle \leq \psi(\langle q \rangle), \tag{5.70}$$

where the equals sign holds if and only if all the q_i are equal for those i's for which $p_i \neq 0$.

Lemma 5.4 If

$$\psi(\langle q \rangle) - \langle \psi(q) \rangle < \varepsilon \tag{5.71}$$

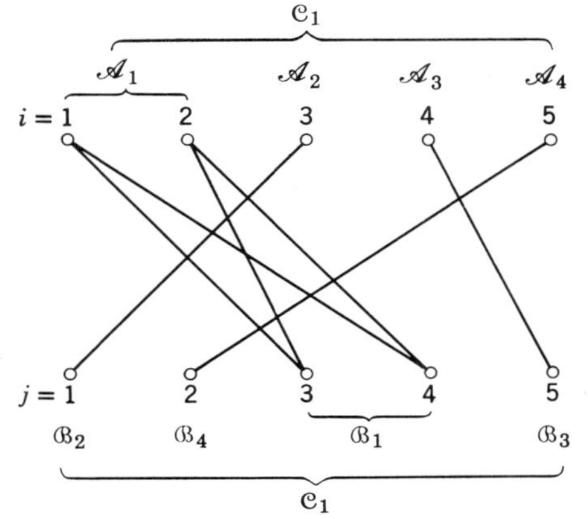

Figure 5.7 This allowed graph has only one $\mathcal{C}$: $z_1 \overset{A}{\sim} z_2$ because $z_1 \overset{T}{\sim} z_2$ see (b), Definition 5.3; $z_3 \overset{A}{\sim} z_4$ because $z_3 \overset{I}{\sim} z_4$ see (c), Definition 5.3; $z_3 \overset{A}{\sim} z_5$ because $z_1 \overset{A}{\sim} z_2$ see (d), Definition 5.3; $z_1 \overset{A}{\sim} z_5$ because $z_3 \overset{A}{\sim} z_4$ see (e), Definition 5.3. Thus all five states are associated see (a), Definition 5.3.

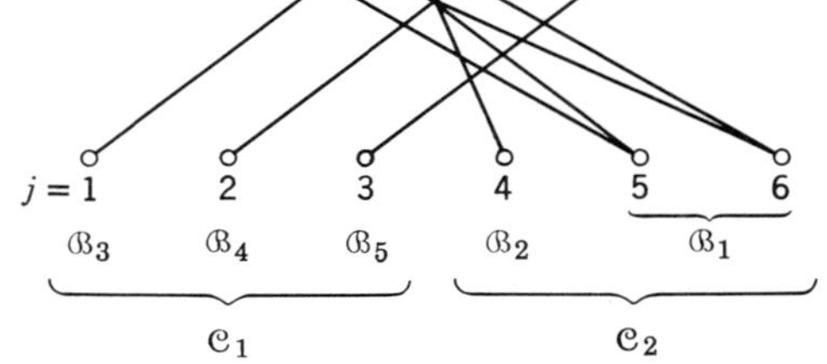

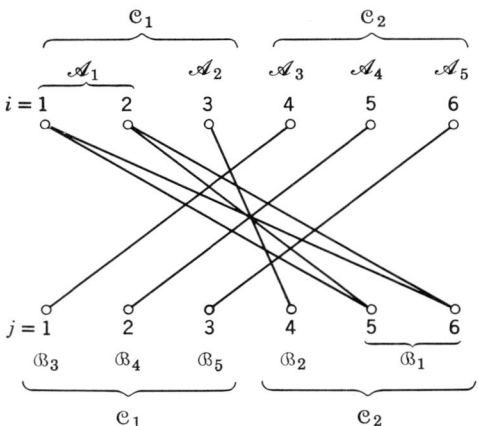

Figure 5.8 This allowed graph has two $\mathcal{C}$'s: $z_5 \overset{A}{\sim} z_6$ because $z_5 \overset{I}{\sim} z_6$; $z_1 \overset{A}{\sim} z_2$ because $z_1 \overset{T}{\sim} z_2$; $z_2 \overset{A}{\sim} z_3$ because $z_5 \overset{A}{\sim} z_6$; $z_4 \overset{A}{\sim} z_5$ because $z_1 \overset{A}{\sim} z_2$.

then

$$\langle \eta^2 \rangle = \sum_{i=1}^{n} p_i \eta_i^2 < 2\varepsilon, \tag{5.72}$$

where

$$\eta_i = q_i - \langle q \rangle. \tag{5.73}$$

Proof of Lemmas 5.3 and *5.4*. The statements of Lemmas 5.3 and 5.4 do not exclude the possibility $\langle q \rangle = 0$. The condition $\langle q \rangle = 0$ means that for any i, p_i or q_i or both must vanish; hence $\langle \psi(q) \rangle = \sum_{i=1}^{n} p_i \psi(q_i) = 0$ and $\psi(\langle q \rangle) = 0$. In this case the conclusion of Lemma 5.3 is certainly upheld, since $q_i = 0$ for all those i's for which $p_i \neq 0$. The conclusion of Lemma 5.4 is also fulfilled because $\sum_{i=1}^{n} p_i \eta_i^2 = \sum_{i=1}^{n} p_i q_i^2 = 0$ in this case.

Consequently we can safely assume $\langle q \rangle \neq 0$ for the rest of the proof. With the help of Taylor's theorem, we then have

$$\psi(q_i) = \psi(\langle q \rangle) - \eta_i(1 + \ln \langle q \rangle) - \frac{\eta_i^2}{2} \frac{1}{\langle q \rangle + \theta_i \eta_i}, \tag{5.74}$$

with $0 < \theta_i < 1$. This formula holds in the entire domain $0 \leq q_i \leq 1$, including the extremity $q_i = 0$, where $\psi(q_i)$ does not have derivatives, but the formula (5.74) still holds with $\theta_i = \frac{1}{2}$. Multiplying (5.74) by p_i and summing over i, we obtain

$$\psi(\langle q \rangle) - \langle \psi(q) \rangle = \frac{1}{2} \sum_{i=1}^{n} p_i \eta_i^2 \frac{1}{\langle q \rangle + \theta_i \eta_i}, \tag{5.75}$$

since $\sum_{i=1}^{n} p_i \eta_i = 0$ as a result of (5.73). $\langle q \rangle + \theta_i \eta_i$ is a value between $\langle q \rangle$ and q_i and hence lies between 0 and 1; that is,

$$\frac{1}{\langle q \rangle + \theta_i \eta_i} \geq 1. \tag{5.76}$$

Consequently

$$\psi(\langle q \rangle) - \langle \psi(q) \rangle \geq \frac{1}{2} \sum_{i=1}^{n} p_i \eta_i^2 \geq 0. \tag{5.77}$$

If the left-hand side vanishes, η_i^2 for $p_i \neq 0$ must vanish. This means, according to (5.73), that $q_i = \langle q \rangle$ provided $p_i \neq 0$ completing the proof of Lemma 5.3. Next, if its left-hand side is less than ε, (5.77) shows that $(1/2) \sum_{i=1}^{n} p_i \eta_i^2 < \varepsilon$. This is Lemma 5.4. We see that the essential features of Lemmas 5.3 and 5.4 will not change if we use some convex function other than ψ.

The condition (5.72) is expressed in terms of p_i and η_i. We should like to bring this condition to a form more easy to handle. First, we exclude from the summation those i's for which $p_i = 0$, and call $p_{\min}$ the smallest among the remaining nonzero p_i's. This $p_{\min}$ must exist, since n is finite. Further let the largest and the smallest among the q_i for which $p_i \neq 0$ be denoted respectively by $q_{\max} = \langle q \rangle + \eta_{\max}$ and $q_{\min} = \langle q \rangle + \eta_{\min}$. Then we have

$$\sum_{i=1}^{n} p_i \eta_i^2 \geq p_{\min} \left(\sum_{i=1}^{n} \eta_i^2 \right)_{p_i \neq 0} \geq p_{\min}(\eta_{\max}^2 + \eta_{\min}^2)$$

$$\geq p_{\min} \tfrac{1}{2}(\eta_{\max} - \eta_{\min})^2 = p_{\min} \tfrac{1}{2}(q_{\max} - q_{\min})^2. \tag{5.78}$$

Thus Lemma 5.5 follows from Lemma 5.4.

Lemma 5.5 If $\psi(\langle q \rangle) - \langle \psi(q) \rangle < \varepsilon$, then

$$q_{\max} - q_{\min} \leq 2 \left(\frac{\varepsilon}{p_{\min}} \right)^{1/2}, \tag{5.79}$$

where $q_{\max} - q_{\min}$ is the largest interval between two q_i for which $p_i \neq 0$ and $p_{\min}$ is the smallest p_i except 0.

Let us now consider n different probability distributions $\{p_i\}$ and label them with an additional index j, $j = 1, 2, \cdots, n$. We thus write p_{ji} for these probability distributions. Corresponding to the normalization of the p's, we have

$$p_{ji} \geq 0, \qquad \sum_{i=1}^{n} p_{ji} = 1. \tag{5.80}$$

Lemma 5.3 can be written for each j; that is,

$$\sum_{i=1}^{n} p_{ji} \psi(q_i) \leq \psi \left(\sum_{i=1}^{n} p_{ji} q_i \right), \qquad j = 1, 2, \cdots, n. \tag{5.81}$$

Further assuming the normalization with respect to the index j,

$$\sum_{j=1}^{n} p_{ji} = 1, \tag{5.82}$$

and summing (5.81) over j, we obtain

$$\sum_{i=1}^{n} \psi(q_i) \leq \sum_{j=1}^{n} \psi \left(\sum_{i=1}^{n} p_{ji} q_i \right). \tag{5.83}$$

Equality in (5.83) holds if and only if equality holds in each of n relations in (5.81).

Now we can interpret p_{ji} as the transition matrix $p_{ji}(t)$ of a doubly stochastic chain, because of the normalizations (5.80) and (5.82). Furthermore, we interpret q_i as the occupancy probability w_i at $t - 1$; then $\sum_{i=1}^{n} p_{ji} q_i$ will be the occupancy probability w_j at t, as in (5.41). Then the left-hand side of (5.83) is the entropy $S(t - 1)$ at $t - 1$, and the right-hand side is the entropy $S(t)$ at t, according to the definition of S in (5.53). We use the natural logarithm unit here.

The equals sign in (5.81) for a given j will hold according to Lemma 5.3 if and only if the $q_i = w_i(t - 1)$ are equal for all the i's for which $p_{ji} \neq 0$, that is, for all those states z_i from which z_j can be reached by nonzero transition probability. Now, in order that the equality in (5.83) may hold, it is necessary and sufficient that the aforesaid be true for all j; that is, the $w_i(t - 1)$ must be equal in each family of terminally connected states. Thus we have the following theorem.

Theorem 5.3 (Prototype H-theorem). We have in a double stochastic Markov chain

$$S(t) \geq S(t - 1), \tag{5.84}$$

where equality holds if and only if the $w_i(t - 1)$ have the same value within each of the families of terminally connected states defined by $p_{ji}(t)$.

It is extremely important to note that the condition mentioned above about the case of equality is meant only for a particular stage from $(t - 1)$ to t. In order to have the equality, the $w_i(t - 1)$ within each set $\mathcal{A}_\alpha$ of terminally connected states must be constant, but this constant can be different from one $\mathcal{A}_\alpha$ to another $\mathcal{A}_\beta$. This guarantees that $S(t) = S(t - 1)$, but not necessarily that $S(t + 1) = S(t)$. Hence, in order to have $S(t) = S(t - 1)$ for consecutive values of t, we have to expect a more restrictive condition. This lemma is important for a case in which we consider only one link of the Markov chain, as in the problem mentioned at the beginning of this section in connection with (5.39). After having introduced Theorem 5.3 which is a direct consequence of Lemma 5.3, we now consider a consequence of Lemmas 5.4 and 5.5.

First, consider the condition

$$S(t) - S(t - 1) < \varepsilon \tag{5.85}$$

or, equivalently,

$$\sum_{j=1}^{n} \psi\left(\sum_{i=1}^{n} p_{ji} q_i\right) - \sum_{i=1}^{n} \psi(q_i) < \varepsilon. \tag{5.86}$$

The left-hand side is the sum over j of the terms

$$\psi\left(\sum_{i=1}^{n} p_{ji} q_i\right) - \sum_{i=1}^{n} p_{ji} \psi(q_i), \qquad j = 1, 2, \cdots, n. \tag{5.87}$$

Since this term is non-negative according to Lemma 5.3, it follows from (5.86) that

$$\psi\left(\sum_{i=1}^{n} p_{ji} q_i\right) - \sum_{i=1}^{n} p_{ji} \psi(q_i) < \varepsilon \tag{5.88}$$

for each j.

What Lemma 5.5 states is that if (5.88) is true, then for a given j, the q_i corresponding to the i's for which $p_{ji} \neq 0$ can vary only within a small domain given by

$$2\left[\frac{\varepsilon}{p_{\min}(j)}\right]^{1/2}. \tag{5.89}$$

Here $p_{\min}(j)$ is the smallest nonzero p_{ji} with the given j. Now consider all the states belonging to a family $\mathcal{A}_\alpha$ of terminally connected states. Then the difference between the largest q_i and the smallest q_i within $\mathcal{A}_\alpha$ must be smaller than the sum of the terms of the type (5.89), where j runs over all the states in $\mathcal{C}(\mathcal{A}_\alpha) = \mathcal{B}_\alpha$. Each term in (5.89) is obviously not larger than $2(\varepsilon/p_0)^{1/2}$, where p_0 is the common lower bound of the $p_{ji}(t)$ explained at the beginning of this section, and the number of j's in $\mathcal{B}_\alpha$ is less than or equal to n. Therefore the maximum variance of q_i within $\mathcal{A}_\alpha$ is less than $2n(\varepsilon/p_0)^{1/2}$. Summarizing these results, we obtain the following lemma.

Lemma 5.6 If we have in a doubly stochastic Markov chain

$$S(t) - S(t-1) < \varepsilon \tag{5.90}$$

then the difference between the largest $w_i(t-1)$ and the smallest $w_i(t-1)$ within a family $\mathcal{A}_\alpha$ of terminally connected sets is bounded by

$$[w_{\max}(t-1) - w_{\min}(t-1)]_{\mathcal{A}_\alpha} < 2n\left(\frac{\varepsilon}{p_0}\right)^{1/2}. \tag{5.91}$$

Let us now consider a temporal sequence

$$S(0), S(1), S(2), \cdots, S(t), \cdots, \tag{5.92}$$

which is a monotonically increasing sequence according to the prototype H-theorem, Theorem 5.3. On the other hand, there exists an upper bound for $S(t)$, namely, $\log n$. Hence we have the previously stated lemma.

Lemma 5.7 If $S(t)$ is defined by (5.53) in a doubly stochastic Markov chain, it converges to a limit as $t \to \infty$.

In other words, for a given $\varepsilon > 0$, there exists an integer T such that

$$S(t) - S(t-1) < \varepsilon \quad \text{for} \quad t \geq T. \tag{5.93}$$

A combined consequence of Lemmas 5.6 and 5.7 is that (5.91) is true for all $t \geq T$. This does not imply that the $w_i(t)$ within an $\mathcal{A}_\alpha$ converge to a limit as $t \to \infty$. It means only that they become close to each other; hence they can, for instance, oscillate in unison with time. However, we can say more than just (5.91) for $t \geq T$. To see that, we introduce Lemma 5.8.

Lemma 5.8 Let it be assumed in a doubly stochastic Markov chain that the difference between the largest $w_{\max}(t-1)$ and the smallest $w_{\min}(t-1)$ occupancy probabilities in each family of terminally connected states is not larger than δ, that is,

$$[w_{\max}(t-1) - w_{\min}(t-1)]_{\mathcal{A}_\alpha} \leq \delta, \tag{5.94}$$

with the same δ for all α. Then

(i) $[w_{\max}(t) - w_{\min}(t)]_{\mathcal{B}_\alpha} \leq \delta$ for all α; (5.95)

(ii) if $|w_k(t) - w_l(t)| \leq \eta$ and $z_i \rightarrow z_k$ and $z_j \rightarrow z_l$, then
$|w_i(t-1) - w_j(t-1)| \leq \eta + 2\delta$; (5.96)

(iii) if $|w_k(t-1) - w_l(t-1)| \leq \eta$ and $z_k \rightarrow z_i$ and $z_l \rightarrow z_j$, then
$|w_i(t) - w_j(t)| \leq \eta + 2\delta$. (5.97)

Proof. (i) If z_i is in $\mathcal{B}_\alpha$ then $w_i(t) = \sum_k^{\mathcal{A}_\alpha} p_{ik}(t)w_k(t-1)$ and $\sum_k^{\mathcal{A}_\alpha} p_{ik}(t) = 1$ because of double stochasticity. Hence

$$[w_{\min}(t-1)]_{\mathcal{A}_\alpha} \leq w_i(t) \leq [w_{\max}(t-1)]_{\mathcal{A}_\alpha}.$$

As a consequence, if both z_i and z_j belong to $\mathcal{B}_\alpha$, $|w_i(t) - w_j(t)| \leq [w_{\max}(t-1) - w_{\min}(t-1)]_{\mathcal{A}_\alpha} \leq \delta$. (ii). Let $z_k \in \mathcal{B}_\alpha$ and $z_l \in \mathcal{B}_\beta$; then $z_i \in \mathcal{A}_\alpha$ and $z_j \in \mathcal{A}_\beta$, where α and β may or may not be the same. Then, z_k being in $\mathcal{B}_\alpha$, $w_k(t)$ will be between $[w_{\min}(t-1)]_{\mathcal{A}_\alpha}$ and $[w_{\max}(t-1)]_{\mathcal{A}_\alpha}$ (see above). Hence $|w_i(t-1) - w_k(t)| \leq \delta$. Similarly, $|z_j(t-1) - z_l(t)| \leq \delta$. These two conditions, combined with the premise $|w_k(t) - w_l(t)| < \eta$, yield $|w_i(t-1) - w_j(t-1)| \leq \eta + 2\delta$. Similarly for (iii).

Lemma 5.9 For a given positive number $\theta > 0$, there exists another positive number $\delta > 0$, such that if we have

$$[w_{\max}(t-1) - w_{\min}(t-1)]_{\mathcal{A}_\alpha} \leq \delta \quad \text{for} \quad t \geq T \tag{5.98}$$

with the same δ for all α in a doubly stochastic Markov chain, we shall have

$$[w_{\max}(t) - w_{\min}(t)]_{\mathcal{C}_\lambda} \leq \theta \quad \text{for} \quad t \geq T \tag{5.99}$$

for all λ, where the left-hand side of (5.99) means the largest difference between two occupancy probabilities at t within each family $\mathcal{C}_\lambda$ of associated states.

Proof. We should recall that there are five alternative (not necessarily disjoint) conditions for membership in a $\mathcal{C}_\lambda$ (see Definition 5.3). Corresponding to them, we have the following five facts. (a) If $|w_i(t) - w_j(t)| \leq \theta_1$ and $|w_j(t) - w_k(t)| \leq \theta_2$, then $|w_i(t) - w_k(t)| \leq \theta_1 + \theta_2$. (b) If $z_i \in \mathcal{A}_\alpha$ and $z_j \in \mathcal{A}_\alpha$, then $|w_i(t) - w_j(t)| \leq \delta$ for $t \geq T - 1$ according to (5.98). (c) If $z_i \in \mathcal{B}_\alpha$ and $z_j \in \mathcal{B}_\alpha$, then $|w_i(t) - w_j(t)| < \delta$ for $t \geq T$ according to (5.95) and (5.98). (d) If $z_i \rightarrow z_k$, $z_j \rightarrow z_l$, and $|z_k(t) - z_l(t)| \leq \theta_1$ for $t \geq T$, then $|z_i(t-1) - z_j(t-1)| \leq \theta_1 + 2\delta$ according to (5.96) for $t \geq T$. (e) If $z_k \rightarrow z_i$, $z_l \rightarrow z_j$, and $|z_k(t-1) - z_l(t-1)| \leq \theta_1$ for $t \geq T$, then $|z_i(t) - z_j(t)| \leq \theta_1 + 2\delta$ for $t \geq T$ according to (5.97). Since there are a finite number of states, a finite number of applications of these five conditions will connect any two states within a $\mathcal{C}_\lambda$. If we make δ sufficiently small, θ_1 and θ_2 appearing in the foregoing formulas will become as small as one wants; hence we can satisfy (5.99).

Now Lemmas 5.6 and 5.7 show that the premise (5.98) of Lemma 5.9 is true for large T. Combining this with Lemma 5.9, we conclude the following.

Lemma 5.10 In a doubly stochastic chain, for a given positive number $\varepsilon > 0$, there exists an integer T such that if $t \geq T$ then

$$[w_{\max}(t) - w_{\min}(t)]_{\mathcal{C}_\lambda} \leq \varepsilon \quad \text{for} \quad t \geq T \tag{5.100}$$

within each family of associated states.

Now from Theorem 5.2, p. 217, we can obviously derive the following lemma.

Lemma 5.11 Suitably labeling the families of associated states in an irreducible Markov chain (doubly stochastic or not), we have

$$\sum_i^{\mathcal{C}_\lambda} w_i(t) = \sum_i^{\mathcal{C}_{\lambda+1}} w_i(t+1);$$

hence

$$\sum_i^{\mathcal{C}_\lambda} w_i(t) = \sum_i^{\mathcal{C}_\lambda} w_i(t+\tau). \tag{5.101}$$

Combination of Lemmas 5.10 and 5.11 yields, in view of (5.66), the next theorem.

Theorem 5.4 (Ensemble-average ergodic theorem—doubly stochastic case). Consider a sequence

$$w_i(k),\ w_i(k+\tau), \qquad w_i(k+2\tau), \cdots, w_i(k+\nu\tau), \cdots, \tag{5.102}$$

where k is an integer such that $0 \leq k < \tau$ and $w_i(t)$ is the occupancy probability at t of the ith state that belongs to the λth family of associated states $\mathcal{C}_\lambda$ in an irreducible doubly stochastic chain with τ families of associated states. Then $w_i(k+\nu\tau)$ converges, as $\nu \to \infty$, to a value dependent only on λ and k,

$$w_\lambda(k) = \frac{\tau \sum_i^{\mathcal{C}_\lambda} w_i(k)}{n}. \tag{5.103}$$

Furthermore, $w_\lambda(k)$ obeys

$$w_{\lambda+1}(k+1) = w_\lambda(k), \tag{5.104}$$

where if $\lambda = \tau$ then $\lambda + 1$ is to be understood as 1, and if $k = \tau - 1$ then $k + 1$ is to be understood as 0.

Proof. From Lemma 5.10 it follows that for a given ε there exists an integer T such that

$$\left| w_i(t) - \frac{\sum_i^{\mathfrak{C}_\lambda} w_i(t)}{N(\mathfrak{C}_\lambda)} \right| \leq \varepsilon \quad \text{for} \quad t \geq T, \tag{5.105}$$

where $z_i \in \mathfrak{C}_\lambda$. Consequently for a given ε there exists an integer ν_0 such that

$$\left| w_i(k + \nu\tau) - \frac{\sum_i^{\mathfrak{C}_\lambda} w_i(k + \nu\tau)}{N(\mathfrak{C}_\lambda)} \right| \leq \varepsilon \quad \text{for} \quad \nu \geq \nu_0. \tag{5.106}$$

But $\sum_i^{\mathfrak{C}_\lambda} w_i(k + \nu\tau)$ is a constant independent of ν according to (5.101) of Lemma 5.11. Hence it is determined by the initial w_i's at $t = 0$. Thus $w_i(k + \nu\tau)$ converges with $\nu \to \infty$ to the limit

$$w_\lambda(k) = \frac{\sum_i^{\mathfrak{C}_\lambda} w_i(k)}{N(\mathfrak{C}_\lambda)} = \frac{\sum_i^{\mathfrak{C}_{\lambda-k}} w_i(0)}{N(\mathfrak{C}_\lambda)}. \tag{5.107}$$

This proves the first statement (5.103) of the theorem in view of (5.66). If $\lambda - k \leq 0$, add τ. Next, according to Lemma 5.10, we can satisfy (5.105) as well as

$$\left| w_j(t + 1) - \frac{\sum_i^{\mathfrak{C}_{\lambda+1}} w_i(t + 1)}{N(\mathfrak{C}_{\lambda+1})} \right| \leq \varepsilon \quad \text{for} \quad t \geq T, \tag{5.108}$$

where $z_j \in \mathfrak{C}_{\lambda+1}$ by choosing a sufficiently large T. According to (5.101) we have $\sum_i^{\mathfrak{C}_\lambda} w_i(t) = \sum_j^{\mathfrak{C}_{\lambda+1}} w_j(t + 1)$, and according to (5.66) we have $N(\mathfrak{C}_\lambda) = N(\mathfrak{C}_{\lambda+1})$ in the doubly stochastic case. Hence

$$|w_i(t) - w_j(t + 1)| < 2\varepsilon \quad \text{for} \quad t > T, \tag{5.109}$$

where $z_i \in \mathfrak{C}_\lambda$ and $z_j \in \mathfrak{C}_{\lambda+1}$. For a sufficiently large ν_0, therefore, we have

$$|w_i(k + \nu\tau) - w_j(k + 1 + \nu\tau)| < 2\varepsilon \quad \text{for} \quad \nu > \nu_0. \tag{5.110}$$

This result, combined with the convergence of $\lim_{\nu\to\infty} w_i(k + \nu\tau)$ to $w_\lambda(k)$, yields (5.104).

Theorem 5.5 (Ensemble-average ergodic theorem—doubly stochastic case with single cyclic class). If the entire set of n states in a doubly stochastic Markov chain consists of a single family of associated states, $w_i(t)$ for any i converges to $1/n$ as $t \to \infty$, and $S(t) \to \log n$.

Proof. This is so because $\sum_i^{\mathfrak{C}_\lambda} w_i(t) = \sum_i^{\mathfrak{Z}} w_i(t) = 1$ and $N(\mathfrak{C}_\lambda) = N(\mathfrak{Z}) = n$ [see (5.103)].

Theorem 5.6 (Time-and-ensemble-average ergodic theorem—doubly stochastic case). No matter how many families of associated states there are in $\mathfrak{Z}$, the time average of $w_i(t)$ in an irreducible doubly stochastic Markov chain converges to $1/n$; that is,

$$\lim_{t\to\infty} \langle w_i(t)\rangle_t = \lim_{t\to\infty} \frac{1}{t}\sum_{t'=0}^{t-1} w_i(t') = \frac{1}{n}. \tag{5.111}$$

Proof. If $t = \nu\tau + k$ with $k < \tau$, we can write

$$\langle w_i(t)\rangle_t = \frac{1}{t}\sum_{t'=0}^{t-1} w_i(t') = \frac{1}{t}\left[\sum_{k'=0}^{\tau-1}\sum_{\nu'=0}^{\nu-1} w_i(k' + \nu'\tau) + \sum_{k'=0}^{k-1} w_i(\nu\tau + k')\right]. \tag{5.112}$$

For large t, that is, for large ν, we may ignore the last k terms and equate t with $\nu\tau$. Hence

$$\lim_{t\to\infty} \langle w_i(t)\rangle_t = \frac{1}{\tau}\sum_{k'=0}^{\tau-1} \lim_{\nu\to\infty}\frac{1}{\nu}\sum_{\nu'=0}^{\nu-1} w_i(k' + \nu'\tau) \tag{5.113}$$

$$= \frac{1}{\tau}\sum_{k'=0}^{\tau-1} w_\lambda(k') = \frac{1}{\tau}\sum_{k'=0}^{\tau-1}\sum_{i}^{\mathfrak{C}_{\lambda-k'}} \frac{w_i(0)}{N(\mathfrak{C}_\lambda)} \tag{5.114}$$

by Theorem 5.4. But this double summation is equivalent to a summation over all the i's, and $N(\mathfrak{C}_\lambda) = N(\mathfrak{Z})/\tau = n/\tau$. Hence $\lim_{t\to\infty}\langle w_i(t)\rangle_t = 1/n$.

Theorem 5.7 (Total connection of states in an ergodic set). In an irreducible doubly stochastic Markov chain, for any two given states z_i and z_j, there exists an integer ν_0 such that $p^t{}_{ji} \neq 0$ with $t = k + \nu\tau$ for $\nu \geq \nu_0$, and k is defined by $k = \mu - \lambda$ with $z_i \in \mathfrak{C}_\lambda$ and $z_j \in \mathfrak{C}_\mu$. The p^t_{ji} is defined by $p^t_{ji} = \sum_a\sum_b \cdots \sum_c\sum_d p_{ja}(t)p_{ab}(t-1)\cdots p_{cd}(2)p_{di}(1)$.

Proof. Let $w_m(0) = 0$ for $m \neq i$ and $w_i(0) = 1$. Then $p^t_{ji} = w_j(t)$. If $z_j \in \mathfrak{C}_\mu$ and $z_i \in \mathfrak{C}_\lambda$, $w_j(k + \nu\tau)$ will be, according to (5.106) and (5.101), arbitrarily close to $\sum_i^{\mathfrak{C}_{\mu-k}} w_i(0)/N(\mathfrak{C}_\lambda)$ for sufficiently large ν, where $\mu - k = \lambda$ and $\sum_i^{\mathfrak{C}_\lambda} w_i(0) = 1$. Hence $w_j(t) = p^t{}_{ji} \neq 0$. There is a finite chain such that $z_i \to z_d \cdots z_a \to z_j$.

Theorem 5.8 (Theorem of recurrence and ubiquitous migration). Let $x(t)$ be the state at time t of a single string belonging to an irreducible doubly stochastic Markov chain. For any given state z_j and for any given time t_1 the probability is 1 that the state $x(t)$ will become z_j for some $t > t_1$, no matter what the starting state $x(0)$ is.

Informal Proof. Consider first the case of one family of associated states. Let T be a length of time such that if $x(t_0) = z_i$ the probability that $x(t_0 + T)$ will be z_j for any j is $(1/n) \pm \varepsilon$. Theorem 5.5 guarantees the existence of such T. Starting at t_0, the probability that $x(t_0 + T)$ will be z_j is approximately $(1/n)$ and the probability that $x(t_0 + T)$ will not be z_j but that

$x(t_0 + 2T)$ will be z_j is approximately $[1 - (1/n)](1/n)$; similarly, the probability that $x(t_0 + \mu T)$ will not be z_j for $\mu < \nu$ but will be z_j for $\mu = \nu$ is approximately $[1 - (1/n)]^{\nu-1}(1/n)$. Hence the probability that $x(t_0 + \mu T)$ will be z_j at some μ is approximately

$$\frac{1}{n}\left[1 + \left(1 - \frac{1}{n}\right) + \left(1 - \frac{1}{n}\right)^2 + \cdots\right] = \frac{1}{n} \cdot \frac{1}{1 - (1 - 1/n)} = 1. \quad (5.115)$$

The total error in this approximation must be finite, because it concerns a definite probability, and this error must tend to zero as $\varepsilon \to 0$ and $T \to \infty$. In the case when there are more than one family of associated states, we should take the first period T equal to $k + \nu\tau$ with some sufficiently large integer ν, and with $k = \mu - \lambda$, where $z_i \in \mathfrak{C}_\lambda$ and $z_j \in \mathfrak{C}_\mu$. The second and later periods T should be $\nu\tau$, with a sufficiently large ν. The probability $1/n$ must be replaced by $1/N(\mathfrak{C}_\lambda) = \tau/n$ [see (5.66)]. The result (5.115) is not affected by this change.

The properties of recurrence and ubiquitous migration being proved, the next step to individual ergodicity (strong law of large numbers) is intuitively almost obvious, but a rigorous mathematical proof would require considerable additional space, which does not seem to warrant its insertion in this section. So far our discussion has been limited to the time-dependent, irreducible, doubly stochastic Markov chain under the assumption of invariant connectivity. We now pass to the case of an irreducible Markov chain that is not necessarily doubly stochastic. As far as the time dependence of transition probabilities is concerned, the assumption of invariant connectivity must be further restricted, as will be explained presently. Of course, the time-independent case is a special case among those allowed by this new restriction. The method we use is what I introduced some time ago under the name of "splitting of states," which consists of reducing the non-doubly-stochastic case to the doubly stochastic case.

When the set of states $\mathfrak{Z} = \{z_i\}$, $i = 1, 2, \cdots, n$, is given, we introduce another set of (split) states $\mathfrak{Z}^* = \{z(i, \alpha_i)\}$ with a double index, $i = 1, 2, \cdots, n$, and $\alpha_i = 1, 2, \cdots, m_i$. When the Markov chain $p_{ji}(t) = p_t(j \mid i)$ is defined on $\mathfrak{Z}$, we introduce a corresponding Markov chain $p_t(j, \alpha_j \mid i, \alpha_i)$ on $\mathfrak{Z}^*$, where

$$p_t(j, \alpha_j \mid i, \alpha_i) = \frac{1}{m_j}\, p_t(j \mid i); \quad (5.116)$$

that is, the transition probabilities do not depend on α_j or an α_i. The factor $1/m_j$ guarantees normalization with regard to the final states for the "split" Markov chain when the original Markov chain satisfies the same normalization:

$$\sum_{j=1}^{n} \sum_{\alpha_j=1}^{m_j} p_t(j, \alpha_j \mid i, \alpha_i) = \sum_{j=1}^{n} p_t(j \mid i) = 1. \quad (5.117)$$

On the other hand, the sum over the initial states is different for the two Markov chains:

$$\sum_{i=1}^{n} \sum_{\alpha_i=1}^{m_i} p_t(j, \alpha_j \mid i, \alpha_i) = \sum_i \frac{m_i}{m_j} p_t(j \mid i). \tag{5.118}$$

The condition that the split Markov chain becomes doubly stochastic is then determined by setting (5.118) = 1; that is,

$$\sum_{i=1}^{n} [p_t(j \mid i) - \delta_{ji}]m_i = 0, \qquad j = 1, 2, \cdots, n. \tag{5.119}$$

If we can solve these n equations for the m_i and get a positive integer value for each m_j, we can reduce the general case to the doubly stochastic case by the use of (5.116). It is clear that (5.119) determines only the ratio of the m_i. If $p_t(j \mid i)$ is already doubly stochastic, the same value for all the m_i will satisfy the equation. To obtain the $w_i(t)$ from the $w_i(0)$ we need to put

$$w_{i,\alpha_i}(0) = \frac{1}{m_i} w_i(0); \tag{5.120}$$

then we obviously obtain, for any t,

$$w_j(t) = m_j w_{j,\alpha_j}(t). \tag{5.121}$$

If the $p_t(j \mid i)$ satisfy the condition of invariant connectivity, then $p_t(j, \alpha_j \mid i, \alpha_i)$ satisfy the same condition. But the m_i's must remain constant to make our correspondence unique. Hence the $p_t(j \mid i)$ can still vary with time insofar as the condition of invariant connectivity is satisfied and the roots of (5.119) remain unchanged. The last condition is also the condition for the connectivity diagram for the split states to remain constant. Hence we call the conjunction of these two conditions "strict condition of invariant connectivity."

The solubility of (5.119) is guaranteed, for

$$\det [p_t(j \mid i) - \delta_{ji}] = 0. \tag{5.122}$$

This last condition is satisfied because the sum of corresponding terms on all the n rows in the determinant gives a row with n zeros:

$$\sum_{j=1}^{n} [p_t(j \mid i) - \delta_{ji}] = 0, \qquad i = 1, 2, \cdots, n. \tag{5.123}$$

What remains to be shown is (a) that the m_i's can all be made positive and (b) that they can all be made integers. The first point can be proven by one of the many theorems attributable to Frobenius [F-10]. We state the theorem without proof. The theorem concerns a simultaneous set of linear algebraic equations of the type

$$\sum_{i=1}^{n} A_{ji} v_i = \lambda v_j, \qquad i, j = 1, 2, \cdots, n, \tag{5.124}$$

where A_{ji} are given and v_j are the unknowns. These equations are soluble if and only if the λ is one of the n roots of the nth degree algebraic equation:

$$\det [A_{ji} - \lambda\delta_{ji}] = 0. \tag{5.125}$$

Lemma 5.12 (Frobenius theorem). If

$$A_{ji} \geq 0, \qquad i, j = 1, 2, \cdots, n \tag{5.126}$$

and the matrix is irreducible, then (5.125) has a positive and single root $\lambda^{(0)}$ such that $\lambda^{(0)}$ is not smaller than the absolute value of any other root of (5.125). The v_i in (5.124) corresponding to $\lambda^{(0)}$ can be taken positive for all i. [This implies, among other things, that $\sum v_i \neq 0$ for $\lambda^{(0)}$.]

In our case the condition (5.126) is, of course, satisfied because A_{ji} are probabilities here. The matrix will be irreducible if we limit the Markov chain to the irreducible cases. As far as λ is concerned, we can easily find the value of $\lambda^{(0)}$ for (5.124) with $A_{ji} = p_t(j \mid i)$. Summation over the j of the left-hand side of (5.124) gives

$$\sum_{j=1}^{n} \sum_{i=1}^{n} A_{ji} v_i = \sum_{i=1}^{n} v_i \tag{5.127}$$

because of the usual normalization. The same summation of the right-hand side gives $\lambda \sum_{j=1}^{n} v_j$; hence $\lambda = 1$, unless $\sum_{j=1}^{n} v_j = 0$, and (5.124) becomes (5.119). Thus $\lambda^{(0)}$ must correspond to $\lambda = 1$, and we are now assured that we can take all m_i's positive. (Other solutions λ must satisfy $\sum_{j=1}^{n} v_j = 0$ and hence cannot be $\lambda^{(0)}$.)

Next we have to show that these m's can all be made integers. For this purpose we note that if the $p_t(j \mid i)$'s are all rational numbers then the m_i's are also all rational, since m_i can be expressed as a quotient of a determinant made from the $p_t(j \mid i)$'s by another determinant made from the $p(j \mid i)$'s. Because all the m_i's can be multiplied by any arbitrary number in (5.119), the rational values of m_i's can be made integers by multiplying by a suitable large number. When the given values of $p_t(j \mid i)$ are irrational numbers, we can always express them in an arbitrarily good approximation by rational numbers, such as decimal numbers of arbitrary length. Consequently, in an arbitrarily good approximation, we can determine the integers m_j to be used in (5.116), and then we can pass to the limit where the $p_t(j \mid i)$ tend to real numbers.

We further note that if two states z_i and z_j are terminally connected, initially connected, or associated in the original chain, any one of the split states (i, α_i) of z_i and any one of the split states (i, α_i) of z_j are respectively terminally connected, initially connected, or associated. The converse is also true. Thus the number of "families" of each kind remains unchanged, whereas the number of states in each family changes, each state z_i splitting into m_i states.

The total number of states, for instance, becomes $\sum_{i=1}^{n} m_i = m$. The reader can also check that a graph that would be forbidden as a doubly stochastic chain is transformed into an allowed one by splitting, insofar as we remain in the irreducible case. A reducible non-doubly-stochastic chain, which allows (5.63), cannot be transformed into a doubly stochastic chain. We see that the proviso of irreducibility in Frobenius' theorem automatically excludes such an eventuality.

Let us now try to extract conclusions regarding non-doubly-stochastic chains by applying the earlier theorems to the doubly stochastic chain obtained by splitting of states. First, the entropy in this latter (designated as S_{DS}) is

$$\begin{aligned} S_{\mathrm{DS}}(t) &= -\sum_{i=1}^{n}\sum_{\alpha_i=1}^{m} w_{i,\alpha_i}(t)\log w_{i,\alpha_i}(t) \\ &= -\sum_{i=1}^{n} w_i(t)\log\frac{w_i(t)}{m_i} \\ &= S^*(t), \end{aligned} \tag{5.128}$$

where $S^*(t)$ is the "relative" entropy function defined in the original non-doubly-stochastic chain, with the *a priori* weight m_i attached to each state z_i. This corresponds to (1.29), in which we put $A = \sum_{i=1}^{n} m_i$, and w_i is replaced by ω_i, given by

$$\omega_i = \frac{m_i}{m} = \frac{m_i}{\sum_{i=1}^{n} m_i}. \tag{5.129}$$

The relative entropy $S^*(t)$ defined in (5.128) differs only by a constant $\log(\sum m_i)$ from another form of the relative entropy $S^{**}(t)$, defined by

$$S^{**}(t) = -\sum w_i(t)\log\frac{w_i(t)}{\omega_i}. \tag{5.130}$$

Because the prototype H-theorem applies to $S_{\mathrm{DS}}(t)$, we obtain the following theorem.

Theorem 5.9 (One-step H-theorem). In an irreducible Markov chain we have

$$S^{**}(t) \geq S^{**}(t-1), \tag{5.131}$$

the equals sign holding if and only if the $w_i(t-1)/\omega_i$ have the same value within each of the families of terminally connected states defined by $p_{ji}(t)$.

The entropy S^{**} in (5.131) is defined by (5.130), and the $m_i = \omega_i m$ are defined by (5.119). The theorem is also true for S^* of (5.128). Because we consider only one step, we need not invoke the assumption of constant connectivity here. In the case of doubly stochastic chains we need not limit ourselves to the irreducible ones, because a reducible chain can be considered as a juxtaposition of more than one independent irreducible chain. But in the case of a non-doubly-stochastic chain we have to limit ourselves to an irreducible chain so that we can use Frobenius' theorem, excluding cases like (5.63).

In usual physics textbooks the result (5.131) is proved on a very restrictive assumption that the $p(j\,|\,i)$ are given so that the following n^2 equations hold, namely,

$$\omega_i p(j\,|\,i) = p(i\,|\,j)\omega_j, \tag{5.132}$$

which is sometimes called the principle of microscopic balance. This is a stronger condition than the n equations obtained from (5.132) by a summation over i [i.e., (5.119) with (5.129)]. Our Theorem 5.9, on the other hand, is based only on the irreducibility of the matrix $p(j\,|\,i)$. The condition (5.132) reduces, in view of (5.116) and (5.129), to

$$p(j\alpha_j \,|\, i\alpha_i) = p(i\alpha_i \,|\, j\alpha_j), \tag{5.133}$$

which is a much more restrictive condition than what we need, namely,

$$\sum_{i=1}^{n} \sum_{\alpha_i=1}^{m} p(j\alpha_j \,|\, i\alpha_i) = 1. \tag{5.134}$$

Although the condition (5.133) plus the usual normalization

$$\sum_{i=1}^{n} \sum_{\alpha_j=1}^{m} p(j\alpha_j \,|\, i\alpha_i) = 1 \tag{5.135}$$

will give (5.134), the combination of (5.134) and (5.135) does not imply (5.133). It is clear that (5.132) automatically excludes the existence of more than one family of associated states $\mathcal{C}$, but within unique $\mathcal{C}$ the condition of irreducibility does not mean any further restriction, whereas the condition (5.132) does.

Applying Theorem 5.4 to the (split) doubly stochastic chain, we obtain the following theorem, in which it is assumed that the "strict" condition of invariant connectivity is satisfied [see the paragraph below (5.79)].

Theorem 5.10 (Ensemble-average ergodic theorem—non-doubly-stochastic case). Consider a sequence

$$w_i(k),\ w_i(k+\tau),\ w_i(k+2\tau), \cdots,\ w_i(k+\nu\tau), \cdots, \tag{5.136}$$

where k is such that $0 \le k < \tau$ and $w_i(t)$ is the occupancy probability at t of the ith state that belongs to $\mathcal{C}_\lambda$ $(1 \le \lambda \le \tau)$ in an irreducible, not-necessarily-doubly-stochastic chain with τ families of associated states. Then $w_i(k+\nu\tau)$ converges, as $\nu \to \infty$, to a value that, if divided by m_i, is dependent only on λ and k; that is, $w_i(k + \nu\tau) \to m_i w_\lambda(k) = m_i \cdot \sum_l^{\mathfrak{C}_\lambda} w_l(k) / \sum_j^{\mathfrak{C}_\lambda} m_j$. This $w_\lambda(k)$ obeys

$$w_{\lambda+1}(k + 1) = w_\lambda(k), \tag{5.137}$$

where if $\lambda = \tau$ then $\lambda + 1 = 1$, and if $k = \tau - 1$ then $k + 1 = 0$.

Similarly, Theorems 5.5–5.8 engender the following four theorems. The proof is not needed.

Theorem 5.11 (Ensemble-average ergodic theorem—single cyclic class). If the entire set of n states in a not-necessarily-doubly-stochastic Markov chain consists of a single family of associated states, $w_i(t)$ converges to $\omega_i = m_i / \sum_j^{\text{all}} m_j = m_i/m$ as $t \to \infty$, hence $S^*(t) \to \log m$, and $S^{**}(t) \to 0$.

Theorem 5.12 (Time-and-ensemble-average ergodic theorem). Independently of the number of families of associated states, the time average of $w_i(t)$ in an irreducible not-necessarily-doubly-stochastic Markov chain converges to $\omega_i = m_i/m$.

Theorem 5.13 (Total connectivity). Theorem 5.7 is also valid in the irreducible, not-necessarily-doubly-stochastic case.

Theorem 5.14 (Recurrence and ubiquitous migration). Theorem 5.8 is valid also in the irreducible, not-necessarily-doubly-stochastic case.

This completes practically all that was intended to be presented in this section. We add two remarks, however, before passing on to the next section, one pertaining to the reducible case and the other to "time reversal" in the Markov chain.

In the doubly stochastic Markov chain the reducible case means that the entire $\mathfrak{Z}$ is divided into subsets of states that are mutually completely isolated. Each subset, if no longer divisible in the same fashion, forms an "ergodic" subset. We can then treat each ergodic subset separately, in the same way as we treat the entire $\mathfrak{Z}$ in the irreducible case. The only difference is that the sum of $w_i(t)$ within a single ergodic subset will not be unity, although it remains constant in time. The entropy defined in the entire $\mathfrak{Z}$ will still increase with time.

In the non-doubly-stochastic Markov chain the reducible case will admit not only completely isolated subsets, but also partly isolated subsets, such as $\mathfrak{C}$ or $\mathfrak{Z} - \mathfrak{C}$ of (5.63). If an influx-free subset contains no smaller efflux-free subsets within it, the entire probability in it will flow out into another subset roughly exponentially with time. If such vanishing subsets exist in $\mathfrak{Z}$, the entropy could in some cases decrease with time until all the vanishing subsets are sufficiently drained out, after which, however, each efflux-free subset, which does not contain a vanishing subset within it, starts to act as an irreducible $\mathfrak{Z}$. The entropy of each such nonvanishing subset will settle to a limiting value dependent on the total probability it has received from the vanishing subsets.

If $\mathfrak{Z}$ consists of one vanishing subset and one nonvanishing subset, the limiting entropy will be the same as in the case in which $\mathfrak{Z}$ consisted only of this nonvanishing subset from the beginning. [The $p_{ji}(t)$ with i in the vanishing subset and j in the nonvanishing subset is then ignored.] When the vanishing subset is very large and the nonvanishing subset is very small, the latter may be called a "sink," because the entire probability is drained into it. In such a case the behavior of the entropy will appear almost reversed and show a conspicuous decreasing tendency, although this decrease may or may not be entirely monotonic. In particular when, the nonvanishing subset consists of only one state, the final entropy will be zero. So much for the reducible Markov chains.

Our discussion of Chapter 3, if applied to a Markov chain, pertains to the equation (3.1) which can be written

$$\begin{aligned} w_i(t-1)p(x_t = z_j \mid x_{t-1} = z_i) &= p(x_{t-1} = z_i, x_t = z_j) \\ &= w_j(t)p(x_{t-1} = z_i \mid x_t = z_j), \end{aligned} \tag{5.138}$$

where $p(x_t = z_j \mid x_{t-1} = z_i) = p_{ji}(t)$ is the transition probability, which is a "predictive" probability, while $p(x_{t-1} = z_i \mid x_t = z_j)$ is not given in the definition of the Markov chain and is to be interpreted as a "retrodictive" probability. The point made in Chapter 3 was that we cannot derive this retrodictive probability from the predictive probability alone because $w_i(t-1)$ is not determined by the predictive probabilities. This is true at each stage of a Markov chain, that is, at any given value of t. See [W-19] and [W-22] for an explanation of the "one-wayness" of physical phenomena from this point of view. The H-theorem (except for the case when the equals sign holds) can be applied in the past-to-future direction but not in the future-to-past direction.

In a Markov chain, however, the final-state occupancy probability $w_j(t)$ at t becomes the initial-state occupancy probability for $t+1$. Thus (5.138) or the formula derived from it,

$$w_j(t) = \sum_{i=1}^{n} w_i(t-1)p(y_t = z_j \mid y_{t-1} = z_i), \tag{5.139}$$

is used *ad infinitum*. And, as we have seen in this section, if there is only one family of associated states, the $w_i(t)$ approach, with $t \to \infty$, values independent of t and independent of the initial condition expressed by $w_i(0)$. These limiting values $\omega_i = w_i(\infty)$ were determined solely by the transition probabilities. This property, which is an aspect of ergodicity, though not in contradiction to the general theorem of impossibility of retrodiction, produces a new interesting situation. Under the strict assumption of invariant connectivity ω_i was determined by

$$\sum_{i=1}^{n} p(x_t = z_j \mid x_{t-1} = z_i)\omega_i = \omega_j, \tag{5.140}$$

with any arbitrary t. In the following discussion we first assume the case of a single family of associated states and later broaden the consideration to cover the irreducible Markov chain with any number of families of associated states.

Let us consider an ensemble of strings belonging to a Markov chain, which all start at $t = 0$. Then, at sufficiently large t, the occupancy probabilities $w_i(t)$ become almost equal to ω_i. Now suppose that we shift the starting points of all these strings as far back into the past as we want. Then the $w_i(t)$ in the finite region of t will have the value ω_i. The chain thus obtained is said to be "aged." Comparison of (5.140) with (5.139) guarantees that if $w_i(t-1) = \omega_i$, $w_i(t) = \omega_i$ too. This creates a situation that may be described in the terminology of Chapter 3 as a uncontrollable system.

Substituting ω_i and ω_j for $w_i(t-1)$ and $w_j(t)$ in (5.138), we can obtain the retrodictive probability in terms of the predictive probability and the ω's, which in turn are determined by the predictive probabilities:

$$p(x_{t-1} = z_i \mid x_t = z_j) = \frac{\omega_i}{\omega_j}\, p(x_t = z_j \mid x_{t-1} = z_i). \tag{5.141}$$

This is normalized with respect to x_{t-1}, as it should be. Equation (5.141) looks somewhat similar to (5.132), but it is entirely different because (5.132) involves only two predictive probabilities and means actually

$$p(x_t = z_i \mid x_{t-1} = z_j) = \frac{\omega_i}{\omega_j}\, p(x_t = z_j \mid x_{t-1} = z_i). \tag{5.142}$$

Equations (5.141) and (5.142) are respectively comparable to (3.39) and (3.42).

The Markov chain under consideration is supposed to have started at a time point in the extremely remote past ($t = -\infty$) with a certain initial condition. But on account of ergodicity the initial condition has been completely forgotten by the time the chain has reached finite values of t. Everything in the finite region of t (and thereafter) is actually determined by the predictive probability $p(x_t = z_j \mid x_{t-1} = z_i)$. Now let us interchange the role of the past and the future and consider a chain that would be produced according to the retrodictive probability $p(x_{t-1} = z_i \mid x_t = z_j)$ of (5.141), starting from an "initial" condition given at a time point in the remote future ($t = +\infty$). The basic equation (5.139) of a Markov chain is now replaced by

$$w_i'(t-1) = \sum_{j=1}^{n} w_j'(t)\, p(x_{t-1} = z_i \mid x_t = z_j), \tag{5.143}$$

where $w_i'(t)$ is the occupancy probability in this reversed Markov chain. The concept of associated states remains unchanged even if we interchange the initial and final states. Therefore, if $\mathfrak{Z}$ consists of one family of associated states $\mathcal{C}$ in the usual interpretation, it also consists of one $\mathcal{C}$ in the reversed interpretation. Of course, if $\mathfrak{Z}$ is irreducible in the usual interpretation, it will be also irreducible in the reverse interpretation. Therefore we can use the same argument as before and the limiting value of $w_i(t)$ will be given, in analogy with (5.140), by

$$\omega_i' = \sum_{j=1}^{n} \omega_j' p(x_{t-1} = z_i \mid x_t = z_j), \tag{5.144}$$

which has only one solution ω_i' except for an arbitrary multiplicative factor (which is fixed by normalization). With the help of (5.141), (5.144) now becomes

$$\omega_i' = \omega_i \sum_{j=1}^{n} \frac{\omega_j'}{\omega_j} p(x_t = z_j \mid x_{t-1} = z_i), \tag{5.145}$$

which is satisfied by

$$\omega_i' = \omega_i, \qquad i = 1, 2, \cdots, n. \tag{5.146}$$

Because (5.144) has only one solution, it must be (5.146). This means that if a Markov chain is ergodic, the reverse Markov chain is also ergodic, and the limiting distribution in the former is the same as that of the latter. Insofar as we are observing the finite region of t, we cannot say which way the Markov chain is running. In the remote past and the remote future, however, the ordinary Markov chain and the reversed Markov chain are generally at variance.

In the case of more than one family of associated states, we could "match" the original chain and the reversed chain by requiring that the sum of $w_i(t)$ in each family of associated states be the same as the sum of the $w_i'(t)$ in the same family at a certain finite value of t. Then the reversed Markov chain would go through the cycle in the opposite direction, keeping the "matching" at least in the finite region of t. But matching depends on the initial condition in this case, unlike the former case, in which matching automatically takes place. The (nonvanishing) retrodictive probability is given by

$$p(x_{t-1} = z_i \mid x_t = z_j) = \frac{w_i(t-1)}{w_j(t)} p(x_t = z_j \mid x_{t-1} = z_i) \tag{5.147}$$

instead of (5.141), where if $z_i \in \mathcal{C}_{\lambda-1}$, $z_j \in \mathcal{C}_\lambda$. If the chain is aged, we have, according to Theorem 5.10, $w_j(t) = m_j w_\lambda(k)$, where $t = k + \nu\tau$ $(\nu \to \infty)$, and $w_i(t-1) = m_i w_{\lambda-1}(k-1)$; but $w_\lambda(k) \doteq w_{\lambda-1}(k-1)$ according to (5.137). Hence, although $w_i(t-1)$ and $w_j(t)$ depend on t, their ratio does not. As a consequence (5.147) becomes

$$p(x_{t-1} = z_i \mid x_t = z_j) = \frac{m_i}{m_j}\, p(x_t = z_j \mid x_{t-1} = z_i), \tag{5.148}$$

which is equivalent to (5.141).

5.3. THE *H*-THEOREM IN DEDUCTION AND OBSERVATION

It is well known, under the name of the second law of thermodynamics, that if one makes two observations at t_0 and t_1 ($>t_0$) on an (adiabatically) isolated system, the value of the thermodynamic entropy at t_1 is larger than, or at least equal to, the value at t_0. Accepting the idea that thermodynamic entropy represents in some sense a degree of lack of information about a system, one notes that the mathematical H-theorem explained in the last section seems to provide a rough model of this thermodynamic phenomenon. However, one is not clear about the kind of information in question here; furthermore, one is puzzled by the role of observation in this increase of entropy. Every time one makes an observation, the entropy seems to increase; that is, our ignorance about the system seems to increase. (A paradigm case: in von Neumann's description of irreversibility of observation [V-3], the act of observation itself is made responsible for an increase in the so-called microscopic entropy.) But is this not contrary to the very notion of observation? After all, we make observation in order to increase information and not in order to increase ignorance. This section and the next will deal with these problems, not from the point of view of theoretical physics, but from the point of view of information theory.

As in the last section we assume that two sets, $\mathcal{A} = \{A_k\}$ and $\mathcal{D} = \{D_i\}$, correspond to two observations of the same kind made at two different times, t_0 and t_1, respectively, thereby A_k and the D_i represent the possible outcomes of the observations. A single hypothesis H_I out of the set $\mathcal{H} = \{H_I\}$ is supposed to be chosen to stand for the "law," that is, for the most highly credited hypothesis. For this reason we no longer mention $\mathcal{H}$ or H_I in this section. The content of H_I expressed by the conditional probability $p(D_i \mid A_k)$, which does not determine the "initial condition" $p(A_k)$. Hence the "final condition" $p(D_i) = \sum_k p(D_i \mid A_k)p(A_k)$ is determined jointly by the law H_I and the initial condition.

$S(\mathcal{A})$ here represents our ignorance about the outcome $\mathcal{A}$ at t_0, and $S(\mathcal{D})$ our ignorance about the outcome $\mathcal{D}$ at t_1; our knowledge about $\mathcal{A}$ is represented by $p(A_k)$, and our knowledge about $\mathcal{D}$ is represented by $p(D_i) = \sum_k p(D_i \mid A_k)p(A_k)$. Suppose that we have actually made an observation at t_0 and obtained a particular result A_k; then the ignorance about $\mathcal{D}$ at t_1 will become $S(\mathcal{D} \mid A_k)$, defined by the probability $p(D_i \mid A_k)$. The quantity $S(\mathcal{D} \mid \mathcal{A})$

is the average of this $S(\mathcal{D} \mid A_k)$. Obviously this average conditional ignorance $S(\mathcal{D} \mid \mathcal{A})$ must be smaller than (at most equal to) the ignorance $S(\mathcal{D})$. In fact, this is guaranteed because $S(\mathcal{D}) - S(\mathcal{D} \mid \mathcal{A})$ is equal to the interrelation $J(\mathcal{D} \otimes \mathcal{A}; \mathcal{D}, \mathcal{A})$ which is non-negative. We may consider, as in Section 5.1, this difference $S(\mathcal{D}) - S(\mathcal{D} \mid \mathcal{A})$ as the information $\inf(\mathcal{D} \mid \mathcal{A})$ about $\mathcal{D}$ provided by the outcome $\mathcal{A}$ at t_0.

From the point of view of ensemble, we can restate the foregoing situation as follows. Take an ensemble $\mathcal{E}_0$ consisting of a large number N of objects, of which $N_k = p(A_k)N$ are to be found in state A_k at t_0. The uncertainty regarding the state of an object at t_0 is then $S(\mathcal{A})$. If the hypothesis H is true—which we assume to be the case—the fraction $p(D_i \mid A_k)$ of N_k objects will be found in state D_i at t_1. This means that, in total, $\sum_k p(D_i \mid A_k)N_k = \sum_k p(D_i \mid A_k)p(A_k)N = p(D_i)N$ will be found in state D_i at t_1. The uncertainty in $\mathcal{E}_0$ about the outcome at t_1 is thus $S(\mathcal{D})$. (In classical physics this conclusion is true whether or not the observation $\mathcal{A}$ is made at t_0 in $\mathcal{E}_0$, but in quantum physics it has to be assumed that the observation is made at t_0 in $\mathcal{E}_0$.) Suppose that we actually perform the observation at t_0 on all the objects in $\mathcal{E}_0$, and classify the objects into families or subensembles $\mathcal{E}_k$ according to the outcome at t_0, such that $\mathcal{E}_k$ consists of the N_k objects that gave the outcome A_k. We now consider each subensemble $\mathcal{E}_k$ separately and ask what is the uncertainty about the outcome of the observation at t_1. This uncertainty is $S(\mathcal{D} \mid A_k)$. Because each subensemble $\mathcal{E}_k$ has N_k members, the average of $S(\mathcal{D} \mid A_k)$ must be given by $\sum_k S(\mathcal{D} \mid A_k)N_k/N$, which is $S(\mathcal{D} \mid \mathcal{A})$. The concept of ensemble is such that the probability can be reinterpreted as a relative frequency in a large collection of similar objects. Thus, when we have made no observation at t_0, the probability distribution of the state of each object at t_0 is $p(A_k)$, which is represented by the ensemble $\mathcal{E}_0$, of which N_k members would give result A_k at t_0. If we perform an observation on an object at t_0 and find it in state A_k, the probability that this particular object is in state A_k is unity. The ensemble corresponding to this new probability distribution is $\mathcal{E}_k$. This means that the ensemble $\mathcal{E}_0$ *contracts* to $\mathcal{E}_k$ as a result of the observation at t_0, which gave result A_k. This contraction of ensemble is an important concept. From the point of view of a set of objects, this contraction can be considered as a polychotomy (see Section 1.4).

From a purely experimental point of view, which is interested only in observational results, a question immediately arises pertaining to the comparison of $S(\mathcal{A})$ and $S(\mathcal{D})$. This means that we keep the same ensemble $\mathcal{E}_0$ and seek to determine the change in time of the uncertainty about the state in which we would find the object. Regarding this problem we have already obtained a theorem (Theorem 5.3) that states the following: if the matrix $p(D_i \mid A_k)$ is such that $\sum_k p(D_i \mid A_k) = 1$, then $S(\mathcal{D}) \geq S(\mathcal{A})$. Since the relation $\sum_i p(D_i \mid A_k) = 1$ is always true, the condition $\sum_k p(D_i \mid A_k) = 1$

means that the matrix is doubly stochastic. Equality $S(\mathcal{D}) = S(\mathcal{A})$ holds if and only if $p(A_k)$ is constant within each of the so-called families of terminally connected states. If $p(D_i \mid A_k)$ determines only one family of terminally connected sets, $S(\mathcal{D}) = S(\mathcal{A})$ will be the case if and only if $p(A_k) = 1/n$. If the matrix is a permutation matrix, that is, if there is one and only one element equal to unity in each column and each row, there will be n families of terminally connected states; as a result, $S(\mathcal{D}) = S(\mathcal{A})$ can be satisfied by any distribution $p(A_k)$. We also know that if $\mathcal{A} \to \mathcal{D}$ is just one link of a Markov chain and if the rest of the chain is also governed by the same transition matrix $p(D_i \mid A_k)$, $S(\mathcal{A}) = S(\mathcal{D})$ will be automatically true if this particular link $\mathcal{A} \to \mathcal{D}$ is located after an infinite number of preceding links of the chain. This is the case for an aged Markov chain. In any event, except in the special case in which $S(\mathcal{D}) = S(\mathcal{A})$ takes place, the uncertainty (ignorance) about the state increases from t_0 to t_1. In other words, we suffer a loss of information $L(\mathcal{A} \to \mathcal{D})$ during the elapse of time from t_0 to t_1, given by

$$L(\mathcal{A} \to \mathcal{D}) = S(\mathcal{D}) - S(\mathcal{A}) \geq 0. \tag{5.149}$$

This relation corresponds to the point of view of the H-theorem, which we may call that of "external" information balance, as distinct from the concept of balance discussed in Section 5.1. The relation (5.149) expresses, so to speak, the degeneration with time of information on $\mathcal{E}_0$. One may then further ask what the information change will be in the temporal development of $\mathcal{E}_k$ from t_0 to t_1 and also what the information change will be in the process of ensemble reduction at t_0 from $\mathcal{E}_0$ to $\mathcal{E}_k$. With regard to the first question, one may note that at t_0 the ensemble $\mathcal{E}_k$ has no uncertainty regarding the observational result, for all the objects in $\mathcal{E}_k$ are to be found in A_k at t_0. That is, the entropy $S(\mathcal{A} \mid A_k) = 0$. Similarly, $S(\mathcal{A} \mid \mathcal{A}) = \sum_k p(A_k) S(\mathcal{A} \mid A_k) = 0$. Regarding the result of an observation made at t_1 on the objects in $\mathcal{E}_k$, however, we have an uncertainty $S(\mathcal{D} \mid A_k)$, which is in general non-negative and in most cases is nonzero. Similarly, $S(\mathcal{D} \mid \mathcal{A})$ is non-negative in general and is nonzero in most cases. Thus the uncertainty also increases from t_0 to t_1 for $\mathcal{E}_k$, fitting the general relation of the type (5.149). Next, in regard to the entropy change in the ensemble reduction, we must note that before the observation at t_0 the uncertainty regarding the states in $\mathcal{A}$ is $S(\mathcal{A})$ and after the observation at t_0 (i.e., right after the ensemble reduction) the uncertainty becomes zero; that is, $S(\mathcal{A} \mid A_k) = 0$. Here a process takes place that involves an entropy decrease, in contrast to (5.149). This must be so, for, after all, an observation brings new information. We come back to this problem a little later in this section and in the next section.

This consideration serves as a simplified model of what we usually encounter in statistical mechanics. We now give some more explanation of the problems as they present themselves in physics, without going into too technical details. Usually the object system under discussion is considered to be in one of the possible quantum states, but the observation can determine the object system to be in a coarsely defined macroscopic state, which consists of many quantum states. It is further assumed that the transition probability from an initial quantum state to a final quantum state can be replaced by an average transition probability that is a function only of the initial macroscopic state and of the final macroscopic state. Let $p(D_i, \alpha_i \mid A_k, \beta_k)$ be the transition probability from the quantum state A_k, β_k, which belongs to the macroscopic state A_k, to the quantum state D_i, α_i, which belongs to the macroscopic state D_i. Let the macroscopic state D_i contain n_i quantum states so that α_i has n_i different values, $1, 2, \cdots, n_i$. Similarly, let the macroscopic state A_k consist of n_k quantum states so that $\beta_k = 1, 2, \cdots, n_k$. It is known that the microscopic transition matrix $p(D_i, \alpha_i \mid A_k, \beta_k)$ in quantum physics is doubly stochastic; that is,

$$\sum_i \sum_{\alpha_i} p(D_i, \alpha_i \mid A_k, \beta_k) = \sum \sum_{\beta_k} p(D_i, \alpha_i \mid A_k, \beta_k) = 1. \tag{5.150}$$

This doubly stochasticity follows from the unitarity of the time development of quantum states [see Section 3.2 as well as (9.163); see also Watanabe [W-7]].

The macroscopic transition probability $p(D_i \mid A_k)$ from A_k to D_i must be obtained from $p(D_i, \alpha_i \mid A_k, \beta_k)$ by averaging it with respect to β_k with an appropriate weight for each β_k and summing it with respect to α_i. But if $p(D_i, \alpha_i \mid A_k, \beta_k)$ depends strongly on β_n, the average of $p(D_i, \alpha_i \mid A_k, \beta_k)$ with respect to β_k will depend on the weight distribution over the β_k and cannot be considered as a function only of D_i and A_k. If, for some physical reason, this weight distribution can be assumed to be uniform (an important physical assumption; see Section 3.2), we can put

$$p(D_i \mid A_k) = \frac{\sum_{\alpha_i} \sum_{\beta_k} p(D_i, \alpha_i \mid A_k, \beta_k)}{n_k}, \tag{5.151}$$

which, of course, satisfies $\sum_i p(D_i \mid A_k) = 1$ on account of the first half of (5.150). On the other hand, the second half of (5.150) yields, with the definition (5.151),

$$\sum_k n_k p(D_i \mid A_k) = n_i. \tag{5.152}$$

This relation, of course, returns to the doubly stochastic condition if all the n's are unity.

According to the one-step H-theorem (Theorem 5.9), we can now derive the following conclusion, provided that the matrix $p(D_i \mid A_k)$ is irreducible.

First define $S(\mathcal{A})$ and $S(\mathcal{D})$ as relative entropies in accordance with (5.128):

$$S(\mathcal{A}) = -\sum_k p(A_k)\log\frac{p(A_k)}{n_k} \tag{5.153}$$

and

$$S(\mathcal{D}) = -\sum_i p(D_i)\log\frac{p(D_i)}{n_i}\,. \tag{5.154}$$

We omit the asterisk from S for simplicity. The H-theorem now states that

$$L(\mathcal{A}\to\mathcal{D}) \equiv S(\mathcal{D}) - S(\mathcal{A}) \geq 0, \tag{5.155}$$

where the equality $S(\mathcal{D}) = S(\mathcal{A})$ holds if $p(A_k)$ is the solution of

$$\sum_k p(D_i \mid A_k)p(A_k) = p(D_i)$$

with $p(A_i) = p(D_i)$. This equation gives, in the light of (5.152), the solution

$$p(A_k) = \frac{n_k}{\sum_k n_k}\,. \tag{5.156}$$

This means that the ignorance about the macroscopic state, as measured by the relative entropy, increases from t_0 to t_1, except for the cases in which $p(A_k) = n_k/\sum_k n_k$, which give the constancy of the entropy. If $\mathcal{A}$ and $\mathcal{D}$ are just one link in a long Markov chain, the entropy increase will cease only after repetition of a large number of such links, that is, only after the Markov chain has become "aged."

The relative entropy used in (5.153) and (5.154) is nothing but what is known as Gibbsian entropy in physics. The relation between Gibbsian entropy and Boltzmannian entropy, which is of the form

$$S_B = \log n_k, \tag{5.157}$$

is, as explained before, such that if the probability distribution sharpens so that $p(A_k) = 1$ in (5.153) for a particular k, then these two entropies will become equal. In other words, if we perform an observation on the object, we can determine to which macroscopic state the object belongs; thus Gibbsian entropy becomes identical with Boltzmannian entropy immediately after each observation. In this sense we speak of "postobservational entropy" to signify the entropy value in the particular situation when both types of entropy coincide.

We can further elaborate this explanation by writing (5.153) as

$$S(\mathcal{A}) = -\sum_k p(A_k)\log p(A_k) + \sum_k p(A_k)\log n_k. \tag{5.158}$$

The first term represents the simple entropy, which measures the uncertainty as to which macroscopic state the object belongs to, and the second term represents the average of the uncertainty (Boltzmannian entropy) regarding the microscopic state, which will remain even after the object has been found in each macroscopic state. If the first term in (5.158) is negligible compared with the second term, $S(\mathcal{A})$ can be interpreted as the expected value of Boltzmannian entropy. This interpretation is allowed irrespective of the actual values of $p(A_k)$ if the number $n_{\min}$ of quantum states in the smallest macroscopic state is still very large compared with the number N of macroscopic states that come into consideration, for the maximum value of the first term in (5.158) is $\log N$ and the minimum value of the second term is $\log n_{\min}$ (This N has nothing to do with the N of p. 238):

$$\log n_{\min} \gg \log N. \tag{5.159}$$

There is good reason to believe that (5.159) is usually the case in thermodynamic observation. Another argument, however, based on grounds different from (5.159), makes it plausible that the first term of (5.158) is negligible compared with the second in practice. Before presenting this argument, we should add a few remarks about the meaning of the first term.

Suppose that we have a correct prediction $p(A_k)$ in the ensemble at t_0 before actually making observation $\mathcal{A}$. We take a member of the ensemble, make the observation $\mathcal{A}$ on it, and obtain a particular value A_k. After the observation $p(A_k)$ becomes 1 for this particular A_k, and 0 for the rest. By this sudden change of $p(A_k)$, the first term changes from $-\sum_k p(A_k) \log p(A_k)$ to zero, decreasing by the amount $-\sum_k p(A_k) \log p(A_k)$, irrespective of which A_k one may obtain by observation. The second term, however, changes to a new value $\log n_k$, which may be larger or smaller than its expected value $\sum_k p(A_k) \log n_k$, depending on the observed A_k. But insofar as the prediction $p(A_k)$ is correct, the average value in the ensemble of the second term does not change at all because of the sudden change in $p(A_k)$ due to the observation. Hence the average change of $S(\mathcal{A})$ at the moment of observation is

$$-(\Delta S)_{\text{obj}} = -\sum_{k=1}^{N} p(A_k) \log p(A_k). \tag{5.160}$$

This is exactly the amount of information obtainable by observation regarding the macroscopic state of the system expressed in terms of simple entropy. We discuss this decrease of entropy by observation in the next section. The subscript "obj" is attached for convenience of discussion in the next section, but it indicates that this entropy refers to the observed object as distinct from the observing system. For the purpose of this paragraph, we note only that the average decrease of entropy (5.160) that results from observation is a negligible amount in the quantity (5.153) whose temporal change is governed by the long-range prediction expressed by (5.155).

Insofar as the entropy is defined by (5.153) and (5.154), the entropy non-decrease law (5.155) is a purely mathematical consequence. However, a question remains as to whether the postobservational entropy at t_1 is actually larger than (or equal to) the postobservational entropy at t_0 and, if so, how this fact is connected with the mathematical theorem (5.155). Suppose that we obtained the result A_k at t_0, making $p(A_k)$ equal to 1 for this particular k so that $S(\mathcal{A})$ became simply $\log n_k$. If we let this system develop with time to the instant t_1, the Gibbsian entropy $S(\mathfrak{D})$ would become

$$S(\mathfrak{D} \mid A_k) = -\sum_i p(D_i \mid A_k) \log p(D_i \mid A_k) + \sum_i p(D_i \mid A_k) \log n_i. \quad (5.161)$$

Although $S(\mathfrak{D} \mid A_k)$ and $S(\mathfrak{D})$ are different entities, they become equal if $\mathcal{E}_0$ is, as we first assumed, equal to $\mathcal{E}_k$ from the beginning. For this reason we write $S(\mathfrak{D})$ for $S(\mathfrak{D} \mid A_k)$ for simplicity in the following. If we perform an observation at t_1, there is probability $p(D_i \mid A_k)$ that we shall obtain the postobservational entropy $\log n_i$. The second term in (5.161) is, as already observed, the expected value of this postobservational entropy. If the condition (5.159) holds, the first term in (5.161) becomes negligible, and the law $S(\mathfrak{D}) \geq S(\mathcal{A})$ (5.155) may be considered as stating that the postobservational entropy at t_1 is on the average larger than (or equal to) the postobservational entropy at t_0. But even in the absence of this condition we can draw a similar conclusion if there is a large number of Markovian links between t_0 and t_1, and if the n_i are considerably larger for a few "privileged" macroscopic states than for the remaining "nonprivileged" macroscopic states. Indeed, if there are many steps of the Markov chain between t_0 and t_1, according to the ensemble-average ergodic theorem (Theorem 5.11) $p(D_i \mid A_k)$ will become approximately $n_i/\sum_i n_i$ (if there is only one family of associated states). As a result, the first term will become of the order of magnitude of the logarithm of the number of "privileged" macroscopic states for which the n_i are appreciably larger than for the rest. The second term will become of the order of magnitude of the logarithm of these large n's. Under these circumstances the first term is also negligibly small compared with the second term. Thus $S(\mathfrak{D})$ again becomes the expected value of the postobservational entropy at t_1. In other words, if $p(D_i \mid A_k)$ becomes exactly $n_i/\sum_i n_i$, then, according to (5.154), $S(\mathfrak{D}) = \log \sum n_i$, where $\sum n_i$ is the total number of quantum states under consideration. (Physicists would say that $\sum n_i$ is the number of quantum states occupied by the micro-canonical ensemble.) But if a few macroscopic states take up a large portion of $\sum n_i$, the $\log n_i$ of those very large macroscopic states will not be much different from $\log \sum n_i$ as far as the order of magnitude is concerned. The above statements will become in fact true if there is one unique macroscopic state (usually called a Maxwell-Boltzmann cell) whose $\log n_i$ is almost equal to $\log \sum n_i$. This shows that the consideration in terms of Gibbsian entropy is closely related to the consideration in terms of postobservational entropy. It is true that if A_k happens to be the Maxwell-Boltzmann cell, $S(\mathfrak{D})$ as calculated by the Markov chain model has to become less than $\log n_k$ even if the difference may be infinitesimal. This is due to the inevitable fluctuations. An object that was once in the Maxwell-Boltzmann cell will be rediscovered in the same cell most of the time, although occasionally it may migrate out of the cell.

The above explanation refers to a special case of (5.153) and (5.154), namely, the case in which $p(A_k) = 1$ and $p(D_i) = p(D_i \mid A_k)$. But the reader can now easily see that a similar explanation is possible in the general case. $S(\mathcal{A})$ of (5.153) is essentially the average of the $\log n_k$ of the states in which the system starts, that is, some of those states with small n's. As far as $S(\mathfrak{D})$ is concerned, it will be about the same as $S(\mathfrak{D} \mid A_k)$ because the above explanation really does not depend on A_k.

So far we have considered the relative entropy of the type (5.153) or (5.154) as the Gibbsian entropy of a physical system, and each state such as A_k or D_i as macroscopic states of the system. But it may be noted in passing that the Boltzmannian entropy of a classical gas can also be written in the form of a relative entropy of the type (5.153) or (5.154), in which the states A_k and D_i are replaced by the macroscopic states of each molecule of the gas.

Let each molecular macroscopic state be labeled with the index σ, and assume it to consist of g_σ microscopic states. The molecules are supposed to have very weak interaction, and the macroscopic state A_k of the gas, consisting of ν molecules, is characterized by the numbers ν_σ of molecules occupying the molecular macroscopic states σ. Then the number of possible ways of distributing ν molecules in different microscopic states in compliance with the prescribed macroscopic distribution ν_σ is

$$n_k = \frac{\nu!}{\nu_1!\,\nu_2!\cdots}\, g_1^{\nu_1} g_2^{\nu_2} \cdots \tag{5.162}$$

according to non-quantum-theoretical statistics. The Boltzmannian entropy of the state A_k will then become, with the use of Sterling's formula,

$$\begin{aligned} S_B(A_k) &= \log n_k = \nu \log \nu - \sum_\sigma \nu_\sigma \log \frac{\nu_\sigma}{g_\sigma} \\ &= \nu\left(-\sum_\sigma \gamma_\sigma \log \frac{\gamma_\sigma}{g_\sigma}\right), \end{aligned} \tag{5.163}$$

where

$$\gamma_\sigma = \frac{\nu_\sigma}{\nu}. \tag{5.164}$$

Thus in this case the Boltzmannian entropy per molecule turns out to be expressed in the form of a relative entropy.

5.4. THERMODYNAMIC COST OF OBSERVATION AND BRILLOUIN'S NEGENTROPY PRINCIPLE

Observation means acquisition of information. Would not this acquisition of information counterbalance the loss of information implied by the H-theorem? Even if this were impossible on the average, could not one take advantage of individual fluctuations from the average to prevent the degeneration of information? If observation revealed that a system (or its constituent part) happens to be in a low-entropy state by accident, one might perhaps be able to change the boundary condition of the system in such a way that it could no longer go back to a high-entropy state. Such a speculation provides the background of the discussion of this section. The name of Brillouin must be mentioned as the one who made in the past the most prominent contribution in this area (see [B-13]).

The first question pertains to whether or not the decrease in entropy (5.160) caused by observation, that is, by ensemble contraction, is in contradiction with the general principle of entropy increase. It should be recalled that this decrease refers to the first term of the expression (5.158) of the Gibbsian entropy. The second term of (5.158) can increase or decrease as a result of observation in an individual case, but on the average it remains unchanged at the moment of observation. Let us concentrate now on the first term, although it may represent a very small quantity compared with the second term.

To discuss intelligibly the problem of observation, it is always advisable to make a clear distinction between what I call "observer's viewpoint" and what I call "bystander's viewpoint." In any observation there are obviously two necessary parties involved: the observed system (object) and the observing system (subject). The latter includes the observing person (who may be a robot) and his apparatus of observation. The observer is concerned with his findings about the observed object, whereas the bystander is concerned with the combined system of the observed and observing systems. For the bystander an observation made by the observer is an interaction between two portions of bystander's combined observed object, and the act of observation by the observer generally causes some change in this combined system. In the observer's view it is quite natural that the simple macroscopic entropy of the observed system represented by the first term of (5.158) decreases as a result of observation, because this entropy measures the degree of his ignorance regarding the macroscopic state of the system. But it is quite conceivable that the act of observation being a physical operation causes an additional increase of entropy in the observing system. This question cannot be raised and discussed by the observer, but only by the bystander. As far as the information about the state of the observed system is concerned, we may assume that the bystander shares the same observational results obtained by the observer's apparatus. (Reading of the observer's meters by the bystander may cause a further additional entropy increase, but this is outside our present consideration.)

The operation of observation, which consists of locating the object in one of N possible cases (macroscopic states), can be considered as a polychotomy in the sense of Section 1.4. Then, by Theorem 1.6, p. 32, we have in general

$$-(\Delta S)_{\text{obj}} \leq \langle C \rangle, \tag{5.165}$$

where $\langle C \rangle$ is the average "cost" of a polychotomy resulting in locating the object in one of N possible cases, for which the probability distribution is given by $p(A_k)$. Now if there takes place a simultaneous increase $(\Delta S)_{\text{obs}}$ of entropy in the observing system, so that the total change $(\Delta S)_{\text{tot}}$ of entropy becomes non-negative, that is, if

$$(\Delta S)_{\text{tot}} = (\Delta S)_{\text{obs}} + (\Delta S)_{\text{obj}} \geq 0, \tag{5.166}$$

the following postulate will suggest itself as a general hypothesis from which (5.166) is derivable.

Postulate 5.1 A polychotomic observation made by an observing system on an object causes an increase of physical entropy in the observing system not less than the average cost of this polychotomy.

It may be noted that in physics entropy is measured in the unit of ergs (per degree centigrade), whereas the cost of polychotomy was defined previously in the unit of bits. Hence the "cost" in Postulate 5.1 must be understood as meaning $k \log_e 2$ times the original definition, where k is the so-called Boltzmann constant. It does not seem to be very easy to derive Postulate 5.1 in its generality from a statistical consideration based on the first principles of physics. It is not hard, however, to make it plausible. Suppose that we perform the polychotomy by a series of repeated dichotomies; then the average number of necessary dichotomies is $\langle C \rangle / k \log_e 2$, since the cost of each dichotomy is 1 bit or $k \log_e 2$ in physical units. To make each dichotomic observation, let us assume that we use photons of energy $h\nu$. If the answer is yes a photon will be absorbed in photomultiplier A, and if the answer is no a photon will be absorbed in photomultiplier B. The energy of the photon is eventually converted into heat ΔQ, which is absorbed by the photomultiplier and connected system at temperature T. Addition of heat causes an increase in the number of available microscopic states, resulting in an increase of physical entropy by the amount $\Delta Q/T$, as is well known in elementary physics. Thus this observation entails an entropy increase of the amount $h\nu/T$ in the observing system of which the photomultiplier is part. In order that the photomultiplier can "see" the photon, that is, in order that the arrival of the photon will be recognizable against the background of thermal agitation (noise), it is necessary that $h\nu \geq kT$. Thus one dichotomic observation results in an entropy increase not less than k. Of course, there is a decrease in entropy because of the emission of the photon at the light source, but this decrease can be made as small as one desires by making the temperature of the light source sufficiently high.

If we need $\langle C \rangle / k \log_e 2$ dichotomies, and each dichotomy results in an entropy increase not less than k, then the total entropy increase of the observing system must be $(\Delta S)_{\text{obs}} \geq \langle C \rangle$. (Note: $\log_e 2$ is of the order of unity and ≤ 1.) Because the average polychotomic cost is $\langle C \rangle$, we can claim that $(\Delta S)_{\text{obs}}$ is not less than the polychotomic cost of the observation. This agrees with, if not proves, Postulate 5.1. This argument is further reinforced by the fact that if an m-chotomy with $m > 2$ is used instead of a dichotomy and if m photons are used for this m-chotomy, the physical entropy increase will be $mh\nu/T \geq mk$ and the cost will be $k \log_e m$ in physical units. In this case the ratio of the entropy increase to the average cost will be larger than in the case of dichotomies, but the result will still agree with Postulate 5.1. We can probably show without much difficulty that an m-chotomy requires at least μ photons to be absorbed at the temperature of the observing system, where μ is an integer satisfying $\log m + 1 > \mu \geq \log m$. If so, we can claim Postulate 5.1 on fairly solid ground. Admittedly, our argument here is correct only in the order of magnitude because the relation $h\nu \geq kT$ is to be understood only in the order of magnitude and also because we have replaced $\log_e 2$ by unity. Probably we can sharpen the argument, as Brillouin did, so that the equality in the "not-less-than" in Postulate 5.1 can be reached under favorable circumstances. The reader may also have noticed that the use of photon $h\nu$ is not an essential part of the argument. If we use any observation that entails energy dissipation ε at temperature T, we should replace $h\nu$ by ε and get the same result. In any event, the results obtained here are in essential agreement with Brillouin's idea, even though the present derivation may appear quite different from his.

In (5.166) the term $(\Delta S)_{\text{obj}}$ had a special meaning referring to (5.160), but Postulate 5.1 can be applied to any polychotomic process of information acquisition. Hence we may give the following theorem as a general consequence of Postulate 5.1 and Theorem 1.6.

Theorem 5.15 Acquisition of information in the amount ΔI about an object by an observing system is accompanied, on the average, by an increase of physical entropy in the observing system in an amount not less than ΔI.

It may be noted that neither Postulates 5.1 nor Theorem 5.15 says anything about the entropy decrease in the observed system. In our foregoing argument we added to the conclusion of Theorem 5.15 another ingredient, which amounts to the statement that an observation that results in a decrease in the simple entropy $(-\Delta S_{obj})$ of the observed system accompanying an ensemble contraction provides the observer with information in the amount $\Delta I = -\Delta S_{obj}$. From Theorem 5.15 it follows that $\Delta I \leq \Delta S_{obs}$, and this result, in conjunction with the above statement, yields $-\Delta S_{obj} = \Delta I \leq S_{obs}$, which was the desired conclusion (5.166).

The main results we have obtained so far may be recapitulated as follows. The Gibbsian entropy consists of two parts: the first part, in the form of simple entropy, expresses the uncertainty regarding the macroscopic state, and the second part represents the expected value of the Boltzmannian entropy. Right after the observation the first term becomes 0, resulting in a decrease in the amount of $(-\Delta S_{obj})$ (5.160). The second term will become one of the possible values of the Boltzmannian entropy, whereas its value before the observation was the expected value of all these possible values. The postobservational value may be larger or smaller than the average in individual cases. But insofar as the theory used in calculating the expectation value is correct, the value of the second term on the average cannot change at the moment of observation. Thus, on the average, there occurs a decrease of $(-\Delta S_{obj})$ in the physical entropy in the observed system as a result of observation. But at each observation there takes place an increase of physical entropy in the observing system in the amount of ΔS_{obs}, which cannot be smaller than $(-\Delta S_{obj})$. Thus we can propose the following theorem.

Theorem 5.16 On the average, there cannot be any over-all physical entropy decrease at the moment of observation in the combined system of object and observer.

All this seems to be unavoidable as long as average behavior is considered. But how about the possible decrease of the second term at the moment of observation in an individual case? In fact, observation may occasionally reveal that the system, by chance fluctuation, happens to be in a small marcoscopic state (i.e., a state with small Boltzmannian entropy). We may then quickly change the boundary condition of the system so that it can no longer go back to a state with higher entropy. It is immediately clear that such a scheme would not work when the observation in question is an observation of the macroscopic state of a large macroscopic object, because a fluctuant migration to a macroscopic state with an appreciably small Boltzmannian entropy practically never happens.

The situation is not the same, however, if the observation concerns a system with a small number of degrees of freedom, such as molecule, because relatively large fluctuations can easily occur in a very small system. Thus, observing the motion of each molecule one after another and changing the boundary condition of its motion according to the observed result, one may be able to prevent the system from going back to a state of larger entropy.

One might thus expect that by repeating this operation, one can eventually decrease the entropy of the entire gas. This is the scheme of Maxwell's demon [B-8, B-10, B-13, D-6, D-7, S-12].

We now show by a simplified model that the demon's scheme does not work either, or, more precisely, that he has to pay for the decrease in entropy of the observed system by a larger increase of entropy in his own observing system. We said at the end of the last section that the Boltzmannian entropy of a classical gas can be written as the sum of the Gibbsian entropy of each molecule:

$$S = \nu \sum_\sigma \left(-\gamma_\sigma \log \frac{\gamma_\sigma}{g_\sigma}\right) = -\sum_\sigma \nu_\sigma \log \left(\frac{\nu_\sigma}{g_\sigma}\right) + \nu \log \nu \quad \text{bits}, \tag{5.167}$$

where the last constant term, $\nu \log \nu$, is inessential. To make it simple, let us assume that there are only two molecular macroscopic cells, $\sigma = 1$ and $\sigma = 2$. We assume that the energy values E_σ of the two cells are the same, so that the molecules can go freely from one cell to another without energy transfer. At equilibrium, ν_σ will become proportional to g_σ, since the energy-dependent factor $\exp(-E_\sigma/kT)$ is the same for both.

The demon's strategy consists of the following: he keeps on observing molecules systematically, and if a molecule happens to be passing from $\sigma = 1$ to $\sigma = 2$ he stops it so that it stays in $\sigma = 1$. If it happens to be passing from $\sigma = 2$ to $\sigma = 1$ he lets it go by. After some time a large number of molecules will be accumulated in $\sigma = 1$, far above the equilibrium value, which is $[g_1/(g_1 + g_2)]\nu$. This will entail a considerable entropy decrease. The change in entropy (5.167) caused by the passage of a single molecule from $\sigma = \beta$ to $\sigma = \alpha$ can be obtained by putting $\delta\nu_\alpha = -\delta\nu_\beta = 1$ in

$$\delta S = -\sum_\sigma \delta\nu_\sigma \left[\log \left(\frac{\nu_\sigma}{g_\sigma}\right) + 1\right]; \tag{5.168}$$

that is,

$$\Delta S_{\beta\to\alpha} = \log \frac{\nu_\beta g_\alpha}{\nu_\alpha g_\beta}. \tag{5.169}$$

The probability of each molecule in $\sigma = \beta$ passing to $\sigma = \alpha$ per unit time may be assumed to be proportional to g_α:

$$\frac{p(\beta \to \alpha)}{p(\alpha \to \beta)} = \frac{g_\alpha}{g_\beta}. \tag{5.170}$$

The number $\Delta\nu_{\beta\to\alpha}$ of molecules passing from $\sigma = \beta$ to $\sigma = \alpha$ per unit time will then be proportional to ν_β and $p(\beta \to \alpha)$:

$$\frac{\Delta\nu_{\beta\to\alpha}}{\Delta\nu_{\alpha\to\beta}} = \frac{\nu_\beta p(\beta \to \alpha)}{\nu_\alpha p(\alpha \to \beta)} = \frac{\nu_\beta g_\alpha}{\nu_\alpha g_\beta}. \tag{5.171}$$

The right-hand side of (5.171) is the argument variable of (5.169). At equilibrium we should have $\Delta\nu_{\beta\to\alpha} = \Delta\nu_{\alpha\to\beta}$, which entails $\Delta S_{\beta\to\alpha} = 0$, satisfying the extremum condition of the entropy.

Without the demon's intervention, molecules pass freely from $\sigma = 1$ to $\sigma = 2$ and from $\sigma = 2$ to $\sigma = 1$, keeping the values ν_α and ν_β in the neighborhood of their averages. Under the demon's control, however, any molecule that has passed from $\sigma = 2$ to $\sigma = 1$ is kept in $\sigma = 1$. This transition is no longer a retrievable fluctuation. The contribution by this molecule to the total Boltzmannian entropy (5.167) will have changed definitely by the amount

$$\Delta S_{\text{obj}} = \Delta S_{2\to1} = \log\left(\frac{\nu_2 g_1}{\nu_1 g_2}\right) \text{ bits.} \tag{5.172}$$

This becomes negative if ν_1/g_1 is larger than ν_2/g_2. This means that by packing more and more molecules in a small cell, the demon can efficiently decrease the entropy.

But the demon has to keep on observing molecules and determine which way they are going, and each observation increases the entropy of his own observing system by $k \log_e 2$ ergs or 1 bit, according to Postulate 5.1. Of the total number of observations, the number of cases in which the molecule turns out to be going from $\sigma = 2$ to $\sigma = 1$ [making (5.172) applicable] will represent the fraction $\Delta\nu_{2\to1}/(\Delta\nu_{1\to2} + \Delta\nu_{2\to1}) = \nu_2 g_1/(\nu_1 g_2 + \nu_2 g_1)$ according to (5.171). This means that in order to succeed in decreasing the entropy of the gas by the amount (5.172), the demon has to increase the entropy of his observing system by

$$\Delta S_{\text{obs}} = \frac{\nu_1 g_2 + \nu_2 g_1}{\nu_2 g_1} \text{ bits.} \tag{5.173}$$

The total change in entropy per intervention will then be

$$\Delta S_{\text{obj}} + \Delta S_{\text{obs}} = \log\left(\frac{\nu_2 g_1}{\nu_1 g_2}\right) + \frac{\nu_1 g_2 + \nu_2 g_1}{\nu_2 g_1}$$

$$= 1 + \frac{1}{x} + \log x, \tag{5.174}$$

with

$$x = \frac{\nu_2 g_1}{\nu_1 g_2}. \tag{5.175}$$

In the measure as the cell $\sigma = 1$ gets more and more crowded, a larger and larger majority of molecules will be found to be passing from cell $\sigma = 1$ to cell $\sigma = 2$, although the decrease of entropy due to a molecule passing from $\sigma = 2$ to $\sigma = 1$ will become larger in magnitude. This situation is reflected in the fact that as x decreases the term $[1 + (1/x)]$ becomes larger, although log x decreases. But $(1/x)$ increases faster than log x decreases with decreasing x, keeping the value of (5.174) positive for the entire positive domain of x. This means that the demon cannot decrease the entropy of the gas by his scheme without increasing the total entropy of the combined system of object and observer.

From the statistical point of view, entropy was defined by probabilities such as $p(A_k)$ or γ_σ, which can be represented by an ensemble. But in the framework of classical thermodynamics, entropy is a quantity determined by the boundary conditions and global thermodynamic variables, such as (uniform) pressure, density, and energy content of the system; in other words, we speak only of the entropy at an equilibrium, which will set in after a certain relaxation time under the given conditions. From this point of view, the ensemble contraction that corresponds to an observation of fluctuation does not result in an entropy decrease. This means that the entropy decrease $(-\Delta S_{\text{obj}})$ is to be considered as taking place not at the moment of observation (as we assumed), but at the moment of the observer's intervention (neglecting the relaxation time), which changes the boundary condition so that the system cannot go back to equilibrium under the old boundary condition. This picture, which is legitimate, allowed Brillouin to interpret the foregoing situation in a rather intriguing fashion. First, we obtain the information ΔI that results in an entropy increase $\Delta S_{\text{obs}}(\geq \Delta I)$ in the observing system, and then the observer, using this information ΔI, changes the boundary condition so that the entropy of the observed system decreases by the amount $(-\Delta S_{\text{obj}}) = \Delta I$. Thus the whole process is divided into two steps. If, according to Brillouin, we introduce the quantity $N = -S_{\text{obs}} - S_{\text{obj}}$, called negentropy, we can say that in the first step we get information by losing negentropy and in the second step we get negentropy by using information. If we write ΔI and ΔN for the gain of information and negentropy, we have $\Delta I + \Delta N \leq 0$ in each of the two steps. It is true that in the second step we had $\Delta I + \Delta N = 0$, but there may be some accompanying entropy increase in the process of intervention; hence it is safer to put the symbol $\leq$ instead of the symbol $=$ in the second step also.

Brillouin's Principle of Negentropy:

Information I can be changed into negentropy N, and negentropy N can be changed into information I. This transformation is governed by the condition that the sum of I and N can never increase:

$$\Delta(I + N) \leq 0. \tag{5.176}$$

The reader may raise some objections to this principle. One of them may be that the original purpose of exorcism of Maxwell's demon was to save the principle of nondecrease of entropy (i.e., $\Delta N \leq 0$) in its generality, but according to Brillouin's picture the physical entropy after an observation and before an intervention is larger than after the intervention; that is, the second step implies a decrease of physical entropy. In our original one-step explanation this difficulty did not exist, but in the two-step explanation it cannot be ignored. The possible answer to this objection is either to admit the fact (which implies that $\Delta N > 0$) and claim that the new principle $\Delta(I + N) \leq 0$ supersedes the old principle $\Delta N \leq 0$, or to maintain that the two steps are separated only in our idea and the intervention has to be made instantaneously before any further fluctuation can take place, so that we should not think that any actual time elapses between the first and the second steps; hence $\Delta N \leq 0$ is also true in the whole process while $\Delta(I + N) \leq 0$ is true in each of the two steps.

Another objection that can be raised is that what we have shown is that $\Delta N + \Delta I \leq 0$, which does not necessarily allow us to write $\Delta(N + I) \leq 0$. Indeed, in classical thermodynamics, we are allowed to write ΔQ for heat exchange but not to write Q, which would imply that there is a state variable expressing the amount of heat possessed by a system. Similarly, we can ask: can we attribute a definite amount of information content to a human brain or to any other system, such as a "magnetic core memory," so that we can write I of which ΔI is supposed to be a change? Of course, in this context the usual notion of information corresponds to ΔI, and this is supposed to be an increment of the total information I possessed by the system. Even if this question is answered affirmatively—that is, if it is meaningful to state that a memory system contains a certain definitive amount of information—it is not at all clear that the use of part of this information actually results in a corresponding decrease of the total stored information. It is true that the second step in Brillouin's description of the demon's operation assumes that the information ΔI is "used" to gain negentropy ΔN. But does this permit us to state that information can be "changed" into negentropy, as Brillouin's negentropy principle explicitly states, and as the expression $\Delta(I + N) \leq 0$ apparently implies?

Let us consider the first part of the question. In an animate system the total amount of accumulated knowledge is well-nigh undefinable. First, memory and habit are not clearly distinguishable. For instance, the response pattern of an animal, inasmuch as it is a result of learning process, may be considered as representing part of its accumulated information; on the other hand, the instinctive response pattern does not seem to be justifiably considered as part of accumulated information, unless we consider the entire process of evolution as a process of information gathering. If this last standpoint is tenable, not only the response pattern but also the entire organic functions of an animal must be considered as part of its accumulated information. Since this problem is much too difficult to tackle, let us consider a simpler case, such as the memory device used in a computer.

We can see immediately that insofar as the memory device is connected in a certain fashion to the main part of the computer, we can meaningfully speak of the information content of the memory device. If we look on the memory device as a physical system isolated from the computer, however, we do not know how to define its memory content. Indeed, a memory device as an isolated system contains all sorts of information other than the information it would supply when connected through a particular write-in and read-out device. The direction of polarization of a magnetic core, which is the carrier of information in ordinary usage, is only one of the many physical properties of the magnetic core. Any other physical property, other than its fixed attributes, can be considered as carrying some information regarding its past history. The intensity (as distinct from the direction) of magnetization (if not perfectly saturated) can be considered as a carrier of information regarding the intensity of electric current that magnetized the core, and so on. In order to include all this information in I, one has to go down to the microscopic state of the system and define its information content by the probability distribution over the possible microscopic states that such a system can take. In this way the information content I will be reduced to some kind of physical negentropy, although it may be different from thermodynamic negentropy. If we come to this extreme, all the attractiveness of Brillouin's principle is lost, and what remains may be something like the old principle of non-decrease of the total entropy. Perhaps the safest way is to define I of a system only relative to a given usage of the system and not talk about the absolute content of information of a system.

Let us now consider the second part of the question: does use of information ΔI really mean a corresponding decrease in the total stored information? At first glance, it seems to be nonsense to claim that the information content of a book is lessened if somebody reads it. But if everybody has learned the book by heart, the information content of the book will have lost its information value. The demon's knowledge ΔI about the physical system becomes worthless the moment he uses it to gain ΔN, because the knowledge does not reflect the reality after his intervention. In a core memory, if care is not taken to re-energize it (which would require negentropy), the information is lost once it is read out. I mention these fragmentary facts, not to prove the claim that use of information ΔI results in a decrease ΔI in the total stored information I, but only to show that this claim is not so absurd as it first looks. In fact, if an atomic state is used as "memory," this claim may be perfectly true.

In conclusion, we have to say that as it is formulated in (5.176) Brillouin's principle does not seem to be rigorously tenable although it reflects a great deal of valid insight into the difficult problem. Postulate 5.1 may be considered a safe minimum claim representing, so to speak, only half of Brillouin's principle. This half, however, may be sufficient to cover a large territory of problems, for in a pair of coupled systems what is "read out" from one is "written into" the other.

Whether or not the reader agrees with Brillouin's principle in the form of (5.176), there is another point Brillouin emphasized that the reader can more readily accept. (As a matter of fact, this point can be derived from Postulate 5.1 only.) It is the thesis that observation of a continuous variable with infinite precision amounts to acquisition of information of infinite amount, and in view of Theorem 5.15, this involves an infinite increase of physical entropy; therefore it is impossible. The first half of this thesis is evident, since, for instance, if a variable x has a uniform probability distribution over a range of length l, the information obtained by an observation that can locate the variable only with inaccuracy Δx will be $\Delta I = \log (l/\Delta x)$, for there are $(l/\Delta x)$ possibilities of equal probability. This becomes infinite as $\Delta x \to 0$. The second half of the thesis is a direct consequence of Theorem 5.15. An increase ΔS_{obs} $(\geq \Delta I)$ of entropy is equivalent to a heat transfer $\Delta Q = T\,\Delta S_{\text{obs}}$ between two bodies whose temperatures are T_1 and T_2, such that $T = T_1 T_2/(T_1 - T_2)$. This means that if $\Delta I \to \infty$, then $\Delta Q \to \infty$, requiring an infinite amount of energy at our disposal. If we do not go to the limit $\Delta I \to \infty$, we can write, with Brillouin,

$$\Delta I = \frac{\Delta E}{T}, \tag{5.177}$$

which gives the upper bound of obtainable information ΔI under the given available energy ΔE. Note, among other things, that $\Delta x = l \exp(-\Delta I \ln 2) \geq l \exp[-(\Delta E/T) \ln 2]$. This relation, together with Heisenberg's uncertainty principle, will completely demolish the dream of Laplace's demon. For details see [B-11–B-14, W-14]. There are, of course, other reasons for which Laplace's demon cannot carry out his scheme. For one thing, he probably would have to have a computer larger than the universe itself. Another more subtle and epistemologically more interesting reason is that he can never gather the information about the state of the entire universe at a given time (see [W-23] for this last point).

6

Induction and Learning—Inverse H-Theorem

6.1. INFORMATION BALANCE IN INDUCTION†

Induction is a process of acquiring knowledge about the underlying general rule from a set of given empirical facts. The source of information is the experimental data ($\mathcal{D}$), and the resulting knowledge concerns relative merits of competing hypotheses ($\mathcal{H}$). In purely passive observation of phenomena, the nature of experience (which determines the set $\mathcal{D}$) as well as the auxiliary condition $A_k \in \mathcal{A}$ is determined by the prevailing circumstances and remains outside the control of the observer. In such a case the information balance that can enter out consideration concerns only the product logical spectrum $\mathcal{D} \otimes \mathcal{H}$ under a given A_k.

In active experimentation on a given set $\mathcal{H}$, the choice of the kind of experiments (which determines the set $\mathcal{D}$) as well as the auxiliarity condition A_k is to a certain extent under the control of the experimenter. During each experiment, however, $\mathcal{D}$ and A_k are fixed, and the information about $\mathcal{H}$ is gained from experimentation under the fixed $\mathcal{D}$ and A_k. The choice of $\mathcal{D}$ and A_k that maximizes the acquired information about $\mathcal{H}$ would be ideal. The choice of $\mathcal{D}$ and A_k such that the uncertainty (entropy) about $\mathcal{H}$ becomes zero by a single observation would indeed provide a "crucial" test.

Formally speaking, when there are more than one possible kind of experiment $\mathcal{D}$ for a given set of $\mathcal{H}$, we could introduce a single set $\mathcal{D}$ of composite experiments (which is a conjunction of these different experiments) and

† Sections 6.1–6.3 are based on [W-13], and Section 6.4 is based on [W-27].

reduce the problem of choice of $\mathfrak{D}$ to a problem of choice of A_k. With regard to the choice of A_k, however, there arise delicate problems, of which we mentioned a few toward the end of Section 4.4. We discuss in a later section some problems relative to the composite experiment. These problems have not yet been perfectly clarified and should be left to a future investigation. In this section, therefore, we limit ourselves to the problem of information balance under given $\mathfrak{D}$ and A_k.

The basic probability distribution that underlies our problem then is one defined over $\mathcal{H} \otimes \mathfrak{D}$ for a given A_k:

$$p_k(H_I \cap D_i) = \frac{p(H_I \cap D_i \cap A_k)}{p(A_k)}. \tag{6.1}$$

In many cases of inductive experimentation the meaning of the probability $p(A_k)$ of A_k is difficult to interpret, but even in those cases the probability $p_k(H_I \cap D_i) = p_k(H_I)p_k(D_i \mid H_I)$ has a clear meaning. The proposition H_I, being a hypothesis, usually determines, on account of its assertion, the probability of each outcome:

$$p_k(D_i \mid H_I) = \frac{p_k(H_I \cap D_i)}{p_k(H_I)} = p(D_i \mid H_I \cap A_k). \tag{6.2}$$

On the other hand, the inverse (inductive) probability

$$p_k(H_I \mid D_i) = \frac{p_k(H_I \cap D_i)}{p_k(D_i)} = p(H_I \mid A_k \cap D_i) \tag{6.3}$$

is not determined by the assertion of the hypothesis. Hence it is more convenient to express it by the Bayes formula, as in (4.28):

$$p_k(H_I \mid D_i) = \frac{p_k(D_i \mid H_I)p_k(H_I)}{\sum_J p_k(D_i \mid H_J)p_k(H_J)}, \tag{6.4}$$

where $p_k(H_I)$ is the prior credibility of H_I before the experimental fact D_i is obtained and $p_k(H_I \mid D_i)$ is the posterior credibility of H_I after the experimental fact D_i has been taken into account. In a repetitive case we should put in (6.4), as (4.35) shows,

$$\begin{aligned} p_k(H_I) &= q^{(\nu-1)}(H_I), \\ p_k(H_I \mid D_i) &= q^{(\nu)}(H_I), \end{aligned} \tag{6.5}$$

provided that D_i is the actual νth experimental outcome.

The information about $\mathcal{H}$ brought about by the experimental fact D_i is the decrease of ignorance about $\mathcal{H}$ caused by D_i,

$$\inf_k(\mathcal{H} \mid D_i) = S_k(\mathcal{H}) - S_k(\mathcal{H} \mid D_i), \tag{6.6}$$

where

$$S_k(H) = -\sum_I p_k(H_I) \log p_k(H_I) \tag{6.7}$$

and

$$S_k(\mathcal{H} \mid D_i) = -\sum_I p_k(H_I \mid D_i) \log p_k(H_I \mid D_i). \tag{6.8}$$

The average of $\text{inf}_k\,(\mathcal{H} \mid D_i)$ with regard to the experimental fact D_i is given by

$$\begin{aligned}\text{inf}_k\,(\mathcal{H} \mid \mathcal{D}) &= \langle \text{inf}_k\,(\mathcal{H} \mid D_i)\rangle_i = S_k(\mathcal{H}) - S_k(\mathcal{H} \mid \mathcal{D}) \\ &= S_k(\mathcal{D}) - S_k(\mathcal{D} \mid \mathcal{H}) \le \min\,[S_k(\mathcal{H}), S_k(\mathcal{D})] \le S_k(\mathcal{D}),\end{aligned} \tag{6.9}$$

where

$$S_k(\mathcal{D}) = -\sum_i p_k(D_i) \log p_k(D_i), \tag{6.10}$$

$$\begin{aligned}S_k(\mathcal{H} \mid \mathcal{D}) &= \sum_i p_k(D_i) S_k(\mathcal{H} \mid D_i) \\ &= -\sum_i \sum_I p_k(H_I \cap D_i) \log \frac{p_k(H_I \cap D_i)}{p_k(D_i)},\end{aligned} \tag{6.11}$$

and

$$S_k(\mathcal{D} \mid \mathcal{H}) = -\sum_i \sum_I p_k(H_I \cap D_i) \log \frac{p_k(H_I \cap D_i)}{p_k(H_I)}. \tag{6.12}$$

Equation (6.9) expresses the information balance (in the sense of Section 3.3) in induction under a given A_k, setting the upper bound to $\text{inf}_k\,(\mathcal{H} \mid \mathcal{D})$ by $\min\,(S_k(\mathcal{H}), S_k(\mathcal{D}))$. The source information is $S_k(\mathcal{D})$, and the loss of information is $S_k(\mathcal{D} \mid \mathcal{H})$. The difference is the gain in information about the hypotheses obtained by this inductive process. If the hypotheses are such that each specifies only one experimental fact D_i, the information loss $S_k(\mathcal{D} \mid \mathcal{H})$ vanishes as it intuitively should.

Equation (6.9) thus allows an interesting picture of inductive learning; namely, the experiment provides the input information $S_k(\mathcal{D})$ and the learner absorbs only a portion of it, $S_k(\mathcal{D}) - S_k(\mathcal{D} \mid \mathcal{H})$. With a correct interpretation this picture certainly helps us to understand the informational structure of inductive learning. But it can also lead to a misconception of the situation. The main source of such a misconception lies in the possible impression that the input information $S_k(\mathcal{D})$ is something entirely objective and independent of the learner, but it is not. This entropy is defined by the probability $p_k(D_i)$, but it is not an objective relative frequency of occurrence of D_i under A_k. It is the predictive probability of D_i based on the observer's credibility distribution over the available hypothesis, $p_k(D_i) = \sum_I p_k(D_i \mid H_I) p_k(H_I)$. Hence it varies with a change in creditation of hypotheses on the part of the observer. In fact, the inductive process can be characterized as adjustment of the $p_k(H_I)$ so as to make $p_k(D_i)$ coincide eventually with the empirical relative frequency of D_i. As a consequence it is not correct to imagine the observer as standing in the midst of a flow of information pouring down on him while he is taking in part of it. Learning is after all not a purely passive function.

In a repetitive experiment, as we discuss in detail in the next section, the credibility distribution usually tends to become sharper and sharper until finally $q^{(\nu)}(H_I)$ becomes 1 for one particular hypothesis whose predictive distribution $p_k(D_i \mid H_I)$ coincides with the empirical frequency distribution. Under such circumstances $S_k(\mathfrak{D})$ and $S_k(\mathfrak{D} \mid \mathcal{H})$ become closer and closer and the observer learns less and less as the repetitive experiment goes on. Hence $S_k(\mathfrak{D})$ represents the source information with particular reference to the learner, whose state of knowledge is represented by $p_k(H_I) = q^{(\nu-1)}(H_I)$. With this understanding (6.9) can be interpreted as expressing the information balance in induction. To make the situation in a repetitive case more explicit, one can substitute (6.5) in (6.6) and obtain

$$\inf{}^{(\nu)} (\mathcal{H} \mid D_i) = U^{(\nu-1)}(\mathcal{H}) - U^{(\nu)}(\mathcal{H}), \tag{6.13}$$

where the inductive entropy $U^{(\nu)}(\mathcal{H})$ is defined by

$$U^{(\nu)}(\mathcal{H}) = -\sum_I q^{(\nu)}(H_I) \log q^{(\nu)}(H_I). \tag{6.14}$$

It should be noted that $q^{(\nu)}(H_I)$ is dependent on the νth outcome D_i. By averaging (6.13) with respect to D_i with the help of

$$p_k(D_i) = \sum_I p_k(D_i \mid H_I) p_k(H_I),$$

which is nothing but the νth predictive probability [see (4.36)],

$$p^{(\nu)}(D_i) = \sum_I q^{(\nu-1)}(H_I) p(D_i \mid H_I), \tag{6.15}$$

we obtain, corresponding to (6.9),

$$\inf{}^{(\nu)} (\mathcal{H} \mid \mathfrak{D}) = U^{(\nu-1)}(\mathcal{H}) - \langle U^{(\nu)}(\mathcal{H}) \rangle \leq S^{(\nu)}(\mathfrak{D}). \tag{6.16}$$

The last term in (6.16) is the deductive entropy $S^{(\nu)}(\mathfrak{D})$, defined by

$$S^{(\nu)}(\mathfrak{D}) = -\sum_i p^{(\nu)}(D_i) \log p^{(\nu)}(D_i), \tag{6.17}$$

where $p^{(\nu)}(D_i)$ is given by (6.15). The source information $S^{(\nu)}(\mathfrak{D})$ is thus dependent on the subjective probability $q^{(\nu-1)}(H_I)$. Toward the end of a series of experiments the upper bound $S^{(\nu)}(\mathfrak{D})$ is usually no longer very pertinent, for $U^{(\nu)}(\mathcal{H})$ converges to a certain value with $\nu \to \infty$, making $\inf^{(\nu)} (\mathcal{H} \mid \mathfrak{D}) \to 0$. Note that the other upper bound $S_k(\mathcal{H})$ is nothing but $U^{(\nu-1)}(\mathcal{H})$.

An important by-product of this consideration is a theorem that may be called the "expected" inverse H-theorem,

$$\langle U^{(\nu)}(\mathcal{H})\rangle \leq U^{(\nu-1)}(\mathcal{H}), \tag{6.18}$$

which follows from (6.16) because

$$\inf{}^{(\nu)}(\mathcal{H} \mid \mathcal{D}) \geq 0. \tag{6.19}$$

This last inequality is a special case of

$$\inf{}_k(\mathcal{H} \mid \mathcal{D}) \geq 0, \tag{6.20}$$

where $\inf_k(\mathcal{H} \mid \mathcal{D})$ is the quantity discussed in relation to (6.9). The reason for (6.20) is that $\inf_k(\mathcal{H} \mid \mathcal{D})$ is the interrelation

$$\inf{}_k(\mathcal{H} \mid \mathcal{D}) = S_k(\mathcal{D}) + S_k(\mathcal{H}) - S_k(\mathcal{D} \otimes \mathcal{H}) \geq 0, \tag{6.21}$$

which is non-negative because of (2.11). We shall come back later to the implications of the expected inverse H-theorem.

Some years ago Ruyer [R-9] made an interesting remark in the nature of an objection to information theory. He said in essence that according to information theory the human mind is nothing but an information transducer, and it is known that information is only diminished or lost in an information transducer; it is never created. Then the total information in the world will only be lost with time. Where, then, is the source of information we need and are actually using in our lives? Although the essence of this question—namely, the problem of ultimate source of information—is still worth investigating, the arguments that led Ruyer to it do not seem to hold tight.

Three kinds of misconceptions are involved in this argument. First, it is tacitly assumed that we can talk about the total amount of information owned by (or assignable to) the world. But we already saw in Section 5.4 that it is dangerous to talk about the information content of any system.

Second, a comparison of the human mind with an information transducer has only limited validity. Ruyer, of course, did not use a formula like (6.9), but this formula permits us to draw a certain analogy with an information channel. In this analogy $\mathcal{H}$ is the input signal and $\mathcal{D}$ is the output signal; the operator receives only the output $\mathcal{D}$ and guesses the input signal $\mathcal{H}$. The basic conditional probability $p_k(D_i \mid H_I)$ gives the error probability. Induction is thus formally comparable to decoding. But in order to speak of the objective amount of input information and output information as in channel theory, we have to assume an objectively fixed input probability distribution $p_k(H_I)$ (ergodic source) and the consequent fixed output probability distribution $p_k(D_i)$. These assumptions, however, do not hold in the case of inductive learning, as we said in the preceding paragraph.

Third, the information about $\mathcal{H}$, expressible as a mathematical quantity like $\inf_k(\mathcal{H})$, has nothing to do with the information that a law (hypothesis that wins out in a successive inductive process) can deliver. If by induction we have established that a particular hypothesis is the right hypothesis (law), each time we use it in deductive prediction we can draw a certain amount of information from the law about the experimental outcomes, as we carefully discussed in the last chapter. But we can use the same law infinitely many times, and hence the amount of information a law can furnish is always infinite. In comparison, the information "about" the law, expressible by $\inf_k(\mathcal{H})$, is finite even if we push a repetitive experiment to the ultimate end (infinite number of trials). It is true that in reality there exist infinitely many possible hypotheses; hence $\inf_k(\mathcal{H})$ can be infinite. However, this infinity and the infinity of information delivery by a single law have different origins. This can be seen by artificially assuming the number of available hypotheses to be finite, as we have done.

For these reasons we can safely say that by inductive learning we do indeed create something from which we can derive an infinite amount of information. It goes without saying that inductive learning involves a deductive-logically impermissible process, as we discussed in Chapter 4. It is the inductive jump (generalization) that creates information. This is obvious, because if no generalization of any kind is allowed, one particular experimental fact cannot provide any predictive information about a similar occasion.

Apart from this logical jump involved in induction itself, there are two phenomena related to induction that exhibit some kind of discontinuity. One may be called "sudden certainty" or "a-ha effect" or "illumination." In this phenomenon a hypothesis, by a sudden bounce, seems to hit credibility unity or at least reach an overwhelmingly large credibility. Since the successive change in degree of confirmation in a repetitive trial is very gradual according to the Bayes rule (6.4), we cannot explicate this phenomenon of sudden certainty by it. There seem to be at least three different kinds of mechanism that result in sudden certainty. One happens in the case when an observer suddenly hits on a new hypothesis that explains the observed facts much better than hitherto existing hypotheses. In terms of our model, this means that the set of hypotheses $\mathcal{H}$ suddenly changes. The source of information, of course, lies in the abduction of the new hypothesis and is purely extra-evidential in the technical use of the word.

In the second kind of sudden certainty, an existing hypothesis receives a sudden boost by some extra-evidential factor. Because $\mathcal{H}$ does not change and the gradual Bayesian evaluation cannot exhibit a sudden jump, the only explication that can be given is that the prior credibility $q^{(0)}(H_I)$ suddenly changes because of new information that is not the experimental fact in $\mathcal{D}$. In the coin-flipping experiment explained under Section 4.3, if the information that the coin came from a psychology professor's pocket reaches the subject in the middle of the experimental sequence, the double-headed coin hypothesis will acquire a sudden certainty. This change can be attributed in our model to a sudden change in the "prior" credibility of this hypothesis. "Prior" does not mean temporally "earlier," but "logically preceding." It can change in the middle of a series of experimentation.

The third kind of sudden certainty seems to originate from the human faculty of guessing the ultimate credibility $q^{(\infty)}(H_I)$ of a hypothesis from the past behavior of the hypotheses of $\mathcal{H}$ up to the νth stage. If, for instance, the credibility $q^{(\nu)}(H_I)$ of a particular hypothesis has grown monotonically from $\nu = 0$ to $\nu = \nu$ while the credibilities of the other hypotheses have gone down monotonically, the observer will extrapolate the trend and decide at a certain stage $\nu = \nu$ that a particular H_I is the winner $[q^{(\infty)}(H_I) = 1]$ even though $q^{(\nu)}(H_I)$ at the present stage is appreciably less than unity. This effect may be called "superinduction," because the extrapolation too is a form of induction. This may be also the reason why comparison of the Bayesian model with a human subject usually reveals that the Bayesian model is much more conservative (less likely to rush into a conclusion) than a human subject.

Another kind of discontinuous effect can be observed in human behavior based on inductive inference. This effect originates from the fact that man often has to choose between taking or not taking a certain action, and this decision has to be a discontinuous function of continuously varying credibilities. Behaviorally, it cannot be decided externally whether it is the action or the credibility of hypotheses that made a jump. Suppose two outcomes D_1 and D_2; if $p(D_1) = 1 - p(D_2) \geq \pi$ a person takes action A_1, and if $p(D_1) < \pi$ he takes action A_2. If he has hypotheses with predictive power over $\mathcal{D}$ so that $p(D_1) = \sum_I p(D_1 \mid H_I)p(H_I)$ (4.36), a gradual change of the $p(H_I)$ can cause a sudden change in his behavior at the point where $p(D_1)$ passes the value π. But the same discontinuous change in behavior could be observed if the $p(H_I)$ made a sudden change, resulting in a sudden increase in $p(D_1)$. In either case he behaves as if he were sure that D_1 would happen. It may also be noted that because of the sudden-certainty phenomenon he would perhaps use in his decision making the value $p(D_1) = p(D_1 \mid H_I)$, where H_I is the sudden winner rather than the average value $p(D_1) = \sum_I p(D_1 \mid H_I)p(H_I)$.

The determination of the threshold π could be attributed to the maximization of utility or minimization of loss according to the now-popular decision theory. Let $V(A_j \mid D_i)$ be the utility (or the negative of loss) that a person acquires (or incurs) by taking action A_j when D_i is the actual outcome. The expected utility of taking action A_1 is $\sum_i V(A_1 \mid D_i)p(D_i)$, and the utility of taking A_2 is $\sum_i V(A_2 \mid D_i)p(D_i)$. In order that the former will not be smaller than the latter, we have to have $[V(A_1 \mid D_1) - V(A_2 \mid D_1) - V(A_1 \mid D_2) + V(A_2 \mid D_2)]p(D_1) \geq V(A_2 \mid D_2) - V(A_1 \mid D_2)$. [Note that $p(D_1) + p(D_2) = 1$.] If $p(D_i)$ satisfying this equality falls in the domain $0 \leq p(D_i) \leq 1$, that will determine the value of π. This method or, more generally, the utilitarian theory of decision making exemplified by this simple illustration is widely adopted. It is certainly a useful approach when the decision should be made to maximize utility. On the other hand, any claim that every human decision is explicable by this model is a gross mistake. In any event, even if one wants to apply the utility consideration, it is important first to have a disinterested, objective evaluation of the probabilities [$p(D_i)$ above].

6.2. MATHEMATICAL PROOF OF THE INVERSE *H*-THEOREM AND THE BAYESIAN CHAIN

A finite Markovian chain (Section 5.2) is characterized by a time parameter $t = 0, 1, 2, \cdots$ and occupancy probabilities $w_i(t)$, $i = 1, 2, \cdots, n$ (n is finite), which are generated by a set of transition probabilities $p_{ji}(t)$ from their initial values $w_i(0)$ according to the rule

$$w_j(t) = \sum_{i=1}^{n} p_{ji}(t) w_i(t-1). \tag{6.22}$$

Because the relation expressing $w_j(t)$ in terms of $w_i(t-1)$ is linear, we can represent the occupancy probabilities $w_i(t)$, $i = 1, 2, \cdots, n$, $t = 0, 1, \cdots$, by an ensemble of strings, each of which assigns a state $i(t)$ to a value of t. In this picture $p_{ji}(t)$ represents the relative frequency of strings with $i(t) = j$ and $i(t-1) = i$ in the population of strings with $i(t-1) = i$. The H-theorem states that $S(t) \geq S(t-1)$, where $S(t)$ is defined by

$$S(t) = -\sum_{i=1}^{n} w_i(t) \log w_i(t).$$

A finite Bayesian chain is characterized by a time parameter $\nu = 0, 1, 2, \cdots$ and the credibilities $q^{(\nu)}(I)$, $I = 1, 2, \cdots, N$ (N is finite), which are generated by a given set of conditional probabilities $p(i \mid I)$, $i = 1, 2, \cdots, n$, from their initial values $q^{(0)}(I)$ according to the rule

$$q^{(\nu)}(I) = \frac{\sum_{i=1}^{n} \delta(i(\nu), i) p(i \mid I) q^{(\nu-1)}(I)}{\sum_{J=1}^{N} \sum_{j=1}^{n} \delta(i(\nu), j) p(j \mid J) q^{(\nu-1)}(J)}, \tag{6.23}$$

where $i(\nu)$, $\nu = 1, 2, \cdots$ is a chain of Bernouilli trials (stochastic chains with range 1) that takes value $i(\nu) = i$ at time ν with a given constant probability γ_i, $i = 1, 2, \cdots, n$. Since $q^{(\nu)}(I)$ is not linear in $q^{(\nu-1)}(I)$, as was the case with $w_i(t)$, we cannot interpret $q^{(\nu)}(I)$ in terms of an ensemble of temporal strings, each of which assigns a state I to a value of ν. Nonetheless it is meaningful to compare $q^{(\nu)}(I)$ to $w_i(t)$. The inverse H-theorem states, for sufficiently large values of ν, $U^{(\nu)} \leq U^{(\nu-1)}$ in a certain sense of average, where $U^{(\nu)} = -\sum_{I=1}^{N} q^{(\nu)}(I) \log q^{(\nu)}(I)$. This is to be compared to the H-theorem which we have proven with regard to the entropy defined in terms of $w_i(t)$. Equation (6.23) is equivalent to (4.35). In the Bayesian model of induction the index i specifies an experimental outcome D_i, a member of $\mathfrak{D}$, and the index I specifies a hypothesis H_I, a member of $\mathfrak{H}$.

Let us take a particular random sequence of length $\nu\{i(1), i(2), \cdots, i(\nu)\}$ and call it $\mathfrak{B}^{(\nu)}$. Let ν_i be the number of i in $\mathfrak{B}^{(\nu)}$ and define the empirical relative frequency $\alpha_i^{(\nu)}$ by (4.37),

$$\alpha_i^{(\nu)} = \frac{\nu_i}{\nu}, \tag{6.24}$$

with

$$\alpha_i^{(\nu)} \geq 0, \qquad \sum_{i=1}^{n} \alpha_i^{(\nu)} = 1. \tag{6.25}$$

Using (6.23) as the recurrence formula, we can obtain, with the help of (6.24),

$$q^{(\nu)}(I) = \frac{\prod_{i=1}^{n} [p(i \mid I)]^{\alpha_i^{(\nu)}\nu} q^{(0)}(I)}{\sum_{J=1}^{N} \prod_{i=1}^{n} [p(j \mid J)]^{\alpha_j^{(\nu)}\nu} q^{(0)}(J)}, \tag{6.26}$$

where 0^0 must be put equal to unity. This $q^{(\nu)}(I)$ corresponds to a particular $\mathfrak{B}^{(\nu)}$, characterized by the $\alpha_i^{(\nu)}$, abstraction having been made of the order of elements of $\mathfrak{B}^{(\nu)}$. Insofar as ν is finite, the $q^{(\nu)}(I)$ calculated by (6.23) with $\mathfrak{B}^{(\nu)}\{i(1), i(2), \cdots, i(\nu)\}$ and the $q^{(\nu)}(I)$ calculated by (6.26) with $\{a_1^{(\nu)}, \alpha_2^{(\nu)}, \cdots, \alpha_n^{(\nu)}\}$ must be the same. If ν becomes infinite, however, we must be a little cautious, (6.23) superseding (6.26) in case of conflict.

Now consider the ensemble $\{\mathfrak{B}^{(\nu)}\}$ of all possible random sequences of length ν, where each $\mathfrak{B}^{(\nu)}$ is a set, $\mathfrak{B}^{(\nu)} = \{i(n), i(2), \cdots, i(\mu), \cdots, i(\nu)\}$. Then at each time μ the relative frequency of $i(\mu) = i$ in the ensemble will be γ_i by definition. Hence the relative frequency of $i(\mu) = i$ for all μ and for all sequences in the domain $1 \leq \mu \leq \nu$ in the ensemble will be also γ_i. This statement (regarding the time and ensemble average) is different from the statement (regarding the time average) of ergodicity: $\alpha_i^{(\nu)} \to \gamma_i$ in each individual sequence $\mathfrak{B}^{(\nu)}$ for $\nu \to \infty$. However, this last statement is also true, of course, because of the strong law of large numbers, which is obvious for a random sequence. At a finite ν, $\alpha_i^{(\nu)}$ and γ_i are different, although the ensemble average of $\alpha_i^{(\nu)}$ is γ_i. For a finite ν we define the "average Bayesian chain" by

$$\bar{q}^{(\nu)}(I) = \frac{\prod_{i=1}^{n} [p(i \mid I)]^{\gamma_i\nu} q^{(0)}(I)}{\sum_{J=1}^{N} \prod_{j=1}^{n} [p(j \mid J)]^{\gamma_j\nu} q^{(0)}(J)}, \tag{6.27}$$

which is equivalent to

$$\bar{q}^{(\nu)}(I) = \frac{\prod_{i=1}^{n} [p(i \mid I)]^{\gamma_i} \bar{q}^{(\nu-1)}(I)}{\sum_{J=1}^{N} \prod_{j=1}^{n} [p(j \mid J)]^{\gamma_j} \bar{q}^{(\nu-1)}(J)}, \tag{6.28}$$

with $\bar{q}^{(0)}(I) = q^{(0)}(I)$. The difference between (6.28) and (6.23) is that (6.28) uses the logarithmic average of $p(i(\nu) \mid I)$ of (6.23).

The deviation of $q^{(\nu)}(I)$ in an individual $\mathcal{B}^{(\nu)}$ from its "average" value $\bar{q}^{(\nu)}(I)$ originates, of course, from the deviation of the $\alpha_i^{(\nu)}$ from their "average" values γ_i. This latter deviation for large ν can be estimated as follows. The fluctuation δ_i of $\nu_i = \nu\alpha_i^{(\nu)}$ from its mean value $\nu\gamma_i$,

$$\delta_i = \nu_i - \nu\gamma_i, \qquad \sum_{i=1}^{n} \delta_i = 0, \tag{6.29}$$

must obey the multinominal distribution, which for large ν becomes

$$\Pr(\delta_1, \delta_2, \cdots, \delta_n) = \left[(2\pi\nu)^{n-1} \prod_{i=1}^{n} \gamma_i\right]^{-\frac{1}{2}} \exp\left(\frac{-\sum_i \delta_i^2}{2\nu\gamma_i}\right). \tag{6.30}$$

The mean value of $\delta_i\delta_j$, under the restriction $\sum_{i=1}^{n} \delta_i = 0$ (6.29), is given by

$$\begin{aligned} \overline{\delta_i\delta_j} &= -\nu\gamma_i\gamma_j \quad \text{for} \quad i \neq j, \\ \overline{\delta_i^{\,2}} &= \nu\gamma_i(1 - \gamma_i). \end{aligned} \tag{6.31}$$

Consequently the root mean square fluctuation of $\alpha_i^{(\nu)}$ from γ_i is

$$\left[\overline{(\alpha_i^{(\nu)} - \gamma_i)^2}\right]^{\frac{1}{2}} = \left[\frac{\gamma_i(1 - \gamma_i)}{\nu}\right]^{\frac{1}{2}}, \tag{6.32}$$

which tends to zero with $\nu \to \infty$, in agreement with what has already been stated. Thus we can write

$$\alpha_i^{(\infty)} = \lim_{\nu \to \infty} \alpha_i^{(\nu)} = \gamma_i, \qquad i = 1, 2, \cdots, n. \tag{6.33}$$

We return later to the fluctuations of $q^{(\nu)}(I)$ from $\bar{q}^{(\nu)}(I)$.

Not unlike our classification of states, $z_i \in \mathfrak{Z}$, in the Markov chain, we now classify the "states" labeled by I in the Bayesian chain, which are hypotheses $H_I \in \mathcal{H}$. In the Markov chain the criteria of classification were based on the transition probabilities $p_{ji}(t)$. In the Bayesian chain the criterion is based on

$$A(I) = \prod_{i=1}^{n} [p(i \mid I)]^{\gamma_i}, \tag{6.34}$$

where $0^0 = 1$. The class $\mathfrak{R}$ is defined as

$$\mathfrak{R} = \{H_I \mid A(I) = 0\}. \tag{6.35}$$

This implies that $H_I \in \mathfrak{R}$ is such that there exists an event D_i whose probability γ_i is not zero and to which H_I gives a zero probability:

$$\mathfrak{R} = \{H_I \mid \text{there exists } D_i \text{ in } \mathfrak{D} \text{ such that } \gamma_i \neq 0 \text{ and } p(i \mid I) = 0\}. \tag{6.36}$$

This means that $\mathfrak{R}$ is a set of "logically refutable hypotheses." Because of ergodicity, $\gamma_i \neq 0$ implies that D_i actually occurs sooner or later in each individual sequence. Hence, according to (6.23), at certain finite ν, $q^{(\nu)}(I)$ will become zero and will remain zero thereafter. Hence $\mathfrak{R}$ is not only the set of "logically refutable hypotheses," but also the set of "logically refuted hypotheses." As far as the average credibility is concerned, $\bar{q}^{(\nu)}(I)$ already vanishes at $\nu = 1$ and thereafter if $H_I \in \mathfrak{R}$.

In contrast to $\mathfrak{R}$, which is the set of minimum $A(I)$, we consider the set of maximum $A(I)$ and call it $\mathfrak{K}$.

$$\mathfrak{K} = \left\{H_I \mid A(I) = \max_J A(J)\right\}. \tag{6.37}$$

Now the set $\mathfrak{H}$ of hypotheses is divided into three disjoint and exhaustive families $\mathfrak{R}$, $\mathfrak{K}$, and $\mathfrak{M} = \neg\mathfrak{R} \wedge \neg\mathfrak{K}$; that is,

$$\mathfrak{M} = \left\{\{H_I \mid A(I) \neq 0, \qquad A(I) \neq \max_J A(J)\right\} \tag{6.38}$$

and

$$\mathfrak{R} \wedge \mathfrak{K} = \emptyset, \quad \mathfrak{K} \wedge \mathfrak{M} = \emptyset, \quad \mathfrak{M} \wedge \mathfrak{R} = \emptyset, \quad \mathfrak{R} \vee \mathfrak{K} \vee \mathfrak{M} = \mathfrak{H}. \tag{6.39}$$

As in (4.53) and (4.54), it is easy to conclude from (6.27) for the average credibility

$$\begin{aligned}
&\bar{q}^{(\nu)}(I) = 0 \quad \text{for} \quad \nu \geq 1, \qquad H_I \in \mathfrak{R},\\
&\bar{q}^{(\nu)}(I) \neq 0 \quad \text{for} \quad \nu \quad \text{finite} \quad \text{and} \quad \bar{q}^{(\nu)}(I) \to 0 \quad \text{as} \quad \nu \to \infty, \qquad H_I \in \mathfrak{M},\\
&\bar{q}^{(\nu)}(I) \neq 0 \quad \text{for} \quad \nu \quad \text{finite} \quad \text{and} \quad \bar{q}^{(\nu)}(I) \to \frac{q^{(0)}(I)}{\sum_{J \in K} q^{(0)}(J)} \quad \text{as} \quad \nu \to \infty,\\
&\qquad\qquad\qquad H_I \in \mathfrak{K}.
\end{aligned} \tag{6.40}$$

Accordingly we may call $\mathfrak{R}$ the refuted class, $\mathfrak{M}$ the vanishing class, and $\mathfrak{K}$ the surviving class. It is also easy to see that the limiting values of $\bar{q}^{(\nu)}(I)$ for $H_I \in \mathfrak{K}$ are never reached at a finite value of ν except in the case $\mathfrak{M} = \emptyset$.

The behavior of the nonaverage credibility $q^{(\nu)}(I)$, which depends on an individual experimental sequence $\mathcal{B}^{(\nu)}$, is not so simple as the average credibility $\bar{q}^{(\nu)}(I)$ in (6.40). If the $q^{(\nu)}(I)$ are known to converge to certain values for $\nu \to \infty$, we can expect their limiting values to be the same as the limiting values of the average counterparts $\bar{q}^{(\nu)}(I)$. But the convergence in each individual sequence may not be guaranteed. In fact, we can immediately show that the mere fact that H_I belongs to the surviving class $\mathcal{K}$ defined by (6.37) is not a sufficient condition for convergence.

Suppose that the $q^{(\nu)}(I)$ of hypotheses belonging to the vanishing class $\mathcal{M}$ converge to 0 and the $q^{(\nu)}(I)$ of hypotheses belonging to the surviving class $\mathcal{K}$ converge to nonvanishing values. The $q^{(\nu)}(I)$ of hypotheses belonging to the refuted class become zero at a finite value of ν in each individual case; hence we can disregard them when we consider the limiting values of the $q^{(\nu)}(I)$ for $\nu \to \infty$. This assumption of convergence implies, in view of (6.23), that

$$C^{(\nu)}(I) \equiv \frac{p(i(\nu) \mid I)}{\sum_J p(i(\nu) \mid J) q^{(\nu-1)}(J)} \to 1, \qquad H_I \in \mathcal{K}. \tag{6.41}$$

This is the same as $[q^{(\nu)}(I) - q^{(\nu-1)}(I)] \to 0$, since (6.23) can be written

$$q^{(\nu)}(I) = C^{(\nu)}(I) q^{(\nu-1)}(I). \tag{6.42}$$

The summation over J in (6.41) can be limited to the surviving class $\mathcal{K}$ because it is assumed that $q^{(\nu-1)}(J) \to 0$ for $H_J \in \mathcal{M}$. The relation (6.41) must hold for any $i(\nu) = D_i$ in $\mathcal{D}$ for which $\gamma_i \neq 0$, since $\gamma_i \neq 0$ implies, because of ergodicity, that D_i is bound to happen after any large ν.

The denominator of (6.41) is the average of the numerator for $H_I \in \mathcal{K}$. In order that the numerator and the denominator become equal at the limit, it is necessary that $p(i(\nu) \mid J)$ have the same value for all H_J in $\mathcal{K}$ and for each $i(\nu) = D_i$ for which $\gamma_i \neq 0$. This means that all hypotheses H_J in $\mathcal{K}$ must have the same (predictive) probability distribution over the D_i for which $\gamma_i \neq 0$.

The conclusion is that in order for the individual $q^{(\nu)}(I)$ for the hypotheses in the surviving class to converge to a finite value, it is necessary that $\mathcal{K}$ consists of hypotheses that not only satisfy (6.37) but also are equivalent in the sense that they assign the same probability to each D_i for which $\gamma_i \neq 0$. For brevity let us call such a $\mathcal{K}$ a "uniform" surviving set. It is easy to see that $\mathcal{K}$ automatically becomes uniform in the case of a perfect match; that is, if $\max_J A(J) = \prod_i \gamma_i^{\gamma_i}$. In the case of a uniform surviving set $\mathcal{K}$, one might as well consider all the members of $\mathcal{K}$ as a single hypothesis and reduce the problem to the simple case in which $\mathcal{K}$ consists of only one member from the beginning. However, the fact that the limiting value of each $q^{(\nu)}(I)$ for $H_I \in \mathcal{K}$ is proportional to its initial value is additional information.

Suppose it is known, for some reason, that all $q^{(\nu)}(I)$ for $H_I \in \mathcal{H}$ converge to values that do not depend on the initial values $q^{(0)}(I)$ of the credibilities. We may speak of an "objective evidential convergence" in this case, since the conclusion reached at the end depends solely on the evidential facts and not on the subjective prior credibilities, which may be influenced by extra-evidential factors. The credibilities of the hypotheses belonging to the refuted class and the vanishing class become zero; hence they do not depend on the prior credibilities of themselves or of other hypotheses. In order that the credibilities in the surviving class $\mathcal{K}$ reach values independent of the prior credibilities, it is necessary that $\mathcal{K}$ consist of a single member, because they are given in general by $q^{(0)}(I)/\sum_{J\in\mathcal{K}} q^{(0)}(J)$, which takes a value independent of the prior credibilities only when it becomes unity. (Note: we have assumed that none of $q^{(0)}(I)$ for $H_I \in \mathcal{H}$ is zero.) It is also to be noted that the individual $q^{(\nu)}(I)$ can never become zero at a finite ν except that H_I is logically refuted [see (6.23)]. Hence the "objective" limiting values of $q^{(\nu)}(I)$ for $\mathcal{M}$ and $\mathcal{K}$ are not reached at a finite ν, except in the special case in which $\mathcal{M} = \emptyset$ and $\mathcal{K}$ consists of one member. Summarizing, we have the following lemma.

Lemma 6.1 Objective evidential convergence of credibilities implies that one of the $q^{(\nu)}(I)$ becomes 1 and all the other $q^{(\nu)}(I)$ become 0 at the limit $\nu \to \infty$. Except in the case when all but one hypothesis are logically refutable, $q^{(\nu)}(I) = 1$ never happens except in the limit $\nu \to \infty$.

This can be stated in the form of an inverse H-theorem.

Theorem 6.1 (Simplest inverse H-theorem). In the case of objective evidential convergence we have always

$$U^{(\nu)} \geq U^{(\infty)} = 0, \tag{6.43}$$

where

$$U^{(\nu)} = -\sum_I q^{(\nu)}(I) \log q^{(\nu)}(I). \tag{6.44}$$

Proof. We have $U^{(\nu)} \geq 0$ for any ν and $U^{(\infty)} = 0$ according to Lemma 6.1. Furthermore, except when all but one hypothesis are logically refuted, one can replace the sign $\geq$ by the sign $>$.

This theorem, however, is almost trivial and not very interesting. Richer in content are two other versions of the inverse H-theorem. One is the "expected" inverse H-theorem (introduced in the last section), and the other is the "average" inverse H-theorem. Usually the two words "expectation" and "average" are used more or less synonymously, but in the present context they suggest the following basic difference in meaning. When the $q^{(\nu-1)}(I)$ are given $q^{(\nu)}(I)$ depends on the νth outcome, $i(\nu) = i$. Objectively, the outcome D_i will happen with probability γ_i, but the observer does not know γ_i.

In fact, his effort is, in a sense, aimed at finding out the values of γ_i. His best evaluation of γ_i at the νth stage is given by

$$p^{(\nu)}(i) = \sum_I p(i \mid I) q^{(\nu-1)}(I). \tag{6.45}$$

The "expected" behavior of the U-function is based on the predictive probability $p^{(\nu)}(i)$. The "average" behavior of the U-function is based on the objective probability γ_i. In the case of the "expected" inverse"H-theorem the entropy value itself, which depends on $i(\nu) = i$, is averaged with the help of $p^{(\nu)}(i)$, whereas in the case of the average inverse H-theorem the relative frequency $\alpha_i^{(\nu)} = \nu_i/\nu$ is consistently replaced by its average value γ_i.

Theorem 6.2 ("Expected" inverse H-theorem). The "expected" value of $U^{(\nu)}$ is never larger than $U^{(\nu-1)}$:

$$U^{(\nu-1)} \geq \langle U^{(\nu)} \rangle. \tag{6.46}$$

Proof. The right-hand side of (6.46) is more precisely

$$\langle U^{(\nu)} \rangle = \sum_{i=1}^{n} p^{(\nu)}(i) U^{(\nu)}, \tag{6.47}$$

where $U^{(\nu)}$ (which depends on the νth outcome i) is obtained from (6.44) by substituting for $q^{(\nu)}(I)$ the value of (6.23) with $i(\nu) = i$. Since the denominator of the expression of $q^{(\nu)}(I)$ of (6.23) is nothing but $p^{(\nu)}(i)$, (6.46) becomes simply

$$0 \geq -\sum_I \sum_i p(i \mid I) q^{(\nu-1)}(I) \log \frac{p(i \mid I) q^{(\nu-1)}(I)}{\left[\sum_J p(i \mid J) q^{(\nu-1)}(J)\right] q^{(\nu-1)}(I)}. \tag{6.48}$$

The right side has the form of

$$-\sum_I \sum_i P(i \cap I) \log \frac{P(i \cap I)}{Q(i \cap I)}, \tag{6.49}$$

where P and Q are some probability distributions over i and I. This expression is obviously nonpositive according to the Gibbs theorem (Theorem 1.1). Q.E.D. For an alternative proof, see the last section, in particular, Equation (6.21).

Theorem 6.3 ("Average" inverse H-theorem). Let the "average" inductive entropy $\bar{U}^{(\nu)}$ at the νth stage be defined by (6.44), in which the individual $q^{(\nu)}(I)$ is replaced by its "average" $\bar{q}^{(\nu)}(I)$ (6.27). Then there exists an integer μ such that

$$\bar{U}^{(\nu-1)} \geq \bar{U}^{(\nu)} \quad \text{for} \quad \nu > \mu. \tag{6.50}$$

Proof. The definition (6.27) of $\bar{q}^{(\nu)}(I)$ can be rewritten with the help of $A(I)$ of (6.34) as

$$\bar{q}^{(\nu)}(I) = \frac{\bar{q}^{(0)}(I)F^{(\nu)}(I)}{\sum_J \bar{q}^{(0)}(J)F^{(\nu)}(J)}, \tag{6.51}$$

with

$$F^{(\nu)}(I) = [A(I)]^\nu. \tag{6.52}$$

In these expressions the time variable ν, which was originally an integer, can also be considered as a continuous variable and $-d\bar{U}^{(\nu)}/d\nu$ represents the information gain per stage regarding the right hypothesis. We demonstrate that for ν larger than a certain lower bound

$$\frac{d\bar{U}^{(\nu)}}{d\nu} \leq 0, \tag{6.53}$$

of which the desired relation (6.50) is a consequence.

It should be remembered that the hypotheses that belong to $\mathcal{R}$ have $A(I) = 0$; hence $\bar{q}^{(\nu)}(I) = 0$ for $\nu \geq 1$. Taking any one H_{I_0} of the hypotheses belonging to $\mathcal{K}$ as standard, we introduce new variables $\beta(I)$ and $\chi(I)$ to characterize any other hypothesis H_I belonging to $\mathcal{M}$ or $\mathcal{K}$:

$$\beta(I) = \frac{\bar{q}^{(0)}(I)}{\bar{q}^{(0)}(I_0)} > 0, \qquad 1 \geq \chi(I) = \frac{A(I)}{A(I_0)} > 0, \qquad H_I \in \mathcal{M} \vee \mathcal{K}. \tag{6.54}$$

The variable $\beta(I)$ is a measure of the *a priori* credibility, and the variable $\chi(I)$ is a measure of the confirmability of H_I. Obviously $\chi(I) = 1$ for H_I belonging to $\mathcal{K}$ (whether $\mathcal{K}$ is uniform or not), and $\chi(I) < 1$ for H_I belonging to $\mathcal{M}$, according to (6.37) and (6.38). Then we can write

$$\bar{U}^{(\nu)} = \frac{Y^{(\nu)}}{X^{(\nu)}}, \tag{6.55}$$

with

$$X^{(\nu)} = \sum_I \beta(I)\chi^\nu(I) \tag{6.56}$$

and

$$Y^{(\nu)} = \sum_I \beta(I)\chi^\nu(I) \log\left[\sum_J \beta(J)\chi^\nu(J)\right] - \sum_I \beta(I)\chi^\nu(I) \log[\beta(I)\chi^\nu(I)], \tag{6.57}$$

where the summation with regard to I and J extends over $\mathcal{M} \vee \mathcal{K}$. Then the derivative

$$\frac{d\bar{U}^{(\nu)}}{d\nu} = \frac{1}{(X^{(\nu)})^2}\left(X^{(\nu)}\frac{dY^{(\nu)}}{d\nu} - \frac{dX^{(\nu)}}{d\nu}Y^{(\nu)}\right) \tag{6.58}$$

becomes

$$\frac{d\bar{U}^{(\nu)}}{d\nu} = \frac{1}{(X^{(\nu)})^2}\sum_I\sum_J \beta(I)\chi^\nu(I)\beta(J)\chi^\nu(J)\log[\beta(I)\chi^\nu(I)]\ln\frac{\chi(J)}{\chi(I)}. \tag{6.59}$$

Adding to this expression another expression obtained from (6.59) by an interchange of I and J, one obtains

$$2(X^{(\nu)})^2 \frac{d\bar{U}^{(\nu)}}{d\nu} = \sum_I \sum_J \beta(I)\chi^\nu(I)\beta(J)\chi^\nu(J) \log \frac{\beta(I)\chi^\nu(I)}{\beta(J)\chi^\nu(J)} \ln \frac{\chi(J)}{\chi(I)}. \quad (6.60)$$

Now if $\log [\beta(I)/\beta(J)]$ and $\log [\chi(I)/\chi(J)]$ have the same sign, we obviously have

$$\log \frac{\beta(I)\chi^\nu(I)}{\beta(J)\chi^\nu(J)} \ln \frac{\chi(J)}{\chi(I)} < 0. \quad (6.61)$$

If $\log [\beta(I)/\beta(J)]$ and $\log [\chi(I)/\chi(J)]$ have opposite signs, we again have (6.61) for ν, which satisfies

$$\nu > -\frac{\log [\beta(I)/\beta(J)]}{\log [\chi(I)/\chi(J)]}. \quad (6.62)$$

If $\log [\beta(I)/\beta(J)] = 0$ and $\log [\chi(I)/\chi(J)] \neq 0$, (6.61) holds for any ν. If $\log [\chi(I)/\chi(J)] = 0$,

$$\log \frac{\beta(I)\chi^\nu(I)}{\beta(J)\chi^\nu(J)} \ln \frac{\chi(J)}{\chi(I)} = 0 \quad (6.63)$$

no matter what value $\log [\beta(I)/\beta(J)]$ may have. Therefore we conclude from (6.60)–(6.62) that

$$\frac{d\bar{U}^{(\nu)}}{d\nu} \leq 0 \quad (6.64)$$

for ν large enough, so that (6.62) is satisfied for those pairs (I, J) that have different χ's. There obviously exists a finite lower bound for such ν, for the number of hypotheses is finite. This completes the proof of the average inverse H-theorem.

The meaning of the condition (6.61) becomes easy to understand if we interpret $\log [\beta(I)/\beta(J)]$ as the *a priori* factor (determined by the prior credibilities) and $\nu \log [\chi(I)/\chi(J)]$ as the *a posteriori* factor (determined by the experimental distribution γ_i). If the *a priori* factor and the experimental factor are in agreement from the beginning, we have $\log [\beta(I)/\beta(J)] \cdot \log [\chi(I)/\chi(J)] > 0$ and (6.61) is satisfied for all ν. If they are not in agreement at the beginning, we have

$$\log [\beta(I)/\beta(J)] \cdot \log [\chi(I)/\chi(J)] < 0,$$

and it takes a certain accumulation of experience to overcome the initial bias, as is expressed by (6.62). After all the initial biases have been corrected to the extent that (6.61) holds for all pairs (I, J), the inductive entropy $U^{(\nu)}$ decreases monotonically. The condition $\log [\beta(I)/\beta(J)] = 0$ means that an equal prior credibility is given to both H_I and H_J; hence the credibility ratio is reduced to the likelihood ratio, and it is not surprising that (6.61) holds for all ν. This leads to the following theorem.

Theorem 6.4 (Average inverse H-theorem: likelihood case). If all the hypotheses are given an equal prior credibility, Theorem 6.3 will hold with $\mu = 1$.

Next we can ask whether the average $\bar{U}^{(\nu)}$ can reach its minimum value at a finite ν. This happens if and only if there is a number ν_1 such that $d\bar{U}^{(\nu)}/d\nu = 0$ for $\nu \geq \nu_1$. To investigate this question one should first note that the factor $\beta(I)\chi^\nu(I)\beta(J)\chi^\nu(J)$ in (6.60) is always nonzero at a finite ν. Therefore $d\bar{U}^{(\nu)}/d\nu$ becomes zero if and only if (6.63) is the case for all pairs (I, J) in $\mathcal{M} \vee \mathcal{K}$. For $I = J$ (6.63) is self-evident. For $I \neq J$ (6.63) holds if and only if $\chi(I) = \chi(J)$ and/or $\beta(I)/\beta(J) = [\chi(J)/\chi(I)]^\nu$. This last equation, for a pair $\chi(I) \neq \chi(J)$, may happen to hold for a particular value of ν, but then it will not hold for any larger value of ν. Therefore, in order that $d\bar{U}^{(\nu)}/d\nu = 0$ may hold for $\nu \geq \nu_1$, it is necessary and sufficient that $\chi(I) = \chi(J)$ for all pairs in $\mathcal{M} \vee \mathcal{K}$. As we have seen, however, the H_I belonging to $\mathcal{K}$ have the largest value of χ. Hence the condition is satisfied if and only if $\mathcal{M}$ is empty. If this is not the case, $d\bar{U}^{(\nu)}/d\nu$ will only tend to zero with $\nu \to \infty$. At this limit $\chi^\nu(I)$ in (6.60) for H_I belonging to $\mathcal{M}$ will vanish because $\chi(I) < 1$, and the terms corresponding to two H's both belonging to $\mathcal{K}$ will vanish, because then $\ln [\chi(J)/\chi(I)] = 0$. Hence $\bar{U}^{(\nu)}$ reaches its minimum value at the limit $\nu \to \infty$ and only at the limit. If $\mathcal{M}$ is empty the $\bar{q}^{(\nu)}(I)$ of $H_I \in \mathcal{K}$ will take their limiting value from $\nu = 1$ on, because all other $\bar{q}^{(\nu)}(I)$ become zero at $\nu = 1$. The average $\bar{U}^{(\nu)}$ takes its limiting value for $\nu \geq 1$ in this case.

Theorem 6.5 (Average inverse H-theorem: supplement). In Theorem 6.3, if $\mathcal{M} \neq \emptyset$, the not-smaller-than sign ($\geq$) in (6.50) can be replaced by the larger-than sign ($>$) and $\lim_{\nu\to\infty} [\bar{U}^{(\nu-1)} - \bar{U}^{(\nu)}] = 0$. If $\mathcal{M} = \emptyset$, the not-smaller-than sign in (6.50) can be replaced by the equality sign ($=$) for $\nu \geq 2$.

If we use a single experimental series $\mathcal{B}^{(\nu)}$ to calculate $U^{(\nu)}$ with the help of the empirical α's (6.24) then the value of $U^{(\nu)}$ is bound to show some fluctuation about the "average" curve $\bar{U}^{(\nu)}$. However, since the fluctuation of the empirical α's decreases with increasing ν, as shown in (6.32), it can be expected, at least within a certain limitation, that the fluctuation of $U^{(\nu)}$ about the smooth curve will also become very small for large ν. As an example we give an estimate of the order of magnitude of the fluctuation of $U^{(\nu)}$ for large ν in the case when $U^{(\nu)} \to 0$ as $\nu \to \infty$, as in Theorem 6.1.

This last condition means that in the expression of $U^{(\nu)}$ (6.44) one of the $q^{(\nu)}(I)$, say, $q^{(\nu)}(I_0)$, becomes very close to unity for larger ν, and all the remaining $q^{(\nu)}(I)$ become close to zero. Because of the nature of the function

$x \log x$, the contribution to $U^{(\nu)}$ from $q^{(\nu)}(I_0)$ becomes negligible compared with the contributions from other $q^{(\nu)}(I)$, $I \neq I_0$. In the same way, in the expression of the small fluctuation $\delta U^{(\nu)}$ of $U^{(\nu)}$,

$$\delta U^{(\nu)} = -\sum_I \delta q^{(\nu)}(I) \log q^{(\nu)}(I), \tag{6.65}$$

we can ignore the term corresponding to I_0. In the denominator of the expression (6.26) of $q^{(\nu)}(I)$ the term corresponding to $J = I_0$ will be very large compared with the other terms. Therefore we can write (6.26) as

$$q^{(\nu)}(I) \approx \frac{q^{(0)}(I)}{q^{(0)}(I_0)} \prod_{i=1}^{n} \left[\frac{p(i \mid I)}{p(i \mid I_0)}\right]^{\nu_i}$$

$$= \frac{q^{(0)}(I)}{q^{(0)}(I_0)} \exp\left[\sum_i \nu_i \ln \frac{p(i \mid I)}{p(i \mid I_0)}\right] \quad \text{for} \quad I \neq I_0, \tag{6.66}$$

from which follows that

$$\frac{\delta q^{(\nu)}(I)}{q^{(\nu)}(I)} \approx \sum_i \delta\nu_i Q_i(I), \tag{6.67}$$

with

$$Q_i(I) = \ln p(i \mid I) - \ln p(i \mid I_0). \tag{6.68}$$

With the aid of (6.31) we derive from (6.67), noting that $\delta\nu_i = \delta_i$,

$$\frac{\overline{\delta q^{(\nu)}(I)\, \delta q^{(\nu)}(J)}}{q^{(\nu)}(I) q^{(\nu)}(J)} \approx \nu D(I, J), \tag{6.69}$$

with

$$D(I, J) = \sum_i \gamma_i Q_i(I) Q_i(J) - \sum_i \gamma_i Q_i(I) \sum_i \gamma_i Q_i(J). \tag{6.70}$$

On the other hand, (6.66) shows that the order of magnitude of $\bar{q}^{(\nu)}(I)$ can be written [putting $\nu_i = \nu\gamma_i$ and assuming $q^{(0)}(I) \approx q^{(0)}(I_0)$]

$$\bar{q}^{(\nu)}(I) \approx e^{-\nu G(I)}, \tag{6.71}$$

with

$$G(I) = -\sum_i \gamma_i Q_i(I) = -\sum_i \gamma_i \ln p(i \mid I) + \sum_i \gamma_i \ln P(i \mid I_0). \tag{6.72}$$

Equation (6.72) shows that $G(I)$ is positive [see (6.34), (6.37), and (6.38)], so that $q^{(\nu)}(I)$ with $I \neq I_0$ becomes very small for large ν. From (6.69) and (6.70) we obtain

$$\overline{\delta q^{(\nu)}(I)\, \delta q^{(\nu)}(J)} \approx D(I, J)\nu \exp\{-\nu[G(I) + G(J)]\}. \tag{6.73}$$

Substituting (6.71) and (6.73) in the square of (6.65), we obtain

$$\overline{(\delta U^{(\nu)})^2} \approx \sum_{I \neq I_0} \sum_{J \neq I_0} D(I, J) G(I) G(J) \nu^3 \exp\{-\nu[G(I) + G(J)]\}. \tag{6.74}$$

The square of the entropy $\bar{U}^{(\nu)}$ itself is, according to (6.71),

$$(\bar{U}^{(\nu)})^2 \approx \sum_{I \neq I_0} \sum_{J \neq I_0} G(I)G(J)\nu^2 \exp\{-\nu[G(I) + G(J)]\}. \tag{6.75}$$

Therefore we see that both $\bar{U}^{(\nu)}$ and the individual fluctuation about $\bar{U}^{(\nu)}$ tend to zero with increasing ν. It is true that $\overline{\delta U^{(\nu)2}}$ vanishes more slowly with increasing ν than $(U^{(\nu)})^2$, but this should not be disconcerting, since the inductive entropy $U^{(\nu)}$, in the same way as the physical entropy, does not change its usefulness if one adds to it an arbitrary constant. For instance, we can use $V^{(\nu)} = (\log N - U^{(\nu)})/\log N$ as a convenient measure of "certainty," where N is the number of hypotheses. Then the fluctuation of $V^{(\nu)}$ will be small compared with $V^{(\nu)}$ itself for large ν. Summarizing this, we may complement Theorem 6.1 with Theorem 6.6.

Theorem 6.6 In the case of objective evidential convergence the average inductive entropy $\bar{U}^{(\nu)}$ and the average fluctuation of an individual entropy $U^{(\nu)}$ about the average $\bar{U}^{(\nu)}$ behave, for sufficiently large values of ν, as

$$\bar{U}^{(\nu)} \propto \nu e^{-a\nu}$$

$$\left[\overline{\delta U^{(\nu)2}}\right]^{1/2} \propto \nu^{3/2} e^{-a\nu}, \tag{6.76}$$

where a is some positive constant.

Before passing to the next section it may be worthwhile to call the reader's attention to the fact that we can regard the polychotomic learning discussed in Section 1.4 as an inductive process and identify the result that $\langle I_\mu^{(1)} \rangle_\mu \geq 0$ [see (1.128)] as a special case of the expected inverse H-theorem. We considered there n possible cases or states for which we have an original probability distribution $p_i^{(0)}$, $i = 1, 2, \cdots, n$. Instead of performing a direct (minute) observation to identify each individual case, we carry out a coarse observation, which allows us to identify a coarse state, $\mu = 1, 2, \cdots, m$. The probability of locating the object in state μ is, of course, given by $p_\mu = \sum_{i \in \mu} p_i^{(0)}$, that is, the sum of the probabilities of the states belonging to the coarse state μ. Since we divide the entire n cases in m coarse cases, we call this operation an m-chotomy. After we have located the object in state μ, our probability distribution for the microscopic states becomes $p_i^{(1)} = p_i^{(0)}/p_\mu$ for $i \in \mu$ and $p_i^{(1)} = 0$ for $i \notin \mu$. Thus the original ignorance $S^{(0)} = -\sum_{i=1}^n p_i^{(0)} \log p_i^{(0)}$ is reduced to $S^{(1)}(\mu) = -\sum_{i=1}^n p_i^{(1)} \log p_i^{(1)}$. The average ignorance after the m-chotomy is then

$$\langle S^{(1)} \rangle = \textstyle\sum_{\mu=1}^m S^{(1)}(\mu) p_\mu = S^{(0)} - (-\sum_{\mu=1}^m p_\mu \log p_\mu).$$

The average information obtained by the m-chotomy is thus $\langle I^{(1)} \rangle = -\sum_{\mu=1}^m p_\mu \log p_\mu \geq 0$. We can reinterpret this process of polychotomic observation as a process of induction. Instead of interpreting $p_i^{(0)}$simply as the probability of state i, we may interpret it as the inductive probability (credibility) of the hypothesis that the object is in state i. There are thus n different hypotheses. The polychotomic observation is the $\mathfrak{D}$ of this chapter (i and n thus correspond respectively to I and N of this chapter, and μ and m correspond to i and n). If hypothesis i is true, the deductive probability that the object will be located in the state μ is $p(\mu \mid i) = 1$ if $i \in \mu$ and $p(\mu \mid i) = 0$ if $i \notin \mu$. The average predictive probability corresponding to (6.45) will be $\sum_i p(\mu \mid i) p_i = p_\mu$. Thus, by the Bayes theorem, we obtain

$$p^{(1)} = \frac{p_i^{(0)} p(\mu \mid i)}{\sum_{i=1}^{n} p_i^{(0)} p(\mu \mid i)} = \frac{p_i^{(0)}}{\sum_{i \in \mu} p_i^{(0)}} = \frac{p_i^{(0)}}{p_\mu}, \tag{6.77}$$

which gives exactly the same result as the previous simple consideration [see (1.123)]. Then obviously $S^{(0)}$, $S^{(1)}(\mu)$, and $\langle S^{(1)} \rangle$ correspond respectively to $U^{(0)}$, $U^{(1)}$, and $\langle U^{(1)} \rangle$ of this section. Thus $\langle I^{(1)} \rangle \geq 0$ corresponds to our expected inverse H-theorem (6.46) for $\nu = 1$.

It may appear a little surprising that in Chapter 1 we compared [see (1.129)] $S^{(0)}$, $\langle S^{(1)} \rangle$, and $S^{(0)} - \langle S^{(1)} \rangle$, respectively, to $S(\mathcal{H} \otimes \mathfrak{D})$, $S(\mathcal{H} \mid \mathfrak{D}$,) and $S(\mathfrak{D})$, whereas in the above explanation we compared them respectively to $S(\mathcal{H})$, $S(\mathcal{H} \mid \mathfrak{D})$, and inf $(\mathcal{H} \mid \mathfrak{D})$. But in the present example these two comparisions are equivalent because $S(\mathfrak{D} \mid \mathcal{H}) = 0$.

6.3. ADDITIONAL REMARKS AND COMPUTER SIMULATION

In this section we add a few remarks on induction that are closely related to the convergence of credibilities in the Bayesian model. Toward the end of the section we adduce an illustration of the computer simulation of credibility convergence and of the inverse H-theorem.

Case of Continuously Varying Hypotheses—Generalized Laplace Law of Succession

In this subsection we consider the case in which a continuous number of hypotheses H_I assign continuously varying probabilities to events D_i, that is, the case in which the probability $p(i \mid I)$ is a continuous function of the continuous argument I. The index i of data, however, is assumed to be discrete. To make the problem intuitively clear we consider a simple concrete example. Two important results should be noted. First, when the prior credibility $p^{(0)}(I)$ is also a continuous function of I, the case turns out to be one of objective evidential convergence. The other interesting result is a generalized formulation of what is known as Laplace's law of succession.

We start with the following problem. It is known that an urn contains a very large number N of balls, of which some are red and the rest are white.

There are $N + 1$ possible hypotheses $H_I (I = 0, 1, 2, \cdots, N)$, namely, H_I states: the urn contains I red balls and $N - I$ white balls.

The experiment consists in taking one ball at random from the urn, determining its color, and placing it back in the urn. There are thus two possible data, or events: D_1—the ball is red, and D_2—the ball is white. Suppose that we perform ν experiments and obtain ν_1 times D_1 and ν_2 times D_2. We have

$$p(1 \mid I) = \frac{I}{N}, \qquad p(2 \mid I) = 1 - \frac{I}{N} \tag{6.78}$$

and

$$q^{(\nu)}(I) = \frac{q^{(0)}(I)\left(\dfrac{I}{N}\right)^{\nu_1}\left(1 - \dfrac{I}{N}\right)^{\nu_2}}{\sum_J q^{(0)}(J)\left(\dfrac{J}{N}\right)^{\nu_1}\left(1 - \dfrac{J}{N}\right)^{\nu_2}}. \tag{6.79}$$

Because N is extremely large, let us put

$$\frac{I}{N} = \chi, \qquad q^{(\nu)}(I) = \frac{1}{N}\, q^{(\nu)}(\chi), \qquad q^{(0)}(I) = \frac{1}{N}\, q^{(0)}(\chi). \tag{6.80}$$

Then $q^{(\nu)}(\chi)\, d\chi$ represents the total sum of those $q^{(\nu)}(I)$ whose argument I lies between I and $I + N\, d\chi$. There are $N\, d\chi$ such I's. The summation over I can be replaced by an integral so that, for instance,

$$\sum_{J=1}^{N} q^{(\nu)}(I) = \int_0^1 q^{(\nu)}(I) N\, d\chi = \int_0^1 q^{(\nu)}(\chi)\, d\chi = 1. \tag{6.81}$$

The Bayes formula (6.79) becomes

$$q^{(\nu)}(\chi) = \frac{q^{(0)}(\chi)\chi^{\nu_1}(1 - \chi)^{\nu_2}}{\int_0^1 q^{(0)}(\chi)\chi^{\nu_1}(1 - \chi)^{\nu_2}\, d\chi}. \tag{6.82}$$

Now the distribution function multiplying $q^{(0)}(\chi)$ in (6.82) is, when normalized,

$$f(\chi) = \frac{(\nu_1 + \nu_2 + 1)!}{\nu_1!\, \nu_2!}\, \chi^{\nu_1}(1 - \chi)^{\nu_2}, \tag{6.83}$$

which has its maximum at

$$\chi = \frac{\nu_1}{\nu_1 + \nu_2}. \tag{6.84}$$

It should be noted that the distribution (6.83) has an appearance somewhat similar to the binomial distribution, but it is not a binomial distribution. In (6.83) the variable χ, but not ν_1 or ν_2, is playing the role of a random variable.

This type of distribution is sometimes called "β-distribution." The average and the higher moments of (6.83) are

$$\bar{\chi} = \int_0^1 \chi f(\chi)\, d\chi = \frac{\nu_1 + 1}{\nu_1 + \nu_2 + 2} \tag{6.85}$$

and

$$\overline{\chi^r} = \int \chi^r f(\chi)\, d\chi = \frac{(\nu_1 + \nu_2 + 1)!}{(\nu_1 + \nu_2 + r + 1)!} \frac{(\nu_1 + r)!}{\nu_1!}. \tag{6.86}$$

For large $\nu_1 \gg 1$

$$\bar{\chi} \to \frac{\nu_1}{\nu_1 + \nu_2}, \qquad \overline{\chi^2} \to \left(\frac{\nu_1}{\nu_1 + \nu_2}\right)^2, \tag{6.87}$$

showing that

$$(\overline{\chi^2} - \bar{\chi}^2) \to 0. \tag{6.88}$$

This means that for a large number of trials ($\nu = \nu_1 + \nu_2 > \nu_1 \gg 1$) the distribution $f(\chi)$ is sharply concentrated about the mean value $\chi = \alpha_1^{(\nu)} = \nu_1/(\nu_1 + \nu_2)$. Therefore, if $q^{(0)}(\chi)$ is continuous in the vicinity of this point or, more generally, if $q^{(0)}(\chi)$ does not change appreciably in the domain of χ of the order of $(\overline{\chi^2} - \bar{\chi}^2)^{1/2}$ that becomes small for large ν, we obtain from (6.82)

$$q^{(\nu)}(\chi) = \frac{(\nu_1 + \nu_2 + 1)!}{\nu_1!\, \nu_1!} \chi^{\nu_1}(1 - \chi)^{\nu_2}, \tag{6.89}$$

which is the same as (6.83). In terms of the variable I this can be written

$$q^{(\nu)}(I) = \frac{I}{N} \frac{(\nu_1 + \nu_2 + 1)!}{\nu_1!\, \nu_2!} \left(\frac{I}{N}\right)^{\nu_1} \left(1 - \frac{I}{N}\right)^{\nu_2}, \tag{6.90}$$

which no longer depends on $q^{(0)}(I)$.

The probability of obtaining the event D_1 in the $(\nu + 1)$st observation according to (6.15) and with the aid of (6.90), is then

$$p^{(\nu+1)}(D_1) = \sum_I q^{(\nu)}(I) p(1 \mid I) = \frac{\nu_1 + 1}{\nu_1 + \nu_2 + 2}, \tag{6.91}$$

which is the same as the mean value given in (6.85). The interesting part of this result is that it is not exactly equal to the past relative frequency, which is $\nu_1/(\nu_1 + \nu_2)$. When $\nu_2 = 0$, that is, when ν observations have consecutively given the same event D_1, the probability of obtaining D_1 in the next observation is, according to (6.91),

$$p^{(\nu+1)}(D_1) = \frac{\nu + 1}{\nu + 2}, \tag{6.92}$$

which is Laplace's law of succession. Equation (6.91) is its generalization.

Coming back to the general case, the distributions (6.89) and (6.90) are functions of ν and $\alpha_1^{(\nu)} = \nu_1/(\nu_1 + \nu_2)$. When ν_1 (hence also ν) tends to infinity, their dependence on ν disappears and the empirical $\alpha_1^{(\nu)}$ can be expected to become a fixed value γ_1. At this limit (6.89) becomes, because of (6.88),

$$q^{(\infty)}(\chi) = \delta(\chi - \gamma_1). \tag{6.93}$$

As far as the prior credibility is concerned, it is sufficient to assume that $q^{(0)}(I)$ is a continuous function of I. When N is not really infinity but only very large, $\delta(\chi - \gamma_1)$ in (6.93) is to be interpreted as zero except in a small vicinity of width $1/N$ of $\chi = \gamma_1$, in which case it takes a value of the order of N. Therefore (6.90) becomes

$$q^{(\infty)}(I) = 1 \tag{6.94}$$

for the particular I that gives the maximum value of

$$\gamma_1 \log \frac{I}{N} + (1 - \gamma_1) \log \left(1 - \frac{I}{N}\right), \tag{6.95}$$

provided that there is only one such I. For the remaining I's we have

$$q^{(\infty)}(I) = 0. \tag{6.96}$$

Maximum Likelihood Method

The Bayesian viewpoint of hypothesis evaluation maintains that the inductive probability of a hypothesis is proportional to the product of the prior credibility and the deductive probability. The classical theory of hypothesis testing advises us to ignore the prior credibility and to base out evaluation solely on the deductive probability. The method suggested by this classical theory is called maximum likelihood method because it tells us to adopt the hypothesis that gives the largest deductive probability (likelihood) to the evidence. From the point of view of the Bayesian model of induction, the maximum likelihood method is, in principle, only partially justifiable and should be considered as a simplifying method of approximation. In this subsection we briefly explain the basis of the maximum likelihood method from this point of view. Other than the maximum likelihood method, there have been different proposals which are claimed to be capable of deriving unique values of prior probabilities from some basic principles, but in the last analysis these principles are found to be even less justified than the questionable but simple principle of equal prior probabilities adopted by the maximum likelihood method.

We can recapitulate the main points of Section 4.4 as follows. Writing $\mathcal{B}^{(\nu)}$ for the experimental sequence $\{i_1, i_2, \cdots, i_\nu\}$ and $p(\mathcal{B}^{(\nu)} \mid I)$ for the deductive probability of $\mathcal{B}^{(\nu)}$ happening on the assumption of H_I, we can express the ratio of credibilities of two hypotheses H_I and H_J according to the Bayesian rule in the form

$$\frac{q^{(\nu)}(I)}{q^{(\nu)}(J)} = \frac{q^{(0)}(I)}{q^{(0)}(J)} \cdot \frac{p(\mathcal{B}^{(\nu)} \mid I)}{p(\mathcal{B}^{(\nu)} \mid J)} \tag{6.97}$$

$$= \frac{q^{(0)}(I)}{q^{(0)}(J)} \exp\left\{- b^{(\nu)}\nu\left[\frac{1}{\Gamma^{(\nu)}(I)} - \frac{1}{\Gamma^{(\nu)}(J)}\right] \ln 2\right\}, \tag{6.98}$$

where

$$b^{(\nu)} = - \sum_i \alpha_i^{(\nu)} \log \alpha_i^{(\nu)} \tag{6.99}$$

and the degree of confirmation $\Gamma^{(\nu)}(I)$ is given by

$$\Gamma^{(\nu)}(I) = \frac{b^{(\nu)}}{-\dfrac{1}{\nu} \log p(\mathcal{B}^{(\nu)} \mid I)}. \tag{6.100}$$

The case $b^{(\nu)} = 0$ happens if and only if the α's are either 0 or 1. For simplicity of argument, we assume in the following that $b^{(\infty)} \neq 0$. Then convergence of $\alpha_i^{(\nu)}$ for $\nu \to \infty$ guarantees the existence of an integer μ such that $b^{(\nu)} \neq 0$ for $\nu > \mu$. If $b^{(\nu)} \neq 0$, $\Gamma^{(\nu)}(I) > \Gamma^{(\nu)}(J)$ is equivalent to $p(\mathcal{B}^{(\nu)} \mid I) > p(\mathcal{B}^{(\nu)} \mid J)$. The logarithm used in (6.99) and (6.100) is assumed to be to the base 2. Equations (6.97) and (6.98) can be derived from (4.34), (4.38), (4.42), and (4.43).

From (6.97) and (6.98) we can derive the following valid statements.

1. If two hypotheses H_I and H_J have the same *a priori* credibility, then the ratio of credibilities $q^{(\nu)}(I)/q^{(\nu)}(J)$ will be $\gtreqless 1$ according to whether the ratio of deductive probabilities $p(\mathcal{B}^{(\nu)} \mid I)/p(\mathcal{B}^{(\nu)} \mid J)$ is $\gtreqless 1$ or, equivalently, according to whether the ratio of confirmabilities $\Gamma^{(\nu)}(I)/\Gamma^{(\nu)}(J)$ is $\gtreqless 1$, provided that $b^{(\nu)} \neq 0$.

2. In the following statement convergence of the quantities involved for $\nu \to \infty$ is assumed and $b^{(\infty)} \neq 0$. For a given ratio of prior credibilities and for a given positive number a there exists an integer μ such that the condition $\Gamma^{(\nu)}(I) - \Gamma^{(\nu)}(J) > a$ for all $\nu > \mu$ implies $q^{(\nu)}(I) > q^{(\nu)}(J)$ for all $\nu > \mu$ and $q^{(\infty)}(I) \geq q^{(\infty)}(J)$.

Statement 1 is a direct consequence of (6.97) and (6.98) and requires no comments. As regards Statement 2, we note that the condition $q^{(\nu)}(I) > q^{(\nu)}(J)$ is equivalent to

$$\nu > \frac{\log\,[q^{(0)}(I)/q^{(0)}(J)]}{b^{(\nu)}\{[1/\Gamma^{(\nu)}(I)] - [1/\Gamma^{(\nu)}(J)]\}} \tag{6.101}$$

according to (6.98) provided that $\Gamma^{(\nu)}(I) > \Gamma^{(\nu)}(J)$. If $[q^{(0)}(I)/q^{(0)}(J)] > 1$, this relation is true for all integers $\nu \geq 0$, provided that $\Gamma^{(\nu)}(I) > \Gamma^{(\nu)}(J)$.

Otherwise this relation will set a certain lower bound for ν. If $\Gamma^{(\nu)}(I) - \Gamma^{(\nu)}(J) > a$ for $\nu > \mu^*$, which also implies $[1/\Gamma^{(\nu)}(J)] - [1/\Gamma^{(\nu)}(I)] > a$ [note: $0 \leq \Gamma^{(\nu)} \leq 1$] for $\nu > \mu^*$, the required condition (6.101) becomes

$$\nu > \frac{\log [q^{(0)}(J)/q^{(0)}(I)]}{b^{(\nu)}a} \quad \text{for} \quad \nu > \mu^*. \tag{6.102}$$

For sufficiently large ν, $b^{(\nu)}$ becomes practically a nonvanishing constant; hence the inequality (6.102) sets a lower bound $\nu > \mu^{**}$. We can then use max (μ^*, μ^{**}) as the lower bound μ of Statement 2. The reason we have to replace $>$ by $\geq$ in the inequality $q^{(\nu)}(I) > q^{(\nu)}(J)$ for $\nu = \infty$ is that $q^{(\nu)}(I)$ and $q^{(\nu)}(J)$ can both converge to zero.

Since the convergence of $\Gamma^{(\nu)}(I)$ is assumed, its limiting value will be the same as the limiting value of its "average" $\bar{\Gamma}^{(\infty)}(I) = [-\sum_i \gamma_i \log \gamma_i]/[-\sum_i \gamma_i \log p(i \mid I)]$. Hence, providing convergence, we can state that if $\bar{\Gamma}^{(\infty)}(I) > \bar{\Gamma}^{(\infty)}(J)$ there will be an integer μ such that $\Gamma^{(\nu)}(I) - \Gamma^{(\nu)}(J) > a$ for some $a > 0$ and for all $\nu > \mu$. As a consequence, if $\Gamma^{(\nu)}(I) - \Gamma^{(\nu)}(J) > c$ for a given c and for a sufficiently large particular value of ν, we can expect to have $\Gamma^{(\nu)}(I) - \Gamma^{(\nu)}(J) > a$ for some a for all subsequent values of ν. The trouble in practice, of course, lies in the fact that we usually do not know "how large" a ν is "sufficiently large" in this context. In any event, combining this result with Statement 2, we can formally state the following.

3. Providing convergence of the quantities involved for $\nu \to \infty$ and $b^{(\infty)} \neq 0$, there exists, for given values of the prior credibilities and for a given positive constant a, an integer μ such that the condition $\Gamma^{(\nu)}(I) - \Gamma^{(\nu)}(J) > a$ [or $p(\mathcal{B}^{(\nu)} \mid I)/p(\mathcal{B}^{(\nu)} \mid J) > a$] for any particular value of ν larger than μ implies $q^{(\nu)}(I) > q^{(\nu)}(J)$ for $\nu > \mu$ and $q^{(\infty)}(I) \geq q^{(\infty)}(J)$.

It may look odd, in view of (6.97), that the empirical factor $p(\mathcal{B}^{(\nu)} \mid I)/p(\mathcal{B}^{(\nu)} \mid J)$ (which is the so-called "likelihood ratio") alone, irrespective of the prior factor $q^{(\nu)}(I)/q^{(0)}(J)$, determines $q^{(\nu)}(I)/q^{(\nu)}(J)$. This is permissible only on account of the restrictive premises of Statement 3: (a) the prior credibilities are fixed; (b) ν is "sufficiently large"; (c) all quantities involved converge; and (d) H_I and H_J are not such that $p(\mathcal{B}^{(\nu)} \mid I) = p(\mathcal{B}^{(\nu)} \mid J)$.

If H_I and H_J both belong to $\mathcal{K}$ and if $\mathcal{K}$ is homogeneous, Statement 3 is not applicable since $\Gamma^{(\nu)}(I) = \Gamma^{(\nu)}(J)$ all the way and the prior factor determines the credibility ratio. If H_I and H_J both belong to $\mathcal{K}$ but $\mathcal{K}$ is not homogeneous, and if H_I and H_J have different predictive probability distributions, $q^{(\nu)}(I)$ and $q^{(\nu)}(J)$ will oscillate at the limit $\nu \to \infty$, and Statement 3 cannot be applied because the assumption of convergence is violated. In such a case $\Gamma^{(\nu)}(I)$ and $\Gamma^{(\nu)}(J)$ become close to each other for large ν, but the difference will change the sign from time to time. Because of the factor ν in the exponent of (6.98), the credibility ratio will show an appreciable oscillation.

If both H_I and H_J belong to $\mathcal{M}$, it can very well happen that $\Gamma^{(\nu)}(I) - \Gamma^{(\nu)}(J) > a$ for all ν beyond a certain lower bound, yet $q^{(\infty)}(I)$ and $q^{(\infty)}(J)$ both become zero. Hence Statement 3 is formally applicable here, but it is not a useful case. If H_I belongs to $\mathcal{K}$ and H_J belongs to $\mathcal{M}$, $\Gamma^{(\nu)}(I) - \Gamma^{(\nu)}(J) > a$ for large ν and $q^{(\infty)}(I) > q^{(\infty)}(I)$. This is the only useful case covered by Statement 3. It may be noted that the condition $H_I \in \mathcal{M}$ or $H_I \in \mathcal{K}$ depends on the other hypotheses in $\mathcal{H}$. Hence we cannot decide $q^{(\infty)}(I) > q^{(\infty)}(J)$ or $q^{(\infty)}(I) = q^{(\infty)}(J)$ by the properties of H_I and H_J only.

Keeping these restrictions on the applicability of Statements 1–3 in mind, let us now introduce the strategy of classical theory of hypothesis testing.

Classical Theory of Hypothesis Testing—Algorithm I. For a given body $\mathcal{B}$ of evidence, hypothesis H_I is to be declared more plausible than hypothesis H_J if the likelihood (predictive probability) of $\mathcal{B}$ on the basis of H_I is larger than that of $\mathcal{B}$ on the basis of H_J.

This is utterly inacceptable from the Bayesian point of view. We can improve the situation somewhat by the following consideration. Since the trichotomy $p(\mathcal{B}^{(\nu)} \mid I)/p(\mathcal{B}^{(\nu)} \mid J) \gtreqless 1$ does not necessarily coincide with the trichotomy $p(\mathcal{B}^{(\infty)} \mid I)/p(\mathcal{B}^{(\infty)} \mid J) \gtreqless 1$, it is not safe to reject H_J in deference to H_I on the ground that $p(\mathcal{B}^{(\nu)} \mid I)/p(\mathcal{B}^{(\nu)} \mid J)$ is larger than the threshold unity, even if we decide to ignore the *a priori* credibilities. We might raise the threshold to some value larger than 1, say, K, so that the probability of rejecting the "right" (on the basis of likelihood) hypothesis becomes less than a certain reasonable value ε, which may be chosen equal to 0.05, for instance. Since we do not know the right hypothesis, all we can do is to estimate, somewhat as in our discussion of fluctuations, the probability of making an error of the kind described above. In our discussion of fluctuations we used the constants γ_i, but if we replace these γ_i by their empirical counterparts $\alpha_i^{(\nu)}$, we obtain an estimation of the fluctuations based only on the empirical data. Similarly, an estimation of the probability of error in question can be made on the basis of the empirical data only. Thus we arrive at the next algorithm.

Classical Theory of Hypothesis Testing—Algorithm II. Hypothesis H_J is to be rejected in deference to hypothesis H_I if $p(\mathcal{B}^{(\nu)} \mid I)/p(\mathcal{B}^{(\nu)} \mid J) > K$, where K is a function of $\mathcal{B}^{(\nu)}$, H_I, H_J, and ε such that the estimated probability that this rejected hypothesis H_J will be the correct hypotheses is less than ε.

One may be tempted to include the *a priori* credibilities in this consideration simply by replacing $p(\mathcal{B}^{(\nu)} \mid I)/p(\mathcal{B}^{(\nu)} \mid J)$ by $[q^{(0)}(I)p(\mathcal{B}^{(\nu)} \mid I)]/[q^{(0)}(J)p(\mathcal{B}^{(\nu)} \mid J)]$ in the above algorithm. But the matter is not so simple in the general case, because K itself may be affected by the values of the *a priori* credibilities.

In human experience the length of observation ν is always finite, and except for superinduction (see Section 6.1) we have to judge hypotheses H_I at the

νth stage only by $q^{(\nu)}(I)$. If ν is finite, no matter how large, the likelihood ratio [the second factor in (6.97)] can always be upset by the prior credibility ratio [the first factor in (6.97)]. From this finitary point of view it may still be considered meaningful to guess the credibility ratio for larger ν, but it is meaningless to guess it for $\nu = \infty$. Insofar as ν is finite, the prior factor can never be ignored. If it is stipulated (against our standpoint) that the prior credibilities must be fixed and cannot be changed in the middle of experience, the prior factor can be finally overcome by the empirical factor. Even then guessing $q^{(\infty)}$ from $q^{(\nu)}$ is a kind of superinduction and introduces additional uncertainty in the inference.

It may be recalled that the *a priori* credibility plays two important roles in induction. For one thing, while ν is small, it has a strong influence on the evaluation of credibilities. Second, when $p(\mathcal{B}^{(\nu)} \mid I) = p(\mathcal{B}^{(\nu)} \mid J)$ the only difference in the credibilities of H_I and H_J for any ν stems from the difference in the *a priori* credibilities. In other words, the credibility ratio at the νth stage is the same as the credibility ratio at the 0th stage in this case, and also remains so in the limit $\nu \to \infty$. The maximum likelihood method cannot say anything in this case. It should not be overlooked either that better-chosen *a priori* credibilities can bring the $q^{(\nu)}$ at a finite ν closer to the ultimate $q^{(\infty)}$.

A Remark on Conjunctive Experimentation

We stated in Chapter 4 that the prior credibility of a hypothesis, while "prior" to a particular experimentation, can reflect the results of other experiences pertinent to the hypothesis, and that in spite of this the prior credibility cannot be determined by experience alone. The following consideration, though limited in its scope, may be regarded as an illustration of these two points of importance.

Let there be two kinds of experiments $\mathfrak{D}^{(1)} = \{D_i^{(1)}\}$ and $\mathfrak{D}^{(2)} = \{D_j^{(2)}\}$ pertinent to $\mathcal{H}$ so that each member H_I can give predictive probabilities $p(D_i^{(1)} \mid I)$ and $p(D_j^{(2)} \mid I)$. Let it be further assumed that the hypotheses satisfies the condition of "predictive separability," that is,

$$p(D_i^{(1)} \cap D_j^{(2)} \mid I) = p(D_i^{(1)} \mid I)p(D_j^{(2)} \mid I) \tag{6.103}$$

for all i, all j, and all I, where $D_i^{(1)} \cap D_j^{(2)}$ is a member of the conjunctive experiment $\mathfrak{D}^{(1)} \otimes \mathfrak{D}^{(2)}$. We can state our result as follows.

Theorem 6.7 Under the assumption of predictive separability the result of evaluation of $\mathcal{H}$ by the conjunctive experiment $\mathfrak{D}^{(1)} \otimes \mathfrak{D}^{(2)}$ as a single test is the same as the result of the same experiment considered as a two-stage evaluation, first by $\mathfrak{D}^{(1)}$ and then by $\mathfrak{D}^{(2)}$, whereby the *a posteriori* credibilities after the test by $\mathfrak{D}^{(1)}$ are used as the prior credibilities before the test by $\mathfrak{D}^{(2)}$.

The proof is as follows. Let ν_{ij}, $\nu_i^{(1)}$, and $\nu_j^{(2)}$ be respectively the number of occurrences of $D_i^{(1)} \cap D_j^{(2)}$, $D_i^{(1)}$, and $D_j^{(2)}$ in the experiment. We obviously have

$$\nu_i^{(1)} = \sum_j \nu_{ij}, \qquad \nu_j^{(2)} = \sum_i \nu_{ij}. \tag{6.104}$$

From the point of view of the conjunctive experiment $\mathfrak{D}^{(1)} \otimes \mathfrak{D}^{(2)}$ we have

$$\begin{aligned} q^{(\nu)}(I) &\propto q^{(0)}(I) \prod_{i,j} [p(D_i^{(1)} \cap D_j^{(2)} \mid I)]^{\nu_{ij}} \\ &= q^{(0)}(I) \prod_{i,j} [p(D_i^{(1)} \mid I)]^{\nu_{ij}} [p(D_j^{(2)} \mid I)]^{\nu_{ij}} \\ &= q^{(0)}(I) \prod_i [p(D_i^{(1)} \mid I)]^{\nu_i^{(1)}} \cdot \prod_j [p(D_{j^{(2)}} \mid I)]^{\nu_j^{(2)}}. \end{aligned} \tag{6.105}$$

In the first line we used $\propto$ instead of $=$ to avoid encumbering the formulas with normalizing factors. The transformation from the first to the second line is allowed by the predictive separability (6.103) and the passage from the second to the third line is made possible by (6.104). The last expression of (6.105) can further be rewritten as

$$q^{(\nu)}(I) \propto q^{*(0)}(I) \prod_j [p(D^{(2)} \mid I)]^{\nu_j^{(2)}}, \tag{6.106}$$

with

$$q^{*(0)}(I) \propto q^{(0)}(I) \prod_i [p(D_i^{(1)} \mid I)]^{\nu_i^{(1)}}. \tag{6.107}$$

These two expressions (6.106) and (6.107) are precisely the description of the experiment according to the second viewpoint, which considers $\mathfrak{D}^{(1)}$ and $\mathfrak{D}^{(2)}$ as successive two-stage tests. The interesting point is that $q^{*(0)}(I)$ has the double role of *a posteriori* credibility after $\mathfrak{D}^{(1)}$ and *a priori* credibility before $\mathfrak{D}^{(2)}$, $q^{(0)}(I)$ of (6.107) being the same as $q^{(0)}(I)$ of the first line of (6.105). We learn from this that the prior credibility in general can reflect the results of other experiments [the second factor in (6.107)], yet cannot free itself from the *a priori* factor entirely [the first factor in (6.107)]. It is to be noted that we have not assumed in the derivation the probabilistic independence

$$\gamma_{ij} = \gamma_i^{(1)} \gamma_j^{(2)}, \tag{6.108}$$

which implies, for large $\nu = \sum_i \sum_j \nu_{ij}$, the empirical separability

$$\frac{\nu_{ij}}{\nu} = \frac{\nu_i^{(1)} \nu_j^{(2)}}{\nu^2}. \tag{6.109}$$

The relations

$$\gamma_i^{(1)} = \sum_j \gamma_{ij} \quad \text{and} \quad \gamma_j^{(2)} = \sum_i \gamma_{ij} \tag{6.110}$$

must always hold, but (6.108) is an extra condition. It is easy, however, to show that if the conjunctive observation $\mathfrak{D}^{(1)} \otimes \mathfrak{D}^{(1)}$ does not satisfy the empirical separability (6.109), any hypothesis that satisfies predictive separability (6.103) cannot have a degree of confirmation of unity.

Theorem 6.7 compares the two viewpoints applied to the same set of experimental data, but if we have the probabilistic independence for large ν we can compare a ν-step test on $\mathfrak{D}^{(1)} \otimes \mathfrak{D}^{(2)}$ with an independent ν-step test on $\mathfrak{D}^{(1)}$ followed by another independent ν-step test on $\mathfrak{D}^{(2)}$. This is so because, for large ν, the three numbers ν_{ij}, $\nu_i^{(1)}$, and $\nu_j^{(2)}$ taken from these independent tests will be approximately equal to $\gamma_{ij}\nu$, $\gamma_i^{(1)}\nu$, and $\gamma_j^{(2)}\nu$, satisfying (6.104) although for smaller ν they may not satisfy (6.104).

This last consideration gives us a clue to a useful algorithm allowing us to introduce uneven weights to different $\mathfrak{D}$'s to which a set $\mathfrak{H}$ of hypotheses is relevant. Suppose that we perform an $\omega^{(1)}\nu$-step test on $\mathfrak{D}^{(1)}$ and an $\omega^{(2)}\nu$-step test on $\mathfrak{D}^{(2)}$, where $\omega^{(1)} \geq 0$, $\omega^{(2)} \geq 0$, and $\omega^{(1)} + \omega^{(2)} = 1$; then from the successive-test viewpoint we have

$$q^{(\nu)}(I) \propto q^{(0)}(I) \prod_i [p(D_i^{(1)} \mid H)]^{\omega^{(1)}\alpha_i^{(1)}\nu} \cdot \prod_j [p(D_j^{(2)} \mid H)]^{\omega^{(2)}\alpha_j^{(2)}\nu} \tag{6.111}$$

to replace the last line of (6.105), where $\alpha_i^{(1)}$ and $\alpha_j^{(2)}$ are the empirical relative frequencies of $D_i^{(1)}$ and $D_j^{(2)}$ in respective tests. When $\omega^{(1)}/\omega^{(2)}$ is a rational number equal, say, to $N^{(1)}/N^{(2)}$ [$N^{(1)}$, $N^{(2)}$ are integers] we can also arrive at (6.111) from the conjunctive test viewpoint applied to $[\mathfrak{D}^{(1)}]^{N^{(1)}} \otimes [\mathfrak{D}^{(2)}]^{N^{(2)}}$, where the power expression should be understood in terms of multiplication in the sense of $\otimes$. We can consider the ω's in (6.111) as the measures of importance of different tests, $\mathfrak{D}$'s.

A Simulated Computer Experiment

The following example is an actual calculation of credibilities $q^{(\nu)}(I)$, confirmabilities $\Gamma^{(\nu)}(I)$, and inductive entropy $U^{(\nu)}$ as functions of ν according to our formulas in a specific problem, in which the random occurrences of observational results are simulated by a computer program using the random number technique (see [W-13] for details).

The problem is as follows. An urn contains 10 balls, of which a certain fraction $I_0/10$ are red and the remaining fraction $(10 - I_0)/10$ are white. The observation consists of taking one ball at random from the urn, determining its color ($n = 2$), and returning it to the urn. The considered hypotheses are 11 in number ($N = 11$) and are of the type,

$H(I)$: I balls out of the 10 are red and the remaining
$(10 - I)$ balls are white, $\quad I = 0, 1, 2, \cdots, 10.$

$H(I_0)$ is the correct hypothesis, that is, the law to be discovered. The process under investigation is one in which the credibilities of hypotheses gradually concentrate on $H(I_0)$ as the number of observations ν increases. If we assign 0 and 1 to red and white, respectively, the deductive probabilities will be

$$p(0 \mid I) = \frac{I}{10}, \qquad p(1 \mid I) = \frac{(10 - I)}{10}. \tag{6.112}$$

The machine simulation consists of producing the numbers 0 and 1 randomly as the ratio of $I_0/(10 - I_0)$. For this purpose a well-tested random number producing program has been utilized. This means in our notation $\gamma_0 = 1 - \gamma_1 = I_0/10$.

The first series of experiments was carried out with $I_0 = 3$. Under this condition $H(0)$ and $H(10)$ must be "logically" refuted sooner or later, since $H(0)$ contradicts the appearance of a single red ball and $H(10)$ contradicts the appearance of a single white ball. Thus $\mathcal{H}$ consists of $H(I)$ with $I = 0, 1, 2, \cdots, 10$, and $\mathcal{M} \vee \mathcal{K} = \mathcal{H} - \mathcal{R}$ consists of $H(I)$ with $I = 1, 2, \cdots, 9$. The $\mathcal{K}$ consists, of course, only of $H(3)$. As regards the *a priori* credibilities, we have tried three different cases: (a) equal *a priori* credibilities are given to all 11 hypotheses; (b) a higher *a priori* credibility is deliberately given to a wrong hypothesis $H(7)$, namely, $q^{(0)}(7) = \frac{12}{22}$ and $q^{(0)}(I) = \frac{1}{22}$ for $I \neq 7$; and (c) a higher *a priori* credibility is given to the right hypothesis, namely, $q^{(0)}(3) = \frac{12}{22}$ and $q^{(0)}(I) = \frac{1}{22}$ for $I \neq 3$. The experiments were continued in each case until the observation number ν became 500 or more. The same sequence of random numbers was used in all three cases. In this sequence $\mathcal{H}$ was retrenched to $\mathcal{M} \vee \mathcal{K}$ with $\nu = 5$.

The smallest number ν_0 for which the inductive entropy satisfied the condition

$$U^{(\nu)} < 0.01 \quad \text{for} \quad \nu \geq \nu_0$$

was found to be $\nu_0 = 330$ in Case (a), $\nu_0 = 330$ in Case (b), and $\nu_0 = 258$ in Case (c). The smallest number ν_1 for which the credibility for the correct hypothesis satisfied the condition

$$q^{(\nu)}(3) > 0.99 \quad \text{for} \quad \nu \geq \nu_1$$

was found to be $\nu_1 = 258$ in (a), $\nu_1 = 258$ in (b), and $\nu = 115$ in (c). Outside of these slight numerical differences, the behavior of the different quantities was the same in all three cases.

The random sequence up to $\nu = 500$ contained 149 red (0's) and 351 white (1's), the ratio of the red to the total number being $\alpha_0^{(\nu)} = 0.298$ instead of $\gamma_0 = 0.3$. The confirmability $\Gamma^{(\nu)}(I)$ of $H(I_0)$, which also serves as a measure of the agreement of $\alpha_0^{(\nu)}$ with γ_0, become 0.999984 at $\nu = 500$.

We describe some of the notable points in the behavior of the credibilities in Case (b) for the correct hypothesis $H(3)$, for the wrong hypothesis $H(7)$ on which a high weight was initially placed, and for the hypothesis $H(2)$, which is an immediate neighbor of the correct hypothesis. Since $p(i \mid 2)$ is numerically close to $p(i \mid 3)$, the chance is high that $q^{(\nu)}(2)$ will remain relatively large compared with other hypotheses whose I is further removed from 3. In our experiment $q^{(\nu)}(3)$, starting from $\frac{1}{22} = 0.04545$ at $\nu = 0$, became definitely larger than 0.9 from $\nu = 203$ on. $q^{(\nu)}(7)$, starting from $\frac{6}{11} = 0.54545$ at $\nu = 0$, became definitively less than 0.1 from $\nu = 6$ on. The credibility $q^{(\nu)}(2)$, starting from 0.04545 at $\nu = 0$, rose to higher values, including a

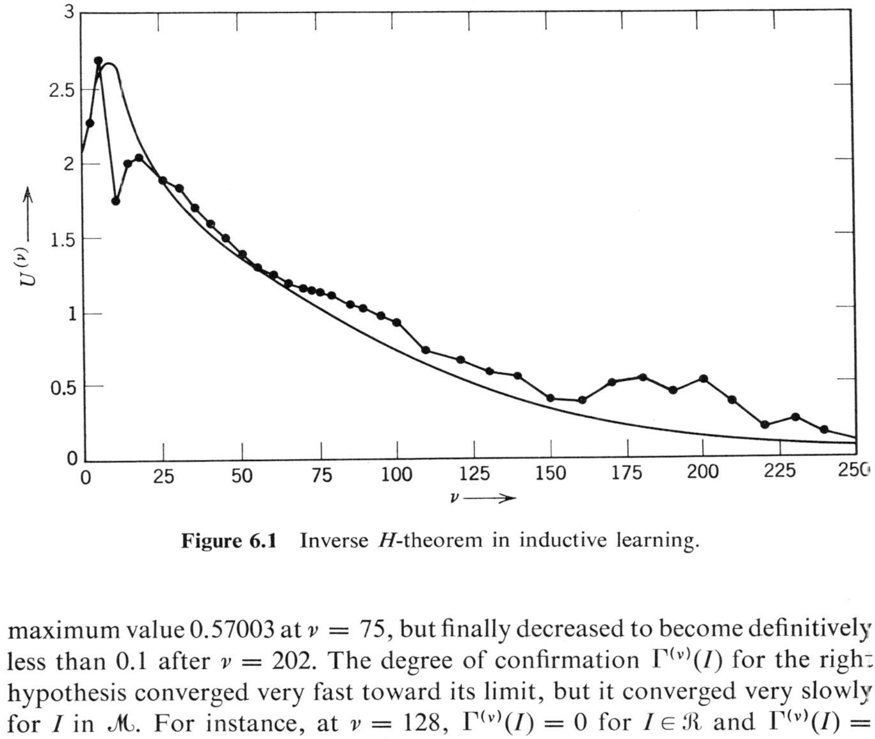

Figure 6.1 Inverse H-theorem in inductive learning.

maximum value 0.57003 at $\nu = 75$, but finally decreased to become definitively less than 0.1 after $\nu = 202$. The degree of confirmation $\Gamma^{(\nu)}(I)$ for the right hypothesis converged very fast toward its limit, but it converged very slowly for I in $\mathcal{M}$. For instance, at $\nu = 128$, $\Gamma^{(\nu)}(I) = 0$ for $I \in \mathfrak{R}$ and $\Gamma^{(\nu)}(I) = 0.999962$ for $I = I_0$, but $\Gamma^{(\nu)}(I) > 0.3$ for all $I \in \mathcal{M}$. Roughly speaking, after $\nu = 200$, everything smoothly tended toward the limiting situation. At $\nu = 500$, $q^{(\nu)}(2) = 0.131090 \times 10^{-5}$, $q^{(\nu)}(3) = 0.999986$, and $q^{(\nu)}(7) < 10^{-30}$.

Figure 6.1 shows the inductive entropy $U^{(\nu)}$ as function of ν. The line connecting the points is the result in Case (b) for one particular random sequence. The smooth curve is the "average" $\bar{U}^{(\nu)}$. Both curves exhibit a small initial rise for a short time but thereafter show a marked "inverse H-theoremic" trend for increasing ν. The initial rise represents the unlearning of the wrong initial bias.

6.4. INVERSE *H*-THEOREM IN BEHAVIORAL LEARNING PROCESSES

Induction is a special type of learning.† It then is "inductively natural" to suspect that the tendency of gradual probability concentration expressed by the inverse *H*-theorem is a common feature of all types of learning. In fact, in any cognitive learning in which a question-and-answer type of test is possible, the process of learning may be characterized by the gradual concentration of weights (probabilities of truth) placed by the subject on different possible answers to a question. Furthermore, even in behavioral learning, in which no question-and-answer tests are possible, we find that the distribution of different responses to a stimulus is gradually concentrated on fewer and fewer alternatives until finally one special response is fixed to the given stimulus. All these tendencies must be expressible by a general law of entropy decrease in learning.

Theorem 6.8 (Generalized inverse *H*-theorem). In learning processes the state of a subject can be characterized by a probability distribution over a set of alternative choices, and the entropy defined by this probability distribution has a general tendency to decrease with time.

The word "theorem" has to be understood in the sense of an empirical law, not in the sense of a mathematical theorem. The expression "has a general tendency to" is used to indicate the possibility of occasional occurrences of temporary reversal of the trend in an individual subject, which, however, can be made to disappear by suitable averaging or suitable smoothing. Since the entropy is non-negative, it is also understood that it approaches a certain limiting value and its rate of decrease has to tend to zero in the limit.

After stating the above theorem we hasten to add that the subject is practically never absolutely neutral (according to the definition of the alternative choices mentioned in the theorem) at the beginning of the process. Hence each subject who learns in the sense of Theorem 6.8 will generally have to go through a period of unlearning to neutralize his pre-existing preferences or bias, which would be characterized by an increase of the entropy in question. Hence we state the following theorem.

Theorem 6.9 (Entropic theorem of learning). The entropy defined in Theorem 6.8 shows a general tendency to increase in the first phase (unlearning) of the process and decrease in the second phase (learning) of the process.

† Compare the case of learning by polychotomic observation, which is not an induction in the usual sense, yet can be forced into the framework of induction and is shown to obey the "expected" inverse *H*-theorem (see the last paragraphs of Section 6.2).

The unlearning phase can have a finite or vanishing duration, while the learning phase occupies the remaining time of the process.

The reader may have noticed that these theorems do not commit themselves as to the nature of "alternative choices" or an observational definition of the probability. Since it is not the purpose of this section to develop a full exposition of learning theory, we have to limit ourselves to a few brief remarks on these questions.

In cognitive cases the alternative choices mean competitive hypotheses or possible answers to a question, and the probability is typically a "subjective" or "personalistic" probability (see Section 7.4). The measurement of this probability has to rely on one of the sophisticated methods developed by psychologists such as Toda and Shuford ([T-5], [S-7] and Appendix A7.1). If one asks the subject which alternative he chooses, he will select the one on which he places the largest probability. Even if one takes an ensemble of similar subjects that have the same probability distribution, the answer about the choice will be unanimous. In order to force the subject to divulge his probability distribution, we have to take advantage of the utility-maximizing nature of the human being and derive the probabilities inductively from the amount he would bet on each alternative when the penalty or reward is cleverly chosen.

In the behavioral cases the alternative choices are usually called response categories. For instance, in a T-maze experiment, left turn, right turn, and no turn are the categories. These categories are in the mind of the experimenter, however, and except for a human subject, to whom the experimenter can tell the definition of the categories, they are entirely unknown to the subject. The problem here is less a practical difficulty in defining the categories than a methodological difficulty in justifying the categories used in practice. Thus the problem is very much like the problem of inductive generalization and class formation (see Chapter 4). To say that the right category of responses is the one associated with the largest positive reinforcement does not seem to say much. First, there is no objective way of evaluating the degree of reinforcement except after many learning experiments, which are based on man-made categories. Second, the subject does not know whether he got the food pellet because he turned left or because he started his journey with the left front paw. In other words, without the notion of a pre-existing category, the subject cannot associate the reinforcement with the category. I should like to add two points in this connection. For one thing, animals seem to have a power of inductive generalization very similar to that of humans. Second, it is tempting to define (though not uniquely) "good" categories as a class of responses such that our entropic theorem becomes true. We should perhaps claim as a postulate that "there exists a set of classes of responses such that Theorem 6.8 or 6.9 is true."

Probability in behavioral cases should probably be understood as a kind of unconscious disposition that can be observed approximately either as the relative frequency of a particular response in a short sequence of responses of a single subject or as the relative frequency of a particular response in an ensemble of subjects of a given kind. This seems to give an observational definition of probability, but it is really not so. On the one hand, an exact measurement of temporal relative frequency requires an infinitely long time, whereas the question we are interested in is the temporal change of the probability. On the other hand, to define an ensemble we have to know exactly what properties of the subject are supposed to be given and what properties are allowed to vary, because no two subjects are identical.

The correct interpretation of the situation would be somewhat as follows. The theory assumes that the state of each subject at each instant is characterizable by the probability distribution, and that the observed quantities are not directly identifiable with these probabilities but are in principle deductively derivable from these. At first glance the probability in the behavioral case is directly observable and the probability in the cognitive case is only indirectly observable. But in reality the probability in both cases is only inductively inferable from observation.

It is interesting that the two modes of manifestation of dispositional probability, one typical of the cognitive case and the other typical of the behavioral case, can sometimes be combined. Suppose that a human subject is exposed to two lamps, left and right. At each trial one or the other lamp is lit instantaneously. The subject is required to predict which one will be lit, and correct predictions are rewarded. The lamps are connected to a switch controlled by a computer so that they are lit randomly, but the ratio of the frequency of the left one being lit to that of the right one being lit is, say, 7:3. In the first stage the subject will also predict left and right randomly, but the ratio of left predictions to right predictions will quickly become close to 7:3. He learns the probability almost instinctively. If the subject is clever enough, he will realize at a certain moment that the optimal strategy is to predict consistently that the left lamp will be lit. By choosing this new strategy he increases the probability of reward from roughly $(7/10)^2 + (3/10)^2$ to $(7/10)$. The first strategy is similar to the behavioral mode of probability manifestation, whereas the second strategy is similar to the cognitive mode.

One more pertinent footnote to Theorem 6.9 is that in a prolonged experiment with the same human subject, the entropy may occasionally suddenly deviate from its minimum value. This is explainable in terms of "boredom," because a human subject gets tired of responding in the same way too many times. But often this kind of deviation from a set pattern is not just a meaningless whim or escape, but serves, consciously or unconsciously, a useful purpose of exploring new possibilities.

After we conjectured the truth of the behavioral inverse H-theorem, Theorems 6.8 and 6.9, we confirmed it with several available data on behavioral learning. We introduce some of the examples at the end of this section. But if the experimental data obey the inverse H-theorem, the various mathematical models of learning that predict the response probabilities must also obey it, if they are good models. On some of the graphs to be introduced later, we have marked both real experimental data and data produced by these models.

The fact that the existing mathematical models of learning give response probabilities that obey the inverse H-theorem prompted a certain psychologist to tell me, "Your inverse H-theorem is nothing more than a corollary of our mathematical models of learning." If he thought that this statement would minimize the value of the inverse H-theorem, he was methodologically very wrong. His claim is similar to saying, "The second law of thermodynamics is a corollary of Ohm's law." It is true that any phenomenon obeying Ohm's law results in an entropy increase and hence obeys the second law, but this does not change the fact that the second law of thermodynamics is a basic law governing all physical phenomena, whereas Ohm's law is a very special law. Similarly, the inverse H-theorem is much more general and basic than individual mathematical models of learning. Keeping this fact in mind, let us glance quickly at some of the so-called mathematical models of learning.

The basic variables in these theories are four time sequences. The time parameter is usually identifiable with the trial number, say, ν. One sequence is the sequence of "conditioned" stimuli $S^{(\nu)}$, which can be any one member of the set $\mathcal{S} = \{S_1, S_2, \cdots, S_m\}$. The second sequence is the sequence of response acts $A^{(\nu)}$, which at each time ν can be any member of the set $\mathcal{A} = \{A_1, A_2, \cdots, A_n\}$. The third is the sequence of reinforcement events $E^{(\nu)}$, which at each ν can be any member of the set $\mathcal{E} = \{E_1, E_2, \cdots, E_r\}$. The experiment is usually so arranged that $E^{(\nu)}$ is a deterministic function of $S^{(\nu)}$ and $A^{(\nu)}$. The fourth is the sequence of probability states $p^{(\nu)}(A_i)$, which is supposed to represent the probability of $A^{(\nu)} = A_i$, that is, the probability that the subject will choose act A_i at the νth trial. The task of the theory is to give $p^{(\nu)}(A_i)$ as a function of the entire past $\Pi^{(\nu)}$, which is the entirety of $S^{(\mu)}$, $A^{(\mu)}$, and $E^{(\mu)}$ for all $\mu < \nu$ and of the present stimulus $S^{(\nu)}$. Since this probability also depends on the individuality σ_k of the subject, we can write

$$p^{(\nu)}(A_i) = \Pr\{A^{(\nu)} = A_i \mid \Pi^{(\nu)}, S^{(\nu)}, \sigma_k\}. \tag{6.113}$$

The response entropy, about which the usual learning theories do not talk, would be, according to (6.113),

$$U^{(\nu)} = -\sum_i p^{(\nu)}(A_i) \log p^{(\nu)}(A_i). \tag{6.114}$$

As explained earlier, a direct and precise observation of the response probability $p^{(\nu)}(A_i)$ is impossible in principle, but its rough experimental estimate can be hoped for. The theoretical values of $p^{(\nu)}(A_i)$ can be compared with such experimental values. Similarly, there can be experimental and theoretical values of the response entropy.

On the ground that the experiment is done with a single subject or subjects of similar backgrounds, the individual parameter σ_k is usually ignored. To make our exposition simple, let us assume that the same stimulus is repeated. On this ground we can eliminate $S^{(\nu)}$ from the right-hand side. By this assumption $E^{(\mu)}$ becomes a given function only of $A^{(\mu)}$. A tremendous simplification of the theory stems from the bold assumption adopted in most present-day learning theories: the effect of the entire past $\Pi^{(\nu)}$ is representable by an n-component vector $\{V_i^{(\nu)}\}$ $i = 1, 2, \cdots, n$, at ν, where n is the number of alternative responses. Then (6.113) becomes

$$p^{(\nu)}(A_i) = \Pr\{A^{(\nu)} = A_i \mid \{V_i^{(\nu)}\}\}. \tag{6.115}$$

The theory is completed if $V_i^{(\nu)}$ is determined as a function of the past $\Pi^{(\nu-1)}$ [which is representable by $\{V_i^{(\nu-1)}\}$] at the $(\nu - 1)$th trial and what happened at the $(\nu - 1)$th trial, which is the choice $A^{(\nu-1)}$ made by the subject at that trial. The theory then consists of giving the functional form of (6.115) and the functional form of

$$\{V_i^{(\nu)}\} = \text{function of } (\{V_i^{(\nu-1)}\}, A^{(\nu-1)}), \tag{6.116}$$

where $A^{(\nu-1)}$ can be any member of $\mathcal{A}$.

Most learning theories further simplify the matter by identifying $p^{(\nu)}(A_i) \equiv p_i^{(\nu)}$ as $V_i^{(\nu)}$. Then the two equations (6.115) and (6.116) can be replaced by one:

$$\{p_i^{(\nu)}\} = \text{function of } (\{p_i^{(\nu-1)}\}, A^{(\nu-1)}). \tag{6.117}$$

The theories usually adopted make another bold assumption that $\{p_i^{(\nu)}\}$ is linear in $\{p_i^{(\nu-1)}\}$:

$$p_i^{(\nu)} = \sum_j a_{ij}^{(k)} p_j^{(\nu-1)} + b_i^{(k)}, \tag{6.118}$$

where $k = 1, 2, \cdots, n$, specifies the response $A^{(\nu-1)} = A_k$ chosen by the subject at the $(\nu - 1)$th trial. There are two versions of theories that assume the linearity (6.117), the homogeneous model and the inhomogeneous model. The homogeneous model, which assumes that $b_i^{(k)} = 0$ for all i and k, that is,

$$p_i^{(\nu)} = \sum_j a_{ij}^{(k)} p_j^{(\nu-1)}, \tag{6.119}$$

is a Markovian model. The inhomogeneous model, which assumes non-vanishing $b_i^{(k)}$ for some pair (i, k), is usually called a linear model. If one speaks of a Markov chain, it would seem to imply that each individual subject corresponds to a string, a member of the Markov ensemble. But in this psychological application each individual is in a state expressible by $p_i^{(\nu)}$ (corresponding to the "occupancy probability"); hence it corresponds to the ensemble, not to a string.

One of the simplest versions of the theories that distinguish between $\{p_i^{(\nu)}\}$ and $\{V_i^{(\nu)}\}$ is provided by what is known as Luce's β-model [L-6]. In this model the $\{V_i^{(\nu)}\}$ change according to a homogeneous linear transformation with a diagonal matrix, that is,

$$V_i^{(\nu)} = c_i^{(k)} V_i^{(\nu-1)}, \tag{6.120}$$

and the probability $p_i^{(\nu)}$ is simply proportional to $V_i^{(\nu)}$:

$$p_i^{(\nu)} = \frac{V_i^{(\nu)}}{\sum_j V_j^{(\nu)}}. \tag{6.121}$$

To make a consistent theoretical picture, the choice of response A_k at the $(\nu - 1)$th stage must be made according to the theoretical prediction, which assigns probability $p_k^{(\nu-1)}$ to A_k. Thus, with the help of a random number-producing program, we can decide in an individual case which k to choose for the coefficient $a_{ij}^{(k)}$ of (6.118) or $c_i^{(k)}$ of (6.120) at each stage. In this way we can produce a chain of values of $p_i^{(\nu)}$. In a later paragraph the expression "statdog" or "statrat" refers to a chain of values of $p_i^{(\nu)}$ produced in such a fashion.

It is easy to see that all three models satisfy our Theorem (6.8) or (6.9.) Leaving a careful examination for the reader's exercise, let us review a few simple cases. The Markovian model usually assumes the existence of a sink (a state whose efflux is zero and to which other states are connected). Then $p_i^{(\nu)}$ of the sink will become 1 as $\nu \to \infty$. In the simplest possible case assume that there are only two states, 1 and 2, which are connected by a constant (independent of k) transition matrix: $a_{11} = a$, $a_{21} = 1 - a$, $a_{12} = 0$, $a_{22} = 1$. This means that State 2 is a sink and transition from State 1 to State 2 takes place with probability $1 - a$. If, for instance, we start with $p_1^{(0)} = 1$ and $p_2^{(0)} = 0$, we shall have $p_1^{(\nu)} = a^\nu$ and $p_2^{(\nu)} = 1 - a^\nu$. The entropy will start with zero and climb until $p_1^{(\nu)}$ and $p_2^{(\nu)}$ become about equal (when the entropy is close to 1 unit) and then will descend gradually until it becomes zero again.

In the case of linear model let us assume again that there are only two states and the coefficients do not depend on k. Because of normalization, we need consider only $p_1^{(\nu)}$ or $p_2^{(\nu)}$ as a variable. For instance, we can write

$$\begin{aligned} p_2^{(\nu)} &= a_{21} p_1^{(\nu-1)} + a_{22} p_2^{(\nu-1)} + b_2 \\ &= (a_{22} - a_{21}) p_2^{(\nu-1)} + (a_{21} + b_2). \end{aligned} \tag{6.122}$$

Hence there are only two independent constants that affect the change of $p_2^{(\nu)}$. Putting $(a_{22} - a_{21}) = (1 - \alpha)$ and $(a_{21} + b_2) = \alpha\lambda$, we obtain from (6.122)

$$p_2^{(\nu)} = p_2^{(\nu-1)} + \alpha(\lambda - p_2^{(\nu-1)}). \tag{6.123}$$

If we assume $0 < \alpha < 1$ and $0 < \lambda \leq 1$, we can see that $p_2^{(\nu)}$ will converge to λ. In fact, the solution is $p_2^{(\nu)} = \lambda - (1 - \alpha)^\nu(\lambda - p_2^{(0)})$, which becomes the same as the Markovian case with State 2 as a sink, if $\lambda = 1$, $a = 1 - \alpha$ and $p_2^{(0)} = 1$.

Finally, in the β-model, the probability itself changes according to

$$p_i^{(\nu)} = \frac{c_i^{(k)} p_i^{(\nu-1)}}{\sum c_j^{(k)} p_j^{(\nu-1)}}. \tag{6.124}$$

This formula has the same mathematical form as the Bayesian model of induction if correspondence is made between $q^{(\nu)}(I)$ and $p_i^{(\nu)}$ and between $p(i \mid I)$ and $c_i^{(k)}$. The difference is that the i and I of $p(i \mid I)$ refer to two different sets $\mathfrak{D}$ and $\mathfrak{K}$, whereas the k and i of $c_i^{(k)}$ both refer to the same set $\mathcal{A}$. The $p(i \mid I)$ is normalized with respect to i, but the $c_i^{(k)}$ is not normalized with respect to k. The D_i of $p(i \mid I)$ happens with probability γ_i, whereas A_k of $c_i^{(k)}$ happens with the response probability whose theoretical value is $p_k^{(\nu-1)}$. In spite of these differences we can guess what happens in the β-model from what we know about the Bayesian model.

If there are only two responses, if $c_i^{(k)}$ does not depend on k, and if $c_2 < c_1$, we can write (6.124) as

$$p_2^{(\nu)} = \frac{p_2^{(\nu-1)}}{p_2^{(\nu-1)} + \beta(1 - p_2^{(\nu-1)})}, \tag{6.125}$$

with

$$\beta = \frac{c_1}{c_2} < 1. \tag{6.126}$$

The repetitive use of (6.125) yields

$$p_2^{(\nu)} = \frac{p_2^{(0)}}{p_2^{(0)} + \beta^\nu(1 - p_2^{(0)})}. \tag{6.127}$$

As in the Bayesian model, if $p_2^{(0)} = 0$, $p_2^{(\nu)} = 0$ for all ν. But insofar as $p_2^{(0)} \neq 0$ the second term in the denominator will become smaller than the first term with time, and at $\nu \to \infty$ we shall have $p_2^{(\nu)} = 1$. The increase of $p_2^{(\nu)}$ is monotonic with ν; hence the entropy curve will show an initial rise insofar as $p_2^{(0)} < \beta^\nu(1 - p_2^{(0)})$, and then it will gradually decrease to zero. The difference in the result of the β-model from the other two models discussed above in this simple case is that at the initial stage $p_1^{(\nu)}$ decreases a little more slowly than in the former two cases. But later on $p_1^{(\nu)}$ decreases more or less exponentially, as it does from the beginning in the other two models. In a more complicated case these models may not agree as much as in the present simple case, and some model may fit an experiment better than others.

In physics, experimental data are usually so abundant and so exact (partly because they are exactly reproducible) that a theory or formula that fits them accurately is very difficult to find, but if it does and if it can be harmonized with the theories of related fields, it immediately acquires a high degree of credibility. In psychology, at least in learning problems, the trouble is the opposite; the experimental data are so few and so inexact that almost any theory can be made to fit them. Any one of the three models mentioned above or any other conceivable model cannot claim much credibility for the reason that its competitors cannot be discredited. Each model says too many things that cannot be put to a crucial test.

As if to aggravate this situation, mathematical psychologists (see Estes [E-2]) introduced a theory, called stimulus sampling theory, that describes the mechanism by which a response is chosen by the subject at each trial. In this theory the stimulus is considered as a collection of so-called stimulus elements, each of which is "conditioned" to a response. The subject is assumed to sample exactly one of the stimulus elements at a time, and his response will be the one to which the sampled element is conditioned. The theory then has to tell how the conditioning changes with time and how the sampling is done. On the one hand, the stimulus sampling theory is a very general framework in the sense that it can engender different kinds of models in agreement with it; on the other hand, it is a very narrow, specific theory because it assumes a detailed mechanism that is not necessarily the unique one resulting in a given model. It is extremely interesting and fertile as a heuristic tool, but as a theory it talks about too many things that cannot be proved or disproved.

Almost at the opposite end of the spectrum of theorization, we can think of a "phenomenological" point of view, in which we are interested only in the observable variable and do not ask about the "mechanism." The point of view of Theorems 6.8 and 6.9 is more or less such a phenomenological one. It is comparable to the standpoint of the second law of thermodynamics as distinct from that of the H-theorem. Granted, in physics, the entropy can be defined in two ways, either in terms of thermodynamic variables ($dS = \Delta Q/T$) or in terms of the probabilities, whereas in the present case we have to define entropy only in terms of response probabilities. A standpoint is possible, however, in which we are not interested in the dynamic law governing the mechanism of temporal change of the probabilities, but only in the dynamic change of the entropy itself. The following model theory may be considered as an attempt along that line. The model is not claimed to be universally valid or to describe the facts accurately in any single case. All that it claims to be is a simple and approximate empirical formula.

We may summarize the underlying thought as follows:

1. The rate of learning $R^{(\nu)}$ is given by the fractional rate of decrease of the entropy,

$$R^{(\nu)} = -\frac{dU^{(\nu)}/d\nu}{U^{(\nu)}}. \tag{6.128}$$

2. The rate of learning $R^{(\nu)}$ is the algebraic sum of the force of pure learning $L^{(\nu)}$ and the force of pure unlearning $F^{(\nu)}$, where the latter includes both the tendency to forget (automatic disorganization) and resistance to learning.

$$R^{(\nu)} = L^{(\nu)} + F^{(\nu)}, \tag{6.129}$$

where

$$L^{(\nu)} \geq 0, \qquad F^{(\nu)} \leq 0. \tag{6.130}$$

3. Insofar as the time variable ν is proportional to the number of trials, the rate of pure learning is constant:

$$L^{(\nu)} = \gamma. \tag{6.131}$$

4. The magnitude of the rate of pure unlearning $F^{(\nu)}$ is a decreasing function of ν.

$$\frac{d\,|F^{(\nu)}|}{d\nu} < 0. \tag{6.132}$$

For simplicity, let us try

$$-F^{(\nu)} = \frac{\beta}{\nu - \nu_0}, \tag{6.133}$$

where β is the strength of unlearning and ν_0 is a certain time point such that the consideration is valid only for $\nu > \nu_0$.† We see presently what to do if $\nu < \nu_0$. Hence, combining (6.129), (6.131), and (6.133), we get the basic dynamic law:

$$R^{(\nu)} \equiv -\frac{dU^{(\nu)}/d\nu}{U^{(\nu)}} = \gamma - \frac{\beta}{\nu - \nu_0} \quad \text{for only} \quad \nu > \nu_0. \tag{6.134}$$

This shows that there are two phases. For sufficiently smaller ν, the force of pure unlearning will be stronger than the force of pure learning and the resultant rate of learning will be negative. This corresponds to the unlearning stage. For sufficiently large ν, the force of pure unlearning becomes weaker than the force of pure learning, making the resultant rate of learning positive, representing the learning stage. But it can hardly be expected of a simple model theory like this that the entire learning curve (the $U^{(\nu)}$-curve) can be

† This β has nothing to do with the β of the β-model.

represented with a formula (6.134) by the same set of constants. Hence we assume the next point.

5. The unlearning and learning stages separately obey (6.134) with different constants (including integration constants), the two curves being connected continuously at the maximum of $U^{(\nu)}$. The solution of (6.134) has the form

$$U^{(\nu)} = \alpha(\nu - \nu_0)^\beta e^{-\gamma\nu} = \alpha'(\nu - \nu_0)^\beta e^{-\gamma(\nu-\nu_0)} \quad \text{for} \quad \nu > \nu_0, \tag{6.135}$$

where α is a non-negative constant and $\alpha' = \alpha e^{-\gamma\nu_0}$. This $U^{(\nu)}$ curve reaches its maximum

$$U_{\text{max}} = \alpha\left(\frac{\beta}{\gamma}\right)^\beta e^{-\gamma\nu_0-\beta} \tag{6.136}$$

at $\nu = \nu_{\text{max}}$, which is

$$\nu_{\text{max}} = \nu_0 + \frac{\beta}{\gamma}. \tag{6.137}$$

The portion $\nu_0 \leq \nu \leq \nu_{\text{max}}$ represents the unlearning stage, and the portion $\nu_{\text{max}} \leq \nu$, the learning stage. It is intuitively understandable that the length of the unlearning period $(\nu_{\text{max}} - \nu_0)$ is determined by the competition of unlearning and learning, namely, by β/γ. In fitting the experimental data with the function of (6.135), however, we propose to allow ourselves to use different sets of constants in the unlearning and learning stages, requiring only that the value of $U^{(\nu)}$ be the same at the connecting point. The slope $dU^{(\nu)}/d\nu$ will be zero on both sides anyhow. This flexible rule also implies the possibility of representing either stage by a constant value of $U^{(\nu)}$, because we can express $U^{(\nu)} = U_0$ for $\nu =$ finite by letting $\nu_0 \to -\infty$, $\gamma \to 0$, and $\alpha \to 0$ in (6.135) while keeping $\alpha(-\nu_0)^\beta = U_0 =$ constant. Furthermore, since the expression of $U^{(\nu)}$ given in (6.135) is meaningless for $\nu < \nu_0$, we extrapolate the curve to $\nu < \nu_0$ continuously.

6. We assume

$$U^{(\nu)} = 0 \quad \text{for} \quad \nu \leq \nu_0. \tag{6.138}$$

Note that we always have $U^{(\nu_0)} = 0$, but $dU^{(\nu)}/d\nu$ is zero at $\nu = \nu_0$ only if $\beta > 1$, and is infinity if $\beta < 1$. If $\nu = 0$ means the actual starting point of the experiment, $\nu_0 > 0$ means, in view of (6.138), a delay in the actual start of the unlearning stage.

The mathematical nature of the curve (6.135) can be studied more easily when it is brought to the standard form by measuring time in the scale of $(1/\gamma)$ and shifting the origin of time to ν_{max}. Equation (6.135) becomes

$$U^{(\xi)} = A(\beta + \xi)^\beta e^{-\xi}, \qquad \xi > -\beta, \tag{6.139}$$

where

$$\xi = \gamma(\nu - \nu_{\text{max}}), \qquad A = \frac{U_{\text{max}}}{\beta^\beta}. \tag{6.140}$$

The point $\nu = \nu_0$ is mapped onto $\xi = -\beta$. In this time scale β is the length of unlearning, representing the strength of prelearned habit. The derivative is given by

$$\frac{dU^{(\xi)}}{d\xi} = -A\xi(\beta + \xi)^{\beta-1}e^{-\xi}. \tag{6.141}$$

The learning model thus obtained is, so to speak, the simplest conceivable one. Yet the agreement with experimental data in various cases is rather remarkable. We now adduce three examples.

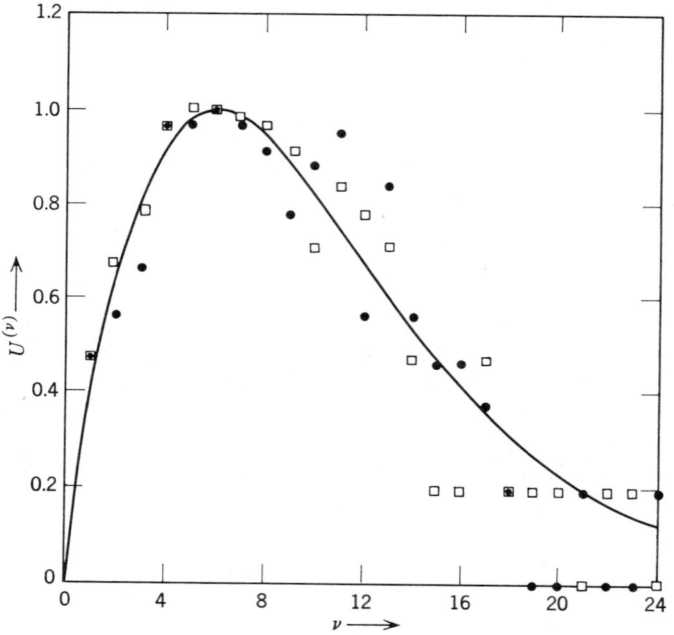

Figure 6.2 Solomon-Wynne shock avoidance experiment.

Figure 6.2 represents the results of the Solomon-Wynne shock avoidance experiment. (See any standard textbook on psychology for the details of the experiment, for instance, Atkinson, Bower, and Crothers [A-7].) The dots are the entropy values calculated from the ensemble average of the avoidance probability of the 30 real dogs used by Solomon and Wynne. The squares are the corresponding values for the 30 statdogs based on the Bush-Mosteller model. The numerical values are taken from Bush-Mosteller's textbook [B-16]. The curve is obtained by our simple formula (6.135). The unlearning part is obtained with $\alpha = 0.454$, $\beta = 1$, $\gamma = 0.167$, and $\nu_0 = 0$. The learning part is obtained with $\alpha = 0.650$, $\beta = 1$, $\gamma = 0.198$, and $\nu_0 = 0.95$. There is no sign of a lag in the start of the learning process. This case is a good example for clarifying the meaning of the word "unlearning" as used in the present discussion. The pre-established habit that is unlearned here is quite a natural one; namely, the dog does not jump the barrier when the light is turned off (which is the conditioned stimulus).

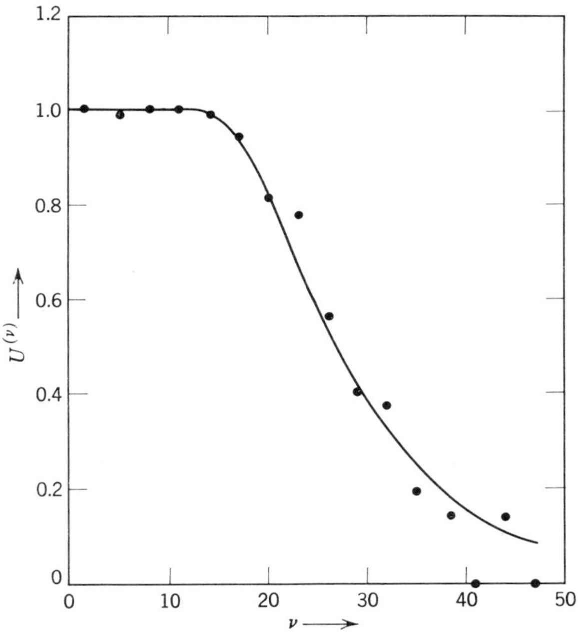

Figure 6.3 Galanter-Bush T-maze experiment, Period 1 of Experiment III.

Figures 6.3 and 6.4 refer to Galanter-Bush's T-maze experiment on rats. (See the article by Galanter and Bush, "Some T-Maze Experiments," in Bush and Estes [B-15].) Figure 6.3 represents the result of what these authors called Period 1 of Experiment III. In this period the rats are exposed to the experiment for the first time. Hence at the beginning approximately as many rats go to the left as go to the right. The dots are the entropy values calculated from the ensemble average of the response frequencies in three-trial blocks. The curve between $\nu = 0$ and $\nu = 13.5$ is $U^{(\nu)} = 1$. The curve for $\nu \geq 13.5$ is obtained from (6.135), with $\alpha = 0.696$, $\beta = 1$, $\gamma = 0.126$, and $\nu_0 = 5.6$. The lag in the start of learning is thus 13.5.

Figure 6.4 represents the result of what Galanter and Bush called Period 2 of Experiment IV. In Period 1 of the same experiment the rats "overlearned"

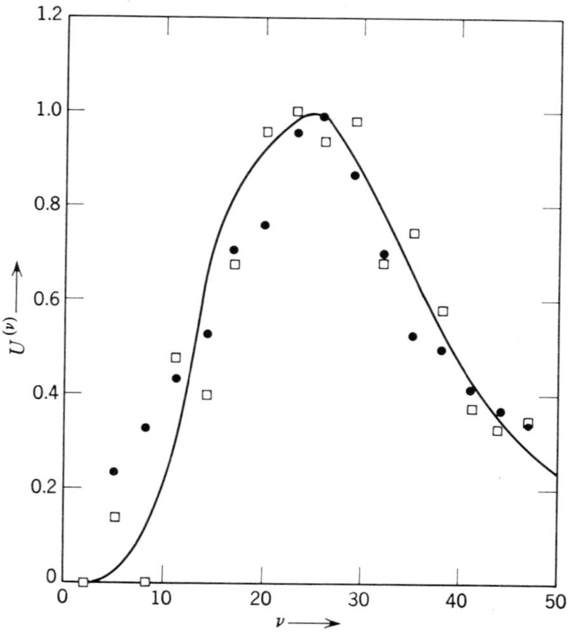

Figure 6.4 Galanter-Bush T-maze experiment, Period 2 of Experiment IV.

to turn to the right (144 trials). In Period 2 the reward was given on the left side. As a result of the initial overlearning, it could be expected that the new learning process encountered a strong resistance. The dots represent the entropy values calculated from the ensemble average of the response frequencies in three-trial blocks for the real rats. The squares are the corresponding values for the statrats based on Luce's β-model. The curve for the unlearning part was computed from (6.135), with $\alpha = 1.84 \times 10^{-5}$, $\beta = 5$, $\gamma = 0.208$, and $\nu_0 = 0$. It is significant that the resistance as represented by the value of β is very high. The learning part was obtained from (6.135), with $\alpha = 0.302$, $\beta = 1$, $\gamma = 0.111$, and $\nu_0 = 15$.

The content of this section is based on [W-27].

7

Logic and Probability

7.1. PROPOSITIONAL LATTICES

It is obviously impossible to give a comprehensive and rigorous exposition of all fundamentals of logic and probability in a single chapter. We aim instead to introduce selectively some basic elements of logic and probability as deemed pertinent to the main purpose of this book, intentionally adopting a rather unorthodox approach to the logical problems. This approach consists first in constructing the notion of a lattice on the basis of implicational relations and then in identifying propositional calculus as a particular interpretation of a special type of lattice called "distributive." This approach is partially motivated by the philosophical view that implication is the most basic form of rational thinking (see the beginning of Section 7.3), but is also necessitated by the purpose of Chapter 9 on non-Boolean (i.e., non-distributive) information theory. To understand the so-called "quantum logic," which is nondistributive, it is necessary to start with the notion of lattices, which is not necessarily distributive. The relationship between the mathematical formulas of distributive lattices and the usual version of propositional calculus is briefly explained in Appendix A7.2.

We first introduce the concept of a lattice without limiting ourselves to its interpretation as a propositional calculus. A lattice $\mathfrak{L}$ is a set of elements $A, B, C, \cdots$ (i.e., $\mathfrak{L} = \{A, B, C, \cdots\}$) that satisfies the following six basic laws.

LAW 7.1. Among some pairs of its elements, say, A and B, there exists a one-way relation "A partakes of B," or "A partakes B," which we write

$$A \rightarrow B. \tag{7.1}$$

We agree that (7.1) and $B \leftarrow A$ mean the same relation. By "one-way relation" is meant that $A \rightarrow B$ does not necessarily imply $B \rightarrow A$. A neutral word "partake" is used instead of "imply" or "entail," to avoid an exclusive commitment to logic.

If there is a two-way relation

$$A \rightarrow B \quad \text{and} \quad B \rightarrow A \tag{7.2}$$

we say that "A and B are equivalent" and write

$$A = B. \tag{7.3}$$

In this case we can consider A and B either as distinct members that are equivalent or as a single element.

LAW 7.2. The relation of partaking is transitive; that is,

$$\text{if } A \rightarrow B \quad \text{and} \quad B \rightarrow C \quad \text{then} \quad A \rightarrow C. \tag{7.4}$$

LAW 7.3. A lattice $\mathfrak{L}$ contains a special element $\square$, called the "whole," such that any arbitrary element X of $\mathfrak{L}$ partakes of it:

$$X \rightarrow \square \qquad (X\text{: arbitrary}). \tag{7.5}$$

LAW 7.4. A lattice $\mathfrak{L}$ contains a special element $\emptyset$, called the "zero," which partakes of every element X of $\mathfrak{L}$:

$$\emptyset \rightarrow X \qquad (X\text{: arbitrary}). \tag{7.6}$$

A set that satisfies Laws 7.1 through 7.4 is a "partially ordered set, with an upper and a lower bound, of which we now mention a few examples. A partially ordered set that further satisfies Laws 7.5 and 7.6 qualifies as a lattice.

Example 7.1. Consider a set of natural numbers starting with 1 and ending with 100, and interpret $\rightarrow$ as $\leq$. Then the transitivity is satisfied, and $\emptyset$ will become the number 1 and $\square$ the number 100.

Example 7.2. If we take as $\mathfrak{L}$ the set of real numbers and interpret $\rightarrow$ as $\leq$, we have to include $-\infty$ and $+\infty$ as additional elements of the set in order to satisfy Laws 7.3 and 7.4. In Examples 7.1 and 7.2 the relation $\rightarrow$ exists in every pair of elements in one direction or the other.

Example 7.3. Take three pebbles, a, b, and c, and consider all possible groups that can be formed with some of the pebbles. There are $3 = \binom{3}{1}$ groups, each containing one pebble, $\{a\}$, $\{b\}$, and $\{c\}$; there are $3 = \binom{3}{2}$ groups, each containing two pebbles, $\{a, b\}$, $\{b, c\}$, and $\{c, a\}$. If we add to this an empty group { } and the group of all pebbles $\{a, b, c\}$, we obtain a set of $8 = (1 + 1)^3 = \sum_{r=0}^{3}\binom{3}{r}$ elements. If we interpret $A \to B$ as meaning that every pebble included in group A is included in group B, and if we interpret the empty group { } as $\emptyset$ and the full group $\{a, b, c\}$ as $\square$, this set of 8 elements satisfies all four laws. The only difficulty may be the assertion that $\emptyset$, which is { }, partakes of any other group. This difficulty may be avoided if we rephrase the meaning of $A \to B$. The above definition of $A \to B$ is equivalent to saying that every pebble not included in B is not included in A either. Then, for instance, $\{\,\} \to \{a, b\}$ holds because the only pebble c that is not included in $\{a, b\}$ is not included in { }, in just the same way as it is not included in $\{a\}$ or in $\{b\}$. Hence $\{\,\} \to \{a, b\}$ as well as $\{a\} \to \{a, b\}$ and $\{b\} \to \{a, b\}$.

Example 7.4. If in a set of propositions we interpret $A \to B$ as standing for "*A* implies *B*" or "if *A* is true then *B* is true," Law 7.3 becomes what is known as syllogism. Let the whole $\square$ stand for a constant truth or tautology such as "5 is 5." Then any proposition A implies a constant truth, because a constant truth is true whether or not A is true and hence is also true if A is true. It follows that all constantly true propositions imply one another; thus they are all equivalent and can be considered as a single element. Let $\emptyset$ stand for a constant absurdity that implies every possible proposition X. This is difficult to understand, but if we paraphrase the meaning of "*A* implies *B*" as "if *B* is false, then *A* is false," it becomes easier to understand. For $\emptyset$ is false whether or not X is false whether or not X is false and hence must also be false in the case when X is false.

Example 7.5. Take a graph that consists of nodes (points), of which some pairs are connected by an arrow or a directed line, in such a way that when one starts from a node and follows the direction of arrows one never comes back to the starting point but finally lands in the same point $\square$, no matter where one has started. Similarly, assume that if one starts from a node and goes against the direction of arrows, one finally comes to the same endpoint $\emptyset$. Then the set of nodes satisfies the four laws if we agree to understand "*A* partakes of *B*" as meaning that node B can be reached from node A by following the arrows (see Figure 7.1).

Example 7.6. A certain category of $\mathfrak{L}$'s can be visualized as a set of all two-dimensional domains taken within a fixed larger two-dimensional domain, D_0. The one-way relation $A \to B$ is then interpreted as "if a point P is in A then P is also in B" (see Figure 7.2). The "whole" $\square$ corresponds to the entire domain D_0, and the "zero" $\emptyset$ corresponds to an "empty" domain, or a domain without any point, which may be considered to be included in

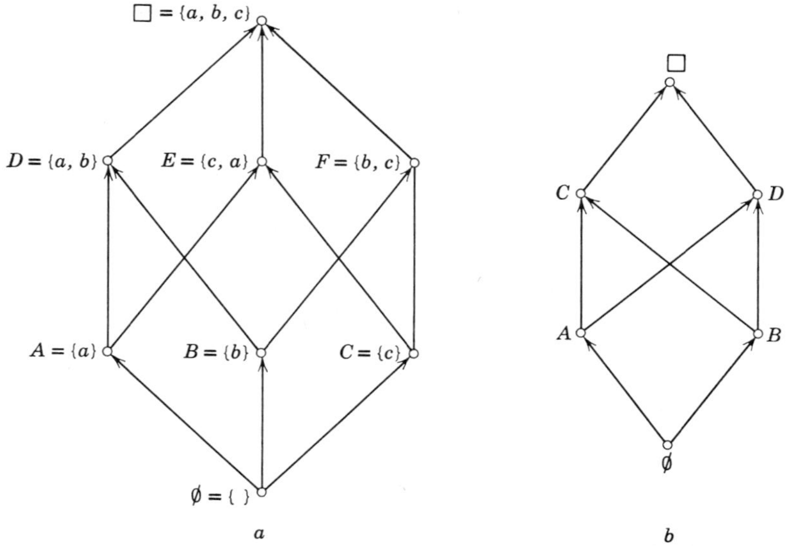

Figure 7.1 Points (nodes) connected by arrows can, under certain conditions, satisfy Laws 7.1–7.4 when the arrow is interpreted as partaking. Graph a is a lattice, but graph b is not.

any arbitrary domain. This correspondence, called Euler's diagram, is of considerable help in an intuitive understanding of Boolean propositional calculus. Too strong reliance on this geometrical model is dangerous, however, because it can lead to a conclusion that is not implied by the axioms of a general lattice but depends on the so-called distributive law, which is peculiar to the Boolean lattice (see a later part of this section). It may be noted in passing that a modification of this model satisfies the four laws also when the

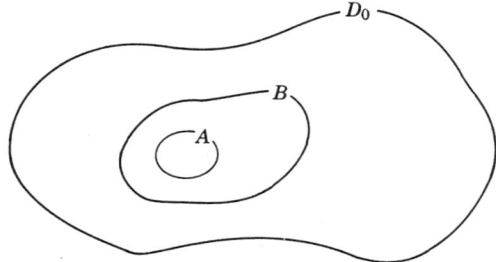

Figure 7.2 Euler's diagram. $A \rightarrow B$ is interpreted as meaning that any point belonging to domain A belongs to domain B; $\emptyset$ is an empty domain, which may be considered to be included in any arbitrary domain. The $\square$, which is D_0, is the total domain, within which other domains are allowed to be taken.

area is quantized in the sense that any allowed member of $\mathfrak{L}$ is composed of a certain number of indivisible elementary areas. In this case the model can be reduced to one similar to Example 7.3.

Example 7.7. In order to avoid misunderstanding as a result of the peculiar nature of the models chosen, we introduce another kind of geometrical model, Figure 7.3. In this correspondence the elements of the set $\mathfrak{L}$ are sets of points

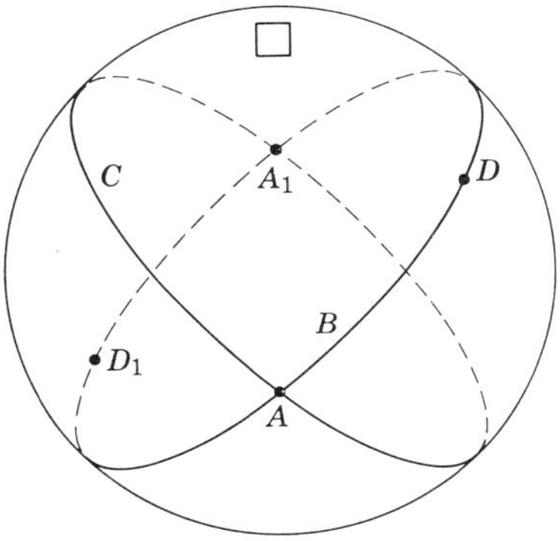

Figure 7.3 The elements are points, great circles, and the entire surface (□) on a sphere, and the zero (∅). Point A and its diametrically opposite point A_1 count as a single point. Point D is on the great circle B; hence $D \to B$. Point A is on the great circles B and C; this means that $A \to B$ and $A \to C$. The "zero" ∅ corresponds to a fictitious entity "included" in every point. There are two X's that satisfy $X \to B$ and $X \to C$ in this example, namely, $X = \emptyset$ and $X = A$; but since $\emptyset \to A$, we have $A = B \cap C$. There are two X's that satisfy $A \to X$ and $D \to X$, namely, $X = B$ and $X = \square$; but since $B \to \square$ we have $B = A \cup D$.

on a sphere, but they can be of different dimensions; that is, they are endpoints of diameters, great circles, and the entire surface of a sphere. No other geometrical domains are admitted as members of $\mathfrak{L}$. A point A and its diametrically opposite point A_1 are considered to be one element and will be referred to as just a point. The arrow relation $A \to B$ is interpreted geometrically by the same principle as in the case of Euler's diagram: if a point P is included in A, it is included in B. If A is a point it is included in itself, and in great circles passing through A and also in the entire surface. If A is a great circle, points on A are included in A and in the entire surface. If A is the entire surface, points on A are included in A. The "zero" is a fictitious entity included in every point (even in this fictitious point itself), and thus in any other higher-dimension elements of the set. The "whole" is the entire spherical surface.

With the help of the notion of partaking, we can define conjunction and disjunction of two elements, A and B, of $\mathfrak{L}$. The "conjunction" C of two given elements A and B is defined as an element of $\mathfrak{L}$ satisfying the following two conditions:

$$\text{(i) } C \to A, \qquad C \to B; \tag{7.7}$$

$$\text{(ii) for any element } X \text{ satisfying } X \to A \text{ and } X \to B, \tag{7.8}$$

$$C \text{ satisfies } X \to C. \tag{7.9}$$

If C satisfies these conditions we write

$$C = A \cap B. \tag{7.10}$$

If we replace the symbol $\to$ by $\leq$, this procedure becomes similar to the one defining the largest lower bound of A and B. The existence of the set of elements X is guaranteed because the element $\emptyset$ always qualifies as an X. But the existence of a unique C in $\mathfrak{L}$ is not always guaranteed. In Figure 7.1*a* any pair, for example, D and F, specifies a unique element B as $D \cap F$. However, the graph of Figure 7.1*b* does not determine $C \cap D$. Thus we add another law as a further requirement of $\mathfrak{L}$.

LAW 7.5. If A and B belong to $\mathfrak{L}$, the unique conjunction $C = A \cap B$ exists and belongs to $\mathfrak{L}$.†

Figure 7.4 gives the geometrical meaning of a conjunction in the Eulerian model. The conjunction of B and C in the spherical model of Figure 7.3 is A. Two X's satisfy $X \to B$ and $X \to C$. One is $\emptyset$ and the other is A. Since $\emptyset \to A$ we have to take A as $B \cap C$.

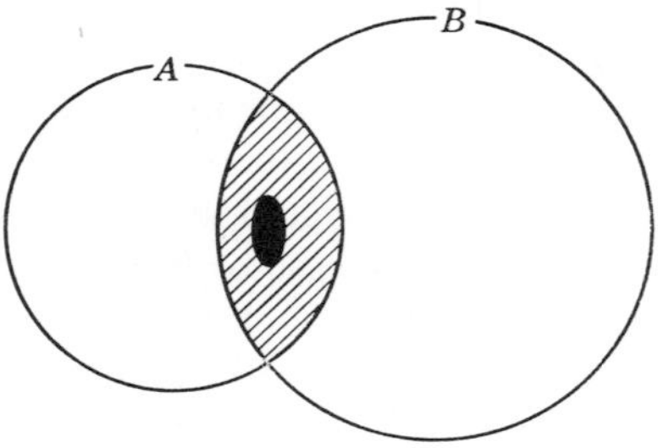

Figure 7.4 The shaded area (if it belongs to $\mathfrak{L}$) is the conjunction of A and B: $C = A \cap B$. The solid black area is an X that satisfies $X \to A$ and $X \to B$; C is the largest of such X's.

† It is easy to see that if two elements C and C' qualify as $A \cap B$, C and C' must be equivalent $C = C'$. In fact, in the definition of C as $A \cap B$, we can use C' as X; then we have, from (7.9), $C' \to C$. Interchanging the roles of C and C', we also get $C \to C'$. In this sense, if $A \cap B$ exists, it is unique.

Similarly, the "disjunction" C of two given elements A and B is defined as an element of $\mathfrak{L}$ satisfying the following two conditions:

$$\text{(i) } A \rightarrow C, \qquad B \rightarrow C; \tag{7.11}$$

$$\text{(ii) for any element } X \text{ satisfying } A \rightarrow X \text{ and } B \rightarrow X, \tag{7.12}$$

$$C \text{ satisfies } C \rightarrow X. \tag{7.13}$$

If C satisfies these conditions we write

$$C = A \cup B. \tag{7.14}$$

The definition of a lattice is now complete with the following law.

LAW 7.6. If A and B belong to $\mathfrak{L}$, the unique disjunction $C = A \cup B$ exists and belongs to $\mathfrak{L}$.

It should be noted that the symbols $\rightarrow$ and $=$ are used to express a relation between members belonging to $\mathfrak{L}$, whereas the symbols $\cap$ and $\cup$ are used to form another member of $\mathfrak{L}$. $\rightarrow$ and $=$ are "relation symbols," whereas $\cap$ and $\cup$ are "connectives." Thus, for instance, (7.13) is a proposition *about* two members C and X that belong to $\mathfrak{L}$. The symbol $\supset$, called a "horseshoe," which we explain later, is different from implication (partaking) ($\rightarrow$) and is not a relation symbol but a connective. In the case of propositions the conjunction and disjunction are called sometimes "and" and "or," respectively. (See Figure 7.5 for the Eulerian representation of the disjunction.)

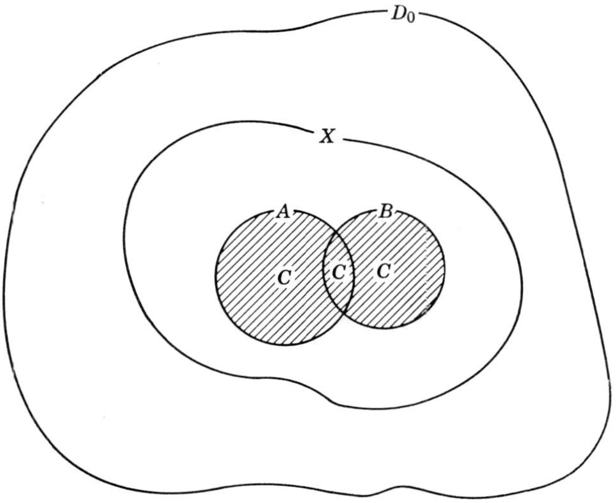

Figure 7.5 The shaded area C (if it is a member of $\mathfrak{L}$) is the disjunction of A and B: $C = A \cup B$. The domain X, defined by the closed curve surrounding A and B, satisfies $A \rightarrow X$ and $B \rightarrow X$; C is the smallest among such X's.

The disjunction in the spherical model has a rather unexpected property. In Figure 7.3 the disjunction of A and D is the great circle B, for no other smaller element in $\mathfrak{L}$ contains both A and D. Similarly, the disjunction of C and D is the entire surface $\square$. In the Eulerian case $A \cup B$ is the set of all points on A or B or on both. In the spherical model, however, the disjunction contains points not in A or B.

Examples 7.3, 7.6, and 7.7, which are all lattices, have one property in common. Each member $A, B \cdots$, of the set $\mathfrak{L}$ is considered again as a set of "items" (pebble and points), and the partaking relation $A \rightarrow B$ is interpreted as meaning that if an "item" belongs to member set A it belongs to member set B. The peculiarity of Example 7.7 stems from the fact that a merger of two member sets of items is not necessarily a member set of $\mathfrak{L}$. This results in a difference in the meaning of the disjunction. In Examples 7.3 and 7.6 the disjunction of A and B is the merger of sets of items, A and B, but in Example 7.7 it is a much larger set than the merger of sets of items of A and B. We see later that Examples 7.3 and 7.6 are "Boolean" lattices, whereas Example 7.7 is not. Examples 7.1 and 7.2 are lattices, but they are not interesting because $A \cap B$ and $A \cup B$ are either A or B themselves.

We can usually reduce a set of propositions to a lattice of a set of sets of items. This is because we can make a one-to-one correspondence between a proposition, say, A, and a set of all possible actual cases in which A is true. If A says, for instance, that all ravens are black, it may be regarded as corresponding to all possible worlds, which are restricted only by the condition that the ravens living in them are black. If A says that object a is red, we can consider all possible worlds in which an object identifiable as a exists and is red. If the object a is not a fixed identifiable item but a variable dummy item, we can also consider a collection of all possible a's, which are, however, under the limitation that they must be red. Each member of $\mathfrak{L}$ is then made to correspond to a set of cases or objects representing cases. This correspondence makes it plausible that the relations among propositions may be like the relations of elements in Examples 7.3 and 7.6 of Boolean lattice.

The precise reason why a non-Boolean case (Example 7.7) cannot be entirely dismissed as a model of lattice of propositions is explained in Chapter 9. For the present I limit myself to explaining how a non-Boolean disjunction (which is larger than the merger of both member sets) can be useful. Suppose that A and B describe the results of some kind of observation. To make it concrete, let us say that A says that "object a is left-handed" and B says that "object a is right-handed," no matter what this handedness refers to. Let us assume that the nature of observation is such that a is found to be either left-handed or right-handed, never in between. If handedness is a constant attribute of object a, it would be sufficient to consider two collections of possible a's, one containing all possible left-handed a's and the other all possible right-handed a's. The first collection corresponds to proposition A, and the second to B. However, there could also be a's that constantly change their handedness infinitely fast and at random (with a certain constant probabilistic ratio between left and right). On observation they will turn out to be left-handed or right-handed, but one cannot say that a particular a is left-handed or right-handed at any instant. We can imagine a case, in which the random decision between left and right takes place only at the moment of observation, so that in principle it is nonsense to say that the object a must be left-handed or right-handed at each time point even though not observationally determined. In a case like this, those vacillating a's do not belong either to the set corresponding to A or to the set corresponding to B, but to a set of a's that unpredictably make A or B true. Hence the set corresponding to this disjunction of A and B is larger than the "merger" of sets corresponding to A and B.

It seems appropriate to mention the following postulate, which underlies Boolean logic. The postulate may be called postulate of fixed predicate-set correspondence or of truth set.

***Postulate** 7.1* (Fixed predicate-set correspondence). Each predicate corresponds one-to-one at each time point to a fixed set of objects that satisfy the predicate.

This postulate does not preclude the use of a time-dependent predicate such as "is red at noon today." According to the postulate there is a fixed set of objects that "are red at noon today." The set is fixed, but the colors of members of this fixed set may be all different at one minute after noon today. On the other hand, the case we discussed above (in which an object will be either red or nonred on observation, but cannot be said to be either red or nonred at a time point if no observation is made) does not satisfy this postulate. Of course, according to the known physical laws, the predicate "is red" does not under normal circumstances allow such a violation of the postulate of fixed predicate-set correspondence, but a predicate like "is linearly polarized in the x direction" (speaking of a photon traveling in the z direction) does entail such a violation. See [W-31, W-31a] for more about this.

From the definitions of the lattice and the operations of conjunction and disjunction, we can derive some important theorems.†

***Theorem** 7.1*

$$A \rightarrow B \quad \text{is equivalent to} \quad A = A \cap B \tag{7.15}$$

† If, we say, as in Section 1.1, that $\cap$ and $\cup$ belong to the object language and $=$ and $\rightarrow$ to the metalanguage, the word "equivalent" in Theorem 7.1 may be said to belong to the meta-metalanguage.

and also to

$$B = A \cup B. \tag{7.16}$$

If it is given that $A \rightarrow B$, all the conditions in (7.7), (7.8), and (7.9) reduce to

$$C \rightarrow A, \tag{7.17}$$

$$X \rightarrow A, \tag{7.18}$$

and

$$X \rightarrow C. \tag{7.19}$$

Because X is required to satisfy only (7.18), it can be A itself. By substituting $X = A$ in (7.19) we obtain

$$A \rightarrow C. \tag{7.20}$$

Then (7.17) and (7.20) imply, because of (7.2) and (7.3), that

$$A = C, \tag{7.21}$$

which is (7.15). Conversely, if (7.15) is true, substitution of A for C in (7.7) yields

$$A \rightarrow A \quad \text{and} \quad A \rightarrow B. \tag{7.22}$$

Hence $A = A \cap B$ implies $A \rightarrow B$. This establishes the equivalence of (7.15) with $A \rightarrow B$. The equivalence of $A \rightarrow B$ and (7.16) can be proven similarly.

Theorem 7.2 The relation

$$D \rightarrow (A \cap B) \tag{7.23}$$

is equivalent to the relations

$$D \rightarrow A \quad \text{and} \quad D \rightarrow B. \tag{7.24}$$

Writing $C = (A \cap B)$, we know from (7.7) that

$$C \rightarrow A \quad \text{and} \quad C \rightarrow B. \tag{7.25}$$

If we assume, as in (7.23), that

$$D \rightarrow C, \tag{7.26}$$

application of the transitivity law (7.4) to (7.25) and (7.26) yields (7.24). Conversely, if we assume (7.24), we can choose X in (7.8) as D. Then (7.9) becomes (7.23). Similarly we can prove the following theorem.

Theorem 7.3 The relation

$$(A \cup B) \rightarrow D \tag{7.27}$$

is equivalent to the relations

$$A \rightarrow D \quad \text{and} \quad B \rightarrow D. \tag{7.28}$$

It is suggested that the student find out for himself what (7.15), (7.23), and (7.27) mean in Euler's diagram and in the spherical model.

Let us assume next in the definition of conjunction $C = A \cap B$ that $B = \emptyset$. Then by (7.7) we have $C \rightarrow \emptyset$. This is possible only if C itself is $\emptyset$, for $C \rightarrow \emptyset \rightarrow C$ [see (7.2) and (7.6)]. On the other hand, if $B = \square$, we can take $X = A$ in (7.8). Then, from (7.7) and (7.9), we get $A \rightarrow C \rightarrow A$, which means $C = A$. We conclude the following theorem, which can also be derived directly from Theorem 7.1 with the help of (7.5) and (7.6).

***Theorem** 7.4*

$$A \cap \emptyset = \emptyset, \qquad A \cap \square = A \quad \text{for any arbitrary} \quad A \tag{7.29}$$

and

$$A \cup \emptyset = A, \qquad A \cup \square = \square \quad \text{for any arbitrary} \quad A. \tag{7.30}$$

The student can also readily verify the following *law of reciprocity*.

***Theorem** 7.5* The system of formulas so far discussed comes back to itself by a simultaneous interchange of

$$\rightarrow \quad \text{with} \quad \leftarrow,$$

$$\cap \quad \text{with} \quad \cup,$$

and

$$\square \quad \text{with} \quad \emptyset. \tag{7.31}$$

We are now prepared to prove the following basic laws of the lattice, which no longer refer to the arrow relation. We presently prove that an alternative definition of a lattice is that it is a set of elements (including $\emptyset$ and $\square$) in which two kinds of operations ($\cap$ and $\cup$) operate in such a way that if A and B are its elements, $A \cap B$ and $A \cup B$ are also its elements, where the operations $\cap$ and $\cup$ satisfy the following rules and (7.29) and (7.30). We refrain from calling the following formulas axioms, because they are not all independent. It is a matter of convenience which rules are selected as the starting axioms.

***Theorem** 7.6* IDEMPOTENT LAW: $A \cap A = A,\ A \cup A = A,$ (7.32)

***Theorem** 7.7* COMMUTATIVE LAW:
$$A \cap B = B \cap A, \qquad A \cup B = B \cup A. \tag{7.33}$$

***Theorem** 7.8* ASSOCIATIVE LAW:
$$A \cap (B \cap C) = (A \cap B) \cap C, \qquad A \cup (B \cup C) = (A \cup B) \cup C. \tag{7.34}$$

***Theorem** 7.9* ABSORPTIVE LAW:
$$A \cap (A \cup B) = A, \qquad A \cup (A \cap B) = A. \tag{7.35}$$

The proof of the idempotent law is as follows. If $A \cap A = C$, $C \rightarrow A$ from (7.7). Next, putting $X = A$ (which is allowed because $A \rightarrow A$) in (7.8) and (7.9), we get $A \to C$. Therefore $A \rightleftarrows C$ or $A = C$. A similar proof can be given for $A \cup A = A$. The proof of the commutative law is hardly necessary, since the definition of conjunction (7.7–7.9) is perfectly symmetrical with respect to A and B. So is the definition of disjunction (7.11–7.13).

The proof of the associative law is as follows. Put $D = A \cap (B \cap C)$. Then, according to the definition of conjunction, we have

$$D \to A \quad \text{and} \quad D \to (B \cap C). \tag{7.36}$$

For any X satisfying

$$X \to A \quad \text{and} \quad X \to (B \cap C) \tag{7.37}$$

we should have

$$X \to D.$$

According to (7.23) and (7.24), $D \to (B \cap C)$ is now equivalent to $D \to B$ and $D \to C$. Hence the two conditions (7.36) and (7.37) can be rewritten as

$$D \to A, \qquad D \to B, \qquad D \to C \tag{7.38}$$

and

$$X \to A, \qquad X \to B, \qquad X \to C. \tag{7.39}$$

This means that the definition of D is completely symmetrical with respect to A, B, and C. Therefore the left-hand side and the right-hand side of (7.34) must be the same. For this reason we can also write

$$A \cap (B \cap C) = (A \cap B) \cap C = (A \cap C) \cap B = A \cap B \cap C. \tag{7.40}$$

Similarly, we can prove the corresponding theorem with the disjunction, and we can also write

$$A \cup (B \cup C) = (A \cup B) \cup C = (A \cup C) \cup B = A \cup B \cup C. \tag{7.41}$$

The absorptive law can be proven as follows. Write $D = A \cap (A \cup B)$. Then according to the definition of conjunction we have

$$D \to A, \qquad D \to (A \cup B), \tag{7.42}$$

and for any X such that

$$X \to A, \qquad X \to (A \cup B) \tag{7.43}$$

we have

$$X \to D. \tag{7.44}$$

According to the definition of disjunction, in particular (7.11), we have

$$A \to (A \cup B). \tag{7.45}$$

Hence we can choose X in (7.43) as A. Equation (7.44) then becomes

$$A \to D. \tag{7.46}$$

We have $D \rightarrow A$ in (7.42) and $A \rightarrow D$ in (7.46). Therefore $A = D$. Q.E.D. The proof becomes simpler if we use Theorem 7.1. This theorem plus (7.45) yields (7.35) immediately.

By the same token we can prove the law with $\cap$ and $\cup$ interchanged.

In the foregoing paragraphs we assumed the properties of Laws 7.1–7.6 and proved Theorem 7.6 through Theorem 7.9, using the definitions of conjunction and disjunction in terms of the participation relation. We can do the opposite and derive the former starting from the latter, proving that the two definitions of lattice are equivalent.

Theorem 7.10 Suppose that $\mathfrak{L}$ is a set of symbols A, B, C, and so on, that is closed for two operations $\cap$ and $\cup$, each of which engenders a third element from a pair of elements, whereby the four laws (7.32)–(7.35) hold. Assume further that $\mathfrak{L}$ contains two elements $\emptyset$ and $\square$ that obey (7.29) and (7.30). Now if a relation $\rightarrow$ is defined so that $A \rightarrow B$ is equivalent to (7.15) or (7.16), $\mathfrak{L}$ satisfies Laws 7.1 through 7.6.

The proof is left to the reader because it is very easy. It may be noted that the equivalence of (7.15) and (7.16) can be shown by the absorptive law, and that the transitive law (7.4) can be proved by the associative law.

Very often, along with the four laws of Theorems 7.6–7.9, another law, called the distributive law, is mentioned. From the viewpoint we adopt in the present section, however, this last law has an entirely different status from the other four.

DISTRIBUTIVE LAW:

$$\begin{aligned} A \cap (B \cup C) &= (A \cap B) \cup (A \cap C), \\ A \cup (B \cap C) &= (A \cup B) \cap (A \cup C). \end{aligned} \tag{7.47}$$

If a lattice satisfies this law, it is called a distributive lattice or a Boolean lattice. To show that the distributive law is not a necessary property of a lattice, it suffices to bring up one example of a lattice in which this law does not hold. This is precisely what we shall presently do, but first we see the meaning of a distributive law in simple examples. If we take a set of ordinary sentences as elements of a lattice, this law indeed seems very natural. For instance, take a very simple case:

A: the animal is good-looking.

B: the animal is a man.

C: the animal is a woman.

Then $A \cap (B \cup C)$ means that "the animal is a good-looking human," where human means "man or woman." On the other hand, $(A \cap B) \cup$

$(A \cap C)$ means that "the animal is a good-looking man or a good-looking woman." Also, in the Eulerian model, we can easily see that the distributive law holds (see Figure 7.6). As a matter of fact, every time we can make a one-to-one correspondence between a proposition and a *fixed* set of items, we can assume the distributive law as an integral part of logic. This is the reason why the postulate of fixed predicate-set correspondence guarantees the validity of Boolean (i.e., distributive) lattices.

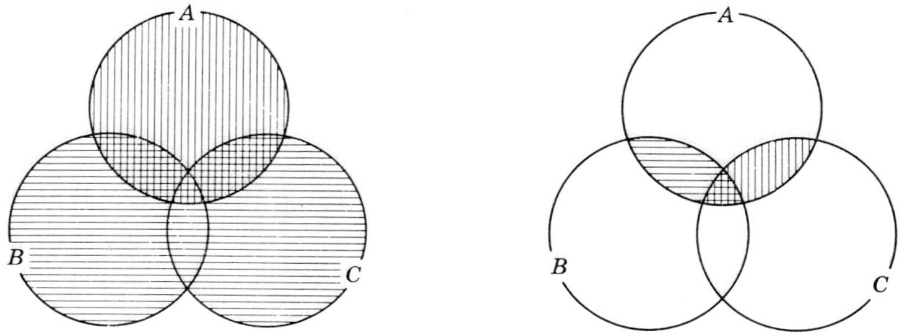

Figure 7.6 On the left the area with horizontal shading is $B \cup C$ and the area with vertical shading is A. The intersection of these two areas, the area with horizontal and vertical shading, is $A \cap (B \cup C)$. On the right the area with horizontal shading is $A \cap B$ and the area with vertical shading is $A \cap C$. The union of these two, that is, the total shaded area, is $(A \cap B) \cup (A \cap C)$. We see that the distributive law $A \cap (B \cup C) = (A \cap B) \cup (A \cap C)$ holds in the Eulerian model. A similar verification of the other distributive law $A \cup (B \cap C) = (A \cup B) \cap (A \cup C)$ is left to the reader as an exercise.

However, that the distributive law does not follow from the definition of a lattice can be seen by the use of the spherical model. Suppose that A is a great circle and B and C are points that are not necessarily on the great circle A. Then $B \cup C$ is another great circle passing B and C. $A \cap (B \cup C)$ will then be the intersection of these two great circles A and $(B \cup C)$. This intersection will be two points D and D_1, which are diametrically opposed. On the other hand, $A \cap B$ must be $\emptyset$ if B is not on A, since there is no single point common to the circle A and the point B. By the same token, $A \cap C$ must also be $\emptyset$ if C is not on A. Then $(A \cap B) \cup (A \cap C)$ becomes $\emptyset \cup \emptyset$, which according to (7.30) is again $\emptyset$. Thus $A \cap (B \cup C) = D$ and $(A \cap B) \cup (A \cap C) = \emptyset$ if neither B nor C is on A. Therefore the distributive law does not hold. Note, however, that the distributive law holds if B or C is on A. See Figure 7.7.

Keeping in mind that the distributive law does not follow from the general definition of conjunction and disjunction, we proceed in this section by occasionally including the distributive law as one of the accepted rules. We shall, however, put an asterisk on the theorems that are proved on the assumption of the distributive law to distinguish them from those theorems that are valid in general. We use the symbol $\mathfrak{L}$ to designate a lattice and the symbol $\mathcal{B}$ to designate a lattice specified as a distributive lattice. From now on, we use more frequently the convenient terminology of logic rather than the neutral terminology that we preferred at the beginning, but this does not necessarily mean that our discussions refer solely to a lattice of propositions.

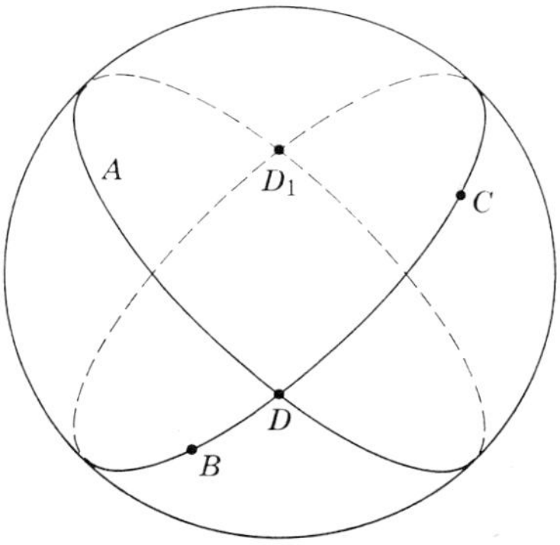

Figure 7.7 A is a great circle; B and C are two points that are not on the circle A. Thus $A \cap B$ and $A \cap C$ are both $\emptyset$. $(B \cup C)$ is a great circle passing B and C. The term $A \cap (B \cup C)$ means the intersection D of these two circles. The term $(A \cap B) \cup (A \cap C)$ is $\emptyset \cup \emptyset$, which is again $\emptyset$. If B or C were on A, $(A \cap B) \cup (A \cap C)$ would be D.

As a next important step, we introduce the complement or negation $\urcorner A$ of a proposition A. If $\mathfrak{L}$ is such that if A belongs to $\mathfrak{L}$ then its negation $\urcorner A$ (to be defined below) also belongs to $\mathfrak{L}$, $\mathfrak{L}$ is called a complemented lattice. Hereafter we always assume that the lattice under consideration is a complemented lattice whether it is Boolean or otherwise. The negation $\urcorner A$ of A is unique when A is given. The operation of negation, or passage from A to $\urcorner A$, is supposed to obey the following basic laws.

LAW OF DOUBLE NEGATION: $\urcorner(\urcorner A) = A$. (7.48)

LAW OF CONTRAPOSITION: if $A \to B$ then $\urcorner B \to \urcorner A$. (7.49)

LAW OF SELF-CONTRADICTION: if $A \to \urcorner A$ then $A = \emptyset$. (7.50)

We consider these laws as the definition of negation. There are other methods of introducing negation, but this set of three formulas is also very convenient for the non-Boolean case.†

† The operation designated by $\urcorner$ must be done before other operations such as $\cap$ and $\cup$ are carried out. Thus, for instance, $\urcorner A \cup B$ is equivalent to $(\urcorner A) \cup B$ but not $\urcorner(A \cup B)$. This is a matter of convention.

From these laws follow several other important rules. First, substituting $\neg B$ and $\neg A$ for A and B, respectively in (7.49), we obtain

$$\text{if} \quad \neg B \rightarrow \neg A \quad \text{then} \quad \neg(\neg A) \rightarrow \neg(\neg B). \tag{7.51}$$

By the law of double negation this becomes

$$\text{if} \quad \neg B \rightarrow \neg A \quad \text{then} \quad A \rightarrow B. \tag{7.52}$$

Combining (7.49) and (7.52), we obtain the following theorem.

Theorem 7.11 $\quad A \rightarrow B$ is equivalent to $\neg B \rightarrow \neg A$. (7.53)

We can also derive the next theorem.

Theorem 7.12 $\quad A = B$ is equivalent to $\neg A = \neg B$, (7.54)

since the premise means that $A \rightarrow B$ and $B \rightarrow A$ and the contrapositions of these are $\neg B \rightarrow \neg A$ and $\neg A \rightarrow \neg B$.

Next take any arbitrary proposition A and its negation $\neg A$. Then, by the very definition of $\emptyset$, we have

$$\emptyset \rightarrow \neg A. \tag{7.55}$$

Now take the contraposition of (7.55) by the use of (7.49); we get

$$\neg(\neg A) \rightarrow \neg\emptyset, \tag{7.56}$$

which, by virtue of the law of double negation, becomes

$$A \rightarrow \neg\emptyset, \tag{7.57}$$

which means that any arbitrary A implies $\neg\emptyset$. This leads, because of the definition of $\square$, to the following theorem.

Theorem 7.13 $\quad \neg\emptyset = \square$, (7.58)

whose negation is

$$\neg\square = \emptyset. \tag{7.59}$$

By definition we have for any A

$$\emptyset \rightarrow A \rightarrow \square, \tag{7.60}$$

which becomes, by virtue of (7.58),

$$\emptyset \rightarrow A \rightarrow \neg\emptyset. \tag{7.61}$$

From this follows, by the use of the transitivity of implication (7.4),

$$\emptyset \rightarrow \neg\emptyset, \tag{7.62}$$

which could also have been obtained by putting $A = \emptyset$ in (7.57). Anyway (7.62) has the form

$$A \rightarrow \neg A. \tag{7.63}$$

What the law of self-contradiction states is that if (7.63) is the case, A is bound to be $\emptyset$. In other words, we have proven the theorem.

Theorem 7.13 The converse of the law of self-contradiction is a consequence of the laws of double negation and of contraposition.

Next we derive the so-called De Morgan law. Consider the definition

$$C = A \cap B. \tag{7.64}$$

We have

$$C \to A \quad \text{and} \quad C \to B, \tag{7.65}$$

and for any arbitrary X satisfying

$$X \to A \quad \text{and} \quad X \to B \tag{7.66}$$

we have

$$X \to C. \tag{7.67}$$

Take the contraposition of all these conditions. We get

$$\neg A \to \neg C \quad \text{and} \quad \neg B \to \neg C, \tag{7.68}$$

and for any arbitrary X satisfying

$$\neg A \to \neg X \quad \text{and} \quad \neg B \to \neg X \tag{7.69}$$

we have

$$\neg C \to \neg X. \tag{7.70}$$

These conditions (7.68)–(7.70) are nothing but those defining $\neg C$ as the disjunction of $\neg A$ and $\neg B$. Hence

$$\neg C = \neg A \cup \neg B. \tag{7.71}$$

Combining (7.64) and (7.71) we obtain the next theorem.

Theorem 7.14

DE MORGAN'S LAW: $$\neg(A \cap B) = \neg A \cup \neg B \tag{7.72}$$

and

DE MORGAN'S LAW: $$\neg(A \cup B) = \neg A \cap \neg B. \tag{7.73}$$

It should be mentioned that the derivation of (7.72) required only the law of contraposition and the definitions of conjunction and disjunction. Similarly, (7.73) can be derived from the same premises. Alternatively, we can derive (7.73) from (7.72) by taking the negation of (7.72) and using (7.54) and (7.48).

Next consider

$$C = A \cap \neg A, \tag{7.74}$$

whose negation, as a result of De Morgan's law and the law of double negation, becomes

$$\neg C = \neg A \cup A. \tag{7.75}$$

By the definition of conjunction and disjunction we have

$$A \cap \neg A \rightarrow A \rightarrow A \cup \neg A, \tag{7.76}$$

which, in consideration of (7.74) and (7.75), entails

$$C \rightarrow \neg C. \tag{7.77}$$

On account of the law of self-contradiction (7.50), this means that $C = \emptyset$. Hence (7.74) and (7.75) can be stated as follows.

Theorem 7.15 For any A we have

LAW OF CONTRADICTION: $A \cap \neg A = \emptyset$ (7.78)

and

LAW OF EXCLUDED MIDDLE: $A \cup \neg A = \square.$ (7.79)

Because of (7.54), (7.58), and De Morgan's law, (7.78) and (7.79) are equivalent.

Sometimes two relations

$$A \cap B = \emptyset \tag{7.80}$$

and

$$A \cup B = \square \tag{7.81}$$

are used as the definition of negation: $B = \neg A$. From our standpoint, however, we know that $B = \neg A$ satisfies (7.80) and (7.81), but it is still an open question whether there is another B that is not $\neg A$ but satisfies (7.80) and (7.81). We see presently that if the distributive law holds then B of (7.80) and (7.81) is bound to be $\neg A$, but otherwise not.

Let us first take (7.80) and form the disjunction of each side with $\neg B$. From the left side we obtain, with the help of the distributive law and (7.79) and (7.30),

$$\begin{aligned} \neg B \cup (A \cap B) &= (\neg B \cup A) \cap (\neg B \cup B) \\ &= (\neg B \cup A) \cap \square \\ &= \neg B \cup A. \end{aligned} \tag{7.82}$$

From the right side of (7.80) we obtain, because of (7.30),

$$\neg B \cup \emptyset = \neg B. \tag{7.83}$$

Combining (7.82) and (7.83), we get

$$\neg B = \neg B \cup A, \tag{7.84}$$

which means, because of (7.16), that

$$A \to \neg B \tag{7.85}$$

or, equivalently,

$$B \to \neg A. \tag{7.86}$$

What we have shown is that if the distributive law holds,

$$\text{if} \quad A \cap B = \emptyset \quad \text{then} \quad A \to \neg B. \tag{7.87)*}$$

It is noteworthy that the converse of this theorem does not require the distributive law. Take (7.85) and express it, according to (7.15), as

$$A = A \cap \neg B, \tag{7.88}$$

which can also be obtained by making the conjunction of each side of (7.84) with A and using the absorptive law. Now make the conjunction of each side of (7.88) with B. The right side becomes, by virtue of the associative law and (7.78) and (7.29),

$$(A \cap \neg B) \cap B = A \cap (B \cap \neg B) = A \cap \emptyset = \emptyset. \tag{7.89}$$

Hence

$$A \cap B = \emptyset. \tag{7.90}$$

Therefore, without the help of the distributive law, we have

$$\text{if} \quad A \to \neg B \quad \text{then} \quad A \cap B = \emptyset. \tag{7.91}$$

Similarly, with the help of the distributive law, we can prove that

$$\text{if} \quad A \cup B = \square \quad \text{then} \quad \neg A \to B. \tag{7.92)*}$$

Without the help of the distributive law we can also prove that

$$\text{if} \quad \neg A \to B \quad \text{then} \quad A \cup B = \square. \tag{7.93}$$

Of course, $\neg A \to B$ in (7.92) is equivalent to $\neg B \to A$. Now, combining (7.87) and (7.92), we conclude the following.

Theorem 7.16 In a distributive lattice $\mathcal{B}$, if A and B satisfy both (7.80) and (7.81), $A = \neg B$.

Renaming $\neg B$ as B in (7.87) and (7.91), and $\neg A$ as A in (7.92) and (7.93), we obtain the following two theorems.

Theorem 7.17 In a general lattice $\mathfrak{L}$,

$$\text{if} \quad A \to B \quad \text{then} \quad A \cap \neg B = \emptyset \tag{7.94}$$

or, equivalently,

$$\text{if} \quad A \to B \quad \text{then} \quad \neg A \cup B = \square. \tag{7.95}$$

Theorem 7.18 In a distributive lattice $\mathscr{B}$,

$$\text{if} \quad A \cap \neg B = \emptyset \quad \text{then} \quad A \to B \tag{7.96}*$$

or, equivalently,

$$\text{if} \quad \neg A \cup B = \Box \quad \text{then} \quad A \to B. \tag{7.97}*$$

In recapitulation, we can state that $A \to B$ is unconditionally equivalent to each of the four expressions

$$A = A \cap B, \qquad \neg A = \neg A \cup \neg B, \qquad B = A \cup B, \qquad \neg B = \neg A \cup \neg B, \tag{7.98}$$

and, if the distributive law holds, equivalent to each of the two expressions

$$A \cap \neg B = \emptyset \quad \text{and} \quad \neg A \cup B = \Box. \tag{7.99}$$

We should point out that the law of reciprocity, Theorem 7.5, can be generalized and replaced by the following theorem.

Theorem 7.19 (Law of reciprocity). The system of laws in a complemented lattice remains invariant for the transformation (7.31), supplemented by the interchange of

$$A \quad \text{with} \quad \neg A \tag{7.100}$$

for every A, including $\emptyset = \neg\Box$ and $\Box = \neg\emptyset$.

In the above discussion pertaining to the formulas with asterisks, such as (7.87), (7.92), (7.96), and (7.97), we qualified the statements by the condition "if the distributive law holds." In the following we show by means of a model consideration that if the distributive law does not hold, there can indeed be cases in which these formulas actually break down. For this purpose we first introduce the concept of negation in the both models in such a way that the laws of negation hold. In Euler's diagram, if A represents a domain D within D_0 then its negation can be represented by the part of D_0 that is not occupied by D (see Figure 7.8). This definition is unique and obviously

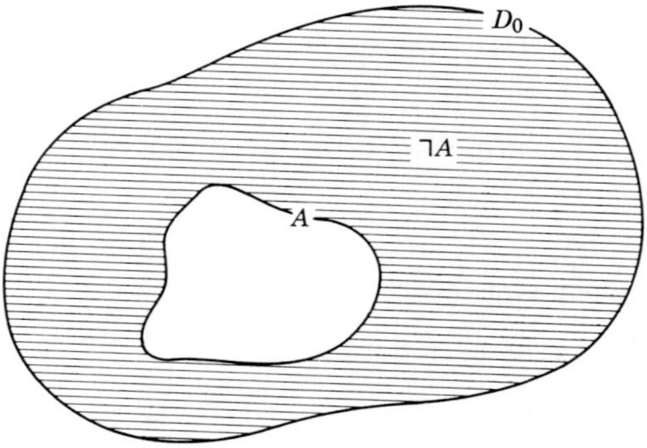

Figure 7.8 If A is the unshaded domain, $\neg A$ is the shaded domain. The one is the remainder of the other in D_0. The law of double negation is obvious.

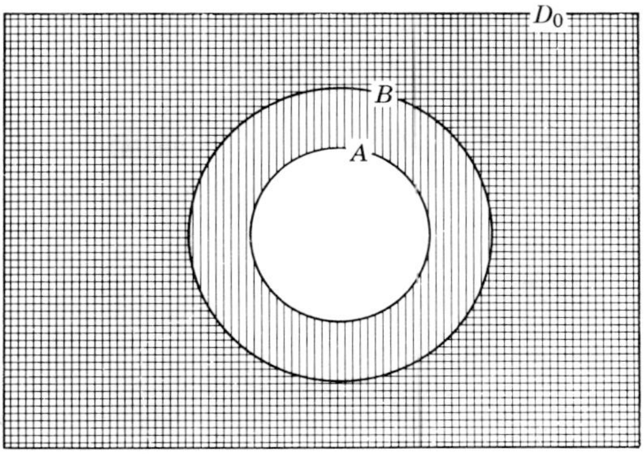

Figure 7.9 The law of contraposition. Assume that A is included in B. Then $\neg B$, which is the horizontally shaded area, is included in $\neg A$, which is the vertically shaded area.

satisfies the law of double negation. The law of contraposition is also very easy to verify. Finally, the law of self-contradiction has to hold, because if A is a finite domain, the domain A is never included in the domain $\neg A$, but if A is $\emptyset$, it is included in its negation, which is D_0 itself (see Figures 7.8 and 7.9).

In the case of a spherical model we can define the negation $\neg A$ of A as follows. If A is $\emptyset$, $\neg A$ is the entire spherical surface. If A is a point, $\neg A$ is the great circle whose normal (i.e., the normal to the plane containing the great circle) cuts the spherical surface at A and A_1 (see Figure 7.10). If A is a great circle, we invert the above definition; namely, $\neg A$ is the point at which the normal to A cuts the sphere. Finally, if A is the entire surface, $\neg A$ is $\emptyset$. With

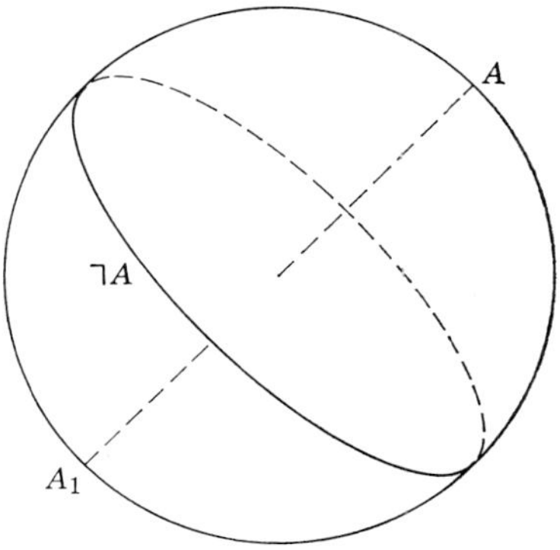

Figure 7.10 If A is a point, $\neg A$ is the great circle perpendicular to the diameter (A, A_1). If A is a great circle, $\neg A$ is the point at which the normal to the plane of the circle intersects the sphere.

this understanding, the uniqueness and the law of double negation are obvious. The law of contraposition applied to the case in which A is a point means the following. In the relation $A \rightarrow B$, B, if different from A, can be one of the great circles passing through A or the entire surface. Let us take the case when B is a great circle passing A. Then the negation $\urcorner B$ of B will be the point perpendicular to the great circle. The negation of A will be a circle $\urcorner A$ perpendicular to A. This circle $\urcorner A$ is bound to pass through the point $\urcorner B$. Thus the law of contraposition is verified in this case (see Figure 7.11).

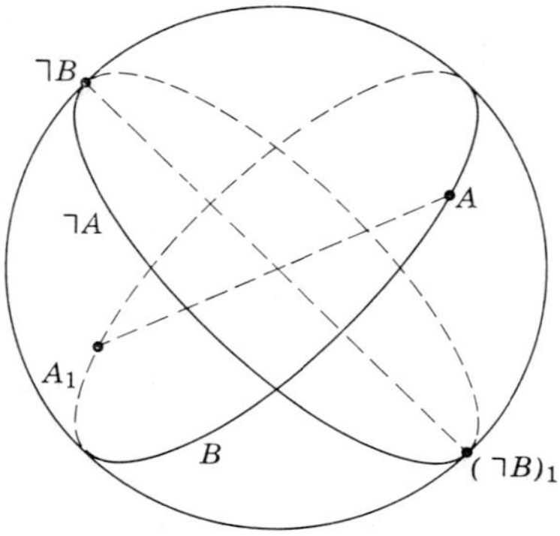

Figure 7.11 A is a point on the great circle B passing A. $A \rightarrow B$. Then $\urcorner A$ is a great circle perpendicular to A and $\urcorner B$ is a point perpendicular to the circle B. Then $\urcorner A$ has to pass $\urcorner B$, since $\urcorner A$ is the locus of all the points perpendicular to A and $\urcorner B$ is perpendicular to any point on B, including A. Thus $\urcorner B \rightarrow \urcorner A$.

Other cases work similarly.

Consider the condition

$$A \cap B = \emptyset \tag{7.101}$$

in Euler's diagram. This means that A and B have no common territory. Therefore the negation of B is bound to cover A completely. This is the meaning of $A \rightarrow \urcorner B$ (see Figure 7.12). In the spherical model, take as A and B two different points as a special example; then (7.101) is satisfied. Assume further that A and B do not subtend a right angle at the center of the sphere. Then the negation $\urcorner B$ of B, which is a great circle perpendicular to the radius to B, will certainly not pass A. This means that A is not included in $\urcorner B$; in other words, A does not imply $\urcorner B$ (see Figure 7.13). This shows that if the distributive law does not hold, there is a case when (7.87) does not hold.

7.2. SPECTRAL DECOMPOSITION, ATOMS, AND TRUTH VALUE

We now introduce the concept of "spectral decomposition" of a (complemented) Boolean lattice, $\mathcal{B}$. We no longer put asterisks on the formulas that

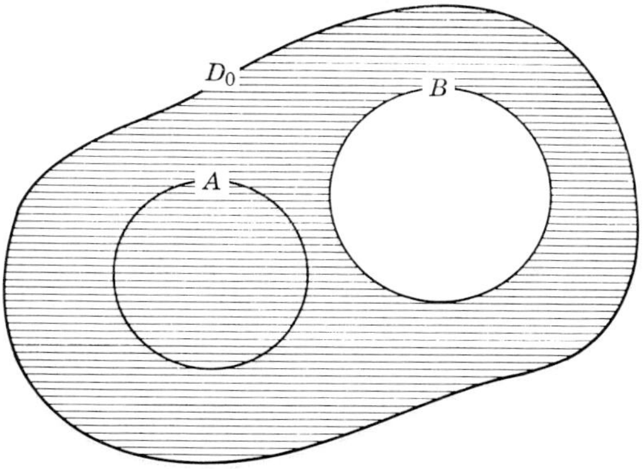

Figure 7.12 If A and B are two domains having no common point, the negation $\neg B$ (shaded area) of B will cover A completely. In this case $A \cap B = 0$ and $A \to \neg B$ are equivalent.

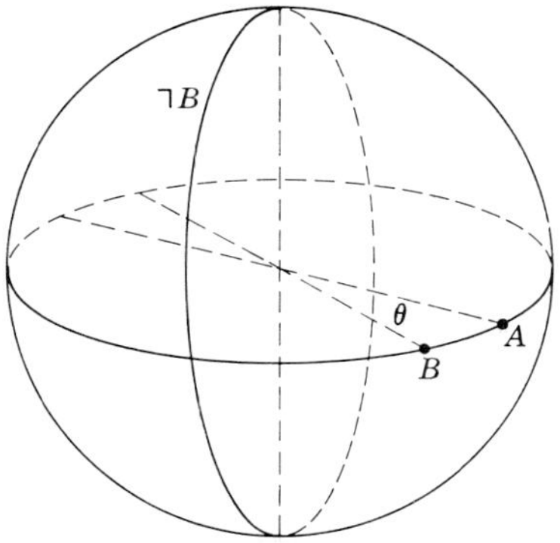

Figure 7.13 A and B are points on the equator, and the angle θ that they subtend at the center of the sphere is not 90°. The angle θ is the difference of the longitude of A and B. The negation $\neg B$ of B is a meridian plane perpendicuar to B. Unless $\theta = 90°$, this meridian does not pass A. Thus $A \cap B = \emptyset$ but $A \to \neg B$ is not true.

refer only to a Boolean lattice. We introduce a new symbol $<$ ("is smaller than") by the following definition:

$$\text{if} \quad A \to B \quad \text{but not} \quad B \to A \quad \text{then} \quad A < B, \quad \text{and conversely.} \tag{7.102}$$

In other words,

$$\text{if} \quad A \to B \quad \text{and} \quad A \neq B \quad \text{then} \quad A < B, \quad \text{and conversely.} \tag{7.103}$$

Now, if a (ordered) group of elements $A_0, A_1, A_2 \cdots, A_{n-1}, A_n$ in $\mathcal{B}$ satisfies the relations

$$\emptyset = A_0 < A_1 < A_2 < \cdots < A_{n-1} < A_n = \square, \tag{7.104}$$

this group of elements is said to form a "chain of inclusion." The existence of such a chain is guaranteed because $\emptyset$ and $\Box$ are necessarily members of $\mathcal{B}$ and $(\emptyset, \Box)$ is already a chain of inclusion. If there is any element X other than $\emptyset$ and $\Box$ in $\mathcal{B}$, $(\emptyset, X, \Box)$ is a chain of inclusion. If X is a member of $\mathcal{B}$, $\neg X$ is also a member; hence $(\emptyset, \neg X, \Box)$ is another chain of inclusion. If X and Y are members of $\mathcal{B}$ different from $\emptyset$ and $\Box$, $(\emptyset, X \cap Y, X, X \cup Y, \Box)$ is one of the chains of inclusion in $\mathcal{B}$, provided that $X \cap Y \neq \emptyset, X \cup Y \neq \Box$, and there is no implication relation between X and Y; that is $X \neq X \cap Y$, $X \neq X \cup Y$.

Now take a pair of consecutive elements, A_{i-1} and A_i, in a chain of inclusion $(A_0, A_1, A_2, \cdots, A_n)$ and form $\neg A_{i-1} \cap A_i = \alpha_i$. This is sometimes written as $A_i - A_{i-1}$, and means the largest element included in A_i and disjoint with A_{i-1}. (A and B are said to be disjoint if $A \cap B = \emptyset$.)

$$\alpha_i = \neg A_{i-1} \cap A_i, \qquad i = 1, 2, \cdots, n. \tag{7.105}$$

Let us note that $\alpha_1 = A_1$. If the lattice is Boolean, as we assume here, we obviously have

$$\alpha_i \neq \emptyset, \qquad i = 1, 2, \cdots, n \tag{7.106}$$

since, if α_i were $\emptyset$, $\emptyset = \neg A_{i-1} \cap A_i$ would imply, because of (7.96), that

$$A_i \rightarrow A_{i-1}, \tag{7.107}$$

which we explicitly excluded in (7.104). Actually, even if the distributive law does not hold, we can derive (7.107) from $\alpha_i = \emptyset$ and $A_{i-1} \rightarrow A_i$ if the lattice allows for the so-called modular law (which we study in Chapter 9). Here, however, we are not interested in this fact.

These n α's have the following important properties:

$$\text{disjointness: } \alpha_i \cap \alpha_j = \emptyset, \qquad i \neq j \tag{7.108}$$

and

$$\text{completeness: } \bigcup_{i=1}^{n} \alpha_i = \alpha_1 \cup \alpha_2 \cup \cdots \cup \alpha_n = \Box. \tag{7.109}$$

We say that $\Box$ is decomposed in the spectrum $\alpha_1, \alpha_2, \cdots, \alpha_n$. What we called a "logical spectrum" in Chapter 1 is precisely this.

The disjointness is obvious, since

$$\alpha_i \cap \alpha_j = \neg A_{i-1} \cap A_i \cap \neg A_{j-1} \cap A_j, \tag{7.110}$$

and if $i < j$ then $A_i \rightarrow A_{j-1}$ (i.e., $A_i = A_i \cap A_{j-1}$); consequently

$$A_i \cap \neg A_{j-1} = A_i \cap A_{j-1} \cap \neg A_{j-1} = A_i \cap \emptyset = \emptyset. \tag{7.111}$$

The completeness follows from the fact that

$$A_{i-1} \cup \alpha_i = A_{i-1} \cup (\neg A_{i-1} \cap A_i) \tag{7.112}$$

by virtue of (7.105) and that this formula (7.112) becomes, by the use of the distributive law,

$$\begin{aligned} A_{i-1} \cup (\neg A_{i-1} \cap A_i) &= (A_{i-1} \cup \neg A_{i-1}) \cap (A_{i-1} \cup A_i) \\ &= \square \cap (A_{i-1} \cup A_i) = A_{i-1} \cup A_i \\ &= A_i \end{aligned} \tag{7.113}$$

because $A_{i-1} \rightarrow A_i$. Combination of (7.112) and (7.113) means that

$$A_{i-1} \cup \alpha_i = A_i \tag{7.114}$$

and repetition of (7.114) yields, because $A_1 = \alpha_1$,

$$A_i = \alpha_1 \cup \alpha_2 \cup \cdots \cup a_i. \tag{7.115}$$

This proves the completeness (7.109), since $A_n = \square$ by definition. Actually the derivation of (7.113) is allowed by the so-called modular law (using the fact that $A_{i-1} \rightarrow A_i$) even in the absence of the distributive law, but not in a general lattice.

Coming back to the chain (7.104), suppose that the A's are such that there is no element B in $\mathcal{B}$ that satisfies

$$A_{i-1} < B < A_i \quad \text{for any} \quad i = 1, 2, \cdots, n. \tag{7.116}$$

Then we say that the chain is "maximal," and the α's obtained from the maximal chain are called "atoms" of $\mathcal{B}$. If the number of members of $\mathcal{B}$ is finite, the existence of atoms is obvious. We consider in this section a finite Boolean lattice, for which n is finite.

The reason why the α's are called atoms (which means indivisible) is that there cannot be any element D in $\mathcal{B}$ such that

$$D < \alpha_k, \qquad D \neq \emptyset \tag{7.117}$$

for some k. We now show that if D satisfying (7.117) existed, then C, defined by

$$C = A_{k-1} \cup D, \tag{7.118}$$

would satisfy (7.116) ($A_{k-1} < C < A_k$), contradicting the assumption that the chain is maximal. In order to show this we have to prove that if D satisfies (7.117), C would satisfy

$$A_{k-1} \rightarrow C \tag{7.119}$$

$$A_{k-1} \neq C \tag{7.120}$$

$$C \rightarrow A_k \tag{7.121}$$

$$C \neq A_k. \tag{7.122}$$

In the first place, (7.119) is obvious by the definition of C in (7.118). Second, as regards (7.120), if A_{k-1} were equivalent to C, it would mean, by virtue of (7.118), that

$$A_{k-1} = A_{k-1} \cup D; \quad \text{that is,} \quad D \to A_{k-1}. \tag{7.123}$$

But this is impossible, for the definition (7.117) implies that

$$D \to \alpha_k = \neg A_{k-1} \cap A_k; \tag{7.124}$$

hence

$$D \to \neg A_{k-1}. \tag{7.125}$$

To make both (7.123) and (7.124) possible, D would have to be $\emptyset$, which was excluded in (7.117). Hence A_{k-1} cannot be equivalent to C. Third, (7.121) is proved by taking the disjunction of each side of (7.118) with A_k and noting that $A_{k-1} \cup A_k = A_k$,

$$C \cup A_k = (A_{k-1} \cup D) \cup A_k = A_k \cup D = A_k, \tag{7.126}$$

since $D \to A_k$ according to (7.124). This (7.126) says that C implies A_k.

Fourth, and finally, to prove (7.122) let us assume that $C = A_k$ and show that this assumption leads to a contradiction. If $C = A_k$, we should have from (7.118)

$$A_k = A_{k-1} \cup D. \tag{7.127}$$

The conjunction of the left side with $\neg A_{k-1}$ gives

$$A_k \cap \neg A_{k-1} = \alpha_k, \tag{7.128}$$

whereas the conjunction of the right side with $\neg A_{k-1}$ gives

$$\begin{aligned} \neg A_{k-1} \cap (A_{k-1} \cup D) &= (\neg A_{k-1} \cap A_{k-1}) \cup (\neg A_{k-1} \cap D) \\ &= \emptyset \cup (\neg A_{k-1} \cap D) = (\neg A_{k-1} \cap D) = D. \end{aligned} \tag{7.129}$$

The last step in (7.129) is justified by (7.125). The two quantities (7.128) and (7.129) must be equivalent, but this is impossible because $D < \alpha_k$, which means that $D \to \alpha_k$ but $D \neq \alpha_k$. Thus we have proved all four relations (7.119)–(7.122). This means that there is no element D in $\mathcal{B}$ satisfying (7.117) if the chain is maximal; that is, the α's are the "smallest" indivisible elements in $\mathcal{B}$ without being $\emptyset$.

The converse of this statement is also obvious: an element α_k such that D satisfying (7.117) does not exist can be considered as a link in a maximal chain. Hence the condition stated with regard to (7.117) can be used as the definition of an atom.

We note that there are more than one maximal chain in a finite $\mathcal{B}$ in general, but it is easy to see that the atoms obtained from these different chains are the same and are uniquely determined by $\mathcal{B}$. This last property is peculiar to a Boolean lattice and is not shared by a general modular lattice. The labeling of the atoms α_i by the index i is, of course, arbitrary, and the two sets of atoms $(\alpha_1, \alpha_2, \cdots, \alpha_i, \cdots, \alpha_n)$ and $(\beta_1, \beta_2, \cdots, \beta_i, \cdots, \beta_m)$ are said to be identical if one set can be obtained from the other by relabeling the index i.

Suppose that there are two sets of atoms $(\alpha_1, \alpha_2, \cdots, \alpha_n)$ and $(\beta_1, \beta_2, \cdots, \beta_m)$ that are not identical in the same $\mathcal{B}$. Then at least one β_k must be different from any one of the α's. Let us make the conjunction of β_k and α_i and call it D_i:

$$D_i = \alpha_i \cap \beta_k, \qquad i = 1, 2, \cdots, n. \tag{7.130}$$

It is impossible that D_i can be $\emptyset$ for all i $(1, 2, \cdots, n)$ because, if it were, we should obtain from the left side

$$\bigcup_{i=1}^{n} D_i = \bigcup \emptyset = \emptyset, \tag{7.131}$$

where $\bigcup_{i=1}^{n}$ means that all n D_i's are connected by disjunction $\cup$, and, from the right side, by the use of the distributive law,

$$\bigcup_{i=1}^{n} (\alpha_i \cap \beta_k) = \left(\bigcup_{i=1}^{n} \alpha_i\right) \cap \beta_k = \square \cap \beta_k = \beta_k. \tag{7.132}$$

The β_k, being an atom, must satisfy (7.106) and cannot be $\emptyset$. Therefore there is at least one D_i that is not $\emptyset$. Such a D_i satisfies

$$D_i \rightarrow \alpha_i, \qquad D_i \neq \emptyset. \tag{7.133}$$

Furthermore, D_i cannot be equivalent to α_i, since, if it were, one would have

$$\alpha_i = \alpha_i \cap \beta_k. \tag{7.134}$$

This would mean $\alpha_i \rightarrow \beta_k$, whereas $\alpha_i \neq \beta_k$ by definition. This leaves the possibility that $\alpha_i < \beta_k$, which also is impossible because it contradicts the fact that β_k is an atom. Consequently D_i must satisfy, according to (7.133),

$$D_i < \alpha_i \quad \text{and} \quad D_i \neq \emptyset. \tag{7.135}$$

This contradicts the fact that α_i is an atom. Therefore we conclude that $(\alpha_1, \alpha_2, \cdots, \alpha_n)$ and $(\beta_1, \beta_2, \cdots, \beta_m)$ must be identical. It may be anticipated here that the role played by the distributive law in (7.132) cannot be replaced by the modular law. Hence the uniqueness of the set of atoms is not a property of a modular lattice.

Now take any arbitrary element X of $\mathcal{B}$; then we have

$$X \cap \alpha_i = \alpha_i \quad \text{or} \quad \emptyset, \tag{7.136}$$

for, if $X \cap \alpha_i$ were neither α_i nor $\emptyset$, the quantity $X \cap \alpha_i$ could be taken as D of (7.117). Hence any X can be written as a disjunction of a certain number of atoms:

$$X = \bigcup \alpha_i, \tag{7.137}$$

where i should be taken for those α_i for which $\alpha_i \cap X = \alpha_i$. Expression (7.137) can be obtained by making the disjunction of (7.136) for all i. The left side gives, by the distributive law,

$$\bigcup_{i=1}^{n} (X \cap \alpha_i) = \left(\bigcup_{i=1}^{n} \alpha_i\right) \cap X = \square \cap X = X, \tag{7.138}$$

whereas the right side gives

$$\bigcup_{i}^{\text{some}} \alpha_i, \tag{7.139}$$

where the index i runs over those for which (7.136) is not $\emptyset$. In a general modular lattice, the "expansion" (7.137) is not always possible for a given X and for a given set of atoms $\{\alpha_i\}$.

It is sometimes helpful to give a systematic label to each member of the Boolean lattice. This can be done by assigning an n-digit binary number to each member such that a 1 occupying the ith digit means that the ith atom is included in the member in the sense of (7.139) and a 0 occupying the ith digit means the opposite. This makes it immediately clear that the total number of members of the lattice with n atoms is 2^n. A permutation of atoms results only in relabeling of members; the lattice remains invariant. (We refer to this fact later as invariance for atomic permutation.)

Let us consider as an exercise the case in which $\mathcal{B}$ contains two elements, A and B. If $\mathcal{B}$ contains A and B it automatically contains all the combinations that can be formed from A and B with the use of three operations: negation, disjunction, and conjunction. Let us assume that $\mathcal{B}$ contains nothing other than these members that can be formed from A and B. The maximum number of such combinations that are not identically equivalent is 16. (We presently see an easy way to prove this fact.) If none of the four possible relations or constraints, $\neg A \cap \neg B = \emptyset$, $A \cap B = \emptyset$, $A \to B$, and $B \to A$, is satisfied, then $\mathcal{B}$ actually contains 16 different members. If one and only one of them is satisfied, $\mathcal{B}$ contains 8 different numbers. If two and only two of them are satisfied, the number becomes 4. If exactly three of them are satisfied, the number becomes 2: $\emptyset$ and $\square$. It is impossible to satisfy all four of them. In the nondegenerate case, that is, the case when $\mathcal{B}$ actually contains the maximum number (16) of different members, there are 24 maximal chains of inclusion. The four members, $A \cap B$, $\neg A \cap \neg B$, $A \cap \neg B = A - B$, and $\neg A \cap B = B - A$, are the atoms. Figure 7.14 represents this nondegenerate case graphically. The symbol $\uplus$ used in the graph is called "exclusive or" and means

$$\begin{aligned} A \uplus B &= (A \cup B) \cap (\neg A \cup \neg B) = (A \cup B) - (A \cap B) \\ &= (A - B) \cup (B - A). \end{aligned} \tag{7.140}$$

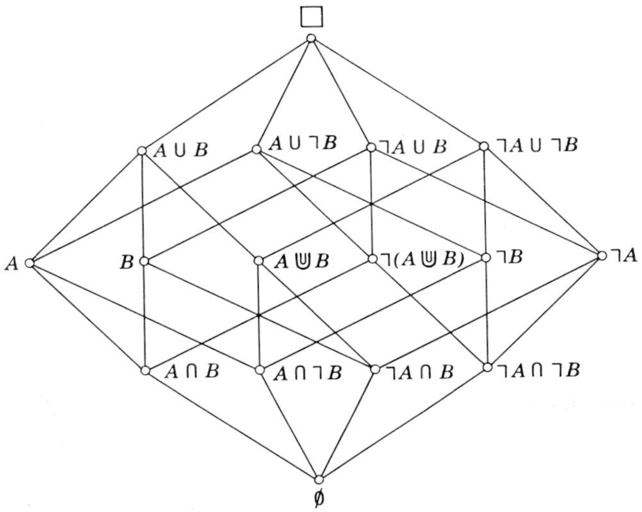

Figure 7.14 Any one of $24 = 4!$ upward paths starting from $\emptyset$ and ending with $\square$ is a maximal chain of inclusion. The atoms are $A \cap B$, $A \cap \neg B$, $\neg A \cap B$, and $\neg A \cap \neg B$. A member, say, X, consists of those atoms through which X can be connected to $\emptyset$. The network has a rotational symmetry for angle 180°. This rotation corresponds to the reciprocal transformation (7.31) and (7.100).

(The last equality holds only in a Boolean lattice.) The disjunction $A \cup B$ is usually read "A or B," but more precisely it means "A or B or both." The "exclusive or" $A \uplus B$ excludes the third possiblity "both." It may be noted as

$$A \uplus B = \neg A \uplus \neg B, \qquad \neg(A \uplus B) = (A \cap B) \cup (\neg A \cap \neg B). \qquad (7.141).$$

The complemented Boolean lattice composed of 16 nonequivalent elements thus formed starting from A and B is called the complemented Boolean lattice "generated" by A and B. Further applications of the three operations on these 16 elements do not produce any new element. Thus the generated lattice is "closed" for the three basic operations. Figure 7.15 shows the case in which there is one constraint, $A \rightarrow B$. In this case we have $A \cap \neg B = \emptyset$, which means that one of the four atoms of the nondegenerate case collapses. As a result, each node in Figure 7.15 corresponds to two nodes in Figure 7.14.

Coming back to the general discussion, we define the "dimension" $D(X)$ of a proposition X as the number of atoms out of which X is built:

$$D(X) = \text{number of } \alpha\text{'s in } X. \qquad (7.142)$$

In other words, $D(X)$ is the number of 1's in the binary labeling scheme mentioned before. The dimension of $\square$ is the total number n of atoms in $\mathfrak{B}$,

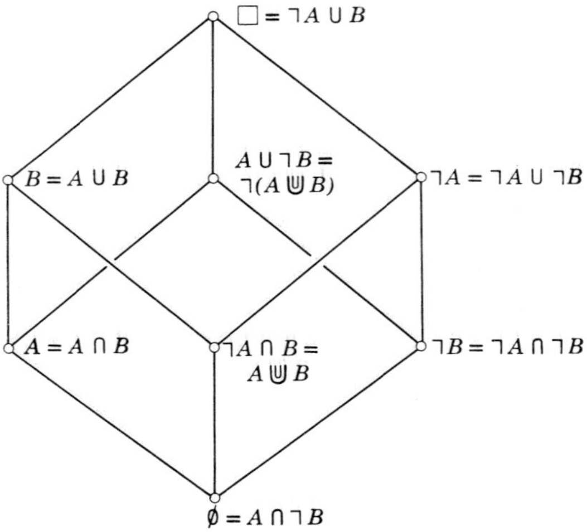

Figure 7.15 If there exists a relation $A \rightarrow B$, that is, $A \cap \neg B = \emptyset$, Figure 7.14 reduces to this simpler diagram, which is structurally identical to Figure 7.1*a*. We can see in Figure 7.2 that there are only three elementary domains in the corresponding Euler diagram (i.e., atoms), A, $B - A$, and $\neg B$, out of which any other domain can be built by disjunction. This means that A, $B - A$, and $\neg B$ are the atoms here. There are $6 = 3!$ maximal chains of inclusion.

and the dimension of $\emptyset$ is zero:

$$D(\square) = n, \qquad D(\emptyset) = 0. \tag{7.143}$$

$D(\square)$ is called the total dimension of the Boolean lattice under consideration. In Figures 7.14 and 7.15 we can see that the dimension corresponds to the height of the level to which an element belongs, calling the lowest level the 0th level. We may also note that $D(X \cap Y)$ must be the number of atoms common to X and Y whereas $D(X \cup Y)$ must be the number of atoms included in either X or Y or both. Hence, if we add these two, we obtain the number of atoms included in X only plus the number of atoms included in Y only plus twice the number of atoms included in both X and Y; that is, the sum of the number of atoms in X plus the number of atoms in Y. Hence

$$D(X) + D(Y) = D(X \cap Y) + D(X \cup Y). \tag{7.144}$$

The dimension function $D(X)$ is not the only function that can satisfy the relation of the type (7.144). In the Euler model of a Boolean lattice, if we can take the area of the domain corresponding to a lattice element, it certainly satisfies the relation (7.144). Even the spherical model can satisfy (7.144) by assigning values 0, 1, 2, and 3, respectively, to $\emptyset$, a pair of diametrically opposite points, a great circle, and the entire spherical surface.

More importantly, the concept of probability is mathematically defined as a non-negative function satisfying (7.144) with two auxiliary conditions $p(\emptyset) = 0$ and $p(\Box) = 1$. In the case of the probability defined on a lattice, the lattice need not necessarily be finite, as was assumed in (7.144) but is usually assumed to be Boolean. The concept of probability in Boolean lattices is discussed in the next section. The necessary modification of the concept of probability in the case of non-Boolean lattices is discussed in Chapter 9.

Coming back to the finite Boolean lattice, we repeat that each proposition makes a one-to-one correspondence with the set of atoms included in the proposition, in such a way that $X \cap Y$ corresponds to the set of atoms included in both X and Y, $X \cup Y$ corresponds to the set of atoms included in either X or Y or both, and $\neg X$ corresponds to the set of atoms not included in X. The relation $X \rightarrow Y$ means that if an atom is included in X it is included in Y. We can generalize this picture a little more by substituting for each atom a fixed collection of objects or points. Then $X \cap Y$ is the set of objects (points) included in both X and Y, and $X \cup Y$ is the set of objects (points) included in either X or Y or both. The implication $X \rightarrow Y$ means that if an object (point) belongs to X it belongs to Y. This is the basis of the set-theoretical interpretation of Boolean logic, as we explained earlier (see Postulate 7.1).

This set-theoretical interpretation makes it easy to introduce the so-called truth values of a proposition. The proposition $\Box$ corresponds to a set of all points P under consideration, and each proposition A corresponds to its subset. The "characteristic function" corresponding to a proposition A is defined as a function of point P, such that its value is 1 if the P belongs to the set that corresponds to A and is 0 otherwise.

$$\begin{aligned} f_A(P) &= 1 \quad \text{if} \quad P \in A \\ &= 0 \quad \text{if} \quad P \notin A. \end{aligned} \tag{7.145}$$

The implication $A \rightarrow B$ then means that

$$\text{if} \quad P \quad \text{is such that} \quad f_A(P) = 1 \quad \text{then} \quad f_B(P) = 1. \tag{7.146}$$

Hence $A = B$ is equivalent to $f_A(P) = f_B(P)$ for all P. It is easy to see from the definition of the characteristic function and from the set-theoretical meaning of the conjunction that the conjunction $A \cap B$ has the characteristic function

$$f_{A \cap B}(P) = f_A(P) \cdot f_B(P). \tag{7.147}$$

Similarly, the disjunction $A \cup B$ has the characteristic function

$$F_{A \cup B}(P) = f_A(P) + f_B(P) - f_A(P) f_B(P), \tag{7.148}$$

and the negation $\neg A$ has the characteristic function

$$f_{\neg A}(P) = 1 - f_A(P). \tag{7.149}$$

One can see that the characteristic function $f_A(P)$ considered as a function of A for a fixed P is a special case of the probability function of A, because we have not only

$$f_A(P) + f_B(P) = f_{A \cap B}(P) + f_{A \cup B}(P), \tag{7.150}$$

satisfying (7.144), but also

$$f_\emptyset(P) = 0 \quad \text{and} \quad f_\square(P) = 1 \tag{7.151}$$

for any arbitrarily fixed point P.

It may be noted that if we use (7.145) for the spherical model, taking P as a point on the sphere, (7.146) and (7.147) can be upheld but (7.148) and (7.149) have to be discarded.

Equations (7.147), (7.148), and (7.149) show that the characteristic function of the resulting proposition obtained by the use of three basic operations, negation, conjunction, and disjunction, from two propositions A and B can be expressed as a function of the characteristic functions of A and B. Therefore the values of the resulting characteristic function are determined by the values of the characteristic functions of A and B. Since each characteristic function f_A and f_B can take two possible values, there are four and only four possible cases. The values of the resulting characteristic function listed according to these four cases are called a truth value table. Tables 7.1 and 7.2 are the truth value tables for $A \cap B$ and $A \cup B$. Very often the value 1 is called "true" in the sense that $f_A(P) = 1$ corresponds to the case in which P is an object (case) that makes proposition A true. Thus a truth value table is often written with the help of T and F instead of 1 and 0, respectively, where T and F stand for true and false. Table 7.1, which shows that $f_{A \cap B} = 1$ if and only if $f_A = 1$ and $f_B = 1$, can be read as "$A \cap B$ is true if and only if A is true and B is true."

Table 7.1. Truth Value Table for $C = A \cap B$

C is true, that is, $f_{A \cap B} = 1$, if and only if A is true, that is, $f_A = 1$, and B is true, that is, $f_B = 1$.

f_A	f_B	$f_{A \cap B}$
0	0	0
0	1	0
1	0	0
1	1	1

Table 7.2. Truth Value Table for $C = A \cup B$

C is true, that is, $f_{A \cup B} = 1$, if and only if either one or both of A and B are true, that is, if $f_A = 1$ and/or $f_B = 1$.

f_A	f_B	$f_{A \cup B}$
0	0	0
0	1	1
1	0	1
1	1	1

These considerations make it easy to derive the result previously used, namely, the fact that no more than 16 different propositions can be built from A and B with the help of negation, disjunction, and conjunction, since, according to the foregoing remark, the characteristic function f of any combination built out of A and B is a function of the characteristic function f_A of A and the characteristic function f_B of B. Hence there are four and only four possible cases, and f can take 0 or 1 in each case. Therefore, according to the choice of 0 and 1 in each of the four cases, there can be $2^4 = 16$ different f's and no more. If two propositions are equivalent, the values of the f's must be the same. If there are more than four cases, the coincidence of the values of f in four cases does not guarantee the equivalence of propositions, but if it is known that there are only four cases then the coincidence of the values of f implies the equivalence of propositions. It is easy to see that what we called atoms are characterized by the fact that there is one and only one entry 1, but three 0's in the corresponding column. More generally, the dimension D of a proposition is the number of 1's in the column of the corresponding truth table. Figure 7.14 gives 16 possible different (i.e., nonequivalent) propositions that can be built from A and B. The particular case we discussed in Figure 7.15 is characterized by the condition $A \to B$, which is equivalent to eight relations: $A \cap \neg B = \emptyset$, $\neg A \cup B = \square$, $A = A \cap B$, $\neg A = \neg A \cup \neg B$, $B = A \cup B$, $\neg B = \neg A \cap \neg B$, $\neg A \cap B = A \uplus B$, and $A \cup \neg B = \neg(A \uplus B)$ [see (7.98) and (7.99)]. This reduces the number of nonequivalent members from 16 to 8.

Among these 16 expressions the combination $\neg A \cup B$ has a special name, "material implication," and is expressed with the help of a connective "horseshoe,"

$$A \supset B = \neg A \cup B. \tag{7.152}$$

This is a proposition built from A and B, not a relation between two propositions. However, if it is equivalent to $\square$, A implies B. More precisely,

$$\begin{array}{llll} \text{if} \quad A \to B & \text{then} \quad A \supset B = \square & \text{unconditionally;} \\ \text{if} \quad A \supset B = \square & \text{then} \quad A \to B & \text{if the distributive law holds} \end{array} \tag{7.153}$$

[see (7.99)]. Table 7.3 gives the truth values of the material implication $A \supset B$. The logicians' favorite notation, $A \supset B$ for $\neg A \cup B$, is highly confusing, however, since the symbol $\supset$ is used in set theory in the sense that $X \supset Y$ means that every member of Y is a member of X. Hence the set-theoretical interpretation of $A \to B$ becomes $A \subset B$, and this is exactly the opposite of the logicians' notation $A \supset B$.

A still more confusing symbol is a connective called equivalence, which is sometimes denoted by $\equiv$, but in order to avoid confusion with the relation

Table 7.3. Truth Value Table for the Material Implication $C = A \supset B$

C is false if and only if A is true and B is false. If you draw Euler's diagram in such a way that there does not exist a domain corresponding to this last case, then you obtain a diagram corresponding to $A \rightarrow B$.

f_A	f_B	$f_{A \supset B}$
0	0	1
0	1	1
1	0	0
1	1	1

symbol $=$, we use $\approx$ here. This connective is defined by

$$A \approx B = (A \supset B) \cap (B \supset A) = \urcorner(A \uplus B), \tag{7.154}$$

and

$$\begin{aligned} &\text{if} \quad A = B \quad \text{then} \quad A \approx B = \square \quad \text{unconditionally;} \\ &\text{if} \quad A \approx B = \square \quad \text{then} \quad A = B \quad \text{if the distributive law holds.} \end{aligned} \tag{7.155}$$

(Actually, in the modular lattice $A \approx B = \square$ also implies $A = B$.) We avoid $\supset$ and $\approx$ as much as possible.

A very interesting connective is what is called the Sheffer stroke, because we can express any other connectives (including negation) in terms of this symbol, showing that it is actually not necessary to use three different symbols to express all the Boolean functions. Its definition is

$$A \mid B = \urcorner(A \cap B), \tag{7.156}$$

from which follows

$$\begin{aligned} \urcorner A &= A \mid A, \\ A \cup B &= (A \mid A) \mid (B \mid B), \\ A \cap B &= (A \mid B) \mid (A \mid B). \end{aligned} \tag{7.157}$$

Another connective in terms of which all others can be expressed is the "neither-nor": $A \parallel B = \urcorner(A \cup B)$. The reader can check this for exercise.

Another source of confusion is that in speaking of the completed Boolean lattice generated by A and B, what we call atoms in accordance with the terminology of lattice theory are $A \cap B$, $\urcorner A \cap B$, $A \cap \urcorner B$, and $\urcorner A \cap \urcorner B$, whereas philosophers often call A and B atoms and term molecules any other expressions containing A, B, and at least one of $\urcorner$, $\cap$, and $\cup$. A further confusion of lesser seriousness stems from the fact that measure theory calls atoms only those set-theoretical atoms whose probability measure is not zero.

The method of characteristic function and truth value table can easily be extended to the case in which the number of starting propositions (in the above we considered A and B) is more than two. If we start with r propositions, there are 2^r different cases, depending on the value of each of the r propositions. Correspondingly, there are 2^r atoms, because each atom has only one entry 1, which could be placed in any one of the 2^r possibilities. There are $\binom{2^r}{2}$ different functions of dimension 2, because this is the number of different ways in which two 1's can be put in 2^r different places. Similarly, $\binom{2^r}{s}$ is the number of functions of dimension s. Hence the total number of members of the lattice generated by r starting propositions is $\sum_{s=0}^{2^r} \binom{2^r}{s} = (1 + 1)^{2^r} = 2^{2^r}$. This, of course, is meant for the case in which there are no constraints among the starting propositions.

7.3. FORMAL CONCEPT OF PROBABILITY

There is a great deal of truth in the assertion maintained by various people, though admittedly on different grounds, that the if-then relation is the most basic form in human thinking. In fact, we showed in Section 7.1 that the entire logical system, except the distributive law, can be derived solely from the notion of implication (provided the unique existence of conjunction and disjunction).† Not only the explicit logical laws, but even the internal structure of a single statement, can be fitted into the form of an implicational relation. Indeed, the sentence "b is Q" can be understood as meaning that if a variable (undefined object) is b then it is Q. C. S. Peirce probably was the first to emphasize this point of view. He wrote, "I have maintained since 1867 that there is one primary and fundamental logical relation, that is illation . . . . A proposition, for me, is but an argument divested of the assertoriness of its premise and conclusion. That makes every proposition a conditional proposition at bottom" [P-2].

In ordinary life, however, one seldom knows with absolute certainty that if A then B. All one can allow oneself as a guide to one's actions is at best a knowledge of the type "if A then probably B," or "if A then probably

† The notion of negation is not derivable from implication either. As we have seen, however, the negation $\urcorner A$ of A is a new name $\urcorner A$ assigned to an element already existing in a certain type of implicational lattice.

not *B*," or "if *A* then more probably *B* than *C*." Thus we are led to consider the notion of conditional probability (or its crude prototype) as the most primary and basic form of thought. It is important in this connection that conditional probability precedes unconditional probability. In fact, the latter can be conceived as a special case of the former, in which the condition is a constant truth or tautology. There is a strong argument for repudiating the notion of unconditional probability. In talking about the probability of a proposition *B*, if we knew (or determined the truth or falsehood of) all the facts relevant to *B*, the occurrence or nonoccurrence of *B* would be already determined by these facts and there would be hardly any room for probabilistic guessing. On the other hand, if nothing relevant were known except logical tautologies, the value of probability could not be decided at all, except perhaps by counting the possible cases the language happens to provide. For instance, if the language provides snow and no snow as the only predicates of weather, and if no factual information is allowed, the best (or the worst, but anyway, the only) thing one could do is attach probability $\frac{1}{2}$ to each of the two linguistically conditioned possibilities, snow and no snow, for any given day. But this kind of probability can hardly be a practical guide in life, and it is questionable whether it deserves the name probability in any sense. The true usefulness of probability resides precisely in the intermediate domain between these two extremes, and the value of probability depends critically on the relevant facts that are selectively taken into account. Hence it is not an exaggeration to say that there is no probability but conditional probability.

To avoid misunderstanding, it should be emphasized that to say that probability depends on the "conditions" is not to say that probability is determined uniquely by the "conditions." This remark is important because an argument similar to the one in the foregoing paragraphs is sometimes used to justify what is often called the Keynesian view of probability (see [K-11], page 11). According to a strict formulation of this view, between any two propositions, taken premise and conclusion, there holds a uniquely determined "probability relation," which can be numerically represented. The argument as developed in the foregoing paragraphs can be used to uphold the importance of conditional probability over the so-called unconditional probability but obviously cannot be used to reinforce the viewpoint that any two propositions automatically determine the numerical value of the conditional probability between them. In fact, one of the main purposes of this book is to demonstrate that conditional probabilities of the type called "retrodictive" are not determined by the two propositions involved.

The foregoing consideration of the importance of conditional probability leads us to speculate on the possibility of a future theory, in which the concept of conditional probability is the starting point from which we can derive the unconditional probability, on the one hand, and logic, on the other, as secondary concepts. From such a perspective it is rather disconcerting that the present-day mathematical theory of probability has to go the opposite way. We have to assume first a set of propositions in which a certain logical structure exists. On this set we define unconditional probabilities, from which conditional probabilities are further derived as a secondary concept. In spite of this undesirable feature, we now briefly review such an orthodox mathematical theory of probability, first from a formal point of view, relegating the problem of interpretation of the formal relations to the next section. For an attempt to derive logic from probability, see [W-31, W-31a].

As we see later, the traditional concept of probability is intimately entwined with the assumed distributive law in the logical substratum. If the logical structure is distributive, it becomes isomorphic with a set of subsets of a collection of points (see Postulate 7.1). For this reason the mathematical theory of probability is based on a set of subsets (Borel field) and is applicable immediately to a Boolean lattice of propositions. If we want to abandon the distributive law in logic and replace it by a weaker law, we have to divorce the foundation of the theory of probability from the point of view of a set of subsets. The concept of probability, as a result, will suffer from a new restriction in usage that did not exist in the Boolean case. We discuss such a non-Boolean case in Chapter 9 in connection with quantum mechanics. The present chapter is strictly Boolean.

The modern theory of probability, which owes its birth to Kolmogorov [K-12], is based on the so-called measure theory, which, so to speak, developed into a whole branch of mathematics by the mid-twentieth century, covering areas other than probabilities, and the average reader of the present book cannot be expected to be familiar with it. The domain in which measure theory becomes really powerful usually involves problems of convergence, differentiation, and integration and requires careful and lengthy preparation before one can enter into any discussion in that area. Fortunately, however, the conceptual status of the quantity "probability" in this theoretical scheme is relatively easy to understand, particularly when the reader is already familiar with lattice theory, because the Borel field on which the probability is defined in measure theory differs essentially from a Boolean lattice only in that an enumerably infinite number of conjunctions and disjunctions can be applied. We try in the following to introduce the concept of probability more or less in accordance with the spirit, if not the letter, of measure theory. (The reader who is seriously interested in measure theory is referred to a textbook like [H-1].) There is some repetition of material from the previous two sections, but we must know that we are dealing with properties of "sets" rather than logic in this section. The reader who is interested in the conceptual basis of probability should not be bothered by details of mathematical argument in the following paragraphs but should try to understand the rough gist of the derivation.

We start with some of the elementary facts in set theory. We consider a set X of objects x, or equivalently an abstract space X consisting of points x. We write $X = \{x\}$. This X corresponds to the $\square$ in the case of logic, and each x corresponds to an object (or a case) that makes some propositions true and others false. The number of x's in X may be finite, countably (enumerably) infinite, or continuously many. X can be the set of pebbles in Example 7.3. A simple example of the continuous case is one in which X is the set of all points lying between 0 and 1. In classical physics the state of a system of N monatomic molecules (like a certain amount of helium gas) can be described by $3N$ space coordinates and $3N$ conjugate momenta. The Euclidean space consisting of these $6N$ coordinates is the phase space of this system (sometimes called Γ-space). Each point in this space represents the microscopic state (theoretically the minutest description) of the system. Each point can then be considered as an x, and the set of all points or the entire Γ-space can be taken as X. In information theory we sometimes consider an infinitely long (to both sides) chain of binary numbers such as

$$\cdots a_{-2}, a_{-1}, a_0, a_1, a_2 \cdots, \tag{7.158}$$

where each a_i can be 0 or 1. Each of such infinitely long chains can be taken as x. Each chain x can then be labeled with a binary number c, defined by

$$c = 0.\ a_0 a_{-1} a_1 a_{-2} a_2 \cdots. \tag{7.159}$$

This c will take any real number in the domain lying between zero and unity. This makes it intuitively clear that the number of x's under consideration is continuous although the number of digit positions is countable. X is then the set of all such x's.

Now we take a subset A of X and write $A \subset X$ to express the fact that if an x belongs to A ($x \in A$) it also belongs to X ($x \in X$). In logic A is a proposition and the x's in A correspond to the individual objects (cases) that make A true.† Similarly, if two subsets, A and B, of X are such that $x \in A$ implies $x \in B$, we write $A \subset B$. We write $A = B$ if $A \subset B$ and $B \subset A$. Obviously, if $A \subset B$ and $B \subset C$, $A \subset C$. X itself is a subset of X, but those subsets of X that are not X itself are called proper subsets of X. Besides X and its proper subsets, we have to consider an empty set $\emptyset$, which is supposed to satisfy $\emptyset \subset A$ for any subset A of X. We are often interested in a set of some subsets A of X, and to make a distinction between a set and a set of sets clear, a set of subsets A is often referred to as a "class," while the subsets A themselves are often called simply "sets."

† As mentioned in Section 7.2, the set-theoretical $A \subset B$ corresponds to the logical $A \supset B = \square$, that is, a horseshoe placed in the opposite direction.

The conjunction (intersection) and disjunction (union) of sets can now be defined as $A \cap B = \{x \mid x \in A \text{ and } x \in B\}$ and $A \cup B = \{x \mid x \in A \text{ or } x \in B\}$. Obviously $A \cap B \subset X$ and $A \cup B \subset X$. It may be noted that in the spherical model discussed in Section 7.1 the definition of $A \cap B$ given here still works whereas the definition of $A \cup B$ given here does not. This is because not all subsets of points on the sphere are allowed to be a member of the class in the case of the spherical model. We can define the "negation" at this stage and then introduce the "difference," or, alternatively, we can reverse the order. The difference (or relative complement of B with respect to A), $A - B$, is defined as $A - B = \{x \mid x \in A \text{ and } x \notin B\}$, where $\notin$ is to be read as "does not belong to."† Of course, $A - B \subset X$. The negation (or complement) $\urcorner A$ of A is then obviously defined as $\urcorner A = X - A$. We have, as before, $A \cup \urcorner A = X$, $A \cap \urcorner A = \emptyset$, and $\urcorner(\urcorner A) = A$ for any A. In general, we can always write $A - B = A \cap \urcorner B$ and $(A - B) \cup (A \cap B) = A$. It may be noted that the second of these two expressions follows from the first only with the help of the distribution law, which is the reason why we did not use the concept of difference in Sections 7.1 and 7.2. When $B \subset A$ the difference $A - B$ is called a proper difference. If $A - B$ is a proper difference, we can use the following symmetrical transcription: $C = A - B$ implies $B = A - C$. The union of two differences $A - B$ and $B - A$, that is, $(A - B) \cup (B - A) = (A \cup B) - (A \cap B)$, is called the symmetrical difference $A \uplus B$, which corresponds to the "exclusive or" in terms of propositions. It may be readily seen that $A \uplus B = \emptyset$ is equivalent to $A = B$.

The main difference between the present case and the Boolean lattice we studied before is that we are allowed to make the conjunction and disjunction of countably many (i.e., a countably infinite number of) sets, and the resulting set is still a subset of X. That is, if $A_1, A_2, A_3 \cdots$ are subsets of X, then $\bigcap_{i=1}^{\infty} A_i \subset X$ and $\bigcup_{i=1}^{\infty} A_i \subset X$. Correspondingly, we have here a generalization of the distributive law and De Morgan's law:

$$\bigcup_{i=1}^{\infty} (B \cap A_i) = B \cap \left(\bigcup_{i=1}^{\infty} A_i\right), \qquad \bigcap_{i=1}^{\infty} (B \cup A_i) = B \cup \left(\bigcap_{i=1}^{\infty} A_i\right) \tag{7.160}$$

and

$$\urcorner\left(\bigcap_{i=1}^{\infty} A_i\right) = \bigcup_{i=1}^{\infty} \urcorner A_i, \qquad \urcorner\left(\bigcup_{i=1}^{\infty} A\right) = \bigcap_{i=1}^{\infty} \urcorner A_i. \tag{7.161}$$

A simple consequence of (7.160) and (7.161) is that

$$\text{if} \quad A_i \subset A \ (i = 1, 2, \cdots) \quad \text{then} \quad \bigcap_{i=1}^{\infty} A_i = A - \bigcup_{i=1}^{\infty} (A - A_i). \tag{7.162}$$

† The negation in this clause belongs to the metalanguage.

Of course one can take in this formula $A = \bigcup_{i=1}^{\infty} A_i$ if one so chooses. The proof of (7.162) is as follows:

$$\bigcup_{i=1}^{\infty} (A - A_i) = \bigcup_{i=1}^{\infty} (A \cap \neg A_i) = A \cap \left(\bigcup_{i=1}^{\infty} \neg A_i\right)$$

$$= A \cap \neg\left(\bigcap_{i=1}^{\infty} A_i\right) = A - \bigcap_{i=1}^{\infty} A_i.$$

But since $A \supset A_i \supset \bigcap_{i=1}^{\infty} A_i$, the last member, $A - \bigcap_{i=1}^{\infty} A_i$, is a proper difference; hence we can interchange $\bigcup_{i=1}^{\infty} (A - A_i)$ of the first member and $\bigcap_{i=1}^{\infty} A_i$ of the last member, which gives (7.162). It may be noted that the conjunction (disjunction) of a finite number of sets can always be written as a conjunction (disjunction) of countably many sets, for $\bigcap_{i=1}^{\infty} A_i = \bigcap_{i=1}^{n} A_i$ if, for instance, $A_j \supset A_n$ for $j > n$.

It is important to distinguish the concept of a "set function" from that of a "point function." If a real value is assigned to each set under consideration (i.e., in a class), we speak of a set function and write $F(A)$ or similarly. If a real value is assigned to each point x in a set, we speak of a point function and write $f(x)$ or similarly. The "characteristic function," $\chi(A; x)$, of a given set A is a point function, which takes the value 1 for $x \in A$ and the value 0 for $x \in \neg A$. If we fix x in $\chi(A; x)$, this quantity is a set function of A. We already used the notion of characteristic function in Section 7.2. We are now prepared to define formally what is known as a Borel field.

DEFINITION 7.1. A Borel field $\mathcal{B}$ is a nonempty class of some subsets of X such that (a) if $B_1, B_2 \in \mathcal{B}$ then $B_1 - B_2 \in \mathcal{B}$; (b) if $B_1, B_2, B_3, \cdots \in \mathcal{B}$ then $\bigcup_{i=1}^{\infty} B_i \in \mathcal{B}$; and (c) $X \in \mathcal{B}$.

In this definition the expression $B_1, B_2 \in \mathcal{B}$, for instance, means $B_1 \in \mathcal{B}$ and $B_2 \in \mathcal{B}$. We sometimes write $(X, \mathcal{B})$ to denote both $\mathcal{B}$ and the entire set X on which $\mathcal{B}$ is defined. From this definition it follows that if B_1, B_2, $B_3, \cdots \in \mathcal{B}$ then $\bigcap_{i=1}^{\infty} B_1 \in \mathcal{B}$.

For the proof we need only recall the formula (7.162) previously obtained, $\bigcap_{i=1}^{\infty} A_i = A - \bigcup_{i=1}^{\infty} (A - A_i)$, where the right-hand side involves only the operations for which $\mathcal{B}$ is "closed" by definition. A class is said to be closed with respect to an operation if the resultant of the operation belongs to the class when the elements on which the operation works are known to belong to the class. It also follows easily from the definition that $\mathcal{B}$ is closed for complementation, because if $B \in \mathcal{B}$ then $\neg B = X - B$; hence $\neg B \in \mathcal{B}$. Since $\emptyset = \neg X$, $\emptyset \in \mathcal{B}$ also. It is easy to see that if a class $\mathcal{B}$ of sets is closed for countable union, countable intersection, and complementation, and if it includes X, then $\mathcal{B}$ is a Borel field. Indeed, $\mathcal{B}$ will then be closed for making the difference, satisfying (a) of Definition 7.1, since the difference can be defined in terms of intersection and complementation.

A Borel field is conceived as a class of sets of points, but if we ignore the points and consider the sets as elements, and equate X with $\square$, it is not different from the complemented Boolean lattice we considered previously except that the number of elements can be infinite and countable union is allowed.

The terms "ring," "algebra," and so on have various meanings, but the following is one of the accepted usages. A ring is a (nonempty) class of sets that is closed for *finite* union and difference formation. If a ring contains X as an element, it is an algebra. The complemented Boolean lattice we discussed in Sections 7.1 and 7.2 is an algebra in this sense if each element is considered as a set of points and $\square$ as X. Furthermore, if a class is closed for *countable* unions and difference formation, it is called a σ-ring, and if it also contains X as an element, it is a Borel field or, equivalently, a σ-algebra. (The prefix σ is an esoteric symbol indicating that the concept has something to do with countable unions.) When X is given there exists an infinite variety of Borel fields that can be defined on X. Of course, the class of all possible subsets of X is a special Borel field.

Suppose that we are given two Borel fields $(X, \mathcal{A})$ and $(X, \mathcal{B})$ defined on X. Consider a class of sets $\mathcal{C}$ of which each element is simultaneously an element of $\mathcal{A}$ and $\mathcal{B}$; that is, $\mathcal{C} = \{C \mid C \in \mathcal{A} \text{ and } C \in \mathcal{B}\}$. Such a class is called the intersection of $\mathcal{A}$ and $\mathcal{B}$. We write $\mathcal{C} = \mathcal{A} \wedge \mathcal{B}$. It is easy to see that $\mathcal{C}$ is another Borel field. In fact, if $C_1, C_2 \in \mathcal{C}$, then $C_1, C_2 \in \mathcal{A}$ and $C_1, C_2 \in \mathcal{B}$; hence $C_1 - C_2 \in \mathcal{A}$ and $C_1 - C_2 \in \mathcal{B}$, which implies that $C_1 - C_2 \in \mathcal{C}$. Similarly, if $C_1, C_2, C_3 \cdots \subset \mathcal{C}$, $\bigcup_{i=1}^{\infty} C_i \in \mathcal{C}$. Finally, X belongs to $\mathcal{C}$ because X is a common element of $\mathcal{A}$ and $\mathcal{B}$. Now consider any arbitrary class $\mathcal{D}$ of sets taken from X, which is not a Borel field. The smallest Borel field that contains $\mathcal{D}$ is called the Borel field $\mathcal{B}(\mathcal{D})$ generated by $\mathcal{D}$. First we note that there exists at least one Borel field that contains the given $\mathcal{D}$, namely, the class of all sets taken in X. Indeed, any element (set) belonging to $\mathcal{D}$ belongs to the class of all subsets of X, which is a Borel field. Now take all the possible Borel fields that contain $\mathcal{D}$ (so that every element of $\mathcal{D}$ is an element of each Borel field) and take the intersection of all these Borel fields. Then this intersection (which exists) is the smallest Borel field $\mathcal{B}(\mathcal{D})$ that contains $\mathcal{D}$. This is the "smallest" in the sense that every element of $\mathcal{B}(\mathcal{D})$ is an element of each one of the Borel fields, whose intersection defines $\mathcal{B}(\mathcal{D})$. The concept of the Borel field $\mathcal{B}(\mathcal{D})$ generated by an arbitrary class $\mathcal{D}$ may be intuitively understood as meaning the following. When an arbitrary class $\mathcal{D}$ is given, it is not guaranteed that the difference between its two elements and the (countable) union of its elements are members of $\mathcal{D}$. If they are not, we add them as new members to $\mathcal{D}$; we continue this procedure of augmenting the class until finally the difference formation and (countable) unions do not produce any new members. The final resulting class produced by this procedure is the Borel field $\mathcal{B}(\mathcal{D})$ generated by $\mathcal{D}$.

Now suppose that we are given two Borel fields $(X, \mathcal{A})$ and $(X, \mathcal{B})$ and consider the class $\mathcal{D}$ of sets, each of which is an element of $\mathcal{A}$ *or* an element of $\mathcal{B}$ (or an element of both $\mathcal{A}$ and $\mathcal{B}$). $\mathcal{D} = \{D \mid D \in \mathcal{A} \text{ or } D \in \mathcal{B}\}$. Thus $\mathcal{D} = \mathcal{A} \cup \mathcal{B}$, where $\cup$ here means a set-theoretical union of classes. Then there is no guarantee that $\mathcal{D}$ is a Borel field. But by the foregoing procedure we can produce the Borel field $\mathcal{C}(\mathcal{D})$ generated by $\mathcal{D}$. Such a Borel field generated by the union of $\mathcal{A}$ and $\mathcal{B}$ is denoted as $\mathcal{C} = \mathcal{A} \vee \mathcal{B}$, in contrast to $\mathcal{D} = \mathcal{A} \cup \mathcal{B}$. We use the same symbols $\cap$ and $\cup$ for sets and classes here, but the symbol $\vee$ applies only to classes, to indicate the generated Borel field.

We are now prepared to introduce the most important concept of "measure" to our consideration. It is helpful to consider this concept as a formal generalization that covers the two familiar notions of volume (length, area, etc.) and probability. First, these quantities are non-negative. Second, if we take two nonoverlapping domains B_1 and B_2 of a space, that is, if $B_1 \cap B_2 = \emptyset$, the volume of the combined domain $(B_1 \cup B_2)$ of the space will be the sum of the volumes of the two separate domains. In the same way, if two propositions B_1 and B_2 are known not to be the case simultaneously (i.e., if $B_1 \cap B_2 = \emptyset$), the probability that B_1 or B_2 will be true is just the sum of two probabilities, one for B_1 and the other for B_2. We considered only two elements, B_1 and B_2, but it may not be far-fetched to consider this situation to be the case even when a countably infinite number of B's are involved. Thus, we have the next definition.

DEFINITION 7.2. A measure $m(B)$ of the Borel field $(X, \mathcal{B})$ is a set function of the members of $\mathcal{B}$ such that (a) $m(B) \geq 0$; (b) if $B_i \in \mathcal{B}$ $(i = 1, 2, \cdots)$ and $B_i \cap B_j = \emptyset$ for $i \neq j$, $m(\bigcup_{i=1}^{\infty} B_i) = \sum_{i=1}^{\infty} m(B_i)$; and (c) $m(\emptyset) = 0$.

When $m(B)$ is given to $(X, \mathcal{B})$ we write $(X, \mathcal{B}, m)$ and call this triple a "measure space." As we have seen, the concept of countable union in the definition includes a finite union too. Any set function $F(B)$ that satisfies Property (b) of Definition 7.2, in which $m(B)$ is replaced by $F(B)$, is said to be σ-additive. [To be more precise, the case when $F(B)$ becomes both $-\infty$ and $+\infty$ must be excluded if $F(B)$ is to be called σ-additive.] A measure is thus a non-negative, σ-additive set function defined on $(X, \mathcal{B})$, with an additional condition that its value be 0 for the empty set $\emptyset$. In probability calculus the quantity probability is usually normalized so that $\Pr(\square) = 1$. For this reason, when $m(X) = 1$, we speak of a probability measure. If, a little more generally, $m(X) < \infty$ (i.e., $m(X)$ is finite), we speak of a finite measure. But any finite measure can be reduced to a probability measure merely by dividing every $m(B)$ by $m(X)$, so that there is no essential difference from the probability measure. However, restriction to a finite measure is often inconvenient, since, for example, we cannot apply this restriction to the concept of volume in an infinite space. In the example of the Γ-space (see above) it is usually assumed in statistical mechanics that the *a priori* probability that the state of a system will be in a certain domain of the Γ-space is *proportional* to the volume of this domain. Such a probability, if the available domain is not restricted by an additional condition, cannot be normalized. If we artificially defined $m(X) = 1$, the measure of all finite (in the usual sense) domains would become zero. To cope with such a situation a less restrictive concept than finite measure is needed, but without letting $m(B)$ be completely arbitrary. The situation here is that any domain (finite or infinite in the intuitive sense) can be covered by the union of countably many domains, each of which has a finite measure. This property, called σ-finiteness, is formally defined as follows: for any $B \in \mathcal{B}$ there exists a sequence $B_1, B_2, B_3, \cdots \in \mathcal{B}$ such that $B \subset \bigcup_{i=1}^{\infty} B_i$ and $m(B_i) < \infty$, $i = 1, 2, \cdots$. The usual concept of volume certainly satisfies the condition of σ-finiteness.

It seems worthwhile to interject an informal remark that shows how important a consequence the restriction on a "countable" union in the definition of probability entails. Suppose that we drop an infinitesimal ball on a line segment whose length is unity, along which the position is labeled by a real number from 0 to 1. Suppose further that the probability distribution along this segment is uniform using this coordinate. Such a probability coincides with what is usually called the Lebesgue measure. The Lebesgue measure of the set of points x satisfying $a \leq x < b$ is given simply by the length $b - a$. Let us write this in the form $m([a, b)) = b - a$, where the left bracket, [, shows that the left endpoint a is included while the right parenthesis,), shows that the right endpoint b is not included; thus $[a, b)$ represents the set of points in question. Then it is obvious that the Lebesgue measure of a set consisting of a single point (whose position is, say, c) is zero, since such a set can be expressed as $[c, c)$ and $m([c, c)) = c - c = 0$. If we drop the ball it has to land somewhere, say, at point c. Thus, an event of probability 0 happens. (In the real physical world the ball has a finite size and the measurement has a finite accuracy. Hence we can avoid this situation if we choose.) This is not very surprising. But consider the ball dropping on any one of the points whose position is expressible as a rational number between 0 and 1. There are infinitely many of them, spread out densely everywhere. But it is known that each rational number between 0 and 1 can be assigned an integer as its name. Let R stand for the set of all rational numbers. Then $m(R)$ can be expressed, according to Item (b) of Definition 7.2, as $\sum_{i=1}^{\infty} m(a_i) = \sum_{i=1}^{\infty} 0 = 0$, where a_i is a rational number whose label is i. Thus the probability (which coincides with the Lebesgue measure here) of all rational numbers is 0.

In other words, the measure of all real numbers between 0 and 1 is the same as the measure of all irrational numbers in the same interval.

Let us now derive some of the useful formulas implied by the definition of measure. First, we show that the famous formula of probability [see (1.9) and (7.144)],

$$m(A) + m(B) = m(A \cap B) + m(A \cup B), \qquad (7.163)$$

follows from Condition (b) of Definition 7.2 of the measure. By the use of the distributive law, we can write $A \cup B$ as a union of disjoint elements in the following three ways: $A \cup B = (A - B) \cup B = (B - A) \cup A = (A - B) \cup (B - A) \cup (A \cap B)$. [Recall the formula $(A - B) \cup (A \cap B) = A$ given earlier.] By the Condition (b) of the definition of measure, we thus obtain three equations: $m(A \cup B) = m(A - B) + m(B)$, $m(A \cup B) = m(B - A) + m(A)$, and $m(A \cup B) = m(A - B) + m(B - A) + m(A \cap B)$. Add the first two and subtract the third; then we obtain (7.163). Thus we obtain (7.163) from Condition (b) with the help of the distributive law. Conversely, if (7.163) is given, with the help of $m(\emptyset) = 0$ we can derive the finite additivity (not σ-additivity) because, if $A \cap B = \emptyset$, (7.163) becomes $m(A \cup B) = m(A) + m(B)$. This derivation does not involve the distributive law. In any event, this makes clear that insofar as the number of elements is finite, we can replace Condition (b) of the definition by (7.163). This is often done in the definition of the concept of probability. When we want to introduce the concept of probability in a nondistributive lattice, as in Chapter 9, it seems particularly advisable to start with (7.163) instead of Condition (b). This certainly could be done in such a case, but the difficulty is not entirely avoided by this choice alone, for a similar difficulty reappears when we come to the definition of conditional probability, and this time the difficulty cannot be avoided by such simple tactics.

From (7.163) follow various useful formulas in the measure (probability) theory. First, putting $B = \neg A$ and noting that $A \cap \neg A = \emptyset$ and $A \cup \neg A = X$, we get $m(\neg A) = m(X) - m(A)$, which equals $1 - m(A)$ if $m(X) = 1$. Next, putting $A = B_1 \cap \neg B_2$ and $B = B_1 \cap B_2$ in (7.163), we obtain $m(B_1) = m(B_1 - B_2) + m(B_1 \cap B_2)$. If $B_1 \supset B_2$, $B_1 \cap B_2 = B_2$ and $m(B_1) = m(B_1 - B_2) + m(B_2)$. Note that $B_1 - B_2$ is a proper difference here. From this follows an important theorem:

$$\text{if} \quad B_1 \supset B_2 \ (\text{i.e., } B_2 \rightarrow B_1) \quad \text{then} \quad m(B_1) \geq m(B_2). \qquad (7.164)$$

Many people consider this fact to be the minimum requirement that can be imposed on a quantity bearing the meaning of what may be called probability. Since $A \cup B \supset A \supset A \cap B$, it follows from (7.164) that $m(A \cap B) \leq m(A)$ and $\leq m(B)$ and that $m(A \cup B) \geq m(A)$ and $\geq m(B)$. It is an interesting problem (not for measure theory, but for a broader consideration) to ask if we can sharpen condition (7.164) a little more, namely, whether it is always possible to introduce a measure m on a given Borel field $(X, \mathcal{B})$ so that

$$\text{if} \quad B_1 \supset B_2 \quad \text{and} \quad B_1 \neq B_2 \quad \text{then} \quad m(B_1) > m(B_2). \tag{7.165}$$

This constitutes, in view of (7.164), the converse of the obvious theorem that directly follows from (7.164):

$$\text{if} \quad B_1 \supset B_2 \quad \text{and} \quad B_2 \supset B_1 \quad \text{then} \quad m(B_1) = m(B_2). \tag{7.166}$$

The condition (7.165), together with (7.166), suggests a possibility of defining the set-theoretical equality and inclusion from the concept of measure: if B_1 and B_2 are such that no measure can be defined violating $m(B_1) \geq m(B_2)$ then $B_1 \supset B_2$, and if B_1 and B_2 are such that no measure can be defined violating $m(B_1) = m(B_2)$ then $B_1 = B_2$. Furthermore, we see in Chapter 9 that the condition (7.165) plays a vital role in defining modular logic.

Let us go back to the standard measure theory and further discuss the properties of the probability measure. Unfortunately, we have to limit ourselves to cases in which the number of members of the Borel field is finite, because otherwise, in order to develop the theory further, we would have to introduce the notion of different types of convergence, integration, differentiation (Radon-Nikodym theorem), and so on, which would be much too lengthy. This means that we have to forgo explaining the most characteristic and most interesting part of measure theory, including the measure-theoretical counterpart of the notion of conditional probability.

We saw in Section 7.2 how we can define "atoms" α_i $(i = 1, 2, \cdots, n)$ when the number of elements in a Boolean algebra is finite, in such a way that any element can be expressed as a *finite* union of such atoms. In other words, in this case we can write $X = \bigcup_{i=1}^{n} \alpha_i$; $\alpha_i \neq \emptyset$; $\alpha_i \cap \alpha_j = \emptyset$ $(i \neq j)$; and for any A, if $A \in \mathcal{B}$, we can write $A = \bigcup_{i}^{\text{some}} \alpha_i$, where the union is taken over some of the i. Set-theoretically this is a partition of X into $\{\alpha_1, \alpha_2, \cdots, \alpha_n\}$. Another obvious way of defining an atom in $\mathcal{B}$ is as follows. Suppose that an element $A \in \mathcal{B}$ is such that if $B \in \mathcal{B}$ and $B \subset A$, $B = A$ or $B \neq \emptyset$. If this is the case and if $A \neq \emptyset$, A is an atom, because there is no element except $\emptyset$ smaller than A within A. All this has been done without reference to the concept of measure. After introducing the measure, we are no longer interested in sets whose measure is zero. Thus we may replace the concept of $\emptyset$ by that of a set of measure 0, and the equality of two sets $A = B$, by $m(A \uplus B) = 0$. A finite measure space $(X, \mathcal{B}, m)$ can be decomposed into a finite number of "atoms" α_i $(i = 1, 2, \cdots, n)$ in such a way that $m(X \uplus \bigcup_{i=1}^{n} \alpha_i) = 0$ and, for any $A \in \mathcal{B}$, $m(A \uplus \bigcup_{i}^{\text{some}} \alpha_i) = 0$. An element A of $\mathcal{B}$ with $m(A) \neq 0$ is called a measure-theoretical "atom" if and only if the conditions $A \in \mathcal{B}$, $B \in \mathcal{B}$, and $B \subset A$ imply $m(B) = m(A)$ or $m(B) = 0$. It is easily seen that a measure-theoretical atom is a set-theoretical atom whose measure is not zero. When a measure space $(X, \mathcal{B}, m)$ is finite it is actually unnecessary to allow each atom to contain many x's. We might as well assign only one x to each atom (which reduces the Borel field to a Boolean lattice), giving a finite measure to this one x, and replace any set of measure 0 by an empty set $\emptyset$. Then X consists of only n x's. An alternative way, of course, is to let each atom contain more than one, possibly an infinite number of, x's and let even a set of measure 0 contain a finite or infinite number of x's. This practice is more common.

Suppose that $\mathcal{B}$ is decomposable into a finite number of set-theoretical atoms, α_i, $i = 1, 2, \cdots, n$; then it follows from the definition of measures that $m(\alpha_i) \geq 0$ and $\sum_{i=1}^{n} m(\alpha_i) = 1$. Conversely, if we have these two conditions and put $m(B) = \sum m(\alpha_i)$ (summation over those $\alpha_i \subset B$) for any $B \in \mathcal{B}$, we can derive the familiar definition of the probability measure. The relation $m(\emptyset) = 0$ can be interpreted as being included in the formula $m(B) = \sum m(\alpha_i)$, where there are no terms on the right-hand side to be added. If we group these α's into a class of disjoint and complete sets, E_i, $i = 1, 2, \cdots, n^*$, we have $E_i \cap E_j = \emptyset$ and $\bigcup_{i=1}^{n^*} E_i = \square$ (the conditions of logical spectrum), and $m(E_i) \geq 0$, $\sum_{i=1}^{n^*} m(E_i) = 1$, and $m(E_i \cup E_j \cup \cdots \cup E_k) = m(E_i) + m(E_j) + \cdots + m(E_k)$ $(i, j, \cdots, k$ are different). The latter are the usual definitions of probabilities, and we also used them extensively in the earlier chapters of this book. Insofar as we are interested in $\emptyset$ and only those sets that can be formed by disjunction of some of the E's, we might as well consider the whole matter as a lattice whose atoms are E's themselves. Then we come back to the first case. Thus we see the relation between the usual notion of probability and the measure-theoretical approach. It may also be added that what is usually called a stochastic variable, say, Q, is a set function of the E's, for it attaches a numerical value $Q(E_i)$ to each of disjoint and complete events E_i. The mean value of Q is then $\langle Q \rangle = \sum_i Q(E_i)m(E_i)$.

It is instructive to consider the structure of $\mathcal{C} = \mathcal{A} \vee \mathcal{B}$, when $\mathcal{A}$ and $\mathcal{B}$ are finite and have respectively set-theoretical atoms α_i $(i = 1, 2, \cdots, n)$ and β_j $(j = 1, 2, \cdots, m)$. The first step to obtain $\mathcal{C}$ is to form the class $\mathcal{D} = \mathcal{A} \cup \mathcal{B} = \{D \mid D \in \mathcal{A} \text{ or } D \in \mathcal{B}\}$. But this $\mathcal{D}$ is not closed to the operations of difference forming and union forming. So, starting from $\mathcal{D}$, we add to the members of the class new members obtained by difference forming and union forming from the present members, until finally no new members are produced by these procedures. Take two D's, D_1 and D_2, which both belong to either $\mathcal{A}$ or $\mathcal{B}$. Then the difference $D_1 - D_2$ belongs again to $\mathcal{A}$ or $\mathcal{B}$ and hence to $\mathcal{D}$. If $D_1 \in \mathcal{A}$ and $D_2 \in \mathcal{B}$, however, the difference $D_1 - D_2 = D_1 \cap \neg D_2$ does not necessarily belong to $\mathcal{A}$ or $\mathcal{B}$. Let D_1 and $\neg D_2$ be expressed as $D_1 = \bigcup_i^{\text{some}} \alpha_i$ and $\neg D_2 = \bigcup_j^{\text{some}} \beta_j$. Then $D_1 - D_2 = (\bigcup_i^{\text{some}} \alpha_i) \cap (\bigcup_j^{\text{some}} \beta_j) = \bigcup_{ij}^{\text{some}} (\alpha_i \cap \beta_j)$. Similarly, we can express

$$D_1 \cup D_2 = \left(\bigcup_i^{\text{some}} \alpha_i\right) \cup \left(\bigcup_j^{\text{some}} \beta_j\right) = \bigcup_{ij}^{\text{some}} (\alpha_i \cup \beta_j).$$

But $\alpha_i \cup \beta_j = (\alpha_i - \beta_j) \cup (\alpha_i \cap \beta_j) \cup (\beta_j - \alpha_i)$, where each term can again be expressed as a union of terms of the type $\alpha_i \cap \beta_j$. Hence $D_1 \cup D_2 = \bigcup_{ij}^{\text{some}} (\alpha_i \cap \beta_j)$. In the same way the union of any number of D's can be expressed as a union of terms of the type $(\alpha_i \cap \beta_j)$. Actually, each $\alpha_i \cap \beta_j$ itself has to be added as a new member, since each $\alpha_i \cap \beta_j$ can be formed by $D_1 - D_2$ with $D_1 = \alpha_i$ and $\neg D_2 = \beta_j$. Once each $(\alpha_i \cap \beta_j)$ is added to the member, all the unions of the type $\bigcup_{ij}^{\text{some}} (\alpha_i \cap \beta_j)$ must of course be added to the membership. But if these are added no new members need be added, because the difference and the union of terms, each of which is of the type $\bigcup_{ij}^{\text{some}} (\alpha_i \cap \beta_j)$ are again of the type $\bigcup_{ij}^{\text{some}} (\alpha_i \cap \beta_j)$. Thus $\mathcal{C} = \mathcal{A} \vee \mathcal{B}$ will consist of members, each of which can be expressed as $\bigcup_{ij}^{\text{some}} (\alpha_i \cap \beta_j)$. This means that the $n \times m$ elements $(\alpha_i \cap \beta_j)$ $(i = 1, 2, \cdots, n)$ $(j = 1, 2, \cdots, m)$ are the set-theoretical atoms of $\mathcal{C} = \mathcal{A} \vee \mathcal{B}$. The measure-theoretical atoms can be obtained simply by dropping those atoms whose measures are zero. We thus see that in the finite case $\mathcal{A} \vee \mathcal{B}$ is nothing but what we denoted by $\mathcal{A} \otimes \mathcal{B}$ in earlier chapters, the only difference being that we characterized each finite Borel field by the set of atoms; thus in the earlier notation we would have written $\mathcal{A} = \{\alpha_1, \alpha_2, \cdots, \alpha_n\}$ in the same way as $\mathcal{E} = \{E_1, E_2, \cdots, E_n\}$.

We now introduce the concept of conditional probability $m(A \mid C)$, which we may call the probability measure of subset A *qua* a subset of the entire set C. In terms of propositions, we call $p(A \mid C)$ the probability of A in the event that C is established to be the case, or simply the probability of A given C. We require three conditions:

1. If $A \subset C$ $(A \to C)$ and $B \subset C$ $(B \to C)$ then

$$\frac{m(A \mid C)}{m(B \mid C)} = \frac{m(A)}{m(B)}, \tag{7.167}$$

which may be interpreted as meaning that the ratio of probabilities of A and B is not affected by the assumed condition C provided that the imposition of C on A and B does not change the meaning of A and B, $A \cap C = A$ and $B \cap C = B$.

2.
$$m(C \mid C) = 1, \tag{7.168}$$

which may be interpreted as meaning that probability of C is unity in the event that C is established to be true.

3. For any A

$$m(A \cap C \mid C) = m(A \mid C), \tag{7.169}$$

which may be interpreted as meaning that once C is assumed to be the case, the probability of A and the probability of $A \cap C$ must be the same. We added the interpretation to the three mathematical relations above, although this section is intended to be neutral toward the different views of the empirical interpretation of probability measure. In fact, the above interpretation of the three relations seems to be rather easily justified from any viewpoint.

For any A and B that may or may not imply C, we have, from (7.167), $m(A \cap C \mid C)/m(B \cap C \mid C) = m(A \cap C)/m(B \cap C)$ because $A \cap C \to C$ and $B \cap C \to C$. This means that

$$m(A \cap C \mid C) = \text{constant} \times m(A \cap C), \tag{7.170}$$

where "constant" is independent of A. Putting $A = C$ in (7.170) and using (7.168), we conclude that constant $= 1/m(C)$. Inserting this value of the constant in (7.170), we obtain, in view of (7.169),

$$m(A \mid C) = \frac{m(A \cap C)}{m(C)}. \tag{7.171}$$

This is the definition of the conditional probability of A given C.

We derived the notion of conditional probability from the unconditional probability, but in a certain sense what we called normalization $m(\square) = 1$, used in the definition of the unconditional probability, can be considered as corresponding to (7.171) with $C = \square$. We have repeated, from Chapter 1 on, that any discussion and argument presupposes a large number of conventional definitions, agreed opinions, and established facts, some of which were explicitly stated and some not even detected. The constant truth $\square$ contains all these presupposed constant truths in conjunction with the logical tautologies. Hence $\square$ varies from one discussion to another. The probabilities must always be renormalized correspondingly, so that the probability of $\square$ in each case becomes unity. What is probably more basic is the ratio of probabilities [(as in (7.167)] rather than their absolute values. This point of view also justifies the use of an infinite σ-finite measure, for in this case we formally have $m(\square) = \infty$, making the absolute unconditional probability meaningless, yet the ratio of finite measures can remain meaningful. This is precisely what physicists are doing in their Γ-space.

Finally we introduce the concept of probabilistic (or statistical) independence. If two subsets or propositions A and B and a measure m are such that

$$m(A \cap B) = m(A) \cdot m(B) \tag{7.172}$$

then A and B are by definition said to be probabilistically independent relative to m. The notion of probabilistic independence is practically the only basic concept that depends critically on normalization, because the left side of (7.172) is linear in m whereas the right side is quadratic. It can be easily shown that if A is such that $m(A) = 0$ or $m(A) = 1$, A is independent of any other B. Indeed, since we have $m(A \cap B) \leq m(A)$ in general, the condition $m(A) = 0$ implies that $m(A \cap B) = 0$, establishing $m(A)m(B) = m(A \cap B)$

for any B. Next, if $m(A) = 1$, $m(\neg A) = 0$; hence $m(B - A) = m(B \cap \neg A) = 0$. From $B = (B - A) \cup (A \cap B)$ we then have $m(B) = m(B - A) - m(A \cap B) = m(A \cap B)$. Since $m(A) = 1$, this last result means that $m(A)m(B) = m(A \cap B)$.

More important is the notion of the probabilistic independence of two lattices $\mathcal{A}$ and $\mathcal{B}$, whose atoms are respectively $\{\alpha_i\}$ and $\{\beta_j\}$. Two sets of propositions $\mathcal{A}$ and $\mathcal{B}$ are said to be probabilistically independent of each other if each atom of one of them is probabilistically independent of every atom of the other; that is, $p(\alpha_i \cap \beta_j) = p(\alpha_i) \cdot p(\beta_j)$ for all i and all j.

In order to show that the conditional probability A given C as defined by (7.171) is in fact a measure of A, we have to show either that for any A and B such that $A \cap B = \emptyset$ we have

$$m(A \cup B \mid C) = m(A \mid C) + m(B \mid C), \qquad (7.173)$$

satisfying Condition (b) of Definition 7.2 (finite case), or else that for any arbitrary A and B we have

$$m(A \mid C) + m(B \mid C) = m(A \cap B \mid C) + m(A \cup B \mid C), \qquad (7.174)$$

satisfying (7.163). To prove (7.173) or (7.174) we have to show at one stage that

$$\begin{aligned} m[(A \cup B) \cap C] &= m[(A \cap C) \cup (B \cap C)] \\ &= m(A \cap C) + m(B \cap C) + m[(A \cap C) \cap (B \cap C)]. \end{aligned}$$

The second equality in this equation is justified as far as the unconditional probability satisfies the definition of a measure. But the first equality requires the distributive law. The last term in this equation, $m[(A \cap C) \cap (B \cap C)]$, can of course be written as $m(A \cap B \cap C)$, and is zero in the case of (7.173).

In Chapter 4 we saw that the relation $A \to B$ requires in its interpretation two different treatments, depending on whether $A \neq \emptyset$ or $A = \emptyset$. For instance, if $A \neq \emptyset$, A can be used as the explanation of B (hypothesis leading to B), but if $A = \emptyset$ this cannot be done. From the probabilistic viewpoint the distinction between the two cases becomes very clear, because if $A \to B$ and $A \neq \emptyset$, (which implies $m(A) \neq 0$ for some measure function), $m(B \mid A) = 1$ whereas if $A \to B$ and $A = \emptyset$, $m(B \mid A) = 0/0 =$ indefinite. If we replace $A \to B$ by $m(B \mid A) = 1$, we can automatically exclude the case $m(A) = 0$, which includes $A = \emptyset$. This shows again how the probabilistic view is closer to our normal way of thinking than the logical view. In fact, everybody knows how difficult it is to explain verbally $\emptyset \to A$ for all A.

7.4. EMPIRICAL INTERPRETATION OF PROBABILITY

The concept of probability as introduced in the last section is purely formal, and no explanation has been given to how and why this concept can be applied in our daily experience and in science. The purpose of this section is to give just such an explanation, though perhaps in a rather sketchy way. This requires that we envisage different traditional schools of thought regarding "what is probability." Our objective in discussing them, however, is not to revive old controversies or take the side of one school or another, but to try to consider each as indicating a possible usage of the mathematical concept of probability, and to determine, above all, the limitations within which each usage can be justified.

The many existing views about probability may be conveniently classified into two major categories, objective and subjective. The former can be subdivided into frequency view and necessary view. Since a purely subjective entity such as confidence or belief or expectation cannot become a direct object of scientific discourse, the subjective view has to rely on a behavioristic method, by which the degree of belief can be measured through a certain protocol. Such a behavioristic version of the subjective view is often called personalistic. Hence we discuss mainly three views: necessary, frequency, and personalistic, in that order. Individual theorists about the nature of probability differ in shades and details, and while some can be placed within one of these three major schools, some others have to be considered as compromisers of more than one school.

All three views have their precursors in the dawning of the theory of probability, which goes back to the beginning of the eighteenth century. The more conspicuous contemporary names connected with the frequency view include Richard von Mises [V-1], and Hans Reichenbach [R-2]. The two most known proponents of the necessary view are John Maynard Keynes [K-7] and Rudolph Carnap [C-2]. The modern personalistic view may be said to have been initiated in 1926 by Frank P. Ramsey whose paper on this subject is included in [R-1], and was given independently thorough mathematical investigation by Bruno de Finetti [D-4]. Lenard J. Savage is probably the best known spokesman for the personalistic view at present [S-1]. In recent years, Masanao Toda developed various ingenious psychological methods to extract subjective probabilities from the subject [T-5, S-7] (see Appendix A7.1).

The necessary view of probability asserts in essence that between any two propositions, taken as the analogues of the logical notions of premise and consequence, there holds, by analogy with logical implication, one and only one relation of a certain kind called probability relation, which corresponds to a numerical value that depends only on the two propositions and not on human opinions or any other element outside these two propositions. This view commits an error that belongs to the same category as the one we discussed in Section 4.2 under the name of Hempel's paradox: namely, starting from the correct and fertile observation that logical relations are a special case of probabilistic relations, one carelessly infers that what was true logically will also be true probabilistically. As was argued in its favor at the beginning of the last section, it is not only a correct but a pregnant thought that the conditional probability $p(B \mid A)$ is an analogue of the logical relation $A \to B$. But the fact that the truth or falsehood of the relation $A \to B$ is a necessary one depending only on the pair (A, B) and not on other elements cannot be extended to the probabilistic relation expressible by $p(B \mid A)$. In fact, Chapter 4 was entirely dedicated to the thesis that when $p(B \mid A)$ (a predictive probability) is determined by the nature of A and B, $p(A \mid B)$ (a retrodictive probability) cannot be determined by the nature of A and B. In this sense, one of the aims of the present book is to refute this "necessary view" of probability. The reader may recall that the inductive probability (which is a retrodictive probability) becomes zero only by "logical" refutation and becomes unity only by an infinitely repeated series of experiments. In these two extreme cases the retrodictive probability may be considered as depending solely on the two argument propositions. But this is not true in general cases. When one limits oneself to the predictive probability, the necessary view has its validity in principle. However, as mentioned in the last section, we have to enumerate, under the "condition" of the conditional probability, all the propositions that are tacitly understood to be true, and this task of "excavating" all the hidden assumptions is practically impossible in reality. In most cases the auxiliary conditions are not specifically mentioned as part of the condition; as a result the conditional probability (even in predictive cases) is not uniquely determined by the two explicitly mentioned propositions.

So much for the necessary view; now for the frequency view. Since this view is supported by the so-called laws of large numbers, we should start with reviewing these laws, first in the customary fashion and then more critically.

Let us consider N logical spectra, each of which consists of two events, $D^{(i)} = \{D_1^{(i)}, D_2^{(i)}\}$, $i = 1, 2, \cdots, N$, with $D_1^{(i)} \cup D_2^{(i)} = \square$ and $D_1^{(i)} \cap \mathfrak{D}_2^{(i)} = \emptyset$, and their product $\mathfrak{D}^{(1)} \otimes \mathfrak{D}^{(2)} \otimes \cdots \otimes \mathfrak{D}^{(N)}$ (measure-theoretically, this is the Borel field generated by the union of the Borel fields corresponding to the $\mathfrak{D}$'s). We apply this to the case of N-fold repetition of a certain event, so that all $D_k^{(i)}$ $(k = 1, 2)$ with different i's and a fixed k refer to the "same" event. We speak of success if $k = 1$ and failure if $k = 2$. Accordingly we may put $p(D_k^{(i)}) = \pi_k$, with $\pi_1 + \pi_2 = 1$, where the π_k do not depend on i. If we repeat the events in such a way that the condition of independent trials is satisfied, we shall have $p(D_{k_1}^{(1)} \cap D_{k_2}^{(2)} \cap \cdots \cap D_{k_N}^{(N)}) = p(D_{k_1}^{(1)}) \cdot p(D_{k_2}^{(2)}) \cdot \cdots p(D_{k_N}^{(N)}) = \pi_{k_1} \cdot \pi_{k_2} \cdot \pi_{k_3} \cdot \cdots \pi_{k_N}$, where each $k_1, k_2, \cdots, k_N$ can be 1 or 2. We can characterize each sequence $(k_1, k_2, \cdots, k_n)$ by the number N_1 of successes $(k_j = 1)$ and the number $N_2 = N - N_1$ of failures $(k_j = 2)$. The probability of having N_1 successes will then obviously be $\binom{N}{N_1}\pi_1^{N_1}\pi_2^{N_2}$. Now the "weak" law of large numbers (for independent trials) is a direct consequence of this distribution and states that for any $\varepsilon > 0$

$$\Pr\left\{\left|\frac{N_1}{N} - \pi_1\right| \geq \varepsilon\right\} \leq \frac{\pi_1\pi_2}{N\varepsilon}, \tag{7.175}$$

from which follows

$$\lim_{N\to\infty} \Pr\left\{\left|\frac{N_1}{N} - \pi_1\right| < \varepsilon\right\} = 1 \tag{7.176}$$

for any given $\varepsilon > 0$.

We can also prove the "strong" law of large numbers, which states that for any given ε there exists N_0 such that

$$\Pr\left\{\left|\frac{N_1}{N} - \pi_1\right| < \varepsilon\right\} = 1 \quad \text{for} \quad N > N_0. \tag{7.177}$$

The usual interpretation of these laws (7.176) and (7.177) is somewhat as follows. Let us assume, for instance, that we throw a die N times. $A_1^{(i)}$ is the proposition that the ith throw will give, say, 1. Let π_1 be $\frac{1}{6}$. N_1/N is the ratio of the number of throws that give 1 to the total number of throws. Let δ_N be the deviation of N_1/N from $\frac{1}{6}$. In the weak law of large numbers, we fix N and ask the probability that δ_N is less in magnitude than ε. This probability as a function of N tends to 1 as $N \to \infty$. But this does not guarantee that if $\delta_N < \varepsilon$ for a certain N, say, N_0, in a particular sequence of throws, the deviation δ_N will remain $< \varepsilon$ for N larger than N_0 if we continue the same sequence of throws. On the other hand, the strong law of large numbers states that there is a certain number N_0 such that δ_N is, and remains, less than ε with probability 1 for all N larger than N_0 when we continue the same sequence of throws. Since probability unity means an undoubtable truth, we say that π_1 must be equal to N_1/N for $N \to \infty$. This, according to the frequency view, indicates that π_1 cannot be arbitrarily assigned and is "objectively" determined by the relative frequency.

We can also apply the theorem to the case of a simultaneous throw of many dice (of a prescribed make) instead of a repeated series of throws of the same die. The proposition to which the tested probability is attached is of the type "the probability of *this* die giving '1' in a (sufficiently violent) throw is $\frac{1}{6}$" or "the probability of dice made according to such and such prescribed manner giving '1' in a throw is $\frac{1}{6}$."

This argument in support of the frequency view of probability contains three major assumptions. First, the proposition whose probability is in question has to be reproducible (sequentially or simultaneously), as the foregoing example clearly shows. Second, the trials are supposed to be statistically independent. Third, we assume $N \to \infty$. These three points require closer attention.

Regarding the first point, we have to point out that most of the propositions to which we usually apply the concept of probability are not reproducible. For instance, it makes perfect sense to talk about the probability of the proposition, "Satosi Watanabe will die in 1969." The holder of the frequency view may say that this probability should be understood as the relative frequency of those who die in 1969 among individuals of the same age and physical constitution as Satosi Watanabe. But this interpretation is fallacious because it presupposes an entirely arbitrary inductive generalization of Satosi Watanabe. The frequency will change every time one selects a group of people who "resemble" him, according to what characteristics are taken into consideration and the extent to which resemblance is required. When we speak of the probability of his dying in 1969, we are thinking solely of this event and nothing else. Admittedly, our estimation of this probability may be influenced by induction based on observation of other people, but the probability itself pertains solely to him and is not changed by the death or survival of the fellow members of an arbitrarily convoked "Watanabe club." Another kind of probability that the frequency view has a hard time in interpreting is the credibility of hypotheses discussed in Chapter 4.

Concerning probabilistic independence, we should recall that it is purely a mathematical notion. To claim the independence $p(A \cap B) = p(A)p(B)$, we have to know $p(A \cap B)$, $p(A)$, and $p(B)$ separately and show that the equality holds. Hence it is circular, to say the least, to use the notion of probabilistic independence in defining the notion of probability in general. The reason why the use of the notion of probabilistic independence prior to the definition of probability itself does not shock many people in the above derivation is that there is a physical situation that may be called "physical independence" of trials: for instance, in the case of die throwing, the throws are either temporally or spatially sufficiently separated. If such physical independence somehow guarantees mathematical independence, that is an empirical physical law, and it should not be silently and uncritically smuggled into the definition of the notion of probability. Thus the frequency view is dependent on a lucky coincidence that a certain physical prescription can be formulated such that compliance with it guarantees the condition of probabilistic independence according to the notion of probability to be defined through this repetitive trial. It is admittedly true that the law of large numbers holds not only in a sequence of independent trials, but also in an ergodic sequence; however, this does not alleviate the difficulties of the frequency view, because we have to use the notion of probability in order to establish ergodicity before defining the notion of probability.

The final and most fundamental objection to the frequency view, particularly to its claim of objectivity, is that we can never realize $N = \infty$ in our human experience. Insofar as N is finite it can always happen that N_1/N

(when $N =$ finite) is quite different from π_1. For instance, if $\pi_1 \neq 0$, a sequence of N events consisting only of A_1 happens with probability $[\pi_1]^N$, which is not zero insofar as $N < \infty$. In this case we have $N_1/N = 1$, which may be quite different from π_1. Therefore we can never logically repudiate an assignment of π_1 different from N_1/N no matter how large N is. We should actually consider the determination of π_1 from N_1/N as a case of Bayesian induction. As we saw in Chapter 4, any assignment of a particular value of probability to an (repetitive) event at a finite value N is strongly influenced by some extra-evidential evaluation (of a hypothesis that the "true" probability is such and such). This includes the case of assignment of a particular value N_1/N to π_1. Hence the so-called objective view is bound to admit a subjective element when it is to be applied realistically, that is, while N is finite. Thus the frequency view is an idealized theory ($N \rightarrow \infty$), applicable only to strictly repetitive propositions, which depend critically on a certain physical law. It has a strength of intuitive appeal, however, particularly to those who are accustomed to dealing with the physical world.

Let us now examine the personalistic view of probability. I shall present a modification of the oldest version of this view, namely, that of Ramsey [R-1]. Admittedly, this version is not so refined and sophisticated as later versions, such as those of de Finetti or Savage, but it has at least three virtues. First, it is simple. Second, it "honestly" brings in the notion of numerically representable "value," which the other versions tend to hide. (See [W-29] for my reservations with regard to the notion of numerically representable value.) Third, it is free from the questionable postulate (which is assumed in various forms by other supporters of this view) that the entire possibility (i.e., $\square$ in logic and X in measure theory) can be considered as a disjunction of arbitrarily many propositions or subsets, each of which has an arbitrarily small probability. (For a modern method of measuring personal probabilities, see Appendix A7.1.)

The great merit of Ramsey lies in his bringing the notion of probability directly into the context of purposive behavior of human beings. There are at least two other places in this book where such behavior is referred to. In Chapter 3 we discussed how probabilistic asymmetry is related to the category of causality and the freedom of purposive action, and in Chapter 6 we saw how the convergence of purposive behavior is characterized by the inverse H-theorem. To have a unified picture of these related aspects of human behavior, let us start with a consideration of general nature, rephrasing the similar consideration introduced in Section 6.4. In a rough description we have to distinguish at least five notions. First, there is a person (or an animal or a robot†) P. In some cases a group of mutually communicating and collaborating persons can be considered as a P. The person P is placed in one of the

† For my view about the difference between man and robot, see [W-28, W-29].

possible situations $\mathcal{S} = \{S_1, S_2, \cdots, S_i, \cdots, S_m\}$. He has a set of available acts to choose from: $\mathcal{A} = \{A_1, A_2, \cdots, A_j, \cdots, A_n\}$. An act A_j is deterministically or probabilistically selected by the situation S_i and the person P. We may symbolically denote this as $A_j(S_i, P)$. As a result of this act there occurs one of the possible consequences $\mathcal{C} = \{C_1, C_2, \cdots, C_k, \cdots, C_q\}$. This consequence usually does not depend directly on P, but on S_i and A_j. Hence $C_k(S_i, A_j)$. This consequence is associated with one of the possible rewards $\mathcal{R} = \{R_1, R_2, \cdots, R_l, \cdots, R_p\}$. In some cases R_l is a desirable or undesirable aspect of C_k itself, but in others R_l is associated with C_k artificially. It is therefore safe to assume $R_l(S_i, C_k, A_j)$. If C_k itself has some desirability, we have to attribute that to the associated R_l, so that the C_k can be considered "neutral."

The person placed in a natural environment where he wants to realize his desired goal proceeds somewhat as follows. In this case R_l is usually determined by C_k alone; hence $R_l(C_k)$. Combining this relation with $C_k(S_i, A_j)$, one gets $R_l(C_k(S_i, A_j))$. The person selects A_j so that the desired value R_l is realized, or R_l is maximized. When A_j is considered as determined by R_l, it is a purposive process. If C_k and hence R_l are determined by A_j and S_i, it is a causal process. In particular, the essence of science is to furnish us with "natural laws" $C_k(S_i, A_j)$. From the person P's point of view, his autonomous selection of an act A_j in the presence of S_i is possible due to the freedom of choice we discussed in Chapter 3. If we look at this process from a third person's viewpoint, i.e., a bystander's viewpoint, we may substitute $A_j(S_i, P)$ in A_j in the above relation and write $R_l(C_k(S_i, A_j(S_i, P)))$. If S_i and P are given, the rest is determined (deterministically or probabilistically). The P often has a more or less correct knowledge of the natural law $C_k(S_i, A_j)$; as a result he tends to select A_j so as to maximize R_l. The knowledge and habits of the person are incorporated in P. The learning process is a change in P with the accumulation of his experience, and the change is observed in his *behavior* $A_j(S_i, P)$. In this context S_i can be called *stimulus*, and A_j, *response*. The reward is called "reinforcement" in psychological experiments, and the function $R_l(C_k)$ is usually controlled by the experimenter. The response as a function of the stimulus is the behavior, and this is controlled by the reinforcement structure.

The gambling situation is characterized by two features that distinguish it from the case of a person in a natural environment. For one thing, the consequence C_k is some uncertain event that is not influenced by the act A_j; thus $C_k(S_i)$. The act is wager and its choice determines the reward under different possible eventualities; thus $R_l(S_i, C_k, A_j)$. Alternatively, each betting act A_j may be described as associating a reward R_l with C_k. The human choice of A_j is determined by the desirability of R_l and the probability (in the sense of expectancy) of C_k. The situation S_i can be fixed and therefore may be placed outside consideration. In Ramsey's construction of the theory, R_l is supposed to be expressed numerically, and the following two basic notions are used. First, each act A_j determines a deterministic function $R_l(C_k)$. For instance, the bet (which is an A_j) may be: if "heads," P gets 5 units of value, and if "tails," he loses 1 unit of value. Second, the person is faced with a certain set of uncertain events $\mathcal{C}$, and given a set of alternative wagers $\mathcal{A}$, he decides on a preferential order of any two alternative bets, $A_j \rightsquigarrow A_k$, which should be interpreted as meaning that he does not prefer A_j to A_k. From this meaning, it is natural to assume that transitivity of $\rightsquigarrow$. If $A_j \overset{\leftsquigarrow}{\rightsquigarrow} A_k$, he is neutral or indifferent with regard to A_j and A_k. This relation $\overset{\rightsquigarrow}{\leftsquigarrow}$ is also transitive. If $A_j \rightsquigarrow A_k$ but not $A_j \overset{\leftsquigarrow}{\rightsquigarrow} A_k$, he will prefer A_k to A_j.

In order to introduce any well-defined, real-valued variable we have to go through four major steps, although their order can be partly changed and some steps can be combined.

1. We consider a set of "points" and have to show the existence of a one-way relation ("not greater than") between *any* two points. (In the case of a lattice we assumed the existence of such a relation between *some* pairs.) From this we can define "equality" as the existence of a two-way relation. All "equal" points are then considered as a single point. This entails the extraction of the particular property reflected in the one-way relation. The set of points becomes immensely retrenched. We can add at this stage a hypothesis of continuity to make handling easier, although the necessity of this hypothesis may be very difficult to prove. By this, we are entitled to assign a real number to each point, but this assignment is completely arbitrary insofar as the greater-than relation is not changed. The concept so far determined is invariant for any rubber-string-like transformation.
2. We have to introduce proportionality. By this is meant that the ratio of two distances is fixed, where distance means the difference between the two real numbers attached to two points. (Hence two distances usually involve *four* points.) This definition still leaves an invariance for multiplication of every value by a constant. Instead of a rubber string, we have, so to speak, a stick that can expand or contract as a whole.
3. Thus we have to introduce the scale unit. This means that a particular real number is assigned to one particular distance (involving *two* points).
4. Finally, we have to decide where the zero point lies. This amounts to assigning a particular real number to *one* particular point. The last two steps can be replaced by assignment of two particular values to two particular points.

In the case of probability there is not much to argue about regarding Steps 3 and 4, for people usually accept as convention that the probability of $\emptyset$ is zero and the probability of $\square$ (or X) is unity. Step 2 is the most difficult part in the case of probability, and every derivation on this point contains a certain degree of artificiality. On the other hand, Step 1 is relatively easy and clearly reveals the basic nature of the quantity we are discussing. Although Ramsey's derivation need not make a clear distinction between Steps 1 and 2, let us discuss Step 1 separately in line with Ramsey's way of concept building. For the purpose of Step 2, Ramsey first introduced proportionality in the numerical expression of value (desirability) and then transplanted it to the numerical expression of probability. In the same way, to introduce the not-greater-than relation in probability, we have to assume the existence of at least two rewards R, one being more desirable than the other. This is simply done by assuming the existence of one reward R, which is more desirable than no reward, which we represent by R_0.

There are two events C_1 and C_2, and there are two optional bets A_1 and A_2, which are defined as follows:

$$\begin{aligned} A_1 &: \quad R \quad \text{if} \quad C_1 \quad \text{and} \quad R_0 \quad \text{if} \quad \neg C_1; \\ A_2 &: \quad R \quad \text{if} \quad C_2 \quad \text{and} \quad R_0 \quad \text{if} \quad \neg C_2. \end{aligned} \tag{7.178}$$

If the person's preference is such that $A_1 \leftsquigarrow A_2$, his personal probability is such that $p(A_1) \geq p(A_2)$. Similarly, if $A_1 \rightsquigarrow A_2$, $p(A_1) \leq p(A_2)$. Suppose that an empirical law says that if $C_1 \to C_2$ then $A_1 \rightsquigarrow A_2$. This entails a corollary, "if $C_1 \to C_2$ then $p(C_1) \leq p(C_2)$," which we mentioned in Section 7.3 as the minimum requirement of the concept of probability. The converse of the above empirical law, "if not $C_1 \to C_2$ then not $A_1 \rightsquigarrow A_2$," is obviously false, but the combination of these two, namely, "if $C_1 \to C_2$ and not $C_2 \to C_1$, then $A_1 \rightsquigarrow A_2$ and not $A_2 \rightsquigarrow A_1$," could be true. This would entail the rule "if $C_1 \to C_2$ and $C_1 \neq C_2$ then $p(C_1) < p(C_2)$" [see (7.165)].

Let us now introduce, according to Ramsey, the metric into the set of points that are "ordered" according to the "not-more-probable-than" relation. Ramsey takes the liberty of assigning, by definition, probability 1/2 to a certain category of events, besides $p(\square) = 1$ and $p(\emptyset) = 0$. This is slightly dangerous, because Steps 3 and 4 permit the assignment of only two values. The third assignment must be such that it does not contradict Step 2 or its equivalent. Events of probability 1/2 are defined as follows: in the foregoing optional bets, we put $C_1 = \neg C_2$. $C_2 = \neg C_1$. If the person has no preference between A_1 and A_2, that is, if $A_1 \underset{\leftsquigarrow}{\rightsquigarrow} A_2$, we say that $p(C_1) = 1/2$. Since this definition is symmetrical, we also have $p(\neg C_1) = 1/2$.

Using the notion of probability 1/2, Ramsey proceeds to define the "proportionality" (Step 2) of the numerical expression of desirability. Of course, it is questionable whether any "value" in life can be expressed as a number. However, material goods such as food or money have a measurable physical quantity or "volume" (such as grams or dollars), and the numerical expression of desirability may be suspected to stand in some functional relation to such a quantity or volume. This function cannot be a linear function, for at least two obvious reasons. For one thing, a negative "volume" (like debt) is extremely painful. Second, beyond a certain amount, an increase in volume gives less pleasure or satisfaction than a similar increase would do if the total volume were still very low. But if we limit the change in volume within a small domain, we may suspect that there could be some proportionality between change in volume and change in numerical desirability. In any event, Ramsey's definition of numerical desirability has an advantage of not being dependent on volume.

Let there be two events C_1 and C_2 of probability 1/2, and four rewards R_1, R_2, R_3, and R_4. The options available to the person are

$$\begin{aligned} A_1&: \quad R_1 \text{ if } C_1 \text{ and } R_4 \text{ if } \neg C_1; \\ A_2&: \quad R_2 \text{ if } C_2 \text{ and } R_3 \text{ if } \neg C_2. \end{aligned} \tag{7.179}$$

According to Ramsey's definition, if $A_1 \underset{\leftsquigarrow}{\rightsquigarrow} A_2$ then $R_1 - R_2 = R_3 - R_4$. This determines the proportionality of numerical desirability in the sense of Step 2, and no further specifications (Steps 3 and 4) about the numerical value are needed for our ultimate purpose of defining probability.

Now we come to the probability proper. We reformulate Ramsey's definition in such a way that we start with the conditional probability. Consider two alternative possibilities of betting A_1 and A_2, which are defined as follows:

$$\begin{aligned} A_1&: \quad R \text{ if } C_1 \text{ and } Q \text{ if } \neg C_1; \\ A_2&: \quad S \text{ if } C_1 \cap C_2, \quad T \text{ if } C_1 \cap \neg C_2, \text{ and } Q \text{ if } \neg C_1. \end{aligned} \tag{7.180}$$

Our definition of the conditional probability of C_2 given C_1 is given by

$$p(C_2 \mid C_1) = \frac{R - T}{S - T} \tag{7.181}$$

if $A_1 \underset{\leftsquigarrow}{\rightsquigarrow} A_2$.

By interchanging the roles of C_2 and $\neg C_2$, we obtain from this definition

$$p(\neg C_2 \mid C_1) = \frac{R - S}{T - S}. \tag{7.182}$$

The reward Q does not appear in (7.181) or (7.182); hence we put

$$Q = T \tag{7.183}$$

for later convenience. From (7.181) and (7.182) it follows that

$$p(C_2 \mid C_1) + p(\neg C_2 \mid C_1) = 1. \tag{7.184}$$

Augmenting (7.180), we introduce a third option,

$$A_3: \quad U \text{ if } \square \text{ and } V \text{ if } \emptyset, \tag{7.185}$$

where V need not even be mentioned, because the person can never get it. U is the value he will get if he chooses option A_3. We adjust U so that $A_1 \rightleftharpoons A_3$, which also entails $A_2 \rightleftharpoons A_3$. Now let us consider first $A_1 \rightleftharpoons A_3$ and apply the definitions (7.180) and (7.181) to this case. All we need is to carry out the following substitution:

$$\begin{gathered} A_3 \text{ for } A_1; \quad A_1 \text{ for } A_2; \\ \Box \text{ for } C_1; \quad \emptyset \text{ for } \neg C_1; \quad C_1 \text{ for } C_2; \\ U \text{ for } R; \quad V \text{ for } Q; \quad R \text{ for } S; \quad Q \text{ for } T. \end{gathered} \tag{7.186}$$

We then obtain, from (7.181),

$$p(C_1 \mid \Box) = \frac{U - Q}{R - Q}. \tag{7.187}$$

Next we apply the definitions (7.180) and (7.181) to $A_2 \rightleftharpoons A_3$. The substitution is

$$\begin{gathered} A_3 \text{ for } A_1; \quad A_2 \text{ for } A_2; \\ \Box \text{ for } C_1; \quad \emptyset \text{ for } \neg C_1; \quad C_1 \cap C_2 \text{ for } C_2; \\ \neg C_1 \cup \neg C_2 \text{ for } \neg C_2; \\ U \text{ for } R; \quad V \text{ for } Q; \quad S \text{ for } S; \quad (T * Q) \text{ for } T. \end{gathered} \tag{7.188}$$

The union of the last two cases, $C_1 \cap \neg C_2$ and $\neg C_1$, of (7.180) is $\neg C_1 \cup \neg C_2$, which now takes the place of $C_1 \cap \neg C_2$ of the original definition. The reward for this case is either T or Q, depending on the case [hence it is expressed as $T * Q$ in (7.188)]. This can be written simply "T or Q" if we agree on (7.183). We then obtain, from (7.181),

$$p(C_1 \cap C_2 \mid \Box) = \frac{U - Q}{S - Q}. \tag{7.189}$$

If we agree to write $p(X \mid \Box)$ as $p(X)$, we obtain from (7.181), (7.187), (7.189) the usual definition of conditional probability:†

$$p(C_2 \mid C_1) = \frac{p(C_1 \cap C_2)}{p(C_1)}. \tag{7.190}$$

With the help of (7.190) we can rewrite (7.184) as

$$p(C_1 \cap C_2) + p(C_1 \cap \neg C_2) = p(C_1). \tag{7.191}$$

This has the form of the basic axiom of probability,

$$p(X) + p(Y) = p(X \cup Y) \quad \text{for} \quad X \cap Y = \emptyset, \tag{7.192}$$

if we put

$$X = C_1 \cap C_2, \qquad Y = C_1 \cap \neg C_2, \qquad X \cup Y = C_1. \tag{7.193}$$

† The X here has nothing to do with the entire set (equivalent to $\Box$) in the sense of measure theory.

One may ask whether any arbitrary pair X, Y satisfying $X \cap Y = \emptyset$ can be fitted into the form (7.193). The answer is yes. For any such given X and Y we can write

$$C_1 = X \cup Y \quad \text{and} \quad C_2 = X \cup (\neg Y \cap Z), \tag{7.194}$$

where Z remains arbitrary. It is left to the reader to prove that (7.193) follows from (7.194). This, together with $p(X) \geq 0$ for any X and $p(\emptyset) = 1 - p(\Box) = 0$, shows that the p-function thus introduced is a probability in the mathematical sense.

The definition (7.181) can be considered as corresponding to Step 2, because it can be written as

$$\frac{p(C_2 \cap C_1) - p(\emptyset)}{p(C_1) - p(\emptyset)} = \frac{R - T}{S - T}. \tag{7.195}$$

The earlier definition of events with probability 1/2 can be rediscovered, with reference to (7.180), as follows. The condition $A_2 \leftrightsquigarrow A_2^*$, where A_2^* is obtained from A_2 by interchange of T and S, entails $p(C_2 \mid C_1) = p(\neg C_2 \mid C_1)$. Combining this with (7.184), we obtain $p(C_2 \mid C_1) = 1/2$ no matter what C_1 is, including $C_1 = \Box$.

Thus we have established that there exists a behaviorally observable way of measuring subjective probability, and the quantities thus measured satisfy the mathematical axioms of probability. Such personal probability is applicable to reproducible as well as nonreproducible events. The question that remains to be answered refers to the meaning personal probability has when applied to reproducible events. Should we always expect agreement between personal probability and frequency-oriented probability? The most natural way to apply personal probability to reproducible events is to assign personal probability (credibility) to hypotheses regarding the relative frequency of occurrence of the events in question. This is essentially what we did in Chapters 4 and 6. What we designated there as q instead of p is our personal confidence placed in a hypothesis. There are actually two aspects to the problem of relation between the personal probability of a hypothesis and the frequency of a certain event. On the one hand, there is a question of prediction of the frequency of occurrence of the event in the future (or in an unknown case), which can be derived from the personal probabilities of hypotheses. On the other hand, there is a question of the influence of the actual frequency of the event in the past on the personal probabilities placed on the hypotheses.

Referring to the notation we used in explaining (7.175), we may introduce a hypothesis $H(\pi)$ with a variable parameter π, which states that the repetitive trials of the event are probabilistically independent and that the probability of each success is π. If this hypothesis $H(\pi)$ is correct, the probability that N_1 out of N will be successes is given by

$$p(N_1 \mid H(\pi)) = \binom{N}{N_1} \pi^{N_1}(1-\pi)^{N-N_1}, \tag{7.196}$$

which inevitably connects π with the frequency N_1/N.

Suppose that we write $q(\pi)\,d\pi$ for our personal probability or credibility that a hypothesis $H(\pi)$ in the domain $(\pi, \pi + d\pi)$ is correct, assuming that at least one of such $H(\pi)$ with some value of π in the domain (0, 1) is correct. Then the personal probability for N_1 successes may be expressed by

$$p(N_1) = \int_0^1 p(N_1 \mid H(\pi))q(\pi)\,d\pi = \binom{N}{N_1}\int_0^1 \pi^{N_1}(1-\pi)^{N-N_1}q(\pi)\,d\pi. \tag{7.197}$$

We saw in Section 6.3 that

$$f(\pi) = \frac{(N+1)!}{N_1!\,(N-N_1)!}\,\pi^{N_1}(1-\pi)^{N-N_1} \tag{7.198}$$

tends to

$$f(\pi) \to \delta\left(\pi - \frac{N_1}{N}\right) \quad \text{as} \quad N \to \infty. \tag{7.199}$$

Note that (6.89) becomes (6.93), where γ_1 is the limiting value of ν_1/ν. Hence (7.197) becomes

$$\begin{aligned} p(N_1)(N+1) &= \int_0^1 \delta\left(\frac{N_1}{N} - \pi\right) q(\pi)\,d\pi \\ &= q\left(\frac{N_1}{N}\right) \quad \text{for} \quad N \to \infty. \end{aligned} \tag{7.200}$$

Now introduce a continuous real-valued variable χ with the domain $(0 \le \chi \le 1)$ in which the rational numbers N_1/N are imbedded, and define $p(\chi)\,d\chi$ as the sum of those probabilities $p(N_1)$, where N_1 lies between $N\chi$ and $N(\chi + d\chi)$. There are $N\,d\chi$ such values of N_1. Consequently (7.200) becomes (replacing $N + 1$ by N)

$$p(\chi)\,d\chi = p(N_1)N\,d\chi = q(\chi)\,d\chi \quad \text{for} \quad N \to \infty. \tag{7.201}$$

This shows that the predictive probability distribution of the frequency of the event (in the future) becomes identical with the personal credibility distribution over the hypotheses.

Next we have to examine the question of the relation between $q(\pi)$ and the frequency of the event in the past. Fortunately, we have already studied this question in detail in Section 6.3. We saw there that no matter what initial credibility distribution $q^{(0)}(\pi)$ we may start with, we end with $q^{(\infty)}(\pi) = \delta(\pi - \gamma_1)$, where γ_1 is given by $\lim_{\nu\to\infty} (\nu_1/\nu) \to \gamma_1$ and ν_1 is the actual number of occurrences of the event in question in ν trials. At the finite number ν of trials, however, we cannot eliminate the effect of $q^{(0)}(\pi)$ altogether, and we can in principle choose $q^{(0)}(\pi)$ in such a way that $q^{(\nu)}(\pi)$ becomes large for values of π appreciably different from ν_1/ν. If we comply with the spirit of the principle of consistent docility and assume that $q^{(0)}(\pi)$ is a slowly changing function of π as compared with the function (6.83) (with χ replaced by π), however, we can eliminate $q^{(0)}(\pi)$ from (6.82) and obtain (6.89).

Thus we see that according to the viewpoint of personal probability, the past record of frequency reflects on the credibility distribution, and the credibility distribution, in turn, reflects on the prediction about the future frequency. The so-called frequency view of probability, so to speak, short-circuits the role of the person who estimates the probability and who makes predictions. From this point of view (6.91) and (6.92) are extremely interesting. This is the result of elimination of the credibility, consistent with the personalistic view, which connects the past frequency directly with the prediction of the future.

In the personalistic view the connection between personal probability and frequency was made possible by the intermediary hypothesis $H(\pi)$. Since this hypothesis asserts that the trials of the event are statistically independent, it is empirical, as we saw in our criticism of the frequency view. To free ourselves as much as possible from an assumption that can be tested only empirically, we may prefer to change the meaning of the hypotheses so that we need not talk about the unclear notion of probabilistic independence. de Finnetti [D-4] showed that our starting equations (7.197) can be obtained from a much less restrictive assumption (called assumption of equivalent events) than the assumption of statistical independence. We consider as before the product $\bigotimes_i \mathfrak{D}^{(i)}$, where each set $\mathfrak{D}^{(i)}$ consists of two events. If the probability of N_1 successes and N_2 failures in the total of N trials depends only on N_1 and N_2 and not on the sequence or order in which N_1 successes and N_2 failures occur, we speak of an "equivalent" case. In such a case we obtain, according to de Finnetti, the formula (7.197), whether or not it is a case of statistical independence. Since the notion of equivalence does not involve any use of numerical values of probabilities and can be defined on the basis of some abstract notion of symmetry, which cannot be regarded as a specific physical condition, the connection between probability and frequency becomes less objectionable.

A final remark to make, in connection with the personalistic view of probability, is that once the existence of numerical probability is assumed then all the axioms used in deriving the numerical probability can be derived from a single principle, so that these axioms are consistent with each other. This single principle is that a subject prefers the alternative to which a higher "expected desirability" is attached. The expected desirability of each event is the product of the probability and the numerical desirability of the reward attached to the event, and the expected desirability of an alternative is the sum of the expected desirability of disjoint and exhaustive events considered in the alternative. For instance, the simplest axiom explained in reference to (7.178) is the following: the expected desirability of alternative A_1 is $\mathrm{ED}(A_1) = Rp(C_1) + R_0 p(\neg C_1) = (R - R_0)p(C_1) + R_0$, and the expected desirability of alternative A_2 is $\mathrm{ED}(A_2) = R_p(C_2) + R_0 p(\neg C_2) = (R - R_0)p(C_2) + R_0$. If $\mathrm{ED}(A_1) \geq \mathrm{ED}(A_2)$, obviously $p(C_1) \geq p(C_2)$, provided that $R - R_0 \geq 0$. The reader is advised to check all the remaining axioms.

An interesting problem was raised and answered by M. Toda [T-5]. The subject is required to assign a non-negative real number r_i to each disjoint and exhaustive event $\{E_i\}$, $\sum_{i=1}^{n} r_i = 1$. The reward R_i that he obtains if E_i happens is a function $R(\{r_i\} \mid E_i)$ of E_i and the set of numbers $\{r_i\}$. The subject, knowing this reward function, then tries to maximize his expected desirability $\mathrm{ED} = \sum_i p_i R(\{r_i\} \mid E_i)$ by adjusting r_i, where p_i is the subjective probability of occurrence of E_i. Toda has shown that by choosing the reward function suitably one can achieve $r_i = p_i$. (See Appendix A7.1 for some of the methods discovered by Toda.)

After reviewing the three basic views on the nature of probability, we should mention the "ensemble" representation of probability. This is sometimes confused with the frequency view of probability, but it does not depend on any specific interpretation of the notion of probability and is compatible with any of the three views discussed above. It is a technique of mental visualization, not a vehicle of any philosophical tenet. If it is confused with the frequency view, it is simply because the ensemble representation takes advantage of our psychological familiarity with the frequency view without subscribing to its philosophy. The idea is very simple. If we have a set $\mathcal{E} = \{E_i\}$ of events with probability $\{p_i\}$, we imagine a collection of a large number N of similar systems in which $\mathcal{E}$ can happen, and we assume that in Np_i systems E_i actually is the case. The basic and important difference of this ensemble representation, from the frequency view, is that these N systems do not coexist in the same space or in the same time dimension. They coexist only in our imagination, and they cannot have any kind of correlation in a physical or mathematical sense. Conditions not specified by the definition of $\mathcal{E}$ must take various possibilities with appropriate numbers of representatives.

That this representation can be applied to a unique event can be seen in the following way. If the probability is 1/2 that Satosi Watanabe will die in 1969, we imagine an immense number N of worlds, in each of which the same Satosi Watanabe is living, and in $1/2N$ of the worlds he will die in 1969. Similarly, if we say that hypothesis H is true with probability 1/2, we imagine a large number of possible universes, of which one half obey the law H. The utility of this ensemble representation was brilliantly exploited by J. W. Gibbs.

7.5. STRUCTURE OF THE OBJECT-PREDICATE TABLE, EXTENSION, INTENSION, AND COMPLEXITY

After having discussed the basic problems pertinent to the notion of probability, we return to the problem of logic, but this time from an "empirical" point of view. This approach has a close relation to computer programs used for different data-processing problems. One can express any empirical datum, that is, the result of an observation, in a proposition of the type "the object possesses a certain property," or its negation.† In some cases the property may concern a relation between two or more individual objects; then one need only consider a pair or a group of objects as a single composite object and the relation as a property of this group. Even if the datum is the result of measurement of a continuous variable, one can "quantify" the value of the variable and reduce the observational result to a set of two-valued propositions, because each experimental measurement inevitably contains a finite error or imprecision; for example, the number 1 or 0 occupying each digit of the approximate binary expression of a real number can be considered as the response, yes (1) or no (0), to a question: Does the object have a certain property? This situation permits us to consider the "object-predicate" table explained below as the general form of an empirical fact of any arbitrary nature. In this section, the postulate of fixed predicate-set correspondence is assumed to hold.

In order to avoid misunderstanding, it may be interjected here that to claim that a certain statement is empirical does not at all imply that it stands for something that is given directly to our senses and is not contaminated by theoretical artifact. On the contrary, only by virtue of our theoretical structure does the empirical determination become possible. If, for instance, the statement in question is "the current is 0.333 amp," we need a quite sophisticated theory of electricity defining "amperes" to dictate the method of measurement before we can determine whether this statement is true or false. Even if the statement is seemingly a nonscientific one such as "this paper is blue," we are using an arbitrary (i.e., not dictated by the outside world or by logic) classification of colors understood and entrenched in our language. As a matter of fact, in the Japanese language, the word roughly corresponding to the English "blue" sometimes includes "green." Classification is undeniably an integral part of theoretical structure (see Section 7.6).

Let $X = \{x_i\}$, $i = 1, 2, \cdots, m$, be the collection of objects under examination and let $Y = \{y_j\}$, $j = 1, 2, \cdots, n$, be a collection of certain predicates

† In a more general formulation, we should state: "the result of a certain test performed on the object system is in the affirmative (negative)." This will avoid any possible misunderstanding that the "property" is necessarily an intrinsic, unalienable attribute of the object. See Chapter 9.

such that the proposition $y_j(x_i)$, which reads "the object x_i satisfies or affirms the predicate y_j," is meaningful for all i and j, whether or not it is true. Under this condition one would have to eliminate from Y, for example, the predicate "is red," if X contains an element that has no color. However, one can introduce a modified predicate that says "has color and, in the affirmative case, is red." This predicate is applicable even to an object that has no color and will be negated by this object. The object-predicate table T is a matrix with m rows and n columns, whose element $T(x_i, y_j)$, or T_{ij}, is equal to 1 or 0 according to whether $y_j(x_i)$ is true or false, that is, according to whether object x_i satisfies (affirms) or does not satisfy (negates) predicate y_j. When a collection of objects X is considered, we do not necessarily mobilize all possible meaningful y's applicable to X. Our choice of Y determines the "*scope of observation*" about X, and this cannot be changed by ulterior transformation of predicates (see below).

In general, if the number n of predicates is given, there can formally be at most 2^n different rows, because each of n elements T_{ij} (i: fixed, $j = 1, 2, \cdots, n$) in a row can be either 1 or 0. Each type of row is called an "object type." In an empirically set up object-predicate table, some of the possible object types may be missing, and some of them may be repeated by more than one member of X. If a table T does not contain two identical rows, we say that T is irreducible with respect to X.

Philosophically minded readers may be inclined to classify into categories the reasons for which certain object types are missing from the table. For instance, if y_1 and y_2 respectively stand for "is red" and "is not red," the absence of any object type that satisfies $T_{i1} = T_{i2}$ (is and is not red simultaneously) will be attributed to an "analytical" ground; in particular, to a "logical" ground. If y_1 and y_2 respectively mean "is a bachelor" and "is not married," the absence of types with $T_{i1} = 1 - T_{i2} = 1$ (is a married bachelor) will also be attributed to an analytical ground, but not a logical ground. It is a semantic constraint. If y_1 and y_2 respectively mean "weighs more than 200 lb" and "is less than 4 ft tall," referring to an X consisting of persons, the absence of types with $T_{i1} = T_{i2} = 1$ will be attributed to an empirical law, that is, to a "synthetic" ground. If y_1 and y_2 respectively mean "weighs more than 200 lb" and "is taller than 6 ft" and if types with $T_{i1} = T_{i2} = 1$ are missing from T, the absence will be attributed to the accidental composition of the collection of samples X. From the purely empirical viewpoint we adopt here, however, the distinctions among these different classes become more or less a matter of degree. In other words, complying with the tacitly understood theoretical structure entrenched in our every-day, or scientific, language, we perform experiments and enter the results in our table. Then we discover that some constraints happen in very few cases and some happen persistently in a great many cases. Leibniz's distinction between truths of reason and truths of fact is often compared with the modern (i.e., since Kant) distinction between analytical truth (grounded in meanings) and synthetic truth (grounded in facts). But Leibniz seems to have been closer to our view when he spoke of the truths of reason as true in all possible worlds (in all possible T's, in our case), if we agree that the entries in the tables are intended to be "empirical." A by-product of the present consideration is that we become ready to accept the view, championed by Quine and White (see Quine [Q-2]), that the distinction between synthetic truth and semantic (analytical but not logical) truth is not as clear-cut as traditional philosophy used to assume. On the other hand, admitting that the distinction is language- and convention-dependent, we can nevertheless uphold this distinction as valid during one discussion or one argument.

Carnap's idea of "state description" [C-1] is somewhat similar to our object-predicate table. He takes all possible "atomic" (in the philosopher's sense, not in the sense of lattice theory) statements in a language, that is, statements that are not compound with the help of logical connectives. This corresponds to our Y. A state description is an assignment of truth values to all these statements. Then, in parallelism with our $\hat{Y}$ (see later paragraphs), he introduces sentences compounded by the use of logical connectives. But the trouble is that he considers all combinatorially possible sets of truth values for the "atomic" statements, ignoring the constraints inherent in the language. Carnap then explains that a statement is analytical if it turns out to be true in every possible description. But by this criterion we shall discover, as Quine pointed out [Q-2], only the logically true statements. There can be nonlogical, analytical truths, which can be discovered only by admitting constraints within Y. The "empirical entry" in compliance with a language or a science allows us, in principle, to discover these analytical truths, which Carnap could not discover.

More important in our view than the distinction between analytic and synthetic is the distinction between operational dependence and operational independence of two observations determining y_1 and y_2. If there are two independent observations such as weight and height, y_1 and y_2 are operationally independent. If the observation determining bachelorhood has to be reduced to examining whether or not a man is married, y_1 and y_2 are not operationally independent. If y_2 is the negation of y_1, we usually do not perform two different observations, and they are operationally dependent. Of course, the operation of observation is dependent on our tacitly understood theoretical structure entrenched in language or science.

Starting from the members y_j of a given Y, we can form a collection $\hat{Y}$ of all possible predicates $\hat{y}_j$, which are some logical combinations of the starting y's. More mathematically speaking, $\hat{Y}$ is the smallest complete Boolean lattice that contains all the y_j of Y. In other words, $\hat{Y}$ can be obtained by augmenting the members of Y by the use of $\cap$, $\cup$, and $\neg$, until finally we can produce nothing new. The $\hat{Y}$ thus obtained will be called the Boolean completion of Y. The matrix T can now be extended to cover the entire $\hat{Y}$. The logical operations can be "defined" in terms of the elements T_{ij} of the extended T. If $\hat{y}_j$ and $\hat{y}_k$ belong to $\hat{Y}$, the third member of $\hat{Y}$, generated by conjunction of $\hat{y}_j$ and $\hat{y}_k$, is "defined" as a column having elements given by

$$T(x_i, \hat{y}_j \cap \hat{y}_k) = T(x_i, \hat{y}_j)T(x_i, \hat{y}_k) \quad \text{for all} \quad i. \tag{7.202}$$

The disjunction of $\hat{y}_j$ and $\hat{y}_k$ is "defined" by

$$\begin{aligned} T(x_i, \hat{y}_j \cup \hat{y}_k) = T(x_i, \hat{y}_j) + T(x_i, \hat{y}_k) \\ - T(x_i, \hat{y}_j)T(x_i, \hat{y}_k) \quad \text{for all} \quad i. \end{aligned} \tag{7.203}$$

The negation of $\hat{y}_j$ is "defined" by

$$T(x_i, \neg\hat{y}_j) = 1 - T(x_i, \hat{y}_j) \quad \text{for all} \quad i. \tag{7.204}$$

The reader will have noticed that these "definitions" are identical with the relations determining the values of the characteristic function (truth values) introduced in (7.147)–(7.149). He can further check for himself that these "definitions" satisfy the axioms introduced in Section 7.1, which are supposed to be obeyed by the operations of disjunction, conjunction, and negation in any complemented Boolean lattice. The constant absurdity and the constant truth are "defined" here by

$$T(x_i, \emptyset) = 0, \qquad T(x_i, \square) = 1 \quad \text{for all} \quad i. \tag{7.205}$$

The equivalence $\hat{y}_j = \hat{y}_k$ may be written as

$$T(x_i, \hat{y}_j) = T(x_i, \hat{y}_k) \quad \text{for all} \quad i. \tag{7.206}$$

The implication $\hat{y}_j \to \hat{y}_k$ (i.e., $y_j \cap \neg y_k = \emptyset$) can obviously be expressed by

$$T(x_i, \hat{y}_j)(1 - T(x_i, \hat{y}_k)) = 0 \quad \text{for all} \quad i. \tag{7.207}$$

It should be recalled that a relation that holds among various columns thus produced in a particular object-predicate table may or may not be attributable to a "logical" relation derivable from the logical expressions (in terms of the logical connectives) corresponding to these columns. They may have a relation peculiar to the data contained in the starting T. Thus, for instance, if rows i satisfying $T_{i1} = 1 - T_{i2} = 1$ are missing from a particular table, the two columns corresponding to $\square$ and $\neg y_1 \cup y_2$ will have exactly the same entries. The implied relation $y_1 \to y_2$ may be or may not have a logical origin. It should also be noted that the extension of T to cover $\hat{Y}$ does not produce any new information not included in the original T, which covered only Y. The "scope of observation" is not changed. On the other hand, any complicated or elevated property that can be extracted from the original Y is found in $\hat{Y}$.

Any hidden property embedded in the scope of observation has its expression in $\hat{Y}$. Among the members of $\hat{Y}$ there must be many that have no intuitive appeal, but some of them could be of unsuspected utility.

Let us assume that the original table T is irreducible; that is, it has m different rows. This m may or may not be equal to its maximum number 2^n, where n is the number of the members in the starting Y. It is easy to see that the extended T will have 2^m different columns (predicate type). First, there will be a column that contains 0's all the way. This column can be produced by $T(x_i, y_j \cap \urcorner y_j) = T_{ij}(1 - T_{ij}) = 0$ for all i. This means that $\emptyset$ is included in $\hat{Y}$. Next there will be a column that has 1 in any given row (say, the kth) and 0 in all the other rows. This column can be produced by the following procedure. For a given x_k let the original Y be divided into two parts $Y^{(1)}$ and $Y^{(2)}$ such that $T_{kj} = 0$ for $y_j \in Y^{(1)}$ and $T_{kj} = 1$ for $y_j \in Y^{(2)}$. Form a member η_k of $\hat{Y}$, which is defined as the conjunction of every y_j belonging to $Y^{(2)}$ and of the negation of every y_j belonging to $Y^{(1)}$. This η_k will certainly have 1 in the kth row, because this element is the product of n terms, each of which is 1. This η_k has to have 0 on any other row, for such an element is the product of n terms, of which at least one must be different from the corresponding value for the kth row. Indeed, all m rows are assumed to be different in the original table. We may write

$$\eta_k = \bigcap_{j=1}^{n} (\urcorner)^{T_{kj}+1} y_j \tag{7.208}$$

with the understanding that $(\urcorner)^1$ is a simple negation and $(\urcorner)^2$ is a double negation.

The operation of disjunction (of $\hat{y}_j$ and $\hat{y}_l$) is putting 1 whenever we have 1 on $\hat{y}_j$ or on $\hat{y}_l$ or on both, and putting 0 otherwise. Hence, unless $\hat{y}_j$ and $\hat{y}_l$ have identical entries or either of them is equivalent to $\emptyset$, the number of 1's on the disjunction is larger than that in either of the columns $\hat{y}_j$ or $\hat{y}_l$. Hence, making a disjunction $\eta_i \cup \eta_k$ of η_i and η_k (respectively corresponding to the ith and kth rows), we obtain a new member of $\hat{Y}$ whose column has two 1's, one on the ith row and the other on the kth row. Repeating the process, we see that any predicate type, that is, any conceivable distribution of 0's and 1's in a column, is to be found in the $\hat{Y}$. Thus the extended T will contain 2^m different columns. The number of 1's in a column is obviously the dimension of the corresponding predicate, as discussed in connection with (7.142)–(7.144).

We can also see that if $\hat{y}_j \to \hat{y}_l$, then, on whichever row where there is 1 in the jth column, there will be 1 in the lth column too. This can be seen from (7.207) or from the relation $\hat{y}_j \cup \hat{y}_l = \hat{y}_l$ with the help of the foregoing paragraph. It follows that a relation $\hat{y}_j \to \eta_k$ means that $\hat{y}_j$ cannot have more than one 1, and if it has one 1 it must coincide with the 1 possessed by η_k.

Hence $\hat{y}_j$ must be equivalent either to $\emptyset$ or to η_k. This shows, in view of what has been said in relation to (7.117) that η_k is an atom (in the lattice-theoretical sense). This means that each row in an irreducible table (or, more generally, each objective type in a reducible table) corresponds one-to-one to an atom of the completed $\hat{Y}$-lattice.

If the original $n \times m$ table is irreducible with respect to X, the completed $\hat{Y}$ will have 2^m elements because $\hat{Y}$ has m atoms and there are 2^m different ways to pick some (0 to m) atoms and form their disjunction. On the other hand, starting from the n given predicate, we can build 2^{2^n} formally different combinations with the help of (7.202)–(7.204). In other words, for a given n there could be 2^n different rows or object types, and by putting $m = 2^n$ we would get a completed $\hat{Y}$ with $2^m = 2^{2^n}$ elements. In a general case we have $m \leq 2^n$, meaning that some possible rows and corresponding atoms may be missing. Starting from a table with a maximum number of rows $m = 2^n$, suppose that one object type, say, x_m, is eliminated. By this omission a pair of two different predicates $\hat{y}_j$ and $\hat{y}_j \cup \eta_m$ will become identical, where $\hat{y}_j$ does not contain η_m. This is because the η_m obtained in the original table with 2^n rows has only 0's in its columns in the new retrenched table with $2^n - 1$ rows. Hence η_m acts as a $\emptyset$ in this modified table. This means that each time we omit one object type, two different predicates are caused to coalesce, reducing the number of elements of $\hat{Y}$ to one-half. This explains the mechanism by which the number of elements of $\hat{Y}$ is reduced to 2^m from its maximum number 2^{2^n} when $m < 2^n$. The reader is urged to examine what has been said in the foregoing paragraphs using Table 7.4 as an example.

It may be noted here that a "permutation of atoms," which we talked about in the last section, corresponds, in the object-predicate table, to a relabeling of objects x_i, $i = 1, 2, \cdots, m$. This may be considered as a consequence of a permutation of corresponding atoms η_k, $k = 1, 2, \cdots, m$. Some of the η_k within Y will be moved outside Y (but of course within $\hat{Y}$), and some of the η_k outside Y will be moved into Y. In another word, Y is not invariant for a permutation of atoms, but $\hat{Y}$ is.

It is often useful to introduce a function of a predicate or of a group of predicates, which we call "dimension" for economy of terminology because it is in a sense a generalization of what was so denoted previously. The dimension $D(\hat{y}_j, \hat{y}_k, \cdots, \hat{y}_l)$ of a group of predicates $(\hat{y}_j, \hat{y}_k, \cdots, \hat{y}_l)$ is defined by

$$D(\hat{y}_j, \hat{y}_k, \cdots, \hat{y}_l) = \sum_{i=1}^{m} T(x_i, \hat{y}_j) T(x_i, \hat{y}_k) \cdots T(x, \hat{y}_l), \qquad (7.209)$$

which is nothing but the number of rows that have 1's in the columns corresponding to $(\hat{y}_j, \hat{y}_k, \cdots, \hat{y}_l)$. Thus $D(\hat{y}_j)$ is the number of 1's in the column of $\hat{y}_j$, that is, the dimension of the predicate y_j, as already used in (7.142). $D(\hat{y}_j, \hat{y}_k)$ is the number of overlapping 1's in the columns of $\hat{y}_j$ and

Table 7.4

		Y							
		Y							
Rank		1	2	0	1	1	2	2	3
	y \ x	$y_1 = y_1 \cap y_2$	$y_2 = y_1 \cup y_2$	$\emptyset_y = y_1 \cap \neg y_2$	$\neg y_2 = \neg y_1 \cap \neg y_2$	$\neg y_1 \cap y_2 = y_1 \uplus y_2$	$\neg y_1 = \neg y_1 \cup \neg y_2$	$y_1 \cup \neg y_2 = \neg(y_1 \uplus y_2)$	$\square_y = \neg y_1 \cup y_2$
X	x_1	0	0	0	1	0	1	1	1
	x_2	0	1	0	0	1	1	0	1
	x_3	1	1	0	0	0	0	1	1

Since there are two original predicates ($n = 2$), there could be four ($2^n = 4$) object types. But one of them is missing ($m = 3$), giving rise to a constraint $y_1 \to y_2$. As a result, each predicate that can be formed in terms of y_1 and y_2 coincides with another, reducing the total number of elements of the completed Boolean lattice $\hat{Y}$ from its maximum value of 16 ($2^{2^n} = 16$) to 8 ($2^m = 8$). The atoms corresponding to x_1, x_2, and x_3 are respectively $\neg y_2$, $\neg y_1 \cap y_2$, and y_1. $\neg y_2 = \eta_1$, $\neg y_1 \cap y_2 = \eta_2$, and $y_1 = \eta_3$. The lattice structure is the same as Figure 7.15 if we replace A and B by y_1 and y_2, respectively.

$\hat{y}_k$. Evidently $D(\hat{y}_j, \hat{y}_k) = D(\hat{y}_j \cap \hat{y}_k)$. The reader can prove, for exercise, that the relation $\hat{y}_j \rightarrow \hat{y}_k$, which can be expressed as (7.207), is also equivalent to

$$D(\hat{y}_j, \hat{y}_k) = D(\hat{y}_j). \tag{7.210}$$

It is interesting that everything done in the foregoing paragraphs can also be done when we interchange the roles of X and Y, because the basic assumption of the entire discussion is that we are given a rectangular matrix with entries that are either 0 or 1. In fact, in later paragraphs we use notions that can be obtained from the ones already explained by an interchange of X and Y. In most cases the interchange of X and Y involves hardly any difficulty from the mathematical point of view. The difficulty, if any, arises mainly from the interpretation. The formal interchangeability of X and Y will be referred to in the following as *object-predicate reciprocity* and has to be distinguished from the well-known logical reciprocity enounced in Theorem 7.19. It is interesting that in some combinatorial problems associated with quantum statistics we often interchange the roles of particle (object) and state (predicates) to facilitate computation. This is related to the fact that in quantum mechanics particles lose their individual identities, but quantum states do not.

By the same method used in extending Y to $\hat{Y}$, we can also construct a completed Boolean lattice $\hat{X}$ starting from a given X. This amounts to adding all object types that were missing from the original table. If the absence of these object types was accidental, their addition does not cause any difficulty in interpretation. But if the absence is such that one would usually attribute it to an analytical cause, we have to exercise a great deal of imagination and mental flexibility to understand the meaning of these "fictitious" objects. Nonetheless, a formal discussion involving $\hat{X}$ proves to be useful and productive. If we take the matrix of Table 7.4 with three rows as the starting point, the total number of elements of the completed $\hat{X}$ could be $2^{2^3} = 256$, because there are three rows giving rise to $2^3 = 8$ possible predicate-types as given by $\hat{Y}$, and the number of possible rows is 2^8. However, in the original X-Y matrix there appear only two of them (y_1 and y_2), implying that there are $8 - 2 = 6$ "constraints" (in the reciprocal sense) reducing the number of elements of $\hat{X}$ from its maximum value 256 to $2^{8-6} = 4$. Thus there is one more line to add to complete the $\hat{X}$, namely, (1, 0) under y_1 and y_2.

It is often stated in traditional philosophy that the notion of concept has two aspects, called extension and intension, which are inversely proportional to each other; extension measures the richness of a concept in terms of the participation of individuals in the class corresponding to the concept, whereas intension measures the richness of the concept in terms of the specification of the properties defining the concept. We give a nonprobabilistic explication of this idea in the following paragraphs, relegating its probabilistic explication or probabilistic generalization to Section 8.3.

Suppose that we are given n starting predicates $y_1, y_2, \cdots, y_n$ and assume first, for simplicity, that the object predicate table shows no constraint, that is, $m = 2^n$, and no repetition (irreducible). The completed $\hat{Y}$ will contain $n = 2^m = 2^{2^n}$ members, but we consider for the moment only predicates of a particular type, namely, those that can be expressed as a conjunction of some factors of which each is either one of the original y's or its negation. (It is easy to see that there are 3^n predicates of this type for a given set of n starting predicates, because each predicate may appear in affirmation or negation or not appear at all.) A predicate of this type can be located on a "dichotomic taxonomy tree," which is defined as follows. First, we take $\square$ as the trunk of the tree. It splits into two branches, y_1 and $\neg y_1$. These two branches are split again into two branches, y_2 and $\neg y_2$, respectively. We thus obtain seven branches $\square$, y_1, $\neg y_1$, $y_1 \cap y_2$, $y_1 \cap \neg y_2$, $\neg y_1 \cap y_2$, and $\neg y_1 \cap \neg y_2$. We continue this process until the last peripheral branches finally become 2^n atoms, given by (7.208). Any predicate of the type specified above can be located as a branch of a tree of this kind, provided that the order of taking the y's is suitably chosen. (For instance, predicates of the allowed type y_2 and $\neg y_2$ are not found on the tree described above, but are found on a tree obtained by interchange of y_1 and y_2.)

Starting from $\square$, as we proceed toward the peripheral branches, the specification of predicates increases. Let us introduce a function of a predicate $R(\hat{y}_j)$ that counts the number of branching points we have to go through to reach the branch $\hat{y}_j$ starting from the trunk $\square$. [We see presently that $R(\hat{y}_j)$ does not depend on the tree when $\hat{y}_j$ is locatable on more than one tree.] This $R(\hat{y}_j)$ may well correspond to the notion of intension. On the other hand, each predicate has a fixed number, dimension $D(\hat{y}_j)$, which counts the number of atoms. Since we assume that the table has no repetition, the number of atoms is the number of objects. Hence $D(\hat{y}_j)$ can be considered as the number of particular objects participating in the class belonging to $\hat{y}_j$. In this sense $D(\hat{y}_j)$ must correspond to extension. Now $R(\hat{y}_j)$ and $D(\hat{y}_j)$ are not independent. In fact, every time one goes through a branching part, the number of atoms becomes one-half. The trunk $\square$ contains m atoms and the peripheral branch contains 1 atom. In general we have

$$D(\hat{y}_j) = m(\tfrac{1}{2})^{R(\hat{y}_j)}. \tag{7.211}$$

[This shows, incidentally, that $R(y_j)$ does not depend on the tree because $D(y_j)$ does not.] If we define the intension and extension of $\hat{y}_j$ by

$$\begin{aligned} \text{Int}\,(\hat{y}_j) &= 2^{R(\hat{y}_j)} \\ \text{Ext}\,(\hat{y}_j) &= D(\hat{y}_j) \end{aligned} \tag{7.212}$$

in accordance with the meaning of D and R as explained above, we get, from (7.211),

$$\text{Ext}\,(\hat{y}_j) \cdot \text{Int}\,(\hat{y}_j) = m, \qquad (7.213)$$

where m is a constant that does not depend on $\hat{y}_j$. This result agrees nicely with the traditional notion of extension and intension. It is sometimes convenient to use what may be called the logarithmic extension ext $(\hat{y}_j)$ and logarithmic intension int $(\hat{y}_j)$, which satisfy

$$\text{ext}\,(\hat{y}_j) + \text{int}\,(\hat{y}_j) = \log_2 m, \qquad (7.214)$$

where

$$\text{ext}\,(\hat{y}_j) = \log_2 \text{Ext}\,(\hat{y}_j) = \log D(\hat{y}_j), \qquad (7.215)$$

$$\text{int}\,(\hat{y}_j) = \log_2 \text{Int}\,(\hat{y}_j) = R(\hat{y}_j).$$

Under the present assumption (no constraint, no repetition) we can replace $\log_2 m$ in (7.214) by n. But for the purpose of generalization it is more desirable to leave it $\log_2 m$.

The first generalization is to define intension and extension for those $\hat{y}_j$ that cannot be located on a branch of a dichotomic taxonomy tree. It may be then natural to use (7.214) inversely to define int $(\hat{y}_j)$, since ext $(\hat{y}_j)$ can be defined without referring to a tree.

$$\text{int}\,(\hat{y}_j) = \log_2 m - \log D(\hat{y}_j). \qquad (7.216)$$

Next we have to consider the case in which there are constraints. One way of looking at this case is, as we did once before, to treat the problem as if there were no constraints, but to remember that the atom corresponding to the missing object is $\emptyset$ in reality. If we apply this approach to the present case, we can use the same tree, but instead of (7.211) we shall have

$$D(\hat{y}_j) \leq 2^n(\tfrac{1}{2})^{R(\hat{y}_j)}. \qquad (7.217)$$

With the use of (7.216) and the first line of (7.215) we have to replace the relation (7.214) by

$$\text{ext}\,(\hat{y}_j) + \text{int}\,(\hat{y}_j) \leq n. \qquad (7.218)$$

This expression includes the possibility that extension will not decrease when intension increases, which is assigned great importance by some writers. This happens simply because a logically possible atom (or some disjunction of atoms) is actually not occupied by a real object. To use Quine's example, the conjunction of "a creature with a heart" and "a creature with kidneys" does not restrict the extension, because the class of creatures with (without) hearts and without (with) kidneys is empty.

Another and perhaps better way of coping with the case with constraints is to abandon the tree picture and resort to a lattice graph, exemplified by Figures 7.14 and 7.15. Here we note that the higher the level, the less specific the characterization, hence the larger the participation of individuals, and that the lower the level, the more specific the characterization, hence the smaller the participation of individuals. Then we can retain (7.214) and (7.215), where $R(\hat{y}_j)$ is now divorced from the tree picture, and we should use (7.216) for the definition of int $(\hat{y}_j)$. We should note that ext $(\hat{y}_j)$ and int $(\hat{y}_j)$ are essentially the altitude and the depth in the lattice graph in a logarithmic scale.

Next we consider a topic that has a certain connection with what some philosophers considered as the essence of "simplicity" of predicates. Their ultimate intention seemed to reduce the entire process of induction to a formally describable prescription, and this involved an attempt to explicate the principle of simplicity in terms of logico-mathematically definable concepts. However, this motivation seems to me to be oriented in the wrong direction. As was made clear in previous chapters, the criterion of simplicity used in induction is admittedly extra-evidential, but this does not mean that it is unempirical. As in most other extra-evidential evaluations, judgement of simplicity is not purely a formal matter. In fact, predicates that exhibit the same degree of simplicity from a formal logico-mathematical point of view do not appeal to our unbiased minds as equally simple. Some concepts seem simple to us merely because they are familiar. Even this kind of primitive, nonformal notion of simplicity plays a useful role in actual induction.

In spite of the misoriented motivation of those who initiated this kind of consideration, it is interesting that a quantity can be defined to characterize a certain aspect of predicates, which may be called "formal simplicity" or "formal complexity." The quantity we introduce will turn out to be in agreement with what Kemeny [K-5] called complexity, although Goodman [G-11] refused to call it by this name. Goodman claimed that certain other quantities were the correct explicata of simplicity or complexity. This controversy appears to be a bit infantile, because neither of these formal concepts corresponds to a universally recognized explicandum, and the controversy degenerates into a question of convenient nomenclature.

Since we consider not only predicates with one argument, but also predicates with more than one argument, we should make a distinction between the set of objects $X = \{x_i\}$, $i = 1, 2, \cdots, m$, which appear in the object-predicate table, and the set of constituent items $\Xi = \{\xi_\rho\}$, $\rho = 1, 2, \cdots, \mu$. For instance, if each object x_i is a pair of two items (ξ_ρ, ξ_σ), each index i can be regarded as standing for a pair of indices (ρ, σ), $\rho, \sigma = 1, 2, \cdots, \mu$. Hence the number of possible pairs is $m = \mu^2$ in this case.† In the following it is important to note that for a given predicate y its ext (y) will depend on the

† This of course permits repetition of the same constituent items, (ρ, ρ) and considers the ordering of items meaningful; that is, pairs (ρ, σ) and (σ, ρ) are considered to be distinct.

choice of Ξ. If y_1 is a one-argument predicate standing for "is red" and if we take only white items in Ξ, ext (y_1) will be zero because in the object set X that coincides with Ξ here, y_1 is equivalent to $\emptyset$. But if we take only red items in Ξ, ext (y_1) will become μ, provided that Y is large enough to distinguish each item in X.

The concept of complexity applies primarily to a "type" of predicate; only secondarily can a predicate be assigned a numerical value of complexity as a member of some type. First, the number of arguments in a predicate is one of the characterizations of a type. The symmetry property with regard to the arguments (when there are more than one) is a second means of characterization. For example, if $x_{\rho\sigma} = (\xi_\rho, \xi_\sigma)$ and if $y_j(x_{\rho\sigma})$ stands for "ξ_ρ and ξ_σ are the same person, or brothers," y_j is a symmetrical two-argument predicate, for $y_j(x_{\sigma\rho})$, which states "ξ_σ and ξ_ρ are the same person, or brothers," is equivalent to $y_j(x_{\rho\sigma})$. If $y_j(x_{\rho\sigma})$ stands for the statement "ξ_ρ is older than ξ_ρ," the y_j is an antisymmetrical, two-argument predicate, because $y_j(x_{\rho\sigma})$ and $y_j(x_{\sigma\rho})$ are the opposite of each other. If $y_j(x_{\rho\sigma})$ stands for "x_σ is purple and x_ρ is yellow," $y_j(x_{\rho\sigma})$ and $y_j(x_{\sigma\rho})$ are disjoint (mutually exclusive) but not the negation of each other. Hence this last y_j is neither symmetrical nor antisymmetrical. Two-argument predicates may be symmetrical or antisymmetrical or neither of them but refer to two items simultaneously.

The complexity of a predicate y as a member of type τ is defined by $C_\tau(y)$, expressed in terms of μ and given by

$$C_\tau(y_j) = \max_{\Xi} \text{Ext}\,(y_j), \qquad y_j \in \tau, \tag{7.219}$$

where it is assumed that Y consists of all members of τ (applicable to X for which y is meaningful). For a given y_j the item set Ξ is supposed to be chosen so as to maximize Ext (y_j). For instance, let y stand for "is red" and τ be the class of general one-argument predicates. Taking only red items, we can then make $C_\tau(y_j) = \mu$, provided that Y is sufficiently detailed to tell μ items apart. Each of the μ items will correspond to an atom, and y_j will act like a $\square$ for this X, including all atoms.

Let $y_j(x_{\rho\sigma})$ stand for "ξ_ρ is the same person as, or a sibling of, ξ_σ" and let τ be a class of symmetrical two-argument predicates. In order to make Ext (y_j) large, we should take μ persons who are all mutually siblings. Y will consist of symmetrical predicates such as "the total weight of ξ_ρ and ξ_σ is over 200 lb." Then X will consist of μ^2 entries, but the rows $x_{\rho\sigma}$ and $x_{\sigma\rho}$ are indistinguishable by the available predicates in Y. This means there are only $\mu(\mu + 1)/2$ different object types (hence atoms), because there are $\mu(\mu - 1)/2$ different rows of the type $\rho \neq \sigma$ and μ different rows of the type $\rho = \sigma$. And in the most favorable case, the column of a y, there will be 1's all the way. Hence the complexity of a symmetrical two-argument predicate is $\mu(\mu + 1)/2$. Similarly, the complexity value of an antisymmetrical two-argument predicate is $\mu(\mu - 1)/2$. This is because there are $\mu(\mu - 1) + 1$ atoms, but each column can include at most $\mu(\mu - 1)/2$ of them.

It should be noted that the same predicate can belong to more than one type and hence have more than one complexity number. Consider, for instance, the predicate $y_j(x_{\rho\sigma})$, which says that "ξ_ρ and ξ_σ are such that either they are the same person or mutually brothers." This y_j, as a general two-argument predicate, has complexity value μ^2, while the same predicate as a symmetrical two-argument predicate, has a complexity value of $\mu(\mu + 1)/2$. The reason is that in the former case we have to introduce all two-argument predicates into Y, as a result of which there will be μ^2 atoms and the rows $x_{\rho\sigma}$ and $x_{\sigma\rho}$ will correspond to separate atoms. Thus the complexity value is peculiar to a type of predicate rather than to an individual predicate. In any event, although it is of some interest that such a complexity value is definable, it is very questionable that this quantity has any deep implication or any practical utility.

Before leaving this section we should mention another effort of formalists who introduced the concept of *a priori* probability of a statement, not because we attach any significance, theoretical or practical, to this concept, but because we should know how dangerous such a formalistic concept can be. Their idea was introduced in connection with the theory of confirmation in induction.

Suppose that a hypothesis H gives a certain deductive probability $p(E \mid H)$ to an experimental event E. If we denote by $\neg H$ the other alternative (which may be the disjunction of all competing hypotheses contradicting H), we could write our Baysian formula in the form

$$q(H \mid E) = \frac{q(H)p(E \mid H)}{q(H)p(E \mid H) + q(\neg H)p(E \mid \neg H)}, \tag{7.220}$$

assigning a suitable value to $p(E \mid \neg H)$. (For an example of such an assignment, see the example of indoor ornithology, Section 6.5.) This credibility $q(H \mid E)$ naturally depends on the extra-evidential probabilities and hence is not a measure of confirmation. A measure of confirmation, as discussed in detail in Chapter 4, must be free from the extra-evidential probabilities $q(H)$ and $q(\neg H)$. One such quantity can be obtained from (7.220) by giving an equal value to $q(H)$ and $q(\neg H)$, that is, by resorting to "blind retrodiction":

$$q_0(H \mid E) = \frac{p(E \mid H)}{p(E \mid H) + p(E \mid H)}. \tag{7.221}$$

We previously noted that a measure of confirmation has to be a monotonically increasing function of likelihood, $p(E \mid H)$, and (7.194) has such a property. What Kemeny and Oppenheim [K-4] called "degree of factual support,"

$f(H, E)$, is formally related to this blind credibility $q_0(H \mid E)$ by a simple relation,

$$f(H, E) = 2q_0(H \mid E) - 1 = \frac{p(E \mid H) - p(E \mid \neg H)}{p(E \mid H) + p(E \mid \neg H)}. \qquad (7.222)$$

An essential difference between our $q_0(H \mid E)$ and their $f(H, E)$, however, stems from the special interpretation they attached to the probability $p(E \mid H)$. This special interpretation is possible because of the special type of statements they considered as H. (An exception is their Example 19.) In our case $p(E \mid H)$ is the predictive probability of E according to the operationally testable contents of H. In their case the probability is reduced to a counting of the *a priori* possible cases logically compatible with H; hence it is a syntactic concept.

Suppose that there are m objects and n predicates. Let $y_j(x_i)$ stand for the proposition that object x_i satisfies the predicate y_j, where $i = 1, 2, \cdots, m$; $j = 1, 2, \cdots, n$. Let $\neg y_j(x_i)$ be its negation Each "possible world" can be described by an object-predicate table $\|T_{ij}\|$ with elements 0 or 1, or by a combined proposition

$$\bigcap_i \bigcap_j (\neg)^{1+T_{ij}} y_j(x_i). \qquad (7.223)$$

The proposition (7.223), which Carnap called "state description," corresponds to the assertion that the world is described by the table T_{ij}. For a given pair (m, n) there can be 2^{mn} different tables, because each element can be 0 or 1. We can now consider a complete lattice whose atoms are given by 2^{mn} propositions of the type (7.223). (The word atom is used in the sense of lattice theory, so that one should not be shocked at seeing that each possible "world" is called an "atom.") Kemeny and Oppenheim gave each of the 2^{mn} atoms an equal *a priori* probability of $(1/2)^{mn}$. Then any proposition in this lattice is a disjunction of some of the atoms of this type (7.223) and receives a probability equal to the number of atoms times $(1/2)^{mn}$. In the Kemeny-Oppeheim theory E and H are limited to propositions in this lattice; hence they could define their probability by this method. For instance, the hypothesis that all objects satisfy a particular predicate y_1 can be written as $\bigcap_i^m y_1(x_i)$ and its probability is $(1/2)^m$, because there are $2^{m(n-1)}$ atoms of the type (7.223) that belong to this hypothesis. The last number can easily be obtained by noting that the matrix T_{ij} must have 1 for $j = 1$ and that it is otherwise arbitrary. They proceed in a similar way for the event E. We have explained here only the case in which the predicates are independent.

The trouble with this kind of probability is its complete dependence on language. If, for instance, the language is such that there is only one predicate to describe the weather, say, snowy, the probability that one arbitrary day will be snowy becomes $\frac{1}{2}$. If the language has rainy and snowy as disjoint predicates, snow will receive probability $\frac{1}{3}$. It is obvious that the conclusions derivable from this kind of probability are unrealistic. Scientific hypotheses are usually so formulated that the predictive probability of an operationally defined event is derivable from them in such a way that the probability does not depend on the whims of the language. Nonetheless, the theory of Kemeny and Oppenheim seems to be about the best one can get along the line of Carnap's idea of induction. It should be remembered, however, that their rather neat results were obtained because they divorced their quantity from the idea of "creditation" and based it solely on what we call confirmation. Thus Carnap's plan of fusing these two ideas was not realized by Kemeny and Oppenheim.

7.6. THEOREM OF THE UGLY DUCKLING

The purposes of this section is to show that from the formal point of view there exists no such thing as a class of similar objects in the world, insofar as all predicates (of the same dimension) have the same importance. Conversely, if we acknowledge the empirical existence of classes of similar objects, it means that we are attaching nonuniform importance to various predicates, and that this weighting has an extralogical origin.

When we employ a concept, we usually understand that there is a group of objects corresponding to this concept that any two members of the group resemble each other more than a member and a nonmember. Two sparrows are very much alike, while a sparrow and a rose are not alike. It is natural to translate the term "to resemble" as "to share many predicates in common." But this interpretation can be shown to lead to a denial of the existence of a class of similar objects by the following theorem, which I have dubbed the theorem of the ugly duckling. The reader will soon understand the reason for referring to the story of Hans Christian Andersen, because this theorem, combined with the foregoing interpretation, would lead to the conclusion that an ugly duckling and a swan are just as similar to each other as are two swans. It is curious that when I have talked about the statement and the proof of the theorem on different occasions since 1961, some people have manifested their surprise and delight, while others grumbled that they knew something like this must be true. But when I asked the latter group of people where they had read or written it, I could get no clear answer [W-18].†

Theorem 7.20 **(*Theorem of the ugly duckling*)** The number of those predicates $\hat{y}_j$ is a completed Boolean lattice $\hat{Y}$ of predicates satisfied simultaneously by two nonidentical objects x_i and x_k (of the list of objects X) is a

† I alluded to this theorem in a lecture in 1961 [W-17] and stated and proved it rigorously in a lecture at Brussels in 1962. The latter lecture is included in [W-18].

fixed constant independent of the choice of the two objects. (See Section 7.5 for the meaning of the symbols.)

By "nonidentical" is meant "belonging to two different object types," that is, "corresponding to two different rows in the object-predicate table." To be able to speak of a completed lattice $\hat{Y}$, we must fix a "scope of observation." But every time we change the scope we get a new $\hat{Y}$, and the theorem holds again in this new $\hat{Y}$.

Proof. Suppose that there are m different rows (object types) in the object-predicate table, which means that there are correspondingly m atoms in the lattice of $\hat{Y}$, and $\hat{Y}$ has 2^m different members. Any predicate $\hat{y}$ in $\hat{Y}$ is a disjunction of a certain number of these atoms. A predicate shared by x_i and x_k is characterized by the fact that it contains the two atoms corresponding to these two object types. It can contain any number of remaining $(m - 2)$ atoms. There are 2^{m-2} different ways of taking some (or none) of the $(m - 2)$ different atoms. Hence 2^{m-2} different predicates are shared by these two objects, and this number, of course, does not depend on the choice of the two objects insofar as they belong to two different rows (nonidentical objects).

More precisely, there are $\binom{m-2}{r}$ different ways of taking r out of the remaining $(m - 2)$ predicates. Hence $\binom{m-2}{r}$ different predicates of dimension $(r + 2)$ are shared by these two objects. Of course the number 2^{m-2} mentioned above is obtained by $\sum_{r=0}^{m-2}\binom{m-2}{r}$. Hence we can assert a theorem a little more precise than Theorem 7.20, namely, the following.

Theorem 7.21 The assertion that is obtained by substituting "those predicates of a given dimension" for "those predicates" in Theorem 7.20.

Since the number of predicates in question above is decided by the different ways of taking some out of the $(m - 2)$ predicates, omitting from consideration the two atoms corresponding to x_i and x_k, the same proof also applies to the following theorem.

Theorem 7.22 The number of predicates (of a given dimension r) that are satisfied by neither x_i nor x_k and the number of predicates [of dimension $(r + 1)$] that are satisfied by x_i but not by x_k are equal to each other and also to the number considered in Theorem 7.20 (7.21) [of dimension $(r + 2)$]. They are all independent of the pair (i, k) chosen.

These three theorems establish that any two objects are equally as similar to each other as any other two objects, and are equally as dissimilar to each other as any other pair, insofar as the number of shared predicates is regarded as a measure of similarity and the number of predicates that are not shared is regarded as an indication of dissimilarity.

We now give a generalization of the above theorems. At first glance it may appear to lack intuitively understandable meaning, but its application will be found in the next chapter. In the foregoing theorems there were two restrictions: the number of objects was two, and the predicates were characterized by the four categories. They were shared (1, 1) or not shared [(0, 1) or (1, 0)], or simultaneously negated (0, 0), where the two numbers indicated here are T_{ij} and T_{kj} (i and k refer to the objects and j refers to the predicate in question). We can also merge two classes (0, 1) and (1, 0); then the number of predicates of this merged class is the number of "unshared" predicates.

Let us consider a group of μ objects $\{x_{i_1}, x_{i_2}, \cdots, x_{i\mu}\}$ and an ordered set of μ numbers, which are 0 or 1: $A^{(\lambda)} = \{\alpha_1^{(\lambda)}, \alpha_2^{(\lambda)}, \cdots, \alpha_\mu^{(\lambda)}\}$. ($\lambda$ is fixed for the time being.) We can characterize a predicate $\hat{y}_j$ by testing the condition that there appear in its column the numbers $(\alpha_1^{(\lambda)}, \alpha_2^{(\lambda)}, \cdots, \alpha_\mu^{(\lambda)})$, respectively, in rows $(i_1, i_2, \cdots, i_\mu)$. In other words, $\hat{y}_j$ can be such that either $T(i_\rho, j) = \alpha_\rho$ $(\rho = 1, 2, \cdots, \mu)$ or not. Let us denote by $A^{(\lambda)}(\hat{y}_j)$ the proposition that $\hat{y}_j$ is such that it satisfies this condition. Now we consider ω sets of numbers $A^{(\lambda)}$, $\lambda = 1, 2, \cdots, \omega$, and we ask if the condition $A^{(\lambda)}(\hat{y}_j)$ with any one of the ω sets of numbers is true for the group of objects $(i_1, i_2, \cdots, i_\mu)$. In other words, we test whether $A(\hat{y}_j) = \bigcup_{\lambda=1}^{\omega} A^{(\lambda)}(\hat{y}_j)$. If we allow all possible combinations of 0 and 1, as the set $A^{(\lambda)}$ of μ numbers (i.e., if $\omega = 2^\mu$), any $\hat{y}_j$ will satisfy $A(\hat{y}_j)$; that is, $A(\hat{y}_j)$ would then become a tautology. The definition of $A(\hat{y}_j)$ is applicable to a group of μ objects $(i_1, i_2, \cdots, i_\mu)$; hence its truth value as a predicate of y_j is also a function of these objects in principle.

To have some mental picture of the notions involved here, let us consider the case of three objects, say, x_1, x_2, and x_3, and characterize a predicate by the condition that it is satisfied by one and denied by the other two. Then we take three sets ($\omega = 3$) of three numbers, $A^{(1)} = \{1, 0, 0\}$, $A^{(2)} = \{0, 1, 0\}$, and $A^{(3)} = \{0, 0, 1\}$, and write $A^{(1)}(\hat{y}_j)$ for the condition that $\hat{y}_j$ is such that $(T_{1j}, T_{2j}, T_{3j}) = (1, 0, 0)$, and so on. The property mentioned above will then be denoted $A(\hat{y}_j) = A^{(1)}(\hat{y}_j) \cup A^{(2)}(\hat{y}_j) \cup A^{(3)}(\hat{y}_j)$. Now the generalized theorem is as follows.

Theorem 7.23 **(*Generalized theorem of the ugly duckling*)** If $A(\hat{y}_j)$ is defined as above, the number of predicates $\hat{y}_j$ satisfying it in the completed predicate lattice $\hat{Y}$ is a constant determined by the set of numbers $A^{(\lambda)}$, $\lambda = 1, 2, \cdots, \omega$, and independent of the choice of the μ objects used in the definition of $A(\hat{y}_i)$.

Proof. The completed $\hat{Y}$ is invariant, as we saw in Section 7.2, for a permutation of its atoms. Suppose that we used μ objects $(i_1^*, i_2^*, \cdots, i_\mu^*)$ instead of the original μ objects $(i_1, i_2, \cdots, i_\mu)$; then there is at least one permutation of m objects $(1, 2, \cdots, m)$, which includes passage from $(i_1, i_2, \cdots, i_\mu)$ to $(i_1^*, i_2^*, \cdots, i_\mu^*)$. Corresponding to this permutation, there will be a permutation of atoms, resulting in a new labeling of each predicate in $\hat{Y}$. Whatever was true in $\hat{Y}$ using the old labeling of predicates will become true in $\hat{Y}$ using the new labeling. Hence the number of predicates in $\hat{Y}$ satisfying a certain property will remain unchanged. The label of a predicate here means an m-digit binary number, as was introduced in Section 7.2, where the ith digit is 1 or 0 according to whether or not the atom corresponding to x_i is included in the predicate. This completes the proof of the generalized theorem of the ugly duckling. In the same way as the simple theorem of the ugly duckling demonstrates the nonexistence of groups of similar objects, this generalized theorem shows the nonexistence of groups of objects that are tied together by a specified combination of affirmation and negation of predicates. An example is given in the next chapter (see the explanation of Table 8.1).

An important point I want to make before leaving this section is that a purely formal (syntactical) discussion must be based on $\hat{Y}$ and should not single out a special subset Y as the set of "primitive predicates," since many subsets Y in $\hat{Y}$ can serve as the set of primitive predicates (which engenders $\hat{Y}$) and selection of a particular one out of these many is a purely nonsyntactical operation. Consider, for instance, the case with two primitive predicates y_1 and y_2, which engender, in the absence of constraints, a lattice of 16 elements (see Section 7.2). The same lattice can be engendered by, for instance, $z_1 = \neg y_1$ and $z_2 = y_1 \uplus y_2 = (\neg y_1 \cap y_2) \cup (y_1 \cap \neg y_2)$. Why should the choice (y_1, y_2) be preferred to (z_1, z_2)? If there is any reason, it must originate from outside the syntactical consideration. [Note that $y_1 = \neg z_1$ and $y_2 = z_1 \uplus z_2 = (\neg z_1 \cap z_2) \cup (z_1 \cap \neg z_2)$.] Any discussion of the cognitive processes (such as deductive and inductive inferences) based on an artificial language(such as Carnap's) that first introduces a set of predicates and then allows logical operations is incapable of making a true distinction between the syntactical and the nonsyntactical because it has already smuggled in a nonsyntactical factor in the form of a chosen set of predicates. Similarly, if we consider y_1 and y_2 as "atomic propositions" (in the philosophers' sense of the word), z_1 and z_2 become "molecular propositions." But if we consider z_1 and z_2 as atomic, y_1 and y_2 become molecular. The distinction between these two views is purely accidental and arbitrary from the logical point of view. To claim that one set is directly given empirically while the other set is not amounts to claiming that one mode of observation is more fundamental than another mode of observation (which can be reduced from the former by a simple additional circuitry in the apparatus), which has no logical significance. Logical atomism is based on a non-logical criterion.

8

Classes and Concepts

8.1. COGNITION AND RECOGNITION

It goes almost without saying that hardly any human thought formulatable as a proposition is possible without the help of "concepts," which are usually supposed to correspond to certain classes of objects or of situations. Even a sentence expressing an immediate perception, such as "I see a red patch here now," is not free from conceptualization. I can distinguish many shades of red, but I say only "red," which includes all the different shades of red; furthermore, I know that what I cannot separate into more than one shade corresponds to many, many different mixtures of electromagnetic waves of different lengths. This chapter presents a model theoretical study of the mental process of class and concept formation, and a brief exposition of computer simulations of the process.

The remark made in the middle of the preceding paragraph points to the important fact that conceptualization starts at the level of sensory organs and continues all the way up to the cortical levels. As the famous experiment of Lettvin shows [L-4], the eyes of a frog send a special signal through the neurons to its brain when a small moving mosquito-like image falls on the eyes. Thus the discrimination of a seemingly complicated concept of "mosquitoness" is carried out by a peripheral organ. Less dramatically, yet in the same vein, we have very sharp discriminatory capabilities for acoustical waves in a certain range of frequency, but we cannot discriminate sounds of too high or too low a pitch. And beyond certain upper and lower limits of frequency we can hear nothing at all. This shows that the sensory signals we receive in our brains are already a result of sifting and molding at lower levels.

In an effort to schematize this situation, we may recognize two levels of conceptualization, roughly corresponding to the precortical level and the cortical level. Conceptualization at the cortical level has to rely on the "raw data" sent to the brain from the sensory organs, where they have already been subjected to a considerable degree of selection and coalescence. In our model theory of cognition and recognition we also have to assume two levels. What we called "scope of observation" in Section 7.5 may be compared to raw data in the above sense. A conscious effort of conceptualization starts from these raw data, which, however, are not as raw and unbiased as people are often led to believe. Actually they are already a result of intense selective and grouping operations. On the other hand, since the exact location of the border between the lower and higher levels is not defined and can be moved arbitrarily, we may conjecture what is going on at the lower level from a study of what is going on at the higher level. For instance, we speak of preselection of variables in the next section, and this preselection already presupposes the raw data provided by what may be called pre-preselection. This latter may have a function and structure somewhat similar to the former.

It is useful to note the existence of two kinds of tasks involved in what may be called class formation or classification. In one of them, which I sometimes call "cognition" for short, we are given a collection of objects with their properties and are asked to group them into classes according to the intensity of intermember relationships. These relationships include, but are not limited to, similarity. Discovery or identification of "clumps" or "clusters" in a vast collection of objects belongs to this category of tasks. In the other category of classification tasks, which I sometimes call "recognition" for short, we are first shown a certain number of sample objects with their properties and their class names. (These samples, which have already been classified, are called "paradigms" in this section.) After this "training period" (during which we are given the paradigms) is over, we are exposed to new samples with their properties, but without their class names, and are asked to place them in existing classes, imitating the classification illustrated by the paradigms. Most so-called problems of "pattern recognition" are of this type.

It is important to know that neither cognition nor recognition is possible from the formal point of view, as shown by the theorems of the ugly duckling. In the case of cognition this is obvious, because any two objects in the collection are as similar to each other as any other two objects. In a collection of m objects there are $2^m - (m + 1)$ possible classes (excluding the empty class and m classes consisting of one object), and each can claim to be a class of strongly similar objects. There is no such thing as a clump or cluster in any collection. Any collection of objects is entirely uniform and homogeneous. This situation does not change even if we are interested in a relation more subtle than similarity, as can be seen from the generalized theorem. In the case of recognition let us assume, for symmetry, that we have been shown k objects of Class I and k objects of Class II. Then we are shown a new object without classification. According to the generalized theorem of the ugly duckling, exactly the same number of predicates is shared by any group of $(k + 1)$ objects, provided that these $(k + 1)$ objects are distinguishable (no two among them are identical). Hence there is no reason to place this newcomer in either Class I or Class II. The exception made above corresponds either to the case where the number of nonidentical paradigms in Class I and Class II is not the same or to the case when the newcomer is identical with a paradigm of one class and not with any paradigm of the other class. In the former case we can take the same number of nonidentical paradigms in each class and repeat the argument. In the latter case recognition is possible in the formal sense, but this is trivial because the newcomer is not a newcomer in reality. In the case of recognition, too, we can invoke a relation more subtle than similarity, but we come to the same conclusion.

In reality, of course, there do exist clusters of similar objects, and there are usually good reasons for placing a new sample in one or the other of the classes that have been indicated by the given paradigms. This can be understood, in the light of the theorem of the ugly duckling, as meaning that the properties shared by similar objects are more important than those shared by nonsimilar objects. This demonstrates that the central problem of cognition and recognition is that of weighting, that is, nonuniform distribution of importance among predicates or, more generally, variables. If there are $n = 2^m$ possible predicates in the lattice $\hat{Y}$, the weight $w(j), j = 1, 2, \cdots, n$, cannot be uniform: $w(j) = 1/n$. The usage of a particular subset Y of $\hat{Y}$, for description of objects, also amounts to a particular type of nonuniform $w(j)$. Logical atomism is a special case of this situation based on a non-logical judgment.

The reader must already have sensed a striking similarity between the impossibility of classification and the formal ambiguity in induction, which we discussed in Chapter 4. In either case a logical handling of empirical data alone leaves the problem unsolved, while an uneven preference placed on predicates originating from an extralogical consideration leads to some useful solution. A hypothesis of the type "all X's are P" (such as "all ravens are black") starts from a "cognitive" act of acknowledging the existence of a group of similar objects that are given the name X. Suppose that $(\mu + \nu)$ predicates are affirmed by all the present members of X and that μ predicates are considered as sufficient to identify an object as a member of the class X. Then we can take the μ predicates as the characterization of X, take any one of the remaining ν predicates as P, and set up a hypothesis, "all X are P." This process requires a selection of predicates at two different levels. First, cognition of the group already implies a special selection of important predicates, which may be (affirmatively or negatively) included in the $(\mu + \nu)$ predicates. At the second stage the core predicates X and one particular P, out of the remaining ν predicates (or their conjunctions), must be selected.

Indeed, for any finite group of objects, the number of predicates that logically qualify as X and P will be enormously large, but only very few of them can be seriously considered for a useful induction. The "recognitive" act is more patently a special type of induction. For from the finite number of paradigms we have to induce a general rule defining the notion of Class I and Class II. The choice of general rules is practically infinite, and the selection is narrowed only by giving importance to a very few predicates. Our preference for certain general rules to other general rules is not based on the evidence (the paradigms), but on an extralogical, extra-evidential consideration. (See [W-30] for more about the relation between pattern recognition and induction.)

If Class I consists of m_{I} nonidentical members and if all these members are fixed and known to us, we could in principle construct the disjunction P_{I} of all the lattice atoms corresponding to these m_{I} objects. This P_{I} will be affirmed by any member of Class I and negated by any nonmember, provided that a member and a nonmember never have the same representation (see Section 7.5). However, the problem never presents itself this way. We know only a small number m_{I} of paradigms of Class I. Let P_{I} be the disjunction of the atoms corresponding to these m_{I} atoms; then the "true" P_{I} could be any one of the predicates implied by P_{I}. In the context of the foregoing paragraph, the "true" P_{I} would correspond to the conjunction of $(\mu + \nu)$ predicates. If we start with n independent predicates, there could be 2^n atoms; hence the number of the candidates for P_{I} would be $2^{2^n - m_{\text{I}}^*}$, which for all practical purposes may be considered as equivalent to infinity. The inductively "good" ones are very few, and buried like a needle in a haystack. This consideration gives some idea of the immensity of the task of induction and recognition, particularly for machines, which do not have a "feeling" of good and bad inductive projections.

Our discussion so far has been based mainly on the assumption that each object (x_i) is represented by the "values" T_{ij} of a finite number (say, n) of predicates y_j, $j = 1, 2, \cdots, n$, each of which can be considered as a two-valued function of a parametric "argument" j. A natural generalization is to consider the case in which each object (x_i) is represented by a continuously valued function $f^{(i)}(\xi)$ of a continuous parametric argument ξ, where ξ takes the place of the predicate label j. The variable ξ can be many-dimensional, but for simplicity we consider the case of one dimension and standardize the discussion by limiting the domain by the condition $0 \leq \xi \leq 1$. Furthermore, to facilitate comparison of functions for more than one object, we occasionally impose the normalization condition

$$\int_0^1 [f^{(i)}(\xi)]^2 w(\xi)\, d\xi = 1, \tag{8.1}$$

where the "weighting" function $w(\xi)$ may also be standardized by

$$\int_0^1 w(\xi)\, d\xi = 1, \qquad w(\xi) \geq 0. \tag{8.2}$$

A "natural" definition of "distance" $D(i, k)$ between two objects x_i and x_k may be

$$[D(i, k)]^2 = \int_0^1 [f^{(i)}(\xi) - f^{(k)}(\xi)]^2 w(\xi)\, d\xi. \tag{8.3}$$

The case in which an object x_i is represented by an n-component vector, $f^{(i)}(j), j = 1, 2, \cdots, n$, can be derived from this formalism regarding ξ as a discrete variable. The distance, for instance, will become here

$$[D(i, k)]^2 = \sum_{j=1}^{n} [f^{(i)}(j) - f^{(k)}(j)]^2 w(j). \tag{8.4}$$

If we ignore the normalization conditions and apply this formalism (8.4) to the binary case, $f^{(i)}(j) = T(i, j)$, with $w(j) = 1$, the total number of predicates that are simultaneously affirmed (i.e., "shared") or simultaneously negated by two objects x_i and x_k becomes $n - [D(i, k)]^2$. This shows that the notion of similarlity, in the sense of the number of shared predicates, is roughly in agreement with the notion of "nearness" in the sense of distance defined in (8.3). It may be noted in passing that the idea of "object-predicate" reciprocity introduced in Section 7.5 in the binary case can readily be extended to the case of continuously valued n-component vectors. In the reciprocal representation, the expressions of distance and normalization involve a summation with respect to "objects" instead of "variables," and the "weight" function will concern the importance of objects.

We already mentioned that the case of a continuously valued function of a continuous argument can be reduced to the case of a finite number of predicates, provided that the range and domain of the variables are finite and there are finite errors or imprecisions. Hence we can apply the theorem of the ugly duckling to the case of continuously valued function of a continuous argument too, and obtain the same conclusion derived in the foregoing paragraphs. However, we can also derive a similar conclusion directly from (8.3). The distance between two objects can be changed by various operations. The simplest is just to change the weight function $w(\xi)$ within the domain $0 \leq \xi \leq 1$ (which corresponds to Y and not to $\hat{Y}$). This alone can give a wide range of flexibility to the notion of distance. The next possibility is to apply a nonunitary linear transformation to each function $f^{(i)}(\xi)$. A third, more drastic transformation is to imitate the extension from Y to $\hat{Y}$ in the case of two-valued variables. This involves an extension of the domain of ξ from the original one (which was assumed to $0 \leq \xi \leq 1$), and the value of the function $f^{(i)}(\xi)$ at each point in the new domain is any functional of the function in the old domain. If we allow these transformations, the structure of distance relation can be altered in any way we want. This consideration serves to show that in the continuous case also there is no "natural class" or "objective" structure of distance and nearness among objects. Conversely, if we take cognizance of a class in our apperception, it implies that there are certain (if not unique) preferred ways of characterizing objects by continuous functions, and the weighting function is no longer entirely arbitrary. We can no longer apply arbitrary transformations on functions that would drastically alter the notion of distance. In summarizing our considerations in the case of predicates and the case of variables, we repeat that cognition and recognition are possible if, and only if, the predicates or variables are selected and weighted.

We spoke of two stages of selection and weighting of predicates and variables: one determines the scope of observation, and the other takes place after the scope of observation has been given. It is useful to make a similar dichotomy again in this latter stage of cognition and recognition; we can distinguish two major directions along which the algorithm of weighting and selection of the given variables can be developed. One of them is applicable both to cognitive and recognitive tasks and is called, for brevity, preselection. The other is applicable especially to recognitive tasks and is called post-evaluation or postselection. Preselection involves production of new variables or functions of the given variables, their evaluation, and the eventual elimination of those judged to be unimportant, and it is assumed that these operations are carried out without preliminary knowledge of the classes that are supposed to be established or recognized. Postselection involves similar operations, but they are guided by the knowledge of at least some paradigms of the classes, so that the importance of a variable can be measured at least partly by its effectiveness in bringing out intraclass cohesion and interclass distinctions. These operations of evaluation and selection, of course, are not independent of the definitions of intermember relations, such as similarity, distance, contiguity, and cohesion, and mutual adjustment of the former (evaluation) and the latter (cohesion) is also important.

The source of value in the process of evaluation is ultimately attributable to the purpose for which the classification is to be used. Such ultimate value reflects itself, perhaps not without deformation, in a scale of importance that can be applied to a variable. In determining such a scale, we can distinguish two kinds of criteria: one "internal" and the other "external." In cognitive problems we are given a collection of objects to be classified. In recognitive problems we are given collections of paradigms of different classes. The criteria derivable from these collections are "internal." In many cases, however, we have criteria that originate from something else than the data furnished by these collections. They are "external" criteria.

A great number of patterns allow for transformations without losing their identities. For instance, an alphabetic character remains invariant for any amount of translation, dilation, a certain extent of rotation, and a slight degree of deformation (topological transformation). If we require the variables describing the characters to be invariant for these transformations, the number of eligible variables will be considerably reduced. Consequencly the invariance consideration of character recognition can be considered a part of preselection, and the implied criterion of importance is external. (See [W-42] for E. Wong's method of extracting invariant features with the help of integral geometry.) However, it is misleading to consider, conversely, any characteristics that are invariant for these transformations as meaningful in character recognition. A great many are not. Many ingenious methods serve the purpose of extracting useful characteristics of the letters of a given alphabet, but usually they are powerful only for a very particular type of classification problem and worthless for other types of problem.† We discuss in this book only those general methods that are applicable to many problems of different fields.

As implied earlier, uneven weight distribution has an aspect of information compression, or economy of description, because the less important data can be ignored. The following is a method of obtaining an uneven weight distribution using only a purely internal criterion. Suppose that we are given a scope of observation and the notion of distance is already defined. There still remains a good deal of room to transform and choose the predicates or variables (without changing the distance relation) in such a way that the unevenness of weights on predicates or variables becomes as conspicuous as possible. It is remarkable that this can be done within the framework of preselection. This process has great utilitarian value because we can economize our description by omitting less important variables. Suppose that each object is described by a two-component vector (f_1, f_2), where each component can take any real number as its value. The distance is defined as the Euclidean distance, defined in the (f_1, f_2) plane. This implies that f_1 and f_2 have the same weight $w(j)$ in (8.4), with $n = 2$. The subscripts 1 and 2 are the j of (8.4). Suppose further that the representative points of the samples are all located very close to the straight line $f_1 - f_2 = 0$ and spread widely along this straight line. If we then rotate the coordinates 45° and use $g_1 = (f_1 + f_2)/\sqrt{2}$ and $g_2 = (f_1 - f_2)/\sqrt{2}$ as the new coordinates, all the representative points have values of g_2 very close to 0, and g_1 alone is enough to identify the different points. Under such circumstances neglect of g_2 really does not hamper our cognitive or recognitive actions. Then the "weight" of g_2 should become

† Many patterns consist of different arrangements of a small fixed number of constituent parts. In such cases we have to make a two-level algorithm, first recognizing the known constituent parts and then recognizing the relationships (such as relative position) among them. This kind of approach, however, is usually possible only in special-purpose algorithms.

practically zero. This is a simple case of what we discuss under preselection in the next section.

In the case of recognition, there is additional information to consider. Using the same two-dimensional representation, it is possible in principle, unless paradigms of two different classes occupy the same geometrical point, to introduce a function (curvilinear coordinate) $h = h(f_1, f_2)$ such that all paradigms of Class I satisfy $h > 0$ and all paradigms of Class II satisfy $h < 0$. Of course the function h is not unique, but we have to admit that such a variable h is more "important" than an arbitrary function that does not have this property. Even if two paradigms of different classes do not fall on the same geometrical point, we may be inclined to discredit an h if it is too complicated a function (i.e., if it gerrymanders the area too much), and we do not give much importance to such an h. But if a simple h, such as a linear function of f_1 and f_2, satisfies the condition and agrees well with the contour suggested by the distribution of paradigm points, we may assign much importance to it. Similarly, the degree of participation (expressible, for example, by a direction cosine between f_i and h) of f_i in the estimation of h measures the importance of f_i. If h is linear in f_1 and f_2 and if $h = 0$ coincides with $f_1 = 0$, the variable f_2 will have no importance as far as discrimination by $h > 0$ and $h < 0$ is concerned. This type of problem, which belongs to postselection, is critically analyzed in Section 8.5.

In the foregoing paragraphs, which are simple illustrations of cognitive and recognitive processes, we can already recognize three major operations at work: (a) weighting and selection of variables, (b) quantitative definition of nearness, and (c) actual placement of objects in classes. It is important to note that these three steps are not independent of one another, and in order to make the entire process work effectively we must adjust and coordinate them correlatively. The notion of nearness can be extended in two ways. One possible generalization, which we do not discuss in this book, introduces a "directed" bilateral relation instead of a symmetrical relation, such as distance or nearness between two objects. Another possible generalization, which we do discuss, is a consideration of multilateral relations that cannot be reduced to a combined effect of bilateral relations. We use the general name "cohesion" to cover all types of multilateral and bilateral relations. Correspondingly, the second step above must now read: (b) quantitative definition of cohesion. The third step (c) may also be paraphrased for brevity as "decision making," because each object has to be placed in one or the other of the classes by our conscious decision.

I stated in the foregoing that what we do consciously by way of cognition and recognition may be useful as a model in understanding its prototype on the subconscious neurophysiological level. Conversely, observation of what the neurophysiological structure performs by way of perception is suggestive of the real role of cognition and recognition in life. More specifically, in view of the formal equivocation of weight (importance) distribution among predicates and variables, one may ask what ultimate factor is, or should be, at work in choosing a particular distribution of weights. In the case of sensory perception, we can unambiguously observe that the decisive factor in determining the particular selection of variables is an empirical relevance to survival and other goals of life. Most animals have a keen discriminatory perception between what is edible and what is poisonous. A constant exposure to electromagnetic waves of the X-ray region would be dangerous to life, but we have no particular sensitivity to them because under usual circumstances the intensity of X-rays in nature is considerably below the danger level. The eyes and ears are sensitive only to the frequencies that are predominant in nature on the earth and that can be used as signals from ambient objects. Frogs are sensitive to mosquitos, and cats are sensitive to small moving objects. We can learn a lesson about perception from these facts and conclude that selection and evaluation of predicates and variables in cognition and recognition can be explicated only in the context of purposive behavior. In higher animals, like human beings, "purposes" are no longer limited to simple survival or reporduction, but we should not forget that without purposes of some sort the world would look entirely "gray," amorphous, and meaningless, as suggests the theorem of the ugly duckling.

In Section 8.2 we discuss in more detail what we called preselection here. In Section 8.3 we discuss the notion of relationships more sophisticated than similarity. Since these higher relationships are properties that emerge only after the number of participant objects becomes more than two, we are automatically led to consider the "cohesion" of a group of any number of objects, rather than just the intensity of bilateral relations, as the basic concept underlying cognition and recognition. From this point of view, similarity and dissimilarity, which are bilateral properties, are only two of the many factors that contribute to the cohesion of a group of objects. Section 8.3 introduces a special way of measuring all kinds of factors contributing to cohesion. Section 8.4 considers one possible algorithm for cognitive tasks, which is applicable to the case when not only similarity and dissimilarity but also multilateral relationships are present. Section 8.5 explains the general nature of recognitive tasks and discusses some of the decision-making algorithms used in recognitive problems. The methods of Section 8.5 are typical of postselection of variables.

8.2. PRESELECTION OF VARIABLES—SELFIC

There are many methods of preselection of variables. The one we explain in this section is distinguished by being entirely automatic (computer-feasible), using only internal criteria, and offering a universal applicability.

It is sometimes referred to as SELFIC (self-featuring information compression). "Self" suggests the internal criteria it uses. "Featuring" suggests its effectiveness in extracting important properties that characterize not only individual objects but also classes of objects [W-21, W-24, W-25].

The rough idea is as follows. Suppose that each object is represented by an n-component vector, of which each component can take any real number. Then the collection of objects can be represented by a crowd of points in the n-dimensional Euclidean space. Suppose further that this crowd has a narrow and elongated shape so that there is one direction (call it L-direction) along which the crowd is spread out considerably, while there is another direction (call it S-direction), perpendicular to L, along which the spread of the crowd is very narrow. We can now rotate the coordinate axes (i.e., transform the variables) so that the L-direction and the S-direction coincide with two of the new n-coordinate axes. Now the new variable corresponding to the S-coordinates is very unimportant, because it has practically the same value for all the objects in the collection. On the other hand, the variable corresponding to the L-coordinate is very important, because it effectively distinguishes the points as well as exposes the internal clusters in the crowd. For certain purposes the S-variable may be neglected completely, to make the description more succinct. Such identification and neglect of unimportant variables may be characterized as a case of information compression in the broad sense.

An important fact to note in this process is that the degree of importance here derives from the collection itself of the points as described by their properties, not from an external criterion or from the classification scheme imposed on them in addition to their known properties. There is no element of "re-cognitive" task involved in this process of weighting. The recognitive task may further introduce a modified weighting of variables, but the process we have sketched in the above paragraph and developed more in detail in this section can be a useful preparation for such further weighting. It is therefore a "preselection" of variables based on internal criteria. In this process of preselection it is assumed, as is obvious in the above illustration, that the notion of distance (which belongs to the second of the three factors mentioned at the very end of Section 8.1) is already determined.

An aspect of what we explained in Section 8.1 may be characterized as follows: cognitive and recognitive tasks presuppose efficient information compression as a first step, since we are always immersed in a sea of information of which the majority is irrelevant and useless. The "flood of information" is, in reality, not a new feature of the modern world. Since the time a living organism was born on the earth, one of the major functions of the nervous system lay in information compression, reducing the number and range of variables to the necessary minimum, and retaining only the very cream of the available information. In a sense, art and science also aim at the same goal: to compress the essence in a nutshell. All this suggests that one bit of information and another bit of information do not have the same value in life. To cope with such a situation we may need a whole new science, which I sometimes refer to as the "theory of valuated information." In any event, information compression presupposes the existence of more important and less important information and, for that reason, is essentially identical with the first of the three tasks mentioned at the end of the last section. Although this definition of information compression entails elimination of redundancy, because redundant information is unimportant information, the elimination of redundancy is not a broad enough definition of information compression. The process we describe in this section is a special case of information compression. The peculiarity of this particular process lies in the fact that the degree of importance is derived from the statistical property of the collection so that the neglect of certain "less important" variables statistically affects the global character of the collection. Of course, there is a danger in this process of losing some of the properties that are important from the "external" point of view.

We explain in the following the method of optimal expansion (SELFIC) and in Appendix A8.5 its relationship to factor analysis. The underlying mathematical methods are well known among applied mathematicians, but our exposition will be different from the usual ones in the sense that we consider SELFIC as a tool of preselection and information compression in the sense explained above. In our explanation of the optimal expansion, we first represent each object by a single-valued function of a continuous argument variable. As explained in the last section, it will be easy to derive from this case, a case in which the objects are represented by n-component vectors. It may not be difficult, either, to generalize the method to a case of more than one argument variable. Since we do not refer to the object-predicate table in this section, we use the symbols x and y in entirely different meanings from the last section.

We consider an ensemble of a large number N of similar objects, of which $Nw^{(\alpha)}$, $\alpha = 1, 2, \cdots, \nu$, belong to type α, whereby the relative frequency $w^{(\alpha)}$ satisfies the probability axioms $w^{(\alpha)} \geq 0$, $\sum_{\alpha=1}^{\nu} w^{(\alpha)} = 1$. Since we usually consider the limiting case $N \to \infty$, we can omit the number N and represent the statistical structure of the ensemble by only $\{w^{(\alpha)}\}$, $\alpha = 1, 2, \cdots, \nu$. Our explanation of mathematical instruments of linear algebra that follows may be unnecessarily detailed for some readers, but it is intended to be useful for other purposes in Chapter 9. We assume that each type specified by a value of the index α is represented by a real square-integrable function $f^{(\alpha)}(\xi)$ in a real domain $a \leq \xi \leq b$, which we normalize by

$$\int_a^b [f^{(\alpha)}(\xi)]^2 \, d\xi = 1. \tag{8.5}$$

We can consider ξ here to be a function of the ξ of (8.1), so that the differential $d\xi$ stands for $w(\xi)\, d\xi$ of (8.1). We can easily generalize the theory to the case in which $f^{(\alpha)}(\xi)$ is a complex function of a real argument ξ, but this generalization is relegated to Appendix A8.1.

A set of functions $\{\psi_i(\xi)\}$, $i = 1, 2, \cdots, \infty$, is said to be an orthonormal (orthogonal and normalized) set in the domain $[a, b]$ if

$$\int_a^b \psi_i(\xi)\psi_j(\xi)\, d\xi = \delta_{ij}. \tag{8.6}$$

The function set $\{\psi_i(\xi)\}$ is said to be complete if any arbitrary function satisfying (8.5) can be expanded in an infinite series of the type

$$f^{(\alpha)}(\xi) = \sum_{i=1}^{\infty} x_i^{(\alpha)} \psi_i(\xi). \tag{8.7}$$

From the orthonormal condition (8.6) we can calculate the coefficients, or components, $x_i^{(\alpha)}$ by

$$x_i^{(\alpha)} = \int_a^b f^{(\alpha)}(\xi)\psi_i(\xi)\, d\xi. \tag{8.8}$$

The normalization condition (8.5) entails, on account of (8.6),

$$\sum_{i=1}^{\infty} [x_i^{(\alpha)}]^2 = 1. \tag{8.9}$$

If we substitute (8.8) in (8.7) and write η for the variable of integration, we obtain

$$f^{(\alpha)}(\xi) = \int_a^b f^{(\alpha)}(\eta)\left[\sum_{i=1}^{\infty} \psi_i(\xi)\psi_i(\eta)\right] d\eta. \tag{8.10}$$

Since this is true for any function $f^{(\alpha)}(\xi)$ because of completeness, we can write the "inverse orthonormal condition" (or the Parseval relation, as it is often called),

$$\sum_{i=1}^{\infty} \psi_i(\xi)\psi_i(\eta) = \delta(\xi - \eta), \qquad a \le \xi \le b, \qquad a \le \eta \le b, \tag{8.11}$$

where $\delta(\xi)$ is Dirac's δ-function.

The ensemble of objects is now fully specified by two sets $\{f^{(\alpha)}(\xi)\}$ and $\{w^{(\alpha)}\}$, $\alpha = 1, 2, \cdots, \nu$. By referring to the "coordinate system" $\{\psi_i(\xi)\}$, $i = 1, 2, \cdots, \infty$, we can equivalently represent the ensemble by the set of "components" $\{x_i^{(\alpha)}\}$ and $\{w^{(\alpha)}\}$. This is already a first step toward information compression, because, while the range of values of each function $f^{(\alpha)}(\xi)$ as well as the domain of its argument ξ is continuous, the range of each component $x_i^{(\alpha)}$ is continuous but the domain of the "argument" i is discontinuous:

$i = 1, 2, \cdots$. The reason this is possible is that we are supposed to know $\{\psi_i(\xi)\}$ besides $\{x_i^{(\alpha)}\}$ and $\{w_i^{(\alpha)}\}$. The main task of SELFIC expansion lies in finding the "optimal" complete orthonormal set $\{\psi_i(\xi)\}$, where "optimal" means, in a certain yet-to-be-specified sense, "most effective in information compression." The reason a function set $\{\psi_i(\xi)\}$ and another function set, say, $\{\varphi_i(\xi)\}$, are not necessarily equally "effective" in information compression is explained presently.

When the objects are represented by n-component vectors $\{v^{(\alpha)}\}$ instead of continuous functions $\{f^{(\alpha)}\}$, we can skip the set $\{f^{(\alpha)}\}$ in the above description and start from the set of components $\{x_i^{(\alpha)}\}$, $i = 1, 2, \cdots, n$, where $x_i^{(\alpha)}$ is now to be considered as the ith component of the vector $v^{(\alpha)}$ in the coordinate system $\{\psi_i\}$, in which ψ_i is no longer a function but a coordinate axis. To include both continuous and discontinuous cases, we may write

$$\sum_{i=1}^{n} [x_i^{(\alpha)}]^2 = 1 \tag{8.12}$$

instead of (8.9), and understand that $n = \infty$ in the case when the objects are continuous functions.

The distance between two functions, $f^{(\alpha)}(\xi)$ and $f^{(\beta)}(\xi)$, is defined, as in (8.3), by

$$[D(\alpha, \beta)]^2 = \int_a^b [f^{(\alpha)}(\xi) - f^{(\beta)}(\xi)]^2 \, d\xi \tag{8.13}$$

$$= 2 - 2\int_a^b f^{(\alpha)}(\xi) f^{(\beta)}(\xi) \, d\xi, \tag{8.14}$$

which, in terms of the components, becomes, by virtue of (8.6) and (8.7),

$$[D(\alpha, \beta)]^2 = \sum_{i=1}^{n} [x_i^{(\alpha)} - x_i^{(\beta)}]^2 \tag{8.15}$$

$$= 2 - 2\sum_{i=1}^{n} x_i^{(\alpha)} x_i^{(\beta)}. \tag{8.16}$$

If the average (with respect to i) of $x_i^{(\alpha)}$ is zero, the second term $\sum_{i=1}^{n} x_i^{(\alpha)} x_i^{(\beta)}$ will acquire the meaning of the correlation between two random variables $x^{(\alpha)}$ and $x^{(\beta)}$. Equation (8.16) shows that the larger the correlation, the smaller the value of $D(\alpha, \beta)$. This fact provides another argument for the definition of distance given in (8.13).

An important fact about the quantity "distance" of the type (8.15) is that the mathematical form of this quantity remains invariant when another orthonormal coordinate system is used instead of $\{\psi_i(\xi)\}$. This is a generalization of the well-known fact that the distance between two points

$(x_1^{(1)}, x_2^{(1)}, x_3^{(1)})$ and $(x_1^{(2)}, x_2^{(2)}, x_3^{(2)})$ in a three-dimensional Euclidian space is expressed in the same (i.e., invariant) form,

$$\{[x_1^{(1)} - x_1^{(2)}]^2 + [x_2^{(1)} - x_2^{(2)}]^2 + [x_3^{(1)} - x_3^{(2)}]^2\}^{1/2},$$

insofar as x_1, x_2, and x_3 represent the components referring to any rectangular coordinate system.

Suppose that we take a second complete orthonormal set $\{\varphi_j(\xi)\}$ and expand $f^{(\alpha)}(\xi)$ as in (8.7), using φ_j for ψ_i, and call the new components $y_j^{(\alpha)}$. Then the $x_i^{(\alpha)}$ and $y_j^{(\alpha)}$ will be related by

$$y_j^{(\alpha)} = \sum_{i=1}^{n} \langle j \mid i \rangle x_i^{(\alpha)} \tag{8.17}$$

with the coefficients

$$\langle j \mid i \rangle = \int_a^b \varphi_j(\xi)\psi_i(\xi)\, d\xi, \tag{8.18}$$

where $\langle j \mid i \rangle$ is a generalization of the cosine of the angle between the i-coordinate and the j-coordinate. Similarly,

$$x_i^{(\alpha)} = \sum_{j=1}^{n} \langle i \mid j \rangle y_j^{(\alpha)} \tag{8.19}$$

and

$$\langle i \mid j \rangle = \int_a^b \psi_i(\xi)\varphi_j(\xi)\, d\xi. \tag{8.20}$$

We then have

$$\langle i \mid j \rangle = \langle j \mid i \rangle. \tag{8.21}$$

By substituting (8.19) in (8.17) and (8.17) in (8.19), we obtain

$$\sum_i \langle j \mid i \rangle \langle i \mid j' \rangle = \delta_{jj'} \quad \text{and} \quad \sum_j \langle i \mid j \rangle \langle j \mid i' \rangle = \delta_{ii'}. \tag{8.22}$$

If we write (8.17) in the matrix form

$$y^{(\alpha)} = Ux^{(\alpha)}, \tag{8.23}$$

we may write (8.19) as

$$x^{(\alpha)} = U^{-1}y^{(\alpha)} \tag{8.24}$$

since UU^{-1} and $U^{-1}U$ are equivalent to multiplication by unity. This notation thus suppresses the need of explicitly writing (8.22). Their matrix elements are

$$\langle j|\, U\, |i\rangle = \langle j \mid i \rangle, \qquad \langle i|\, U^{-1}\, |j\rangle = \langle i \mid j \rangle. \tag{8.25}$$

By introducing the transpose matrix U^T by

$$\langle i|\, U^T\, |j\rangle = \langle j|\, U\, |i\rangle, \tag{8.26}$$

the relation (8.21) becomes

$$U^T = U^{-1}, \tag{8.27}$$

which is the succinct characterization of an orthogonal transformation. The orthogonal transformation (8.27) leaves the scalar product of two vectors invariant, for

$$\sum_{i=1}^{n} x_i^{(\alpha)} x_i^{(\beta)} = \sum_{i,j,j'} \langle i \mid j \rangle \langle i \mid j' \rangle y_j^{(\alpha)} y_{j'}^{(\beta)} = \sum_{j}^{n} y_j^{(\alpha)} y_j^{(\beta)} \tag{8.28}$$

because of (8.17), (8.21), and (8.22). This entails that the distance, which is expressible as (8.16), is also invariant for the orthogonal transformation.

The same formalism explained with regard to the transformation between $x_i^{(\alpha)}$ and $y_j^{(\alpha)}$ can be applied to the transformation between $f^{(\alpha)}(\xi)$ and $x_i^{(\alpha)}$. By introducing

$$\begin{aligned} \langle i \mid \xi \rangle &= \langle i | \, U \, | \xi \rangle = \psi_i(\xi), \\ \langle \xi \mid i \rangle &= \langle \xi | \, U^{-1} \, | i \rangle = \psi_i(\xi), \end{aligned} \tag{8.29}$$

we can write (8.8) as

$$x_i^{(\alpha)} = \int_a^b \langle i \mid \xi \rangle f^{(\alpha)}(\xi) \, d\xi, \tag{8.30}$$

which can be compared to (8.17). All the basic formulas (8.21), (8.22), and (8.27) remain unchanged if the summation is replaced by integration wherever necessary. From this point of view (8.14) and (8.16) have exactly the same form, showing an invariance of the distance for the transformation from ξ to i.

The magnitude $x_i^{(\alpha)}$ or, more conveniently, the squared magnitude $[x_i^{(\alpha)}]^2$ of the component along the ith axis ψ_i of the coordinate system $\{\psi_i\}$ in the expansion of the function $f^{(\alpha)}(\xi)$, or of the vector $v^{(\alpha)}$, may be considered as a good measure of the usefulness or importance of the ith coordinate axis ψ_i in representing the given function $f^{(\alpha)}(\xi)$ or vector $v^{(\alpha)}$. Those coordinate axes ψ_i along which the object $f^{(\alpha)}$ or $v^{(\alpha)}$ has small components can be ignored without altering $f^{(\alpha)}$ or $v^{(\alpha)}$ appreciably. Since we have $f^{(\alpha)}$ or $v^{(\alpha)}$ with relative frequency $w^{(\alpha)}$ in the ensemble, the average importance of the axis ψ_i in the ensemble is given by

$$\rho_i = \sum_{\alpha=1}^{\nu} w^{(\alpha)} [x_i^{(\alpha)}]^2. \tag{8.31}$$

The quantity ρ_i is convenient because it has, from (8.12), the properties of a probability measure,

$$\rho_i \geq 0, \qquad \sum_{i=1}^{n} \rho_i = 1. \tag{8.32}$$

For a given ensemble of objects $\{f^{(\alpha)}\}$, $\{w^{(\alpha)}\}$ or $\{v^{(\alpha)}\}$, $\{w^{(\alpha)}\}$, the values of the ρ's vary depending on the choice of the coordinate system $\{\psi_i\}$. If we use $\{\varphi_j\}$ as the frame of reference, the ρ's will become $\sum_{\alpha=1}^{\nu} w^{(\alpha)} \{y_j^{(\alpha)}\}$, where the $y_j^{(\alpha)}$ are derivable from the $x_i^{(\alpha)}$ by (8.17). From the point of view of information compression, it is desirable that the ρ's be concentrated on a few axes instead of being widely spread over many axes. To formulate this idea mathematically it would be convenient to introduce the entropy function in terms of the ρ's,

$$S(\{\psi_i\}) = -\sum_{i=1}^{n} \rho_i \log \rho_i, \tag{8.33}$$

which depends not only on $\{\psi_i\}$ but also on the ensemble $\{\psi^{(\alpha)}\}$, $\{w^{(\alpha)}\}$ or $\{v^{(\alpha)}\}$, $\{w^{(\alpha)}\}$, although the ensemble is kept unchanged in the present discussion. The desirable choice of the coordinate system $\{\psi_i\}$ will then be characterized as one that minimizes S. If there are more than one such optimal coordinate system, we need select any one of them. If $\{\phi_i\}$ is one of the optimal coordinate systems, we have

$$S(\{\phi_i\}) = \min_{\{\psi_i\}} S(\{\psi_i\}). \tag{8.34}$$

One of the main results asserted in this section is the following theorem.

Theorem 8.1 A coordinate system is optimal in the sense of (8.34) if, and only if, it is a Karhunen-Loève (K-L) coordinate system.

The full proof of this theorem is given later in this section and in Appendix A8.2, but to understand the proof we have to know what the K-L expansion and K-L coordinate systems are. A real function G with two argument variables ξ and ξ' is called autocorrelation function for a given ensemble $\{w^{(\alpha)}\}$ and $\{f^{(\alpha)}(\xi)\}$ when it is defined by

$$\langle \xi | \, G \, | \xi' \rangle = \sum_{\alpha} w^{(\alpha)} f^{(\alpha)}(\xi) f^{(\alpha)}(\xi'). \tag{8.35}$$

This quantity, when extended to the complex case, is a familiar tool in quantum mechanics under the name of "density matrix." This G can be considered a symmetrical matrix with two continuous indices ξ and ξ',

$$\langle \xi' | \, G \, | \xi \rangle = \langle \xi | \, G \, | \xi' \rangle. \tag{8.36}$$

By the use of expansions (8.7) and (8.30), we can write (8.35) as

$$\langle \xi | \, G \, | \xi' \rangle = \sum_{i} \sum_{i'} \langle \xi \mid i \rangle \langle i | \, G \, | i' \rangle \langle i' \mid \xi' \rangle, \tag{8.37}$$

where

$$\langle i | \, G \, | i' \rangle = \sum_{\alpha} w^{(\alpha)} x_i^{(\alpha)} x_{i'}^{(\alpha)} = \int_a^b \int_a^b \langle i \mid \xi \rangle \langle \xi | \, G \, | \xi' \rangle \langle \xi' \mid i' \rangle \tag{8.38}$$

and $\langle \xi \mid i \rangle$ and $\langle i' \mid \xi' \rangle$ are given by (8.29). This new expression $\langle i | \, G \, | i' \rangle$ can be considered as equivalent to the original $\langle \xi | \, G \, | \xi' \rangle$, except that it refers to another coordinate system. By the same token, if we use $\{\varphi_j\}$ instead of $\{\psi_i\}$, we obtain, instead of (8.38),

$$\langle j|\, G\,|j'\rangle = \sum_{\alpha} w^{(\alpha)} y_j^{(\alpha)} y_{j'}^{(\alpha)}, \tag{8.39}$$

which is related to (8.38) by the orthogonal transformation U of (8.25), that is, by

$$\langle j|\, G\,|j'\rangle = \sum_{i}\sum_{i'} \langle j \mid i\rangle\langle i|\, G\,|i'\rangle\langle i' \mid j'\rangle. \tag{8.40}$$

We speak of the i-representation or ψ_i-representation of the "same" matrix G when we use (8.38). In the discrete representation also, the matrix G is symmetrical:

$$\langle i|\, G\,|i'\rangle = \langle i'\, |G|\, i\rangle. \tag{8.41}$$

It may be noted that the ρ's introduced in (8.31) are nothing but the diagonal elements of the matrix (8.38):

$$\rho_i = \langle i|\, G\,|i\rangle. \tag{8.42}$$

Now consider a function $\phi_k(\xi)$ (k is fixed for a while) such that there exists a constant λ_k that satisfies the equation

$$\int_a^b \langle \xi|\, G\,|\xi'\rangle \phi_k(\xi')\, d\xi' = \lambda_k \phi_k(\xi). \tag{8.43}$$

If we expand $\phi_k(\xi)$ in the $\{\psi_i\}$ coordinate as

$$\phi_k(\xi) = \sum_i \psi_i(\xi)\langle i \mid k\rangle, \qquad \langle i \mid k\rangle = \int_a^b \phi_k(\xi)\psi_i(\xi)\, d\xi, \tag{8.44}$$

the equation (8.43) becomes, in terms of (8.38),

$$\sum_{i'} \langle i|\, G\,|i'\rangle\langle i' \mid k\rangle = \lambda_k \langle i \mid k\rangle. \tag{8.45}$$

We can consider this as a set of homogeneous linear equations for the n unknowns $\langle i \mid k\rangle$ $i = 1, 2, \cdots, n$ (k is fixed for the present). The condition for existence of a set of solutions is the vanishing of the determinant made out of the coefficients of the unknowns:

$$\|\langle i|\, G\,|i'\rangle - \delta_{ii'}\lambda_k\| = 0. \tag{8.46}$$

If the space is n-dimensional, this equation is an equation of n degrees for λ_k; hence there are n roots if we count multiple roots separately. To each of them we can assign an integral value of k; we obtain $\lambda_1, \lambda_2, \cdots, \lambda_n$, and correspondingly also $\phi_1, \phi_2, \cdots, \phi_n$. Each λ_k is called an eigenvalue of matrix G, and ϕ_k is its eigenfunction or eigenvector.

Write an equation similar to (8.45) for k' instead of k, and rewrite it in the form

$$\sum_{i'} \langle k' \mid i'\rangle\langle i'|\, G\,|i\rangle = \lambda_{k'}\langle k' \mid i\rangle \tag{8.47}$$

using the symmetry (8.41) and the symmetry (8.21). Multiplying (8.45) by $\langle k' \mid i \rangle$ and summing over i, we obtain

$$\sum_{i,i'} \langle k' \mid i \rangle \langle i | G | i' \rangle \langle i' \mid k \rangle = \lambda_k \sum_i \langle k' \mid i \rangle \langle i \mid k \rangle. \tag{8.48}$$

Similarly, by multiplying (8.47) by $\langle i \mid k \rangle$ and summing over i, we obtain (after interchanging the dummy labels i and i')

$$\sum_{i,i'} \langle k' \mid i \rangle \langle i | G | i' \rangle \langle i' \mid k \rangle = \lambda_{k'} \sum_{i'} \langle k' \mid i' \rangle \langle i' \mid k \rangle. \tag{8.49}$$

Subtraction of (8.49) from (8.48) yields

$$0 = (\lambda_k - \lambda_{k'}) \sum_{i'} \langle k' \mid i' \rangle \langle i' \mid k \rangle. \tag{8.50}$$

This shows that the two eigenfunctions ϕ_k and $\phi_{k'}$, corresponding to two different eigenvalues $\lambda_k \neq \lambda_{k'}$, are orthogonal.

$$\int_a^b \phi_k(\xi)\phi_{k'}(\xi)\, d\xi = 0, \tag{8.51}$$

which follows, by virtue of (8.8) and (8.11), from the vectorial relation of orthogonality,

$$\sum_i \langle k \mid i \rangle \langle i \mid k' \rangle = 0. \tag{8.52}$$

It is also easy to show that in a "degenerate" case, when a λ is an s-ple root, we can take s functions (vectors) that are mutually orthogonal and are orthogonal to all other functions (vectors) corresponding to different values of λ. These s functions (vectors) are indeterminate only in the sense that we can apply to them any s-dimensional orthogonal transformation and obtain still another set of s functions that possess the same properties as the original s functions. But the s-dimensional subspace is determined.

For this reason the set of functions, or vectors, $\{\phi_k\}$, $k = 1, 2, \cdots, n$ (n is ∞ in the continuous case), can be used as another orthogonal coordinate system like $\{\psi_i\}$ and $\{\varphi_j\}$, because we are now entitled to put

$$\sum_i \langle k \mid i \rangle \langle i \mid k' \rangle = \delta_{kk'} \tag{8.53}$$

for $k, k' = 1, 2, \cdots, n$. The G matrix in ϕ_k-representation will then have, according to the definition (8.40), the elements given by (8.48) or (8.49), that is,

$$\langle k | G | k' \rangle = \lambda_k \delta_{kk'}, \tag{8.54}$$

because of (8.53). This shows that the G-matrix is diagonal in the ϕ_k-representation and the λ_k are the diagonal elements. In other words, the λ_k are a special case of the ρ's of (8.42) and (8.31). This means that if we expand the function $f^{(\alpha)}(\xi)$ as

$$f^{(\alpha)}(\xi) = \sum_k c_k^{(\alpha)} \phi_k(\xi), \tag{8.55}$$

with

$$c_k^{(\alpha)} = \int \phi_k(\xi) f^{(\alpha)}(\xi)\, d\xi = \sum_i x_i^{(\alpha)} \langle k \mid i \rangle, \tag{8.56}$$

then (8.38) becomes, as a result of (8.54)

$$\sum_\alpha w^{(\alpha)} c_k^{(\alpha)} c_{k'}^{(\alpha)} = \lambda_k \delta_{kk'}. \tag{8.57}$$

As was the case in (8.32), we also have

$$\lambda_k \geq 0 \quad \text{and} \quad \sum_k \lambda_k = 1. \tag{8.58}$$

This shows that the G-matrix is a symmetrical matrix with non-negative eigenvalues that add up to unity. From (8.57) it also follows that

$$\sum_{\alpha=1}^{\nu} w^{(\alpha)} c_k^{(\alpha)} c_{k'}^{(\alpha)} = 0, \qquad k \neq k'. \tag{8.59}$$

This allows the following interpretation. If we consider $c_k^{(\alpha)}$ with a given k as a random variable that takes a particular value $c_k^{(\alpha)}$ with relative frequency $w^{(\alpha)}$, two such random variables $c_k^{(\alpha)}$ and $c_{k'}^{(\alpha)}$ are statistically uncorrelated if their mean values are set equal to zero.

At this point we should interject a remark about the meaning of the multiple eigenvalues equal to zero. Among the ν original functions $f^{(\alpha)}(\xi)$ there could be a relation of linear dependence of the type

$$\sum_{\alpha=1}^{\nu} a_\alpha f^{(\alpha)}(\xi) = 0, \tag{8.60}$$

where not all the ν coefficients a_α are supposed to be zero. Then we can express one of the ν original functions as a linear combination of the remaining $(\nu - 1)$ functions. For instance, if $a_\nu \neq 0$,

$$f^{(\nu)}(\xi) = -\frac{\sum_{\alpha=1}^{\nu-1} a_\alpha f^{(\alpha)}(\xi)}{a_\nu}. \tag{8.61}$$

Among these $(\nu - 1)$ functions there could be again a relation of the type (8.60), in which ν is replaced by $\nu - 1$. We can again represent one of the $(\nu - 1)$ functions as a linear combination of $(\nu - 2)$ functions of the type, (8.61). We can continue this process until we come to, say, μ functions. among which there no longer exists any linear dependence of the type (8.60). This number μ is independent of the way in which we eliminate the "dependent" functions, and is determined by the original set of functions

$\{f^{(\alpha)}(\xi)\}$, provided that none of the w's is zero. These μ functions determine a μ-dimensional space, in which the μ orthogonal function $\phi_k, k = 1, 2, \cdots, \mu$, corresponding to μ nonzero eigenvalues, can be taken. In order that the function set $\{\phi_k\}$ may be complete, k has to run to n, which is infinity here. All those functions for which $k > \mu$ are not determined by the G except that they have to be orthogonal to the first μ functions. These functions can be considered as eigenfunctions corresponding to the infinitely degenerate eigenvalue zero.

In the case of n-component vectors there can be no more than n linearly independent vectors in the collection $\{v^{(\alpha)}\}$, $\alpha = 1, 2, \cdots, \nu$. If the number of linearly independent vectors in $\{v^{(\alpha)}\}$ is μ, which is less than n, μ will be the number of nonzero eigenvalues determining a μ-dimensional subspace. The vectors in the remaining $n - \mu$ dimensional subspace correspond to the eigenvalue 0.

The coordinate system $\{\phi_k(\xi)\}$ defined by (8.43) or (8.45) is a Karhunen-Loève coordinate system (the indefinite article is used because of ambiguity due to the degeneracy) determined by the ensemble $\{f^{(\alpha)}(\xi)\}$, $\{w^{(\alpha)}\}$. The expansion of $f^{(\alpha)}(\xi)$ or $v^{(\alpha)}$ (8.55) in terms of this coordinate system is a Karhunen-Loève expansion. As far as Theorem 8.1 is concerned this definition is sufficient, but for the purpose of Theorem 8.2, which follows, it is convenient to arrange the functions (vectors) ϕ_k in such a way that the series of inequalities

$$\lambda_1 \geq \lambda_2 \geq \cdots \geq \lambda_{k-1} \geq \lambda_k \geq \cdots \tag{8.62}$$

holds.

The diagonal elements of the G in the K-L coordinate system are the eigenvalues λ_k of G. The diagonal elements ρ_i of G in another coordinate system can be derived from λ_k by the use of the general transformation (8.40). The matrix elements of G in the coordinate system $\{\psi_i\}$ will then be given by

$$\langle i|\, G\,|i'\rangle = \sum_{k,k'} \langle i|\, U\,|k\rangle\langle k|\, G\,|k'\rangle\langle k'|\, U^{-1}\,|i'\rangle, \tag{8.63}$$

where $\langle i|\, U\,|k\rangle$ is nothing but $\langle i \mid k\rangle$ given by (8.44). Since $\langle k|\, G\,|k'\rangle$ is diagonal, as given by (8.54), the $\rho_i = \langle i|\, G\,|i\rangle$ will be given by

$$\rho_i = \sum_k A_{ik}\lambda_k, \tag{8.64}$$

with

$$A_{ik} = \langle i \mid k\rangle\langle k \mid i\rangle = |\langle i \mid k\rangle|^2. \tag{8.65}$$

Because the transformation is orthonormal, we have (8.22), which results in the double stochasticity of the matrix

$$\sum_k A_{ik} = \sum_i A_{ik} = 1, \qquad A_{ik} \geq 0. \tag{8.66}$$

Theorem 8.1 now consists of two parts, one associated with the "if" and the other associated with the "only if" in the statement of the theorem. The "if" part is then easily proven by the well-known simplest version of the H-theorem (see Theorem 5.3).

Theorem 8.2 (Prototype H-theorem). If two probability distributions $\{\lambda_k\}$, $k = 1, 2, \cdots$ and $\{\rho_i\}$, $i = 1, 2, \cdots$ are related by (8.64) with a doubly stochastic matrix (8.66),

$$-\sum_k \lambda_k \log \lambda_k \leq -\sum_i \rho_i \log \rho_i. \tag{8.67}$$

The "only-if" part of Theorem 8.1 is proven in Appendix A8.2. The Karhunen-Loève expansion is usually known as the "error-minimizing" expansion. We now explain why it is so characterized and give in Appendix A8.3 a proof (different from the usually known one) of the underlying theorem.

Insofar as the function set $\{\psi_i(\xi)\}$ is complete, any function of reasonable behavior can be expressed by an infinite series as in (8.7) without committing error. But if we use only a finite number, say, m, of terms in this series, we of course commit some errors in expressing the function. In the case of n-component vectors the errors are, in general, inevitable if $m < n$. The mean squared errors of this kind for a given m in the ensemble of objects $\{w^{(\alpha)}\}$, $\{f^{(\alpha)}\}$ is

$$E(\{\psi_i\}, m) = \sum_\alpha w^{(\alpha)} \int_a^b \left| f^{(\alpha)}(\xi) - \sum_{i=1}^m x_i^{(\alpha)} \psi_i(\xi) \right|^2 d\xi \tag{8.68}$$

$$= \sum_\alpha w^{(\alpha)} \sum_{i=m+1}^n |x_i^{(\alpha)}|^2 \tag{8.69}$$

$$= 1 - \sum_{i=1}^m \rho_i, \tag{8.70}$$

where $x_i^{(\alpha)}$ is given by

$$x_i^{(\alpha)} = \int_a^b \psi_i(\xi) f^{(\alpha)}(\xi)\, d\xi, \tag{8.71}$$

as in (8.8). In the case of n-dimensional vectors we should consider $x_i^{(\alpha)}$ as their components in a coordinate system that can be varied later. The K-L expansion can now be characterized† by Theorem 8.3.

Theorem 8.3 The K-L coordinate system $\{\phi_k\}$ defined by (8.43) or (8.45) with the additional condition (8.62) is such that

$$E(\{\phi_k\}, m) = \min_{\{\psi_i\}} E(\{\psi_i\}, m) \tag{8.72}$$

for every integer $m > 0$ ($m < n$ in the vector case).

† While Theorem 8.3 is well-known (see [L-5]), Theorem 8.1 is an original contribution by the author (see [W-36]). The idea of using the K-L expansion for the analysis of continuous signals on the ground of Theorem 8.3 was suggested to the author by Eugene Wong who also extended the K-L expansion to the 2-dimensional case (see [W-39]).

One may suspect that when $\{\psi_i\}$ and m are given one can better approximate an $f^{(\alpha)}(\xi)$ by using some coefficients $x_i^{(\alpha)\prime}$, which are different from those $x_i^{(\alpha)}$ given by (8.71). But it is easy to see that this is not the case, because

$$\int_a^b |f^{(\alpha)}(\xi) - \sum_{i=1}^{m} x_i^{(\alpha)\prime}\psi_i(\xi)|^2 d\xi = \sum_{i=1}^{m} |x_i^{(\alpha)} - x_i^{(\alpha)\prime}|^2 + \sum_{i=m+1}^{n} |x_i^{(\alpha)}|^2. \quad (8.73)$$

In view of the fact that the ρ's become λ's in the ϕ-system, Theorem 8.3 must be, because of (8.70), equivalent to the following theorem.

Theorem 8.4 If two probability distributions λ_k, $k = 1, 2, \cdots$, and ρ_i, $i = 1, 2, \cdots$, are related by (8.64) with a doubly stochastic matrix (8.66) and if the labeling of λ_k is in descending order in the sense of (8.62), then

$$\sum_{k=1}^{m} \lambda_k \geq \sum_{i=1}^{m} \rho_i \quad (8.74)$$

for any given integer $m > 0$.

I conjectured without proof the validity of Theorem 8.4 and told Professor Shizuo Kakutani about it; he immediately provided me with a proof, which I reproduce in Appendix A8.3 in my own language. There could be a simpler proof, but Kakutani's proof is intuitively clear and very instructive.

When I explained Theorems 8.1 and 8.3 to my students, Jay Luck, came up with a theorem that proves the consequence of Theorem 8.1 from the consequence of Theorem 8.3.

Theorem 8.5 If two probability distributions λ_k, $k = 1, 2, \cdots$, and ρ_i, $i = 1, 2, \cdots$, are such that

$$\lambda_{k-1} \geq \lambda_k, \qquad \rho_{i-1} \geq \rho_i \quad \text{for all} \quad k \quad \text{and all} \quad i \quad (8.75)$$

and

$$\sum_{k=1}^{m} \lambda_k \geq \sum_{i=1}^{m} \rho_i \quad (8.76)$$

for every integer $m > 0$, then

$$-\sum_k \lambda_k \log \lambda_k \leq -\sum_i \rho_i \log \rho_i, \quad (8.77)$$

with the equality holding in (8.77) if, and only if,

$$\sum_{k=1}^{m} \lambda_k = \sum_{i=1}^{m} \rho_i \quad (8.78)$$

for every integer $m > 0$. The summation in (8.77) runs up to n or to ∞.

A proof of this theorem is given in Appendix A8.4. This theorem, combined with Theorem 8.4, thus provides a proof of Theorem 8.1 without the help of the H-theorem.

As noted earlier, (8.59) shows that the K-L coefficients c_k with different k's are statistically uncorrelated if α is regarded as specifying a particular value and $w^{(\alpha)}$ is the statistical weight of occurrence of this value. This provides another argument for using K-L coefficients as variables to represent the members of the ensemble economically. It would be desirable, inasmuch as information compression means elimination of redundancy, to use variables that are statistically independent, but in the absence of such variables, statistically uncorrelated variables may be the next best. Furthermore, if each of the c_k's has a Gaussian distribution, they become statistically independent.

As practical advice, it may be added that for the purpose of discrimination it is advisable to subtract the average function $\sum_\alpha w^{(\alpha)} f^{(\alpha)}(\xi)$ from each function $f^{(\alpha)}(\xi)$ before applying the K-L method, for if this average is not zero the first eigenfunction tends to represent it, and it will have no discriminating power.† It should be emphasized again that the method explained in this section is a method of "preselection" as suggested solely by the statistical property of the ensemble. It is quite possible that the variable that is unimportant in this sense will become important again from the point of view of "postselection" of variables. For the rest of this section I mention a successful case of application of the present method to a practical problem.

As we have seen, ρ_i can be considered as a measure of the degree to which the object vectors of the ensemble belong to the coordinate ψ_i. Similarly, $\sum_{i=1}^{m} \rho_i$ is the degree to which they belong to the subspace subtended by the m coordinates $\{\psi_i\}$, $i = 1, 2, \cdots, m$. For a given m this degree becomes maximal when we use the optimal coordinate system $\{\phi_k\}$. We can realize "$100 \times \sigma\%$ faithful" representation by retrenching the n-dimensional representation space to its m-dimensional subspace subtended by $\{\phi_k\}$, $k = 1, 2, \cdots, m$, if m and σ are linked by

$$\sum_{k=1}^{m} \lambda_k \geq \sigma > \sum_{k=1}^{m-1} \lambda_k. \tag{8.79}$$

† As we see in Section 8.5, in recognitive problems (as distinct from cognitive problems) we can also obtain the optimal coordinate systems for the paradigms of each class separately. In such a case we should not subtract the average of the paradigms of the class, because this average may be the most conspicuous feature of the class [W-25]. If we want the average of all functions to vanish exactly, we should replace $f^{(\alpha)}(\xi)$ with $f^{(\alpha)*}(\xi)$, which is given by the following condition, instead of just subtracting the average as described above:

$$f^{(\alpha)*}(\xi) = \frac{f^{(\alpha)}(\xi) - f^{(0)}(\xi)}{\int [f^{(\alpha)}(\xi) - f^{(0)}(\xi)]^2} \, d\xi,$$

where the constant function $f^{(0)}(\xi)$ is to be determined by the condition that

$$\sum_{\alpha=1}^{\nu} \quad w^{(\alpha)} f^{(\alpha)*}(\xi) = 0.$$

In many applications we have found that we can attain a high fidelity (such as $\sigma = 0.9$) by a surprisingly small value of m ($m \ll n$). This m-dimensional space may be called "feature space."

Figure 8.1 represents results obtained by applying the K-L method to the deviation from the average of the power spectra (obtained by 36 bandpass filters) of 12 different vowels (i, I, ε, a, e, o, ʌ, u, U, ɔ, ɚ, æ) spoken by 19 different persons (male and female). Each record is a 36-component vector. The center frequencies of the filters range from 100 to 10,000 cycles/sec, and their bandwidths, from 50 to 1200 cycles/sec. Figure 8.1 represents 12 × 19 points in the space defined by "longitude" $\tan^{-1}(c_2/c_1)$ and "latitude" $\tan^{-1}[c_3/(c_1^2 + c_2^2)^{1/2}]$, where c_1, c_2, and c_3 are the first, second, and third K-L coefficients as defined in (8.56). The subscript α in this case identifies each of the 12 × 19 points. Figure 8.1 also shows a rough determination of nine disjoint regions, where intrusion of some "alien" elements in each territory is purposely tolerated in order to avoid excessive gerrymandering. These zones have been drawn by a human hand to facilitate inspection. The o and the ɔ occupy a common region, and this may lie in the nature of things. The distinction between I and e is so small that confusion between them may be forgiven at this stage. The most troublesome one is ɚ (er, ir, or ur in the usual spelling), which intrudes into several regions. However, the histogram of Figure 8.2, which represents the distribution along c_4, shows that ɚ forms almost a separate island in the c_4 space. It is to be expected that considerable overlapping of regions is inevitable in speech recognition. We should probably be surprised that the extent of overlapping is not more than seen in Figure 8.1. Note that the concept of "formant" is not used in our method. The determination of zones and the location of a new arrival in one of the ready-made zones are a problem of decision procedure discussed in Section 8.5.

8.3. ENTROPIC MEASURE OF SIMILARITY AND COHESION

The operation of cognition and recognition involves three main factors: (a) selection and weighting of predicates and variables, (b) determination of intensity of similarity and of other interobject relationships, and (c) placement of objects in classes. As explained in Section 8.1, these three factors do not necessarily represent successive steps that take place in the simple sequence given, but mutually influence one another. For instance, the special kind of preselection discussed in Section 8.2 is a case in which the concept of distance [which belongs to (b) here] is given first, and we look for the most efficient set of variables [which is part of (a) here] to express the average interobject distance within a given collection of objects. The problem of postselection is usually of the following type: from a given (b) and given illustrations of (c), we derive a suitable (a). In this section we discuss a special type of operation

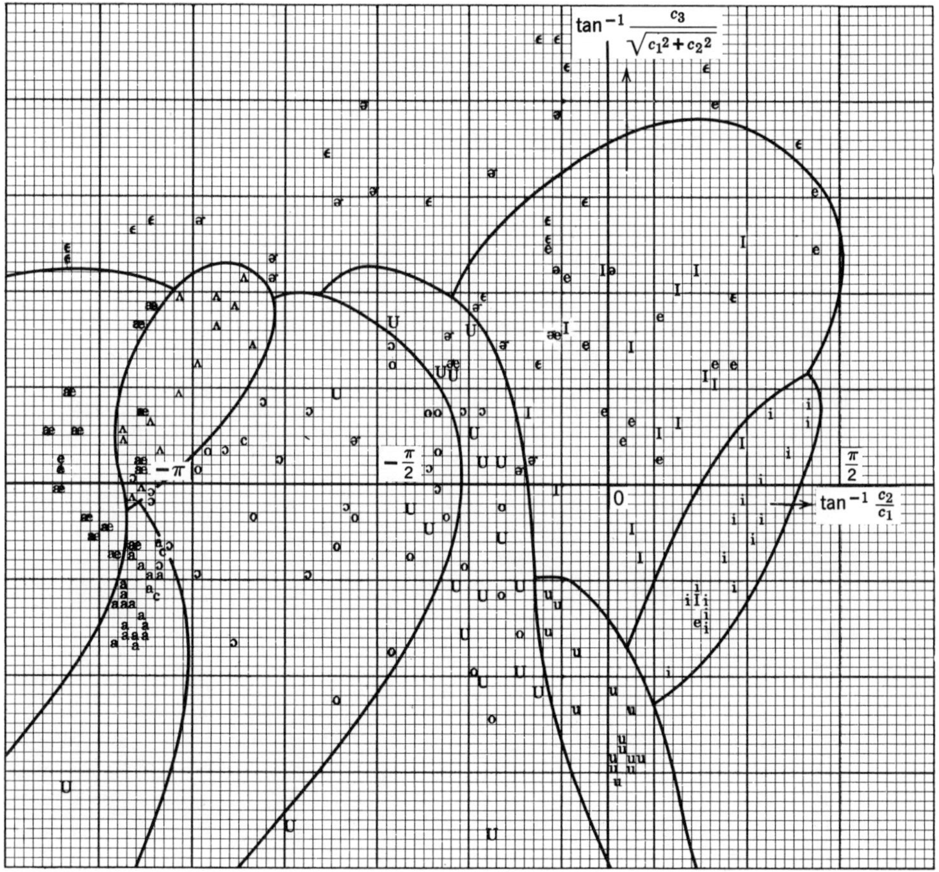

Figure 8.1 Two variables are sufficient to distinguish nine zones discriminating English vowels. We can thus bypass the formant analysis. The experimental data were provided by R. Bakis of IBM Research Laboratory. The programming on the IBM 7094 was carried out by

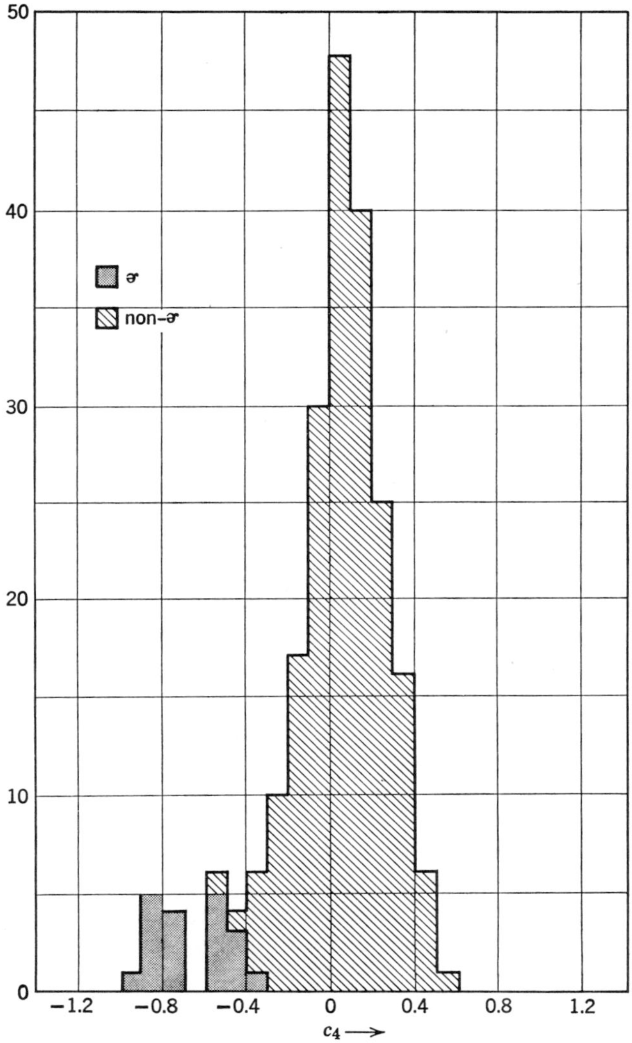

Figure 8.2 The fourth coefficient in the optimal expansion discriminates "ɚ".

belonging to (b). First, we consider the case in which operation (a) has already been done, but later we discuss how (c) can influence (a) through the intermediary of the adopted step (b) in recognitive tasks.

What follows in this section assumes that each object is represented by a binary vector of n components. In other words, we are given an object-predicate table $T(x_i, y_j)$, $i = 1, 2, \cdots, m$, and $j = 1, 2, \cdots, n$. It is assumed that the set of predicates $Y = \{y_j\}$ is not a completed Boolean lattice. If it were, and if all the predicates had the same weight, we could not discover (as shown by the theorem of the ugly duckling) any pair or subset of objects in $X = \{x_i\}$ that are tied together by similarity or other bonds more firmly than another pair or subset. However, selection of a particular subset Y from the completed set $\hat{Y}$ already amounts to giving uneven weights to the members of $\hat{Y}$, namely, nonzero weights to the members of the subset Y and zero weight to the nonmembers of Y. That is why we can define similarity and other types of bonds in the present case, even if we give equal importance to each of the adopted predicates (members of Y).

There are many ways of defining a measure of similarity and other bonds in a table of this sort, and the entropic measure we introduce in this section is only one possibility (see [W-11, W-12, W-15]). However, the entropic measure has a large number of advantages, among which is that it leads to a measure of cohesion that is a function of every group of objects and that is additive in the following two senses. First, the terms due to similarity and the different terms due to more sophisticated kinds of relationship can be simply added on the same footing, contributing to the over-all cohesion. Second, if we juxtapose two unrelated groups of objects, the measure of cohesion of the combined system is simply the sum of the measures of cohesion of the partial systems.

Let us start with the idea of similarity. If we limit ourselves to a proper subset Y of "more important" predicates taken from the completed lattice $\hat{Y}$, the number of predicates simultaneously affirmed by two objects can be used as a mark of their similarity. We shall agree here that a predicate y_j and its negation $\neg y_j$ are equally important, and we shall count a simultaneous negation of a predicate by two objects just as significant a mark of their similarity as its simultaneous affirmation by them.

Let us take two rows corresponding to x_i and x_k in the object-predicate table, in which Y is a subset of important predicates. If two rows are approximately the same, that is, if

$$T(x_i, y_j) = T(x_k, y_j) \tag{8.80}$$

for a "large number" of y's, nobody will disagree that the two objects x_i and x_k resemble each other. But what should we understand by a "large number"? Suppose that there are $n_1(x_i)$ 1's and $n_0(x_i)$ 0's in the row of x_i, and $n_1(x_k)$ 1's and $n_0(x_k)$ 0's in the row of x_j. If these 1's and 0's in the two lines are distributed completely at random, one would not say that the two rows are alike. If, on the contrary, the 1's (0's) of one row do attract the 1's (0's) of the other row to the same places (to the same columns), one would say that the two rows resemble each other. Suppose that $n_{11}(x_i, x_k)$ 1's coincide on the two rows and $n_{00}(x_i, x_k)$ 0's coincide on the two rows. If the 1's on the row of x_i and the 1's on the row of x_k are distributed at random, we have

$$\frac{n_{11}(x_i, x_k)}{n} = \frac{n_1(x_i)}{n} \cdot \frac{n_1(x_k)}{n}. \tag{8.81}$$

But if the 1's in the row of x_i and the 1's in the row of x_k have a tendency to coincide more frequently than by accident, we have

$$\frac{n_{11}(x_i, x_k)}{n} > \frac{n_1(x_i)}{n} \cdot \frac{n_1(x_k)}{n}, \tag{8.82}$$

which may be regarded as a sign of similarity. By the same token, the condition

$$\frac{n_{00}(x_i, x_k)}{n} > \frac{n_0(x_i)}{n} \cdot \frac{n_0(x_k)}{n} \tag{8.83}$$

is another indication of similarity. These last two inequalities may be taken as an explication of the term "large number" above.

To simplify the notation let us introduce

$$\begin{aligned} \alpha &= \frac{n_{11}(x_i, x_k)}{n}, \qquad & \beta &= \frac{n_{10}(x_i, x_k)}{n}, \\ \gamma &= \frac{n_{01}(x_i, x_k)}{n}, \qquad & \delta &= \frac{n_{00}(x_i, x_k)}{n}, \end{aligned} \tag{8.84}$$

where $n_{rs}(x_i, x_k)$ is the number of columns on which the row x_i has the value r and the row x_k has the value s. We have, of course,

$$\begin{aligned} \alpha + \beta &= \frac{n_1(x_i)}{n}, \qquad & \gamma + \delta &= \frac{n_0(x_i)}{n}, \\ \alpha + \gamma &= \frac{n_1(x_k)}{n}, \qquad & \beta + \delta &= \frac{n_0(x_k)}{n}. \end{aligned} \tag{8.85}$$

The inequalities of similarity (8.82) and (8.83) become

$$\alpha > (\alpha + \beta)(\alpha + \gamma) \quad \text{and} \quad \delta > (\beta + \delta)(\gamma + \delta). \tag{8.86}$$

Inspection of (8.86) leads one to agree that the values of $\alpha/[(\alpha + \beta)(\alpha + \gamma)]$ and $\delta/[(\beta + \delta)(\gamma + \delta)]$ beyond the threshold of unity measure the degree of similarity. Taking the logarithms of these two fractions and joining them with relative weights α and δ, we obtain a measure of similarity:

$$C^+ = \alpha \log \frac{\alpha}{(\alpha + \beta)(\alpha + \gamma)} + \delta \log \frac{\delta}{(\delta + \beta)(\delta + \gamma)} \tag{8.87}$$

The use of the logarithm ensures the additivity mentioned above, and also because the threshold mentioned above can be expressed by $C^+ = 0$. The relative weights α and δ are justified by the fact that they are the actual relative frequencies of two types of coincidences (1, 1) and (0, 0).

By an analogous argument we can put

$$\beta > (\alpha + \beta)(\beta + \delta) \quad \text{and} \quad \gamma > (\gamma + \delta)(\alpha + \gamma) \tag{8.88}$$

as the conditions of dissimilarity and define a measure of dissimilarity by

$$C^- = \beta \log \frac{\beta}{(\beta + \alpha)(\beta + \delta)} + \gamma \log \frac{\gamma}{(\gamma + \delta)(\gamma + \alpha)}. \tag{8.89}$$

Although we can invent all kinds of measures of similarity and dissimilarity, I know of none more convenient than C^+ and C^-. The only other one that may be worth mentioning is the following. Suppose that we multiply the left sides of the two inequalities of (8.86) by $\alpha + \beta + \gamma + \delta = 1$. Then we obtain $\alpha\delta - \beta\gamma > 0$ from both inequalities. This shows that the two inequalities of (8.86) are actually equivalent to each other and also to

$$\theta > 0, \tag{8.90}$$

with

$$\theta \equiv \alpha\delta - \beta\gamma. \tag{8.91}$$

Similarly, the two inequalities of (8.88) are equivalent to each other and to

$$\theta < 0. \tag{8.92}$$

Thus θ is a useful measure of similarity. From these facts it follows that if $\theta > 0$ then $C^+ > 0$ and $C^- \leq 0$, and that if $\theta < 0$ then $C^+ \leq 0$ and $C^- > 0$. The anomalous case $[C^+ > 0,\ C^- = 0]$ occurs when $\beta = \gamma = 0$. Similarly, the case $[C^+ = 0,\ C^- > 0]$ occurs when $\alpha = \delta = 0$. The cases $[C^+ > 0,$ $C^- < 0]$ and $[C^+ < 0,\ C^- > 0]$ never happen.

The absence of all similarity and all dissimilarity means in our interpretation the probabilistic independence of the two "random variables" x_i and x_k, which can take the values 0 and 1, as can be seen, for example, in (8.81). This neutrality means the four equalities that can be obtained by replacing the "larger-than" symbol by the equality symbol in (8.86) and (8.88). These four equations are actually equivalent to one another and to $\theta = 0$. This absence of similarity or dissimilarity entails $C^+ = 0$ and $C^- = 0$. It should be noted however, that $C^+ = 0$ alone, or $C^- = 0$ alone, does not conversely imply the absence of similarity and dissimilarity expressed by $\theta = 0$. For $C^+ = 0$ implies either $C^- = 0$ or else $C^- > 0$ and $\alpha = \delta = 0$. By the same token $C^- = 0$ implies either $C^+ = 0$ or else $C^+ > 0$ and $\beta = \gamma = 0$. If $C^+ = C^- = 0$ simultaneously, $\theta = 0$.

It is interesting to introduce the sum of C^+ and C^-,

$$C = C^+ + C^-. \tag{8.93}$$

This C, as we shall soon see, is nothing but what we called interdependence between two random variables x_i and x_k in Chapter 2. Because of Gibb's theorem C is non-negative, and the case $C = 0$ happens if, and only if, the variables are probabilistically independent, that is, if $\theta = 0$. C measures the strength of the deviation from independence, either toward similarity or toward dissimilarity.

Since our formalism is perfectly symmetrical with respect to 0 and 1, if we replace all the predicates in Y by their negations the results will remain unchanged. If we replace only some of the predicates by their negations, however, the numerical values of various qualities considered above will change, although the general trends will not change too easily. (By interchange of a predicate with its negation, $\alpha + \delta$ and $\beta + \gamma$ remain unchanged, but $\alpha\delta$ and $\beta\gamma$ change.) To avoid this ambiguity and to be faithful to the idea that y_j and $\neg y_j$ are equally important, the best practical way is to agree always to list both y_j and $\neg y_j$ pairwise in the list of Y. This entails

$$\alpha = \delta \quad \text{and} \quad \beta = \gamma, \tag{8.94}$$

which makes the entire analysis very simple. Since we then have the restriction

$$\alpha + \beta = \tfrac{1}{2}, \tag{8.95}$$

there is only one degree of freedom left, which can be determined by the parameter θ. Figure 8.3 shows the value of C^+, C^-, and θ $(= \alpha - \frac{1}{4})$ as functions of $2\alpha = 1 - 2\beta$. This convention has another advantage of avoiding the difficulty we encounter in defining similarity and dissimilarity when there are too many 1's or too many 0's. In fact, if we adopt this convention, there will be an equal number of 1's and 0's in each row.

Let us now attempt to extend the method we developed for two objects to cases of more than two objects. In the case of two objects, when two rows are compared, there are only four combinations [(0, 0), (0, 1), (1, 0), and (1, 1)] of values that can appear in a column. We attached to these four the "probabilities" determined by the relative frequencies with which each of the four combinations appears in the object-predicate table. If we take a subset X_ρ of X, consisting of r objects $(x_1^{(\rho)}, x_2^{(\rho)}, \cdots, x_r^{(\rho)})$, each column can be described by a set of numbers $(a_1^{(\rho)}, a_2^{(\rho)}, \cdots, a_r^{(\rho)})$ that appear in the r rows corresponding to these objects. Since each $a_i^{(\rho)}$ can be either 0 or 1, there are 2^r possible combinations of values. The relative frequency with which each combination $(a_1^{(\rho)}, a_2^{(\rho)}, \cdots, a_r^{(\rho)})$ appears in the object-predicate table is given by

$$p(a_1^{(\rho)}, a_2^{(\rho)}, \cdots, a_r^{(\rho)}) = \sum_{j=1}^{n} \frac{1}{n} \prod_{i=j}^{r} \delta[a_i^{(\rho)}, T(x_i^{(\rho)}, y_j)], \tag{8.96}$$

where $\delta[\xi, \eta]$ is 1 if $\xi = \eta$ and is 0 if $\xi \neq \eta$. In this expression the factor $1/n$ can be considered as a uniform weight attached to each of the n predicates

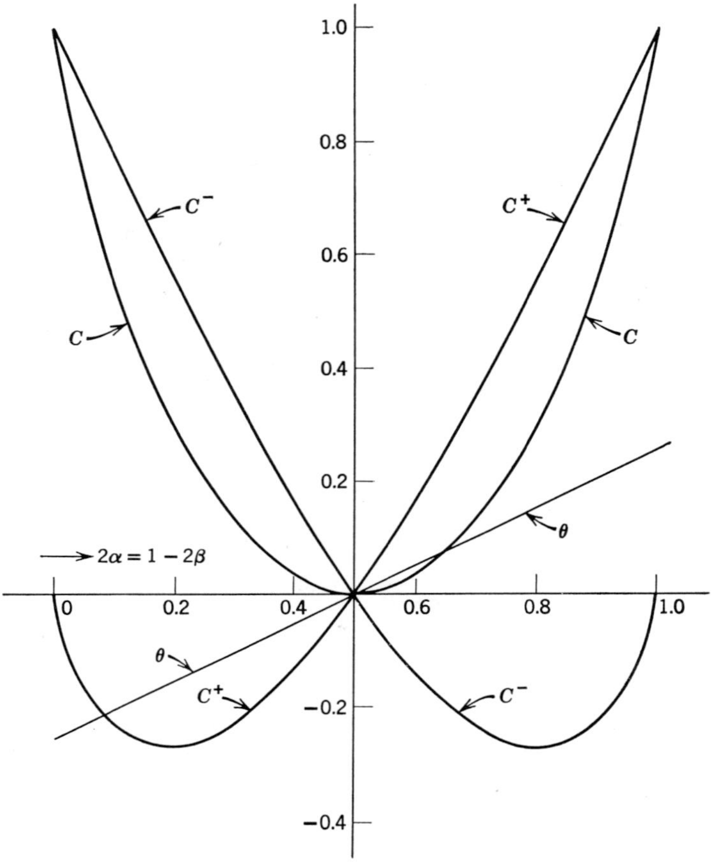

Figure 8.3 Relations among various variables measuring similarity and dissimilarity.

included in Y. As suggested by this interpretation, we can generalize this formula so as to make it applicable to the completed Boolean lattice $\hat{Y}$ by attaching weight w_j to each of the $\hat{n}$ predicates $\hat{y}_j$ of $\hat{Y}$:

$$p(a_1^{(\rho)}, a_2^{(\rho)}, \cdots, a_r^{(\rho)}) = \sum_{j=1}^{\hat{n}} w_j \prod_{i=j}^{r} \delta[a_i^{(\rho)}, T(x_i^{(\rho)}, \hat{y}_j)], \tag{8.97}$$

where $w_j \geq 0$ and $\sum_{j=1}^{\hat{n}} w_j = 1$. If we put $w_j = 1/\hat{n}$, we get the same value of $p(a_1^{(\rho)}, a_2^{(\rho)}, \cdots, a_r^{(\rho)})$ for all possible subsets $X^{(\rho)}$ consisting of r objects, insofar as the set of numbers $(a_1^{(\rho)}, a_2^{(\rho)}, \cdots, a_r^{(\rho)})$ is fixed, according to the generalized theorem of the ugly duckling (Theorem 7.23). This will destroy any kind of structure in the entire set X. By choosing an uneven weight, we start to see the structure. We obtain (8.96) from (8.97) by putting $w_j = 0$ for those predicates not included in the retrenched set Y and giving an equal weight $w_j = 1/n$ to each member of Y. We can, of course, give $w_j = 0$ to all nonmembers of a retrenched Y and uneven weights to the members of Y. This is a useful model.

With the help of (8.96) or (8.97) we can define the entropy of the subset X_ρ by

$$S(X_\rho) = -\sum_{a_1^{(\rho)}} \sum_{a_2^{(\rho)}} \cdots \sum_{a_r^{(\rho)}} p(a_1^{(\rho)}, a_2^{(\rho)}, \cdots, a_r^{(\rho)}) \log p(a_1^{(\rho)}, a_2^{(\rho)}, \cdots, a_r^{(\rho)}). \tag{8.98}$$

This entropy becomes minimal when the group of objects appear to have the same relationship [represented by a particular set of numbers $a_i^{(\rho)}$] independent of the predicate referred to, and increases in the measure as it exhibits diverse relations depending on different predicates. The expression (8.98) suggests that each set X_ρ of objects is comparable to a set of interdependent random variables and that each predicate is comparable to an occurrence of a possible set of values of these random variables. (This viewpoint is the "reciprocal" of the viewpoint of factor analysis.)

Once we have obtained the expression of entropy for each set of objects, our whole arsenal of interdependence analysis developed in Chapter 2 will become instrumental in unraveling the mutual relationships among subgroups of objects. The basic quantity is what we called "interdependence" J, which was defined as follows. Suppose that we have a subset X_ρ of objects, which we now divide into τ sub-subsets, $X_{\rho 1}, X_{\rho 2}, \cdots, X_{\rho\tau}$. Then the interdependence among these sub-subsets constituting X_ρ is

$$J(X_\rho; X_{\rho 1}, X_{\rho 2}, \cdots, X_{\rho\tau}) = \sum_{\alpha=1}^{\tau} S(X_{\rho\alpha}) - S(X_\rho) \tag{8.99}$$

in accordance with (2.15). To simplify the expression, we sometimes omit the merged subset X_ρ from the argument of J; thus we often write the left-hand side of (8.99) simply $J(X_{\rho 1}, X_{\rho 2}, \cdots, X_{\rho\tau})$. When X_ρ consists of two objects x_i and x_k, and if we divide X_ρ into two sub-subsets, each of which consists of a single object, namely, x_i or x_k, then J becomes C of (8.93):

$$J(x_i, x_k) = C = C^+ + C^-. \tag{8.100}$$

The reader is urged to restudy, at this point, the main results of the method of interdependence analysis (IDA) explained in Chapter 2.

In Chapter 2 we said that the method of IDA can extricate different kinds of interdependence of stochastic variables. Now we are in a position to apply this powerful method to the object-predicate table to analyze the structural dependence of a collection of objects, using (8.97), which allows us to consider the objects as stochastic variables. This OPIDA method, when applied to any two objects, can discover the bilateral relation consisting of their similarity and dissimilarity. The J of a group of more than two objects can be reduced in some cases to pairwise similarity and/or pairwise dissimilarity within the group, but more often it represents some "emergent" properties of the group, which cannot be reduced to bilateral relations. It is not claimed that all properties that have been called "emergent" by philosophers and psychologists can be rediscovered by OPIDA, but the following example will convince the reader that the method is indeed capable of exposing very subtle multilateral properties.

Suppose that four girl students live in a dormitory. Three of them are bound by a peculiar mixture of friendship and jealousy, so that none of the group wants to sit alone in the lounge without another member, yet none wants to

Table 8.1

x \ y	y_1	y_2	y_3	y_4	y_5	y_6	y_7	y_8
x_1	1	1	0	0	1	1	0	0
x_2	1	1	1	1	0	0	0	0
x_3	0	0	1	1	1	1	0	0
x_4	1	0	1	0	1	0	1	0

sit there with both of the remaining two because she cannot stand seeing the evidence of friendship between these latter two. The fourth girl is entirely neutral to these three and sits in the lounge no matter who else may or may not be sitting there; reciprocally these three pay no attention to the fourth girl. Suppose that x_1, x_2, x_3, and x_4 represent these four girls, and let y_j stand for the predicate "is sitting in the lounge at the jth observation." Because of the relationship among the girls, there can be only eight types of columns in the object-predicate table thus obtained. If we write each of the eight possible column types only once in the table, we obtain Table 8.1. The reader should compare this situation with Language D of Section 2.3, Figure 2.2.

In each row there are four 1's and four 0's; hence, using $w_j = \frac{1}{8}$ in (8.97) for $j = 1, 2, \cdots, 8$, we obtain $p(0) = p(1) = \frac{1}{2}$ for each girl, meaning that the probability that she will or will not be sitting in the lounge is $\frac{1}{2}$. If there is a tendency for any pair of girls to sit together more often than by accident, C^+ becomes positive, showing "similarity." However, for any of the six pairs we can take from the four girls, we obtain $p(0, 0) = p(0, 1) = p(1, 0) = p(1, 1) = \frac{1}{4}$, which is equal to $p(0) \cdot p(0) = p(1) \cdot p(1) = p(0) \cdot p(1)$. Hence not only C^+ but also C^- and $C = C^+ + C^-$ all disappear for any pair. This shows that we cannot detect any special relation insofar as we observe the girls pairwise. Now if we take a group of three girls, something new happens. Consider first a group of three consisting of x_2, x_3, and x_4; then all eight possible column types appear with an equal frequency: $p(0, 0, 0) = p(0, 0, 1) = p(0, 1, 0) = p(0, 1, 1) = p(1, 0, 0) = p(1, 0, 1) = p(1, 1, 0) = p(1, 1, 1) = \frac{1}{8}$. This makes $J_{tot}(x_2 \cup x_3 \cup x_4) = J(x_2, x_3, x_4)$ for this group disappear, because $S^{(1)}$ for each girl is 1 and $S^{(3)}$ of the group of three is 3. The same situation also prevails for the groups (x_1, x_3, x_4) and (x_1, x_2, x_4), but not for the special group (x_1, x_2, x_3). For this group the column types (1, 0, 0), (0, 1, 0), (0, 0, 1), and (1, 1, 1) are missing, the first three corresponding to "sitting without a friend" and the fourth corresponding to "sitting with both friends." The result is that $J_{tot}(x_1 \cup x_2 \cup x_3) = J(x_1, x_2, x_3) = 1$, because $S^{(3)}$ of this group of three is only 2. This indicates that this particular group of three (x_1, x_2, x_3) has a special multilateral structure. If we take J_{tot} of the entire four girls, we obtain $J_{tot}(x_1 \cup x_2 \cup x_3 \cup x_4) = 1$, which is the same as $J_{tot}(x_1 \cup x_2 \cup x_3)$. In the light of Corollary 2.2, this means that there is no relation between the group of three (x_1, x_2, x_3) and the fourth girl, because $J(x_4, x_1 \cup x_2 \cup x_3) = 0$.

This example shows that the OPIDA method can discover and quantitatively describe a complicated group property of objects as reflected in their behavior, although of course it will never penetrate into the causation or motivation of the behavior. It is interesting that none of the other mathematical methods so far proposed seems to be able to analyze a table as simple as Table 8.1.

Since we have become familiar with the structure of Table 8.1, we may as well use it to illustrate the generalized theorem of the ugly duckling (Theorem 7.23). The special group (x_1, x_2, x_3) is characterized by four column types, (0, 0, 0), (0, 1, 1), (1, 0, 1), and (1, 1, 0), which may now be considered respectively to be $A^{(1)}$, $A^{(2)}$, $A^{(3)}$, and $A^{(4)}$ in the statement of the theorem, with $\omega = 4$. We can see from Table 8.1 and the definition of $A^{(\lambda)}(\hat{y}_j)$ that for the group (x_1, x_2, x_3) we have $A^{(1)}(y_7)$, $A^{(1)}(y_8)$, $A^{(2)}(y_3)$, $A^{(2)}(y_4)$, and so on, where, for instance, $A^{(2)}(y_3)$ is the statement that the column of y_3 for the group (x_1, x_2, x_3) is (1, 1, 0). Now $A(\hat{y}_j) = \bigcup_{\lambda=1}^{4} A^{(\lambda)}(\hat{y}_j)$ is true of any of the eight predicates considered in Table 8.1 and for the group of objects (x_1, x_2, x_3). But if we take another object group, say, (x_2, x_3, x_4), $A(\hat{y}_j)$ is true only for y_1, y_4, y_5, and y_8 and is not true for y_2, y_3, y_6, and y_7. Now suppose that we enlarge Y to its Boolean completion $\hat{Y}$. Then $\hat{Y}$ will contain 16 predicates in all. None of the eight newcomers will satisfy $A(\hat{y}_j)$ for (x_1, x_2, x_3), but four of them will satisfy $A(\hat{y}_j)$ for (x_2, x_3, x_4). As a result, there are exactly eight predicates in $\hat{Y}$ that satisfy $A(\hat{y}_j)$ for any group of three objects. This will formally cause the special structure of the group (x_1, x_2, x_3) to disappear. But of course we should not forget that the newcomers in $\hat{Y}$ would require a different interpretation.

It is interesting to note in passing that A is a predicate characterizing a predicate $\hat{y}_j$, so that $A(\hat{y}_j)$ is true or false depending on $\hat{y}_j$ but also depends on the object on which the property is tested. As a result, A can also be considered a predicate for three objects, say, x_i, x_k, x_l. The sentence $A(\hat{y}_j; x_i,$

x_k, x_l) can be true or false depending on all four variables. But if we fix $\hat{y}_j$ and consider (x_i, x_k, x_l) as the argument, it amounts to considering A as a predicate for a group of three objects. That is a group property. This predicate turns out to be true for (x_1, x_2, x_3) no matter which predicate one selects from the original Y.

In summary, we can say that $J_{\text{tot}}(X_\rho)$ is an excellent measure of the cohesion existing among the individual members x_i belonging to X_ρ, where the cohesion may stem from bilateral relations of similarity and dissimilarity as well as from more complicated multilateral relations. Similarly $J(X_{\rho 1}, X_{\rho 2}, \cdots, X_{\rho\tau})$ is a good measure of cohesion existing among the subgroups $X_{\rho\mu}$ belonging to X_ρ. Such a measure can be expected to find many applications in cognitive and recognitive problems. Relegating more elaborate discussions to later sections, we mention a few simple applications here. Suppose that in a collection X of m objects, for an integer r $(r < m)$, we are asked to locate the subgroup of r objects that are most strongly tied together. An answer would be the subset X_ρ of r objects that maximizes $J_{\text{tot}}(X_\rho)$. Suppose that we are given a collection X of objects and an integer τ and are asked to classify X into τ classes in such a way that the members of each class are strongly tied together. An answer will be a division of X into $X_1, X_2, \cdots, X_\tau$ that minimizes $J(X_1, X_2, \cdots, X_\tau)$, because, by the fundamental theorem of IDA, this will maximize $\sum_{\mu=1}^{\tau} J_{\text{tot}}(X_\mu)$ [see, in particular, (2.30)].

OPIDA also allows us to address ourselves to an inverse problem. The problem of classification aims usually at determination of classes such as $X_1, X_2, \cdots, X_\tau$ when the weights of preference w_j [in (8.97)] are given. The inverse problem would be to determine the weights w_j when the classes $X_1, X_2, \cdots, X_\tau$ are given. This is applicable to the "training period" of a recognitive task. A formal answer would be to maximize $J(X_1, X_2, \cdots, X_\tau)/J_{\text{tot}}(X)$ by varying the w_j for the given division $X = X_1 \cup X_2 \cup \cdots X_\tau$. This is a case of what we called reaction of Step (c) on (a) at the beginning of this section. Of course, each cognitive or recognitive problem has its own peculiar aspect, purpose, and source of complications, which must be taken into consideration in addition to the general orientation given above. Furthermore, as we see later, a major trouble in executing on a computer the simplest algorithms based on the above consideration originates from the fact that they require a tremendous storage space and a tremendous number of computations. This is mainly because in a set of m objects there are $2^m - 2$ nonempty proper subsets, which can easily become an enormous number.

I mentioned in Section 7.5 that the mathematical symmetry between X and Y in the object-predicate table is so obvious that whatever can be mathematically formulated for X in its relation to Y can also be immediately reformulated for Y in its relation to X. This reciprocity, of course, also applies to the use of the OPIDA method. All that is required is to introduce the entropy of a subset of predicates Y_ρ consisting of, say, $y_1^{(\rho)}, y_2^{(\rho)}, \cdots, y_r^{(\rho)}$ in analogy to (8.98):

$$S(Y_\rho) = -\sum_{b_1^{(\rho)}} \sum_{b_2^{(\rho)}} \cdots \sum_{b_r^{(\rho)}} p(b_1^{(\rho)}, b_2^{(\rho)}, \cdots, b_r^{(\rho)}) \log p(b_1^{(\rho)}, b_2^{(\rho)}, \cdots, b_r^{(\rho)}). \tag{8.101}$$

The "probability" $p(b_1^{(\rho)}, b_2^{(\rho)}, \cdots, b_r^{(\rho)})$ that a horizontal row will have the values $(b_1^{(\rho)}, b_2^{(\rho)}, \cdots, b_r^{(\rho)})$ at the positions defined by $(y_1^{(\rho)}, y_2^{(\rho)}, \cdots, y_r^{(\rho)})$ is given by

$$p(b_1^{(\rho)}, b_2^{(\rho)}, \cdots, b_r^{(\rho)}) = \sum_{i=1}^{m} v_i \prod_{k=1}^{r} \delta[b_k^{(\rho)}, T(x_i, y_k^{(\rho)})], \tag{8.102}$$

where the summation with respect to i extends over all the objects included in the table, and the weight v_i is a measure of importance of each sample object in determining the property and relation of predicates. We assume $v_i \geq 0$ and $\sum_{i=1}^m v_i = 1$. When Y_ρ is Y itself, the vector $(b_1^{(\rho)}, b_2^{(\rho)}, \cdots, b_r^{(\rho)})$ becomes an "object type."

With the help of $S(Y_\rho)$ we can introduce the entropic measure of cohesion of predicates. As in the case of objects, the total cohesion (total interdependence) of predicates within the subset Y_ρ is $J_{\text{tot}}(Y_\rho) = \sum_{k=1}^r S(y_k^{(\rho)}) - S(Y_\rho)$, which is a special case of the cohesion among the sub-subsets of predicates, $Y_{\rho 1}, Y_{\rho 2}, \cdots, Y_{\rho\tau}$, constituting Y_ρ:

$$J(Y_{\rho 1}, Y_{\rho 2}, \cdots, Y_{\rho\tau}) = \sum_{\alpha=1}^{\tau} S(Y_{\rho\alpha}) - S(Y_\rho). \tag{8.103}$$

These quantities can be used for various useful purposes. In a very simple example, those predicates that show strong mutual similarity, having a large value of C^+/C close to unity for each pair, may be grouped together as a single predicate when a rough first-approximation description is required. Another interesting example of inversion of X and Y (object-predicate reciprocation) can be found in a probabilistic consideration of the ideas of intention and extension of a predicate. We are also going to see presently a quantitative generalization of logical relations by the use of predicate entropies.

In close accord with the usually accepted notion of intension of a predicate, we may consider intension as the richness of information carried by the knowledge K that an object satisfies the predicate in question. On the other hand, extension must represent the measure of indeterminacy about individuals still left after acquisition of knowledge K.

Before acquisition of K, our ignorance about the object type of an individual is given by $S(Y)$, which is defined by the probabilities $p(b_1, b_2, \cdots, b_n)$, where Y consists of all the predicates in the table, $y_1, y_2, \cdots, y_n$. After acquisition of the knowledge K that the object satisfies a certain predicate, say, y_1, this probability changes to

$$p^*(Y) = p^*(b_1, b_2, \cdots, b_n) = 0 \quad \text{for} \quad b_1 = 0,$$
$$p^*(Y) = p^*(b_1, b_2, \cdots, b_n) = \frac{1}{p(1)}\, p(1, b_2, \cdots, b_n) \quad \text{for} \quad b_1 = 1, \tag{8.104}$$

where p(1) = $\Pr(b_1 = 1)$ is a normalizing denominator. Thus the ignorance about the object type after acquisition of knowledge K is $S^*(Y)$, which is defined by $p^*(Y)$ with the help of the usual definition of S-function (8.101). Then the information brought about by knowledge K is the decrease in ignorance,

$$\text{int}\,(y_1) = S(Y) - S^*(Y), \tag{8.105}$$

which must correspond to the intension of predicate y_1. On the other hand, the extension must be given by the degree of indeterminacy of the object type of an object after the acquisition of knowledge K,

$$\text{ext}\,(y_1) = S^*(Y). \tag{8.106}$$

Similar definitions, of course, can be applied to any predicate $\hat{y}_j$ in the Boolean completion $\hat{Y}$.

The intension and extension thus defined satisfy the relation int (y_1) + ext (y_1) = constant, which they should satisfy in the logarithmic scale according to the inverse proportionality law. The constant is defined by the object-predicate table and does not depend on the particular predicate y_1 under consideration.

It is easy to see that this information-theoretical explication of intension and extension reduces to our former explication when each of the possible object types receives a constant probability or zero probability. The summation in the (8.101) extends, in this case, with $Y_\rho = Y$, over all possible object types, which are 2^n in all. However, those object types that do not appear in the object-predicate table have zero probability and hence make no contribution to $S(Y)$. If m is the number of different object types that appear in the object-predicate table, the maximum of $S(Y)$ under a fixed value of m is $\log m$, which is reached when each of the m object types receives probability $1/m$. Suppose this to be the case; then $S^*(Y)$ becomes $\log D(y_1)$, where $D(y_1)$ is the number of object types that have $b_1 = 1$. Under these circumstances (8.105) and (8.106) become

$$\text{int}\,(y_1) = \log m - \log D(y_1) \quad \text{and} \quad \text{ext}\,(y_1) = \log D(y_1), \tag{8.107}$$

which are exactly equivalent to (7.214) and (7.216).

Since the problem of the so-called "complexity" can be reduced to a problem of extension, as we saw in Section 7.5, its probabilistic generalization is immediate and is not described here.

In Section 7.5 we saw how we can discover the "empirical logical structure" among predicates from a given object-predicate table [see (7.202)–(7.207)]. It is not only instructive but also useful to project the problem of logical structure against the continuous background provided by probability and entropy. In so doing we shall be able to talk about an "approximate" implication, and "approximate" negation, and so on, which are certainly very convenient notions. This attempt is interesting also from the point of view, hinted at in Section 7.3, that probability is prior to logic, in spite of the usual definition of the concept of probability, which presupposes logic. Once an object-predicate table is given, it is easy to extend it to its Boolean completion and we can consider the probability distribution for all the predicates in the completed lattice. For this reason what we denote by y_j can be a $\hat{y}_j$ of our former notation. See [W-31, W-31a] for another line of attempt to derive logic from probability.

We can translate the logical constraints in terms of probabilities as follows. The relation $y_j = \emptyset$ is equivalent† to

$$p(b_j) = 0 \tag{8.108}$$

for $b_j = 1$. The v_i are supposed to be nonzero in (8.102). The relation $y_j = \Box$ is equivalent to (8.108) for $b_j = 0$. The implication $y_j \to y_k$ is equivalent to

$$p(b_j, b_k) = 0 \tag{8.109}$$

for $b_j = 1 - b_k = 1$. The equivalence relation $y_j = y_k$ is equivalent to (8.109) for $b_j = 1 - b_k$. The predicate y_k is the negation of y_j if and only if (8.109) is true for $b_j = b_k$. The conjunctive relation $y_l = y_j \cap y_k$ is equivalent to

$$p(b_j, b_k, b_l) = 0 \tag{8.110}$$

for $b_j b_k = 1 - b_l$. The disjunctive relation $y_l = y_j \cup y_k$ is equivalent to (8.110) for $(1 - b_k)(1 - b_j) = b_l$. By replacing "$= 0$" in (8.108)–(8.110) by "$\ll 1$," we get the notion of approximate logical relations.

Passing to the notion of entropy, we mention some of the simple theorems. The reader will be able to prove them by introducing four parameters corresponding to α, β, γ, and δ of (8.84) and checking case by case.

Theorem 8.6 If $y_l \to y_k$ then $C^+(y_l, y_k) \geq 0$ and $C^-(y_l, y_k) \leq 0$. Under the condition $y_l \to y_k$ the equality $C^+(y_l, y_k) = 0$ happens if $y_l = \emptyset$ or $y_k = \Box$, and the equality $C^-(y_l, y_k) = 0$ happens if $y_l = \emptyset$ or $y_k = \Box$ or $y_l = y_k$.

According to its usual definition, the average conditional entropy for y_k on the hypothesis of y_l is given by

$$S(y_k \mid y_l) = -\sum_{b_l} \sum_{b_k} p(b_l, b_k) \log \frac{p(b_l, b_k)}{p(b_l)}. \tag{8.111}$$

† According to the "empirical definition" based on the given object-predicate table.

We can decompose this $S(y_k \mid y_l)$ into two terms,

$$S^+(y_k \mid y_l) = -\sum_{b_l}\sum_{b_k} \delta[b_l, 1]\, p(b_l, b_k) \log \frac{p(b_l, b_k)}{p(b_l)} \tag{8.112}$$

and

$$S^-(y_k \mid y_l) = -\sum_{b_l}\sum_{b_k} \delta[b_l, 0]\, p(b_l, b_k) \log \frac{p(b_l, b_k)}{p(b_l)} \tag{8.113}$$

so that

$$S(y_k \mid y_l) = S^+(y_k \mid y_l) + S^-(y_k \mid y_l). \tag{8.114}$$

Theorem 8.7 If $y_l \to y_k$, then $S^+(y_k \mid y_l) = 0$. If $S^+(y_k \mid y_l) = 0$, then $y_l \to y_k$ or $y_l \to \neg y_k$.

Combining the foregoing two theorems, we obtain Corollary 8.1.

Corollary 8.1 If $S^+(y_k \mid y_l) = 0$ and $C^+(y_l, y_k) > 0$ (strictly positive) then $y_l \neq \emptyset$, $y_k \neq \square$, and $y_l \to y_k$.

Proof. From $S^+(y_k \mid y_l) = 0$ follow two possibilities, $y_l \to y_k$ and $y_l \to \neg y_k$. In the former case we have $y_l \neq \emptyset$ and $y_k \neq \square$ according to Theorem 8.6, because $C^+ \neq 0$. In the second case the relation $C^-(y_l, \neg y_k) = C^+(y_l, y_k) \leq 0$, which follows from Theorem 8.6, contradicts the premise $C^+(y_l, y_k) > 0$. Hence the second case is excluded. This establishes a probabilistic definition of implication that eliminates the undesirable trivial cases. (See the end of Section 7.3) Because it is probabilistic we can extend it to an "approximate" notion by giving allowance to "$= 0$" and "> 0."

8.4. ALGORITHM OF CLUSTERING

In Section 8.1 we pointed out that the problem of clustering is insolvable, because it depends crucially on the two unknown factors: the notion of similarity and the weight distribution among the variables. This does not imply, however, that the problem becomes easily solvable when these two factors are clearly defined. In the foregoing sections we talked about the three ingredients of cognitive and recognitive tasks: selection and weighting of variables (predicates), definition of cohesion (which includes similarity), and actual placement of the objects into classes. In this section we assume that the first two factors are already clarified and we concentrate on the third aspect of the task. Even under these restricted circumstances the problem of clustering is still laden with ambiguities and difficulties.

One trouble is that we do not know whether we should allow an object to belong to only one class (of the same level) or more than one. Taxonomic classification usually presupposes the first alternative, while Roget's Thesaurus makes a good use of the second alternative. How should we define in terms of cohesion the qualifying conditions of membership of a class? The results would depend greatly on the nature and stricture of these conditions. About how many classes do we want to have? Do we allow the existence of extremely large classes and extremely small classes, or do we want the classes to be more or less of the same size? Do we want the classes to be fixed once and for all? Can we not devise a method that automatically gives different kinds of classes and groupings as the purpose varies and our experience grows? No matter what choice we may make in the presence of these alternatives, there is one further common difficulty that we have to face when we want to use a computer to carry out the task: the memory space and the computing time we need in a clustering job usually become prohibitively large.† For analogy between clustering and abduction, see [W-30].

There can be an unlimited number of strategies to carry out the "clustering" task; none of those known at present is really satisfactory, and all of them impose an enormous burden on the present-day computer. What follows is an algorithm of clustering that I have been developing in recent years and may be claimed to be one of the better ones. The merit of this approach is that it gives a unique solution based on theory and, as shown later, it can be reconciled to some extent with computer economy. One feature of the present algorithm that is not shared by other known algorithms is that it is applicable to cases when cohesion is not due solely to bilateral relations. To fix our idea, we shall orient our explanation toward a taxonomic classification, and our end product will be a taxonomic tree in which each peripheral branch contains one object (or a group of extremely similar objects). If we have to satisfy some conditions on the number of classes or on the number of members in each class, we "trim" the tree accordingly. Although the ideas behind our algorithm of taxonomy can be adapted for other, more sophisticated types of classification, we do not intend to discuss those possibilities.

First, we should examine the basic nature of the idea of cohesion. Suppose that we take a lump C of matter (collection of objects) and tear it apart into two parts, A and B. By this tearing action we destroy the cohesion between A and B, but within both A and B there still remains internal cohesion. This permits a mental picture of cohesion satisfying the condition that the total cohesion within C is always larger than the sum of the cohesion in A and the cohesion in B, the difference being the portion of cohesion destroyed by the tearing operation. A student of physics can visualize this by comparing cohesion to the (negative) potential energy between molecules. Suppose that for a group of people $X = \{x_1, x_2, \cdots, x_m\}$ we introduce a quantity "acquaintance" $A(i, j)$, which is equal to 1 if persons i and j know each other and equal to 0 if not. If we define the "cohesion" of a group of people by the

† The deepest reason for the inherent difficulties of clustering is that clustering is comparable to abduction if pattern recognition is comparable to induction. (See Chapter 4 for the definition of abduction, and see [W-30] for more about this matter.)

sum of $A(i, j)$ for all pairs that can be taken in the group, this cohesion certainly satisfies the condition mentioned above. The same is true if $A(i, j)$ is allowed to take any non-negative values not necessarily equal to 0 or 1. The entropic cohesion J_{tot} defined by (8.99) is another quantity that satisfies this condition. This is a direct consequence of Theorem 2.1, p. 59. We now consider the problem from a more formal viewpoint.

Let $X = \{x_1, x_2, \cdots, x_m\}$ be the collection of objects (or object types) we are interested in clustering. Let us designate by X_ρ a subset of X, that is, $X_\rho \subset X$, which means that if $x_i \in X_\rho$ then $x_i \in X$. Consider the collection $\mathcal{B} = \{X_\rho\}$ of all such subsets. (ρ will run from 1 to 2^m, including the empty set $\emptyset$ and the whole set X.) This $\mathcal{B}$ forms a Boolean lattice (finite Borel field) if we interpret the conjunction and disjunction in the usual set-theoretical way and the negation in the sense of a complementary set. The x's are the atoms of this lattice (see Section 7.2). Note also that each x could represent a set of objects (points) that are identical in the adopted mode of description. We can now give a definition of supra-additive, additive, and subadditive measures as follows.

DEFINITION 8.1. A set function $\mu(X_\rho)$ is called a measure if (a) $\mu(\emptyset) = 0$ (b) $\mu(X_\rho) \geq 0$ for all $X_\rho \in \mathcal{B}$. A measure is called supra-additive, additive, or subadditive according to whether, for all pairs $X_1 \in \mathcal{B}$ and $X_2 \in \mathcal{B}$ with $X_1 \cap X_2 = \emptyset$, (c) $\mu(X_1 \cup X_2) \geq \mu(X_1) + \mu(X_2)$, (d) $\mu(X_1 \cup X_2) = \mu(X_1) + \mu(X_2)$, or (e) $\mu(X_1 \cup X_2) \leq \mu(X_1) + \mu(X_2)$.

What was called "measure" in Definition 7.2 is called "additive measure" here. A measure that is simultaneously supra-additive and subadditive is an additive measure. A supra-additive measure that is not additive, that is, a supra-additive μ-function that satisfies (c) with "$>$" instead of "$\geq$" for at least one pair (X_1, X_2), is called a properly supra-additive measure. Similarly, a subadditive measure that is not additive is a properly subadditive measure. Either condition (c) or (d) implies (f) that if $X_1 \in \mathcal{B}$, $X_2 \in \mathcal{B}$, and $X_1 \subset X_2$ then $\mu(X_1) \leq \mu(X_2)$. But condition (e) does not imply (f). Hence, if (f) is required, it would be an additional condition in the case of a subadditive measure. The reader should notice that condition (f) is equivalent to (7.164). In the case of a subadditive measure it is advisable to assume that there exists at least one atom x_i for which $\mu(x_i) \neq 0$; otherwise the measure of every subset will become zero. We try in this section the assumption that the cohesion of a group is a supra-additive function of the group. We introduce the following lemma, which is so obvious that it does not require proof.

Lemma 8.1 Let $\mu_+(X_\rho)$, $\mu(X_\rho)$, and $\mu_-(X_\rho)$ be three measures defined on $\mathcal{B}$ such that $\mu_+(X_\rho) + \mu_-(X_\rho) = \mu(X_\rho)$ for all X_ρ and $\mu(X_\rho)$ is an additive measure. Then the statement that $\mu_+(X_\rho)$ is (properly) supra-additive is equivalent to the statement that $\mu_-(X_\rho)$ is (properly) subadditive. (μ_+ and μ_- are said to be conjugate to each other with respect to μ.)

Theorem 8.8 The entropy function $S(X_\rho)$ defined in (8.98) is a subadditive measure on $\mathcal{B}$, and the total cohesion function $C(X_\rho) = J_{\text{tot}}(X_\rho)$ derivable from it by (2.28) is a supra-additive measure on $\mathcal{B}$. $\mu_-(X_\rho) = S(X_\rho)$ and $\mu_+(X_\rho) = C(X_\rho)$ are conjugate to each other with respect to an additive measure $\mu(X_\rho) = \sum_{x_i \in X_\rho} S(x_i)$.

Proof. In the definition of C,

$$C(X_\rho) = \sum_{x_i \in X_\rho} S(x_i) - S(X_\rho), \tag{8.115}$$

the first term on the right, $\mu(X_\rho) = \sum_{x_i \in X_\rho} S(x_i)$, is additive because it is a sum with respect to the members of X_ρ, and the second term, $\mu_-(X_\rho) = S(X_\rho)$, is subadditive because of the basic property of the entropy function introduced in Theorem 1.3. Hence $\mu_+(X_\rho) = C(X_\rho)$ is supra-additive. The reader can easily check that $\mu_- = S$ satisfies condition (f). [Note that (2.7) is an expression of condition (f). See Theorem 9.15 for breakdown of this condition.]

In the case of "acquaintance" $A(i, j)$, the sum of the $A(i, j)$ for all the pairs that exist in X_ρ is obviously supra-additive:

$$\mu_+(X_\rho) = \sum_{\text{pair } (i,j) \in X_\rho} A(i, j). \tag{8.116}$$

In this case $\mu_+(x_i)$ must be zero for each individual member. Assuming $A = 0$ or 1, we can introduce its conjugate subadditive measure $\mu_-(X_\rho)$ by

$$\mu_-(X_\rho) = \frac{m_\rho(m - 1)}{2} - \mu_+(X_\rho), \tag{8.117}$$

where m and m_ρ are respectively the total number of members of X and that of X_ρ. The first term, $m_\rho(m - 1)/2$, is so chosen that it is additive and is never smaller than the second term, $\mu_+(X_\rho)$, whose maximum value is $m_\rho(m_\rho - 1)/2$. The reader can easily produce a case in which condition (f) is violated by μ_- of (8.117). The definitions (8.115) and (8.116) are good examples of the idea that the cohesion can be expressed by a supra-additive set function.

Now let us pass to the consideration of a taxonomic tree. At each branching point a subset X_ρ is divided into disjoint sub-subsets $X_{\rho\alpha}$, $\alpha = 1, 2, \cdots, \tau$, such that $X_{\rho 1} \cup X_{\rho 2} \cup \cdots \cup X_{\rho\tau} = X_\rho$ and $X_{\rho\alpha} \cap X_{\rho\beta} = \emptyset$ for $\alpha \neq \beta$. We introduce the "branching cost" Q at such a branching point by

$$\begin{aligned} Q(X_\rho \mid X_{\rho 1}, X_{\rho 2}, \cdots, X_{\rho\tau}) &= \sum_{\alpha=1}^{\tau} \mu_-(X_{\rho\alpha}) - \mu_-(X_\rho) \\ &= \mu_+(X_\rho) - \sum_{\alpha=1}^{\tau} \mu_+(X_{\rho\alpha}), \end{aligned} \tag{8.118}$$

where μ_+ is a general cohesion function and μ_- is its conjugate. If we take $\mu_- = S$ and $\mu_+ = C$, this Q becomes equal to $J(X_\rho; X_{\rho 1}, X_{\rho 2}, \cdots, X_{\rho\tau})$ of (8.99).

Theorem 8.9 Given X and the cohesion functions, the sum of the branching costs at all the branching points in a taxonomic tree is constant independent of the way the taxonomic tree is taken in X.

Proof

$$\sum_{\substack{\text{all branching}\\ \text{points}}} Q(X_\rho \mid X_{\rho 1}, X_{\rho 2}, \cdots, X_{\rho\tau}) = \sum^{\text{all}} \mu_-(x_i) - \mu_-(X)$$

$$= \mu_+(X) - \sum^{\text{all}} \mu_+(x_i). \qquad (8.119)$$

Except for the trunk and the peripheral branches (the individual x's), each branch appears once as an X_ρ and another time as an $X_{\rho\alpha}$ on the left-hand side of (8.119). In the two examples we used, the $\mu_+(x_i)$ are all zero, and the sum becomes simply $\mu_+(X)$. Theorem 2.1 is a special case of the present theorem.

The problem we face is as follows. A taxonomic tree is determined by a collection of branching points of the type $(X_\rho \mid X_{\rho 1}, \cdots, X_{\rho\tau})$, starting from the trunk X and ending with the peripheral branches x_i, with the in-between branches appearing once as X_ρ and another time as $X_{\rho\alpha}$. We want to establish an algorithm to select a unique tree among many possible ones by imposing a certain rule on the distribution of the branching costs at the branching points, while the sum total of these branching costs is fixed according to Theorem 8.9. Since a taxonomic tree is a successive repetition of branching, if we can apply a unique strategy at each branching point we shall end up with a unique taxonomic tree.

With regard to a single branching point, we made some suggestions toward the end of the last section. Since $\mu_+(X_\rho)$ and $\sum_\alpha \mu_+(X_{\rho\alpha})$ represent respectively the cohesion of the stem and the cohesion of the new branches, the difference, which is the branching cost, is the loss of cohesion caused by the partition of a group X_ρ into τ smaller groups $X_{\rho\alpha}$. Since the purpose of a classification is to form groups within which members cohere intensely, the most desirable partition must be the one that minimizes the branching cost. This minimization is easily (at least mathematically) defined if τ (which corresponds to r of the last section) is fixed; but when τ is not fixed the same idea of minimization of cost would suggest that the smaller the τ the better it is, because the branching cost will become smaller in general with a smaller τ. Of course, $\tau = 1$ does not serve any purpose because it amounts to doing nothing. This leads us to conclude (at least provisionally) that a dichotomy is better than a partition into more than two branches. But do we really always want to have a dichotomy?

Consider a case of trichotomy: $(X_\rho \mid X_{\rho1}, X_{\rho2}, X_{\rho3})$. Under what condition would we prefer this trichotomy to dichotomies, such as $(X_\rho \mid X_{\rho1} \cup X_{\rho2}, X_{\rho3})$ and $(X_{\rho1} \cup X_{\rho2} \mid X_{\rho1}, X_{\rho2})$? The condition is that there would be no reason to combine $X_{\rho1}$ and $X_{\rho2}$ in preference to $X_{\rho1}$ and $X_{\rho3}$ or $X_{\rho2}$ and $X_{\rho3}$. In other words, the trichotomy is preferred if the dichotomies $(X_\rho \mid X_{\rho1} \cup X_{\rho2}, X_{\rho2})$, $(X_\rho \mid X_{\rho1} \cup X_{\rho3}, X_{\rho2})$, and $(X_\rho \mid X_{\rho1}, X_{\rho2} \cup X_{\rho3})$ are more or less equally desirable. By generalizing this idea, we can formulate a rule that allows us to decide between dichotomies and polychotomies.

Our algorithm consists of three principles that determine a unique taxonomic tree. It is assumed that a general cohesion function μ_+ is defined on $\mathcal{B}$; hence the branching cost is calculable at each possible branching point.†

STRATEGY 8.1. Strategies 8.2 and 8.3 give a prescription for determining a partition $(X_\rho \mid X_{\rho1}, \cdots, X_{\rho\tau})$ when X_ρ is given. We apply this prescription first to the entire collection (trunk) X as X_ρ, and then to each of the resulting branches $X_{\rho\alpha}$ (i.e., $X_{\rho\alpha}$ of the first stage will be the X_ρ of the second stage). We continue this process until the branches $X_{\rho\alpha}$ become all x's.

STRATEGY 8.2. Given X_ρ, consider all possible dichotomies $(X_\rho \mid X_{\rho1}, X_{\rho2})$. If there is only one particular pair $(X^*_{\rho1}, X^*_{\rho2})$ such that

$$Q(X_\rho \mid X^*_{\rho1}, X^*_{\rho2}) = \min_{(X_{\rho1}, X_{\rho2})} Q(X_\rho \mid X_{\rho1}, X_{\rho2}), \tag{8.120}$$

the prescribed branching is the dichotomy $(X_\rho \mid X^*_{\rho1}, X^*_{\rho2})$.

STRATEGY 8.3. If there are more than one particular pair $(X^*_{\rho2}, X^*_{\rho2})$, take the collection $\mathcal{C}$ of all subsets C of objects that can appear either as $X^*_{\rho1}$ or as $X^*_{\rho1}$ in the (8.120). Consider the lattice $\overline{\mathcal{C}}$ generated by this collection $\mathcal{C}$ and let $\gamma_1, \gamma_2, \cdots, \gamma_\tau$ be the atoms of $\overline{\mathcal{C}}$. Then the prescribed branching is $(X_\rho \mid \gamma_1, \gamma_2, \cdots, \gamma_\tau)$.

The γ's can also be defined as follows. If a subset C of X_ρ is a member of the $\mathcal{C}$ defined above, its complement $\urcorner' C = X_\rho \cap \urcorner C$ is also a member. Let us keep either C or $\urcorner' C$ and eliminate the other from the collection. Let us designate by $\mathcal{C}^*$ this new collection with half the members, and denote its members as C^*_i, $i = 1, 2, \cdots, k$. Consider 2^k possible combinations: $\beta(i_1, i_2, \cdots, i_k) = (\urcorner')^{i1} C^*_1 \cap (\urcorner')^{i2} C^*_2 \cap \cdots, \cap (\urcorner')^{ik} C^*_k$, where each i can be either 0 or 1 and $(\urcorner')^0$ means the absence of $\urcorner'$ and $(\urcorner')^1$ means the presence of $\urcorner'$. Those β's that are not $\emptyset$ are γ's.

† These strategies were proposed by me around 1961 in an internal publication of the IBM Research Center. Francis W. Dauer tested its effectiveness in a linguistic problem [D-1], and Casimir Kulikowski tested it in a character recognition problem (1965, unpublished). The strategies were presented at an international conference in 1962, whose proceedings were published in 1965 [W-18]. See also [K-2a].

The leading idea of this strategy can be characterized as a "principle of parsimony" in the sense that it advises us to spend as little as possible of the branching cost Q at the beginning so that we may have as much as possible to spend toward the end, the total sum of spending being constrained to be constant by Theorem 8.9.

To convince himself that this algorithm is fairly reasonable, the reader can try it on the following examples given by the graphs of Figure 8.4. The cohesion is the sum of bilateral relations of the type $A(i, j)$, which in this case

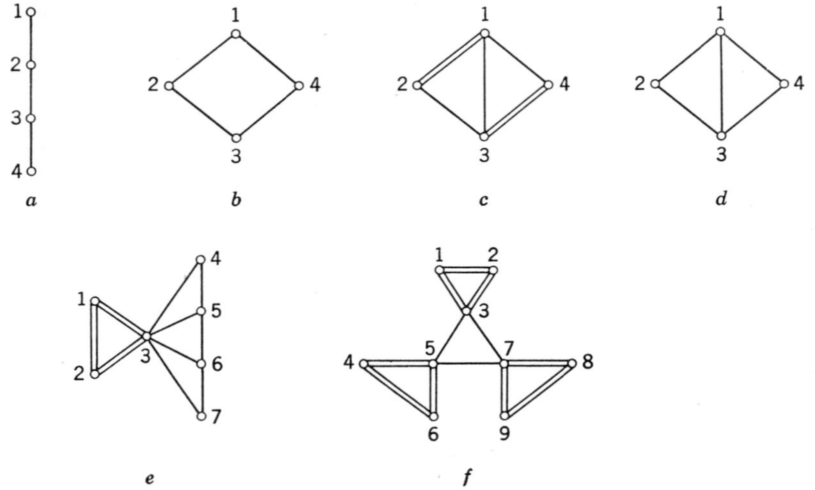

Figure 8.4 Objects bound by bilateral bonds of different strengths.

can take any non-negative integers 0, 1, 2, ... In the Figure 8.4, each circle represents an object and the number of bonds between two circles represents the integer values of $A(i, j)$. The reader will note that the strategy tells us to cut the least number of bonds. As a result, for instance, in Figure 8.4*d*, the pair (1, 3) survives the first trichotomy, while the pair (1, 2) is separated. It may be argued that (1, 3) is strongly adhered directly as well as indirectly through 2 and 4.

To show that the proposed algorithm also works reasonably well when the cohesion is not due to a bilateral relationship, let us consider the example of dormitory girls illustrated in Table 8.1. The X consists of the four girls, x_1, x_2, x_3, and x_4. The first-step dichotomy of X gives a division of X into (x_1, x_2, x_3) and x_4, because this is the only division whose branching cost is zero. The second-step dichotomy must be effected on (x_1, x_2, x_3). Strategy 8.3 dictates that (x_1, x_2, x_3) must be divided into three parts, x_1, x_2, and x_3, because the lattice in question is the one generated by x_1 and (x_2, x_3); x_2 and (x_3, x_1); and x_3 and (x_1, x_2). The atoms of this lattice are obviously x_1, x_2, and

x_3. This classification (Figure 8.5*a*) is perfectly satisfactory. We have the total cohesion in the amount of 1 unit available. In the taxonomic tree *a* we "spend" 0 unit at the first stage and 1 unit at the second stage. In the taxonomic tree *b* we spend 1 unit at the first stage and 0 unit at the second stage.

In spite of its obvious strong points, our strategy as it stands is not free from shortcomings. We mention four major shortcomings and show how they can be overcome.

First, if we take the condition (8.120) too seriously, we shall seldom have the opportunity to make polychotomies other than dichotomies. We should

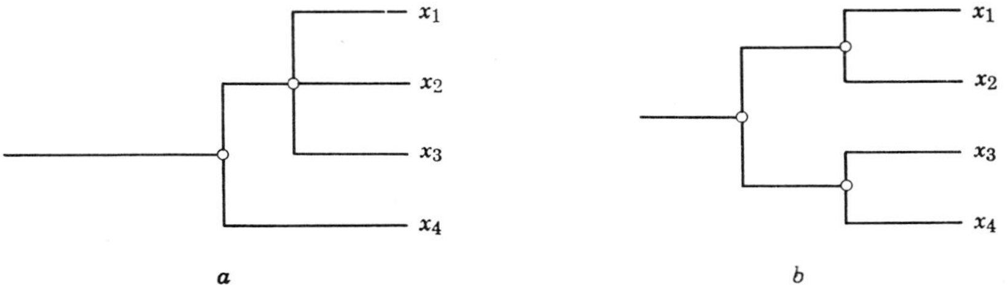

Figure 8.5 Clustering of four girls in dormitory.

therefore introduce a suitable constant ε and replace (8.120) by

$$Q(X_\rho \mid X^*_{\rho 1}, X^*_{\rho 2}) - \min_{(X_{\rho 1}, X_{\rho 2})} Q(X_\rho \mid X_{\rho 1}, X_{\rho 2}) \leq \varepsilon \, \langle Q(X_\rho \mid X_{\rho 1}, X_{\rho 2}) \rangle_{(X_{\rho 1}, X_{\rho 2})}, \tag{8.121}$$

where the angle brackets indicate the average dichotomic branching cost.

Second, the strategy as introduced above divides the entire set until each branch becomes a subset with a single member. An actual taxonomy, however, usually must finish with a subset of a certain size. We explain the problem, limiting ourselves to the case of classification based on similarity. It is obvious that the maximum of uniformity within a class is reached when the subset consists only of one member, but such a subset cannot be called a class of similar objects. The taxonomic tree obtainable according to our strategies must be trimmed at a certain point. We can introduce different conditions governing the trimming process. The simplest condition is an upper bound to the number of classes and a lower bound to the size of each class. Next, by the help of some measure of similarity, we can require that the process of branching be stopped when the intraclass similarity becomes appreciably larger than the interclass similarity. There are obviously many different ways to implement this idea. For instance, "intraclass similarity" could mean the minimum similarity in the class or the average similarity in the class. Similarly, "interclass similarity" could mean the maximum, or the average, or something more complicated. The term "appreciably larger" can also be given different interpretations. Another factor that can be used in trimming is the fact that the degree of uniformity (definable by the aid of some measure of similarity) within a branch changes suddenly when we pass from the trunk toward the periphery of the tree. The branch that has become uniform after a branching point is a good candidate to become the end of division. These factors must be flexibly incorporated in the algorithm according to the nature of the problem.

Third, our primary idea of the division of a subset into two parts such that the sum of the "cohesion" within each of the two parts is maximal may be somewhat simple minded. Should "cohesion" be understood, as in the foregoing, in the sense of the total cohesion of each part? Should it not be understood in the sense of "cohesion per member," or in the sense of "cohesion per binary bond"? According to the results of various experiments, we cannot make a single rigid rule. To give flexibility in this respect to our strategy, we can replace the definition of branching cost (8.118) (for $\tau = 2$) by

$$Q(X_\rho \mid X_{\rho 1}, X_{\rho 2}) = A[\mu_+ (X_\rho) - B_1\mu_+ (X_{\rho 1}) - B_2\mu_+ (X_{\rho 2})], \qquad (8.122)$$

where A is a function of $m(X_\rho)$, $m(X_{\rho 1})$, and $m(X_{\rho 2})$, while B_1 and B_2 are functions of $m(X_{\rho 1})$ and $m(X_{\rho 2})$, respectively. The symbol m means "the number of members of."

The maximum cohesion within the resultant parts implies minimal cohesion between the two parts. Hence our strategy may be characterized as a search for the easiest cleavage. Suppose that the μ_+-function is the sum of the bilateral bonds within the group. By cleaving the group of $m(X_\rho)$ members into a group of $m(X_{\rho 1})$ members and a group of $m(X_{\rho 2})$ members, we have to cut $m(X_{\rho 1}) \cdot m(X_{\rho 2})$ bonds of different strengths (including those of zero strength). If we want $Q(X_\rho \mid X_{\rho 1}, X_{\rho 2})$ to represent the average strength of the bonds to be cut, we should put in (8.122)†

$$A = \frac{1}{[m(X_{\rho 1}) \cdot m(X_{\rho 2})]}, \qquad B_1 = B_2 = 1. \qquad (8.123)$$

If the bond represents similarity, this Q will represent the average interclass similarity.

Unless A is a function only of X_ρ, (AB_1) is a function only of $X_{\rho 1}$, (AB_2) is a function only of $X_{\rho 2}$, and these three functions are the same function, Theorem 8.9 will no longer hold. Strategies 8.1, 8.2, and 8.3, however, will still make sense and remain effective in this general case.

Fourth, while our strategies are applicable in principle to any number of objects, there is a severe limitation in practice to the number imposed by the memory capacity and speed of present-day computers. Thanks to Strategy 8.3, we do not need to check all possible taxonomic trees at one stroke, but

† This choice was suggested by Tadao Takekawa.

needed check only dichotomies at each stage. For instance, if there are $m = 4$ objects, there are actually 26 different taxonomic trees but only seven dichotomies. This ratio grows very fast with m. In some cases we can place a lower limit on the size of each part of the dichotomies, thus lowering the number of dichotomies to be examined. But if the number of objects (m) becomes more than, say, 20, the number of dichotomies ($2^{m-1} - 1$) becomes prohibitive. This is the price we pay for having a unique solution unlike other approaches.

We can, however, cope with cases of large m by the following preprocessing strategy. We first select a small number (say, 20 or less) of "representative" objects from among the given large number m of objects and apply our clustering Strategies 8.1, 8.2, and 8.3 to these representatives. After we have obtained the classes of representative objects, the other objects that have been provisionally omitted from consideration will be reinstated and each object will be placed in the class to which the "nearest" representative belongs. This strategy can be characterized as the reduction of a clustering problem to a pattern-recognition problem. In fact, by preprocessing we obtain a small number of representative objects whose class is determined. Hence these representative objects can be considered as the paradigms of pattern recognition (see Section 8.1), and the other objects are to be classified in simulation of these paradigms. See [W-32a] and [W-32b].

I can suggest two methods of extracting the representative objects from the given collection. In one, called REPREX (for representative extraction), each object is given the degree of importance that measures essentially the degree to which it shares the important properties in the sense of SELFIC, and those objects that are important in this sense are used as representatives (see [W-32]). In the other method, which may be called "preclustering," we first group those objects that are strongly similar and use one of them (or their average) as a group representative. In many cases we can adjust the criterion of similarity so that we get relatively a small number of groups of similar objects. It is true that preclustering is itself a case of clustering, because it divides the given collection of objects into classes of similar objects. The reason why this preclustering can be done with a large number of objects (whereas the aforementioned clustering algorithm could not) is that in the present case we simplify cohesion and reduce it to a binary, bilateral relation, "similar or dissimilar."

It is important to note that after the representatives are chosen, either by REPREX or by preclustering, we can apply a cohesion function of a sophisticated kind in the main clustering program. The resulting classification scheme may be based on nonbilateral as well as bilateral relations.

REPREX works as follows. We apply the SELFIC method explained in Section 8.2 to the entire collection of objects at hand, after subtracting the average vector from each vector. Let $\{\phi_k\}$ be the optimal coordinate system, and let $c_k^{(\alpha)}$ be the component of the object vector α in the direction of ϕ_k. For a given value of σ we can obtain a feature space of n^* dimensions with $100 \times \sigma\%$ fidelity by requiring

$$\sum_{k=1}^{n^*} \lambda_k \geq \sigma > \sum_{k=1}^{n^*-1} \lambda_k, \tag{8.124}$$

where

$$\lambda_k = \sum_{\alpha=1}^{\nu} w^{(\alpha)} [c_k^{(\alpha)}]^2, \tag{8.125}$$

as in (8.57). By attaching the meaning of "important properties" or features to the vector components in this feature subspace, we can define the importance of each vector α by

$$I^{(\alpha)} = \sum_{k=1}^{n^*} [c_k^{(\alpha)}]^2. \tag{8.126}$$

Indeed, those objects that have the major part of their components outside the feature space may not be considered as representative vectors. We can now rearrange the object labels α so that

$$I^{(1)} \geq I^{(2)} \geq \cdots \geq I^{(\nu-1)} \geq I^{(\nu)}. \tag{8.127}$$

If we want to take m^* representative objects, we need take only $I^{(1)}, I^{(2)}, \cdots, I^{(m^*)}$. If we set m^* in the neighborhood of 20 or so, we can apply our main clustering algorithm to them and obtain the desired number of classes with class paradigms. The remaining job of placing the other $m - m^*$ objects into these classes is a typical pattern-recognition problem. For this purpose we can use the recognition methods explained in the next section. The simplest way is to place each object in the class of the paradigm most similar to the object.

The REPREX method, combined with the idea of object-predicate reciprocity, opens a new possibility for coping with the problem in which the number of predicates is very large. As is obvious from the explanation of Section 8.2, the SELFIC method can be used for a limited number (say, 100 or less) of predicate variables, but for a practically unlimited number of objects. When the number of objects is limited but the number of predicate variables is very large, we can interchange the roles of object and predicate variable by object-predicate reciprocation and apply SELFIC-REPREX. As a result we can obtain a smaller number of representative predicates in lieu of representative objects. After thus retrenching the number of predicate variables, we can once more apply SELFIC in the ordinary way. When both the number of objects and the number of predicate variables are large, it is often advisable to pick 100 or so objects at random and apply SELFIC-REPREX in the reciprocated fashion, first to decrease the number of predicate variables and then to reinstate the provisionally discarded objects.

In the rest of this section we explain a special clustering algorithm applicable to the case when cohesion is bilateral and symmetrical (nondirectional) and the bilateral relation has only two values, yes-no or 0-1. This method can be used independently or as a preclustering for a bigger problem.The bilateral relation could be, for instance, "person i and person k know each other," or "object i is similar to object k." The special feature of this algorithm is that although the relation itself is binary, it introduces another expression of the intensity of relation. For instance, person i and person k may not know each other, but there may be a third person l who knows and is known by person i and person k. Then i and k may be said to be acquaintances of the second degree. In the same way, two persons can be linked by $(s - 1)$ intermediate persons but no fewer than $(s - 1)$ intermediate persons; then we may speak of acquaintance of the sth degree, and so forth. This kind of intensity is incorporated in the following considerations.

The purpose of the algorithm is to divide the entire set of objects into classes so that any two members of a class are mutually linked by a bond of a degree not larger than s, and a member and a nonmember are not linked at all or are linked by a bond of a degree larger than s. If the cohesion is due to a bilateral relation but the relation has continuous degree, we should first determine a threshold and reduce the relation to a binary characterization. When this bilateral binary clustering program is used as preclustering for the main clustering algorithm introduced earlier, we have to go through the binarization process first, but later, when we use the main clustering algorithm, we can restore the nonbilateral, gradated relationship among the objects.

In the nonbinary, bilateral case we can define an object-object "proximity" table, P_{ik}, $i, k = 1, 2, \cdots, m$, where P_{ik} represents the strength of the bilateral relation between object i and object k. When the bilateral relation consists of "similarity," P_{ik} could be the entropic measure $C^+(x_i, x_k)$, $C^+(x_i, x_k) - C^-(x_i, x_k)$, and so on, of Section 8.3, which is derived from the object-predicate table $T(x_i, y_j)$. The parameter θ of Section 8.3 also qualifies as P_{ik}. If we consider only the coincidence of 1's (yes to a predicate) as indicative of similarity, we can also consider $\sum_{j=1}^{n} T(x_i, y_j)T^{-1}(y_j, x_k)$ [$=\alpha n$ to use the α of (8.84)] as a measure of proximity. Tanimoto's favorite measure of similarity can be written, with the help of (8.84), as $t = (\alpha - \beta - \gamma)/(\alpha + \beta + \gamma)$, which is deliberately asymmetrical with respect to 1 and 0 [T-6]. The symmetrical version of the same idea is $s = (\alpha + \delta - \beta - \gamma)$, which was considered in Section 8.3. In a symmetrical table (symmetrized with respect to 1 and 0), this s becomes $2\alpha - 2\beta = 4\alpha - 1$. That was the reason why α was used as a measure of similarity in Figure 8.3. The s and t can be useful as P_{ik}, depending on the purpose. P_{ik} can, of course, represent the intensity of any bilateral, symmetrical relation other than similarity between two objects i and k. For brevity, however, we shall use the word proximity to denote any bilateral relations. We assume throughout this section the symmetry

$$P_{ik} = P_{ki}. \tag{8.128}$$

From this continuously valued proximity table P_{ik} we derive a binary proximity table Q_{ik} with a suitable threshold τ:

$$\begin{aligned} Q_{ik} &= 1 \quad \text{if} \quad P_{ik} \geq \tau \\ &= 0 \quad \text{if} \quad P_{ik} < \tau. \end{aligned} \tag{8.129}$$

For definiteness, we assume

$$Q_{ik} = Q_{ki}, \qquad i, k = 1, 2, \cdots, m, \tag{8.130}$$

$$Q_{ii} = 0, \qquad i = 1, 2, \cdots, m. \tag{8.131}$$

We sometimes write $Q(i, k)$ for Q_{ik} for convenience.

We speak of first-degree proximity between object i and object k if we have $Q_{ik} = 1$, and of second-degree proximity between them if we have $Q_{ik} = 0$ and if for some j, $Q_{ij}Q_{jk} = 1$. Consider an r-unit-long sequence of object labels $(i = j_0, j_1, \cdots, j_r = k)$ such that $Q(j_0, j_1)Q(j_1, j_2) \cdots Q(j_{r-1}, j_r) = 1$. If for a given pair (i, k) r is the length of the shortest sequence connecting them, having this property, we speak of rth-degree proximity. The degree of proximity and the value of P_{ik} are not necessarily related, but in general, by increasing the threshold τ, we pass to a larger degree of proximity between the same pair, and for a given threshold value a smaller degree of proximity often reflects a higher value of the P's. If we denote the rth power of the matrix Q by Q^r, the rth-degree proximity between i and k is equivalent to the condition $Q^s_{ik} = 0$ for $s < r$ and $Q^r_{ik} = 1$. It is particularly convenient to introduce a matrix

$$\begin{aligned} &(Q^{r*})_{ik} = (Q^1)_{ik} \dotplus (Q^2)_{ik} \dotplus \cdots \dotplus (Q^r)_{ik} \quad \text{for nondiagonal elements,} \\ &(Q^{r*})_{ii} = 0 \quad \text{for diagonal elements,} \end{aligned} \tag{8.132}$$

where the "addition" ($\dotplus$) means $(A \dotplus B)_{ik} = 1$ if $A_{ik} \neq 0$ or $B_{ik} \neq 0$ and $(A \dotplus B)_{ik} = 0$ if $A_{ik} = 0$ and $B_{ik} = 0$. Then $Q^{r*}_{ik} = 1$ means that the proximity between i and k is of the rth degree "or better" (better means smaller). By this artifice, proximity of the rth degree or better is mathematically expressed by Q^{r*} in the same way as first-degree proximity is mathematically expressed by Q. The vanishing of diagonal elements is assumed for this purpose.

We define first a "closed family of proximity" $\mathcal{F}$ by rather strict conditions: (a) a member i of $\mathcal{F}$ has proximity with any other member k of $\mathcal{F}$, and (b) it has no proximity with any nonmember k of $\mathcal{F}$. This means that $Q_{ik} = 1$ if $i \in \mathcal{F}$ and $k \in \mathcal{F}$, and $Q_{ik} = 0$ if $i \in \mathcal{F}$ and $k \notin \mathcal{F}$ or $i \notin \mathcal{F}$ and $k \in \mathcal{F}$. We can liberalize these conditions and define a closed family of rth-degree-or-better proximity using Q_{ik}^{r*} in lieu of Q_{ik}. Figure 8.6 shows Q_{ik}^{r*} graphically; the first column is for $r = 1$, the second, for $r = 2$, and so on, and a bond is used whenever $Q_{ik}^{r*} = 1$. The four points in the first row constitute a closed family of first-degree proximity. Three points (1, 2, 3) of the second row do

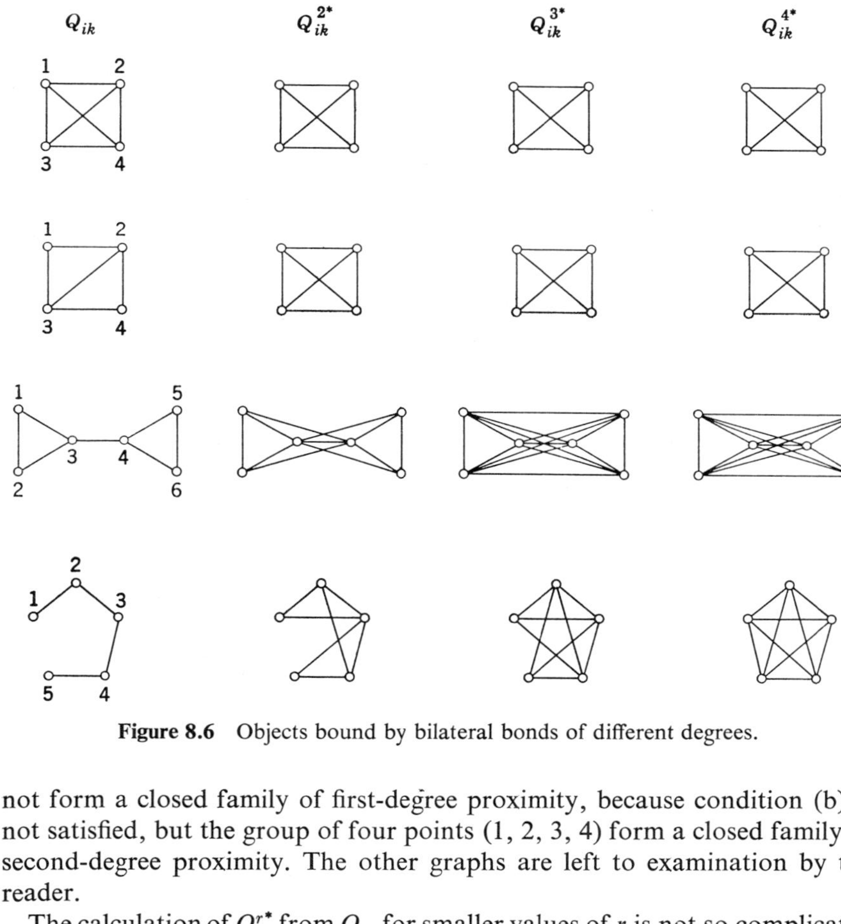

Figure 8.6 Objects bound by bilateral bonds of different degrees.

not form a closed family of first-degree proximity, because condition (b) is not satisfied, but the group of four points (1, 2, 3, 4) form a closed family of second-degree proximity. The other graphs are left to examination by the reader.

The calculation of Q_{ik}^{r*} from Q_{ik} for smaller values of r is not so complicated on a computer. After we have obtained Q_{ik}^{r*} for a chosen value of r, the further algorithm will be the same for any value of r. For brevity we use Q_{ik} in our discussion, but the same thing can be done for any Q_{ik}^{r*}. The purpose of the algorithm is to find the closed family of proximity to which each object belongs. If we use Q the closed family is one of the first degree, but if we use

Q^{r*} in lieu of Q what follows will refer to a closed family of rth degree or better.

First, if an object is similar to no other object, it has to be represented as a one-member family. Such an object x_i, and only such an object, will satisfy the condition

$$\sum_{k=1}^{m} Q_{ik} = 0. \tag{8.133}$$

Next, if an object x_i and another form a family of two members, we have, as a necessary condition,

$$\sum_{j=1}^{m} Q_{ij} = 1. \tag{8.134}$$

Third, if an object x_i is a member of a closed family with p members, we should have, as a necessary condition (the addition here is the usual one)

$$\sum_{j=1}^{m} Q_{ij} = p - 1 \tag{8.135}$$

and

$$(Q^3)_{ii} = 2\binom{p-1}{2} = (p-1)(p-2). \tag{8.136}$$

The first condition means that there are $(p - 1)$ partners in the family, and the second condition means that, starting from x_i, one can come back to x_i by any of $\binom{p-1}{2}$ triangular paths passing through two of the $(p - 1)$ partnerpoints. The factor 2 means that the two opposite directions go over a given triangle. Q^3 means $(Q^{r*})^3$ if Q^{r*} is used for Q. Actually, the condition (8.134) can be considered as a special case of (8.135) and (8.136) for $p = 2$. The condition (8.136) becomes unnecessary for $p = 2$, because (8.134) implies that $(Q^3)_{ii} = 0$. Similarly, (8.133) is a special case of (8.135) for $p = 1$, (8.136) again being automatically satisfied. Let us say that an object x_i is "saturated" if it satisfies the conditions (8.135) and (8.136) for some p. (According to this definition, a one-man family is saturated, but this will be soon reclassified as a case of nonfamily member.) We can now mention two obvious theorems.

Theorem 8.10 An unsaturated object belongs to no closed family of proximity.

Note that the converse of this theorem is not necessarily true. For instance, point 1 of the second row of Figure 8.6 is a saturated point with $p = 3$, but does not belong to a closed family as far as Q is concerned.

Theorem 8.11 Objects $x(i_1), x(i_2), \cdots, x(i_p)$ form a closed family with p members if, and only if, $Q(i_1, i_k) = 1$ only for $k = 2, \cdots, p$, and each object is saturated.

Proof. The "only if" part is obvious. The proof of the "if" part is as follows. Point $x(i_1)$ is connected with $(p - 1)$ other points, and according to (8.136) these $(p - 1)$ points are connected pairwise. Hence each of these $(p - 1)$ points is connected with $x(i_1)$ and with all the remaining $(p - 2)$ points, but not with other points. This last condition follows from the saturation of every point and is important. For instance, points 1, 2, and 3 of the third row (first column) satisfy the criteria of this theorem with $p = 3$ except that point 3 is not saturated, not satisfying (8.135) and (8.136) simultaneously by the same p.

Now the first step in our algorithm is clear. Check each row of the table of the matrix Q and determine whether it is unsaturated or find the integer p with regard to which it is saturated. The unsaturated rows and the saturated rows with $p = 1$ will be classified as "nonfamily" members.

In the second step we start from the first row again and proceed downward. If a row i_1 is unsaturated, remove this row from the table and note that i_1 is a nonfamily member. If i_1 is saturated with p, determine the $(p - 1)$ "partners" for which $Q(i_1, i_k) = 1$. Check to see if all these $(p - 1)$ partners are saturated (with p). If yes, then note that $(i_1, i_2, \cdots, i_p)$ form a closed family, and remove all these rows from the table. If any of the $(p - 1)$ prospective partners is unsaturated or absent from the table as a result of earlier deletion, remove all these existing rows referred to from consideration and classify such objects as "nonfamily" members.

As a result of these two steps, we obtain a class of nonfamily members and a certain number of families. Each of the nonfamily members must be represented by itself. Thus we complete the task of (binary, bilateral) clustering. As stated before, we can use it as preclustering and apply another clustering program with nonbilateral, nonbinary cohesion, treating each family (and each nonfamily member) as an object because their number is smaller than the number of all the objects. Each family can be represented either by one of the family members or by the average vector of the family. These will be the input for the earlier algorithm of taxonomy. After the taxonomy is determined, all the members of a family will be reinstated in the branch in which the family representative is located.

In carrying out the present algorithm of family formation, we have to determine the threshold τ and the r of the Q^{r*} to use in place of Q. The larger the τ and the smaller the r, the more restrictive the condition of family membership; as a result, the larger will be the number of nonfamily fragments and the smaller the number of families. The number of items to be subjected to the earlier taxonomic algorithm is precisely the number of nonfamily members plus the number of families. We should adjust τ and r so that a combination of the two algorithms becomes feasible on the available computer.

8.5. DECISION PROCEDURES IN RECOGNITION

Let each object be represented by a vector $(x_1, x_2, \cdots, x_n)$ and let the classes to which they are supposed to belong be denoted by I, II, $\cdots$. There are two kinds of conditional probabilities to consider: $p(x \mid X)$ and $p(X \mid x)$, where x stands for an n-component vector and X for any one of the classes I, II, $\cdots$. In the case when each component x_i is a continuous variable, $p(x \mid X)$ will be considered as a probability "density" in the x-space. If the intension of class X (i.e., the definition of the class by the predicate variables) were unambiguously definable in terms of the variables x_i, the probability $p(X \mid x)$ would become 0 or 1. But in the problem of recognition the intension of classes is not only unknown but also often indeterminate. Furthermore, the vector x does not usually contain all possible information about the object (which may be relevant to the intension), and the observed values of x are not free from distortion and external noise. For these reasons $p(X \mid x)$ is not 0 or 1. On the other hand, $p(x \mid X)$ is not 0 or 1 (or a δ-function) either, because a class X contains members that correspond to many different points.

Under these circumstances it is important to determine which one of the two probabilities, $p(x \mid X)$ or $p(X \mid x)$, is determined by the nature of the problem and not affected by the contingent conditions (see Chapter 3). In most recognition problems it is more natural to consider $p(x \mid X)$ as determined by the nature of the problem. Indeed, an empirical frequency distribution of members of a class in the x-space is always conceivable. If so, the inverse probability $p(X \mid x)$ can be given only indirectly through the Bayes formula $p(X \mid x) = p(X)p(x \mid X)/\sum_Y p(Y)p(x \mid Y)$. In this case if, and only if, the "prior" probability $p(X)$ is given and fixed (which is not always the case), $p(X \mid x)$ will acquire a unique value at each point x for each class X.† If, on the other hand, $p(X \mid x)$ is determined directly by the nature of the problem without

† At first glance $p(X)$ seems to mean the probability that an object will belong to class X when its properties (x) are not specified and hence can be replaced by the relative frequency of objects of class X in the entire population of objects. But this is an oversimplification, since X actually can be considered as $H \cap A$, where hypothesis H specifies the probability distribution in x for an object of class X and A states that the object at hand is an object of class X. If there are more than one competing H, the prior probability $p(X) = p(H \cap A)$ depends on the credibility of H other than the relative frequency: $p(A)$. If, however, we take one particular population of objects of different classes and interpret probabilities as frequencies in this population, then $p(X)$ may be equated with $p(A)$ because only one H is valid, namely, the one that gives the actual frequency distribution of class samples. See [W-30] for a more rigorous discussion of this matter.

calling for the use of the Bayesian formula, we need not worry about the fixity of $p(X)$ as an additional assumption.

In any event, the conditional probability $p(X \mid x)$ is, at least conceptually, the basic quantity in the recognition problem, because the task here is to decide the class X to which a given x should be assigned. For this reason we often refer to $p(X \mid x)$ as "decision probability." The problem then is reduced to estimation of the function $p(X \mid x)$ on the basis of the information provided by the paradigms that have been given. A straightforward procedure for determining class affiliation after the decision probability has been obtained would be to derive a "decision function" $X(x)$, such that this X as a function of x is the one that maximizes $p(X \mid x)$ at point x. If the decision function $X(x)$ is such that the set of points x that give the same value of X constitutes an n-dimensional domain in the n-dimensional x-space, it provides a justification of the idea of "zones," according to which the entire x-space is divided into the zones corresponding to the classes. Theoretically, the best way to determine the zones $X(x)$ is to derive them from the probability $p(X \mid x)$, as explained above, but the zoning algorithms usually employed in practice bypass the probability $p(X \mid x)$ and try to determine the zones directly from the paradigms. We return later to the problem of zoning. It can also happen that the function $X(x)$ derived from the probability $p(X \mid x)$ is such that the set of points x that give the same value of X constitutes, at least in a certain approximation, a less-then-n-dimensional subspace rather than an n-dimensional domain (zone). In this case an approach that I call "subspace" approach (explained later) becomes more appropriate.

In some practical problems it can happen that the straightforward idea of defining the decision function $X(x)$ by the most probable X according to the conditional probability $p(X \mid x)$ turns out to be inadequate. This occurs, for instance, if the damage or loss caused by a "misclassification" of an object depends critically on the class to which it should be assigned and on the class into which it is "mistakenly" placed. In such a case a reasonable strategy is to determine the decision function $X(x)$ in such a way as to minimize the expected loss. Suppose, for instance, that there are two classes, I and II, and let $L(\text{I}, \text{II})$ and $L(\text{II}, \text{I})$ be two constants representing, respectively, the loss due to an object of Class II being misplaced in Class I and that due to an object of Class I being misplaced in Class II. Let the decision function $X(x)$ be such that it assigns x to Class I if $p(\text{I} \mid x) = 1 - p(\text{II} \mid x) \geq \pi$, and let $p(x)\, dx$ be the probability that an arbitrary object will be in the small volume dx at x in the n-dimensional x-space. Then the expected loss is

$$\int p(x)\, dx \{D[p(\text{I} \mid x) - \pi] p(\text{II} \mid x) \cdot L(\text{I}, \text{II})$$
$$+ D[\pi - p(\text{I} \mid x)] \cdot p(\text{I} \mid x) \cdot L(\text{II}, \text{I})\},$$

where D is the step function such that $D(q) = 1$ if $q \geq 0$ and $D(q) = 0$ if $q < 0$. When $p(\text{I} \mid x)$, $p(x)$, $L(\text{I, II})$, and $L(\text{II, I})$ are all known, the minimization of this expected loss will determine the parameter π and consequently the decision function $X(x)$. If $L(\text{I, II}) = L(\text{II, I})$, this minimization procedure reduces to our earlier procedure, namely, to the case $\pi = \frac{1}{2}$. If $L(\text{I, II}) > L(\text{II, I})$, π will move to a larger value. Since this generalization, which incorporates the loss (or utility) under consideration, adds only a fairly obvious complication irrelevant to the problem of recognition *per se*, we no longer pursue it in this section.

As is always the case with cognitive and recognitive tasks, the first and most significant step is a choice of variables, $x_1, x_2, \cdots, x_n$. If we take too many x's, the density $p(x \mid X)$ becomes so thin that the limited number of available paradigms gives no reasonable guidance in determining the probability $p(X \mid x)$. What is important is to use as few variables as possible that are in fact pertinent to class distinction. Objections are often raised against the Bayesian method on the ground that the probability $p(x \mid X)$ is factorized as $p(x_1 \mid X), p(x_2 \mid X), \cdots, p(x_n \mid X)$ in practice, as if the x's were all independent. This, however, has nothing to do with the Bayesian method as such, and the main reason why many researchers have resorted to this unwarranted simplification lies in the fact that they have too many variables x, and $p(x_1, x_2, \cdots, x_n \mid X)$ becomes too thin to be inferred from the available paradigms.

As emphasized in Section 8.1, reduction of the number of variables in the case of recognition can be done in two steps. First, we can use a preselection of variables in the sense of Section 8.2, which can be done independently of the paradigms. Thereafter we can apply a postselection or postevaluation of variables in such a way that the class distinction as suggested by the paradigms becomes as conspicuous as possible. We gave an example of postevaluation toward the end of Section 8.3 (determination of w_j by the class members). The general idea of postselection may be described as follows. Starting from n variables x_i, $i = 1, 2, \cdots, n$, we introduce n^* ($\leq n$) variables, $\xi_i = f_i(x_1, x_2, \cdots, x_n)$, $i = 1, 2, \cdots, n^*$, in which we aim to minimize the number n^* and minimize the ambiguity regarding class affiliation. A good measure of this ambiguity is the average conditional entropy,

$$\bar{S}(X \mid \xi) = -\sum_X \int p(\xi)\, d\xi\, p(X \mid \xi) \log p(X \mid \xi), \tag{8.137}$$

where the integral is actually an n^*-ple integral because ξ stands for an n^*-component vector. If $p(X \mid \xi)$ is such that at each point ξ there is only one X that makes $p(X \mid \xi) \neq 0$ (hence $=1$), this entropy takes its minimum value, which is zero. In such a case the zoning concept is fully guaranteed to be successful and we may speak of perfect separation, but we should not expect this to happen too often. It is interesting that this entropy (8.137) is invariant for the transformation from n^* variables ξ to the same number n^* of new variables η, such that $p(\eta)\,d\eta = p(\xi)\,d\xi$. A remark in Section 1.3 shows that $S(X) - \bar{S}(X \mid \xi)$ is invariant, but in the present case $S(X)$ is also invariant; hence $\bar{S}(X \mid \xi)$ is invariant.† This shows that if we introduce a new set of n^* η's from a given set of n^* ξ's, we gain nothing. When the set of n x's is given, however, if we introduce a new set of n^* ξ's from the x's ($n^* < n$), the value of $\bar{S}(X \mid \xi)$ will depend on how we define the ξ's as functions of the x's. If we reduce the number of n^*, the decisional ambiguity $\bar{S}(X \mid \xi)$ increases in general. Hence it presents a meaningful problem to reduce n^* as much as possible without increasing $\bar{S}(X \mid \xi)$. This can be considered as the aim of the postselection of variables. In the literature we sometimes encounter attempts to stretch the x-space in certain directions and contract it in certain other directions in the hope that the class distinction will become more conspicuous, but except for a certain convenience arising from the possibility of making the distribution of paradigms of one class roughly spherical, the practical gain from such a distortion method usually turns out to be minimal. The reason is obvious because (8.137) is invariant for such a transformation. In the following we no longer make a distinction between the x's and ξ's.‡ If we compare X and x respectively to the "hypothesis" and "evidence" of the inductive process, the entropy $\bar{S}(X \mid x)$ can be identified as the average of what I introduced under the name of "inductive entropy" in my 1960 paper [W-13]. Kamentsky and Liu rediscovered and used it for pattern recognition later [K-2]. (See Chapter 5 also.) It measures, in a sense, the negative of the power of evidence.

The entire reasoning in the foregoing paragraphs is based on the assumption that we can make a *reasonable* estimate of the probability $p(X \mid x)$ from a limited number of given paradigms. The value of $\bar{S}(X \mid x)$, however, may depend strongly on the method of this estimation. This estimation is essentially a "smoothing process," because we have to obtain continuous functions $p(X \mid x)$ from discrete paradigm points. If a certain caution is not taken in this smoothing process, we may, at one extreme, be easily led to an unrealistic estimate of $p(X \mid x)$ that allows a perfect separation under any circumstances, which is not real. At the other extreme we may be led to a useless estimate $p(X \mid x)$, which does not depend on x at all. We illustrate this point with the following simple example.

One of the simplest ways of estimating $p(X \mid x) \propto p(x \mid X)p(X)$ from the given paradigms is to mark all the known paradigms in the x-space, making sure that the total number of representative points of each class X is roughly proportional to the *a priori* probability $p(X)$ of the unknown newcomers,

† $S(X) - \bar{S}(X \mid \xi)$ corresponds to an interrelation J in (1.80).

‡ We return to the problem of postselection at the end of this section.

which will be classified later. This last condition is more or less automatically satisfied if paradigms, as well as newcomers, are taken randomly out of the same population and if all known paradigms are marked. At each point x we consider an n-dimensional sphere with the center at x such that it contains a fixed number ν of paradigms, and the value of $p(X \mid x)$ is determined by the relative frequency of the paradigm points of class X in the sphere. If we increase ν indefinitely, the value of $p(X \mid x)$ will become independent of x and equal to $p(X)$, because the radius of the sphere has to become indefinitely large. If we decrease ν, the dependence of $p(X \mid x)$ on x becomes more and more marked, and at the same time the distribution of $p(X \mid x)$ over different X's becomes less widely spread. Finally, at $\nu = 1$, we get $p(X \mid x) = 0$ or 1 and the entropy $\bar{S}(X \mid x)$ of (8.137) becomes zero, giving rise to a case of apparently perfect separation. As a result, the zoning method will formally become 100% guaranteed to be successful. But of course the estimate based on small values of ν may be influenced strongly by local fluctuations of paradigms, and can hardly be trusted. A guideline for a reasonable value of ν may be the following: for a given density of paradigm points, the value ν must be sufficiently large so that a linear shift of the center x by a distance comparable to the average distance between two neighboring paradigm points does not appreciably change the values of $p(X \mid x)$ determined by the relative frequency. The density of paradigms is often outside the control of the classifying agent, but it is desirable that the density be large enough so that a further increase in it will not continue to alter values of $p(X \mid x)$ determined with proper caution. In the adaptive case, the radius and the ν should be gradually adjusted with the increasing density. If the density is large enough, $\nu = 1$ (the nearest neighbor method) will become as good as anything else.

The following method of determining the decision probability $p(X \mid x)$ uses a mathematical theorem developed by Kirillov, Chentsov, Frolov, and Aizerman (see [A-4]), but is adapted to incorporate the idea explained in the foregoing paragraph. We consider a set of orthonormal functions $\{\psi_i(x)\}$ in the n-dimensional x-space, where the index i is supposed to be such that the higher its value, the more rapidly the function changes its value with x. Let N be the smallest number such that $\psi_i(x)$ for $i > N$ changes its value appreciably for a shift of x by a distance comparable to l, where l indicates the limit of distance below which we do not want to consider any further detailed change of probabilities. We may roughly equate l to the order of magnitude of the average distance between neighboring paradigms. (If the density of the paradigm points varies drastically from one region of the space to another, we have to know that the result will be unreliable in the region where the points are sparse.) We thus expand the decision probability as (N: finite)

$$p(X \mid x)p(x) = p(x \mid X)p(X) = \sum_{i=1}^{N} c_i(X)\psi_i(x). \tag{8.138}$$

Let $\{x^{(\alpha)}\}$, $\alpha = 1, 2, \cdots, m$, be the set of m available paradigms, of which a subset $\{x_X^{(\alpha)}\}$, $\alpha = 1, 2, \cdots, m_X$, belongs to class X. If we increase m indefinitely, the ratio m_X/m will tend to $p(X)$ and the density of the points $x_X^{(1)}, x_X^{(2)}, x_X^{(3)}, \cdots$ will become proportional to $p(x \mid X)p(X)$. Let K be the density matrix† whose eigenfunctions are $\psi_i(x)$, $i = 1, 2, \cdots, N$, with equal eigenvalues:

$$(x|\, K\, |x') = \sum_{i=1}^{N} \psi_i(x)\, \psi_i(x'). \tag{8.139}$$

Note that $(x|\, K\, |x')$ would become an n-dimensional δ-function if $N \to \infty$ [see (8.11)]. If we stop at a finite $i = N$, this K becomes appreciably different from zero only if $|x - x'|$ is comparable to l or less. This K may intuitively be interpreted as representing the "influence" of point x' at point x, whereby the range of influence is of the order of l. Consider

$$P_m{}^X(x) = \frac{1}{m} \sum_{\alpha=1}^{m_X} (x|\, K\, |x_X^{(\alpha)}), \tag{8.140}$$

where the summation refers only to the paradigms of X and the divisor m is the number of all paradigms. This $P_m{}^X(x)$ may be interpreted as representing the proportionate "influence" of the m_X paradigms of class X at point x. The theorem of the Russian researchers states that

$$P_m{}^X(x) \to p(X \mid x)\, p(x) \quad \text{as} \quad m \to \infty, \tag{8.141}$$

"in probability," meaning that for any $\varepsilon > 0$ the probability that $|P_m{}^X(x) - p(X \mid x)p(x)|$ will be smaller than ε becomes unity as $m \to \infty$.

A simplified version of the proof is as follows. Since the expression of (8.140) means the average of $(x|\, K\, |x_X^{(\alpha)})$, we shall have

$$P_m{}^X(x) \to \int (x|\, K\, |x')\, p(x' \mid X)\, p(X)\, dx' \quad \text{as} \quad m \to \infty, \tag{8.142}$$

in probability.‡ If N were ∞, $(x|\, K\, |x')$ would be a δ-function, and the last integral expression of (8.142) would immediately become $p(X \mid x)p(x) = p(x \mid X)p(X)$, as desired. But even if N is finite, as in the present case, we can obtain the same result as follows. Substituting (8.138) and (8.139) in (8.142), we obtain

$$\begin{aligned} P_m{}^X(x) &\to \int \sum_{i=1}^{N} \psi_i(x)\, \psi_i(x') \sum_{j=1}^{N} c_j(X) \psi_j(x')\, dx' \\ &= \sum_{i=1}^{N} \sum_{j=1}^{N} \psi_i(x)\, c_j(X)\, \delta_{ij} = p(X \mid x)\, p(x) \quad \text{as} \quad m \to \infty, \end{aligned} \tag{8.143}$$

where the orthonormality of the $\{\psi_i(x)\}$ is used. Since $p(x)$ is easily obtained by

† Note that if $f^{(\alpha)}(\xi)$ used in (8.35) are orthogonal, they themselves become the $\phi_k(\xi)$ of (8.43).

‡ See the remark four lines above (8.139).

$p(x) = \sum_X p(X \mid x)p(x)$ [i.e., the sum of (8.143) with respect to X], we can calculate $p(X \mid x)$ from (8.143). An interesting thing about this theorem is that the result does not depend on N at all, provided that we can make $m \to \infty$. The choice of N determines the smoothness of the function (8.138).†

In practice we can never realize $m \to \infty$, but we may use (8.140) as a good approximate estimate of $p(X \mid x)p(x)$. One example of the function set that could be used for the present purpose may be Someya-Shannon's sampling functions applied to each coordinate x. We can automatically eliminate wave numbers larger than K by using the sampling function $\psi_n(x) = \sqrt{K/\pi} \cdot \sin(Kx - n\pi)/(Kn - n\pi)$; that is, any change of the value of a function taking place in a domain of x smaller than the order of π/K is automatically smoothed out. It would be safe to take π/K approximately equal to several times the average distance between neighboring paradigms. The index n refers to the "sampling point"; hence, if the domain of x is finite, we need only a finite number of functions.

The method of determining the decision probability $p(X \mid x)$ sketched above, combined with an approximate prescription for deriving the decision function $X(x)$ from $p(X \mid x)$ [such as $p(X(x) \mid x) = \max_X p(X \mid x)$], seems to provide the most logical way of performing the task of recognition, once the variables x's have been chosen. This method, however, is not always the most practical one. For this reason the popular zoning techniques that aim at producing a decision function $X(x)$ directly from the paradigms without the intermediary of decision probability $p(X \mid x)$ cannot be ignored. A common defect of almost all those zoning techniques is that they do not pay sufficient attention to the delicate problem of smoothing. As a result, they are liable either to end with meaningless gerrymandering of zones or to force excessively "flat" border surfaces, which do not conform with a visibly curved distribution of paradigms. The simplest yet quite effective algorithm of zoning is what is known as the "nearest neighbor method." This coincides with the case $\nu = 1$ in the probabilistic consideration of p. 435, and it is certainly a good method when the density is high, but it is not so certain when this is not the case. This method is adaptive in the sense that the zones are modified each time a new paradigm arrives, but the burden on the memory can be severe. See, for instance, [C — 7] for a theoretical justification of this method.

Let us review the philosophy behind the usual zoning techniques, assuming, for simplicity, that there are only two classes, I and II. (When there are more than two classes, the required classification can be considered as a superposition of several dichotomies, where each dichotomy can be taken between two groups of classes.) The idea amounts to introducing a new variable

† The arbitrariness of N is a special case of the arbitrariness inherent in induction. In practice, N should probably be chosen so that l will be larger than the average distance between neighboring paradigms.

$\xi = \xi(x_1, x_2, \cdots, x_n)$. such that ξ takes values belonging to a set of numbers, say, A, if the point x is a paradigm of Class I and ξ takes values belonging to another set, say, B, if the point x is a paradigm of Class II, where A and B do not overlap ($A \cap B = \emptyset$). For instance, we may require $\xi(x) > 0$ if $x \in \text{I}$ and $\xi(x) < 0$ if $x \in \text{II}$. Of course, this condition is supposed to be satisfied by all the paradigms, and it is hoped that any newcomer will satisfy it too.

It is obvious that there always exists such a function ξ for a given set of paradigms, provided that no two paradigms of two different classes have exactly the same representation point. The basic defect of this idea is, as mentioned above, that it completely ignores the necessity of reasonable smoothing even at the expense of increasing the decisional ambiguity. As a consequence of this basic defect, the method is usually infected with the following practical troubles: it turns out that there actually exist infinitely many possible ξ's satisfying the conditions, and some of the possible border surfaces $\xi = 0$ exhibit too little or too much contouring around paradigms. We may then have to resort to a certain external judgment to choose a "good" function ξ. But what is the criterion of goodness of a variable? The answer cannot be unique. As a matter of fact, if the probabilistic derivation is correct, a good ξ may be such that some paradigms are allowed to invade the zones of other classes.

It may be noted that a good variable ξ corresponds to the "projectible" predicate in the inductive generalization discussed in Chapter 4. As was the case in the classical problem of induction (which in a way becomes a special case of "recognition"), the variable of inductive generalization is never uniquely determined by the paradigms (which correspond to the "evidence" in the classical case). A choice of the variable ξ has to depend essentially on what we called an "extra-evidential" consideration in Chapter 4. The requirement, often used in practice, that ξ be a linear function of x (the method of linear decision function) is a crude and often unjustified application of the principle of simplicity. Logically speaking, the reasonable smoothing used previously finds its justification, not in the paradigms themselves, but in our broader judgment; hence it too is a kind of extra-evidential consideration. However, it seems to be a sounder guiding principle than forcing a certain functional form on the $X(x)$. The nearest neighbor method too has its own hidden extra-evidential philosophy.

The method of linear decision function certainly limits the set of possible ξ's and eliminates any possibility of excessive gerrymandering. It has its own difficulties, however. First, there often does not exist a hyperplane (linear ξ) satisfying the condition. Second, if one does exist, there are usually infinitely many. Third, even if there exists a plane separating paradigms of Class I and paradigms of Class II, the shape of the distribution of paradigms often suggests—to human eyes—that a curved border surface may be more appropriate.

One possible (but very dubious) way to avoid these difficulties, at least on the surface, is to introduce a single representative point $x_X{}^R$ of each class X, and place planes at a "well-balanced" position between class representatives. The representative point $x_X{}^R$ of class X may be defined as the average of the paradigms of class X: $x_X{}^R = \sum_{\alpha=1}^{m_X} x_X^{(\alpha)}/m_X$ (to be understood vectorially). Suppose we decide that a newcomer x should be placed in the class whose representative is the nearest to x. That would mean $x \in Y$ if $(x - x_Y{}^R)^2 = \min_X (x - x_X{}^R)^2$. This amounts to seeking the X that minimizes $-\sum_i x_i x_{X_i}^R + (x_X{}^R)^2$, which is linear in x_i. If there are only two classes, for instance, the border will be given by a hyperplane: $\sum_i (x_{\mathrm{I}}{}^R - x_{\mathrm{II}}^R)_i x_i - (x_{\mathrm{I}}{}^R)^2 + (x_{\mathrm{II}}^R)^2 = 0$. It is easy to attach some weighting factors to each representative. There are many variants of this type of linear method, with or without hardware realization. A device developed by the Karlsruhe group is one of them [S-9]. This kind of method works if the density distribution of paradigms of each class is of the same type and has a roughly spherical symmetry. In the absence of this last condition (which is usually the case), the success of the method is fortuitous. Conversely, it can also be argued that the preselection of variables should be such that these linear methods work successfully. Variable selection and decision are interdependent. See. Nilssen's book [N-1] for various recognition techniques similar to the linear methods discussed in this paragraph.

An interesting method of reducing ambiguity in the definition of the separating hyperplane in the separable case (i.e., the case in which such a hyperplane exists at all) is as follows. Consider always a pair of parallel planes that are both allowed separation planes. The volume between the two planes is a dead zone, a "no man's land." Obtain the pair that maximizes the distance between them; then they will have an orientation that is fairly reasonable with reference to the given groups of paradigms. The actual decision function may be defined by a third parallel plane suitably located between the first two. This method was developed by my student, P. F. Lambert, who showed that this maximization operation can be done on a computer [L-2, W-25]. He called the method DZM (dead zone maximization). The interesting part of this algorithm is that it can be applied, with some modification, to a nonseparable case. Consider a pair of parallel planes such that one of them has all the paradigms of Class I on its positive (suitably defined) side and the other has all the paradigms of Class II on the negative side. In the separable case the second plane can lie on the negative side of the first plane, and the algorithm maximizes the distance between the two planes. In the nonseparable case the second plane will lie on the positive side of the first plane, and the algorithm minimizes the distance. The maximization principle becomes a minimization principle for the absolute value of the negative distance between the planes. The only trouble is that in the nonseparable case the solution is not always unique.

Various actual algorithms for determining zones by paradigms can be divided into those that are "adaptive" and that those are "static." In a static method all the available paradigms are supposed to have been chosen once and for all and the zones are determined on the basis of these fixed data. In an adaptive method the paradigms are supposed to come in one by one, and the zones are revised every time such a revision is required by the arrival of a new paradigm. The distinction between the two kinds of methods defined this way is not really essential, because each static method can, in principle, be converted into an adaptive method, provided that all the past paradigms are recorded somewhere in the machine. In fact, with the arrival of a new paradigm, the machine can calculate all over again new "fixed" zones on the basis of all the past paradigms plus the new one. The more important distinction between the static and adaptive methods from a practical point of view lies in the ease with which the zones can be modified by the arrival of a new paradigm. If the method determining the zones is such that the machine does not need to keep a record of all the past paradigms, in particular, if the modified zone is determined solely by the present zone and the new paradigm, it may be called genuinely adaptive. As the speed and memory capacity of computers increase, however, the distinction between "genuinely adaptive" and the "adaptive" method based on a static method will become unimportant. Because of the obvious superficial resemblance of an "adaptive" system to a living intelligence, excessive emphasis was placed on this feature in earlier days, but now that all the mystery is gone we need not view the so-called adaptive systems with any special sense of awe.

It may also be mentioned, as a shortcoming of adaptive methods in general, that they usually do not fully exploit the information provided by the paradigms, because the existing zone is not revised if the next newcomer is "correctly" classified by the existing zone. Actually the position of all the newcomers could be used to bring the zone to a better-balanced position, but this is not done by the known adaptive method. This remark also applies to some of the adaptive methods designed to estimate the decision probability $p(X \mid x)$ (see [A-4]).

After making these reservations regarding the adaptive zoning method, I hasten to add that no exposition on pattern recognition can be complete without an explanation of how this method works. It is unfortunate that the publicity that accompanied the early successes of the "perceptron" and the "adaline," and the subsequent reaction to this excessive popularity of the linear devices, obscured the real issue. The method has many defects, as explained above, but, combined with an adequate preselection of variables it can be quite useful. In the following I briefly explain the gist of the adaptive linear zoning method, incorporating some of the ideas of Aizerman's non-linear generalization. The proof of the so-called perceptron convergence theorem is relegated to Appendix A8.6.

As we have seen, the purpose here is to obtain a function $\xi = \xi(x_1, x_2, \cdots, x_n)$ such that all the paradigms of Class I satisfy $\xi > \varepsilon$ and those of Class II satisfy $\xi < -\varepsilon$, where ε could be zero or a small positive number. The algorithm consists of a process of gradual modification of a tentative function ξ, whereby the function is revised if a new paradigm is incorrectly classified by the present function ξ, and is not revised in the opposite case. However, this corrective revision at each stage is not necessarily effective enough to make the modified ξ capable of classifying correctly all the past paradigms and the newcomer which was the cause of the correction. It only modifies the zone in the "right direction," in the sense that the ξ-value at the position of the new paradigm changes toward a correct value. To assure that this gradual adaptation finally reaches a ξ-function satisfying the desired condition, the theory assumes the following. (a) Of the set of all possible objects, probably infinite in number, of both classes, a subset P_{I} of a limited number of paradigms of Class I and a subset P_{II} of a limited number of paradigms of Class II are selected and fixed. (b) The sequence of paradigms shown to the recognizer during the training period is assumed to be such that each paradigm out of $P_{\mathrm{I}} \cup P_{\mathrm{II}}$ comes back an indefinite number of times in the sequence. Given sufficient repetition of each member of the fixed finite subset $P_{\mathrm{I}} \cup P_{\mathrm{II}}$, the adaptive algorithm is guaranteed to revise ξ so that it finally classifies all the members of $P_{\mathrm{I}} \cup P_{\mathrm{II}}$ correctly.

If the function ξ is square-integrable, it may be expanded in the form

$$\xi(x) = \sum_{i=1}^{N} c_i \varphi_i(x), \tag{8.144}$$

where x stands for $(x_1, x_2, \cdots, x_n)$ and $\{\varphi_i(x)\}$ is a set of orthonormal functions in which larger values of index i indicates faster change of the values. The upper limit N of the summation with respect to i should be determined by the consideration of allowable gerrymandering of contours and the desired degree of smoothing. If $N \to \infty$, we could in principle choose the c_i's to express any surface by the equation $\xi = 0$ and to satisfy (with $\varepsilon = 0$) any arbitrary paradigm subsets P_{I} and P_{II}, provided that they contain no self-contradiction (points of different classes occupying the same point). We can also consider (8.144) as a power expansion; in the simple case, if we take only the first order terms $\varphi_i(x) = x_i$ and a constant term, we obtain a hyperplane and the method becomes a linear case to which the perceptron and adaline belong. If we include second-order expressions in (8.144) we can obtain n-dimensional conical sections. In general, if N is finite, either in the case of orthonormal expansion or in the case of power expansions, the existence of ξ satisfying (with $\varepsilon = 0$) the paradigm subsets is not guaranteed. The existence of at least one ξ in the linear case is what is usually called "linear separability."

In order to reduce the general case (8.144) to a linear one, at least in form, we can introduce an N-dimensional representation vector $(y_1, y_2, \cdots, y_N)$ instead of the n-dimensional representation vector $(x_1, x_2, \cdots, x_n)$ for each object, where $y_i = \psi_i(x)$. The function system $\{\psi_i(x)\}$ may or may not contain a constant term as one of the ψ's. But we can always write a constant term separately outside the summation; hence abbreviating $\xi(x(y))$ as $\xi(y)$, we get

$$\xi(y) = \sum_{i=1}^{N} c_i y_i - p. \tag{8.145}$$

Since we are interested mainly in the sign of $\xi(y)$, we can, when it is more convenient, assume that the vector c_i is normalized so that

$$\sum_{i=1}^{N} c_i^2 = 1. \tag{8.146}$$

Then $\xi(y) = 0$ represents a hyperplane in the N-dimensional space whose normal is given by the unit vector $\{c_i\}$, and the perpendicular distance from the origin is p. The same plane has two equivalent equations in general, because a simultaneous transformation of $\{c_i\}$ and p into $\{-c_i\}$ and $-p$ leaves $\xi(y) = 0$ invariant. But in the present context, if the paradigms of Class I are on the side of the origin of the plane p must be negative, and if the paradigms of Class II are on the side of the origin p must be positive, because $\xi > \varepsilon$ and $\xi < -\varepsilon$ are supposed to correspond to Class I and Class II, respectively. The region between two parallel planes $\xi = \varepsilon$ and $\xi = -\varepsilon$ is called "dead zone." If the c's are normalized in the sense of (8.146), the constant ε will represent half the thickness of the dead zone. To explain the adaptive procedure, however, it is more convenient not to assume the normalization (8.146) because the adaptive procedure that changes the c's gradually does not leave the normalization condition invariant.

To make our description clearer, let us put a superscript α on each vector y to distinguish each paradigm. The desired function is such that (putting $p = -c_{N+1}$)

$$\begin{aligned} \xi(y^{(\alpha)}) &= \sum_{i=1}^{N} c_i y_i^{(\alpha)} + c_{N+1} > \varepsilon \qquad \text{if} \quad \alpha \in \mathrm{I} \\ &= \sum_{i=1}^{N} c_i y_i^{(\alpha)} + c_{N+1} < -\varepsilon \quad \text{if} \quad \alpha \in \mathrm{II}. \end{aligned} \tag{8.147}$$

We can start the adaptive procedure with any arbitrary set of initial values, $c_1, c_2, \cdots, c_{N+1}, \varepsilon$. Although the final function ξ depends on the choice of these initial values, the final ξ will satisfy (8.147) in each case. The prescription of a gradual change of the c's is as follows. If a newly arrived paradigm is correctly classified, that is, if it satisfies (8.147) with the present values of the c's, they should be left as they are. If (8.147) is violated by a new paradigm, say, $y^{(\gamma)}$, of Class I, that is, if $\xi(y^{(\gamma)}) < \varepsilon$, $\gamma \varepsilon$ I, the c_i should be changed to c_i^*, given by

$$\begin{aligned} c_i^* &= c_i + \lambda y_i^{(\gamma)}, & i &= 1, 2, \cdots, N \\ c_{N+1}^* &= c_{N+1} + \lambda, & \lambda &> 0. \end{aligned} \tag{8.148}$$

If (8.147) is violated by a new paradigm $y^{(\gamma)}$ of Class II, that is, if $\xi(y^{(\gamma)}) > \varepsilon$, $\gamma \in$ II, the c_i should be changed to c_i^*, given by

$$\begin{aligned} c_i^* &= c_i - \lambda y_i^{(\gamma)}, & i &= 1, 2, \cdots, N \\ c_{N+1}^* &= c_{N+1} - \lambda & \lambda &> 0. \end{aligned} \tag{8.149}$$

The positive parameter λ is usually kept constant during the whole process of adaptation. In Appendix A8.6 we suggest a way to vary λ during the process to bring the final plane to a better-balanced position.

It is evident that the revision process (8.148) and (8.149) will bring the plane to a "better" position as far as the paradigm $y^{(\gamma)}$ is concerned, because $\xi^*(y^{(\gamma)}) = \sum_{i=1}^N c_i^* y_i^{(\gamma)} + c_{N+1}^* = \xi(y^{(\gamma)}) + \lambda \sum (y_i^{(\gamma)})^2 + \lambda > \xi(y^{(\gamma)})$ if $\gamma \in$ I and, similarly, $\xi^*(y^{(\gamma)}) < \xi(y^{(\gamma)})$ if $\gamma \in$ II. However, this does not guarantee $\xi^*(y^{(\gamma)}) > \varepsilon$ for $\gamma \in$ I or $\xi^*(y^{(\gamma)}) < -\varepsilon$ for $\gamma \in$ II, and it is not obvious that the new plane will continue to satisfy those previous paradigms that the old ξ satisfied. However, the so-called perceptron convergence theorem guarantees that the necessity for correction of the c's will stop after a finite number of steps, if $P_{\text{I}} \cup P_{\text{II}}$ is finite, if the case is linearly separable at all in the y's, and if we arrange the sequence of paradigms to be shown to the machine in such a way that, for any given integer M, each paradigm type reappears at least once (hence arbitrarily many times) after the Mth position in the sequence. Termination of the correction procedure means, of course, that the final set of c's satisfies all the paradigms of $P_{\text{I}} \cup P_{\text{II}}$, because each paradigm comes back again and will no longer cause any revision. The term "convergence" is misleading, however, because the final set of values of the c's is not uniquely determined by the subset $P_{\text{I}} \cup P_{\text{II}}$. It depends on the initial values of the c's, ε and λ, as well as on the order in which members of P_{I} and P_{II} are shown. Novikoff's proof of this theorem is given in Appendix A8.6.

We should not leave this chapter without mentioning a new algorithm of recognition, which may be called a "subspace method," as distinct from the "zoning method" (see [W-25]). The subspace method may be characterized as embodying most faithfully the idea of postselection, because it is based on extraction of the most conspicuous properties of each class as represented by a body of paradigms. The method is called CLAFIC (class-featuring information compression).† Let $x_i^{(\alpha)}$, $i = 1, 2, \cdots, n$, $\alpha = 1, 2, \cdots, \mu$, be the normalized representative vectors of paradigms of Class I, α denoting different paradigm types. Each variable x_i can be the result of a direct observation or a quantity indirectly derived from the observation such as the y_i of the foregoing paragraphs, or even the results of SELFIC. We may attach a weight $w^{(\alpha)}$ to each paradigm type α with the hope that $w^{(\alpha)}$ faithfully reflects the relative frequency of type $x^{(\alpha)}$ of Class I not only in the available body of paradigms, but also in the entire population from which the newcomers will be drawn after the training period is over. Allowing ourselves only an orthogonal linear transformation,

$$\xi_j^{(\alpha)} = \sum_i T_{ji} x_i^{(\alpha)}, \qquad T^T = T^{-1}, \tag{8.150}$$

we want to require that the new components $\{\xi_j\}$ will extract the conspicuous group properties of the body of paradigms of Class I most efficiently. This requirement may be interpreted as follows. The transformation T (8.150) can be understood as representing passage from one coordinate system, say, $\{\mathbf{e}^{(i)}\}, i = 1, 2, \cdots, n$, to which the components x_i refer, to another coordinate system $\{\boldsymbol{\eta}^{(j)}\}$, $j = 1, 2, \cdots, n$, to which the components ξ_j refer, where $\boldsymbol{\eta}^{(j)} = \sum_j \mathbf{e}^{(i)} T_{ij}^{-1}$ or $\eta_i^{(j)} = T_{ij}^{-1}$. Then $|\xi_j^{(\alpha)}|^2$ is the squared component of the vector α along the axis $\boldsymbol{\eta}^{(j)}$. If this is large, the $\boldsymbol{\eta}^{(j)}$ will represent a major property or feature of the item α. Thus its average in the body of paradigms of Class I, namely,

$$\rho_j = \sum_\alpha w^{(\alpha)} \, |\xi_j^{(\alpha)}|^2, \tag{8.151}$$

will be the measure of importance of the property $\boldsymbol{\eta}^{(j)}$ in representing the features of the paradigms of Class I. Note that $\rho_j \geq 0$ and $\sum \rho_j = 1$. The desirable coordinate system $\{\boldsymbol{\eta}^{(i)}\}$ is such that the distribution of ρ's is concentrated on very few coordinates. If, for instance, $\rho_1 = 1$ and $\rho_j = 0$, $j \neq 1$, $\eta^{(1)}$ will be the unique characteristic feature of Class I as represented by the paradigms. Such an optimum coordinate system was discussed in detail in Section 8.2 and can be obtained as the eigenvectors of the equation

$$\sum_{i'} G_{ii'} \eta_{i'}^{(j)} = \lambda_j \eta_i^{(j)}, \tag{8.152}$$

where

$$G_{ii'} = \sum_\alpha w^{(\alpha)} x_i^{(\alpha)} x_{i'}^{(\alpha)}. \tag{8.153}$$

In contrast to SELFIC we apply the method to the paradigms of each class separately and should not subtract the average vector (of the class) from each vector as we did in Section 8.2. If we arrange the eigenvalues so that

$$\lambda_1 \geq \lambda_2 \geq \cdots, \tag{8.154}$$

† The idea of CLAFIC proposed by me was first successfully tested by C. Kulikowski.

we get the desired coordinates $\boldsymbol{\eta}^{(1)}, \boldsymbol{\eta}^{(2)}, \cdots$ in descending order of importance. The coordinates $\boldsymbol{\eta}^{(j)}$ whose ρ_j are small may be considered as corresponding to fluctuation and noise within the class. If we choose a constant θ $(0 \le \theta \le 1)$ as a measure of fluctuations that should be suppressed, we should take only σ "feature coordinates" $\eta^{(1)}, \eta^{(2)}, \cdots, \eta^{(\sigma)}$, where σ is the smallest integer satisfying

$$\sum_{j=1}^{\sigma} \lambda_j \ge (1 - \theta). \tag{8.155}$$

In most of the cases, with a reasonable choice of θ, this σ will be considerably smaller than the original number n of variables.

Now consider the projection operator (matrix)

$$(\mathcal{S}_{\mathrm{I}})_{ii'} = \sum_{j=1}^{\sigma} \eta_i^{(j)} \eta_{i'}^{(j)}; \tag{8.156}$$

then the result of application of $\mathcal{S}_{\mathrm{I}}$ on any vector x_i, that is,

$$x_i^* = \sum_{i'} (\mathcal{S}_{\mathrm{I}})_{ii'} x_{i'}, \tag{8.157}$$

will become the projection of x on the subspace subtended by the feature variables of Class I. (See Section 9.2 for details of the concept of projection operator.) If we write

$$(x\mathcal{S}_{\mathrm{I}}x) = \sum_i \sum_{i'} x_i (\mathcal{S}_{\mathrm{I}})_{ii'} x_{i'}, \tag{8.158}$$

this $(x\mathcal{S}_{\mathrm{I}}x)$ will be equal to the squared cosine of the angle between the vector x and the feature subspace.

Now we should do the same thing for Class II and obtain $\mathcal{S}_{\mathrm{II}}$. We may then assign to the probabilities that x will be a member of I and a member of II the values given by

$$p(\mathrm{I} \mid x) = \frac{(x\mathcal{S}_{\mathrm{I}}x)}{(x\mathcal{S}_{\mathrm{I}}x) + (x\mathcal{S}_{\mathrm{II}}x)},$$
$$p(\mathrm{II} \mid x) = \frac{(x\mathcal{S}_{\mathrm{II}}x)}{(x\mathcal{S}_{\mathrm{I}}x) + (x\mathcal{S}_{\mathrm{II}}x)}. \tag{8.159}$$

The two subspaces $\mathcal{S}_{\mathrm{I}}$ and $\mathcal{S}_{\mathrm{II}}$ may have a common subspace $\mathcal{S}_{\mathrm{I}\cap\mathrm{II}}$, which may be obtained by multiplying $\mathcal{S}_{\mathrm{I}}$ and $\mathcal{S}_{\mathrm{II}}$ alternately a sufficiently large number of times. The subspace that is peculiar to Class I may be depicted as $\mathcal{S}_I^* = \mathcal{S}_{\mathrm{I}} - \mathcal{S}_{\mathrm{I}\cap\mathrm{II}}$, which may or may not be perpendicular to the subspace $\mathcal{S}_{\mathrm{II}}^* = \mathcal{S}_{\mathrm{II}} - \mathcal{S}_{\mathrm{I}\cap\mathrm{II}}$ Peculiar to Class II. The case in which $\mathcal{S}_{\mathrm{I}}^*$ and $\mathcal{S}_{\mathrm{II}}^*$ are not perpendicular is discussed in detail in Appendix A9.2.

9
Non-Boolean Information Theory

9.1. MODULAR LATTICE

This chapter is devoted to an attempt to reconstruct probability theory and information theory on a logical basis broader than the Boolean logic to which the customary version of these theories is intimately committed. It has long been recognized that many patterns of human inference that can be considered valid or useful do not fit into the usual Boolean logic. Although many of these complaints have remained on an imprecise, verbal level and have not been formulated in terms of symbolic logic, a few points of view have been successfully formulated as a formal logic broader than the usual Boolean logic. Examples of such broader formalisms are (a) intuitionistic logic, which claims that double negation is not equivalent to the affirmation, (b) modal logic, which makes a distinction between necessary truth and contingent truth, and (c) three-valued logic, which assumes that a proposition can be true, false, or "indefinite."

The "modular" logic that we explain in this section, which provides the basis for the entire chapter, is different from the three well-known forms of broader logic and is based on the following heuristic idea: we should make as broad a formalism as possible, abiding, however, with two principles—A, the if-then relation is the unalterable basis of logical thinking (Peirce's principle); and B, the propositions must be such that they can be assigned "probabilities."† The reader may note that the observations made in the beginning paragraphs of Section 7.3 are in general agreement with these two points, A and B. (It must be admitted, however, that the concept of "conditional" probabilities will be revived after a long detour.) As we saw in

† For a slightly different approach to this problem, see [W-31], [W-31a].

Section 7.1, the if-then relation (together with the existence of negation, conjunction, and disjunction) is sufficient to engender the formulas of the usual propositional calculus except the distributive law. This suggests that a generalization of logic must start with questioning the validity of the distributive law. But complete abandonment of this law makes the notion of probability impossible, thus contradicting the second principle, *B*. The resolution of this dilemma achieved by modular logic is a restricted admittance of the distributive law in a special way just sufficient to allow the notion of probability to survive. The notions of logic and probability thus modified turn out to be precisely those used in quantum mechanics. (Historically speaking, this last statement must be reversed. Birkoff and von Neumann worked out modular logic as an interpretation of quantum mechanics [B-2].)

Before explaining the theory of modular lattice, we should add two remarks about the relation between the usual Boolean logic and any proposed broader form of logic. First, we have to acknowledge that there exists a domain of inference where the usual logic is undoubtedly valid. For this reason any new logic must, when applied to certain restricted areas, reduce to the usual logic. Therefore a "broader" logic must be a "weaker" logic in the sense that with an additional stronger restriction it engenders the usual logic. Second, since we are accustomed to the usual logic, the new logic must be explained in a language based on the usual logic. In other words, "metalogic," according to which the formalism of the new logic is handled, has to be Boolean logic. This second point is obviously justifiable only if the first point is true, because otherwise the Boolean "metalogic" would become groundless. The situation is somewhat reminiscent of explaining the geometry of a non-Euclidean space by that of a Euclidean space of higher dimensions in which it is embedded, whereas conceptually a Euclidean space is a special case of the non-Euclidean space.

We saw in Sections 7.2 and 7.3 that in a Boolean lattice we can assign a weight function $w(X)$ to each proposition X such that

$$w(X) + w(Y) = w(X \cap Y) + w(X \cup Y). \tag{9.1}$$

[See (7.144) and (7.163).]. The guiding idea in this section consists of requiring that each proposition in a lattice (which is not necessarily Boolean) can be assigned at least one weight function $w(X)$ to satisfy condition (9.1) instead of requiring general validity of the distributive law.† Actually, in addition to (9.1), we require a second condition on $w(X)$, namely, that

$$\text{if} \quad X < Y \quad \text{then} \quad w(X) < w(Y), \tag{9.2}$$

† It should be recalled, however, that if the formula of the type (9.1), as was shown towards the end of Section 7.3, does not guarantee the existence of conditional probabilities in the absence of the distributive law.

where $X < Y$ of course means that $X \to Y$ but not $Y \to X$. This is (7.165). In order to obtain some feeling for the matter, let us try to understand what (9.2) means in the case of a Boolean lattice, although we are now interested in a more general kind of lattice. In a Boolean lattice we have $Y = X \cup (Y \cap \neg X)$ for X and Y satisfying $X \to Y$, where X and $Y \cap \neg X$ are disjoint. [Actually the decomposition $Y = X \cup (Y \cap \neg X)$ for $X \to Y$ is also possible in the modular lattice.] Hence $w(Y) = w(X) + w(Y \cap \neg X)$. But the condition that Y does not imply X means $Y \cap \neg X \neq \emptyset$. Hence, if it is true as a general rule that $w(A) \neq 0$ for any $A \neq \emptyset$, $w(Y) > w(X)$. One of the axioms of probability states that $w(\emptyset) = 0$, but it does not state the converse, that $w(A) = 0$ implies $A = \emptyset$. What we need in the Boolean case in order to obtain (9.2) is exactly this last property. This property, although not required by the axioms of probability, is not terribly unnatural, since if some proposition is not identically false one may assign to it some nonzero probability no matter how small, particularly when the number of propositions under consideration is finite. Furthermore, we are not going to require (9.2) of all probability assignments; we require only the existence of one w-function satisfying (9.2).

Now we have an important theorem due to Dedekind (see [D-3]).

Theorem 9.1 If a lattice $\mathfrak{L}$ is such that each element $X \in \mathfrak{L}$ can be assigned a value $w(X)$ in such a way that (9.1) and (9.2) are satisfied, the lattice $\mathfrak{L}$ satisfies the modular law, which states that if $X, Y, Z \in \mathfrak{L}$ and if $X \to Z$ then

$$X \cup (Y \cap Z) = (X \cup Y) \cap Z. \tag{9.3}$$

(The lattice $\mathfrak{L}$ is then called a "modular lattice." For a mnemotechnical purpose, note that the formula has the appearance of an associative law.)

Proof. The proof consists of two steps. First we prove without the premise of this theorem the following conclusion, which is a little weaker than (9.3), namely, that

$$\text{if} \quad X \to Z \quad \text{then} \quad X \cup (Y \cap Z) \to (X \cup Y) \cap Z. \tag{9.4}$$

This follows directly from the laws of general lattices and will be separately proved as Lemma 9.1. The second step is to derive, with the premise of the present theorem, the desired result (9.3) from (9.4). An immediate consequence of the property (9.2), which states that $A \to B$ with $A \neq B$ implies $w(A) < w(B)$, is that if $A \to B$ and $w(A) = w(B)$ then $A = B$. For when $A \to B$ there are two possibilities, $A \neq B$ and $A = B$; in the former case we shall have $w(A) \neq w(B)$. Therefore, in order to derive (9.3) from (9.4), it suffices to show that

$$w[X \cup (Y \cap Z)] = w[(X \cup Y) \cap Z]. \tag{9.5}$$

This can be proved as follows, with the help of the property (9.1) and the premise $X \rightarrow Z$. The weight function on the left side of (9.5) can be transformed by the use of (9.1) to

$$w[X \cup (Y \cap Z)] = w(X) + w(Y \cap Z) - w(X \cap Y \cap Z). \tag{9.6}$$

The second term on the right side of (9.6) can be written with the help of (9.1) as $w(Y) + w(Z) - w(Y \cup Z)$. As far as the third term in (9.6) is concerned, we have $X \cap Y \cap Z = X \cap Y$, since the premise $X \rightarrow Z$ means $X = X \cap Z$. This is true even in the non-Boolean case [see (7.98)]. Hence the third term becomes $w(X \cap Y) = w(X) + w(Y) - w(X \cup Y)$. Substituting these expressions for the second and third terms of (9.6), we obtain

$$w[X \cup (Y \cap Z)] = w(Z) - w(Y \cup Z) + w(X \cup Y). \tag{9.7}$$

The right side of (9.5) becomes, because of (9.1) and the relation $X \cup Y \cup Z = Y \cup Z$, which is implied by $X \rightarrow Z$,

$$\begin{aligned} w[(X \cup Y) \cap Z] &= w(X \cup Y) + w(Z) - w(X \cup Y \cup Z) \\ &= w(X \cup Y) + w(Z) - w(Y \cup Z). \end{aligned} \tag{9.8}$$

Comparison of (9.7) and (9.8) establishes the desired conclusion, (9.5), hence (9.3).

The remaining task is to prove (9.4), which can be formulated as the following lemma.

Lemma 9.1 If $X, Y, Z \in \mathfrak{L}$, where $\mathfrak{L}$ is any lattice of a general kind, and $X \rightarrow Z$, then

$$X \cup (Y \cap Z) \rightarrow (X \cup Y) \cap Z.$$

Proof. Let us write $A = X \cup (Y \cap Z)$ and $B = (X \cup Y) \cap Z$. Then the desired result $A \rightarrow B$ can be proved by showing $A \cup B = B$, which is equivalent to $A \rightarrow B$ even in a non-Boolean lattice [see (7.98)]. We now show that $A \cup B$ can be transformed into B by the following steps. First, because of the associative law (7.34), we have

$$\begin{aligned} A \cup B &= [X \cup (Y \cap Z)] \cup [(X \cup Y) \cap Z] \\ &= X \cup \{(Y \cap Z) \cup [(X \cup Y) \cap Z]\}. \end{aligned} \tag{9.9}$$

Within the braces we have two terms, $C = Z \cap Y$ and $B = Z \cap (X \cup Y)$. It is easy to see that $C \rightarrow B$ because $C \cap B = Z \cap Y \cap Z \cap (X \cup Y) = Z \cap Y \cap (X \cup Y) = Z \cap Y = C$, where the last step is allowed by the absorptive law, or because $Y \rightarrow X \cup Y$ [see (7.35) and (7.11)]. Since $C \rightarrow B$, we have $C \cup B = B$, which means that the quantity in the braces is equivalent to the second factor $(X \cup Y) \cap Z$. By writing the factor X outside the braces as $X \cap Z$ (which is allowed because $X \rightarrow Z$), we can transcribe (9.9) as

$$A \cup B = [X \cap Z] \cup [(X \cup Y) \cap Z]. \tag{9.10}$$

Now, by the same argument (with X and Y interchanged) that allowed us to reduce $C \cup B$ to B, we can reduce the right side of (9.10) to $[(X \cup Y) \cap Z]$, because $X \to X \cup Y$. Thus (9.10) becomes $A \cup B = B$. Q.E.D.

Now that we have proved the modular law (9.3), let us examine its implications. Consider the distributive law

$$X \cup (Y \cap Z) = (X \cup Y) \cap (X \cup Z). \tag{9.11}$$

If we use the condition $X \to Z$, the last factor $(X \cup Z)$ becomes simply Z in any lattice. Hence the distributive law takes the form of the modular law. This means that the modular law amounts to permitting the distributive law (9.11) within a limited domain of applicability imposed by the condition $X \to Z$. As a consequence any distributive (or Boolean) lattice that claims general validity of (9.11) for any X, Y, or Z is a special case of a modular lattice.

If we take the negation of both sides of (9.11), we obtain, by the use of De Morgan's law (which is also valid in a nondistributive lattice if we introduce the negation as we did in Chapter 7),

$$\neg X \cap (\neg Y \cup \neg Z) = (\neg X \cap \neg Y) \cup (\neg X \cap \neg Z). \tag{9.12}$$

If we rename $\neg X$, $\neg Y$, and $\neg Z$ as X, Y, and Z, we obtain the other distributive law,

$$X \cap (Y \cup Z) = (X \cap Y) \cup (X \cap Z). \tag{9.13}$$

If the lattice allows the modular law, (9.11) and (9.12) should be valid under the condition $X \to Z$; that is, $\neg Z = \neg X \cap \neg Z$. This means that (9.13) should be valid under the condition $Z = X \cap Z$; that is, $Z \to X$. If we substitute Z for $X \cap Z$ in (9.13), we obtain

$$X \cap (Y \cup Z) = (X \cap Y) \cup Z, \tag{9.14}$$

which should be valid under the condition that $Z \to X$. But this is exactly the same as (9.3) with X and Z interchanged. This means that the modular law is self-reciprocal (invariant for the reciprocal transformation considered under Theorems 7.5 and 7.19), whereas the distributive laws pass from one form to the other by the reciprocal transformation.

We thus see that the distributive law of the type (9.11) is valid when $X \to Z$ and the distributive law of the type (9.13) is valid when $Z \to X$. This does not mean that (9.11) and (9.13) are valid in a modular lattice only under these conditions. In fact, it is easy to see that (9.11) must also be valid when $X \to Y$ and (9.13) must also be valid when $Y \to X$, since the roles of Y and Z are entirely symmetrical in these formulas. Along this line let us note some more facts about the validity of the distributive law in a modular lattice. The simplest fact is the following theorem.

Theorem 9.2 Both distributive laws (9.11) and (9.13) hold for three elements in any nondistributive lattice (a) if two of them are equivalent or (b) if one of the three is either $\emptyset$ or $\square$.

Proof. In Case (a) the distributive laws reduce to the law of reflexivity of equivalence ($A = A$) or to the absorptive law (7.35) because of the idempotent law (7.32). In Case (b), the same is true because of the laws governing $\emptyset$ and $\square$ (7.29) and (7.30). All the laws involved here are valid in any nondistributive (not necessarily modular) lattice.

Theorem 9.3 In a modular lattice the distributive laws hold for three elements if one of them implies one of the remaining two.

Proof. Since there are three elements, X, Y, and Z, there are three different pairs, and there could be six different implication relations. Because the roles of Y and Z are symmetrical, however, we need consider only three distinct cases: (i) $X \to Y$, (ii) $Y \to X$, and (iii) $Y \to Z$. Case (i) has already been dealt with. In Case (ii) we note that $Y \cap Z \to Y$ by the definition of conjunction and $Y \to X$ by assumption; hence $Y \cap Z \to X$. This makes the left side of (9.11) equivalent to X. The right side of (9.11) becomes, by the assumption, $X \cap (X \cup Z)$, which is equivalent to X by the absorptive law. This proves the validity of (9.11) in Case (ii). In Case (iii), the left side of (9.11) becomes $X \cup Y$ by the assumption. On the right side we note that the first factor $(X \cup Y)$ implies the second factor $(X \cup Z)$, because $(X \cup Y) \cup (X \cup Z) = X \cup Y \cup Z = X \cup Z$ by the use of commutativity and idempotence and by the assumption that $Y \cup Z = Z$. Hence the right side becomes equivalent to the first factor $(X \cup Y)$ [cf. (7.98)]. This proves (9.11) in Case (iii). Thus we have proven the distributive law (9.11) for each of six possible implications. If we apply the reciprocity transformations and call $\neg X$, $\neg Y$, and $\neg Z$ simply X, Y, and Z, the collection of six possible implications becomes the original collection and the distributive law (9.11) becomes the other distributive law (9.13). Hence we have proven the theorem for this version of the distributive law too. It should be noted that the modular law was invoked only in Case (i), and not in Cases (ii) and (iii). Theorem 9.2 is included as a special case of Theorem 9.3.

Theorem 9.4 The distributive laws hold for three elements in any lattice if they satisfy either one of the following two conditions: (a) the disjunction of any two elements is equivalent to the negation of the third element; (b) the conjunction of any two elements is equivalent to the negation of the third element.

Proof. Let us first take Case (a) and prove (9.11). The premise implies that $X \cup Y \cup Z = \square$ as well as $X \cup Y = \neg Z$, $Y \cup Z = \neg X$, and $Z \cup X = \neg Y$, from which also follow $\neg X \cap \neg Y \cap \neg Z = \emptyset$, $X = \neg Y \cap \neg Z$, $Y = \neg Z \cap \neg X$, and $Z = \neg X \cap \neg Y$ according to De Morgan's law. From these last four equations it follows further that $X \cap Y = Y \cap Z = Z \cap X = \emptyset$. Hence the left side of (9.11) becomes $X \cup (Y \cap Z) = X \cup \emptyset = X$. The right side becomes $(X \cup Y) \cap (X \cup Z) = \neg Z \cap \neg Y = X$. Let us next take Case (b) and prove (9.11). The premise means that $X \cap Y \cap Z = \emptyset$, $X \cap Y = \neg Z$, $Y \cap Z = \neg X$, and $Z \cap X = Y$, from which follow $\neg X \cup \neg Y \cup \neg Z = \Box$, $X = \neg Y \cup \neg Z$, $Y = \neg Z \cup \neg X$, and $Z = \neg X \cup \neg Y$. These last four equations entail further that $X \cup Y = Y \cup Z = Z \cup X = \Box$. The left side of (9.11) becomes $X \cup (Y \cap Z) = X \cup \neg X = \Box$. The right side becomes $(X \cup Y) \cap (X \cup Z) = \Box \cap \Box = \Box$. This completes the proof for (9.11). The proof for the other distributive law is unnecessary, since the two premises (a) and (b) transform into each other by reciprocity transformation and renaming of affirmed and negated propositions.

We can introduce a theorem that imposes still less restrictive conditions on X, Y, and Z than Theorem 9.4, but we are more interested in the concept of "compatibility" than the mere fact that X, Y, and Z satisfy the distributive law. For this reason we now pass to a consideration of sublattices within a given (modular) lattice. First, a sublattice $\mathfrak{L}_1$ of any lattice $\mathfrak{L}$ is defined by the conditions that if L_1 is a member of $\mathfrak{L}_1$ then L_1 is a member of $\mathfrak{L}$ (we denote this fact as $\mathfrak{L}_1 \subset \mathfrak{L}$) and that the members of $\mathfrak{L}_1$ are closed for the operations of disjunction and conjunction, obeying Theorems 7.6–7.9, p. 309, Thus any two elements X and Y of $\mathfrak{L}$, together with their disjunction $X \cup Y$ and conjunction $X \cap Y$, form a sublattice $\mathfrak{L}_1$, $X \cup Y$ and $X \cap Y$ playing the roles of $\Box$ and $\emptyset$, respectively, of $\mathfrak{L}_1$. In general, the $\Box$ and $\emptyset$ of a sublattice $\mathfrak{L}_1$ may be defined as the disjunction and conjunction of all members of $\mathfrak{L}_1$. The negation in $\mathfrak{L}_1$ is the complementation with respect to the new $\Box$ of $\mathfrak{L}_1$, and laws (7.48)–(7.50) will be obeyed.

More frequently used than the notion of sublattice in our discussion is what we call a "partial" lattice of $\mathfrak{L}$. A partial lattice $\mathfrak{L}_1$ of a lattice $\mathfrak{L}$ is a sublattice of $\mathfrak{L}$ that shares the same $\emptyset$ and $\Box$ with $\mathfrak{L}$. If $\mathfrak{L}_1$ does not include all the members of $\mathfrak{L}$, $\mathfrak{L}_1$ is said to be a proper partial lattice. The nonproper partial lattice of $\mathfrak{L}$ is $\mathfrak{L}$ itself. A first simple, but important, theorem is that any modular lattice $\mathcal{M}$ has a Boolean partial lattice $\mathcal{B}$. The proof is that $\mathcal{M}$, being a lattice, has $\emptyset$ and $\Box$ as its members. But the set $\mathcal{B}_0 = \{\emptyset, \Box\}$ already forms a lattice, and this lattice is Boolean. In fact, because of Theorem 9.2, the distributive law holds if any one of the propositions involved is $\emptyset$ or $\Box$; hence $\{\emptyset, \Box\}$ is Boolean. This is sufficient as a proof of the theorem, but we can go one step further. Suppose that $\mathcal{M}$ contains an element A that is different from $\emptyset$ and $\Box$. Then $\mathcal{B}_1 = \{\emptyset, A, \neg A, \Box\}$ is a partial lattice of $\mathcal{M}$ and is Boolean. The latter property is guaranteed by Theorem 9.2, since if we want to take three propositions out of these given four, we have to take some member twice or take at least $\emptyset$ or $\Box$ once. In either case the condition of Theorem 9.2 is satisfied, which makes the distributive law hold among any three elements of $\mathcal{B}_1$.

We write $\mathcal{B}_0 < \mathcal{B}_1 \subset \mathcal{M}$ to express that any member of $\mathcal{B}_0$ is a member of $\mathcal{B}_1$ and $\mathcal{B}_0 \neq \mathcal{B}_1$. It may be quite possible that there is another partial lattice, say, $\mathcal{B}_2$, of $\mathcal{M}$ such that $\mathcal{B}_1 < \mathcal{B}_2 \subset \mathcal{M}$ and $\mathcal{B}_2$ is Boolean. We can continue this procedure until finally we reach a Boolean partial lattice, say, $\mathcal{B}_C$, such that there is no larger Boolean partial lattice $\mathcal{B}_D$ in $\mathcal{M}$, and $\mathcal{B}_C < \mathcal{B}_D \subset \mathcal{M}$. Such a Boolean partial lattice $\mathcal{B}_C$ of $\mathcal{M}$ is called a complete Boolean partial lattice of $\mathcal{M}$. There can be many such complete Boolean partial lattices. In fact, the definition of $\mathcal{B}_1$ is dependent on the choice of A, which can be any member of $\mathcal{M}$. When we make $\mathcal{B}_2$ we cannot arbitrarily choose new members

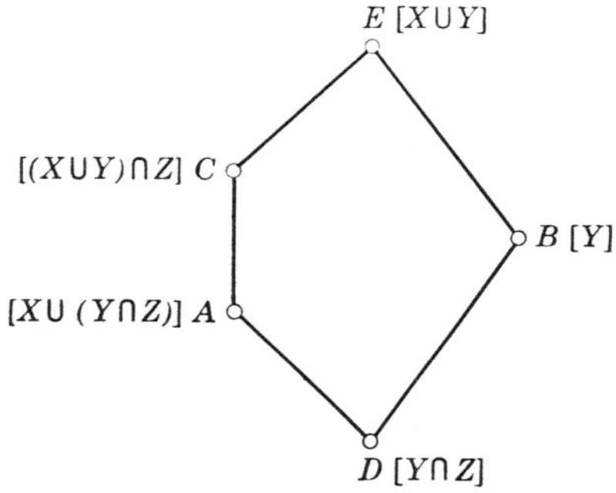

Figure 9.1 Non-existence of pentagonal sublattice is equivalent to modularity of lattice.

to be added to $\mathcal{B}_1$; yet, of course, this choice is not unique in general. Furthermore, there is no guarantee that the final product of this successive enlargement of $\mathcal{B}$'s will be unique. Actually, in the case we discuss most frequently in the following, each Boolean partial lattice has only a finite number of members, but there are continuously many different Boolean partial lattices in all, so that $\mathcal{M}$ itself has continuously many members. In fact it is very common for an element to belong to many different Boolean partial lattices. A question then arises as to the dimension function of an element that can be defined by (7.142) with respect to each Boolean partial lattice. Do the dimensions defined in different Boolean partial lattices coincide or not? To answer this question we shall have to study more closely the implications of the modular law (9.3).

We have to recall that in any lattice Lemma 9.1 is true; that is, (9.4) holds, whereas (9.3) holds (for any X, Y, and Z) only in a modular lattice. That means that in a "nonmodular" lattice there is at least one set of three members X, Y, and Z for which (9.3) does not hold. It is interesting that this non-modularity of a lattice $\mathfrak{L}$ is equivalent to the existence of at least one sublattice in $\mathfrak{L}$ of the type depicted in Figure 9.1. As usual, the upper end of a line segment is "partaken" (implied) by, but does not partake (imply) the lower end of the line segment, and, besides that, we understand that no line satisfying these conditions is omitted from the graph.

First, we show that if $\mathfrak{L}$ has a sublattice of the pentagonal form (Figure 9.1), $\mathfrak{L}$ is not modular. Equate X, Y, and Z of (9.4) with A, B, and C of Figure 9.1, because $A \to C$ and $X \to Z$. Then $X \cup (Y \cap Z)$ in the figure becomes $A \cup (B \cap C) = A \cup D = A$, and $(X \cup Y) \cap Z$ becomes $(A \cup B) \cap C = E \cap C = C$. Since $A \to C$ and not $A = C$, we see that (9.4) is satisfied but not (9.3), that is, $\mathfrak{L}$ is nonmodular. The relation $B \cap C = D$, which we used in this derivation, follows from the definition of conjunction as the largest element that partakes both elements involved, and similarly for $A \cup B = E$.† Second, we can see, conversely, that any nonmodular lattice has a sublattice of the pentagonal type (Figure 9.1). Let X, Y, $Z \in \mathfrak{L}$ be such that $X \to Z$ and (9.4) holds, but not (9.3). Consider a set of five members, $Y \cap Z$, Y, $X \cup Y$, $X \cup (Y \cap Z)$, and $(X \cup Y) \cap Z$. It is easy to see that these five form a sublattice of the type depicted in Figure 9.1 if we equate these five to D, B, E, A, and C, respectively. The relation $A \to C$ and not $A = C$ is what the assumption of nonmodularity means. The four partaking relations $D \to B$, $B \to E$, $D \to A$, and $A \to E$ are trivial. The most problematic point is that both $A \cup B$ and $B \cup C$ become E, and both $A \cap B$ and $B \cap C$ become D. First, $A \cup B = [X \cup (Y \cap Z)] \cup Y = X \cup [(Y \cap Z) \cup Y] = X \cup Y = E$ (by the associative and absorptive laws). Next, as far as $B \cup C$ is concerned, since $X \to X \cup (Y \cap Z) \to (X \cup Y) \cap Z = C$ we have $B \cup X = X \cup Y \to B \cup C$. On the other hand, since $C \to X \cup Y$ and $B = Y$, we have $B \cup C \to X \cup Y$. Hence $B \cup C = E$ too. Similarly for $A \cap B = B \cap C = D$. This completes the proof that the existence of a pentagonal sublattice is equivalent to the nonmodularity of the lattice. From this follows Theorem 9.5.

Theorem 9.5 In a modular lattice, if an element A immediately partakes (is immediately partaken by) X and Y, then $X \cup Y$ ($X \cap Y$) is immediately partaken by (immediately partakes) both X and Y.

Proof. U is said to partake V immediately if $U < V$ and if there exists no W such that $U < W < V$, where $U < V$ means $U \to V$ and $U \neq V$. The maximal chain [see (7.116)] is such that each term immediately partakes the next term. If $X \cup Y$ is not immediately partaken by X, let an element between X and $X \cup Y$ be called Z; then the five elements A, X, Y, Z, and $X \cup Y$ would form a sublattice of the forbidden kind. This gives rise to the following important theorem.

† It is easy to see that the hexagon obtained by putting one more node point between E and B or between B and D is also nonmodular. Note: the original pentagon is a sublattice of this hexagon. A similar statement can be made about a polygon of still higher rank.

Theorem 9.6 In a modular lattice the length of a maximal chain connecting $\emptyset$ to a given element, say, A, is determined by A and is independent of the chain.

Proof. The length of a maximal chain

$$\emptyset = A_0 < A_1 < A_2 < \cdots < A_n = A \tag{9.15}$$

is defined as n. Let us state the theorem in the form: if one maximal chain ending with an element has length n then all other maximal chains ending with the same element have also length n. This theorem is obviously true for $n = 1$.

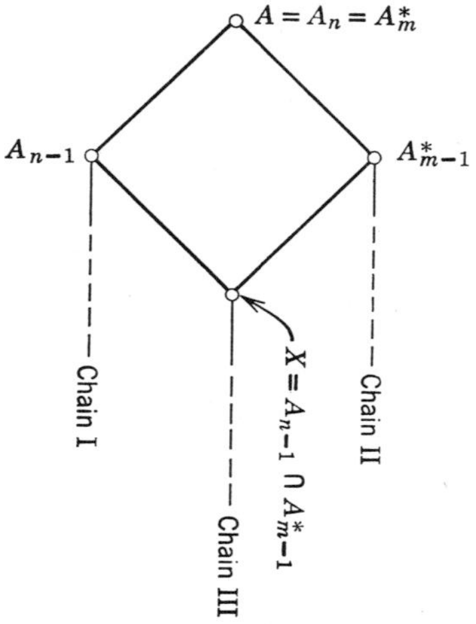

Figure 9.2 Length of maximal chain connecting $\emptyset$ and A is determined solely by A.

Suppose now that it is true for $n - 1$, and let us demonstrate that it will then be true for n. Suppose that two maximal chains end with A; one of them (Chain I) is given by (9.15) and the other (Chain II) is given by

$$\emptyset = A_0^* < A_1^* < A_2^* < \cdots, < A_m^* = A, \tag{9.16}$$

where m may or may not be equal to n (see Figure 9.2). If $A^*_{m-1} = A_{n-1}$, we can apply the theorem for $n - 1$ to the chains ending with A_{n-1} and the proof is immediate. If $A^*_{m-1} \neq A_{n-1}$, consider $X = A^*_{m-1} \cap A_{n-1}$; then, applying theorem 9.5 to $A = A_n = A^*_m$, A_{n-1}, A^*_{m-1}, and X, we conclude that X immediately partakes A^*_{m-1} as well as A_{n-1}. We call any one maximal chain ending with X Chain III. Now consider the portion of Chain I ending with A_{n-1} (of length $n - 1$) and compare it with the extension of Chain III ending with A_{n-1}. Since this extension is also maximal, it must, by the inductive assumption, have length $n - 1$; that is, Chain III itself must have the length $n - 2$. Next compare the portion of Chain II ending with A^*_{m-1} with the extension of Chain III ending with A^*_{m-1}. Since this latter has length $n - 1$, the former must, by the inductive assumption, have length $n - 1$ also. This means that the entire length of Chain II ending with $A^*_m = A$ must be of length n (see Figure 9.2). This completes the proof, which is taken from Birkoff [B-3].

We can thus define a function $D(X)$ by the length between $\emptyset$ and X, which is uniquely determined by the element X in a modular lattice. If we are given two elements X and Y such that $Y \to X$, we can make a maximal chain ending with X through Y; the portion of the chain between $\emptyset$ and Y will be a maximal chain for Y, and the portion between X and Y will be maximal in the sense that we cannot squeeze any more elements in between. It is then obvious that $Y \to X$ implies $D(Y) \leq D(X)$, and unless $Y = X$, we have $D(Y) < D(X)$. If Y immediately partakes X, $D(Y) + 1 = D(X)$. If $Y \to X$, we can speak of the length of the maximal chain $L(Y, X)$ between Y and X, which will be $L(Y, X) = D(X) - D(Y)$. Now we can prove the next theorem.

Theorem 9.7 The D-function defined in the preceding paragraph is such that for any two elements X and Y in a modular lattice we have

$$D(X) + D(Y) = D(X \cap Y) + D(X \cup Y). \tag{9.17}$$

Proof. Note that a maximal chain for $X \cup Y$ can pass through $\emptyset \to X \cap Y \to X \to X \cup Y$ and another one can pass through $\emptyset \to X \cap Y \to Y \to X \cup Y$. Consider the portion $X \cap Y \to Y$. Let a maximal chain be

$$X \cap Y = Y_0 < Y_1 < \cdots < Y_m = Y \tag{9.18}$$

and similarly let a maximal chain for the portion $X \cap Y \to X$ be

$$X \cap Y = X_0 < X_1 < \cdots < X_n = X. \tag{9.19}$$

This means that

$$D(Y) = D(X \cap Y) + m \tag{9.20}$$

and

$$D(X) = D(X \cap Y) + n. \tag{9.21}$$

Next consider a chain (see Figure 9.3)

$$Y = X_0 \cup Y \to X_1 \cup Y \to \cdots \to X_n \cup Y = X \cup Y. \tag{9.22}$$

An open question is whether or not this is a maximal chain. As shown presently, the distance (difference in the D-values) between two consecutive terms in (9.22) is actually 0 or 1. Hence the length

$$L(Y, X \cup Y) = D(X \cup Y) - D(Y) \leq n. \tag{9.23}$$

Similarly,

$$L(X, X \cup Y) = D(X \cup Y) - D(X) \leq m. \tag{9.24}$$

Substituting for m and n in (9.23) and (9.24) the values obtained from (9.20) and (9.21) and adding these two equations, we obtain

$$D(X \cup Y) + D(X \cap Y) - D(X) - D(Y) \leq 0. \tag{9.25}$$

Reversing the argument, suppose that the distance between $X \cup Y$ and X is m and the distance between $X \cup Y$ and Y is n; that is,

$$D(X \cup Y) = D(X) + m \tag{9.26}$$

and

$$D(X \cup Y) = D(Y) + n. \tag{9.27}$$

The distance between X and $X \cap Y$ then cannot be larger than n:

$$L(X \cap Y. X) = D(X) - D(X \cap Y) \leq n, \tag{9.28}$$

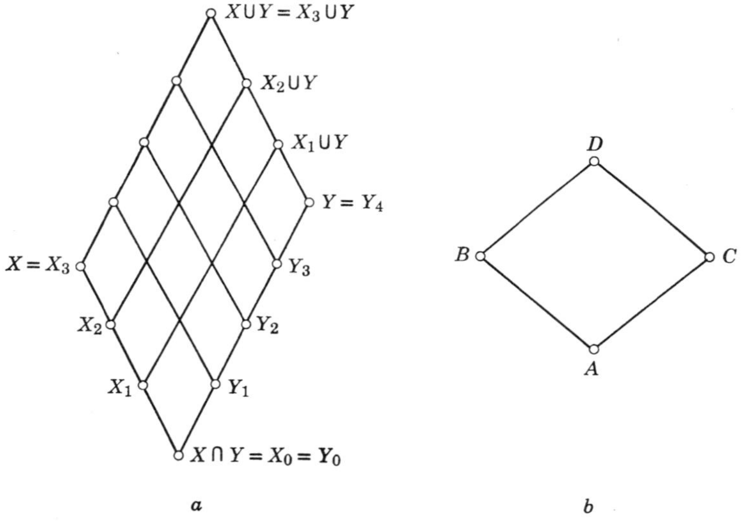

Figure 9.3 The D-function satisfies the axiom of probability (9.17).

and similarly

$$L(X \cap Y, Y) = D(Y) - D(X \cap Y) \leq m. \tag{9.29}$$

Eliminating m and n from these four equations, we get

$$D(X) + D(Y) - D(X \cap Y) - D(X \cup Y) \leq 0. \tag{9.30}$$

Hence, combining (9.25) and (9.30), we get the desired result (9.17). The remainder is to prove (9.23), (9.24), (9.28), and (9.29). Let us consider (9.23) as an example. To have a mental picture let us depict the lattice as in Figure 9.3*a* for a simple case of $m = 4$ and $n = 3$. Let each small diamond in Figure 9.3*a* be represented as A, B, C, D, as in Figure 9.3*b*: $A \rightarrow B$, $A \rightarrow C$, $D = B \cup C$. The point is that if $L(A, B) \leq 1$ and $L(A, C) \leq 1$, then because of Theorem 9.5, $L(B, D) \leq 1$ and $L(C, D) \leq 1$. The case $L(A, B) = 0$ or

$L(A, C) = 0$ is trivial. But if $L(A, B) = L(A, C) = 1$, we usually have $L(B, D) = 1$ and $L(C, D) = 1$ as in Theorem 9.5, although we could also have $L(B, D) = 0$ or $L(C, D) = 0$. This last eventuality can happen because B can coincide with C. Either way, however, $L(B, D) \leq 1$ and $L(C, D) \leq 1$. Thus, starting at the lowest diamond, $A = X \cap Y$, $B = X_1$, $C = Y_1$, $D = X_1 \cup Y_2$, we may have $L(C, D) = 0$ or 1. Continuing this argument gradually upward, we wind up with $L(Y, X_1 \cup Y) \leq 1$, $L(X_1 \cup Y, X_2 \cup Y) \leq 1$, $L(X_2 \cup Y, X_3 \cup Y) \leq 1$. Hence $L(Y, X \cup Y) \leq 3 = n$. This is (9.23).The other three cases work the same way. In the last two cases (9.28), (9.29), we have to use the second version (the one in parentheses) of Theorem 9.5.

We have thus come back to the starting point (9.1). We first derived the modular law from the existence of a probability-like function, and then, in the next stage, assuming the modular law, we derived constructively a probability-like function.† Now take a maximal chain in a modular lattice from $\emptyset$ to $\square$,

$$\emptyset = A_0 < A_1 < \cdots < A_n = \square, \tag{9.31}$$

and suppose that we apply the argument of Section 7.2 in deriving atoms. There are four places where we used the distributive law, namely, in deriving (7.107), (7.113), (7.132), and (7.138). To derive (7.107) from $\emptyset = \urcorner A_{i-1} \cap A_i$, we used the distributive law in the following way: $A_{i-1} = A_{i-1} \cup (\urcorner A_{i-1} \cap A_i) = (A_{i-1} \cup \urcorner A_{i-1}) \cap (A_{i-1} \cup A_i) = A_{i-1} \cup A_i$. This is perfectly legitimate in a modular lattice too, in the light of Theorem 9.3, because it is understood that $A_{i-1} \rightarrow A_i$. The use of the distributive law in (7.113) is also legitimate according to Theorem 9.3, because $A_{i-1} \rightarrow A_i$. This means that we can derive from (9.31) n atoms $(\alpha_1, \alpha_2, \cdots, \alpha_n)$ satisfying the conditions of disjointness (7.108), completeness (7.109), and atomicity (7.117).

The basic difference is that in a modular lattice we cannot derive (7.132). This means that the two sets of atoms $\{\alpha_1, \alpha_2, \cdots, \alpha_n\}$ and $\{\beta_1, \beta_2, \cdots, \beta_n\}$ are not necessarily equivalent. Similarly, the use of the distributive law in (7.138) is not always legitimate in a modular law. This means that an arbitrary element of the modular lattice cannot always be expressed as a disjunction of a certain number of atoms taken from a given set of atoms $\{\alpha_1, \alpha_2, \cdots, \alpha_n\}$.

On the other hand, if the system $\{\alpha_1, \alpha_2, \cdots, \alpha_n\}$ is derived from the maximal chain of A_i (9.31), any A_i is, by definition, expressible as a disjunction of α's, namely, $A_i = \alpha_1 \cup \alpha_2 \cup \cdots \cup \alpha_i$. Now consider the collection (lattice) of all elements that can be obtained from the A's by conjunction, disjunction, and negation. Two things become clear. First, each member of this collection can be expressed as a disjunction of α's; second, the distributive law is allowed in this collection. The reason is that each A_i (hence also each member of the collection) corresponds to a subset of α's, and

† See Section 9.3 (S-63 and sequel) for more about the probability in a modular lattice.

the disjunction and conjunction become simply set-theoretical operations of disjunction and conjunction. We have repeatedly shown that set-theoretical disjunction and conjunction allow the distributive law. The negation of an element then corresponds to the set-theoretical complementation in the set $\{\alpha_i\}$. Hence the collection is a Boolean lattice. This is what we called a Boolean partial lattice before, and is defined each time a maximal chain of inclusion is given.

This consideration makes it clear that the D-function of Section 7.2 and the D-function of this section must have the same value. Theorem 9.6 shows that the value of $D(X)$ is independent of the Boolean partial lattice in which the maximal chain (connecting $\emptyset$ to X) is taken to evaluate $D(X)$.

We now introduce an important concept of compatibility of propositions.

DEFINITION 9.1. Two propositions A and B in a modular lattice $\mathcal{M}$ are said to be compatible if there exists a Boolean sublattice $\mathcal{B}$ of $\mathcal{M}$ such that $A \in \mathcal{B}$ and $B \in \mathcal{B}$.

By the nature of this definition, there is no guarantee that compatibility is transitive. If we write $A \sim B$ for compatibility of A and B, we obviously have $A \sim A$ as well as

$$A \sim B \quad \text{is equivalent to} \quad B \sim A. \tag{9.32}$$

But

$$A \sim B \quad \text{and} \quad B \sim C \quad \text{do not imply} \quad A \sim C. \tag{9.33}$$

It should be noted that $A \sim B$ is not guaranteed by the existence of a C such that A, B, and C satisfy a distributive law. For $A \sim B$ it is necessary that any three taken from all the logical combinations made from A and B satisfy the distributive laws. We have the following two theorems.

Theorem 9.8 If $A = \neg B$ then $A \sim B$.

Proof. A partial lattice $\{\emptyset, A, \neg A, \square\}$ is obviously Boolean. We can "complete" this Boolean lattice, if we want, and the resulting completed Boolean partial lattice will include A and $\neg A$.

Theorem 9.9 If $A \rightarrow B$ then $A \sim B$.

Proof. We can take a maximal chain (9.31) so that it goes through A and B.

From this it obviously follows that $A \sim (A \cap B)$ and $A \sim (A \cup B)$, because $A \cap B \rightarrow A \rightarrow A \cup B$. If $A \sim B$ then $A \sim \neg B$. This is because a Boolean partial lattice containing A and B has to contain $\neg A$ and $\neg B$ also. One may ask why it was possible to define a lattice of general nature by using the concept of implication, which is valid only between compatible propositions. The best way to understand the situation is to go back to the definition of conjunction (7.7)–(7.10). In (7.7) we have $C \sim A$ and $C \sim B$, but this does not limit the discussion to the case $A \sim B$. Similarly, we have $X \sim A$ and $X \sim B$, but not necessarily $A \sim B$. Furthermore, there are many X's, and they need not be compatible in pairs; C is compatible with all those X's because of (7.9), but this does not at all imply that any two X's will be compatible. The C is a particular one among X because C is compatible with any other X.

It is worth noting that in a modular lattice $\neg A \cup B = \square$ (A horseshoe $B = \square$) or $A \cap \neg B = \emptyset$ does not imply $A \rightarrow B$, but does imply $A \rightarrow B$ if A and B belong to a Boolean lattice [see (7.99)]. Similarly, the condition of the excluded middle and the condition of contradiction between A and B imply that they are mutual negations if A and B belong to a Boolean lattice. Hence we have the following theorem:

Theorem 9.10 If $\neg A \cup B = \square$ and $A \sim B$ or if $A \cap \neg B = \emptyset$ and $A \sim B$, then $A \rightarrow B$. If $A \cup B = \square$, $A \cap B = \emptyset$, and $A \sim B$, then $A = \neg B$.

So much for a formal consideration; we now introduce a geometric model that has all the characteristic features of a modular lattice. Relegating more mathematical treatment to the next section, let us limit ourselves to a more intuitive explanation of the model. The model consists essentially of making a one-to-one correspondence between a proposition and a subspace passing the origin in an n-dimensional Euclidean space. If $n = 3$, the possible subspaces are zero-dimensional, one-dimensional, two-dimensional, and three-dimensional. The zero-dimensional subspace is unique and is the origin of the space. Any straight line passing through the origin is a one-dimensional subspace. Any plane passing the origin is a two-dimensional subspace. The three-dimensional subspace is unique and is the entire space. $\emptyset$ is made to correspond to the origin, and $\square$ is made to correspond to the entire space. The implication $A \rightarrow B$ is interpreted as meaning that the subspace corresponding to A "lies in" (i.e., is a subspace of) the subspace corresponding to B. Thus, for instance, if A is a line and B is a plane, $A \rightarrow B$ means that A is on the plane B. The conjunction $A \cap B$ then should mean, according to the definition, the largest subspace lying both in A and in B. Hence $A \cap B$ should be the intersection of A and B. The disjunction $A \cup B$ should, according to the definition, be the smallest subspace in which both subspace A and subspace B lie. Hence we can picture $A \cup B$ as the subspace defined by all the vectors that can be formed by the sum $a\mathbf{v}_A + b\mathbf{v}_B$ of two vectors, one of which, $\mathbf{v}_A$, lies in subspace A and the other, $\mathbf{v}_B$, lies in subspace B. Thus, for instance, if A and B are two different lines, $A \cup B$ must be a plane on which both A and B lie. The reader may have already noticed that the so-called spherical model in Chapter 7 is nothing but the present model, the only difference being that each subspace is represented by its intersection with a sphere.

Figure 7.7, which illustrates a breakdown of the distributive law $A \cap (B \cup C) = (A \cap B) \cup (A \cap C)$, can be used as an example of the modular law. In fact, if $B \rightarrow A$, this distributive law reduces to the modular law $B \cup (C \cap A) = (B \cup C) \cap A$, and we have already noted that if point B lies on circle A the distributive law holds. Figure 7.13 shows that $A \cap B = \emptyset$ does not imply $A \rightarrow \neg B$, except in the case when $\theta = 90°$. On the other hand, we have shown that $A \cap B = \emptyset$ plus $A \sim B$ implies $A \rightarrow \neg B$. This is a special case of the general rule that $A \sim B$ means that A and B belong to the same family of orthogonal subspaces. This last statement means that, given A and B, there exist three orthogonal directions $\mathbf{x}$, $\mathbf{y}$, and $\mathbf{z}$ such that A and B are some subspaces built with the help of $\mathbf{x}$, $\mathbf{y}$, and $\mathbf{z}$. The family of orthogonal subspaces built with the help of $\mathbf{x}$, $\mathbf{y}$, and $\mathbf{z}$ consists of $\emptyset$, x-axis, y-axis, z-axis, xy-plane, yz-plane, zx-plane, and $\square$. A complete Boolean partial lattice in the present case consists of the eight elements of a family of orthogonal subspaces. The reader can translate into geometric language what we have done with a maximal chain of inclusion. The dimension $D(A)$ is exactly the dimension of the subspace corresponding to A.

Regarding the "probability" function w of (9.1), we have to make three important remarks. First, we required the existence of one function w satisfying (9.1) and showed that the D-function is just such a function. The condition $w(\square) = 1$ can be satisfied by defining $w(X) = D(X)/D(\square)$, but $w(X)$ could be different from $D(X)/D(\square)$. We can see that in the Boolean case we can change the values of w very easily, because there is only one set of atoms and we can arbitrarily choose the probability of each atom, $w(\alpha_i)$. But in the non-Boolean case we have to consider different sets of atoms at once, because $w(X)$ must have a unique value determined by X independent of the set of atoms used to express X as their disjunction.

Second, we should remember that a function that satisfies (9.1) or, equivalently, (7.163) automatically satisfies the condition

$$w(A \cup B) = w(A) + w(B) \quad \text{for} \quad A \cap B = \emptyset, \tag{9.34}$$

which is Condition (b) of Definition 7.2. But a w-function defined by (9.34) does not necessarily satisfy (9.1) unless the distributive law is guaranteed. This means that the w-function defined by (9.1) is a more restrictive notion than the w-function defined by (9.34) in the absence of the distributive law as a general rule. [See the discussion following (7.163).]

Third, a w-function satisfying (9.1) is not like the usual probability in the absence of the distributive law, because it does not allow us to derive a conditional probability by the usual formula,

$$w(B \mid A) = \frac{w(A \cap B)}{w(A)}. \tag{9.35}$$

In order to use this quantity for the purpose it is intended to serve, one has to show that $w(B \mid A)$ as a probability of B satisfies the axioms of probability, particularly the law (9.1), which takes the form

$$w(B \mid A) + w(C \mid A) - w(B \cap C \mid A) = w(B \cup C \mid A). \qquad (9.36)$$

Of course, we are hereby allowed to consider the unconditional probabilities $w(X)$ as satisfying the axioms. Hence we can write

$$\begin{aligned} w(B \cap C \mid A) &= \frac{w[A \cap (B \cap C)]}{w(A)} \\ &= \frac{w[(A \cap B) \cap (A \cap C)]}{w(A)} \\ &= \frac{w(A \cap B) + w(A \cap C) - w[(A \cap B) \cup (A \cap C)]}{w(A)}. \end{aligned} \qquad (9.37)$$

With the help of this result, the left side of (9.36) becomes

$$\text{left side of (9.36)} = \frac{w[(A \cap B) \cup (A \cap C)]}{w(A)}, \qquad (9.38)$$

whereas

$$\text{right side of (9.36)} = \frac{w[A \cap (B \cup C)]}{w(A)}. \qquad (9.39)$$

In a general or a modular lattice the two propositions $(A \cap B) \cup (A \cap C)$ in (9.38) and $A \cap (B \cup C)$ in (9.39) are not necessarily equivalent. Therefore $w[(A \cap B) \cup (A \cap C)]$ and $w[A \cap (B \cup C)]$ can have different values. Only with the help of the distributive law can we assume these two values to be equal. This means that in the absence of the distributive law the so-called conditional probability defined by (9.35) is not guaranteed to be a probability. The notion of conditional probability, which is obviously the most important concept in any science (Section 7.3), is revived in Section 9.3 for the non-Boolean case.

9.2. PROJECTION OPERATORS

In this section we investigate more carefully the geometric model of a modular lattice of propositions introduced briefly in the last section, which consists of assigning to each proposition A a subspace $\mathfrak{M}(A)$ passing the origin in an n-dimensional real Euclidean space. In quantum physics we actually have to use a complex space to accomodate the complexity (the so-called phase relation) of physical phenomena, besides an additional difficulty due to the fact that n can become continuously many. We deliberately use a real space of finite dimensions, however, because the use of a real space makes it intuitively easier to visualise the situation, and also because it is possible that the application of non-Boolean logic to fields other than physics may require only a real space [W-16, W-31a]). Further, it is mathematically very simple to pass from a real space to a complex space. As far as the number n of dimensions of the space in quantum physics is concerned, if the spatial extension of the physical system under consideration and the total available energy are finite, the number of available quantum states becomes finite, which allows a model with finite n. For this reason what is characteristic of quantum physics seems to have nothing to do with the finiteness of n. (See [V-5] for a space with continuously many dimensions.) The main mathematical tool introduced and extensively employed in this and following sections is the so-called projection operator $\mathfrak{P}[\mathfrak{M}(A)]$, which corresponds one to one to each subspace, $\mathfrak{M}(A)$. Thus each proposition is represented by a projection operator $\mathfrak{P}$. The reader may be surprised to see that the mathematical structure used in Chapter 8 in connection with the CLAFIC method in pattern recognition reappears in this section.

For the basic operations of the lattice we use the following geometric interpretations. The subspace corresponding to the conjunction of two propositions A and B is defined by

$$\mathfrak{M}(A \cap B) = \{\mathbf{x} \mid \mathbf{x} \in \mathfrak{M}(A) \quad \text{and} \quad \mathbf{x} \in \mathfrak{M}(B)\}, \tag{9.40}$$

which means that $\mathfrak{M}(A \cap B)$ is the set of all vectors that lie in both $\mathfrak{M}(A)$ and $\mathfrak{M}(B)$. By "vectors," of course, are meant vectors attached to the origin. For disjunction we have

$$\mathfrak{M}(A \cup B) = \{\mathbf{x} \mid \mathbf{x} = c_1\mathbf{x}_1 + c_2\mathbf{x}_2; \quad \mathbf{x}_1 \in \mathfrak{M}(A); \quad \mathbf{x}_2 \in \mathfrak{M}(B); \quad c_1, c_2 = \text{real}\}, \tag{9.41}$$

which means that $\mathfrak{M}(A \cup B)$ is the set of vectors that are linear combinations of a vector in $\mathfrak{M}(A)$ and a vector in $\mathfrak{M}(B)$. The limitation of the c's to real numbers will be lifted when we discuss quantum physics. A subspace in general is by definition a collection of vectors that is closed with respect to the operation of linear combination. This can be seen by putting $A = B$ in (9.41), which makes the left side equal to $\mathfrak{M}(A)$.

By equating $\mathfrak{M}(A)$ with $\mathfrak{M}(A \cap B)$ of (9.40) or by equating $\mathfrak{M}(B)$ with $\mathfrak{M}(A \cup B)$ of (9.41), we can immediately see that $A \to B$ is equivalent to the statement

$$\text{if} \quad \mathbf{x} \in \mathfrak{M}(A) \quad \text{then} \quad \mathbf{x} \in \mathfrak{M}(B). \tag{9.42}$$

We denote this relation by $\mathfrak{M}(A) \subset \mathfrak{M}(B)$. The symbol $\emptyset$ corresponds to the origin, that is, an empty set of vectors. The symbol $\square$ corresponds to the set of all vectors of the entire n-dimensional space. The negation $\neg A$ of A

corresponds to the subspace defined by

$$\mathfrak{M}(\neg A) = \{\mathbf{x} \mid \mathbf{x} \perp \mathbf{y} \quad \text{for all} \quad \mathbf{y} \in \mathfrak{M}(A)\}; \tag{9.43}$$

that is, $\mathfrak{M}(\neg A)$ is the set of vectors of which each is perpendicular (orthogonal) to all the vectors in $\mathfrak{M}(A)$.

In order to justify this correspondence (between subspaces and elements of a modular lattice), we can show that these definitions satisfy the basic laws of general lattices (7.29)–(7.30), (7.32)–(7.35), as well as the law of syllogism (7.4), the definition of $\cap$ and $\cup$ from $\rightarrow$ (7.7)–(7.14), the three laws of negation (7.48)–(7.50), and the modular law (9.3). Since we have already given such a justification in the simple case of a spherical model (three-dimensional space) and since such a justification in the general case is also extremely simple, we do not include it here. Perhaps the only one that is not so easy to prove is the modular law, and we shall come back to it a little later.

As far as compatibility is concerned, its interpretation is as follows. Suppose first that two propositions X and Y are compatible and $X \cap Y = \emptyset$. This implies $Y \rightarrow \neg X$ (see Theorem 9.10). Hence every vector in $\mathcal{M}(Y)$ is perpendicular to all vectors in $\mathcal{M}(X)$, because any vector in $\mathcal{M}(Y)$ is in $\mathcal{M}(\neg X)$, which is the set of vectors in which each is perpendicular to all vectors in $\mathcal{M}(X)$. Thus $\mathcal{M}(X) \perp \mathcal{M}(Y)$. Now consider two vectors A and B that are compatible: $A \sim B$. We can use the distributive law and obtain $B = (A \cap B) \cup (\neg A \cap B)$, where $(A \cap B) \cap (\neg A \cap B) = \emptyset$. Since A and B are in the same Boolean lattice, $X = (A \cap B)$ and $Y = (\neg A \cap B)$ are also in the same Boolean lattice; that is, X and Y are compatible and $X \cap Y = \emptyset$. Hence every vector in Y is perpendicular to all vectors in X. Since $B = X \cup Y$ this means that

$$\mathcal{M}(B) = \{\mathbf{v} \mid \mathbf{v} = a\mathbf{x} + b\mathbf{y}, \quad \mathbf{x} \in \mathcal{M}(X), \quad \mathbf{y} \in \mathcal{M}(Y)\}, \tag{9.44}$$

where $\mathbf{x} \perp \mathbf{y}$, i.e., $(X) \perp \mathcal{M}(Y)$. By a similar argument (interchanging the roles of A and B),

$$\mathcal{M}(A) = \{\mathbf{u} \mid \mathbf{u} = a\mathbf{x} + b\mathbf{z}, \quad \mathbf{x} \in \mathcal{M}(X), \quad \mathbf{z} \in \mathcal{M}(Z)\}, \tag{9.45}$$

where $X = A \cap B$ as before, and $Z = A \cap \neg B$ and $\mathcal{M}(X) \perp \mathcal{M}(Z)$. Finally, $Y = \neg A \cap B$ and $Z = A \cap \neg B$ belong to the same Boolean lattice as A and B; hence $Y \sim Z$. Since $Y \cap Z = \emptyset$, we have $\mathcal{M}(Y) \perp \mathcal{M}(Z)$. As a consequence we can take n orthogonal coordinates in the entire space in such a way that a certain group of them define the subspace $\mathcal{M}(X)$, another group define $\mathcal{M}(Y)$, and another group define $\mathcal{M}(Z)$, the remaining ones defining a space orthogonal to $\mathcal{M}(X)$, $\mathcal{M}(Y)$, and $\mathcal{M}(Z)$. Then the space $\mathcal{M}(A)$ is defined by the coordinates of $\mathcal{M}(X)$ and $\mathcal{M}(Y)$, while the space $\mathcal{M}(B)$ is defined by the coordinates of $\mathcal{M}(X)$ and $\mathcal{M}(Z)$. In other words, $\mathcal{M}(A)$ and $\mathcal{M}(B)$ are subspaces defined by some coordinates belonging to the same orthogonal coordinate system. Hence $A \sim B$ implies that $\mathcal{M}(A)$ and $\mathcal{M}(B)$ belong to the same family of orthogonal subspaces. It is easy to show, conversely, that if, $\mathcal{M}(A)$, $\mathcal{M}(B)$, and $\mathcal{M}(C)$ belong to the same family of orthogonal subspaces, $\mathcal{M}[A \cup (B \cap C)] = \mathcal{M}[(A \cup B) \cap (A \cup C)]$, which, in view of the one-to one correspondence of subspace and proposition, means that the distributive law holds for A, B, and C. This is because, insofar as a family of orthogonal subspaces is concerned, the problem reduces to a set-theoretical consideration of coordinates. This fact shows that if $\mathcal{M}(A)$ and $\mathcal{M}(B)$ belong to the same family of orthogonal subspaces, $A \sim B$. But to prove the general modular law $A \cup (B \cap C) = (A \cup B) \cap C$ for $A \rightarrow C$ in our model without a guarantee of $A \sim B$ and $B \sim C$, we need a more subtle consideration.

So far we have not explicitly used the coordinates, but from this point on it becomes easier to refer to them. With the help of a system of n orthogonal unit vectors, $\mathbf{e}_i$, $i = 1, 2, \cdots, n$, we can express any vector $\mathbf{x}$ in the space as

$$\mathbf{x} = x_i \mathbf{e}_i, \tag{9.46}$$

Where the summation symbol $\sum_{i=1}^{n}$ is dropped for simplicity. Any time the same index, such as i, appears twice in a formula, it means that the sum $(\sum_{i=1}^{n})$ with respect to the index is to be taken, except when the contrary is explicitly stated. The real numbers, x_i, $i = 1, 2, \cdots, n$, in (9.46) are called components of the vector $\mathbf{x}$ with reference to the coordinate system $\mathbf{e}_i$. The relation $\mathbf{x} = 0$ is equivalent to $x_i = 0$ for all $i = 1, 2, \cdots, n$. The sum of two vectors is a vector whose components are the sum of their corresponding components. The product of a vector by a number is a vector whose components are this number times the components of the original vector. The scalar product of two vectors $\mathbf{x}$ and $\mathbf{y}$ is defined by

$$(\mathbf{x}, \mathbf{y}) = x_i y_i = (\mathbf{y}, \mathbf{x}). \tag{9.47}$$

If $\mathbf{x} = \mathbf{y}$ then the scalar product (9.47) becomes the square of the length $|\mathbf{x}|$ of the vector $\mathbf{x}$. If $|\mathbf{x}| \neq 0$, $|\mathbf{y}| \neq 0$, and $(\mathbf{x}, \mathbf{y}) = 0$, we say that $\mathbf{x}$ and $\mathbf{y}$ are orthogonal. The fact that the coordinate vectors $\mathbf{e}_i$ are normalized (i.e., have length unity) and are mutually orthogonal can be written as

$$\begin{aligned}(\mathbf{e}_i, \mathbf{e}_j) = \delta_{ij} &= 1, \qquad i = j,\\ &= 0, \qquad i \neq j.\end{aligned} \tag{9.48}$$

Let us now introduce another system of orthogonal unit vectors, $\mathbf{e}_i$, and write

$$\mathbf{e}_i' = \mathbf{e}_j T_{ji}. \tag{9.49}$$

Now let us consider a new vector $\mathbf{x}'$, whose components, with reference to the primed system $\mathbf{e}_i'$, are the same as the components of the vector $\mathbf{x}$ with reference to the original system $\mathbf{e}_i$. Then

$$\mathbf{x}' = x_i\mathbf{e}_i' = \mathbf{e}_j T_{ji} x_i \equiv \mathbf{e}_i x_i' \tag{9.50}$$

where x_i' is the component of $\mathbf{x}'$ with reference to the original coordinate system $\mathbf{e}_i$. The last equality in (9.50) yields, because of (9.48),

$$x' = Tx, \tag{9.51}$$

which is the matrix expression for $x_i' = T_{ij}x_j$. In order that the condition that the new coordinate vectors $\mathbf{e}_i'$ are orthogonal and normalized, that is, that the condition

$$(\mathbf{e}_i' \cdot \mathbf{e}_j') = \delta_{ij} \tag{9.52}$$

follows from the condition that the old coordinate vectors $\mathbf{e}_i$ are orthogonal and normalized, it is necessary and sufficient to have

$$TT^T = T^TT = I, \tag{9.53}$$

where the matrices I and T^T mean $(I)_{ij} = \delta_{ij}$ and $(T^T)_{ij} = T_{ji}$, I is called the identity matrix, and T^T is called the transpose of T. If two matrices T and T^{-1} satisfy

$$TT^{-1} = T^{-1}T = I \tag{9.54}$$

we say that T and T^{-1} are mutually inverse. Equation (9.53) shows that

$$T^{-1} = T^T, \tag{9.55}$$

which is the condition for an orthogonal transformation [see (8.27)]. With the help of this T^{-1}, we can solve (9.49) for $\mathbf{e}_j$ and write

$$\mathbf{e}_i = \mathbf{e}_j' T_{ji}^{-1} \tag{9.56}$$

and also solve (9.51) for x' and write

$$x = T^{-1}x'. \tag{9.57}$$

In the above explanation we moved the vector and considered how its components change with reference to the fixed coordinates. Alternatively, we can consider the vector as fixed in the space and ask how its components change when referred to a new coördinate system. The only difference in these two viewpoints is that the roles of T and T^{-1} are interchanged.

Now we define the operation "projection of a vector $\mathbf{x}$ on another vector $\boldsymbol{\xi}$." The result of this operation, in the familiar three-dimensional case, would be a vector in the direction of $\boldsymbol{\xi}$, and its length would be that of $\mathbf{x}$ times the cosine between $\mathbf{x}$ and $\boldsymbol{\xi}$. More generally, the resulting vector of the projection is given by

$$\mathbf{y} = \frac{(\mathbf{x}, \boldsymbol{\xi})\boldsymbol{\xi}}{(\xi^2)} = \mathcal{P}[\boldsymbol{\xi}]\mathbf{x}, \tag{9.58}$$

where the "projection operator" $\mathcal{T}[\boldsymbol{\xi}]$ stands symbolically for the operation of obtaining $\mathbf{y}$ from $\mathbf{x}$. We note first that this operation is independent of the length of $\boldsymbol{\xi}$, for a multiplication of ξ by any real number does not alter $\mathbf{y}$. This remark, of course, includes the fact that $\mathbf{y}$ remains unchanged by an interchange of $\boldsymbol{\xi}$ by $-\boldsymbol{\xi}$. If we take $|\boldsymbol{\xi}| = 1$ then (9.58) becomes simpler. Second, $\mathbf{y}$ is a vector in the direction of $\boldsymbol{\xi}$ (or $-\boldsymbol{\xi}$). The quantity $(\mathbf{x}, \boldsymbol{\xi})/|\mathbf{x}| \cdot |\boldsymbol{\xi}|$ is the cosine between $\mathbf{x}$ and $\boldsymbol{\xi}$.

If the vector $\mathbf{x}$ is in the direction of $\boldsymbol{\xi}$, the operation $\mathcal{T}[\boldsymbol{\xi}]$ leaves the vector $\mathbf{x}$ unchanged. In the general case, when $\mathbf{x}$ is not in the direction of $\boldsymbol{\xi}$, the first application of $\mathcal{T}[\boldsymbol{\xi}]$ on $\mathbf{x}$ yields a vector in the direction of $\boldsymbol{\xi}$, but the second application of $\mathcal{T}[\boldsymbol{\xi}]$ on this resulting vector no longer alters it. Thus we obtain an idempotent law,

$$\mathcal{T}[\boldsymbol{\xi}]\,\mathcal{T}[\boldsymbol{\xi}] = \mathcal{T}[\boldsymbol{\xi}], \tag{9.59}$$

where we have dropped $\mathbf{x}$, since the formula applies to any arbitrary vector. The condition: $\mathcal{T}[\boldsymbol{\xi}]\mathbf{x} = 0$ means $(\boldsymbol{\xi}, \mathbf{x}) = 0$, which in turn means either that $\mathbf{x} = 0$ or that $\mathbf{x}$ and $\boldsymbol{\xi}$ are orthogonal. [It is assumed, of course, that $\boldsymbol{\xi} \neq 0$, since if $\boldsymbol{\xi} = 0$ then (9.58) would become meaningless.]

When $\mathbf{x}$ is given, one can of course always write, with the help of a suitable vector $\mathbf{z}$,

$$\mathbf{x} = \mathcal{T}[\boldsymbol{\xi}]\mathbf{x} + \mathbf{z} = \mathbf{y} + \mathbf{z}. \tag{9.60}$$

Applying $\mathcal{T}[\boldsymbol{\xi}]$ on this expression and taking account of (9.59), we obtain

$$\mathcal{T}[\boldsymbol{\xi}]\mathbf{z} = 0 \quad \text{or} \quad (\boldsymbol{\xi}, \mathbf{z}) = 0. \tag{9.61}$$

In the light of (9.61), the relation (9.60) can be interpreted as expressing the fact that any vector $\mathbf{x}$ can be decomposed into two (uniquely defined) parts, one in the direction of $\boldsymbol{\xi}$ and the other perpendicular to $\boldsymbol{\xi}$.

Equation (9.58) can be written, in terms of the components of the vectors involved, as

$$y_i = \frac{(\xi_i, \xi_j)x_j}{|\xi|^2}. \tag{9.62}$$

Therefore we can write $\mathbf{y} = \mathcal{T}[\boldsymbol{\xi}]\mathbf{x}$ as

$$y_i = (\mathcal{T}[\boldsymbol{\xi}])_{ij}x_j \tag{9.63}$$

with the help of

$$(\mathcal{T}[\boldsymbol{\xi}])_{ij} = \frac{\xi_i \xi_j}{(\xi_k, \xi_k)}. \tag{9.64}$$

Thus we can interpret $\mathcal{T}[\boldsymbol{\xi}]$ as either an operation or a matrix. We note immediately the following two important properties of the projection matrix:

$$(\mathcal{T}[\boldsymbol{\xi}])^T = \mathcal{T}[\boldsymbol{\xi}]; \tag{9.65}$$

$$\text{trace}\,\mathcal{T}[\boldsymbol{\xi}] = 1. \tag{9.66}$$

The "trace" of a matrix, say, A, means the sum of the diagonal elements of A, that is,

$$\text{trace } A = A_{ii} \qquad \text{(summation over } i\text{).} \tag{9.67}$$

A useful property of the diagonal sum is that for any two matrices, say, A and B, we have

$$\text{trace } (AB) = \text{trace } (BA) = A_{ij}B_{ji} \qquad \text{(summation over } i \text{ and } j\text{)} \tag{9.68}$$

even though AB as a matrix may or may not be equal to BA as a matrix. The idempotent law (9.65) also holds, of course, as a matrix relation and can be proven with the help of the definition (9.64).

If every vector is transformed as

$$x \to Tx, \tag{9.69}$$

as in (9.51), $\mathcal{T}[\boldsymbol{\xi}]$ obviously will be transformed as

$$\mathcal{T}[\boldsymbol{\xi}] \to T\mathcal{T}[\boldsymbol{\xi}]T^{-1}. \tag{9.70}$$

It is very easy to show that the three basic properties (9.59), (9.65), and (9.66) remain unchanged by the transformation (9.70). In this proof the relations (9.55) and (9.68) have to be used.

The foregoing discussion pertains to the projection of $\mathbf{x}$ on the direction $\boldsymbol{\xi}$, that is, on the one-dimensional subspace defined by $\boldsymbol{\xi}$. We have to extend this concept to that of projection on a more-than-one-dimensional subspace. As the first step we consider the projection of a vector on a two-dimensional subspace. If we are given two nonzero vectors, $\boldsymbol{\xi}$ and $\boldsymbol{\eta}$ ($\boldsymbol{\xi} \neq 0$ and $\boldsymbol{\eta} \neq 0$) such that we cannot satisfy a relation

$$a\boldsymbol{\xi} + b\boldsymbol{\eta} = 0 \tag{9.71}$$

by any real numbers a and b except for $a = b = 0$, we say that $\boldsymbol{\xi}$ and $\boldsymbol{\eta}$ are "linearly independent." In the present simple case linear independence simply means that the two vectors have different directions. (The terminology here is such that two antiparallel vectors are also in the "same" direction.) Then the set of all vectors $\mathbf{x}$ that can be written as

$$\mathbf{x} = a\boldsymbol{\xi} + b\boldsymbol{\eta} \tag{9.72}$$

with arbitrary a and b forms a two-dimensional subspace. We can take two orthogonal coordinate vectors in such a subspace. Let us take $\boldsymbol{\xi}_1 = \boldsymbol{\xi}/|\boldsymbol{\xi}|$ as one of the coordinates. Then, because of (9.60), $\boldsymbol{\omega}$, defined by

$$\boldsymbol{\eta} = \mathcal{T}[\boldsymbol{\xi}]\boldsymbol{\eta} + \boldsymbol{\omega}, \tag{9.73}$$

is orthogonal to $\boldsymbol{\xi}$ (i.e., $\boldsymbol{\xi}_1$). Hence we can take as a second coordinate vector $\boldsymbol{\xi}_2 = \boldsymbol{\omega}/|\boldsymbol{\omega}|$. Then any vector $\mathbf{x}$ expressed as in (9.72) can be written as

$$\mathbf{x} = a'\boldsymbol{\xi}_1 + b'\boldsymbol{\xi}_2. \tag{9.74}$$

Conversely, any vector written as (9.74) can be written as (9.72). Therefore two spaces, one defined by (9.72) and the other by (9.74), are identical. Now we can define the projection operator corresponding to this subspace by

$$\mathfrak{P}[\mathfrak{M}] = \mathfrak{P}[\boldsymbol{\xi}_1] + \mathfrak{P}[\boldsymbol{\xi}_2]. \tag{9.75}$$

This definition seems to depend on the choice of the orthogonal coordinates $\boldsymbol{\xi}_1$ and $\boldsymbol{\xi}_2$ taken arbitrarily in the subspace, but it becomes clear by the following consideration that the operator (9.75) is actually determined by the subspace itself.

Suppose that we take n orthogonal coordinates $\mathbf{e}_i$ in such a way that two of them, say, $\mathbf{e}_1$ and $\mathbf{e}_2$, coincide with $\boldsymbol{\xi}_1$ and $\boldsymbol{\xi}_2$. In this particular coordinate system the matrix corresponding to $\mathfrak{P}[\mathfrak{M}]$ will, according to (9.64), have the form

$$\begin{array}{c} \\ 1 \\ 2 \\ 3 \\ \vdots \\ n \end{array} \begin{array}{c} \begin{array}{cccccc} 1 & 2 & 3 & \cdots & n \end{array} \\ \left(\begin{array}{cc|c} 1 & 0 & \\ 0 & 1 & \mathbf{0} \\ \hline & & \\ \multicolumn{2}{c|}{\mathbf{0}} & \mathbf{0} \end{array}\right) \end{array}, \tag{9.76}$$

where $\mathbf{0}$ stands for a rectangular or a square matrix with 0's as its elements. If we now apply an arbitrary orthogonal transformation T of (9.51), $\mathfrak{P}[\mathfrak{M}]$ will lose the particular form (9.76). However, if the transformation T that brings $\mathbf{e}_i$ to $\mathbf{e}_i'$ is such that the subspace defined by $\{\mathbf{e}_1, \mathbf{e}_2\}$ and the subspace defined by $\{\mathbf{e}_3, \mathbf{e}_4, \cdots, \mathbf{e}_n\}$ are transformed separately into themselves, T will have the form

$$\begin{array}{c} \\ 1 \\ 2 \\ 3 \\ \vdots \\ n \end{array} \begin{array}{c} \begin{array}{cccccc} 1 & 2 & 3 & \cdots & n \end{array} \\ \left(\begin{array}{c|c} T_{(1)} & \mathbf{0} \\ \hline \mathbf{0} & T_{(2)} \end{array}\right) \end{array}, \tag{9.77}$$

where $T_{(1)}$ is the well-known orthogonal transformation in a plane, which can be written as

$$\begin{pmatrix} \cos\theta & \sin\theta \\ -\sin\theta & \cos\theta \end{pmatrix}. \tag{9.78}$$

If we apply the transformation T of the form (9.77) on the matrix (9.76), the latter will remain unchanged. Since this original form (9.76) is always the same irrespective of the initial choice of $\boldsymbol{\xi}_1$ and $\boldsymbol{\xi}_2$ and remains unchanged by the transformation T, we can conclude that it is a property of the subspace independent of the choice of $\boldsymbol{\xi}_1$ and $\boldsymbol{\xi}_2$.

The above consideration refers to a two-dimensional subspace. Let us next consider an m-dimensional subspace, $m \leq n$. An m-dimensional subspace is defined as a set of linear combinations

$$\mathbf{x} = \sum_{i=1}^{m} a_i \boldsymbol{\eta}^{(i)} \tag{9.79}$$

with arbitrary real coefficients a_i $(i = 1, 2, \cdots, m)$, where the m vectors $\boldsymbol{\eta}^{(i)}$ are linearly independent, that is, the $\boldsymbol{\eta}^{(i)}$ are such that a relation of the form

$$\sum_{i=1}^{m} b_i \boldsymbol{\eta}^{(i)} = 0 \tag{9.80}$$

cannot be satisfied by any real b's except $b_1 = b_2 = \cdots = b_m = 0$. We can take m orthogonal coordinates within this subspace, in such a way that any linear combination of the $\boldsymbol{\eta}$'s (9.79) can be expressed as a linear combination of these coordinates and vice versa. One way of defining such m coordinates is as follows. First take $\boldsymbol{\eta}^{(1)}$ and $\boldsymbol{\eta}^{(2)}$, define two orthogonal coordinates $\boldsymbol{\xi}_1$ and $\boldsymbol{\xi}_2$ by the method explained before, and denote the projection operator defined in (9.75) by $\mathcal{T}[\mathfrak{M}^{(2)}]$. Take the third vector $\boldsymbol{\eta}^{(3)}$ and define $\boldsymbol{\omega}^{(3)}$ by

$$\boldsymbol{\eta}^{(3)} = \mathcal{T}[\mathfrak{M}^{(2)}]\boldsymbol{\eta}^{(3)} + \boldsymbol{\omega}^{(3)}. \tag{9.81}$$

If we apply $\mathcal{T}[\boldsymbol{\xi}_1]$ on (9.81) from the left, we obtain $\mathcal{T}[\boldsymbol{\xi}_1]\boldsymbol{\omega}^{(3)} = 0$ since $\mathcal{T}[\boldsymbol{\xi}_1]\,\mathcal{T}[\mathfrak{M}^{(2)}] = \mathcal{T}[\boldsymbol{\xi}_1]$, which follows directly from (9.75) because $\boldsymbol{\xi}_1 \perp \boldsymbol{\xi}_2$. This means that $\boldsymbol{\omega}^{(3)} \perp \boldsymbol{\xi}_1$. Similarly, $\boldsymbol{\omega}^{(3)} \perp \boldsymbol{\xi}_2$. Further $\boldsymbol{\omega}^{(3)}$'s cannot be 0, for if $\boldsymbol{\omega}^{(3)}$ were 0 it would imply that $\boldsymbol{\eta}^{(3)}$ can be expressed as a linear combination of $\boldsymbol{\xi}_1$ and $\boldsymbol{\xi}_2$ and therefore as a linear combination of $\boldsymbol{\eta}^{(1)}$ and $\boldsymbol{\eta}^{(2)}$, which contradicts the assumption that the $\boldsymbol{\eta}$'s are linearly independent. Hence we can introduce a third coordinate $\boldsymbol{\xi}_3$ by $\boldsymbol{\xi}_3 = \boldsymbol{\omega}_3/|\omega_3|$, so that $\boldsymbol{\xi}_1$, $\boldsymbol{\xi}_2$, and $\boldsymbol{\xi}_3$ are mutually orthogonal (and normalized). We can repeat this procedure until finally we obtain m orthogonal and normalized coordinates $\boldsymbol{\xi}_i$ $(i = 1, 2, \cdots, m)$. It is easy to show that the set of linear combinations of the $\boldsymbol{\xi}$'s and that of the $\boldsymbol{\eta}$'s are identical. We can define the projection operator corresponding to this subspace $\mathfrak{M}$ by

$$\mathcal{T}[\mathfrak{M}] = \sum_{i=1}^{m} \mathcal{T}[\boldsymbol{\xi}_i]. \tag{9.82}$$

It is also easy to show that this operator is uniquely determined by the subspace $\mathfrak{M}$ itself. In the coordinate system of which m coordinates subtend the subspace $\mathfrak{M}$, the matrix $\mathcal{T}[\mathfrak{M}]$ becomes a diagonal, having m diagonal elements equal to unity corresponding to these m coordinates and all other elements equal to zero. This form remains unchanged insofar as the transformation brings the subspace $\mathfrak{M}$ into itself. By a more general transformation $\mathcal{T}[\mathfrak{M}]$ becomes a nondiagonal symmetrical matrix.

Using the definition (9.82) and the properties of the one-dimensional projection operator $\mathcal{T}[\boldsymbol{\xi}_i]$, the reader can easily check

$$\mathcal{T}[\mathfrak{M}]\,\mathcal{T}[\mathfrak{M}] = \mathcal{T}[\mathfrak{M}], \tag{9.83}$$

$$(\mathcal{T}[\mathfrak{M}])^T = \mathcal{T}[\mathfrak{M}], \tag{9.84}$$

and

$$\text{trace}\ \mathcal{T}[\mathfrak{M}] = m. \tag{9.85}$$

If we write

$$\mathbf{x} = \mathcal{T}[\mathfrak{M}]\mathbf{x} + \mathbf{y} \tag{9.86}$$

for any given $\mathbf{x}$, $\mathbf{y}$ is either 0 (meaning that $\mathbf{x}$ lies in $\mathfrak{M}$) or a vector perpendicular to all vectors of $\mathfrak{M}$, that is, perpendicular to $\mathfrak{M}$. The vector $\mathcal{T}[\mathfrak{M}]\mathbf{x}$ is always a vector in $\mathfrak{M}$ no matter where $\mathbf{x}$ lies.

If we are given two subspaces $\mathfrak{M}_1$ and $\mathfrak{M}_2$ such that $\mathbf{x} \in \mathfrak{M}_1$ implies $\mathbf{x} \in \mathfrak{M}_2$ (i.e., $\mathfrak{M}_1 \subset \mathfrak{M}_2$ corresponding to $A_1 \to A_2$), it is obvious that

$$\mathcal{T}[\mathfrak{M}_1]\,\mathcal{T}[\mathfrak{M}_2] = \mathcal{T}[\mathfrak{M}_2]\,\mathcal{T}[\mathfrak{M}_1] = \mathcal{T}[\mathfrak{M}_1] \quad (\mathfrak{M}_1 \subset \mathfrak{M}_2), \tag{9.87}$$

since we can take m_2 coordinates in $\mathfrak{M}_2$ in such a way that m_1 ($\leq m_2$) of them are the coordinates of $\mathfrak{M}_1$. If $\mathfrak{M}_1$ and $\mathfrak{M}_2$ are such that $\mathbf{x}_1 \in \mathfrak{M}_1$ and $\mathbf{x}_2 \in \mathfrak{M}_2$ imply $\mathbf{x}_1 \perp \mathbf{x}_2$, that is, if $\mathfrak{M}_1 \perp \mathfrak{M}_2$, it is evident that

$$\mathcal{T}[\mathfrak{M}_1]\,\mathcal{T}[\mathfrak{M}_2] = \mathcal{T}[\mathfrak{M}_2]\,\mathcal{T}[\mathfrak{M}_1] = 0 \quad (\mathfrak{M}_1 \perp \mathfrak{M}_2), \tag{9.88}$$

for we can take the m_1 coordinates in $\mathfrak{M}_1$ and m_2 coordinates in $\mathfrak{M}_2$ in such a way that they constitute part of the n orthogonal coordinates. We have seen that the compatibility $A_1 \sim A_2$ is equivalent to the condition that $\mathfrak{M}(A_1) = \mathfrak{M}_1$ and $\mathfrak{M}(A_2) = \mathfrak{M}_2$ belong to the same family of orthogonal subspaces. This last condition means that there exists a system of n orthogonal coordinates such that $\mathfrak{M}_1$ is defined by a certain group of them and $\mathfrak{M}_2$ is defined by another group. If we write

$$\mathcal{T}[\mathfrak{M}_1] = \sum_{i \in \mathfrak{M}_1} \mathcal{T}[\boldsymbol{\xi}_i] \quad \text{and} \quad \mathcal{T}[\mathfrak{M}_2] = \sum_{i \in \mathfrak{M}_2} \mathcal{T}[\boldsymbol{\xi}_i],$$

the product $\mathcal{T}[\mathfrak{M}_1]\,\mathcal{T}[\mathfrak{M}_2]$ as well as $\mathcal{T}[\mathfrak{M}_2]\,\mathcal{T}[\mathfrak{M}_1]$ will become the sum of $\mathcal{T}[\boldsymbol{\xi}_i]$ such that $\boldsymbol{\xi}_i$ is contained in both $\mathfrak{M}_1$ and $\mathfrak{M}_2$. Hence we have

$$\mathcal{T}[\mathfrak{M}_1]\,\mathcal{T}[\mathfrak{M}_2] = \mathcal{T}[\mathfrak{M}_2]\,\mathcal{T}[\mathfrak{M}_1] = \mathcal{T}[\mathfrak{M}_3], \qquad A_1 \sim A_2, \tag{9.89}$$

where $\mathfrak{M}_3$ is the space defined by the coordinates common to $\mathfrak{M}_1$ and $\mathfrak{M}_2$ and obviously corresponds to the proposition $A_1 \cap A_2$. We should not forget that we are assuming $A_1 \sim A_2$. An important conclusion we can draw is that $A_1 \sim A_2$ implies the commutativity of $\mathcal{T}[\mathfrak{M}_1]$ and $\mathcal{T}[\mathfrak{M}_2]$:

$$\mathcal{T}[\mathfrak{M}_1]\,\mathcal{T}[\mathfrak{M}_2] - \mathcal{T}[\mathfrak{M}_2]\,\mathcal{T}[\mathfrak{M}_1] = 0 \quad \text{if} \quad A_1 \sim A_2. \tag{9.90}$$

Obviously (9.87) and (9.88) are special cases of (9.89), since $\mathfrak{M}_1 \subset \mathfrak{M}_2$ ($A_1 \rightarrow A_2$) as well as $\mathfrak{M}_1 \perp \mathfrak{M}_2$ ($A_1 \cap A_2 = \emptyset$ and $A_1 \sim A_2$) implies $A_1 \sim A_2$. It should be noted that commutativity is invariant for any orthogonal transformation (actually for any nonsingular linear transformation). This we can see by noticing

$$(T\mathcal{T}_1T^{-1})(T\mathcal{T}_2T^{-1}) - (T\mathcal{T}_2T^{-1})(T\mathcal{T}_1T^{-1}) = T(\mathcal{T}_1\mathcal{T}_2 - \mathcal{T}_2\mathcal{T}_1)T^{-1} = TOT^{-1} = 0 \tag{9.91}$$

where $\mathcal{T}_1 = \mathcal{T}[\mathfrak{M}_1]$ and $\mathcal{T}_2 = \mathcal{T}[\mathfrak{M}_2]$.

The next step is to demonstrate the converse of (9.90), namely, that if $\mathcal{T}[\mathfrak{M}_1]$ and $\mathcal{T}[\mathfrak{M}_2]$ commute with each other, $\mathfrak{M}_1$ and $\mathfrak{M}_2$ belong to a family of orthogonal subspaces; hence $A_1 \sim A_2$. The proof is given in Appendix A9.1. Thus we have established that compatibility of two propositions A_1 and A_2, membership of $\mathfrak{M}(A_1)$ and $\mathfrak{M}(A_2)$ in the same family of orthogonal subspaces, and commutativity of $\mathcal{T}[\mathfrak{M}(A_1)]$ and $\mathcal{T}[\mathfrak{M}(A_2)]$ are equivalent to one another.

Since the projection operator corresponds one to one to a subspace and hence to a proposition, the projection operator corresponding to the negation $\neg A$ of a proposition A is also uniquely determined. From (9.43)

$$\mathcal{T}[\mathfrak{M}(\neg A)] = I - \mathcal{T}[\mathfrak{M}(A)], \tag{9.92}$$

where I is an $n \times n$ diagonal matrix having only 1's on the diagonal. In their diagonal form (they can be simultaneously diagonalized since $A \sim \neg A$), wherever $\mathfrak{M}(A)$ has 1 on the diagonal $\mathfrak{M}(\neg A)$ has 0, and wherever $\mathfrak{M}(A)$ has 0 on the diagonal $\mathfrak{M}(\neg A)$ has 1. The constantly true proposition $\Box$ corresponds to the entire space and the constantly false proposition $\emptyset$ corresponds to an empty set of vectors; hence

$$\mathcal{T}[\mathfrak{M}(\Box)] = I \quad \text{and} \quad \mathcal{T}[\mathfrak{M}(\emptyset)] = 0. \tag{9.93}$$

When two propositions A_1 and A_2 are compatible, we have already seen [in a remark below (9.89)] that the product of $\mathcal{T}[\mathfrak{M}(A_1)]$ and $\mathcal{T}[\mathfrak{M}(A_2)]$ is equal to $\mathcal{T}[\mathfrak{M}(A_1 \cap A_2)]$, where the last matrix will have 1's where, and only where, both $\mathcal{T}[\mathfrak{M}(A_1)]$ and $\mathcal{T}[\mathfrak{M}(A_2)]$ have 1's when these matrices are simultaneously diagonalized. Using (9.92) and De Morgan's law, we obtain

$$\begin{aligned}\mathcal{T}[\mathfrak{M}(A_1 \cup A_2)] &= I - \mathcal{T}[\mathfrak{M}(\neg A_1 \cap \neg A_2)]\\ &= I - \mathcal{T}[\mathfrak{M}(\neg A_1)]\mathcal{T}[\mathfrak{M}(\neg A_2)]\\ &= I - (I - \mathcal{T}[\mathfrak{M}(A_1)])(I - \mathcal{T}[\mathfrak{M}(A_2)])\\ &= \mathcal{T}[\mathfrak{M}(A_1)] + \mathcal{T}[\mathfrak{M}(A_2)] - \mathcal{T}[\mathfrak{M}(A_1 \cap A_2)] \quad \text{if } A_1 \sim A_2,\end{aligned} \tag{9.94}$$

using the fact that if $A_1 \sim A_2$ then $\urcorner A_1 \sim \urcorner A_2$. In the diagonal form (9.94) means simply that $\mathcal{T}[\mathfrak{M}(A_1 \cup A_2)]$ has 1's where, and only where, $\mathcal{T}[\mathfrak{M}(A_1)]$ or $\mathcal{T}[\mathfrak{M}(A_2)]$ or both have 1's. It is very interesting to see that (9.93) and (9.94) formally satisfy the basic axioms of probability functions of propositions. The only difference is that these relations use matrix relations instead of numerical relations. The only missing relation to complete the axioms of probability is $\mathcal{T}[\mathfrak{M}(A)] \geq 0$. This last relation also holds in this case in the sense that the eigenvalues (i.e., the values of the diagonal elements when the matrix is diagonalized) are 0 or 1, hence ≥ 0. It should not be forgotten, however, that (9.94) has been proven for $A_1 \sim A_2$. In fact, in the general case we cannot prove

$$\mathcal{T}[\mathfrak{M}(A_1)] + \mathcal{T}[\mathfrak{M}(A_2)] = \mathcal{T}[\mathfrak{M}(A_1 \cup A_2)] + \mathcal{T}[\mathfrak{M}(A_1 \cap A_2)] \quad (9.95)$$

but we can prove

$$\text{trace}\,\mathcal{T}[\mathfrak{M}(A_1)] + \text{trace}\,\mathcal{T}[\mathfrak{M}(A_2)] = \text{trace}\,\mathcal{T}[\mathfrak{M}(A_1 \cap A_2)] + \text{trace}\,\mathcal{T}[\mathfrak{M}(A_1 \cup A_2)]. \quad (9.96)$$

satisfying (9.1). Of course, when we have (9.95), we can derive (9.96) from it, but we cannot go the other way around. The proof is given in Appendix A9.2.†

There still remains one law of modular logic that we have not established in our geometric model, namely, the modular law, which states that if $A_1 \rightarrow A_3$ then $A_1 \cup (A_2 \cap A_3) = (A_1 \cup A_2) \cap A_3$. Introducing three subspaces $\mathfrak{M}_1 = \mathfrak{M}(A_1)$, $\mathfrak{M}_2 = \mathfrak{M}(A_2)$, and $\mathfrak{M}_3 = \mathfrak{M}(A_3)$ such that $\mathfrak{M}_1 \subset \mathfrak{M}_3$, we could demonstrate $\mathfrak{M}[(A_1 \cup (A_2 \cap A_3)] = \mathfrak{M}[(A_1 \cup A_2) \cap A_3]$ or, equivalently, $\mathcal{T}[\mathfrak{M}\{A_1 \cup (A_2 \cap A_3)\}] = \mathcal{T}[\mathfrak{M}\{(A_1 \cup A_2) \cap A_3\}]$ by a method similar to the one we use in Appendix A9.2. A still simpler method of proving the modular law in the geometric model is to resort to Theorem 9.1 and show that in fact (9.1) and (9.2) hold in the geometric model, because we have already established that the subspaces obey the laws of a general complemented lattice. To carry out this program it suffices to show that to the geometric entity $\mathfrak{M}(X)$ or $\mathcal{T}[\mathfrak{M}(X)]$ corresponding to a proposition X can be assigned a numerical value $w[\mathfrak{M}(X)]$ or $w[\mathcal{T}[\mathfrak{M}(X)]]$ such that the two relations (9.1) and (9.2) hold. However, (9.96), which we have just introduced and is proven in Appendix A9.2, interpreted as (9.1) by identifying

$$w(X) = \text{trace}\,\mathcal{T}[\mathfrak{M}(X)]. \quad (9.97)$$

All that remains to be proven is that if $X \rightarrow Y$ and $X \neq Y$ then $w(X) < w(Y)$, which becomes in the geometric model

$$\text{if} \quad \mathfrak{M}(X) \subset \mathfrak{M}(Y) \quad \text{and} \quad \mathfrak{M}(X) \neq \mathfrak{M}(Y)$$
$$\text{then} \quad \text{trace}\,\mathcal{T}[\mathfrak{M}(X)] < \text{trace}\,\mathcal{T}[\mathfrak{M}(Y)]. \quad (9.98)$$

This is obvious because the premise means that $\mathfrak{M}(X)$ is a *proper* subspace of $\mathfrak{M}(Y)$, entailing that the number of dimensions of the subspace $\mathfrak{M}(X)$ is

† See S-63 of the next section and the sequel for more about the probability in a non-distributive lattice.

less than the number of dimensions of the subspace $\mathfrak{M}(Y)$, because if they were equal $\mathfrak{M}(X)$ would become identical with $\mathfrak{M}(Y)$. This result is exactly the conclusion of (9.98), for trace $\mathcal{T}[\mathfrak{M}]$ is the number of dimensions of $\mathfrak{M}$. This completes the proof for the modular law in the geometric model.

If we want to make $w(X)$ satisfy all the axioms of probability, we need to put

$$w(X) = \frac{\text{trace } \mathcal{T}[\mathfrak{M}(X)]}{n}, \tag{9.99}$$

where n stands for trace $\mathcal{T}[\mathfrak{M}(\square)]$. This guarantees that $w(\square) = 1$. We have to be careful, however, because we cannot define a conditional probability. Furthermore, we should note that this probability $w(X)$ of a proposition depends *only* on X, which is not realistic if the present formalism is to be applied to an empirical science. We usually have some preliminary information about the system under scrutiny. Hence a genuine probability of a proposition must be some kind of conditional probability that depends not only on X but also on other elements, such as the result of an observation previously performed on the system. For this reason the $w(X)$ of (9.99) should be called pure a priori probability and it is not sufficiently versatile a tool to express the probabilistic situation in an empirical science. We discuss this problem in later sections.

Figure 9.4 is an illustration showing that (9.95) as well as (9.89) breaks down when A_1 and A_2 are not compatible, but that (9.96) still holds. In a plane $\mathfrak{M}_1$ is assumed to be the x-axis while $\mathfrak{M}_2$ is supposed to be a vector coinciding with neither the x-axis nor the y-axis. $\mathfrak{M}_{1\cap 2} = \emptyset$ and $\mathfrak{M}_{1\cup 2} = \square$, which is the xy-plane. Hence, if we apply the operator of the right side of

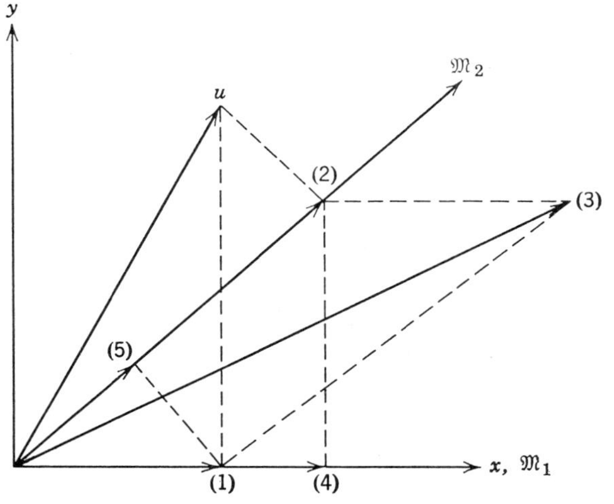

Figure 9.4 $\mathcal{T}[\mathfrak{M}_1]\mathbf{u} = (1)$, $\mathcal{T}[\mathfrak{M}_2]\mathbf{u} = (2)$, $\mathcal{T}[\mathfrak{M}_{1\cap 2}]\mathbf{u} = 0$, $\mathcal{T}[\mathfrak{M}_{1\cup 2}]\mathbf{u} = \mathbf{u}$. $\mathcal{T}[\mathfrak{M}_1] \cdot \mathcal{T}[\mathfrak{M}_2]\mathbf{u} = (4)$, $\mathcal{T}[\mathfrak{M}_2]\mathcal{T}[\mathfrak{M}_1]\mathbf{u} = (5)$, $(\mathcal{T}[\mathfrak{M}_1] + \mathcal{T}[\mathfrak{M}_2])\mathbf{u} = (3)$.

(9.95) on any arbitrary vector **u**, the result will be **u** again. If we apply the operator of the left side of (9.95), however, the first term will produce a vector in the direction of $\mathfrak{M}_1$ and the second term will produce a vector in the direction of $\mathfrak{M}_2$. The vector sum of these two will not be **u** unless $\mathfrak{M}_1$ and $\mathfrak{M}_2$ are perpendicular (unless $A_1 \sim A_2$). Thus (9.95) breaks down in the general case. Furthermore, $\mathfrak{T}[\mathfrak{M}_{1\cap 2}]\mathbf{u} = 0$ for $\mathfrak{M}_{1\cap 2} = \emptyset$, whereas $\mathfrak{T}[\mathfrak{M}_1]\,\mathfrak{T}[\mathfrak{M}_2]\mathbf{u}$ is a vector along the x-axis and $\mathfrak{T}[\mathfrak{M}_2]\,\mathfrak{T}[\mathfrak{M}_1]\mathbf{u}$ is along the direction of $\mathfrak{M}_2$; hence these three vectors are all different. Thus the three projection operators in (9.89) are all different. Finally, trace $\mathfrak{T}[\mathfrak{M}_1]$ = trace $\mathfrak{T}[\mathfrak{M}_2] = 1$, trace $\mathfrak{T}[\mathfrak{M}_2] = 1$, trace $\mathfrak{T}[\mathfrak{M}_{1\cup 2}] = 2$, and trace $\mathfrak{T}[\mathfrak{M}_{1\cap 2}] = 0$; therefore (9.96) still holds in this case.

9.3. EMPIRICAL BACKGROUND FOR NON-BOOLEAN LOGIC

It was made reasonably plausible in Section 9.1 that modular logic may be the first natural step to take if one embarks on a quest for a framework of thinking broader than the usual distributive or Boolean logic. But in Chapter 7 it was pointed out that the distributive law is deeply rooted in the set-theoretical classification of objects (postulate of fixed truth set). The moment we interpret an observational proposition (see Sections 4.1 and 4.2) such as "this is red" as meaning that this object is a member of the set of red objects, we are already trapped in Boolean logic. Is there, then, any alternative interpretation for such an observational proposition? Indeed there is, and this alternative seems to be a more basic one, from which the usual set-theoretical interpretation is derived on an assumption that is not always justified. (See also Section 7.1).

Obviously, as shown in Chapter 4, the immediate content of the observational proposition "this is red" is that "if we make a redness test on this object, the observational result will be affirmative." This does not necessarily mean that there are two fixed boxes such that each colored object has to be placed in one or the other. (We are not talking about the continuous gradation between red and nonred, which can be overcome by some convenient agreement). It is quite possible that an object by itself has no fixed color, but haphazardly sometimes appears red and sometimes nonred in such a way that the selection between the two cannot be determined by any other factor.† Under such circumstances an observational proposition that makes perfect sense in its primary meaning cannot be interpreted in any set-theoretical manner. As we shall see, the precaution of distinguishing between the primary and secondary meanings of an observational proposition pays off

† This might suggest a breakdown of the law of the excluded middle, but the law is upheld in the nondistributive logic we study. For an intuitive understanding of this fact, it may help the reader to recall the broader meaning of disjunction. For an interesting remark by Birkoff and von Neumann on the law of the excluded middle, see [B-2]. In the type of logic we study, $A \cup B = \square$ and $A \cap B = \emptyset$ are equivalent to $A = \neg B$, only if $A \sim B$.

in physics. The reader may speculate that a similar idea may be true for propositions with psychological connotations, such as "he is a good man" or "he is happy." In fact, a man is usually neither good nor bad, and a man can be both happy and sad at the same time; only his observed behavior can be good and he may behave like a man who is completely happy. In any event, there is good reason to doubt the universal validity of the distributive law, and, combined with this doubt, the argument of Section 9.1 encourages us to investigate modular logic as the basis on which to rebuild the whole structure of our logical and probabilistic theories.† If this is the right orientation, the first thing to try is to assume that an observational proposition is represented by an element in a modular lattice, hence by a projection operator in an orthogonal space. This is exactly what has happened in quantum mechanics through a long detour of theoretical groping. In the case of atomic physics, of course there is a more direct, compelling experimental reason for adopting a non-Boolean logic. The aim of this and the next sections is to present the theoretical skeleton of quantum mechanics from the point of view of a logical analysis of the observational propositions involved in atomic phenomena. We shall start from a study of the nature of probabilistic predictive propositions ("probabilistic experiential propositions" of Chapter 4) and derive therefrom the properties of observational propositions, showing that these propositions form a modular lattice.‡ This will naturally lead us to the use of projection operators as a basic tool. After this point the basic theorems of quantum mechanics will be stated in terms of projection operators, not always with sufficient justification. In a word, we try to sketch a possible way of constructing quantum mechanics on the basis of the logical structure of the propositions involved. In so doing we shall not attempt to make every theoretical postulate of quantum physics "necessary" on the ground of experimental facts. We shall be satisfied with pointing out a few simple experimental facts that support some of the more important postulates of the theory. After all, construction of a theory is an inductive process and not every postulate of it can be made logically necessary. All the main statements in the folloing are numbered for easy reference. This might

† For applications of the non-distributive logic to problems outside physics, see [W-16], [W-31] and [W-31a]. If the CLAFIC model of class (predicate) introduced in Chapter 8 were universally correct, then the modular logic would become a universal logic for all predicates, since in this model each predicate corresponds to a subspace. The fact is that this model is only approximately correct, in particular, in connection with the concept of disjunction, but it implies, on the other hand, non-universal validity of the distributive logic.

‡ Birkoff and von Neumann [B-2] derived "quantum logic" from the structure of the Hilbert space used in quantum mechanics. Husimi [H-18] tried to establish the axioms of modular logic directly on empirical facts. The present work takes a direction similar to Husimi's pioneer work, but our formulation is quite different from his, See [W-23] for a slightly different version of the present work.

give the appearance of an axiomatic system to our exposition but it is not meant to be one. Our exposition, as in many textbooks on quantum mechanics, is a process, subject to logical gaps, intended to make the theory reasonably acceptable. Some of the numbered statements are indeed "postulates" that are supported, if not proven, by empirical facts. But some of them are simply rules of the game whose acceptance is at the mercy of the individual's discretion. Some are just definitions; some are conclusions or summaries of antecedent statements. Some are just statements, in our terminologies, of the basic accepted theorems in quantum mechanics whose justification is far beyond the scope of this section. It must be expected from the beginning that a logical standpoint will not be able to yield the entire substance of an empirical science. In fact, all we can derive is an empty framework of quantum mechanics, in which we have to put more detailed empirical facts in order to make it become a science.

The following three statements seem to be almost inevitable for any theory that can serve as a tool of prediction, which is the main function of any science. They are in agreement with our consideration of Chapter 4.

S-1. An "observational proposition" has the form: if one performs a certain well-defined operation $\mathfrak{D}$ of observation on an object system, he obtains result D_i. (In general case there can be continuously many different D_i's, but we limit ourselves to the case in which this number is finite.) For simplicity we write $\mathfrak{D} = D_i$ for this proposition. We use the first letters of the alphabet, A, B, C, $\cdots$, to denote this type of proposition. The numerical value attached to each possibility does not matter in the following discussion. It is possible that an observational proposition, means a disjunction of the type $\mathfrak{D} = D_i$ or D_j, and also a negative proposition, such as it is not the case that $\mathfrak{D} = D_i$. These can be brought to the standard form by relabeling the experimental results. This $\mathfrak{D}$ corresponds to what was denoted by the same symbol in Chapter 4.

S-2. An object system is described by a proposition of the type: the object system is in state condition G_k. We use the last letters of the alphabet, $\cdots$, X, Y, Z, to denote this type of propositions.

S-3. A theory in physics is supposed to produce an "experiential proposition" of the type: the probability that A will be true on "condition" Z, $p(A \mid Z)$, has such-and-such. If A means $\mathfrak{D} = D_i$ and if Z corresponds to state condition G_k, there is probability $p(A \mid Z)$ of obtaining result D_i.†

It is very important to realize first that the "object system" in S-2 is not

† In the ordinary predicate calculus $A(a)$ is a proposition that object a satisfies predicate A. The assertion $A(a) = \square$ or $= \emptyset$ can be expressed by the characteristic functions $f(A \mid a) = 1$ or $= 0$. Our $p(A \mid Z)$ can be considered as a natural generalization to a continuous case of the characteristic function $f(A \mid a)$, Z playing the role of a. Our derivation of logic from the "probability" $p(A \mid Z)$ may be considered as a realization of the ideal mentioned in Chapter 7. See [W-31a].

necessarily a chunk of matter as in classical physics. It could be a certain portion of the vacuum, or a train of waves of some sort. Second, the "state condition" of a system is almost, but not entirely, synonymous with the "state" or "situation" of the system. If we used the word "state," we would be on the verge of falling into the pitfall of the postulate of a fixed truth set. A "state condition" is a collection of recorded observed facts about the system, hence serves as the identification of the system. It may be a collection of the past observational propositions proven to be true about the system. In some cases, however, there could be included in G_k a fact such as "the photon (the observed system) has been let go through such and such a slit." A record of this sort constitutes part of the preparation of the system, and is somewhat different in nature from a customary standard observational proposition, but by an appropriate broadening of definition it could be considered an observational proposition too. We discuss later, with more clarity, the nature of a Z or G_k. For the time being we should understand that G_k represents our existing "knowledge" about the system based on observation or operationally defined preparation. If the reader does not like the subjective flavor attached to the word "knowledge," he may interpret the "knowledge" as the "condition" to which the system fits. The reader should recall that a similar remark was made in Chapter 7 with regard to the notion of "condition" in conditional probability.

In the context of Chapter 4, the theory itself and the method of measurement must also be included in the condition of conditional probability, but we do not specifically mention this here, for we have in mind only one theory to construct, and the measurement method is implied by the proposition A. Although these three statements seem to be almost self-evident, the fact is that they, in particular S-1 and S-2, leave room for further investigation. Wigner [W-37] and Yanase [Y-2] have in fact shown that, except for a few special physical quantities, a "nondestructive" way of performing an observation of the type S-1 is possible only in an approximate sense, in contradiction to von Neumann's earlier conclusion [V-3]. By a nondestructive method is meant an observation after which the observed object system remains in an observable state, to which the observation just made is relevant. An example of a destructive observation is determination of the linear polarization of a photon by letting it pass through a Polaroid filter and letting it fall on a photomultiplier. The observation is done in conformity with S-1, but after the observation the photon is destroyed. Similarly, a measurement of the momentum of an election in a photographic emulsion under a magnetic field allows the election to be absorbed and to stay in a state that has nothing to do with the measured momentum. In practice it is sometimes difficult to perform a nondestructive observation corresponding to A, but for simplicity of argument we assume that there always exists a means of performing a nondestructive observation corresponding to A. Later we give an example in which a destructive observation is changed into a nondestructive one. It is understood that the object after a nondestructive observation A on state condition Z is in a certain state condition determined by Z and the result of A. This is because Z usually includes past observations on the system, and the observation A, after it has been done, becomes part of the past record of the system.

The reader may remember that we saw that conditional probability of the type (9.35) in the non-Boolean case cannot be manipulated as freely as probability in the usual sense. We shall return to this problem after we have studied more about the nature of the "conditional" probability $p(A \mid Z)$. For the time being it can be understood in terms of the relative frequency of A in systems that have the same record Z.

For each given observational proposition A we can think of a set $\mathfrak{Z}(A)$ of state conditions such that the conditional proposition "A on a system in Z" is meaningful if $Z \in \mathfrak{Z}(A)$. Similarly, for each state condition Z we can think of a set $\mathcal{A}(Z)$ of observational propositions such that the conditional proposition "A on a system in Z" is meaningful if $A \in \mathcal{A}(Z)$.

S-4. There exists a set $\mathcal{A}$ of observational propositions and a set $\mathfrak{Z}$ of state conditions such that if $A \in \mathcal{A}$ then $\mathfrak{Z}(A) = \mathfrak{Z}$ and if $Z \in \mathfrak{Z}$ then $\mathcal{A}(Z) = \mathcal{A}$. In the following, our universe of discourse refers to a pair, $\mathcal{A}$ and $\mathfrak{Z}$, that stand in such a mutual relationship.

S-5. Two observational propositions A and B (in $\mathcal{A}$) are said to be equivalent ($A = B$) if $p(A \mid Z) = p(B \mid Z)$ for all Z (in the corresponding $\mathfrak{Z}$).

S-6. Two state conditions X and Y (in $\mathfrak{Z}$) are said to be equivalent to each other if $p(A \mid X) = p(A \mid Y)$ for all A (in $\mathcal{A}$).

S-7. There is an element (or a group of equivalent elements) in $\mathcal{A}$, denoted by $\emptyset$, that satisfies $p(\emptyset \mid Z) = 0$ for all Z in $\mathcal{A}$.

S-8. There is an element (or a group of equivalent elements) in $\mathfrak{Z}$, denoted by $\square$, that satisfies $p(\square \mid Z) = 1$ for all Z in $\mathfrak{Z}$.

For instance, when we are talking about photons traveling in the z-direction, we can put a polaroid filter perpendicular to the z-direction and place a photomultiplier behind it. This would correspond to an destructive observation about the linear polarization of the photon. But if we use an opaque plate instead of the polaroid plate, the observation corresponds to $\emptyset$, for there is no chance of a photon being detected by the photomultiplier. On the other hand, if we place a transparent, nonpolarizing glass in front of the photomultiplier, or, for that matter, if we do not place anything in front of the photomultiplier, the observation will correspond to $\square$, for every time a photon arrives it will be detected.

S-9. Let A stand for observational proposition $\mathfrak{D} = D_i$ and B for $\mathfrak{F} = F_j$. Then we write BA (or $B \cdot A$) to express the proposition "if one performs the observational operation $\mathfrak{D}$ in a nondestructive way and immediately thereafter the operation $\mathfrak{F}$ in the same object system, one will obtain D_i and F_j, respectively."

This type of observation will be called a composite observation. The opposite concept is a simple observation that obeys the next rule, S-10. If no specification regarding "composite" or "simple" is made, an observation will mean

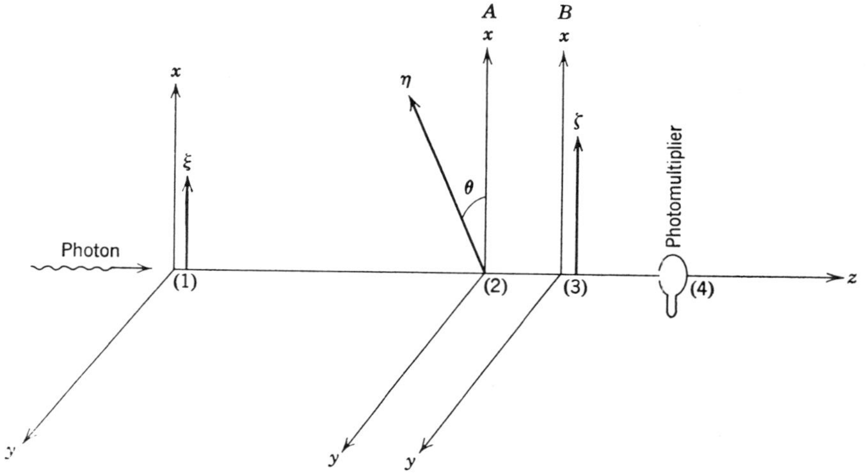

Figure 9.5 The directions ξ, η, and ζ are the directions of linear polarization caused by three polaroid plates placed at positions (1), (2), and (3). The intensity of light after (3) is $\cos^4 \theta$ times the intensity of light after (1).

a simple observation. If we repeat the same observation $\mathfrak{D}$, the result of the second observation is the same as the first result.

S-10. For any state condition Z and for any simple observational proposition A, $p(AA \mid Z) = p(A \mid Z)$. A composite observational proposition $C = BA$ may or may not satisfy the condition $p(CC \mid Z) = p(C \mid Z)$. For a simple but destructive observation A, the condition will also break down, but we exclude from now all destructive observations from consideration.

S-11. There are pairs (A, B) in $\mathcal{A}$ such that $p(AB \mid Z) = p(BA \mid Z)$ for all Z in $\mathfrak{Z}$. In this case A and B are said to be compatible and we write $A \sim B$.

The physical background for these last statements is as follows. Suppose that electromagnetic waves of visible frequency are traveling in the z-direction and go through a polarizer at position (1) (Figure 9.5), so that the waves are linearly polarized in the ξ-direction, which coincides with the x-direction. The electromagnetic waves or photons after this polarizer are considered as Z. The first observation A consists of determining at position (2) whether or not the photon is polarized in the η-direction in the xy-plane, which makes an angle θ with the x-direction. This can be done by placing the polaroid in the η-direction and placing a photomultiplier immediately after it. The second observation, B (without intervention of A), consists of determining at position (3) whether or not the photon is polarized in the ζ-direction, which coincides with the x-direction. This can be performed by placing a polaroid in the ζ-direction and a photomultiplier after it. Now the observation corresponding to BA can be performed by placing the polaroid in the η-direction at (2) and another polaroid in the ζ-direction at (3) and putting a photomultiplier at (4), where positions (2), (3), and (4) follow immediately after one another along the increasing direction of the z-axis. Elimination of the photomultiplier between (2) and (3) may be argued as follows. (This is a trick by which a destructive observation is changed to a non-destructive one.) Suppose that there exists an imaginary instrument that can detect passage of a photon between (2) and (3) without destroying the photon and without affecting its polarization. Then the case "BA = true" corresponds to the case in which the imaginary photon detector detects the photon and the photomultiplier (4) detects the photon, and the case "BA = false" corresponds necessarily to any one of the following three cases: (a) the imaginary photomultiplier detects no photon and the photomultiplier (4) detects a photon, (b) the imaginary photomultiplier detects a photon but the photomultiplier (4) detects no photon, or (c) the imaginary photomultiplier detects no photon and the photomultiplier (4) detects no photon. Among these three, however, (a) is excluded, for in order for a photon to exist after polarizer (3) it must have existed before (3). [An opaque shield is assumed to be provided so that no photon can bypass (2) before reaching (3).] This shows that "BA = true" and "BA = false" can be determined simply by the photomultiplier (4) alone.

According to Maxwell's electromagnetic theory, the intensity of light waves after the polarizer (3) is $\cos^4 \theta$ times the intensity of the light waves before the polarizer (2). In the particle picture this implies that $p(BA \mid Z) = \cos^4 \theta$. Now let us interchange positions (2) and (3), keeping the directions of the polaroids attached to them unchanged. This corresponds to observation AB. The electromagnetic theory tells that the intensity of the light after both polarizers is $\cos^2 \theta$ times the original intensity. This implies that $p(AB \mid Z) = \cos^2 \theta$. Hence $p(AB \mid Z) \neq p(BA \mid Z)$ except in the case $\theta = 0$ or $\pi/2$. This shows the existence of two observations that are not compatible. On the other hand, if $\theta = 0$ (or equivalently $\theta = \pi$), the two polarizers have the same direction of polarization and the effects of the two are equivalent to only one of them: $p(AB \mid Z) = p(BA \mid Z) = p(A \mid Z)$. In the particular Z we are considering, the values of these three probabilities are all equal to 1, but equality of these three probabilities (whatever their values) obviously holds for any kind of light traveling in the z-direction. Hence A and B are compatible. By noting that A and B are equivalent ($A = B$), we can rewrite the relation as $p(AA \mid Z) = p(A \mid Z)$. This is a special case of S-10. Next, if $\theta = \pi/2$ (or equivalently $\theta = 3\pi/2$), $p(AB \mid Z) = p(BA \mid Z) = 0$. This holds not only for the particular Z we are considering but also for any Z. Hence $AB = BA$. In the present simple example, compatibility $AB = BA$ reduced to $A = B$ or $AB = \emptyset$, but there are cases in which $AB = BA$ holds without $A = B$ or $AB = \emptyset$.

It is important to know that the combination of two arbitrary observations A and B is not a simple observational proposition, since $ABAB$ is not equivalent to AB, as the reader can easily check this for the case $\theta \neq 0$, $\theta \neq \pi/2$. If $A \sim B$ then $ABAB = AB$, and AB becomes equivalent to a simple observational proposition.

S-12. Among compatible pairs $A \sim B$ there are some that satisfy the relation $p(AB \mid Z) = p(BA \mid Z) = p(A \mid Z)$ for all Z. In this case we say that A implies B and write $A \rightarrow B$.

Roughly speaking, if $A \rightarrow B$ (and not $B \rightarrow A$), A is a more detailed observation than B. The reader can convince himself that according to this definition the proposition "the photon has polarization in the y-direction" implies the proposition "the photon has polarization in the xy-plane." (A vector meson can have its polarization in all three directions when it is moving in one direction, whereas a photon can have polarization only in a direction perpendicular to the motion.)

S-13. If, and only if, $A \rightarrow B$ and $B \rightarrow A$, then $A = B$. If $A \rightarrow B$ and $B \rightarrow C$ then $A \rightarrow C$.

From S-12 it follows that if $A \rightarrow B$ and $B \rightarrow A$ then $p(A \mid Z) = p(B \mid Z)$ for all Z, which in view of S-5 implies $A = B$. Conversely, if $A = B$, then, on account of S-10, we have $p(AB \mid Z) = p(BA \mid Z) = p(A \mid Z) = p(B \mid Z)$ for all Z, implying $A \rightarrow B$ and $B \rightarrow A$. The transitivity law of implication is proved if we can derive $A = AC = CA$ from $A = AB = BA$ and $B = BC = CB$. This can be done easily by noticing $AC = (AB)C = A(BC) = AB = A$ and $CA = C(BA) = (CB)A = BA = A$. In writing these equations we have used the associative law of the type $(AB)C = A(BC)$. This is obviously justifiable if we translate this formula in terms of experimental procedure. The inference $AC = (AB)C$ from $A = AB$ is allowed because of the assumption that the system in state Z enters, after any observation C, a certain state, say Z_1, that is uniquely determined by Z and the results of C.

S-14. For any arbitrary A and B and for all Z we have $p(BA \mid Z) \leq p(A \mid Z)$ but not necessarily $p(BA \mid Z) \leq p(B \mid Z)$.

By definition, $p(BA \mid Z)$ is the relative frequency of the joint event A on state condition Z, and then B on state condition Z' that ensues from observation of A on Z. The probability of the first component of the joint event is $p(A \mid Z)$. Hence $p(BA \mid Z) \leq p(A \mid Z)$. On the other hand, the second component of the joint event is not the same as the event $(B \mid Z)$, for state condition Z' after A may be different from state condition Z. If, in fact, in the experimental setup of Figure 9.5 the observation B uses the ζ-direction parallel to the y-axis, $p(BA \mid Z) = \cos^2 \theta \sin^2 \theta$ while $p(B \mid Z) = 0$.

S-15. $A \rightarrow B$ implies $p(A \mid Z) \leq p(B \mid Z)$ for all Z.

The proof is immediate, since, from S-12 and S-14, $p(A \mid Z) = p(BA \mid Z) = p(AB \mid Z) \leq p(B \mid Z)$.

Next, by the use of the definition of implication, we can define conjunction and disjunction following the general prescription (7.7)–(7.14). Thus we can state the following.

S-16. For given A and B let C be (i) such that $C \rightarrow A$ and $C \rightarrow B$, that is, $p(C \mid Z) = p(CA \mid Z) = p(AC \mid Z) = p(CB \mid Z) = p(BC \mid Z)$ for all Z, and (ii) such that $D \rightarrow C$, that is $p(D \mid Z) = p(DC \mid Z) = p(CD \mid Z)$ for all Z and for all those D that satisfy $D \rightarrow A$ and $D \rightarrow B$; that is, $p(D \mid Z) = p(DA \mid Z) = p(AD \mid Z) = p(DB \mid Z) = p(BD \mid Z)$ for all Z. Then C is called conjunction of A and B and denoted as $C = A \cap B$. For given A and B there exists C. (If C exists it is unique because of its definition, but its existence is nontrivial.)

S-17. For given A and B let C be (i) such that $p(A \mid Z) = p(CA \mid Z) = p(AC \mid Z)$ for all Z and $p(B \mid Z) = p(CB \mid Z) = p(BC \mid Z)$ and (ii) such that $p(C \mid Z) = p(DC \mid Z) = p(CD \mid Z)$ for all Z and for all those D that satisfy $p(A \mid Z) = p(DA \mid Z) = p(AD \mid Z)$ for all Z and $p(B \mid Z) = p(DB \mid Z) = p(BD \mid Z)$ for all Z. Then C is called disjunction of A and B and denoted as $C = A \cup B$. For a given A and B there exists C.

The existence of a unique conjunction and a unique disjunction in S-16 and S-17 must be supported by experiments. Applying A and B of Figure 9.5 to S-16, the reader will find that $C = A \cap B$ must be $\emptyset$ except when $\theta = 0$, which entails $A = B = C$ because there is no other observation that would imply both A and B. Similarly, $A \cup B$ can be nothing but $\square$ for the photon traveling in the z-direction, for the only observational proposition implied by both A and B is the statement that the polarization is in the xy-plane. In the case of conjunction it should be noted that in general $p(AB \mid Z) \neq 0$ and $p(BA \mid Z) \neq 0$, while $p(A \cap B \mid Z) = 0$, implying $A \cap B \neq AB$ and $A \cap B \neq BA$ except in the cases $\theta = 0$ and $\theta = \pi/2$. This is a special case of S-18, which follows. In the case of $A \cup B$ the reader will note that $A \cup B$ includes (is implied by) not only all linear polarizations in the xy-plane, but also all the elliptical polarizations in the xy-plane. It may be noted that since the definitions S-16 and S-17 are in agreement with the general definitions (7.7)–(7.14), all the laws of lattice (7.32)–(7.35) ensue from them. See the end of this section for the experimental meaning of the conjunction $A \cap B$ in the general case.

S-18. If $A \sim B$ then $A \cap B = AB = BA$.

First, we should note that since $A \sim B$, the compound observation $X = AB$ is equivalent to a simple observation in the sense of S-10. Next, since $C = A \cap B \rightarrow A$ and $C \rightarrow B$, $C \sim A$ and $C \sim B$. In order to show that $X = C$ it suffices to show two things. First, $X \rightarrow A$ and $X \rightarrow B$. This can be shown by noticing that the equations of (i) of S-16 are satisfied when C there is replaced by X because $A \sim B$. By the definition of $C = A \cap B$, C is implied by any simple observation that imples both A and B: $X \rightarrow C$. Second, we should show $C \rightarrow X$. Since $C \rightarrow A$, we have $p(C \mid Z) = p(CA \mid Z)$ for all Z. Let Z be the state condition right after an observation of B in Z. Then $p(CB \mid Z) = p(CAB \mid Z)$ for all Z. Since $C \rightarrow B$, we have $p(C \mid Z) = p(CB \mid Z)$ for all Z. These two together engender $p(C \mid Z) = p(CAB \mid Z)$. Since $C \sim A$ and $C \sim B$, we can also write $p(C \mid Z) = p(CX \mid Z) = p(XC \mid Z)$ for all Z. This means $C \rightarrow X$. As we have both $C \rightarrow X$ and $X \rightarrow C$, we conclude that $C = X$; that is, $A \cap B = AB = BA$.

S-19. We have for any $A \in \mathcal{A}$, $\emptyset \rightarrow A$ and $A \rightarrow \square$.

From S-7 we have $p(\emptyset \mid Z) = 0$ for all Z. Applying this to the state right after observation of A on an arbitrary Z, we conclude that $p(\emptyset A \mid Z)$ must be zero times whatever the value $p(A \mid Z)$ may have. Hence $p(\emptyset A \mid Z) = 0$. Next, by S-14, we have $p(A\emptyset \mid Z) \leq p(\emptyset \mid Z)$ for all Z. Since $p(\emptyset \mid Z) = 0$, this means that $p(A\emptyset \mid Z) = 0$. Thus we get $p(\emptyset \cdot A \mid Z) = p(A \cdot \emptyset \mid Z) = p(\emptyset \mid Z)$, which means $\emptyset \rightarrow A$. As far as $\square$ is concerned, we should remember that the destructive observation corresponding to $\square$ is simply the test of existence (in no matter what state) of the object system, and the nondestructive observation corresponding to $\square$ amounts to doing nothing. Hence $p(\square \cdot A \mid Z) = p(A \mid Z)$ and $p(A \cdot \square \mid Z) = p(A \mid Z)$, implying $A \rightarrow \square$. Summarizing the foregoing, we state the following conclusion.

S-20. The set $\mathcal{A}$ of observational propositions as defined in S-4 is a lattice in which the relation implication is defined by S-12, and the operations conjunction and disjunction are defined by S-16 and S-17.

We are now prepared to introduce the negation $\urcorner A$ of an observational proposition A. If A means that the result of an observation $\mathfrak{D}$ is D_i, it will be natural to define $\urcorner A$ as meaning that the result of the observation $\mathfrak{D}$ is other than D_i. This is in agreement with the notion of negation in the ordinary language, but it is very important that A and $\urcorner A$ refer to the same operation of observation. For instance, if A is the A of Figure 9.5, there are two possible experimental results: the photon is linearly polarized either in the θ-direction (A) or in the direction perpendicular to it ($\urcorner A$). First, we should note that if we carry out the operation $\mathfrak{D}$, the result must be D_i or some of the remaining possibilities D_j, $j \neq i$. Hence we can make the following statement.

S-21. $p(A \mid Z) + p(\urcorner A \mid Z) = 1$ for any Z.

Next, suppose that we observe $\mathfrak{D}$ and get D_i and then repeat the same observation $\mathfrak{D}$. Then the result must be D_i again according to S-10. This leads to the following statement.

S-22. $p(\urcorner A \cdot A \mid Z) = p(A \cdot \urcorner A \mid Z) = 0$ for all Z, implying also that $A \sim \urcorner A$.

Let us remember that an observational proposition A in the foregoing definition can have a disjunctive meaning, such as "$\mathfrak{D} = D_1$ or D_2," in the same way as its negation can. Keeping this in mind, the reader will find in the foregoing definition of the notion of negation that the relation between A and $\urcorner A$ is entirely symmetrical; that is, if A and B are such that $B = \urcorner A$, then $A = \urcorner B$. This can be seen in the basic relations of the negation S-21 and S-22. Consequently we conclude the following.

S-23. $A = \urcorner(\urcorner A)$ (law of double negation).†

By applying S-12 to an assumed relation $A \rightarrow \urcorner A$, we obtain $p(A \mid Z) = p(A\urcorner A \mid Z\cdot)$, of which the right side, according to S-22, is zero. Hence, by virtue of the definition S-7, the following statement can be made.

S-24. If $A \rightarrow \urcorner A$ then $A = \emptyset$ (law of self-contradiction),

Next consider two observational propositions, A: $\mathfrak{D} = D_i$ and B: $\mathfrak{F} = F_j$. If $A \sim B$, it means that it does not make any difference whether we do $\mathfrak{D}$ first and then $\mathfrak{F}$ next, or vice versa. It is physically to be expected that this property (indifference to the order of the two) belongs to the operations $\mathfrak{D}$ and $\mathfrak{F}$, and not to the results D_i and F_j. In fact, this is confirmed by experiment. This entails the next statement.

S-25. If $A \sim B$ then $A \sim \urcorner B$, $\urcorner A \sim B$, and $\urcorner A \sim \urcorner B$.

This may be considered a limited admission of transitivity of compatibility, since it states $A \sim \urcorner B$ when it is by definition true that $A \sim B$ and $B \sim \urcorner B$. But transitivity is not a general rule for compatibility. The reader can easily test S-25 in the example of polarization of photons by adding another polarizer corresponding to $\urcorner B$. Note that if B means a certain linear polarization, $\urcorner B$

† From S-21, we also have $p(\urcorner A \mid Z) + p(\urcorner\urcorner A \mid Z) = 1$. Hence $p(A \mid Z) = p(\urcorner\urcorner A \mid Z)$ for all Z. Combining this result with S-5, we get S-23.

means a polarization perpendicular to that of B. We shall presently develop various consequences of S-25, which include the law of contraposition.

S-26. For any A and B and for all Z it is true that

$$p(BA \mid Z) + p(\neg B \cdot A \mid Z) = p(A \mid Z).$$

According to our earlier explanation, the object system after observation A is supposed to be in a state condition, say, Z_1, determined by Z and A. Hence we shall have $p(BA \mid Z) = p(B \mid Z_1)p(A \mid Z)$ and $p(\neg B \cdot A \mid Z) = p(\neg B \mid Z_1)p(A \mid Z)$. But according to S-21 we have $p(B \mid Z_1) + p(\neg B \mid Z_1) = 1$. This leads to S-26. It may be noted that S-14 is a corollary of S-26.

S-27. If $A \sim B$ then $p(BA \mid Z) + p(B\neg A \mid Z) = p(B \mid Z)$ for all Z.
The proof follows immediately from S-25 and S-26. The latter becomes, by interchanging the names of A and B, $p(A \cdot B \mid Z) + p(\neg A \cdot B \mid Z) = p(B \mid Z)$. The first term on the left is equal to $p(BA \mid Z)$ because of the compatibility of A and B, and the second is equal to $p(B \cdot \neg A \mid Z)$ by virtue of S-25.

S-28. The relation $A \rightarrow B$ is equivalent to $p(A \cdot \neg B \mid Z) = p(\neg B \cdot A \mid Z) = 0$ for all Z.
For any A and B we have, from S-26, $p(A \mid Z) = p(B \cdot A \mid Z) + p(\neg B \cdot A \mid Z)$. If $A \rightarrow B$, then, according to S-12, $p(A \mid Z) = p(BA \mid Z)$; hence $p(\neg BA \mid Z) = 0$. If $A \rightarrow B$, then, according to S-12, $A \sim B$; hence, according to S-25, $A \sim \neg B$. As a result we also have $p(A \cdot \neg B \mid Z) = 0$. Conversely, if $p(A \cdot \neg B \mid Z) = p(\neg B \cdot A \mid Z) = 0$ for all Z, we have $A \sim \neg B$ and hence $A \sim B$, from S-25. Consequently we have not only $p(A \mid Z) = p(B \cdot A \mid Z) + p(\neg B \cdot A \mid Z) = p(BA \mid Z)$ but also $p(A \mid Z) = p(AB \mid Z) + p(A \cdot \neg B \mid Z) = p(AB \mid Z)$. Hence $A \rightarrow B$ because of S-12.
It may be noted that the relation $p(A \cdot \neg B \mid Z) = p(\neg B \cdot A \mid Z) = 0$ for all Z implies $A \cap \neg B = A \cdot \neg B = \neg B \cdot A = \emptyset$ according to S-7, S-11, and S-18. But the converse is not true; that is, the condition $A \cap \neg B = \emptyset$ alone does not entail $A \cdot \neg B = \neg B \cdot A = \emptyset$. The condition $A \cap \neg B = \emptyset$ in conjunction with $A \sim B$ entails $A \cdot \neg B = \neg B \cdot A = \emptyset$, which in turn is equivalent to $A \rightarrow B$. This is in agreement with Theorem 9.10 and is of great importance.

S-29. If $A \rightarrow B$ then $\neg B \rightarrow \neg A$ (law of contraposition).
The proof is as follows. By S-28, the premise $A \rightarrow B$ means $p(A \cdot \neg B \mid Z) = p(\neg B \cdot A \mid Z) = 0$ for all Z. By putting $C = \neg A$ and $D = \neg B$, we obtain, because of the law of double negation, $p(\neg C \cdot D \mid Z) = p(D \cdot \neg C \mid Z) = 0$ for all Z. Consequently, by using S-28 we obtain $D \rightarrow C$, which is equivalent to $\neg B \rightarrow \neg A$. We have thus derived all the basic laws concerning negation.

S-30. The lattice $\mathcal{A}$ is a complemented lattice.

We can considerably simplify our formalism if we introduce another axiom, which has experimental backing. It is left to the reader to test the meaning of S-31 by the example of photon polarization.

S-31. If $p(AB \mid Z) = 0$ for all Z then $p(BA \mid Z) = 0$ for all Z.

A simple corollary to S-31 is that if $p(AB \mid Z) = 0$ for all Z then $A \sim B$. It may be noted that if $p(AB \mid Z) = 0$ for all Z then $AB = BA = A \cap B = \emptyset$. But $A \cap B = \emptyset$ does not necessarily imply $AB = BA = \emptyset$.

S-32. It suffices to define $A \rightarrow B$ by $p(A \mid Z) = p(BA \mid Z)$ for all Z.

By the use of S-26, we can infer from $p(A \mid Z) = p(BA \mid Z)$ for all Z that $p(\neg B \cdot A \mid Z) = 0$ for all Z. This establishes $A \sim \neg B$ because of S-31; hence $A \sim B$. This allows us to write $p(A \mid Z) = p(BA \mid Z) = p(AB \mid Z)$ for all Z, justifying S-32. As another consequence of S-31, we can derive the basic relations of negation S-21 and S-22 for $B = \neg A$ from $p(AB \mid Z) = 0$ and $p(A \mid Z) + p(B \mid Z) = 1$ for all Z. Note $A \cap B = \emptyset$ and $A \cup B = \square$ do not imply $A = \neg B$.

S-33. If $A \sim B$ and $A \cap B = \emptyset$ [or, equivalently, if $p(AB \mid Z) = 0$ for all Z], $p(A \mid Z) + p(B \mid Z) = p(A \cup B \mid Z)$ for all Z.

This formula re-establishes one form of the probability axiom for the p-function within the compatible propositions. We can connect this formula with the ones we are already familiar with in the following way. Because of S-30, we know that De Morgan's law can be applied to the negation as we introduced it. Hence we can write, with the help of S-21, $p(A \cup B \mid Z) = 1 - p(\neg A \cap \neg B \mid Z)$. Since $A \sim B$, we also have $\neg A \sim \neg B$ because of S-25. With the aid of S-18, we can then rewrite $p(\neg A \cap \neg B \mid Z) = p(\neg A \cdot \neg B \mid Z)$. Now $\neg A \cdot \neg B$ means that we do not get results B in the first test, nor do we get results A in the second test. The remaining possibilities are either that we do not get B first and we do get A or else that we do get B in the first test. Hence $p(\neg A \cdot \neg B \mid Z) + p(A \cdot \neg B \mid Z) + p(B \mid Z) = 1$. Combining these, we get

$$\begin{aligned} p(A \cup B \mid Z) &= 1 - p(\neg A \neg B \mid Z) \\ &= 1 - [1 - p(A \cdot \neg B \mid Z) - p(B \mid Z)] \\ &= p(A \cdot \neg B \mid Z) + p(B \mid Z) \end{aligned}$$

for all Z. Since $A \sim \neg B$, from S-26 we get $p(A \cdot \neg B \mid Z) = p(\neg B \cdot A \mid Z) = p(A \mid Z) - p(BA \mid Z)$, where $p(BA \mid Z) = 0$ because $A \sim B$ and $A \cap B = \emptyset$. This establishes S-33. The reader can obtain an experimental insight into S-33 by the illustration explained just after S-41. The earlier formulas S-26 and S-21 can be regarded as special cases of S-33.

Our next step is to show that the set of observational propositions $\mathcal{A}$ is a nondistributive modular lattice. Since a modular lattice can be Boolean (distributive), we have to show a case in which the distributive law is violated.

To show this, let us remember that if the lattice is Boolean, $\neg A \cap B = \emptyset$ implies $B \rightarrow A$. Let us assume that these A and B correspond to A and B of Figure 9.5. Then A means linear polarization in θ-direction and $\neg A$ means linear polarization in the direction $\theta + \pi/2$. Hence the relation $\neg A \cap B = \emptyset$ is true except in the case $\theta = \pi/2$ according to our explanation below S-17. $B \rightarrow A$ is not true, however, because B could imply the same linear polarization as A but it could also imply an entirely unspecified polarization in the xy-plane. This shows that $\neg A \cap B = \emptyset$ does not imply $B \rightarrow A$, violating the distributive law.†

S-34. The lattice $\mathcal{A}$ is nondistributive.

We have to establish now that the lattice $\mathcal{A}$ satisfies the modular law (7.3). We have seen that this can be done by showing that we can assign a weight function $w(A)$ to each proposition A satisfying (7.1) and (7.2). We have also seen that $w(A)$ can be identified with the number of "atoms" included in A. Those who are familiar with quantum mechanics may already guess that this atom must correspond to a "quantum state."

S-35. An observational proposition α in the lattice is called "atomic" if $\alpha \neq \emptyset$ and if there exists no observational proposition A in the lattice such that $A \rightarrow \alpha$ and $\emptyset \neq A \neq \alpha$.

Of course, we can interpret all these conditions in terms of probabilities. For instance, $\alpha \neq \emptyset$ means that there exists some Z such that $p(\alpha \mid Z) \neq 0$. More explanation of this statement is given a little later. Remembering the definition of implication (S-12), the reader will understand that S-35 states the existence of a maximal description (observational result) of a system in the sense that there cannot be more detailed description; that is, if we add a further description it will contradict the former. We introduce some concrete examples after S-40.

S-36. There exists a "complete" set $\{\alpha_i\}$, $i = 1, 2, \cdots, n$, of atoms in the lattice such that $\alpha_i \sim \alpha_j$ and $\alpha_i \cap \alpha_j = \emptyset$ for $i \neq j$ and $i, j = 1, 2, \cdots, n$, and that $\alpha_1 \cup \alpha_2 \cup \cdots \cup \alpha_n = \square$. (Two complete sets of atoms are considered to be identical if we can pass from one set to another just by relabeling the indices i.)

Probabilistically, disjointness compatibility and completeness mean $p(\alpha_i\alpha_j \mid Z) = 0$ for $i \neq j$ and for all Z, by virtue of S-31, and $\sum_{i=1}^{n} p(\alpha_i \mid Z) = 1$ for all Z because of S-33.

† Both $A \cap B = \emptyset$ and $\neg A \cap B = \emptyset$ (where $B \neq \emptyset$) can be true, yet this does not imply a breakdown of the law of the excluded middle. In fact, we still have $A \cup \neg A = \square$, as well as $A \cup B = \square$, $\neg A \cup B = \square$, $A \cup \neg B = \square$ and $\neg A \cup \neg B = \square$. This may appear strange, but the meaning of disjunction is different from what we intuitively understand under the name of disjunction.

S-37. There exists in general more than one complete set of atoms.

S-38. For any observational proposition A that is not $\emptyset$, there exists at least one (usually many) complete set $\{\alpha_i\}$ such that $A = \alpha_1 \cup \alpha_2 \cup \cdots \cup \alpha_m$; that is, because of S-33, $p(A \mid Z) = \sum_{i=1}^{m} p(\alpha_i \mid Z)$ for all Z. This implies $\alpha_i \sim A$.

A given observational proposition A can be expressed as in S-38, but this can be done with the help of different sets of atoms. For instance, it is quite possible that $A = \alpha_1 \cup \alpha_2 = \beta_1 \cup \beta_2$, where the α's and β's are not compatible. This can happen because $\alpha_1 \sim A$, $\alpha_2 \sim A$, $\beta_1 \sim A$, $\beta_2 \sim A$, $\alpha_1 \sim \alpha_2$, $\beta_1 \sim \beta_2$ do not imply $\alpha_i \sim \beta_j$, $i, j, = 1, 2$. Let us now introduce a weight function of the type (7.2) and see what kind of experimental backing we can give to it.

S-39. The number of atoms required to express a given A as their disjunction is independent of the used set of atoms and is a constant characteristic of A. We denote this constant by $w(A)$.

S-40. If $A \rightarrow B$ and $A \neq B$ then $w(A) < w(B)$.

S-41. For any A and B, which are not necessarily compatible, $w(A) + w(B) = w(A \cup B) + w(A \cap B)$.

Let us consider whether S-35 through S-39 can be substantiated in our example of photon polarization. The example is actually too simple to check all the details of these statements, but it can give support to most of the main points involved. As far as the polarization of a photon traveling in the z-direction is concerned, there is no description more precise than "linearly polarized in the θ-direction," "left-circularly polarized," or "right-elliptically polarized with the major axis in the θ-direction and ellipticity ε." These are the atoms. This is an experimental fact and the basis of existence of the so-called "quantum state," which we introduce presently. Now take two directions, θ and $\theta + \pi/2$, and consider two atoms, α_1 and α_2, which mean, respectively, "the photon is linearly polarized in the θ-direction" and "the photon is linearly polarized in the $(\theta + \pi/2)$-direction." We have already seen that the conditions $p(\alpha_1\alpha_2 \mid Z) = 0$ and $p(\alpha_1 \mid Z) + p(\alpha_2 \mid Z) = 1$ for all Z entail $\alpha_1 \sim \alpha_2$, $\alpha_1 \cap \alpha_2 = \alpha_1\alpha_2 = \alpha_2\alpha_1 = \emptyset$, and $\alpha_1 \cup \alpha_2 = \square$. Hence α_1 and α_2 constitute a complete set of atoms. Similarly, two atoms β_1 and β_2, standing for "the photon is left-circularly polarized" and "the photon is right-circularly polarized," satisfy $p(\beta_1\beta_2 \mid Z) = 0$ and $p(\beta_1 \mid Z) + p(\beta_2 \mid Z) = 1$ for all Z. Hence β_1 and β_2 constitute a complete set of atoms. It may be noted that the observation corresponding to β_1 or β_2 can be performed with the help of a $\frac{1}{4}$-wavelength plate and linear polarizer. More generally, we can take two elliptical polarizations γ_1 and γ_2 characterized by the descriptions $\gamma_1 = [E_x = a \cos \omega t, E_y = b \cos (\omega t + \theta)]$ and $\gamma_2 = [E_x = b \sin (-\omega t),$

$E_y = -a \sin(-\omega t - \theta)]$, where E_x and E_y are the components of the electric field in the x-direction and the y-direction. The observation corresponding to γ_1 (or γ_2) can be performed by a plate of doubly refracting material (calcite cut parallel to the optical axis) of various thicknesses and a linear polarizer. These two, γ_1 and γ_2, again form a complete set of atoms, since $p(\gamma_1\gamma_2 \mid Z) = 0$ for all Z and $p(\gamma_1 \mid Z) + p(\gamma_2 \mid Z) = 1$ for all Z. Of course $\{\alpha_1, \alpha_2\}$ and $\{\beta_1, \beta_2\}$ are special cases of $\{\gamma_1, \gamma_2)$. It may be noted that each set of elliptical polarization is characterized by two parameters, (a/b) and θ. Hence there are actually as many complete sets of atoms as values of two real parameters or of one complex parameter. These facts substantiate S-35 through S-37. As far as S-38 is concerned, our example verifies it for $A = \square$, where $\square$ means that the polarization is within the xy-plane. As regards S-37, we can note that if we take $A = \square$ then $w(A) = 2$ no matter which complete set of atoms we may take.

We stated in Section 9.1, and repeated at the beginning of the present section, that it is an unwarranted inference to attribute the result of an observation of an object system as one of the properties of the system independent of the observation. By now we have developed sufficient material to substantiate this point by a concrete example. Suppose that an observation gives two alternatives, A and its negation $\neg A$, when the observation is made on an object, say, O. Then we usually automatically allow ourselves to state that O must *be* either A or $\neg A$, as if A and $\neg A$ were attributes of O. But this is a dangerous habit of thinking.† Suppose that A is the observational statement that the polarization of a photon is found to be linear and in the direction θ, and $\neg A$ is the observational statement that the polarization of the photon is found to be linear and in the direction $\theta + \pi/2$. These two are genuinely negations of each other, for it is impossible for A and $\neg A$ to be true simultaneously, and the probability of A and the probability of $\neg A$ add up to unity. It is entirely false, however, to say that any photon must be either linearly polarized in the θ-direction or linearly polarized in the $(\theta + \pi/2)$-direction. In fact, the photon may be linearly polarized in a direction different from both θ and $\theta + \pi/2$, or it may be rather elliptically polarized. A skeptical reader may contend that I am after all using the same kind of predicate for the result of an observation and for the object itself, because I stated that the photon *may actually be* polarized in an angle different from θ or $\theta + \pi/2$. The answer to this criticism is simple. When I say that the photon "is found to be" polarized in the θ-direction, I mean the observational procedure explained before. When I say that the photon "is" actually

† This may be used as an argument for abandoning the law of the excluded middle. But our approach is such that we broaden the notion of disjunction so that this law is still valid This makes the entire system much more elegant than abandoning the law of the excluded middle. In our system $A \cup B$ includes elements that were in neither A nor B.

polarized in the θ-direction, I mean that the photon is "in such a state that there is probability unity of finding it linearly polarized in the direction θ." More formally, the observational result may be A or $\neg A$, but the *appropriate* predicate for an object must be of the type "has probability 1 of giving A" or "has probability 1 of giving $\neg A$." The finding that the observation showed A does not imply that the object was in a state such that the probability of finding it in A is unity. When there is no need to make this distinction, we can use the postulate of fixed predicate-set correspondence, hence also the usual distributive logic.

If there were only one complete set of atomic observational propositions, no confusion would arise from using observational predicates for the object predicates (attributes). But in physics, and perhaps in other actual situations in life, there are more than one complete set of atoms. Thus S-37 is one of the conspicuous discrepancies from the ordinary logic resulting from the broader point of view adopted at the beginning of this chapter.

Next, with regard to S-40, let us take as A a linear polarization in the direction θ. Then among all possible polarizations (including all elliptic polarizations), the B that satisfies $A \rightarrow B$, that is, $p(BA \mid Z) = p(A \mid Z)$ for all Z, is A itself. But this is excluded by the premise of S-40. Then $p(A \mid Z) = p(BA \mid Z)$ automatically determines B as $\square$, which means in this case that the polarization is in the xy-plane. Then $w(A) = 1$ and $w(B) = 2$, verifying S-40. Next, in connection with S-41, let us take as A and B two linear polarizations in the directions θ_A and θ_B. Let us first assume that $\theta_A \neq \theta_B$ (and possibly $\theta_A \neq \theta_B + \pi$). Then $C = A \cap B$ must be $\emptyset$, because the necessary condition $C \rightarrow A$ implies $C = \emptyset$ or $C = A$; similarly, $C \rightarrow B$ implies $C = \emptyset$ or $C = B$. Hence $C = A \cap B = \emptyset$ on condition that $A \neq B$. Next $D = A \cup B$ must be $\square$, because the condition $A \rightarrow D$ limits D to A or $\square$, and similarly $B \rightarrow D$ limits D to B or $\square$. Hence, on condition $A \neq B$, we have $A \cup B = \square$. As a consequence we have $w(A) = w(B) = 1$, $w(A \cap B) = 0$, and $w(A \cup B) = 2$, verifying S-41. If $A = B$ then we have $A = B = A \cap B = A \cup B$; hence $w(A) = w(B) = w(A \cap B) = w(A \cup B) = 1$, again verifying S-41.

It is true that we have shown that S-40 and S-41 hold in a simple example. If S-40 and S-41 are general rules, we can conclude that the set of observational propositions forms a modular lattice. The fact that there are in fact more than one nonidentical complete set of atomic propositions (S-37), which we have demonstrated by our example, is sufficient to conclude that the modular lattice under consideration does not reduce to a Boolean lattice even in the simplest case.

Birkoff and von Neumann raised in their original paper on quantum logic the following question (see [B-2] and also Husimi's answer [H-18]): What experimental meaning can one attach to the meet and join of two given experimental propositions? Formally speaking, if the implication is defined in experimental terms (S-12), the conjunction (join) is defined in terms of implication (S-16) and hence is indirectly defined in experimental terms. However, we cannot visualize what kind of experiment corresponds to the conjunction at all. The following consideration is my answer to Birkoff and von Neumann's question.

Let $\Gamma^{(n)}$ be a sequential observational proposition defined by

$$BABA \cdots BA,$$

where n A's and n B's are placed alternately with an A at the right extremity. There is no doubt that there is a limit $p(\Gamma^{(\infty)} \mid Z)$ for each Z, because $0 \leq p(\Gamma^{(n+1)} \mid Z) \leq p(\Gamma^{(n)} \mid Z)$ from S-14. Since $\Gamma^{(n)}\Gamma^{(m)} = \Gamma^{(n+m)}$, we have $p(\Gamma^{(\infty)}\Gamma^{(\infty)} \mid Z) = p(\Gamma^{(\infty)} \mid Z)$, showing that $\Gamma^{(\infty)}$ is a simple observational proposition in the sense of S-10 and hence belongs to $\mathcal{A}$.

Let $\Delta^{(n)}$, $\Theta^{(n)}$, and $\Lambda^{(n)}$ be defined respectively as $A\Gamma^{(n-1)}B$, $A\Gamma^{(n)}$, and $\Gamma^{(n)}B$. Then $\Gamma^{(\infty)} = \Delta^{(\infty)} = \Theta^{(\infty)} = \Lambda^{(\infty)}$. Proof for $\Delta^{(\infty)}$: note first that $\Delta^{(\infty)}\Delta^{(\infty)} = \Delta^{(\infty)}$. Hence $\Delta^{(\infty)} \in \mathcal{A}$. Next

$$p(\Gamma^{(n)}\,\Delta^{(m)} \mid Z) = p(B\,\Delta^{(n+m-1)} \mid Z) \leq p(\Delta^{(n+m-1)} \mid Z),$$

because of S-14, and $p(\Gamma^{(n)}\,\Delta^{(m)} \mid Z) \geq p(A\Gamma^{(n)}\,\Delta^{(m)} \mid Z) = p(\Delta^{(n+m)} \mid Z)$. Hence $p(\Gamma^{(\infty)}\,\Delta^{(\infty)} \mid Z) \geq p(\Delta^{(\infty)} \mid Z) \geq p(\Gamma^{(\infty)}\,\Delta^{(\infty)} \mid Z)$; that is,

$$p(\Gamma^{(\infty)}\,\Delta^{(\infty)} \mid Z) = p(\Delta^{(\infty)} \mid Z).$$

In view of S-32, we conclude that $\Delta^{(\infty)} \to \Gamma^{(\infty)}$. Interchanging the roles of A and B and hence also the roles of $\Delta^{(n)}$ and $\Gamma^{(n)}$, we get $\Gamma^{(\infty)} \to \Delta^{(\infty)}$. This completes the proof that $\Delta^{(\infty)} = \Gamma^{(\infty)}$. The other three cases can be proven in a similar way. Since $\Gamma^{(\infty)}$, $\Delta^{(\infty)}$, $\Theta^{(\infty)}$, and $\Lambda^{(\infty)}$ all belong to $\mathcal{A}$ and are equivalent to each other, we use a single symbol S to denote any one of them.

We can now prove that this S satisfies the definition of $C = A \cap B$. First, we have to show $S \to A$ and $S \to B$. Proof: $\Gamma^{(n)}A = \Gamma^{(n)}$ because $AA = A$. Hence $SA = S$. Similarly, $A\Gamma^{(n)} = \Theta^{(n)}$. Hence $AS = S$. This completes the proof of $S \to A$, because of S-12 or S-32. For the proof of $S \to B$, note that $\Gamma^{(n)}B = \Lambda^{(n)}$ and $B\Gamma^{(n)} = \Gamma^{(n)}$. Next we have to show that if $D \to A$ and $D \to B$ then $D \to S$. The proof is as follows. The definition of D is $DA = AD = DB = BD = D$. If we multiply the sequence of $\Gamma^{(n)}$ by D from the left, we can reduce the length of the sequence of $\Gamma^{(n)}$ step by step, using $DA = D$ and $DB = D$. Thus $D\Gamma^{(n)} = D\Theta^{(n-1)} = D\Gamma^{(n-1)} = \cdots = DA = D$. Similarly, by multiplying $\Gamma^{(n)}$ by D from the right and using $AD = D$ and $BD = D$, we obtain $\Gamma^{(n)}D = \Lambda^{(n-1)}D = \Gamma^{(n-1)}D = \cdots = BD = D$. Hence $D\Gamma^{(n)} = \Gamma^{(n)}D = D$, entailing $DS = SD = D$; that is, $D \to S$.

In this section, we showed that the observational propositions in quantum physics form a modular lattice, relying mainly on experimental facts in atomic physics. But, in reality, we can come to a similar conclusion without relying on atomic physics, indicating that the necessity of a modular logic is not limited to physics, see [W-31a].

9.4. THEORETICAL SCAFFOLD OF QUANTUM MECHANICS

Once it is established that the set of observational propositions forms a modular lattice, it is natural, according to Section 9.2, to expect that each observational proposition can be made to correspond to the projection operator of a subspace in an orthogonal space. (Note a strange coincidence with CLAFIC, p. 447.) Furthermore, this correspondence suggests that $w(A)$ may be identified with the number of dimensions of the subspace corresponding to A.

S-42. Each observational proposition A of weight $w(A)$ corresponds to a subspace $\mathfrak{M}(A)$ of dimensions $w(A)$ in an orthogonal space that corresponds to $\square$. The proposition $\emptyset$ corresponds to a subspace of zero dimension.

S-43. The relation $A \to B$ between two observational propositions A and B is translated as meaning that if any vector ψ belongs to $\mathfrak{M}(A)$, $\psi \in \mathfrak{M}(A)$, then ψ belongs to $\mathfrak{M}(B)$, $\psi \in \mathfrak{M}(B)$.

Of course, S-43 is suggested by the analysis of Section 9.2, and our acceptance of S-43 leads to the following statement.

S-44. If $\mathfrak{M}(A)$ and $\mathfrak{M}(B)$ correspond to A and B, respectively, the subspace $\mathfrak{M}(A \cap B)$, corresponding to $A \cap B$, is given by $\mathfrak{M}(A \cap B) = \{\psi \mid \psi \in \mathfrak{M}(A) \text{ and } \psi \in \mathfrak{M}(B)\}$. Similarly, the subspace $\mathfrak{M}(A \cup B)$, corresponding to $A \cup B$, is given by $\mathfrak{M}(A \cup B) = \{\psi \mid \psi = a\psi_1 + b\psi_2, \psi_1 \in \mathfrak{M}(A), \psi_2 \in \mathfrak{M}(B); a, b\text{: complex numbers}\}$.

In our analysis of Section 9.2 we assumed a and b to be real, but in the present case we have to admit complex coefficients. The reason is as follows.

First, we have to know that in the definition of a subspace as a set of vectors the absolute value of the vector ψ is immaterial, since if ψ belongs to the subspace then $a\psi$ does also. Hence in the expression $\psi = a\psi_1 + b\psi_2$ what is meaningful is a/b, and this was assumed to be a real number in Section 9.3. But this is not broad enough. Suppose that we take two linear polarizations θ_1 and $(\theta_1 + \pi/2)$ as two atoms α_1 and α_2. Then there will be two vectors $\psi(\alpha_1)$ and $\psi(\alpha_2)$. (Note that according to the definitions, these ψ-vectors have nothing to do with the vectors of the electromagnetic field in the xy-plane, but because of a situation that cannot be explained here, we can use the electric field or the vector potential as the ψ-vectors in this case if we allow them to become complex. See [W-8], for instance.) Then $\alpha_1 \cup \alpha_2 = \square$ (where $\square$ corresponds to the observational proposition that the polarization is in the xy-plane) must correspond to $\mathfrak{M}(\alpha_1 \cup \alpha_2) = \{\psi \mid \psi = a\psi(\alpha_1) + b\psi(\alpha_2)\}$. Next take two mutually exclusive elliptical polarizations γ_1 and γ_2. Then there will be two vectors $\psi(\gamma_1)$ and $\psi(\gamma_2)$. Hence $\mathfrak{M}(\alpha_1 \cup \alpha_2) = \mathfrak{M}(\gamma_1 \cup \gamma_2) = \{\psi \mid \psi = c\psi(\gamma_1) + d\psi(\gamma_2)\}$. This means that a vector of this type $c\psi(\gamma_1) + d\psi(\gamma_2)$ can also be expressed as $a\psi(\alpha_1) + b\psi(\alpha_2)$. But this is impossible if a/b and c/d are real. Suppose that we put $d = 0$ and consider only $\psi(\gamma_1)$, which can be any elliptical polarization. But it is well known, and was shown in our previous explanations, that two parameters or one complex parameter are required to express an arbitrary elliptical polarization in terms of two linear polarizations. This cannot be done by one real parameter a/b.

S-45. The orthogonal space required in physics is a complex space.

This entails a slight modification in our mathematical tools developed in Section 9.2. Thus the components of a vector become complex, and the scalar product of two vectors **x** and **y** will be defined by

$$(\mathbf{xy}) = x_i^* y_i = (\mathbf{yx})^*, \tag{9.100}$$

where x_i and y_i are the components of **x** and **y**, respectively, and the summation is taken with respect to i. The asterisk (*) means the complex conjugate. The orthogonal transformation (9.55), which leaves (9.47) invariant, is replaced by the unitary transformation

$$\bar{T} = T^{-1}, \tag{9.101}$$

which leaves (9.100) invariant. [See (A8.1.4), where a similar transition to complex space is done.] Here a matrix $\bar{T}$ is the so-called Hermitian conjugate of T and is defined by

$$(\bar{T})_{ij} = (T^T)_{ij}^* = T_{ji}^*. \tag{9.102}$$

Any matrix equal to its Hermitian conjugate is called a Hermitian matrix. Among the three properties of a projector operator, the idempotent law (9.83) and the law of dimensionality (9.85) remain unchanged. The symmetry of a projection operator (9.84) is replaced by "hermiticity":

$$\bar{\mathfrak{I}}[\mathfrak{M}] = \mathfrak{I}[\mathfrak{M}]. \tag{9.103}$$

For instance, a one-dimensional projection operator is expressed instead of (9.64), by

$$(\mathfrak{I}[\boldsymbol{\xi}])_{ij} = \frac{\xi_i \xi_j^*}{\xi_k \xi_k^*}. \tag{9.104}$$

Since each subspace $\mathfrak{M}$ and its projection operator correspond to each other one to one, there is no doubt about the next statement.

S-46. The probability $p(A \mid Z)$ must be a function $F(\mathfrak{I}[\mathfrak{M}(A)], Z)$ of $\mathfrak{I}[\mathfrak{M}(A)]$ and Z, no matter what mathematical expression may be given to Z.

Let us now assume that A is a union of r atoms $\alpha_1, \alpha_2, \cdots, \alpha_r$ (see S-38),

$A = \alpha_1 \cup \alpha_2 \cup \cdots \cup \alpha_r$; then we have (since $\alpha_i \cap \alpha_j = \emptyset$)

$$p(\alpha_1 \cup \alpha_2 \cup \cdots \cup \alpha_r \mid Z) = \sum_{i=1}^{r} p(\alpha_i \mid Z) \quad \text{for all} \quad Z. \tag{9.105}$$

On the other hand, since $\mathfrak{M}(A)$ is the set of linear combinations of vectors $\mathfrak{M}(\alpha_i)$, we have from (9.82)

$$\mathfrak{T}[\mathfrak{M}(\alpha_1 \cup \alpha_2 \cup \cdots \cup \alpha_r)] = \sum_{i=1}^{r} \mathfrak{T}[\mathfrak{M}(\alpha_i)]. \tag{9.106}$$

Because $p(\alpha_i \mid Z) = F(\mathfrak{T}[\mathfrak{M}(\alpha_i)], Z)$, we obtain from (9.105) and (9.106)

$$F\left(\sum_{i=1}^{r} \mathfrak{T}[\mathfrak{M}(\alpha_i)], Z\right) = \sum_{i=1}^{r} F(\mathfrak{T}[\mathfrak{M}(\alpha_i)], Z), \tag{9.107}$$

showing a kind of linear dependence of F on the argument $\mathfrak{T}[\mathfrak{M}]$. Thus we may also state the following.

S-47. If A_i $(i = 1, 2, \cdots)$ are all mutually exclusive, that is, if $A_i \cap A_j = \emptyset$ $(i \neq j)$, then

$$F(\mathfrak{T}[\mathfrak{M}(A_1 \cup A_2 \cup A_3 \cdots)], Z) = F\left(\sum_i \mathfrak{T}[\mathfrak{M}(A_i)], Z\right) = \sum_i F(\mathfrak{T}[\mathfrak{M}(A_i)], Z).$$

The linear dependence in this sense does not necessarily imply, but does suggest, a more general linear dependence of F on its first argument. Thus we shall see whether a satisfactory theory can be built on the following assumption.

S-48. If A_i $(i = 1, 2, \cdots)$ are all mutually exclusive, that is, if $A_i \cap A_j = \emptyset$ $(i \neq j)$, and if q_i $(i = 1, 2, \cdots)$ are arbitrary (real) numbers, then

$$F\left(\sum_i q_i \mathfrak{T}[\mathfrak{M}(A_i)], Z\right) = \sum_i q_i F(\mathfrak{T}[\mathfrak{M}(A_i)], Z).$$

We considered in Chapter 4 a set of mutually exclusive and complete experimental results D_i $(i = 1, 2, \cdots)$ that can be obtained by an observational operation $\mathfrak{D}$, and we expressed by $\mathfrak{D} = D_i$ an observational proposition that one obtains D_i as a result of the observation $\mathfrak{D}$. The so-called physical quantities fit in this framework in the sense that there is a well-defined operation corresponding to each physical quantity and the possible results of each observation are mutually exclusive and complete. To make clear that we are dealing with a physical quantity that is a special case of $\mathfrak{D}$, we hereafter use Q and q_μ instead of $\mathfrak{D}$ and D_i, and write A_μ for $Q = q_\mu$. The q's will be called "eigenvalues" of Q. Then we have $p(A_\mu A_\nu \mid Z) = 0$ for all Z if $\mu \neq \nu$, and $\sum_\mu p(A_\mu \mid Z) = 1$ for all Z (see S-21, S-31, and S-33). The probability $p(A_\mu \mid Z)$ of obtaining q_μ as a result of Q on state Z is, according to

S-45, given by $F(\mathfrak{T}[\mathfrak{M}(A_\mu)], Z)$. The anticipated property S-48 makes it very convenient to express the average (or expected) value $\langle Q \mid Z\rangle$ or Q on state Z. If we now introduce a quantity $Q = \sum_\mu q_\mu \mathfrak{T}[\mathfrak{M}(A_\mu)]$, then $F(Q, Z) = F(\sum q_\mu \mathfrak{T}[\mathfrak{M}(A_\mu)], Z) = \sum_\mu q_\mu F(\mathfrak{T}[\mathfrak{M}(A_\mu)], Z) = \sum q_\mu p(A_\mu \mid Z) = \langle Q \mid Z\rangle$. This leads to the next statement.

S-49. The physical quantity Q, which has eigenvalues q_μ, will be given a mathematical expression,

$$Q = \sum_\mu q_\mu \mathfrak{T}[\mathfrak{M}(A_\mu)],$$

where A_μ is the proposition $Q = q_\mu$.

S-50. The average or expected value $\langle Q \mid Z\rangle$ of the q's on a system in state condition Z is given by

$$\langle Q \mid Z\rangle = F(Q, Z),$$

where F is defined in S-46 and S-48.

The A_μ's that appear in the same Q in S-49 are mutually exclusive (which implies compatibility) and complete. The compatibility, mutual exclusion, and completeness are expressed, in terms of the corresponding projection operators $\mathfrak{M}_\mu = \mathfrak{M}(A_\mu)$, respectively, as

$$\begin{aligned} \mathfrak{T}[\mathfrak{M}_\mu]\mathfrak{T}[\mathfrak{M}_\nu] - \mathfrak{T}[\mathfrak{M}_\nu]\mathfrak{T}[\mathfrak{M}_\mu] &= 0, \\ \mathfrak{T}[\mathfrak{M}_\mu]\mathfrak{T}[\mathfrak{M}_\nu] &= 0 \quad \text{for} \quad \mu \neq \nu, \\ \sum_\mu \mathfrak{T}[\mathfrak{M}_\mu] &= 1. \end{aligned} \tag{9.108}$$

[See (9.90), (9.88), and (9.93).]

Since all $\mathfrak{T}[\mathfrak{M}_\mu]$ commute, the $\mathfrak{M}_\mu$ belong to a family of orthogonal subspaces. Hence we can take one orthogonal coordinate system ψ_i ($i = 1, 2, \cdots, n$) and express $\mathfrak{T}[\mathfrak{M}_\mu]$ as a sum of some of the $\mathfrak{T}[\psi_i]$. Because of mutual exclusiveness [the second line of (9.108)], if a ψ_i belongs to a $\mathfrak{M}_\mu$ it does not belong to any other $\mathfrak{M}_\nu$ ($\nu \neq \mu$). When $\mathfrak{M}[A_\mu]$ reduces to one dimension, that is, to a vector ψ_i, the proposition A_μ becomes what was called atomic proposition α.

S-51. For a given physical quantity there exists a coordinate system $\{\psi_i\}$ ($i = 1, 2, \cdots$) such that

$$Q = \sum_i q_i \mathfrak{T}[\psi_i],$$

where $q_i = q_\mu$ if $\psi_i \in \mathfrak{M}(A_\mu)$. If $\psi_i \in \mathfrak{M}(A_\mu)$, the value of trace $\mathfrak{T}[\mathfrak{M}(A_\mu)]$ is the number of those q's equal to q_μ. If trace $\mathfrak{T}[\mathfrak{M}(A_\mu)] > 1$, the eigenvalue $q_\mu = q_i$ is said to be degenerate.

S-52. The coordinate system $\{\psi_i\}$ is uniquely defined by Q if each $\mathfrak{M}(A_\mu)$ is one-dimensional; otherwise there remains ambiguity because ψ_i can be subjected to any unitary transformation that maps each $\mathfrak{M}(A_\mu)$ into itself.

Since each $\mathfrak{I}[\mathfrak{M}(A_\mu)]$ is a Hermitian operator and since the q's are real, we have the following statement.

S-53. The mathematical expression of a physical quantity is a Hermitian operator.

In older textbooks on quantum mechanics, it was sometimes assumed that the converse of S-53 was also true. But it has now become clear that not all Hermitian operators correspond to physical quantities (see [W-4, W-8, W-35]).

In macroscopic physics all physical quantities are compatible; hence we need only one set of coordinates. Suppose that we are given two physical quantities Q with q_μ and R with r_ν and that all propositions A_μ $(Q = q_\mu)$ and B_ν $(R = r_\mu)$ are compatible. Then all $\mathfrak{I}[\mathfrak{M}(A_\mu)]$ and $\mathfrak{I}[\mathfrak{M}(B_\nu)]$ commute. Hence $Q = \sum_\mu q_\mu \mathfrak{I}[\mathfrak{M}(A_\mu)]$ and $R = \sum_\nu r_\nu \mathfrak{I}[\mathfrak{M}(B_\nu)]$ commute. It is not difficult to prove the converse either.

S-54. Given two quantities Q and R, which are expressed as $Q = \sum_\mu q_\mu \mathfrak{I}[\mathfrak{M}(A_\mu)]$ and $R = \sum_\nu r_\nu \mathfrak{I}[\mathfrak{M}(B_\nu)]$, the compatibility of A_μ and B_ν for all μ and ν (for which q_μ and r_ν are not zero) is equivalent to the commutability:

$$QR - RQ = 0.$$

In such a case there exists at least one set of coordinates that can be used to express both Q and R in the form of S-51 using the same $\{\psi_i\}$. Similarly, if we are given a set of many quantities that all commute pairwise, there exists at least one set of common coordinates. We usually assume that if we take the set of *all* commuting quantities, it uniquely determines one system of coordinates.

So far we have introduced mathematical expressions for observational propositions and physical quantities. We now pass to the mathematical expression of state conditions Z. Since we have already utilized considerable space in this section, the rest of the section has to be sketchy, pointing out only the main established theorems in quantum mechanics without giving sufficient justification to each statement.

S-55. A state condition Z that gives $p(A \mid Z) = 1$ for an observational proposition A is said to be an "eigenstate" corresponding to A, and symbolically we write $Z = E(A)$, which does not necessarily mean that the condition $p(A \mid Z) = 1$ defines Z uniquely by A.

When A is an atom α, $\mathfrak{M}(A)$ is a vector $\boldsymbol{\psi}_\alpha$. Then it may be natural to assume that the eigenstate $E(\alpha)$ is also uniquely determined by $\boldsymbol{\psi}_\alpha$. Hence let us make the following assumption.

S-56. The eigenstate corresponding to an atomic observational proposition α is called a "quantum state" and is mathematically represented by $E(\alpha) = \mathcal{T}[\boldsymbol{\psi}_\alpha]$.

Let us consider two sets of atomic propositions $\{\alpha_i\}$ and $\{\beta_j\}$. Assume that Z specifies the eigenstate corresponding to one of the α_i, say, α_1. $Z = E(\alpha_1)$. Then, since $\bigcup_j^{\text{all}} \beta_j = \square$, we have, from S-8 and (9.105), $\sum_{j=1}^n p(\beta_j \mid E(\alpha_1)) = 1$. On the other hand, let us consider the product of two projection operators, $\mathcal{T}[\boldsymbol{\psi}]$ and $\mathcal{T}[\boldsymbol{\varphi}]$, where $\boldsymbol{\psi}$ and $\boldsymbol{\varphi}$ are vectors corresponding to one of the α's and to one of the β's, respectively. From (9.104) we have

$$(\mathcal{T}[\boldsymbol{\psi}]\,\mathcal{T}[\boldsymbol{\psi}])_{ab} = \psi_a \overset{*}{\psi_c} \varphi_c \overset{*}{\varphi_b} \tag{9.109}$$

if the vectors are normalized: $\psi_a \psi_a^* = \varphi_a \varphi_a^* = 1$. If we take the diagonal sum of (9.109) we obtain

$$\text{trace}\,(\mathcal{T}[\boldsymbol{\psi}]\,\mathcal{T}[\boldsymbol{\varphi}]) = \psi_a \overset{*}{\varphi_a} \overset{*}{\psi_b} \varphi_b = |\psi_a \overset{*}{\varphi_a}|^2 \geq 0. \tag{9.110}$$

Now let us put $\boldsymbol{\psi} = \boldsymbol{\psi}(\alpha_1)$ and $\boldsymbol{\varphi} = \boldsymbol{\varphi}(\beta_j)$ in (9.110) and sum over j:

$$\sum_j (\psi(\alpha_1))_a (\psi(\alpha_1))_b^* (\varphi(\beta_j))_b (\varphi(\beta_j))_a^* . \tag{9.111}$$

The factor $\sum_j (\varphi(\beta_j))_b (\varphi(\beta_j))_a^*$ is equal to δ_{ba}. This is a consequence of (9.101), because $(\varphi(\beta_j))_b$ can be considered as a matrix element of the unitary transformation connecting the coordinate system $\{\boldsymbol{\psi}(\beta_1), \boldsymbol{\psi}(\beta_2), \cdots\}$ to the coordinate system with respect to which the components labeled with $a, b, c, \cdots$ are taken. Therefore the sum in (9.111) is equal to unity. This means that

$$\text{trace}\,(\mathcal{T}[\boldsymbol{\psi}(\alpha_1)]\,\mathcal{T}[\boldsymbol{\varphi}(\beta_j)]) \geq 0 \tag{9.112}$$

and

$$\sum_{j=1}^{n} \text{trace}\,(\mathcal{T}[\boldsymbol{\psi}(\alpha_1)]\,\mathcal{T}[\boldsymbol{\varphi}(\beta_j)]) = 1, \tag{9.113}$$

which satisfy the basic laws of probability of disjoint and complete events with respect to the index j. Comparison of these two equations with the required relation $p(\beta_j \mid E(\alpha_1)) \geq 0$ and the previously established relation $\sum_{j=1}^n p(\beta_j \mid E(\alpha_1)) = 1$, with the understanding expressed in S-55, makes it almost inescapable to put

$$p(\beta \mid E(\alpha)) = \text{trace}\,(\mathcal{T}[\boldsymbol{\psi}(\alpha)]\,\mathcal{T}[\boldsymbol{\varphi}\,(\beta)]). \tag{9.114}$$

So far we have considered the case in which the object system is in a quantum state $E(\alpha_1) = \mathcal{T}[\boldsymbol{\psi}(\alpha_1)]$. Suppose that we take an ensemble of N

similar object systems of which $N_i = \omega_i N$ are in quantum state $\psi(\alpha_i)$. $\omega_i \geq 0$ and $\sum_{i=1}^{n} \omega_i = 1$. If the state condition Z of the system corresponds to such an ensemble, the probability that β will be true on Z must be given by

$$p(\beta \mid Z) = \sum_i \omega_i p(\beta \mid E(\alpha_i)) = \text{trace}\,(\mathfrak{T}[\boldsymbol{\varphi}(\beta)] \sum_{i=1}^{n} \omega_i \mathfrak{T}[\boldsymbol{\psi}(\alpha_i)]. \quad (9.115)$$

Therefore we can incorporate (9.114) and (9.115) in a single relation

$$\rho(\beta \mid Z) = \text{trace}\,(\mathfrak{T}[\boldsymbol{\varphi}(\beta)]\,Z), \quad (9.116)$$

with

$$Z = \sum_{i=1}^{n} \omega_i \mathfrak{T}[\boldsymbol{\psi}(\alpha_i)]. \quad (9.117)$$

The ω_i's are the eigenvalues of Z. The most common name used by physicists for this mathematical entity Z is "density matrix." The correspondence we make between a "state condition" and an "ensemble" may be perfectly justifiable, because a "condition" is a statistical collection of the past data on the system. Such a correspondence between state and ensemble is customary even in classical physics. For instance, when we speak of a thermodynamic "state," we mean in fact an ensemble, not a microscopically well-defined state. It is a surprizing coincidence that the "autocorrelation function" G, which is useful in pattern recognition, is mathematically the same entity as Z. [See (8.35) and also Appendix A9.5.]

Next, if an observational proposition A is given by $\mathfrak{T}[\mathfrak{M}(A)] = \sum_j^{\text{some}} \mathfrak{T}[\boldsymbol{\varphi}(\beta_j)]$, the probability of A being true is the sum of the probabilities of β_j included in A being true. Hence we obtain from (9.116)

$$p(A \mid Z) = \text{trace}\,(\mathfrak{T}[\mathfrak{M}(A)]\,Z). \quad (9.118)$$

This is an embodiment of $F(\mathfrak{T}[\mathfrak{M}(A)], Z)$ of S-46, satisfying the anticipated property S-48. If a physical quantity Q is given by $Q = \sum q_\mu \mathfrak{T}[\mathfrak{M}(A_\mu)]$, then $\langle Q \mid Z \rangle = F(Q, Z)$ of S-50 must be $\sum_\mu q_\mu p(A_\mu \mid Z)$, which is

$$\langle Q \mid Z \rangle = F(Q, Z) = \text{trace}\,(Q\,Z). \quad (9.119)$$

Repeating the main points, we can state the following.

S-57. The state of an object system is represented by a matrix that is Hermitian, that is,

$$\tilde{Z} = Z; \quad (9.120)$$

whose diagonal sum is 1, that is,

$$\text{trace}\, Z = 1; \quad (9.121)$$

and whose eigenvalues $\omega_1, \omega_2, \cdots, \omega_n$ are non-negative. If $\{\boldsymbol{\psi}_i\}$ are the coordinates that diagonalize Z, then Z can be written in the form (9.117).

S-58. . an observational proposition corresponding to the subspace $\mathfrak{M}(A)$, the probability that A will be found to be true on a system in state condition Z is given by (9.118).

S-59. If a physical quantity Q is expressed as in S-49, the expected (average) value of Q in Z is given by (9.119).

S-60. A "state" Z that is idempotent, that is,

$$Z^2 = Z, \tag{9.122}$$

is called a quantum state or pure state. Otherwise Z is called a mixed state.

This is because if Z is expressed as (9.117), $Z^2 = \sum_i \omega_i^2 \mathscr{T}[\psi(\alpha_i)]$ because of the orthogonality of the α and the idempotent law of the $\mathscr{T}$. Hence $Z^2 = Z$ is possible if and only if $\omega_i^2 = 0$ or 1.

S-61. Let $\mathfrak{N}_\nu$ $(\nu = 1, 2, \cdots, \lambda)$ be mutually exclusive and complete subspaces defined by the coordinate system $\psi(\alpha_i)$. If the ω_i of (9.117) are constant within each $\mathfrak{N}_\mu$, (9.117) can be written

$$Z = \sum_{\nu=1}^{\lambda} \frac{\omega_\nu}{\text{trace } \mathscr{T}[\mathfrak{N}_\nu]} \mathscr{T}[\mathfrak{N}_\nu], \tag{9.123}$$

where ω_ν is the sum of ω_i's corresponding to the subspace $\mathfrak{N}_\nu$.

S-62. A state Z, that gives an equal probability $\omega_i = 1/n$ to all $\psi(\alpha_i)$ is called a virgin state and is expressed, according to (9.93) and (9.123), by

$$Z_0 = \frac{1}{n} I. \tag{9.124}$$

The probability that a proposition A will be true on Z_0 is, according to (9.118), equal to the *a priori* probability of A, given by (9.99).

If the energy of the system is limited to a small domain, say between E and $E + \Delta E$, the virgin state corresponds to what is known as the "microcanonical ensemble."

Let us now examine if the probability (9.118) obeys the axioms of probabilities. The non-negativity of (9.118) derives directly from (9.112) and (9.117). $p(\emptyset \mid Z) = 0$ and $p(\square \mid Z) = 1$ are obvious because $\mathscr{T}[\mathfrak{M}(\emptyset)] = 0$ and $\mathscr{T}[\mathfrak{M}(\square)] = I$. The only remaining point of interest is to determine whether or not

$$p(A \cap B \mid Z) + p(A \cup B \mid Z) = p(A \mid Z) + p(B \mid Z) \tag{9.125}$$

holds for all Z, where A and B are not necessarily compatible. The relation (9.125) is, according to (9.118), equivalent to

$$\text{trace } (\{\mathscr{T}[\mathfrak{M}(A)] + \mathscr{T}[\mathfrak{M}(B)] - \mathscr{T}[\mathfrak{M}(A \cap B)] - \mathscr{T}[\mathfrak{M}(A \cup B)]\} \cdot Z) = 0. \tag{9.126}$$

If A and B are compatible, we have (9.94); hence (9.126) holds. If A and B are not compatible, however, (9.126) does not hold as a general rule. A counterexample to (9.126) can easily be presented. Suppose that A, B, and Z are all one-dimensional and $\mathcal{T}[\mathfrak{M}(A)] = Z = \mathcal{T}[\boldsymbol{\psi}]$ and $\mathcal{T}[\mathfrak{M}(B)] = \mathcal{T}[\boldsymbol{\varphi}]$ with the condition that A and B are neither equivalent nor mutually exclusive; that is, $\mathcal{T}[\boldsymbol{\varphi}]\mathcal{T}[\boldsymbol{\psi}] \neq 0$ and $\boldsymbol{\psi} \neq \boldsymbol{\varphi}$. Then we have

$$p(A \mid Z) = \text{trace}\,(\mathcal{T}[\boldsymbol{\psi}] \cdot \mathcal{T}[\boldsymbol{\psi}]) = 1,$$
$$p(B \mid Z) = \text{trace}\,(\mathcal{T}[\boldsymbol{\psi}] \cdot \mathcal{T}[\boldsymbol{\varphi}]) \neq 0,$$
$$p(A \cap B \mid Z) = \text{trace}\,(\mathcal{T}[\boldsymbol{\psi} \cap \boldsymbol{\varphi}] \cdot \mathcal{T}[\boldsymbol{\psi}]) = \text{trace}\,(0 \cdot \mathcal{T}[\boldsymbol{\psi}]) = 0,$$

and

$$p(A \cup B \mid Z) = \text{trace}\,(\mathcal{T}[\boldsymbol{\psi} \cup \boldsymbol{\varphi}] \cdot \mathcal{T}[\boldsymbol{\psi}]) = \text{trace}\,(\mathcal{T}[\boldsymbol{\psi}]) = 1.$$

Hence (9.126) does not hold.

S-63. For any given Z, the quantity $p(A \mid Z)$ can be considered as a probability for A in the sense of (9.125) insofar as the A's are limited within a family of compatible propositions. If the A's are allowed to be taken from a set of observational propositions that are not all compatible, the quantity does not behave as a usual probability of A in the sense of (9.125). Thus, the quantity $p(A \mid Z)$ defined for all $A \in \mathcal{A}$ and all $Z \in \mathfrak{Z}$ is a generalization of the notion of probability.

S-64. For all $A \in \mathcal{A}$, the quantity $p(A \mid Z)$ behaves as a probability in the sense of (9.125) regardless of compatibility if Z is the virgin state Z_0. However, if the A's involved are not all compatible, the conditional probabilities cannot be derived from $p(A \mid Z)$.

If $Z = Z_0$, (9.125) reduces to (9.96), each of the four probabilities involved becoming an *a priori* probability. Even in this case, as was shown at the end of Section 7.3, the quantity $p(A \cap B \mid Z)/p(B \mid Z)$ does not behave like a probability measure of B unless the distributive law is valid. This trouble does not exist in S-63 if all the A's are compatible.

The next important question is under what conditions we can reestablish the postulate of fixed predicate-set correspondence. Unless this postulate is upheld, the probabilities—even if they satisfy the mathematical axioms of probability—cannot be interpreted in the customary way. The postulate requires that $p(A \mid Z) = 1 - p(\neg A \mid Z)$ can be interpreted as the relative frequency of those object systems in the ensemble which make the proposition A true with certainty. In fact, the postulate assumes that each single sample of the object system has either to belong to or not belong to the truth set of A without any intrinsic ambiguity. According to the interpretation of Z as an ensemble mentioned in the foregoing, this requirement can be expressed as $Z = pE(A) + (1 - p)E(\neg A)$, where $p = p(A \mid Z)$ and $E(A)$ is an eigenstate of A. This entails that $\mathfrak{T}[\mathfrak{M}(A)]$ and Z are commutable. In order for this to be the case for all A's and all Z's under consideration, this commutability must be a universal rule for all A's and all Z's. This means, according to S-66 through S-71 which follow, that the observational propositions determining these Z have to be compatible with the A's. Conversely, it is obvious that if this last is the case, the postulate of fixed predicate-set correspondence becomes tenable.

S-65. In order that the postulate of fixed predicate-set correspondence and the usual logic and the usual meaning of probability based on this postulate become valid, it is necessary and sufficient that not only all A's but also all the observational propositions determining the state conditions Z's are all mutally compatible.

S-58 and S-59 give the prescription for deriving the prediction of an observational result provided that we know how to determine Z. Thus the remaining task is to give a formalism to determine Z at t from the result of an observation made at t_0 ($<t$). This process of determination breaks down to two steps: (a) determination of Z right after the observation at t_0 and (b) determination of the change in Z from t_0 to t. There are processes of determining a state that seem to be different from the one just described, but actually they are usually special cases of it. For instance, it is often said that if an electron is left alone for a long time in a fixed potential, it takes the lowest accessible energy state in that potential. Thus the state is determined without previous observation. This process can be considered as a special case of (b) if all the interacting physical entities are taken into consideration and if $(t - t_0) \to \infty$. Under the special circumstances of this case the state at t_0 within a certain limit does not affect the Z at t. Similarly, there is a method of preparing a desired state Z at t by applying a short-lived pulselike potential before t [L-1]. This is also a special case of Process (b), where a shocklike intervention takes place between t_0 and t and as a result the state Z at t_0 affects the state Z at t only very little. In any event, if we determine the basic processes (a) and (b), the formalism can be applied to various special cases. First we go through all the pertinent facts regarding Process (a).

S-66. If one makes an observation of the quantity $Q = \sum_\mu q_\mu \mathfrak{T}_\mu$ ($\mathfrak{T}_\mu = \mathfrak{T}[\mathfrak{M}(A_\mu)]$) on a system in state Z and if one obtains a particular result q_μ, the state immediately after the observation becomes

$$Z'_\mu = \frac{\mathfrak{T}_\mu Z \mathfrak{T}_\mu}{\text{trace}\,(\mathfrak{T}_\mu Z)}. \tag{9.127}$$

It is possible that the reader cannot find this formula in standard textbooks on quantum mechanics, but it is in perfect accord with the accepted usage of quantum mechanics. The reader can easily check that $Z' = \bar{Z}'$ and trace $Z' = 1$ and that Z' has non-negative eigenvalues.

Suppose that an extremely large number N of systems is in state Z. If we carry out observation $Q = \sum_\mu q_\mu \mathcal{T}_\mu$ on all of them, then, according to S-57, N trace $(\mathcal{T}_\mu Z)$ $(\mu = 1, 2, \cdots)$ will give the result q_μ. If we consider only those systems that give a particular result q_μ, the ensemble is represented by Z'_μ. The reader should note that what happens here corresponds to an "ensemble contraction" of earlier chapters. If we do not limit ourselves to those systems that give a particular q_μ but consider all the systems, the ensemble after the observation becomes the sum over μ of Z'_μ times trace $(\mathcal{T}_\mu Z)$. Hence we make the following statement.

S-67. If we perform an observation of a quantity $Q = \sum_\mu q_\mu \mathcal{T}_\mu$ on a system in Z, the expected state Z'' after observation is

$$Z'' = \sum_\mu \text{trace}\,(\mathcal{T}_\mu \cdot Z)\, Z'_\mu = \sum_\mu \mathcal{T}_\mu\, Z\, \mathcal{T}_\mu. \tag{9.128}$$

There is a fairly widespread misunderstanding that observation of Q on Z implies a transition from Z to Z''. But this is only the expected behavior of Z at observation. If we know the particular result q_μ, the state passes from Z to Z'_μ. The next two statements are corollaries of S-66 and S-67 for a special case in which $\mathcal{T}_\mu$ is one-dimensional. The reader can derive them from S-66 and S-67 by a simple computation.

S-68. If the eigenvalue q_μ is nondegenerate in the sense of S-51, the state after observation that gave $q_\mu = q_i$ is

$$Z'_i = \mathcal{T}[\psi_i] \tag{9.129}$$

irrespective of the anterior state Z. [If trace $(\mathcal{T}[\psi_i] \cdot Z) = 0$, the observer will never obtain result q_i.]

S-69. If all the eigenvalues of Q are nondegenerate in the sense of S-51, the expected state after an observation of Q is

$$Z'' = \sum_i \{\text{trace}\,(\mathcal{T}[\psi_i] \cdot Z)\}\mathcal{T}[\psi_i]. \tag{9.130}$$

Theorem S-68 is usually used in the explanation of the preparation of a state, because it does not depend on Z. There is another case in which the state after observation can be expressed only in terms of the operators defining the observation.

S-70. If the anterior state Z is the virgin state, the state after the observational result q_μ is

$$Z'_\mu = \frac{\mathcal{T}_\mu}{\text{trace}\,\mathcal{T}_\mu}. \tag{9.131}$$

The expected state after the observation of Q is again Z_0, since

$$Z'' = \sum_{\mu} \frac{1}{n} (\text{trace } \mathfrak{T}_\mu) Z'_\mu = \frac{1}{n} \sum_{\mu} \mathfrak{T}_\mu = Z_0. \tag{9.132}$$

Another simple corollary of S-66 and S-67 is S-71.

S-71. The states Z'_μ and Z'' after the observation of Q commute with Q.

$$\begin{aligned} Z'_\mu Q - QZ'_\mu &= 0, \\ Z''Q - QZ'' &= 0. \end{aligned} \tag{9.133}$$

So much for the determination of the state immediately after an observation. The next necessary ingredient of a predictive theory must concern how the state Z develops with time, say, from t_0 to t, so that one can make a prediction of the result of an observation at t on the basis of a previous observation at t_0. We can differentiate two cases regarding the change of state Z of a system from t_0 and t. In one case the system can be considered as "isolated" during the period from t_0 to t, in the sense that the interaction between the system and the rest of the universe is so weak that we can ideally neglect it completely. In the other case the system is open in the sense that such an interaction cannot be neglected. The latter case is divided into three subcases. In one, we can include part or all of the remaining world in the system so that the new combined system becomes almost isolated. In this subcase the method applicable to the isolated system can be used for the combined system. In the second subcase the interaction can be so formulated that we actually need not worry about the dynamic situation of the rest of the world. This is possible when, for instance, the effect of the external world can be represented in terms of a constant or a time-dependent but well-defined potential. In the third subcase the effect of the external world is so unpredictable that we have to use a statistical consideration with regard to the situation. Except for this third subcase, the following formalism is directly applicable. The following gives the basic postulates regarding the development of Z with time.

S-72. The state $Z(t)$ at t and the state $Z(t_0)$ at t, are related by a unitary transformation

$$Z(t) = TZ(t_0)T^{-1} = TZ(t_0)\bar{T}, \tag{9.134}$$

provided that no observation is made between t_0 and t.

On account of this property of time development, it is guaranteed that if trace $Z(t_0) = 1$ then trace $Z(t) = 1$, that if $Z(t_0)$ is Hermitian then so is $Z(t)$, and that the eigenvalues of $Z(t)$ are the same as those of $Z(t_0)$. This last property becomes obvious if we remember that the eigenvalues of a state Z (for that matter, of any Hermitian operator) are defined as the diagonal elements when the matrix is brought to a diagonal form by a suitable unitary transformation. Hence if $Z(t_0)$ is a state matrix (in the sense of S-57) then so is $Z(t)$. One may further note that if $Z(t_0)$ is the virgin state Z_0 then so is $Z(t)$ (Liouville's theorem). Also, if $Z(t_0)$ is a quantum state [i.e., if $Z^2(t_0) = Z(t_0)$], so is $Z(t)$.

It is evident from the definition of T in (9.134) that T has to depend in general on t and t_0; $T = T(t, t_0)$. It is also evident that

$$T(t, t_0) = 1 \quad \text{if} \quad t = t_0. \tag{9.135}$$

Suppose next that we increase t by an infinitesimal amount dt. Then we can expect in general that the difference between $T(t + dt, t_0)$ and $T(t, t_0)$ has a part proportional to dt. Thus we may write

$$T(t + dt, t_0) = (1 + K\,dt)T(t, t_0), \tag{9.136}$$

where K is a yet-undetermined operator. Neglecting any quantity proportional to the higher powers of dt, we can derive from (9.136)

$$T^{-1}(t + dt, t_0) = T^{-1}(t, t_0)(1 - K\,dt). \tag{9.137}$$

On the other hand, if we take the Hermitian conjugate [see (9.102)] of (9.136), we obtain

$$\bar{T}(t + dt, t_0) = \bar{T}(t, t_0)(1 + \bar{K}\,dt). \tag{9.138}$$

But since T is a unitary transformation,

$$T^{-1}(t, t_0) = \bar{T}(t, t_0)$$

and

$$T^{-1}(t + dt, t_0) = \bar{T}(t + dt, t_0).$$

Hence comparison of (9.137) and (9.138) yields

$$K = -\bar{K} \tag{9.139}$$

or

$$H = \bar{H} \tag{9.140}$$

if we define

$$H = iK. \tag{9.141}$$

If we use this H, (9.136) and (9.137) can be rewritten as

$$\frac{dT(t, t_0)}{dt} = -iHT(t, t_0) \tag{9.142}$$

and

$$\frac{dT^{-1}(t, t_0)}{dt} = +iT^{-1}(t, t_0)H. \tag{9.143}$$

Using these properties of T, we can rewrite the dynamic law (9.134) in the form of a differential equation,

$$\frac{dZ(t)}{dt} = \frac{dT}{dt} Z(t_0)T^{-1} + TZ(t_0)\frac{dT^{-1}}{dt} = -i[HZ(t) - Z(t)H]. \quad (9.144)$$

If $Z(t)$ is a quantum state, we can write $(Z(t))_{ij} = (\mathcal{T}[\psi(t)])_{ij} = \psi_i(t)\psi_j^*(t)$. Hence (9.144) becomes

$$\left(\frac{d\psi_i}{dt} + iH_{ik}\psi_k\right)\psi_j^* + \psi_i\left(\frac{d\psi_i^*}{dt} - \psi_j^* H_{jk}\right) = 0. \quad (9.145)$$

This can be satisfied by

$$\frac{d\psi_i}{dt} = -iH_{ik}\psi_k. \quad (9.146)$$

Obviously (9.146) is a sufficient condition and not a necessary condition for (9.145). This stems from the fact that $\mathcal{T}[\psi(t)]$ is not affected by multiplying $\psi(t)$ by a "phase factor" $e^{i\alpha(t)}$, where $\alpha(t)$ is a real-valued function of t. The choice of (9.146) among other possibilities amounts to fixing $\alpha(t)$ in a suitable fashion. When we consider a process of superposition $\psi = a\psi_1 + b\psi_2$, the ratio a/b has a real meaning; hence the phases of ψ_1 and ψ_2 are important. But as far as we consider the evolution of $\psi(t)$ in time, the phase of $\psi(t)$ can be arbitrarily chosen. Thus we state the following.

S-73. The time development of a quantum state $\psi(t)$ is governed by an equation of the type

$$\frac{d\psi}{dt} = -iH\psi, \quad (9.147)$$

where H is a Hermitian operator. This type of equation is called a Schrödinger equation, and the H (multiplied by a constant) is called Hamiltonian.

In (9.136) K was introduced to represent a small change of $T(t, t_0)$ at t; hence it is natural to assume that K depends on t but not on t_0. Therefore H is in general a function of t. This possibility allows us to include in H the effect of the external world, which is a function of t. In the so-called "Schrödinger picture," H is supposed not to depend on t except to represent a time-dependent external effect. In this picture all the Q's are supposed to be independent of t.

Now the expected value of a physical quantity Q in $Z(t)$ is given by

$$\langle Q \mid Z(t)\rangle = \text{trace } (Q \cdot Z(t)), \quad (9.148)$$

which, in view of (9.134), can also be written as

$$\begin{aligned}\langle Q \mid Z(t)\rangle &= \text{trace } (Q \cdot TZ(t_0)T^{-1}) \\ &= \text{trace } (T^{-1}QT \cdot Z(t_0)) \\ &= \langle T^{-1}QT \mid Z(t_0)\rangle.\end{aligned} \quad (9.149)$$

This last expression is the expected value of the quantity $T^{-1}QT$ in the initial state $Z(t_0)$. As a consequence, to obtain the expected value of a quantity at t, we can alternatively adopt a point of view that the state remains unchanged in time but the quantity changes with time according to the rule

$$Q(t) = T^{-1}Q(t_0)T. \tag{9.150}$$

This point of view is called the "Heisenberg picture."

In solving practical problems, the so-called "interaction picture" (see [T-3], for instance) is often used. This picture is somewhere between the Schrödinger picture and the Heisenberg picture and is particularly suitable for solving the problem in which the Hamiltonian H consists of a main part H_0 and a small "perturbation" H_1. The idea is as follows. Suppose that

$$H = H_0 + H_1 \tag{9.151}$$

and let us try to separate T into two factors,

$$T = T_0 T_1, \tag{9.152}$$

in such a way that T_0 is essentially governed by H_0 and T_1 is essentially governed by H_1. Equation (9.142) becomes, by insertion of (9.151) and (9.152),

$$\frac{dT_0}{dt} T_1 + T_0 \frac{dT_1}{dt} = -iH_0T_0T_1 - iH_1T_0T_1, \tag{9.153}$$

which can be solved if we put

$$\frac{dT_0}{dt} = -iH_0T_0 \quad \text{and} \quad \frac{dT_1}{dt} = -iH_1'T_1 \quad \text{with} \quad H_1' = T_0^{-1}H_1T_0. \tag{9.154}$$

Then the expected value of quantity Q at t can be written

$$\begin{aligned} \langle Q \mid Z(t)\rangle &= \text{trace}\,(Q \cdot T_0T_1Z(t_0)T_1^{-1}T_0^{-1}) \\ &= \text{trace}\,(T_0^{-1}QT_0 \cdot T_1Z(t_0)T_1^{-1}), \end{aligned} \tag{9.155}$$

which suggests an interpretation that the quantity Q varies with time according to

$$Q(t) = T_0^{-1}Q(t_0)T_0 \tag{9.156}$$

and the state evolves with time according to

$$Z(t) = T_1Z(t_0)T_1^{-1}. \tag{9.157}$$

This means that physical quantities change with time as in the Heisenberg picture with Hamiltonian H_0, whereas the state changes with time as in the Schrödinger picture with Hamiltonian $H_1' = T_0^{-1}H_1T_0$. It may be noted that H_1' and H_1 are connected in the definition (9.154) by the same relation as $Q(t)$ and $Q(t_0)$ in (9.156). This point of view is called an interaction picture, because usually H_1 is chosen to represent the Hamiltonian corresponding to the interaction between constituent entities of the system under consideration or the interaction of the system with the exterior system.

Coming back to the Schrödinger picture, let us spend a few paragraphs in deriving the so-called "transition probability," which has a close relation to our earlier discussion of temporal sequences. In the following the summation convention is suspended; that is, even if the same index appears twice in the same expression, the summation with respect to this index is not carried out except when it is explicitly stated. Let $\{\boldsymbol{\varphi}_1, \boldsymbol{\varphi}_2, \cdots, \boldsymbol{\varphi}_a, \cdots\}$ and $\{\boldsymbol{\psi}_1, \boldsymbol{\psi}_2, \cdots, \boldsymbol{\psi}_i, \cdots\}$ be two coordinate systems, and assume that we determined the state of a system at t_0 to be $\boldsymbol{\varphi}_c$ according to S-68. The transition probability in question is the probability of finding the system in $\boldsymbol{\psi}_k$ at t. This probability, $p(c \to k)$, must be given by (9.119), in which Q is replaced by $\mathcal{T}[\psi_k]$ and Z is replaced by $Z(t)$ of (9.134). Thus

$$p(c \to k) = \text{trace}\,(\mathcal{T}[\boldsymbol{\psi}_k] \cdot T\mathcal{T}[\boldsymbol{\varphi}_c]T^{-1}). \tag{9.158}$$

Let us write this expression in terms of matrix elements in the coordinate system $\{\boldsymbol{\psi}_i\}$. Then $(\mathcal{T}[\boldsymbol{\psi}_k])_{ij}$ is obviously $\delta_{ik}\,\delta_{ij}$, because $\mathcal{T}[\boldsymbol{\psi}_k]$ is diagonal in the coordinate system $\{\boldsymbol{\psi}_i\}$ and has 1 at the position corresponding to $\boldsymbol{\psi}_k$. If we introduce a unitary transformation U such that

$$\boldsymbol{\varphi}_a = \sum_a \boldsymbol{\psi}_i U_{ia} \tag{9.159}$$

as in (9.49), $(\mathcal{T}[\boldsymbol{\varphi}_c])_{ij}$ must be given by

$$(\mathcal{T}[\boldsymbol{\varphi}_c])_{ij} = \sum_a \sum_b U_{ia}(\mathcal{T}[\boldsymbol{\varphi}_c])_{ab}U_{bj}^{-1} = U_{ic}U_{cj}^{-1} \tag{9.160}$$

as in (9.70). Then (9.158) becomes

$$p(c \to k) = \sum_l \sum_m T_{kl}U_{lc}U_{cm}^{-1}\;T_{mk}^{-1}. \tag{9.161}$$

Now we sum this probability over all the possible final states ψ_k, and, because $T^{-1}T = 1$ and $U^{-1}U = 1$, we obtain

$$\sum_k p(c \to k) = \sum_l U_{lc}U_{cl}^{-1} = 1. \tag{9.162}$$

Similarly, if we sum (9.161) over all the possible initial states, we obtain, because $TT^{-1} = 1$ and $UU^{-1} = 1$,

$$\sum_c p(c \to k) = \sum_l T_{kl}T_{lk}^{-1} = 1. \tag{9.163}$$

The first relation (9.162) is necessary because $p(c \to k)$ is a probability distribution over k. The second relation (9.163) is an additional property that may be called inverse normalization. Both relations represent simply what was called "double stochasticity" in earlier chapters. It is usually claimed that double stochasticity in quantum mechanics is a consequence of the unitarity of T. But our derivation used only linearity of T and U (see [V-3, S-10, W-7]). We used $UU^{-1} = U^{-1}U = TT^{-1} = T^{-1}T = 1$, but not $U\bar{U} = \bar{U}U = T\bar{T} = \bar{T}T = 1$. The reason is that the usual derivation, which uses the state vectors ψ_n and φ_c directly instead of $\mathcal{T}[\psi_n]$ and $\mathcal{T}[\varphi_c]$, yields the expression

$$p(c \to k) = \left| \sum_l T_{kl} U_{lc} \right|^2 \tag{9.164}$$

instead of (9.161). In our present derivation, in order to prove $p(c \to k) \geq 0$, we have to replace U_{cm}^{-1} and T_{km}^{-1} by $\bar{U}_{cm} = U_{mc}^*$ and $\bar{T}_{mk} = T_{km}^*$ in (9.161), using the unitarity of U and T. Then (9.161) becomes (9.164) and there is no essential discrepancy.

The fact that we can define the transition probability may seem to provide a solid ground on which to derive the H-theorem using a Markov chain model. Double stochasticity in particular is such a convenient property that we may be tempted to jump to the conclusion that quantum mechanics provides a proof that the simple entropy $S = -\sum p(k) \log p(k)$ unconditionally increases with time. But this is too hasty, for the basic assumption of the Markov chain,

$$p(i \to j, 2t) = \sum_k p(i \to k, t) p(k \to j, t), \tag{9.165}$$

is not guaranteed in quantum mechanics, where $p(i \to k, t)$ is the transition probability from i to k during the time period t and $p(i \to j, 2t)$ is the transition probability from i to j during the time period $2t$. In quantum mechanics (9.165) would be true if observations were made every t seconds (and the average were taken with respect to these observational results), but if $p(i \to j, 2t)$ means the transition probability during $2t$ without an intermediate observation, (9.165) is not true. In order to re-establish the Markov chain condition in quantum mechanics, we have to use a coarse-grained observation and impose a fairly severe condition on the time interval that underlies the Markov chain (see [W-10] for details).

The fundamental framework so far constructed shows a great similarity to quantum mechanics, and has indeed proven to be capable of yielding some of the important consequences of quantum mechanics. This theoretical framework will remain a frame without a picture, however, unless and until we introduce Planck's constant and specify the unitary transformations connecting various physical quantities, even if we overlook the basic limitation inherent in our framework as a result of the assumed finiteness of the dimensions of the orthogonal space. On the other hand, the present skeleton theory can serve as a guiding model according to which one can build a full-fledged quantum mechanics. I do not undertake such an attempt in sufficient detail here, but try to give some explanation about the relation between the present framework and the real physical theory. The main purpose is to make it easy for the reader to understand the correspondence between the present exposition and the conventional ones. It may be well to anticipate from the beginning the role Planck's constant should play in quantum theory. In usual physical units, such as centimeter, gram, and second, Planck's constant, $\hbar$, which incidentally has the dimension of action (i.e., square centimeters per gram per second), is a very small number, $\approx 10^{-27}$ cgs. Hence it is negligible if we are dealing with large-scale physical phenomena. In complete parallelism with this situation, we know that the "non-Booleanity" of propositions does not let itself be felt when we are dealing with the large-scale phenomena, whereas it inevitably becomes conspicuous when we start to talk about microscopic phenomena involving atomic entities such as photons or electrons. Non-Booleanity is expressed, in terms of mathematical representations, as noncommutativity of some physical quantities. Hence it can be expected that Planck's constant is a measure of the noncommutativity of physical quantities and therefore of non-Booleanity of propositions. Thus, for instance, the basic noncommutative expression (9.144) is actually proportional to Planck's constant.

S-74. Boolean logic is not an infallible set of rules in properly handling all propositions in human experience. As far as observational propositions in physics are concerned, a measure of deviation from Boolean logic is given by Planck's constant. The use of Boolean logic in handling observational propositions about phenomena is permitted approximately in the measure to which the "action" involved in the phenomena becomes appreciably large compared with Planck's constant.

Readers who want more elucidation of this matter are referred to Appendix A9.3, where the connection between non-Booleanity and Planck's constant is discussed more closely with respect to the basic commutation relation of quantum mechanics.

9.5. NON-BOOLEAN INFORMATION THEORY IN QUANTUM PHYSICS

The main purpose of this section is to reformulate in quantum-mechanical language some of the basic relations useful in the information-theoretical dependence analysis presented in earlier chapters. This amounts to a generalization of the usual information theory on the basis of non-Boolean logic. Before doing this, however, I want to show the relationship between the formalism introduced in the foregoing sections and the customary

(Boolean) method used in statistical handling of observational data. This will also make it easier to see the relationship between quantum-mechanical information theory and the usual one.

In view of the fact that Boolean logic is a special case of modular logic and, moreover, that any modular lattice contains a Boolean lattice as a partial lattice, the method applicable to the modular lattice must reduce to the one used in ordinary statistics by imposing a certain additional restriction on the method. The restriction is that all physical quantities Q's and all states (density matrices) Z's under consideration are limited to those that commute with one another. Under this restriction all the projection operators $\mathfrak{P}$ that appear in the theory can be diagonalized by using a single orthogonal coordinate system, say, $\{\psi_i\}$, $i = 1, 2, \cdots$. Then each $\mathfrak{P}$ is characterized by its diagonal elements. Thus we do not need a double index (i, j) to specify matrix elements, but a single index i. For instance, if $\mathfrak{P}[\psi_k]$ is a one-dimensional projection operator corresponding to vector ψ_k, its (i, i)-element, or simply its ith element, is given by

$$(\mathfrak{P}[\psi_k])_i = \delta_{ik}. \tag{9.166}$$

Similarly, if $\mathfrak{P}[\mathfrak{M}]$ is a projection operator corresponding to a subspace $\mathfrak{M}$,

$$\begin{aligned}(\mathfrak{P}[\mathfrak{M}])_i &= 1 \quad \text{if} \quad \psi_i \in \mathfrak{M} \\ &= 0 \quad \text{if} \quad \psi_i \perp \mathfrak{M}.\end{aligned} \tag{9.167}$$

The idea of a subspace itself becomes very simple when we are—as now—actually dealing with subspaces belonging to a single family of orthogonal spaces. For such a subspace corresponds uniquely to a subset of $\{\psi_i\}$, and the operations of implication, conjunction, and disjunction reduce simply to the corresponding set-theoretical operations applied to subsets of $\{\psi_i\}$. The notion of a subspace is actually not necessary. All we need is simply the notion of subsets of the set of objects $\{\psi_i\}$, $i = 1, 2, \cdots$. Thus in the Boolean case a subspace $\mathfrak{M}$ is to be understood as a subset of ψ's. As a consequence the relation $\psi_i \perp \mathfrak{M}$ in (9.167) simply means that ψ_i belongs to a set that is the complement of the subset $\mathfrak{M}$ with respect to the entire set. Conversely, as we explained at length in Chapter 7 and in the earlier sections of the present chapter, the moment we can apply the concept of set to propositions (truth set), the distributive law becomes valid and the logic becomes Boolean.

The quantity Q in S-51 becomes

$$(Q)_i = \left(\sum_j q_j \mathfrak{P}[\psi_j]\right)_i = \sum_j q_j \delta_{ij} = q_i \tag{9.168}$$

and the state Z of (9.117) becomes

$$(Z)_i = \left(\sum_k \omega_k \mathfrak{P}[\psi_k]\right)_i = \sum_k \omega_k \delta_{ik} = \omega_i. \tag{9.169}$$

The expression (9.169) means that there is probability ω_k that the object system will be in state ψ_k. The expression (9.168) implies that Q is a quantity that has the value q_j if the object system is in state ψ_j. The basic relation (9.119) that gives the average value of Q in Z becomes, by the use of (9.168) and (9.169),

$$\langle Q \mid Z \rangle = \text{trace}\,(Q \cdot Z) = \sum_i Q_i Z_i = \sum_i q_i \omega_i, \tag{9.170}$$

which is nothing but the usual procedure of obtaining the average of q_i when there is probability ω_i for each q_i.

The procedures we used in determining the state by an observational procedure also have a simple interpretation. Suppose that we make an observation of a quantity $Q = \sum_\mu q_\mu \mathfrak{J}[\mathfrak{M}_\mu]$ and obtain the value q_μ, which corresponds to a subset $\mathfrak{M}_\mu$ defined by some ψ's. Then (9.127) becomes

$$(Z'_\mu)_i = \frac{(\mathfrak{J}_\mu)_i Z_i (\mathfrak{J}_\mu)_i}{\sum_i (\mathfrak{J}_\mu)_i Z_i}. \tag{9.171}$$

Using (9.167) and (9.169), we can rewrite this as

$$\begin{aligned} (Z'_\mu)_i &= \frac{\omega_i}{\sum_{j \in \mathfrak{M}_\mu} \omega_j} \quad \text{if} \quad \psi_i \in \mathfrak{M}_\mu \\ &= 0 \quad \text{if} \quad \psi_i \in \mathfrak{M}_\mu. \end{aligned} \tag{9.172}$$

This is exactly what we have consistently done in the past. For instance, (1.123) expresses exactly the same procedure. This is essentially a process of ensemble reduction, which we discussed in Chapter 5.

The only part of the non-Boolean formalism that cannot be directly translated into ordinary language by the same method is the dynamic law of the type (9.134) or (9.147). In fact, if all operators involved commute with one another, T and Z will also commute with each other, and (9.134) will lead to $Z(t) = Z(t_0)$. This should not be cause for surprise, because there is no reason why the true microscopic Hamiltonian H and the true microscopic unitary transformation T used in (9.134) or (9.147) should commute with the quantities Q we measure and with the state operator Z. In fact, in the case of mechanics, we can derive classical Newtonian equations of motion from Schrödinger equations by assuming that the state ψ is not the eigenstate of the Hamiltonian.

Similarly, we can derive the basic notion of transition probabilities, necessary for statistical mechanics, from quantum mechanics by assuming that the Hamiltonian or unitary transformation that governs the time development of the state does not commute with the thermodynamic quantities, and noting that state Z right after an observation made with thermodynamic quantities commutes with these thermodynamic quantities. This last point is justifiable for the following reason. Suppose that $Q^{(1)}$, $Q^{(2)}, \cdots Q^{(v)}$ are thermodynamic quantities, which commute with one another. Then we can express them as $Q^{(s)} = \sum_{\mu(s)} q_{\mu(s)} \mathcal{P}_{\mu(s)}$, where $\mathcal{P}_{\mu(s)}$ is the sum of the one-dimensional projection operators of the type $\mathcal{P}[\psi_i]$ corresponding to a subspace $\mathfrak{M}_{\mu(s)}$, in which $Q^{(s)}$ has the value $q_{\mu(s)}$. We can use a single coordinate system $\{\psi_i\}$ for different $Q^{(s)}$ because all $Q^{(s)}$ commute. Then we can divide the entire set $\{\psi_i\}$ into disjoint and complete subsets $\mathfrak{M}_1, \mathfrak{M}_2, \cdots$, in such a way that within a subset $\mathfrak{M}_\lambda$ each $Q^{(s)}$ has a constant value and for $\mathfrak{M}_\lambda$ and $\mathfrak{M}_\kappa$ $(\lambda \neq \kappa)$ there is at least one $Q^{(s)}$ that has different values in $\mathfrak{M}_\lambda$ and $\mathfrak{M}_\kappa$. In other words, $\mathfrak{M}_\lambda$ $(\lambda = 1, 2, \cdots)$ represents a state that cannot be divided into more than one part by thermodynamic means. These subspaces are called "thermodynamic cells." Let $\mathcal{P}_\lambda$ $(\lambda = 1, 2, \cdots)$ be the projector operators corresponding to these $\mathfrak{M}_\lambda$: $\mathcal{P}_\lambda = \mathcal{P}[\mathfrak{M}_\lambda]$. Then, after an observation that determined the object system in macroscopic cells μ, the state will become $Z'_\lambda = \mathcal{P}_\lambda Z \mathcal{P}_\lambda / \text{trace}\,(\mathcal{P}_\lambda Z)$, according to (9.127). We can see immediately that $\mathcal{P}_\kappa Z' - Z' \mathcal{P}_\kappa = 0$ for any κ whether or not the original state Z commutes with the $\mathcal{P}_\lambda$.

The transition probability from state (or thermodynamic cell) λ at $t = t_0$ to state κ at $t = t$ is given, according to (9.134) and (9.118), as

$$p(\lambda \to \kappa, t - t_0) = \text{trace}\,(\mathcal{P}_\kappa T(t, t_0) Z'_\lambda T^{-1}(t, t_0)), \qquad (9.173)$$

where Z'_λ is the state immediately after the first observation and should be, according to the above explanation,

$$Z'_\lambda = \frac{\mathcal{P}_\lambda Z \mathcal{P}_\lambda}{\text{trace}\,(\mathcal{P}_\lambda Z)}. \qquad (9.174)$$

where Z is the state just before the first observation. It may be noted that even though the state right after the first observation, Z', commutes with the $\mathcal{P}$'s, the state right before the second observation, $TZ'T^{-1}$, generally does not commute with the $\mathcal{P}$'s. We now note that the coordinate system $\{\psi_i\}$ is not uniquely determined by the thermodynamic quantities Q. We can indeed apply unitary transformations that map each $\mathfrak{M}_\lambda$ into itself. Since Z'_λ commutes with all $\mathcal{P}$'s, we can take advantage of this freedom and choose $\{\psi_i\}$ so that Z'_λ is also diagonalized. Using such vectors $\{\psi_i\}$, Z'_λ will be expressed simply as $Z'_\lambda = \sum_{i \in \lambda} \omega'_i \mathcal{P}[\psi_i]$, where the sum over the ψ's is limited to those ψ's included in $\mathcal{P}_\lambda = \sum_{i \in \lambda} \mathcal{P}[\psi_i]$. This last fact can be seen by noting that $\mathcal{P}_\lambda Z'_\lambda = Z'_\lambda \mathcal{P}_\lambda = Z'_\lambda$ and $\mathcal{P}_\kappa Z'_\lambda = Z'_\lambda \mathcal{P}_\kappa = 0$ for $\kappa \neq \lambda$. Thus the transition probability (9.173) can be written as

$$p(\lambda \to \kappa, t - t_0) = \sum_{i \in \lambda} \omega'_i \,\text{trace}\,(\mathcal{P}_\kappa \cdot T(t, t_0) \mathcal{P}[\psi_i] T^{-1}(t, t_0)). \qquad (9.175)$$

Usually at this stage we introduce a simplifying assumption, either that trace $(\mathcal{S}_\kappa T(t, t_0)\mathcal{S}[\psi_i]T^{-1}(t, t_0))$ does not depend strongly on the ψ's insofar as they are limited within a thermodynamic cell, or that ω_i' is a constant within $\mathcal{S}_\lambda$. We did the same thing in (3.34). If so (9.175) becomes approximately equal to

$$p(\lambda \to \kappa, t - t_0) = \frac{\text{trace}\,(\mathcal{S}_\kappa T(t, t_0)\mathcal{S}_\lambda T^{-1}(t, t_0))}{\text{trace}\,\mathcal{S}_\lambda}, \tag{9.176}$$

which depends only on λ, κ, and $t - t_0$. If we are allowed to use this expression, we can also replace Z_λ' of (9.174) simply by $Z_\lambda' = \mathcal{S}_\lambda/\text{trace}\,\mathcal{S}_\lambda$. This is why we often loosely state that if an observation of the quantity $Q = \sum_\mu q_\mu \mathcal{S}_\mu$ gives the result q_μ, the state becomes $\mathcal{S}_\mu/\text{trace}\,\mathcal{S}_\mu$. But this is rigorously true only if $\mathcal{S}_\mu$ is one-dimensional. Justification of this procedure for the more general case must be based on a consideration like the one just given.

Now let us define the entropy functions for a system in state Z. To do this we have to look into the nature of probability of the type (9.118), namely, $p(A \mid Z) = \text{trace}\,(\mathcal{S}[\mathfrak{M}(A)] \cdot Z)$. In order to have the basic rule of probability

$$p(A \mid Z) + p(B \mid Z) = p(A \cap B \mid Z) + p(A \cup B \mid Z) \tag{9.177}$$

for any Z, we have to have

$$\mathcal{S}[\mathfrak{M}(A)] + \mathcal{S}[\mathfrak{M}(B)] = \mathcal{S}[\mathfrak{M}(A \cap B)] + \mathcal{S}[\mathfrak{M}(A \cup B)]. \tag{9.178}$$

This is possible in general only if A and B commute [see (9.96) and (9.125)]. An exception is the case when Z in (9.177) is the virgin state Z_0 of (9.124), making (9.177) hold for any arbitrary A and B (see S-64). Hence, in order to define the entropy functions, we have to use a set of commuting quantities. Let $\mathcal{S}_\mu$, $\mu = 1, 2, \cdots$, be mutually commuting projection operators such that $\mathcal{S}_\mu \mathcal{S}_\nu = \delta_{\mu\nu}\mathcal{S}_\nu$, which entails $\mathcal{S}_\mu\mathcal{S}_\nu - \mathcal{S}_\nu\mathcal{S}_\mu = 0$ for $\mu \neq \nu$. The dimension of $\mathcal{S}_\mu$ will be denoted by D_μ: trace $\mathcal{S}_\mu = D_\mu$. These $\mathcal{S}$'s may or may not represent macroscopic cells in the thermodynamic sense. Let $\mathcal{S}_A$, $\mathcal{S}_B$, and so on be sums of some of the $\mathcal{S}_\mu$; that is, each of A, B, and so on logically corresponds to a disjunction of the μ's involved (see S-33). We then have (9.178) and we can apply the usual rules of probabilities. Furthermore, the reader can easily check, in the present case of mutually compatible propositions, that the conditional probability of A on condition C in Z given by $p(A/C \mid Z) = \text{trace}\,(\mathcal{S}_A\mathcal{S}_C Z)/\text{trace}\,(\mathcal{S}_C Z)$ satisfies the rule

$$p(A/C \mid Z) + p(B/C \mid Z) = p(A \cap B/C \mid Z) + p(A \cup B/C \mid Z). \tag{9.179}$$

The situation here is entirely different from the one discussed with reference to (9.35) through (9.39); here A, B, and C all commute.

Thus we can define the simple entropy (1.25) by

$$S(Q, Z) = -\sum_\mu \text{trace}\,(\mathcal{S}_\mu Z) \log \text{trace}\,(\mathcal{S}_\mu Z), \tag{9.180}$$

where the symbol Q is inserted in S as an argument, because we assume that we perform an observation of a quantity that can distinguish among the $\mathfrak{I}_\mu$'s. Such a quantity can be expressed as $Q = \sum_\mu q_\mu \mathfrak{I}_\mu$, with $q_\mu \neq q_\nu$ for $\mu \neq \nu$. A relative entropy of the type (1.29) and (1.38) can now be written

$$S(Q, Z) = -\sum_\mu \text{trace}\,(\mathfrak{I}_\mu Z) \log \frac{\text{trace}\,(\mathfrak{I}_\mu Z)}{D_\mu}, \tag{9.181}$$

with $D\ = \text{trace}\,\mathfrak{I}_\mu$. This is a quantum mechanical analogue of the classical Gibbsian entropy. von Neumann's famous quantum-mechanical H-theorem discusses the behavior of the quantity (9.181) in time development, putting $Z = Z(t) = T(t, t_0)Z(t_0)T^{-1}(t, t_0)$ and assuming $[Z(t_0)]^2 = Z(t_0)$, hence also $[Z(t)]^2 = Z(t)$ (quantum state) (see [V-2]). For a survey of the more recent developments of the discussion started by von Neumann on this problem, see for instance, [R-6].

A simple fact that can be mentioned here is that if Z at $t = t_0$ is given by $Z = \sum_\mu \omega_\mu \mathfrak{I}_\mu/\text{trace}\,\mathfrak{I}_\mu$ then $\text{trace}\,(\mathfrak{I}_\mu Z(t_0)) = \omega_\mu$ and $\text{trace}\,(\mathfrak{I}_\mu Z(t)) = \omega_{\mu'} = \sum_\lambda P_{\mu\lambda}\omega_\lambda$, where the transition probability $P_{\mu\lambda}$ is given by (9.176). Since this transition probability satisfies the condition $\sum_\lambda P_{\mu\lambda} D_\lambda = D_\mu$ along with $\sum_\mu P_{\mu\lambda} = 1$, we can conclude the H-theorem: $S(Q, Z(t_0)) \leq S(Q, Z(t))$, where $S(Q, Z(t))$ should be understood in the sense of (9.181) (see Theorem 5.9).

If the observation Q is maximal, that is, if $Q = \sum_i q_i \mathfrak{I}[\psi_i]$ with $q_i \neq q_j$ for $i \neq j$, then both entropies (9.180) and (9.181) reduce to

$$S(Q, Z) = -\sum_i \text{trace}\,(\mathfrak{I}[\psi_i], Z) \log \text{trace}\,(\mathfrak{I}[\psi_i], Z). \tag{9.182}$$

This quantity of course depends on Q, but if Z and Q commute with each other, that is, $Z = \sum_i \omega_i \mathfrak{I}[\psi_i]$ and $Q = \sum q_i \mathfrak{I}[\psi_i]$ with the same $\{\psi_i\}$, (9.182) becomes

$$S(Z) = -\sum_i \omega_i \log \omega_i = -\text{trace}\,(Z \log Z). \tag{9.183}$$

This quantity is determined solely by Z and coincides with $S(Q, Z)$ of (9.182) under the conditions stated above. The quantity $-\text{trace}\,(Z \log Z)$, which is called "microscopic entropy" of Z, was first introduced by von Neumann [V-3] and extensively used by Watanabe [W-2]. von Neumann used this quantity more or less as a simplified facsimile of the thermodynamic entropy, whereas I attached a new dependence-analytical meaning to it. This is an extremely convenient quantity because it is independent of Q and invariant for any unitary transformation applied to Z.

The second of the two expressions involved in (9.183) requires some explanation, because it involves a function of an operator Z. A function $f(Z)$ of an operator Z can be defined in two alternative ways. One way is to use the power series. For instance, by analogy with the numerical expansion

$$-\log x = \sum_{r=i}^{\infty} \frac{1}{r} (1 - x)^r, \tag{9.184}$$

which is valid for $0 < x \leq 2$, we define Z by

$$-\log Z = \sum_{r=1}^{\infty} \frac{1}{r} (1 - Z)^r \tag{9.185}$$

except when the vector $\boldsymbol{\varphi}$ on which the operator $\log Z$ is applied is such that $Z\boldsymbol{\varphi} = 0$. We are using the natural logarithm here, which of course does not impair the generality of the conclusion. On the right side of (9.185) we have only multiplication and addition of operators, whose meaning we already know. Now because $\mathcal{T}[\boldsymbol{\psi}_i]\mathcal{T}[\boldsymbol{\psi}_j] = \delta_{ij}\mathcal{T}[\boldsymbol{\psi}_i]$ and $\sum_i^{\text{all}} \mathcal{T}[\boldsymbol{\psi}_i] = 1$, we have

$$(1 - Z)^r = \left(1 - \sum_i^{\text{all}} \omega_i \mathcal{T}[\boldsymbol{\psi}_i]\right)^r = \left(\sum_i^{\text{all}} (1 - \omega_i)\mathcal{T}[\boldsymbol{\psi}_i]\right)^r = \sum_i^{\text{all}} (1 - \omega_i)^r \mathcal{T}[\boldsymbol{\psi}_i]. \tag{9.186}$$

Hence (9.185) becomes

$$-\log Z = -\sum_i^{\text{all}} (\log \omega_i)\mathcal{T}[\boldsymbol{\psi}_i]. \tag{9.187}$$

This result is a special case of the alternative definition of a function $f(Z)$ of an operator $Z = \sum_i^{\text{all}} \omega_i \mathcal{T}[\boldsymbol{\psi}_i]$:

$$f(Z) = \sum_i f(\omega_i)\mathcal{T}[\boldsymbol{\psi}_i]. \tag{9.188}$$

In the case when $\boldsymbol{\varphi}$ is such that $Z\boldsymbol{\varphi} = 0$, one may consider the result of application of $-(\log Z)$ on such a $\boldsymbol{\varphi}$ as the limit of $c\boldsymbol{\varphi}$ (c is positive) for $c \to +\infty$. As regards the operator $-Z_1 \log Z_2$, which we presently encounter, we should require an additional condition that the result of application of this operator on $\boldsymbol{\varphi}$ is zero if $Z_2\boldsymbol{\varphi} = 0$ and $Z_1\boldsymbol{\varphi} = 0$; this is the generalization of the usual convention $0 \log 0 = 0$.

In the following we discuss the properties of the microscopic entropy $S(Z)$ of (9.183) as a tool for analysis of interdependence existing in a quantum-mechanical system. In the usual (Boolean) case, all the basic formulas were derived from the Gibbs theorem, $-\sum_i p_i \log q_i \geq -\sum_i p_i \log p_i$ (see Theorem 1.1). We therefore mention its quantum-mechanical analog, which I introduced and proved in [W-18].

Theorem 9.11 For any two state conditions (density matrices) Z_1 and Z_2 we have

$$-\text{trace}\,(Z_1 \log Z_2) \geq -\text{trace}\,(Z_1 \log Z_1) \tag{9.189}$$

whether or not Z_1 and Z_2 commute. The equality in (9.189) holds if, and only if, $Z_1 = Z_2 \cdot$ (Z_1 is obviously assumed to be such that $-\text{trace}\, Z_1 \log Z_1 < \infty$.)

Proof. See Appendix A9.4.

Once we have established the generalized Gibbs theorem, we can derive various useful results from it. The most important is the quantum (non-Boolean) analogue of Theorem 1.2, regarding the additivity of entropy and probabilistic independence. To introduce such a theorem, we have to find out what should be the quantum analogue of the concept of "probabilistic independence." Let us take any physical system that can be described by two groups of variables, x and y. If, for instance, the system is an elementary particle, x may stand for the spatial coordinates and y for the spin and isospin. If the system is many-bodied consisting of N particles, x may stand for all the spatial and spin coordinates of a group of N_1 ($<N$) particles taken from the N particles, and y may stand for all the spatial and spin coordinates of the remaining $N - N_1$ particles. There can be infinitely many coordinate systems to represent the degrees of freedom referring to x, but for simplicity let us take one coordinate system $\{\psi_i(x)\}$, $i = 1, 2, \cdots$, and use i as the label of a quantum state referring to x. Similarly, let us use $\{\psi_j(y)\}$, $j = 1, 2, \cdots$, as the coordinate system for the degrees of freedom represented by y, and use j as the label of a quantum state referring to y. Then the state Z of the system may be represented in the matrix form

$$\langle i, j|\, Z\, |i', j'\rangle. \tag{9.190}$$

Now consider a physical quantity Q_x, which refers to x and can be written as $Q_x = \sum_i q_i \mathcal{T}[\psi_i(x)]$. Similarly, we consider a physical quantity R_y, which refers to y and can be written as $R_y = \sum_j r_j \mathcal{T}[\varphi_j(y)]$. We assume that Q_x and R_y are nondegenerate. The probability of getting the value q_i by observation of Q_x in Z is obviously given by

$$\Pr\{q_i \mid Z\} = \sum_j \langle i, j|\, Z\, |i, j\rangle = \langle i|\, Z_x\, |i\rangle, \tag{9.191}$$

where Z_x is defined by

$$\langle i|\, Z_x\, |i'\rangle = \sum_j \langle i, j|\, Z\, |i', j\rangle, \tag{9.192}$$

which means that the trace is taken with respect to y only. Similarly, the probability of obtaining result r_j by the observation of R_y in Z is given by

$$\Pr\{r_j \mid Z\} = \sum_i \langle i, j|\, Z\, |i, j\rangle = \langle j|\, Z_y\, |j\rangle, \tag{9.193}$$

where

$$\langle j|\, Z_y\, |j'\rangle = \sum_i \langle i, j|\, Z\, |i, j'\rangle. \tag{9.194}$$

Finally, we ask the probability of getting q_i by Q_x and r_j by R_y simultaneously. We may reinterpret this as the probability of obtaining q_{ij} by observing the quantity $Q_{xy} = \sum_{ij} q_{ij} \mathcal{T}[\psi_i(x)]\mathcal{T}[\varphi_j(y)]$, where the q_{ij} can be arbitrarily chosen under the condition that they are not degenerate. This probability is obviously

$$\Pr\{q_i, r_j \mid Z\} = \langle i, j| Z |i, j\rangle \tag{9.195}$$

Now the condition of independence,

$$\Pr\{q_i, r_j \mid Z\} = \Pr\{q_i \mid Z\} \Pr\{r_j \mid Z\}, \tag{9.196}$$

can be reinterpreted in view of (9.191), (9.193), and (9.195) as

$$\langle i, j| Z |i, j\rangle = \langle i, j| Z' |i, j\rangle, \tag{9.197}$$

where Z' is defined by

$$\langle i, j| Z' |i'j'\rangle = \langle i| Z_x |i'\rangle\langle j| Z_y |j'\rangle. \tag{9.198}$$

The reader can easily convince himself that this Z' is a possible state matrix for the object system, by noting that Z' can be written

$$Z' = \sum_{a,b} \omega_a\omega_b \mathfrak{T}[\xi_a(x)]\mathfrak{T}[\eta_b(y)] \tag{9.199}$$

where Z_x and Z_y are $Z_x = \sum_a \omega_a \mathfrak{T}[\xi_a(x)]$ and $Z_y = \sum_b \omega_b \mathfrak{T}[\eta_b(y)]$. Insofar as we consider unitary transformations that do not mix up the x-part and y-part, the product form (9.198) will remain unchanged and (9.197) can be considered the condition of independence for any pair of observables Q_x and R_y. Hence the complete independence of the x-part and the y-part in Z can be characterized by

$$Z = Z'. \tag{9.200}$$

Let us consider the microscopic entropies for Z, Z', Z_x, and Z_y, which are

$$S(Z) = -\text{trace}_{x,y}\,(Z \log Z),$$

$$S(Z') = -\text{trace}_{x,y}\,(Z' \log Z'),$$

$$S(Z_x) = -\text{trace}_x\,(Z_x \log Z_x),$$

and

$$S(Z_y) = -\text{trace}_y\,(Z_y \log Z_y).$$

We are particularly interested in the fact that the mixed expression $-\text{trace}$ $(Z \log Z')$ can be written

$$-\text{trace}\,(Z \log Z') = S(Z_x) + S(Z_y). \tag{9.201}$$

To derive this one need note only that

$$\langle i, j| \log Z' |i', j'\rangle = \langle i| \log Z_x |i'\rangle\delta(j, j') + \delta(i, i')\langle j| \log Z_y |j'\rangle, \tag{9.202}$$

which is easily proved with the help of (9.187) and the expression (9.199).

We can now enunciate a generalized version of Theorem 1.2.

Theorem 9.12 If Z_x and Z_y are derived from Z by (9.192) and (9.194) then

$$S(Z_x) + S(Z_y) \geq S(Z). \tag{9.203}$$

Equality holds in (9.203) if and only if, $Z = Z_xZ_y$, that is, if, and only if, the result of any observation in the x-part and the result of any observation in the y-part in Z are completely independent.

Proof. Substitute Z and Z' for Z_1 and Z_2 of Theorem 9.11. The condition for equality $Z_1 = Z_2$ becomes $Z = Z' = Z_xZ_y$.

We may now introduce

$$J(x, y \mid Z) = S(Z_x) + S(Z_y) - S(Z) \geq 0 \tag{9.204}$$

as a measure of the extent to which x and y in Z deviate from the independent situation.

DEFINITION 9.2. The quantity (9.204) is called interdependence between and x-part and the y-part in Z.

We can sometimes split the x-part (and for that matter the y-part) into two parts and apply Theorem 9.12 to this splitting. Repeating this process, we can derive a generalization of Theorem 1.3. Let the system under consideration be described by N groups of variables $x^{(1)}, x^{(2)}, \cdots, x^{(N)}$, each of which has a coordinate system $\{\psi_{i(I)}(x^{(I)})\}$, $i = 1, 2, \cdots, I = 1, 2, \cdots, N$. The index $i(I)$ labels a quantum state in the Ith group of variables. Then the state Z for this system can be expressed as a matrix of the form

$$\langle i(1), i(2), \cdots, i(N)|\, Z\, |i'(1), i'(2), \cdots, i'(N)\rangle. \tag{9.205}$$

Let us define

$$\begin{aligned}&\langle i(I)|\, Z^{(I)}\, |i'(I)\rangle\\ &= (\textstyle\sum)^{N-1}\langle i(1), \cdots, i(I), \cdots, i(N)|\, Z\, |i(1), \cdots, i'(I), \cdots, i(N)\rangle,\end{aligned} \tag{9.206}$$

where $(\sum)^{N-1}$ means the summation with respect to $i(1), \cdots, i(N)$, except for $i(I)$. In other words, $Z^{(I)}$ is obtained from Z by taking trace with respect to the indices except the one corresponding to $x^{(I)}$. Then we have the following from Theorem 9.12.

Theorem 9.13 If the $Z^{(I)}$, $I = 1, 2, \cdots, N$, are derived from Z by (9.206) then

$$\sum_{I=1}^{N} S(Z^{(I)}) \geq S(Z). \tag{9.207}$$

Equality holds in (9.207) if, and only if, $Z = \prod_{I=1}^{N} Z^{(I)}$, that is, if, and only if, all observational results with respect to different parts $x^{(I)}$, $I = 1, 2, \cdots, N$, are independent.

DEFINITION 9.3. The amount of interdependence among the parts $x^{(I)}$, $I = 1, 2, \cdots N$, in Z is given by

$$J(x^{(1)}, x^{(2)}, \cdots, x^{(N)} \mid Z) = \sum_{I=1}^{N} S(Z^{(I)}) - S(Z) \geq 0. \tag{9.208}$$

The fundamental theorem of interdependence analysis, Theorem 2.1, obviously also has its quantum-mechanical analogue.

Theorem 9.14 Suppose that we divide the entire set of x's in a successive polychotomic fashion (in a taxomic tree). Then the sum of interdependences of the type (9.208) taken at each branching point is a constant independent of the way the polychotomic tree is made, provided that the last stage division $[x^{(1)}, x^{(2)}, \cdots, x^{(N)}]$ is kept fixed.

Coming back to the case of the dichotomy of Z into Z_x and Z_y, let us consider whether we can push the analogy of the Boolean case still further in our non-Boolean case. Of course, the analogy of the formulas of Section 2.2 with our present case lies in the correspondence of $S(\mathcal{E})$, $S(\mathcal{F})$, and $S(\mathcal{G})$ with $S(Z_x)$, $S(Z_y)$, and $S(Z)$, respectively. In Chapter 2 we could show $S(\mathcal{F} \mid \mathcal{E}) = S(\mathcal{G}) - S(\mathcal{E})$. This allowed us to rewrite $J = S(\mathcal{E}) + S(\mathcal{F}) - S(\mathcal{G})$ as $J = S(\mathcal{E}) - S(\mathcal{E} \mid \mathcal{F})$ and $J = S(\mathcal{F}) - S(\mathcal{F} \mid \mathcal{E})$. Then, using the fact that $S(\mathcal{E} \mid \mathcal{F}) \geq 0$ and $S(\mathcal{F} \mid \mathcal{E}) \geq 0$, we concluded $J \leq \min\,(S(\mathcal{E}), S(\mathcal{F}))$. The maximum value of J for a given $S(\mathcal{E})$ or a given $S(\mathcal{F})$ was reached when $S(\mathcal{E} \mid \mathcal{F}) = 0$ or $S(\mathcal{F} \mid \mathcal{E}) = 0$, depending on the case. $S(\mathcal{E} \mid \mathcal{F}) = 0$ meant complete dependence of $\mathcal{E}$ on $\mathcal{F}$ and $S(\mathcal{F} \mid \mathcal{E}) = 0$ meant complete dependence of $\mathcal{F}$ on $\mathcal{E}$. This fact was used as a further argument for calling J interdependence, because it takes its minimum value when $\mathcal{E}$ and $\mathcal{F}$ are independent and its maximum value when either one depends completely on the other. We now show that we can reproduce these facts only partially in the non-Boolean case.

According to the interpretation given to (9.191), (9.193), and (9.195), the conditional probability of obtaining r_j on the condition of obtaining q_i must be given by (*Note*. Q_x and R_y are compatible)

$$\Pr\{r_j/q_i \mid Z\} = \frac{\Pr\{q_i, r_j \mid Z\}}{\Pr\{q_i \mid Z\}} = \frac{\langle i, j|\, Z\, |i, j\rangle}{\langle i|\, Z_x\, |i\rangle}. \tag{9.209}$$

Hence we can define the average conditional entropy of the y-part as measured by $R_y = \sum r_j \mathcal{P}[\varphi_j(y)]$, given the x-part as measured by $Q_x = \sum q_i \mathcal{P}[\psi_i(x)]$, by analogy with (1.66) by,

$$\begin{aligned} S(R_y/Q_x \mid Z) &= -\sum_i \sum_j \langle i, j|\, Z\, |i, j\rangle \log \frac{\langle i, j|\, Z\, |i, j\rangle}{\langle i|\, Z_x\, |i\rangle} \\ &= -\sum_i \sum_j \langle i, j|\, Z\, |i, j\rangle \log \langle i, j|\, Z\, |i, j\rangle \\ &\quad + \sum_i \langle i|\, Z_x\, |i\rangle \log \langle i|\, Z_x\, |i\rangle \geq 0. \end{aligned} \tag{9.210}$$

The last inequality ensues from the fact that $0 \leq p\{r_j/q_i \mid Z\} \leq 1$. The parallelism with the Boolean case mentioned above would suggest that (9.210) may be equated to $S(Z) - S(Z_x)$, but this is not always possible, because the quantity (9.210) is not invariant for an arbitrary unitary transformation. In order that the first term in (9.210) may be equated with $S(Z) = -\sum_i \sum_j \langle i, j|\, Z \log Z\, |i, j\rangle$, it is necessary that Z be diagonal in the

ψ_i-φ_j-representation. Similarly, in order that the second term in (9.210) may be equated with $S(Z_x) = -\sum_i \langle i|\, Z_x \log Z_x |\, i\rangle$, it is necessary that Z_x be diagonal in the ψ_i-representation. Thus we can conclude the following.

Theorem 9.15 If Z_x is the x-part of Z in the sense of (9.192) and if there exists a coordinate system $\{\psi_i(x)\}$ for the x-part and a coordinate system $\{\varphi_j(y)\}$ for the y-part such that Z_x becomes diagonal in the ψ_i-representation and Z becomes diagonal in the ψ_i-φ_j-representation, $S(Z) - S(Z_x)$ can be shown to be non-negative and interpreted as the average conditional entropy of the y-part as measured by R_y given the x-part as measured by Q_x in Z, where R_y is nondegenerate and diagonal in the φ_j-representation and Q_x is nondegenerate and diagonal in the ψ_i-representation. If the diagonalization condition about Z and Z_x is not satisfied, $S(Z) - S(Z_x)$ may become negative.

The last statement will presently be shown to be true by a simple example. This situation, $S(Z) - S(Z_x) < 0$, is contrary to what we had in the Boolean case, $S(\mathcal{G}) - S(\mathcal{E}) = S(\mathcal{F} \mid \mathcal{E}) \geq 0$, and we must be careful in handling the formulas involving conditional entropies in quantum-mechanical information theory. $S(Z) - S(Z_x)$ becomes non-negative, however, and can be interpreted as an average conditional entropy under the above given restrictive condition; hence the relation $S(Z) - S(Z_x) = 0$ under this condition means the complete dependence of the y-part on the x-part in a certain sense. This fact provides a further argument for considering $J = S(Z_y) - [S(Z) - S(Z_x)]$ as a measure of interdependence in the non-Boolean case too. In fact, if Z is such that there exist $\{\Psi_i\}$ and $\{\Psi_j\}$ that simultaneously make Z, Z_x, and Z_y diagonal, $J \leq \min\,(S(Z_x), S(Z_y))$ and the maximum value of J is reached if the x-part is completely dependent on the y-part or vice versa, where the term "complete dependence" is to be understood as vanishing of the conditional entropy in the sense of Theorem 9.15.

The diagonalization condition in Theorem 9.15 is readily satisfied if the x-part and the y-part are entirely different in nature. For instance, if the y-part represents an atomic object under examination and the x-part represents a macroscopic instrument of observation, the condition may be assumed to be approximately fulfilled. In fact, the possibility of observation of an atomic object by a macroscopic instrument, as assumed in von Neumann's theory of observation [V-3], hinges on the kind of dependence we discussed in Theorem 9.15. However, if the x-part and y-part are of the same kind—if, for instance, both x and y stand for groups of electrons or photons—then the diagonalization condition usually cannot be satisfied, owing to the basic indistinguishability of the particles. In the following we discuss the relation between the entropy of the whole system and the entropies of its constituent particles in the case of an assembly of fermions, such as electrons and nucleons.

Let x_I $(I = 1, 2, \cdots, N)$ denote the spatial coordinates and internal variables such as spin and isotopic spin of the Ith fermion. The state function of this N-fermion system $\Psi^{(N)}$ $(x_1, x_2, \cdots, x_N)$ must satisfy

$$\Psi^{(N)}(x_1, x_2, \cdots, x_N) = \rho(\pi)\Psi^{(N)}[\pi(x_1, x_2, \cdots, x_N)], \qquad (9.211)$$

where $\pi(x_1, x_2, \cdots, x_N)$ means any permutation π of the variables $(x_1, x_2, \cdots, x_N)$, and $\rho(\pi) = +1$ if π is an even permutation and $\rho(\pi) = -1$ if π is an odd permutation. Equation (9.211), if it is to be satisfied for every possible permutation π, represents the Pauli principle. The state matrix $Z^{(N)}$ corresponding to this $\Psi^{(N)}$ is, of course, given by

$$\langle x_1, x_2, \cdots, x_N| Z^{(N)} |x_1', x_2', \cdots, x_N'\rangle$$
$$= \Psi^{(N)}(x_1, x_2, \cdots, x_N)\Psi^{(N)*}(x_1', x_2', \cdots, x_N'). \qquad (9.212)$$

This $Z^{(N)}$ is multiplied by $\rho(\pi)$ if we apply a permutation π to either of the two sets of variables $(x_1, \cdots, x_N)$ and $(x_1', \cdots, x_N')$, but it remains unchanged if we apply the permutation π to both sets of variables. We now take any assembly of $M \leq N$ fermions within this N-fermion assembly and ask its state. This state will be given by the M-fermion state matrix $Z^{(M)}$, which is derived from $Z^{(N)}$ by the following recurrence formula:

$$\langle x_1, \cdots, x_M| Z^{(M)} |x_1', \cdots, x_M'\rangle$$
$$= \int dx_{M+1}\langle x_1, \cdots, x_M, x_{M+1}| Z^{(M+1)} |x_1', \cdots, x_M', {}_{M+1}\rangle \qquad (9.213)$$

where the integral also implies a summation with respect to the spin and isospin coordinates. Equation (9.213) means that the trace is taken with respect to x_{M+1}, but since $Z^{(M+1)}$ remains unchanged by the simultaneous permutations of $(x_1, \cdots, x_{M+1})$ and $(x_1', \cdots, x_{(M+1)}')$, it does not make any difference with which coordinate the trace is taken. $Z^{(M)}$ is antisymmetrical for an odd permutation of the arguments $(x_1, \cdots, x_M)$ and also for an odd permutation of the arguments $(x_1', \cdots, x_M')$. Since we assume $\Psi^{(N)}$ to be normalized, we have

$$Z^{(0)} = 1, \qquad (9.214)$$

which implies trace $Z^{(M)} = 1$ for any M.

Now let us make our representation more specific by introducing an orthogonal and normalized complete function system $\{\varphi_i(x_I)\}$, $i = 1, 2, \cdots$, for the Ith particle [$\varphi_i(x_I)$ is the x_I-representation of a vector $\boldsymbol{\varphi}_i$]. Since the particles are all identical, we can use the same functions φ for all particles. Take a set μ of M functions $\mu = \{\varphi_{\mu 1}, \varphi_{\mu 2}, \cdots, \varphi_{\mu M}\}$ out of the set $\{\varphi_i\}$ and form a "Slater function."

$$\Psi_\mu^{(M)}(x_1, \cdots, x_M) = \frac{1}{\sqrt{M!}} \begin{vmatrix} \psi_{\mu 1}(x_1) & \psi_{\mu 2}(x_1) & \cdots & \psi_{\mu M}(x_1) \\ \psi_{\mu 1}(x_2) & \psi_{\mu 2}(x_2) & \cdots & \psi_{\mu M}(x_2) \\ \cdots & \cdots & \cdots & \cdots \\ \psi_{\mu 1}(x_M) & \psi_{\mu 2}(x_M) & \cdots & \psi_{\mu M}(x_M) \end{vmatrix}. \tag{9.215}$$

If we want to define the sign of $\Psi_\mu^{(M)}$ uniquely, we can agree that the sequence of ψ's in $\mu = \{\psi_{\mu 1}, \psi_{\mu 2}, \cdots \psi_{\mu M}\}$ is taken in the ascending order of the i's in the $\{\psi_i\}$. Taking all possible combinations of M functions from $\{\psi_i\}$, we can form an orthogonal and normalized complete coordinate system $\{\Psi_\mu^{(M)}\}$, $\mu = 1, 2, \cdots$, for the M-fermion systems. To designate each such function we use the symbol $\mu^{(M)}$ when it is necessary to indicate how many particles are involved. Each such function $\mu^{(M)}$ can also be expressed in terms of the coordinates $\{\psi_i\}$; that is, in terms of Dirac's notation the function (9.125) can be expressed as

$$\langle i_1, i_2, \cdots, i_M \mid \mu^{(M)} \rangle = \frac{1}{\sqrt{M!}}\, \rho(\pi) \tag{9.216}$$

if there is a parameter π such that

$$\pi(i_1, i_2, \cdots, i_M) = (\mu_1, \mu_2, \cdots, \mu_M), \tag{9.217}$$

and as zero otherwise.

Using the N-particle Slater functions, we can write any $\Psi^{(N)}$ of (9.211) as

$$\Psi^{(N)} = \sum_\mu a_\mu \Psi_\mu^{(N)}, \tag{9.218}$$

with

$$\sum_\mu |a_\mu|^2 = 1. \tag{9.219}$$

If we express $Z^{(N)}$ in terms of the coordinate system $\{\Psi_\mu^{(N)}\}$, writing $\mu^{(N)}$ to denote $\Psi_\mu^{(N)}$, we get

$$\langle \mu^{(N)} | \, Z^{(N)} \, | \mu^{(N)\prime} \rangle = a_\mu a_{\mu'}^*, \tag{9.220}$$

where a_μ is the coefficient of $\Psi_\mu^{(N)}$ in (9.218). With the aid of the recurrence formula (9.213), we can derive $Z^{(M)}$ for $M < N$ by simple calculation. The result, expressed in the coordinate system of $\{\Psi_\mu^{(M)}\}$, will be of the form

$$\langle \mu^{(M)} | \, Z^{(M)} \, | \mu^{(M)\prime} \rangle = \binom{N}{M}^{-1} A(\mu^{(M)}, \mu^{(M)\prime}), \tag{9.221}$$

where $A(\mu^{(M)}, \mu^{(M)\prime})$ is a sum of terms of the type $a_\mu a_{\mu'}^*$ with appropriate signs. The pairs a_μ and $a_{\mu'}^*$ that appear in this sum must obviously satisfy the following condition: The two sequences of N functions, $\mu^{(N)}$ and $\mu^{(N)\prime}$, corresponding to a_μ and $a_{\mu'}$, contain, respectively, the sequences of M functions $\mu^{(M)}$ and $\mu^{(M)'}$, and the remaining $(N - M)$ functions in $\mu^{(N)}$ and $\mu^{(N)'}$ coincide. In particular, it is easy to see that the diagonal elements of $Z^{(M)}$ in the representation of $\{\Psi_\mu^{(M)}\}$ will be

$$\langle\mu^{(M)}|\, Z^{(M)}\, |\mu^{(M)}\rangle = \binom{N}{M}^{-1} \sum_\mu a_\mu a_\mu^*, \tag{9.222}$$

where the summation should be extended over all those $\mu^{(N)}$ that contain $\mu^{(M)}$. Since we have the normalization (9.219), each of the diagonal elements of the matrix $Z^{(M)}$ in the $\mu^{(M)}$-representation is $\leq \left[1 \Big/ \binom{N}{M}\right]$:

$$\langle\mu^{(M)}|\, Z^{(M)}\, |\mu^{(M)}\rangle \leq \binom{N}{M}^{-1} \tag{9.223}$$

The equality in (9.223) obviously will hold if, and only if, all $\mu^{(N)}$ involved in the expansion (9.218) contain the given $\mu^{(M)}$. A special case of (9.223) for $M = 1$ is

$$\langle i|\, Z^{(1)}\, |i\rangle \leq \frac{1}{N}, \tag{9.224}$$

where i refers to ψ_i, because $\Psi_\mu^{(M)}$ of (9.215) for $M = 1$ is nothing but one of the ψ's. The equality in (9.224) will hold if, and only if, all $\mu^{(N)}$ contain the given function ψ_i.

If we use an arbitrary one-particle function system $\{\psi_i\}$ in forming an N-particle function system $\{\Psi_\mu^{(N)}\}$, there is no guarantee that $Z^{(1)}$ will become diagonal in the ψ_i-representation. But for a given $Z^{(N)}$ there will be a one-particle function system, say, $\{\varphi_j\}$, that diagonalizes the derived $Z^{(1)}$. We could also have used this $\{\varphi_j\}$ to build $\{\Psi_\mu^{(N)}\}$ from the beginning. Hence we obtain the following theorem.

Theorem 9.16 The one-particle entropy $S(Z^{(1)})$ in an N-fermion system is equal to or larger than $\log N$:

$$S(Z^{(1)}) \geq \log N. \tag{9.225}$$

The equality in (9.225) holds if, and only if, the original N-particle system can be expressed as a single Slater matrix.

Proof. If we use the one-particle function $\{\varphi_j\}$ that diagonalizes $Z^{(1)}$, we can write $S(Z^{(1)}) = -\sum_j \langle j|\, Z^{(1)}\, |j\rangle \log \langle j|\, Z^{(1)}\, |j\rangle$, where $\langle j|\, Z^{(1)}\, |j\rangle \leq 1/N$ according to (9.224). Writing $p_j = \langle j|\, Z^{(1)}\, |j\rangle$, we have $S(Z^{(1)}) = -\sum_j p_j \log p_j \geq -\sum_j p_j \log (1/N) = \log N$. Since $S(Z^{(1)})$ is invariant for a unitary transformation, this result does not depend on the function system $\{\varphi_j\}$. This proof also shows that the equality in (9.225) will hold if, and only if,

$p_j = 1/N$ for N different φ_j's and is zero for all the rest. This means that the equality in (9.224) holds for exactly N functions φ_j; that is, $\langle j| Z^{(1)} |j\rangle = 1/N$ for these N φ_j's and $\langle j| Z^{(1)} |j\rangle = 0$ for all other φ_j's. This in turn means that there should be only one $\mu^{(N)}$ (containing N φ_j's) in the expansion (9.218). This Slater function is a determinant made with the help of N eigenfunctions of $Z^{(1)}$. Theorem 9.16 was first published in 1939, see [W-2].

The preceding discussion is based on the assumption that the N-fermion system is in a pure quantum state $[(Z^{(N)})^2 = Z^{(N)}]$; however, the result of Theorem 9.16 remains unchanged even if the state of the N-fermion system is a mixed state $[(Z^{(N)})^2 \neq Z^{(N)}]$. The reason is that $Z^{(N)}$ will then be an average of more than one $Z^{(N)}$, each representing a pure quantum state, and consequently $\langle i| Z^{(1)} |i\rangle$ will also become an average of those satisfying (9.224); hence it will itself satisfy (9.224). The equality in (9.224) will hold if, and only if, $Z^{(N)}$ is a pure quantum state expressible by a single Slater matrix.

We have derived our result (9.224) concerning $Z^{(1)}$ on the assumption that the N-particle system expressed by $Z^{(N)}$ satisfies the Pauli principle; that is, (9.224) was proven to be a necessary condition for $Z^{(N)}$ to satisfy the Pauli principle. It is also a sufficient condition? Intuitively, it seems obviously to be so. This conjecture and other related matters are explained in [W-3]. At my request, and under an IBM contract, Dr. H. Kuhn gave a rigorous mathematical proof (which I do not duplicate here) for the sufficiency of (9.224). [K-15]. Thus we have the following theorem.

Theorem 9.17 There exists an N-body density matrix $Z^{(N)}$ satisfying the Pauli principle, from which a given one-body density matrix $Z^{(1)}$ can be derived by the successive integration (9.213) if, and only if, the eigenvalues of the $Z^{(1)}$ are not larger than $1/N$.

It should be noted, however, that when a given $Z^{(M)}$ $(1 < M \leq N)$ is such that the $Z^{(1)}$ derived from it has eigenvalues not larger than $1/N$, there does not necessarily exist a $Z^{(N)}$ satisfying the Pauli principle from which this given $Z^{(M)}$ can be derived [if $M = N$, $Z^{(N)} = Z^{(M)}$]. The above condition guarantees only that there exists a Pauli-principle-satisfying $Z^{(N)}$ from which the same $Z^{(1)}$ can be derived as from the given $Z^{(M)}$. It should also be noted that the condition (9.225) is weaker than the condition (9.224) and is only a necessary condition for the Pauli principle.

Returning to information theory, let us first show by a simple example that $S(Z) - S(Z_x)$, considered in Theorem 9.15, is not necessarily non-negative. Suppose that we have a 2-fermion system, which is in the quantum state

$$\Psi^{(2)}(x_1, x_2) = \frac{1}{\sqrt{2}} \begin{vmatrix} \psi_1(x_1) & \psi_2(x_1) \\ \psi_1(x_2) & \psi_2(x_2) \end{vmatrix}. \tag{9.226}$$

The $\mu^{(2)}$-representation [corresponding to (9.220)] of $Z^{(2)}$, defined by (9.212) with the help of (9.226), is diagonal and has eigenvalue 1 for the position corresponding to the function (9.226). On the other hand, the ψ_i-representation [corresponding to (9.190)] of $Z^{(2)}$,

$$\langle i_1, i_2 | \, Z^{(2)} \, | i_1', i_2' \rangle, \tag{9.227}$$

is not diagonal. For instance, the off-diagonal element $\langle 1, 2 | \, Z^{(2)} \, | 2, 1 \rangle$ is $-(1/2)$. The one-body state matrix $Z^{(1)}$ in the ψ_i-representation,

$$\langle i_1 | \, Z^{(1)} \, | i_1' \rangle, \tag{9.228}$$

now has only the diagonal elements $\langle 1 | \, Z^{(1)} \, | 1 \rangle = \langle 2 | \, Z^{(1)} \, | 2 \rangle = \frac{1}{2}$. This shows that the diagonalization condition of Theorem 9.15 is not satisfied. In fact, we have

$$S(Z^{(2)}) = 0 \quad \text{and} \quad S(Z^{(1)}) = 1, \tag{9.229}$$

which entails $S(Z^{(2)}) - S(Z^{(1)}) < 0$, as anticipated in Theorem 9.15. On the other hand, the total correlation (9.208) in $Z^{(2)}$ is $J = S(Z^{(1)}) + S(Z^{(1)}) - S(Z^{(2)}) = 2$.

If the N-fermion system is isolated from the rest of the world, we can assume that it is in a quantum state. Hence we may assume that

$$S(Z^{(N)}) = 0. \tag{9.230}$$

This is a direct consequence of $Z^{(N)} = \Psi^{(N)}\Psi^{(N)*}$ or, equivalently, $(Z^{(N)})^2 = Z^{(N)}$, and has nothing to do with the number of Slater functions needed to express $\Psi^{(N)}$.

Theorem 9.18 The total interdependence among particles in an isolated N-fermion system is

$$J_{\text{tot}} = N S(Z^{(1)}) \geq N \log N. \tag{9.231}$$

Proof. The total system is divided into N partial systems, each of which has entropy $S(Z^{(1)})$; hence the total interdependence J_{tot} should be $J_{\text{tot}} = NS(Z^{(1)}) - S(Z^{(N)})$, which becomes $NS(Z^{(1)})$ because of (9.230). The second relation in (9.231) is a consequence of (9.225).

Thus we may introduce

$$V = S^{(1)} - \log N \geq 0 \tag{9.232}$$

as a measure of the interdependence (per particle) among fermions over and above the interdependence due to the Pauli constraint.

The Pauli principle itself imposes a certain dependence among particles, for if one particle occupies a state ψ_i no other particle can occupy ψ_i. The lower bounds in (9.231) and (9.232) are obtained by imposing only the Pauli principle. Hence it is natural to consider any excess over these lower bounds to represent dependence originating from a situation other than the constraints due to the Pauli principle. In fact, the lower bounds are reached when the N-particle system is expressible by a single Slater function; hence $V > 0$ measures in a sense the extent to which the state of the N-fermion system deviates from a single Slater function. But a Slater function represents the simplest structure that an N-fermion system can possess. In fact, apart from the Pauli condition, the Slater function (9.215) means that each particle has its own state independent of the state of the other particles. This argument is reinforced by looking at the situation from the point of view of perturbation theory in the Hartree-Fock model, in which the difference between the real two-body interaction Hamiltonian and the approximate one-body interaction Hamiltonian plays the role of a perturbation. The 0th approximation, which is the Hartree-Fock model, is represented by a single Slater function. If we switch on the perturbation, quantum jumps start to occur; when these jumps correspond to a simultaneou change of states of two particles instead of one, the resulting perturbed state function becomes a sum of many Slater functions. The stronger this cooperative effect is, the more Slater functions become involved with appreciable amplitudes. From (9.222) we can see that this tends to spread the probabilities on more and more states ψ_i, entailing larger and larger values of $S(Z^{(1)})$. This consideration makes the interpretation of the quantity (9.231) as a measure of organization even more convincing.

Thus we have come back to the thesis that $S^{(1)}$ is not only a measure of uncertainty of the state of a particle but also a measure of correlation of particles, a thesis which I arrived at in 1939 and which marked the starting point of my series of studies in information theory (see [W-2]). In this paper I obtained an estimate of $S^{(1)} = \log N + 0.73$ (in the natural logarithmic unit) for nucleons in heavy nuclei.

In Appendix A9.5 we give another example of the use of quantum-mechanical entropy.

Appendices

A1.1. USE OF RELATIVE ENTROPY TO AVOID NEGATIVE INFORMATION†

Suppose that two logical spectra $\mathcal{E} = \{E_i\}$, $i = 1, 2, \cdots, n$, and $\mathcal{F} = \{F_j\}$, $j = 1, 2, \cdots, m$, are not probabilistically independent, so that $p(E_i)$ and $p(E_i \mid F_j)$ are not equal for some pair (i, j). Under these circumstances the information about $\mathcal{E}$ brought about by the observation of $\mathcal{F}$, which gave an outcome F_j, is not zero and is given by

$$\begin{aligned} K(\mathcal{E} \mid F_j) &= S(\mathcal{E}) - S(\mathcal{E} \mid F_j) \\ &= -\sum_i p(E_i) \log p(E_i) + \sum_i p(E_i \mid F_j) \log p(E_i \mid F_j), \end{aligned} \quad \text{(A1.1.1)}$$

where the first term represents the ignorance about $\mathcal{E}$ before the observation and the second term represents the ignorance about $\mathcal{E}$ after the observation of F_j. As remarked following (1.126) and also after (3.17), the quantity $K(\mathcal{E} \mid F_j)$ of (A1.1.1) is not necessarily non-negative. One might consider this fact disturbing on the ground that the observational result F_j must improve our knowledge about $\mathcal{E}$ as compared with before the observation; hence the information brought in about $\mathcal{E}$ by F_j must be positive.

This complaint, however, is based on a confusion due to semantic ambiguity surrounding the expression "improve our knowledge" or, for that matter, the notion of "good" knowledge. Probabilistic knowledge expressible by a wide probability distribution is "bad" in the sense that it is far removed from deterministic knowledge, yet it may be "better," at least in some cases, than probabilistic knowledge expressible by a sharp probabilistic distribution, in the sense that it may correspond better to reality or to the ensemble at hand. In the first sense, goodness means lack of indeterminacy in the contents of a

† The first digit of each appendix number corresponds to the number of the chapter to which the appendix applies. There are no appendices for Chapters 2 and 6.

probabilistic statement or of a prediction, and in the second, it means reliability of the statement or the prediction itself. Information theory is primarily concerned with the first aspect, and it can handle the second aspect only through a complicated procedure such as used in our Chapters 4 and 5. Thus the statement based on $p(E_i \mid F_j)$ is better in the second sense than the one based on $p(E_i)$, but could be worse in the first sense. This is why $K(\mathcal{E} \mid F_j)$ could be negative. On the other hand, it could not be denied that in most cases the more side information or auxiliary indirect information about $\mathcal{E}$ we have, the sharper (i.e., the closer to a deterministic case) our prediction becomes. This fact is reflected in the mathematical relation $K(\mathcal{E} \mid \mathcal{F}) \geq 0$, which means that $K(\mathcal{E} \mid F_j)$ is on the average non-negative.

Apart from the confusion mentioned above, there could be a more simple-minded yet not unfounded complaint that a negative amount of information is very difficult to accept according to our ordinary usage of the word "information." The following is one way of avoiding negative information without modifying the main line of argument developed in the text.

We use the notion of relative entropy introduced in (1.29). Suppose that there is an underlying probability w_i for each event E_i of $\mathcal{E}$; then the ignorance, expressed by a probability distribution $p^{(1)}(E_i)$, will be

$$S^{*(1)}(\mathcal{E}) = -\sum_i p^{(1)}(E_i) \log \frac{p^{(1)}(E_i)}{w_i A}, \qquad \text{(A1.1.2)}$$

where the arbitrary constant A can be equated to n, the number of E_i's. If, by aquisition of some observational knowledge, $p^{(1)}(E_i)$ changes to $p^{(2)}(E_i)$, the information about $\mathcal{E}$ due to this change will be

$$\text{nf*}\,(p^{(1)} \to p^{(2)}) = S^{*(1)}(\mathcal{E}) - S^{*(2)}(\mathcal{E}) = -\sum_i p^{(1)}(E_i) \log \frac{p^{(1)}(E_i)}{w_i} + \sum_i p^{(2)}(E_i) \log \frac{p^{(2)}(E_i)}{w_i}. \qquad \text{(A1.1.3)}$$

Applying this formula to our present problem, if we put $p^{(1)}(E_i) = p(E_i)$ and $p^{(2)}(E_i) = p(E_i \mid F_j) = p(E_i \cap F_j)/p(F_j)$ and also adopt $p(E_i)$ as our standard probability distribution w_i, we obtain as the information defined by (A1.1.3)

$$K^*(\mathcal{E} \mid F_j) = -\sum_i p(E_i \mid F_j) \log \frac{p(E_i)}{p(E_i \mid F_j)}, \qquad \text{(A1.1.4)}$$

which is non-negative because of the simple Gibbs theorem. The interesting fact is that the average of (A1.1.4) is exactly the same as the average of (A1.1.1):

$$K(\mathcal{E} \mid \mathcal{F}) = \sum_j p(F_j)\, K(\mathcal{E} \mid F_j) = \sum_j p(F_j)\, K^*(\mathcal{E} \mid F_j). \qquad \text{(A1.1.5)}$$

In view of this fact, one might be tempted to define in general the information provided by the change from $p^{(1)}(E_i)$ to $p^{(2)}(E_i)$ by

$$\text{inf}^{**}\,(p^{(1)} \to p^{(2)}) = -\sum p^{(2)}(E_i) \log \frac{p^{(1)}(E_i)}{p^{(2)}(E_i)}. \tag{A1.1.6}$$

However, this is likely to lead to some confusion, because (A1.1.6) does not satisfy the transitive additivity satisfied by (A1.1.3):

$$\text{inf}^*\,(p^{(1)} \to p^{(3)}) = \text{inf}^*\,(p^{(1)} \to p^{(2)}) + \text{inf}^*\,(p^{(2)} \to p^{(3)}). \tag{A1.1.7}$$

A3.1. REMARKS ON CONDITIONAL PROBABILITIES

In Section 3.1 we discussed the problem of under what circumstances we can derive the conditional probabilities $p(E_i \mid F_j)$ from the conditional probabilities $p(F_j \mid E_i)$, when these two sets of conditional probabilities are related to each other by (3.1). We can elucidate the situation by envisaging the problem of what freedom is left when the conditional probabilities $p(F_j \mid E_i)$ are determined. We can summarize the answer to the second question in the following way.

Theorem A3.1.1 When $p(F_j \mid E_i)$ for all i and all j are given (determined by nature or by the nature of the problem investigated), we have the following freedom and constraints regarding the choice of values of the other probabilities involved in (3.1).

1. When only the $p(F_j \mid E_i)$ for all i and all j are given, we can still choose arbitrarily the values of the $p(E_i)$ for all i.† If the $p(F_j \mid E_i)$ and $p(E_i)$ for all i and j are given, all the remaining probabilities are determined by them. When the $p(F_j \mid E_i)$ for all i and all j are given, we cannot freely choose the $p(E_i \mid F_j)$ for all i and all j because these can contradict $p(F_j \mid E_i)$. If there is no contradiction, the $p(F_j \mid E_i)$ and $p(E_i \mid F_j)$ determine all the other probabilities, since

$$p(E_i) = \left[\sum_j \frac{p(F_j \mid E_i)}{p(E_i \mid F_j)}\right]^{-1}. \tag{A3.1.1}$$

2. When the $p(F_j \mid E_i)$ for all i and all j are given, we cannot choose the values of the $p(F_j)$ entirely arbitrarily. Even when the $p(F_j \mid E_i)$ and $p(F_j)$ do not involve contradiction, these two sets of probabilities may not lead to definite values of the remaining probabilities. This happens when we cannot solve $p(F_j) = \sum_i p(F_j \mid E_i) \cdot p(E_i)$ for $p(E_i)$.
3. When the $p(F_j \mid E_i)$ for all i and all j are given, the other conditional probabilities $p(E_i \mid F_j)$ can be derived from the former if, in addition, (a) the $p(E_i)$ are (arbitrarily) given or (*b*) the $p(F_j)$ are given in a such a way as not to

† To simplify, we shall not usually mention obvious constraints of the type $p(E_i) \geq 0$ and $\sum_i p(E_i) = 1$.

contradict $p(F_j \mid E_i)$ and if the simultaneous linear equations involving $p(F_j \mid E_i)$, mentioned above, are solvable. (c) Except in those cases when, as in these last two, all the probabilities are determined, there is a very special case in which the $p(E_i \mid F_j)$ are determined directly by the $p(F_j \mid E_i)$ while the unconditional probabilities remain undetermined. This happens when the values of $p(F_j \mid E_i)$ are either 0 or 1 and for a given F_j there is only one E_i for which $p(F_j \mid E_i)$ is 1. (We call this case a bilaterally deterministic case because the matrix in this case is a permutation; i.e., there is only one 1 in each row and in each column.) In the bilaterally deterministic case the ratio $p(E_i)/p(F_j)$ is determined (equal to unity) for a pair for which $p(F_j \mid E_i) = 1$, but $p(E_i)$ and $p(F_j)$ are not determined separately. We have $p(E_i \mid F_j) = p(F_j \mid E_i)$ here.

Proof. The argument given in Section 3.1 may be used as proofs of most of the assertions made in this theorem. Equation A3.1.1 can be obtained by dividing (3.1) by $p(E_i \mid F_j)$ and summing with respect to F_j. The contradiction mentioned under Item 1 can happen if the $p(E_i \mid F_j)$ are given in such a way that the sum with respect to E_i of the right side of (A3.1.1) does not become unity. This can easily happen even if the conditions $p(E_i \mid F_j) \geq 0$ and $\sum_i p(E_i \mid F_j) = 1$ are satisfied. The contradiction mentioned under Item 2 can happen, for instance, if $p(F_j) < \min_k p(F_k \mid E_i)$ for some j. Then, even if the equations $p(F_j) = \sum_i p(F_j \mid E_i)p(E_i)$ are soluble for the $p(E_i)$, some of the resulting $p(E_i)$ will become >1 or <0, satisfying $\sum_i p(E_i) = 1$ all the same.

A4.1. CHURCH'S OBJECTION TO AYER

Church [C-6] showed that if we define a verifiable proposition according to Ayer [A-8] by (4.1) and (4.2), any proposition or its negation or both will become verifiable. Church's proof can be formulated as follows. Let $\mathcal{T}$ be the set of all possible propositions, let $\mathcal{V}$ be a subset of $\mathcal{T}$, and let $\mathcal{E}$ be a subset of $\mathcal{V}$, whereby the relation between $\mathcal{V}$ and $\mathcal{E}$ is as defined in (4.1) and (4.2). Furthermore, Church assumed that $\mathcal{E}$ contains at least three propositions E_i, $i = 1, 2, 3$, such that $E_1 \not\rightarrow E_3$ and $E_3 \not\rightarrow E_2$. Then he demonstrated that $X \in \mathcal{T}$ implies either $X \in \mathcal{V}$ or $\neg X \in \mathcal{V}$, or maybe both. His proof consisted of two steps. First, he considered a proposition C defined by $C = (\neg E_1 \cap E_2) \cup (E_3 \cap \neg X)$ and showed that $C \in \mathcal{V}$. Next he showed that if $C \in \mathcal{V}$ then $X \in \mathcal{V}$ or $\neg X \in \mathcal{V}$. The first step is done as follows. Make the conjunction of C with E_1; then this conjunction can be shown to imply E_3, meaning, according to the definition of $\mathcal{V}$, that $C \in \mathcal{V}$ unless $E_1 \rightarrow E_3$, which is excluded by the premise. Indeed,

$$\begin{aligned} C \cap E_1 &= [(\neg E_1 \cap E_2) \cup (E_3 \cap \neg X)] \cap E_1 \\ &= (\neg E_1 \cap E_2 \cap E_1) \cup (E_3 \cap \neg X \cap E_1) \quad \text{(by the distributive law)} \\ &= \emptyset \cup (E_3 \cap \neg X \cap E_1) = E_3 \cap (\neg X \cap E_1) \rightarrow E_3 \qquad \text{(A4.1.1)} \end{aligned}$$

by virtue of the rules $\emptyset \cup \alpha = \alpha$ for any α, and $\alpha \cap \beta \rightarrow \alpha$ for any α and β. Hence $C \in \mathcal{V}$. At the second step, make the conjunction $C \cap X$ of C and X; then we can show $C \cap X \rightarrow E_2$, just interchanging the roles of $\neg X$ and $\neg E_1$ and the roles of E_2 and E_3 in (A4.1 1). This means, since $C \in \mathcal{V}$, that $X \in \mathcal{V}$ provided that $C \not\rightarrow E_2$. (Note here that if $C \rightarrow E_2$, we cannot conclude that $X \in \mathcal{V}$, but this does not necessarily exclude $X \in \mathcal{V}$.) If $C \rightarrow E_2$, it follows from the definition of C that $(\neg E_1 \cap E_2) \rightarrow E_2$ (which is a tautology) as well as $(\neg X \cap E_3) \rightarrow E_2$, because $\alpha \cup \beta \rightarrow \gamma$ implies both $\alpha \rightarrow \gamma$ and $\beta \rightarrow \gamma$. Hence $\neg X \cap E_3 \rightarrow E_2$, which means $\neg X \in \mathcal{V}$, since $E_3 \not\rightarrow E_2$. There are two cases: (a) if $C \rightarrow E_2$ then $\neg X \in \mathcal{V}$ and X may or may not belong to $\mathcal{V}$; (b) if $C \not\rightarrow E_2$ then $X \in \mathcal{V}$ and $\neg X$ may or may not belong to $\mathcal{V}$. Hence the possibility that both $X \notin \mathcal{V}$ and $\neg X \notin \mathcal{V}$ is precluded.

If we use our revised definition of $\mathcal{V}$ [(i), (ii), (iii) of Section 4.1] instead of Ayer's, Church's proof no longer holds. In fact, C can be written

$$\begin{aligned} C &= [(\neg E_1 \cap E_2)\neg \cup E_3] \cap [(\neg E_1 \cap E_2) \cup \neg X] \\ &= (\neg E_1 \cup E_3) \cap (E_2 \cup E_3) \cap (\neg E_1 \cup \neg X) \cap (E_2 \cup \neg X) \qquad \text{(A4.1.2)} \end{aligned}$$

and Church's proof $C \cap E_1 \rightarrow E_3$ depends only on the first term $(\neg E_1 \cup E_3)$. To see that C cannot be shown to be verifiable, one needs only to put in (4.5), $V_1 = C$, $V_1^* = \neg E_1 \cup E_3$, $V_2 = E_1$, $E = E_3$. Hence $C \cap E_1 \rightarrow E_3$ does not prove that $C \in \mathcal{V}$. What we can prove to be a verifiable proposition in this case with the help of V_2 and E, according to Condition (ii) of our definition, is that $\neg V_2 \cup E = \neg E_1 \cup E_3 = V_1^*$. But this proof is not necessary, since V^* here, being a combination of experiential propositions, is an experiential proposition and therefore verifiable according to Condition (i). Furthermore, if we want to invoke Condition (iii), we have to prove that $(\neg E_1 \cup \neg X)$ and $(E_2 \cup \neg X)$ are verifiable propositions. There is no guarantee that we can do this. It can be seen in a similar way that the second step, $X \cap C \rightarrow E_2$, does not yield $C \rightarrow E_2$ or $X \in \mathcal{V}$, since X is not $E_2 \cup \neg C$ [as required by (ii)] but only satisfies $X \rightarrow E_2 \cup \neg C$ [satisfying (4.1)]. Furthermore, $C \rightarrow E_2$ does not imply $\neg X \in \mathcal{V}$ here.

A5.1. PROOF OF THEOREM 5.1

If b_r in (5.24) is given, it is obvious that a single-valued $f(x)$, such that $0 \le f(x) \le P - 1$, is determined by the equation. The crucial point is whether the b_r ($r = 0, 1, \cdots, P - 1$) are conversely and uniquely determined when the function $y = f(x)$ is given. We have already seen that any given function can be written in the form (5.23). Therefore our task is to show first that by adjusting the additive constant involved in each derivative, we can make $f^{(r)}(0)/r!$ an integer, and second that this integer $f^{(r)}(0)/r!$ is uniquely determined (modulo P) by $f(x)$.

The first step is to show that for a given $f^{(r)}(0)$ there exists an integer n such that $f^{(r)}(0) + np$ is divisible by $r!$. One way of proving this is to invoke the so-called Chinese remainder theorem, which asserts the following. Let $\xi_1, \xi_2, \cdots$ be given non-negative integers and $\eta_1, \eta_2, \cdots$ be given non-negative integers which are relatively prime in pairs. Then there exists a non-negative integer μ such that $\mu = \xi_i$ (modulo η_i). To apply this theorem, we note first that P and $r!$ are relatively prime, since P is a prime number and $r!$ is a product of numbers all less than P. Next let n_0 be such that $f^{(r)}(0) + n_0P$ becomes non-negative. Also put $\xi_1 = f^{(r)}(0) + n_0P$, $\xi_2 = 0$, $\eta_1 = P$, and $\eta_2 = r!$. Then the theorem says that there exists a number μ such that $\mu = f^{(r)}(0) + n_0P$ (modulo P) and $\mu = 0$ (modulo $r!$). This means that $\mu = f^{(r)}(0) + n_0P + n'P$, with some integer n', and $\mu = n''r!$, with some integer n''. In other words, $f^{(r)} + (n_0 + n')P$ is divisible by $r!$. This is what we wanted. The next step is to show that if $f^{(r)}(0) + n_1P$ and $f^{(r)}(0) + n_2P$ are both divisible by $r!$ with $n_1 > n_2$, the quotients $q_1 = (f^{(r)}(0) + n_1P)/r!$ and $q_2 = (f^{(r)}(0) + n_2P)/r!$ are equal to each other (modulo P). Let us subtract $q_2r! = f^{(r)}(0) + n_2P$ from $q_1r! = f^{(r)}(0) + n_1P$, side by side; then we obtain $r!\,(q_1 - q_2) = (n_1 - n_2)P$. Since P is a prime number, the left side $r!\,(q_1 - q_2)$ must contain P as a factor (fundamental theorem of arithmetics). But $r!$ does not contain P, because $r < P$. Hence $(q_1 - q_2)$ must contain P as a factor; that is, $q_1 - q_2 = 0$ (modulo P). This completes the proof. It is easy to see that if a pair (n, q) is such that $q = (f^{(r)}(0) + nP)/r!$, the pair $(n + r!, q + P)$ also satisfied the same equation. The mathematical gist of this proof can be formulated as the following theorem. Let P be a prime number and s and r be integers, with the restriction $0 \leq r < P$. Then there are infinitely many integers n_i such that $s + n_iP$ is divisible by $r!$ and the quotients $q_i = (s + n_iP)/r!$ are all equal (modulo P). A simple corollary of Theorem 5.1 is that $y(x) = 0$ for all x, $0 \leq x \leq P - 1$, corresponds one-to-one to $b_r = 0$, $r = 0, 1, \cdots, P - 1$. Conversely, we can also derive the theorem from this corollary.

A7.1. TODA'S METHODS OF MEASUREMENT OF SUBJECTIVE PROBABILITIES

In the early 1950s Masanao Toda of Hokkaido University was experimenting with various methods ingeniously devised to induce a subject to divulge his subjective probabilities on a set of alternatives [T-5]. Some of his methods have since been rediscovered independently by other researchers [D-5, R-4, S-7].

The subject is faced with a set of n alternatives t_i, $i = 1, 2, \cdots, n$, of which only one is supposed to be "correct." An easy example is the case of a multiple-choice examination. The student has a subjective probability distribution p_i,

$i = 1, 2, \cdots, n$, on these alternatives and is asked to assign a positive number r_i to each alternative. This number r_i may be compared to the amount he would bet on the ith alternative only in the sense that the more probable the ith alternative seems to him to be "correct," the higher a value of r_i he is advised to put. But the reward he gets when he hits the "correct" alternative is not necessarily proportional to r_i. When the subject has assigned r-values to all the alternatives, the experimenter (computer) tells him the list of payoffs $w(t_j, \mathbf{r})$, $j = 1, 2, \cdots, n$, which is the amount of reward the subject will receive if the jth alternative happens to be the correct one, given that the subject has "bet" $\mathbf{r}$ $(r_1, r_2, \cdots, r_n)$.

It is now assumed that each human subject behaves so as to maximize the expected utility, although he himself may not be able to tell the exact values of his subjective probabilities. This is a serious assumption (adopted since the time of Ramsey; see Section 7.4), but it has proved to be fairly close to reality, at least in the sense that the values of subjective probabilities obtained by different methods based on this assumption are approximately the same. If this assumption is accepted, the subject will adjust the "bet" $\mathbf{r}$ so as to maximize the expected payoff

$$E_w(\mathbf{p}, \mathbf{r}) = \sum_{j=1}^{n} p_j w(t_j, \mathbf{r}). \qquad \text{(A7.1.1)}$$

This value E will depend on two vectors $\mathbf{r}$ and $\mathbf{p}$ and the payoff function w.

Toda's problem is to find a payoff function w that satisfies the "matching condition"

$$\max_{\mathbf{r}} E_w(\mathbf{p}, \mathbf{r}) = E_w(\mathbf{p}, k\mathbf{p}); \qquad \text{(A7.1.2)}$$

that is, a w-function such that E becomes maximum when $\mathbf{r}$ is proportional to $\mathbf{p}$. If this is achieved, the experimenter will only have to read the bet $\mathbf{r}$ and it will give him the subjective probabilities directly. This condition can be satisfied if we have

$$\frac{\partial E}{\partial r_i} = 0, \qquad \frac{\partial^2 E}{\partial r_i^{\,2}} < 0, \qquad i = 1, 2, \cdots, n \qquad \text{(A7.1.3)}$$

for $\mathbf{r} = k\mathbf{p}$. It is easy to see the validity of the following theorem derived by Toda.

Toda's Theorem of Linear Combination. If each of $w_1, w_2, \cdots, w_m$ is a payoff function satisfying the matching condition, then $w = \sum_{i=1}^{m} a_i w_i + b$, $a_i > 0$, $i = 1, 2, \cdots, m$, is another payoff function satisfying the matching condition.

Toda proposed to use this theorem to correct a systematic error peculiar to each matching payoff function by superimposing more than one matching payoff function.

No general solution of Toda's problem (A7.1.2) is known to date. But we mention some of the known matching payoff functions. The proof is left to the reader.

Spherical Gain Function

$$w(t_j, \mathbf{r}) = \frac{r_j}{\left(\sum_{j=1}^{n} r_j^2\right)^{1/2}}. \tag{A7.1.4}$$

Logarithmic Loss Function

$$w(t_j, \mathbf{r}) = k \log r_j - \sum_{j=1}^{n} r_j, \tag{A7.1.5}$$

where k is a positive constant.

Quadratic Loss Function. This is usually used for the case $n = 2$ (true-or-false question instead of multiple choice) and under the restriction $\sum_{i=1}^{2} r_i =$ fixed (total bet is a constant). (See de Finetti [D-5] for a generalization for $n > 2$.) The function is

$$w(t_j, \mathbf{r}) = -r_j^2. \tag{A7.1.6}$$

The first two functions can also be used under the restriction $\sum_{i=1}^{n} r_i =$ fixed.

Toda proposed another interesting way of measuring subjective probability. It can be applied to the discrete case, as above, but is more easily explained in the continuous case. Let us assume that the "correct" point lies somewhere in the domain $0 \leq t \leq 1$. The subject has a subjective probability density $p(t)$. Although the method works in a more general case, let us assume that the distribution $p(t)$ is unimodular. The "bet" that the subject can make consists of specifying a domain $[a, b]$, $0 \leq a < b \leq 1$; that is, he bets that the correct point is in this domain. Of course, the reward must be devised so that if the subject "bets" a large domain his gain will become small even if he is right. The payoff function $w(t, a, b)$ is defined as the amount of reward the subject gets when he bets the domain $[a, b]$ and the correct point is at t. The function suggested by Toda is

$$\begin{aligned} w_k(t, a, b) &= k - d \quad \text{if} \quad a \leq t \leq b, \\ &= -d \quad \text{otherwise}, \end{aligned} \tag{A7.1.7}$$

where $d = b - a$ is the size of the domain and k is a positive parameter. The experiment is repeated with different values of k. The experimental situation is somewhat like this. The subject has to pay an admission fee d, which is proportional to the size of the domain $d = b - a$. In each experiment the experimenter tells the subject the value of k, which is the prize if the subject answers correctly. Obviously, if the prize k is large compared with d, the subject will tend to increase the domain d, because he does not mind a relatively large (but small compared wlth k) admission fee. If the prize k is small, he would rather decrease the admission fee d.

The average payoff is then

$$E_k(a, b) = (k - d)\int_a^b p(t)\,dt + (-d)\left[1 - \int_a^b p(t)\,dt\right]$$
$$= kF(a, b) - (b - a), \tag{A7.1.8}$$

where

$$F(a, b) = \int_a^b p(t)\,dt. \tag{A7.1.9}$$

Since the subject will choose a and b to maximize $E_k(a, b)$ for the given k, we should have

$$\begin{aligned} 0 &= \frac{\partial E_k(a, b)}{\partial a} = -kp(a) + 1, \\ 0 &= \frac{\partial E_k(a, b)}{\partial b} = kp(b) - 1, \\ 0 &> \frac{\partial^2 E_k(a, b)}{\partial a^2} = -kp'(a), \\ 0 &> \frac{\partial^2 E_k(a, b)}{\partial b^2} = kp'(b), \end{aligned} \tag{A7.1.10}$$

where p' means $dp(t)/dt$. If the curve is unimodular, the last two conditions are automatically satisfied. The first two conditions give

$$p(a) = p(b) = \frac{1}{k}, \tag{A7.1.11}$$

meaning that a and b are the abscissas (t-coordinates) of the points of intersection between the curve of the subjective probability distribution and the horizontal line $p = 1/k$. Repeating the experiment at different values of k, the experimenter can reconstruct the p-curve. If the p-curve is known to contain only two parameters to be determined by experiment, then one experiment with one value of k should in principle be sufficient. It should be warned that if k is very large, the subject will stop thinking hard to determine the domain $[a, b]$ and bet on the entire domain $[0, 1]$ in order to be certain of obtaining the prize k. But this corresponds to small p-values, which are usually not interesting.

In the discrete case we may interpret (A7.1.8) as applicable to the integer values of a and b, $a = 1, 2, \cdots, n$, $b = 1, 2, \cdots, n$, and $F(a, b) = \sum_{i=a}^{b} p(t)$, where $p(t)$, $t = 1, 2, \cdots, n$, is the subjective probability. Then, provided that $p(t)$ is unimodular in t, the optimal domain $[a, b]$ is given by

$$p(a-1) \leq \frac{1}{k} \leq p(a),$$
$$p(b) \geq \frac{1}{k} \geq p(b+1). \qquad \text{(A7.1.12)}$$

The usefulness of Toda's different methods of measuring subjective probability in educational evaluation is obvious. According to the usual method, two students will obtain the same grade if, for instance, in the case of a true-or-false examination ($n = 2$) one student has $p(1) = 0.51$ and $p(2) = 0.49$ and the other student has $p(1) = 1.00$ and $p(2) = 0.00$. If the first alternative is the correct answer, the second student is by far a better student than the first, but they will still get the same grade. Probability distribution in general reflects infinitely more information about the student's state of knowledge than grading according to the usual true-or-false or multiple-choice method. Shuford and collaborators have done considerable experimental and theoretical work along this line [S-7].

A7.2. PROPOSITION CALCULUS

In this appendix we show informally that the formalism of logic in terms of the lattice used in this book and the usual propositional calculus are equivalent. Let us first start from our formalism. In our formalism there are two kinds of formulas: those that contain neither $=$ nor $\rightarrow$, and those that contain either of these two signs. Let us call the former "formulas" and the latter "relations." Relations do not appear as such in propositional calculus. The former correspond to what Kleene calls formulas [K-10], Fitch calls propositions [F-7], and Quine calls schemata [Q-1]. In principle, each formula can be either true or false. In our formalism any formula standing alone (i.e., without such things as $=\square$, $=\emptyset$) may or may not be true. In ordinary propositional calculus we must be a little more cautious. If a formula is standing in a series of formulas called "deduction," each formula is meant to be true. In our formalism we write $A = \square$ to show that A is meant to be true. Of course, $A = \emptyset$ is equivalent to $\neg A = \square$.

In any practical use of logic, certain formulas are assumed to be true during the duration of an argument; they are usually called hypotheses. Sometimes the axioms of logic are counted as hypotheses, but not necessarily. From these hypotheses we can derive logical consequences. These consequences are also true within the context when the adopted hypotheses are assumed to be true; for instance, if A and $\neg A \cup B$ (usually denoted as $A \supset B$) are hypotheses, we can derive B as the consequence [with the help of the distributive law; see (7.97)]. This is usually taken as an axiom, but put in the form of a deduction. This may be written, in Kleene's notation, A, $A \supset B \vdash B$. In Fitch's notation it may be written

1	A	hypothesis
2	$A \supset B$	hypothesis
3	B	1, 2 modus ponens.

In Quine's notation it may be written

*(1)	A	
**(2)	$A \supset B$	
**(3)	B	(1), (2).

All three formulas, A, $A \supset B$, and B, are true within this argument. Thus in our notation we can say that $A = \square$ and $\neg A \cup B = \square$; hence $B = \square$.

In our formalism $A \rightarrow B$ means that there exists a logical deduction such that if A is assumed to be true then B becomes true. Thus we can also say $A = \square$ and $A \rightarrow B$; hence $B = \square$. $A \rightarrow B$ is under the assumption of the distributive law, equivalent to the relation $\neg A \cup B = \square$, that is, the formula $A \supset B$ standing in a deduction. Our $A = B$ is by definition equivalent to $A \rightarrow B$ and $B \rightarrow A$ and hence, in the distributive logic, is equivalent to the formula $(A \supset B) \cap (B \supset A)$ standing in a deduction.

As far as the axioms or postulates are concerned, we may mention as an example those adopted by Kleene:

1*a*. $A \supset (B \supset A)$, 1*b*. $(A \supset B) \supset ((A \supset (B \supset C)) \supset (A \supset C)$

2. $A, A \supset B \vdash B$

3. $A \supset (B \supset (A \cap B))$

4*a*. $A \cap B \supset A$, 4*b*. $A \cap B \supset B$

5*a*. $A \supset (A \cup B)$, 5*b*. $B \supset (A \cup B)$

6. $(A \supset C) \supset ((B \supset C) \supset ((A \cup B) \supset C))$

7. $(A \supset B) \supset ((A \supset \neg B) \supset \neg A)$

8. $\neg\neg A \supset A$.

Starting from our lattice laws (7.32)–(7.35), the distributive laws (7.47), and the laws concerning negation (7.48)–(7.50), we can prove all these axioms. Thereby we must put $= \square$ to each of the formulas except Axiom 2 and interpret $A \supset B$ as $\neg A \cup B$. For instance, Axiom 1*a* is proved as follows: $\neg A \cup (\neg B \cup A) = (\neg A \cup A) \cup \neg B = \square \cup \neg B = \square$.

Conversely, starting from Axiom 1*a* through Axiom 8, we can derive all our laws mentioned above. In fact, in every textbook of propositional calculus, we can find our laws mentioned as theorems derivable from the axioms; for instance, see Section 27, Chapter VI, of Kleene's text.

A8.1. COMPLEX LINEAR ALGEBRA

The linear algebra explained in Section 8.2 can be generalized easily to cover cases in which the functions $f^{(\alpha)}(\xi)$ considered are not necessarily real-valued but can be complex-valued. The argument ξ, however, has to be assumed to be real. This generalization is precisely what is needed again in Chapter 9 to introduce quantum-mechanical theory from a logical consideration. I do not repeat all the formulae for the complex case, but mention only the major differences that have to be introduced. In the normalization conditions such as (8.5), (8.9), and (8.12) the brackets ([ ]) have to be replaced by absolute-value symbols (| |). The scalar product must be formed from a product of a function and the complex conjugate of another function. Hence the orthonormal conditions (8.6) and their inverse (8.11) will become

$$\int_a^b \psi_i^*(\xi)\psi_j(\xi)\, d\xi = \delta_{ij}, \qquad \sum_i \psi_i^*(\xi)\psi_i(\eta) = \delta(\xi - \eta), \tag{A8.1.1}$$

where the asterisks denote the complex conjugates. In the distance expressions (8.13)–(8.16), the brackets have to be replaced by absolute-value symbols. Correspondingly, in (8.14) and (8.16), the scalar products must be changed to a mixed product in the same fashion as in (A8.1.1).

As a result of (A8.1.1) the expansion coefficient of the type (8.18) must be changed to

$$\langle j \mid i\rangle = \int_a^b \varphi_j^*(\xi)\psi_i(\xi)\, d\xi. \tag{A8.1.2}$$

This entails an important change in (8.21); namely, we now have

$$\langle i \mid j\rangle = \langle j \mid i\rangle^*, \tag{A8.1.3}$$

although the orthonormal relation (8.22) remains formally unchanged. As a result of the change (A8.1.3), the relation (8.27) has to be replaced by

$$\bar{U} = U^{-1}, \tag{A8.1.4}$$

where $\bar{U}$ is the Hermitian conjugate of U and is defined, instead of (8.26), by

$$\langle i|\, \bar{U}\, |j\,\rangle = \langle j|\, U\, |i\rangle^*. \tag{A8.1.5}$$

The unitary transformation defined by (A8.1.4) does not leave (8.28) invariant, but does leave the mixed (Hermitian) scalar product $\sum_i x_i^{(\alpha)*}x_i^{(\beta)}$ invariant.

The ρ_i as defined by (8.31)will no longer be real, but its modified definition is

$$\rho_i = \sum_{\alpha=1}^{\nu} w^{(\alpha)} \, |x_i^{(\alpha)}|^2. \tag{A8.1.6}$$

The autocorrelation function or density matrix G will be defined by

$$\langle \xi | \, G \, | \xi' \rangle = \sum w^{(\alpha)} f^{(\alpha)}(\xi) f^{(\alpha)*}(\xi') \tag{A8.1.7}$$

instead of (8.35). The G will not necessarily be symmetrical ($G^T = G$), as in (8.41), but will be "Hermitian": $\bar{G} = G$. This guarantees that its diagonal elements (and hence also eigenvalues) will all be real. Furthermore, because of the special definition of G (A8.1.7), which entails

$$\langle j | \, G \, | j' \rangle = \sum_{\alpha} w^{(\alpha)} y_j^{(\alpha)} y_j^{(\alpha)*}, \tag{A8.1.8}$$

the diagonal elements ρ_i and λ_k are non-negative. Most of the formulas we used in the discussion of the property of G remain formally unchanged, although many of them involve complex quantities instead of real quantities. In the product expressions in (8.56), (8.57), and (8.59) we have to put an asterisk on the first factor. Other than these small changes, the major conclusions of the section remain entirely intact.

A8.2. PROOF OF THEOREM 8.1 FOR A DEGENERATE CASE

The "if" part of Theorem 8.1 was proved in Section 8.2. The "only-if" part of the theorem means that an optimal coordinate system is bound to be a Karhunen-Loève coordinate system. In other words, if $\{\phi_k\}$ and $\{\psi_i\}$ are such that the equality holds in (8.67), $\{\psi_i\}$ must be another K-L coordinate system, which is equally allowed because of degeneracy of the G matrix. To prove this, we invoke a detailed version of the prototype H-theorem, which in conjunction with Theorem 8.2, constitutes the contents of Theorem 5.3.

Lemma A8.2.1 Equality in (8.67) holds if, and only if, the λ_k's whose indices belong to a "terminally connected" family have the same value.

DEFINITION. Two indices k and l are said to be "terminally connected" (a term borrowed from a consideration of Markov chains, which therefore may sound a little awkward here) if there exists a chain of indices $(k, \cdots, m, n, \cdots, l)$ in which each pair of consecutive indices, say, m and n, are such that there is at least one index i that satisfies both $A_{im} \neq 0$ and $A_{in} \neq 0$. A family of terminally connected indices is such that any two members of the family are terminally connected and a member and a nonmember of the family are not terminally connected.

Applying this idea to our problem, let us assume that the λ's and ρ's are such that equality in (8.67) holds. The lemma then tells us that if ϕ_k and ϕ_l belong to a terminally connected family, λ_k and λ_l must be equal. Hence each nondegenerate eigenfunction ϕ_k constitutes a family with only one member. On the other hand, two eigenfunctions corresponding to a multiple eigenvalue may or may not belong to a single terminally connected family. Now $A_{ik} = 0$ means that ϕ_k and ψ_i are orthogonal and $A_{ik} \neq 0$ means that they have mutually nonvanishing projections. Hence, if a subspace subtended by a certain number, say, d, of ϕ's is also subtended by d ψ's in such a way that no subspace of lesser dimension within it is shared by ϕ's and ψ's, then, and only then, are the d ϕ's "terminally connected." From this we conclude that $\{\phi_k\}$ and $\{\psi_i\}$ yield the same entropy value if, and only if, they are systems of eigenfunctions of the same matrix G and differ only through the arbitrariness allowed by the degeneracy of the eigenvalues.

A8.3. PROOF OF THEOREM 8.4

The following proof of Theorem 8.4 was given orally to me by Professor S. Kakutani.

Lemma A8.3.1 For any given integer m and for a given doubly stochastic matrix A_{ij} $(i, j = 1, 2, \cdots, \infty)$ there exists another doubly stochastic matrix B_{ij}, such that $B_{ij} = 0$ for $i > m$ or $j > m$ and $B_{ij} \geq A_{ij}$ for $1 \leq i \leq m$ and $1 \leq j \leq m$.

Proof. Let D denote the class of $m \times m$ matrices with non-negative elements C_{ij} $(i, j = 1, 2, \cdots, m)$ that satisfy $\tau_j \equiv \sum_{i=1}^m C_{ij} \leq 1$ and $\sigma_i \equiv \sum_{j=1}^m C_{ij} \leq 1$. A row (column) whose row sum σ_i (column sum τ_j) is less than unity is said to be deficient. The total number δ of deficient rows and columns is called the degree of deficiency. The degree zero, $\delta = 0$, means that the $m \times m$ matrix is doubly stochastic. A simple fact to note about the deficient rows and columns is that if there are one or more deficient rows (columns) then there is at least one deficient column (row). This can be seen easily if one notes that the sum of all row sums is equal to the sum of all column sums:

$$\sum_{i=1}^{m} \sigma_i = \sum_{i=1}^{m} \tau_j.$$

We now prove the following statement by induction with respect to δ. If $\|C_{ij}\|$ is a matrix belonging to D and has degree of deficiency δ, there exists another matrix $\|C'_{ij}\|$ belonging to D with degree $\delta - 1$ or $\delta - 2$ such that $C'_{ij} \geq C_{ij}$. Among δ deficient row sums and column sums of $\|C_{ij}\|$, let the row sum σ_k be (one of) the largest. (The argument goes exactly the same way when a column sum is the largest.) Because of the fact mentioned above, there is at least one column, say, the lth, that is deficient, since there is a deficient row $\sigma_k < 1$. Increase the element C_{kl} until the kth row becomes nondeficient. This will not make the lth column larger than unity. The matrix thus obtained will be denoted by $\|C'_{ij}\|$. This matrix $\|C'_{ij}\|$ differs from $\|C_{ij}\|$ only by one element at the intersection of the kth row and the lth column, and $C'_{kl} > C_{kl}$; hence we can write in general $C'_{ij} \geq C_{ij}$ $(i, j = 1, 2, \cdots, m)$. The degree of deficiency of $\|C'_{ij}\|$ will be either $\delta - 1$ or $\delta - 2$ according to whether $\sigma_k > \tau_l$ or $\sigma_k = \tau_l$. The case $\sigma_k < \tau_l$ is excluded, since σ_k is the largest nonunity sum. We can continue this process until the degree of deficiency is 2, and then these two deficient rows will be eliminated by one stroke. (The case $\delta = 1$ cannot happen.) Coming back to the statement of the lemma, let the matrix $\|A'_{ij}\|$ be defined as an $m \times m$ matrix such that $A'_{ij} = A_{ij}$ for $1 \leq i \leq m$ and $1 \leq j \leq m$. Then this $\|A'_{ij}\|$ is a member of D, and the above proof shows that there exists a nondeficient (doubly stochastic) matrix $\|B'_{ij}\|$ belonging to D, such that $B'_{ij} \geq A'_{ij} = A_{ij}$ for $1 \leq i \leq m$ and $1 \leq j \leq m$. The B_{ij} of the lemma can be defined by $B_{ij} = B'_{ij}$ for $1 \leq i \leq m$ and $1 \leq j \leq m$, and by $B_{ij} = 0$ otherwise. This completes the proof of the lemma.

Proof of Theorem 8.4. Let ε_{ij} $(i, j = 1, 2, \cdots, m)$ be defined by

$$B_{ij} = A_{ij} + \varepsilon_{ij}, \tag{A8.3.1}$$

where A_{ij} and B_{ij} $(i, j = 1, 2, \cdots, \infty)$ are related as in Lemma A8.3.1. Since the matrices A_{ij} and B_{ij} are doubly stochastic, we have

$$\sum_{j=1}^{m} \varepsilon_{ij} = \sum_{j=m+1}^{\infty} A_{ij}. \tag{A8.3.2}$$

Then we have

$$\begin{aligned}
\sum_{j=1}^{m} \lambda_j &= \sum_{i=1}^{m} \sum_{j=1}^{m} B_{ij}\lambda_j && \left[\text{since} \sum_{i=1}^{m} B_{ij} = 1\right] \\
&\geq \sum_{i=1}^{m} \sum_{j=1}^{m} A_{ij}\lambda_j + \sum_{i=1}^{m} \sum_{j=1}^{m} \varepsilon_{ij}\lambda_{m+1} && [\text{from (8.62) and (A8.3.1)}] \\
&\geq \sum_{i=1}^{m} \sum_{j=1}^{m} A_{ij}\lambda_j + \sum_{i=1}^{m} \sum_{j=m+1}^{\infty} A_{ij}\lambda_j && [\text{from (8.62) and (A8.3.2)}] \\
&= \sum_{i=1}^{m} p_i. && [\text{from (8.64)}]
\end{aligned}$$

This completes the proof.

A8.4. PROOF OF THEOREM 8.5

Consider the entropy S defined by the probability distribution p_i, $i = 1, 2, \cdots$,

$$S = -\sum_{i=1}^{\infty} p_i \log p_i, \tag{A8.4.1}$$

under the condition that

$$p_i \geq p_{i+1}, \qquad i = 1, 2, \cdots . \tag{A8.4.2}$$

Introducing a new set of variables

$$\sigma_r = \sum_{i=1}^{r} p_i, \tag{A8.4.3}$$

we can rewrite S as

$$S = -\sum_{r=1}^{\infty} (\sigma_r - \sigma_{r-1}) \log (\sigma_r - \sigma_{r-1}) \tag{A8.4.4}$$

with the obvious provisions that

$$\sigma_0 = 0, \qquad \lim_{r\to\infty} \sigma_r = 1 \tag{A8.4.5}$$

and

$$\sigma_{r+1} - \sigma_r \geq 0, \qquad r = 1, 2, \cdots . \tag{A8.4.6}$$

The additional condition (A8.4.2) is now rewritten

$$\sigma_{r+1} - 2\sigma_r + \sigma_{r-1} \leq 0, \qquad r = 1, 2, \cdots . \tag{A8.4.7}$$

Now the partial derivatives of S with respect to σ_r is

$$\frac{\partial S}{\partial \sigma_r} = \log \frac{\sigma_{r+1} - \sigma_r}{\sigma_r - \sigma_{r-1}} = \log \frac{p_{r+1}}{p_r} . \tag{A8.4.8}$$

The condition (A8.4.7) shows that

$$\frac{\partial S}{\partial \sigma_r} \leq 0 \tag{A8.4.9}$$

in the domain allowed by (A8.4.6).† Suppose that we are given two probability distributions $\{p_i\}$ and $\{p'_i\}$ or, equivalently $\{\sigma_r\}$ and $\{\sigma'_r\}$, both obeying (A8.4.2) or, equivalently, (A8.4.7), such that

$$\sigma_r \geq \sigma'_r, \qquad r = 1, 2, \cdots . \tag{A8.4.10}$$

These two probability distributions (two points in the σ-space) can be connected continuously by a straight line that lies entirely within the allowed domain. This is because each of the conditions (A8.4.6) and (A8.4.7) means that one side of a hyperplane is allowed. Then our result (A8.4.9) entail that

$$S(\{\sigma_r\}) \leq S(\{\sigma'_r\}). \tag{A8.4.11}$$

In order for the equality to hold in (A8.4.11) we must have, for each r, either (a) $(\partial S/\partial \sigma_r) = 0$ all along the path or (b) $\sigma_r = \sigma'_r$. The former case (a)

† $(\partial S/\partial \sigma_r)$ becomes indefinite if $p_r = p_{r+1} = 0$, but we can define the function S so that $(\partial S/\partial \sigma_r) = 0$ at such a point: $\sigma_{r-1} = \sigma_r = \sigma_{r+1}$.

implies $p_{r+1} = p_r$ and $p'_{r+1} = p'_r$. Suppose that $\sigma_1 = \sigma'_1$, $\sigma_2 = \sigma'_2, \cdots$, $\sigma_{r-1} = \sigma'_{r-1}$, and $\sigma_r \neq \sigma'_r$ (which implies $p_r \neq p'_r$). Then (a) must hold for r and $\sigma_{r+1} \neq \sigma'_{r+1}$, because $\sigma_{r+1} = \sigma_{r-1} + 2p_r$ and $\sigma'_{r+1} = \sigma'_{r-1} + 2p'_r$. Hence the option (a) must continue to hold for all larger values of r. This in turn means that $p_r = p_{r+1} = p_{r+2} = \cdots$, as well as $p'_r = p'_{r+1} = p'_{r+2} = \cdots$. Since $p_r \neq p'_r$ for a value of r, one of them must be nonzero and this would imply either $\sum_{i=0}^{\infty} p_{r+i} = \infty$ or $\sum_{i=0}^{\infty} p'_{r+i} = \infty$. This is impossible; hence, we have to adopt the option (b) for all r. This means that the equality in (A8.4.11) holds if, and only if, $\sigma_r = \sigma'_r$ for all r.

A8.5. FACTOR ANALYSIS

This appendix is added in order to explain how the so-called factor analysis is related to the SELFIC method explained in Section 8.2.

The method of factor analysis, as now practiced, is laden with divergencies and complications, partly because of historical accident and partly because of the personal preferences of the large number of researchers in the field. As a result, the brief schematized account of factor analysis given below may not be regarded as a standard version of the method by many researchers (see, for instance, [H-3, T-4, H-14, K-3] for classical descriptions of factor analysis). But since the purpose of the present appendix is, in a sense, to introduce another modified version of factor analysis, which will be explained later, it really does not matter where we start.

For the purpose of this appendix we introduce various notations entirely anew. Hence it is advisable to forget, for the time being, the meanings we attached to various symbols and letters in Section 8.2.

We consider n real random variables, x_i, $i = 1, 2, \cdots, n$, which may be in general mutually correlated. We assume that if we made observations of these random variables in the entire population of corresponding systems, we would obtain a set of values $(x_1^{(\alpha)}, \cdots, x_n^{(\alpha)})$, $\alpha = 1, 2, \cdots, \nu$, with relative frequencies $\gamma^{(\alpha)}$. We agree to take the origin of each x_i so that its average becomes zero:

$$\langle x_i^{(\alpha)} \rangle_\alpha = \sum_{\alpha=1}^{n} \gamma^{(\alpha)} x_i^{(\alpha)} = 0. \tag{A8.5.1}$$

Similarly, we use a suitable scaling so that its standard deviations become unity:

$$\langle x_i^{(\alpha)2} \rangle_\alpha = \sum_{\alpha=1}^{\gamma} \gamma^{(\alpha)} x_i^{(\alpha)2} = 1. \tag{A8.5.2}$$

The central idea of factor analysis is to consider the existing correlation among the x's as stemming from a relatively small number of common variables on which the x's are linearly dependent. Thus the analysis starts with writing the random variables x_i as

$$x_i = \sum_{j=1}^{m} a_{ij} y_j + b_i z_i, \qquad i = 1, 2, \cdots, n, \tag{A8.5.3}$$

with

$$m \leq n, \tag{A8.5.4}$$

where the y's are the hidden common variables (which are called "factors") and the z's are introduced as a necessary compromise because the variation of x_i cannot be attributed entirely to the common factors. At this stage we can already point out the basic dilemma underlying factor analysis; namely, in an effort to attribute the multifariousness of the apparent data (x_i) to fewer hidden "causes" (y_j), we are forced to introduce more new variables than we had at the beginning ($m + n$ versus n variables). The enormous degree of arbitrariness brought into the problem by this fact is partly eliminated by requirement that all the $m + n$ variables (y_j and z_i) be statistically independent. Yet arbitrariness is a characteristic feature, if not a necessary evil, inherent in the method of factor analysis. It is possible, of course, to utilize this arbitrariness to enhance the "effectiveness" of factor analysis in some sense. In this connection, it should be kept in mind that the factorization would become more "effective" by using as small a number as possible as the integer m and by making the magnitudes of the b's as small as possible.

It is customary to introduce the standardizing conditions

$$\begin{aligned} \langle y_j^{(\alpha)} \rangle_\alpha &= 0, \qquad \langle y_j^{(\alpha)} \rangle_\alpha^{\ 2} = 1, \\ \langle z_i^{(\alpha)} \rangle_\alpha &= 0, \qquad \langle z_i^{(\alpha)} \rangle_\alpha^{\ 2} = 1, \end{aligned} \tag{A8.5.5}$$

as well as the "independence" assumption (which means uncorrelatedness),

$$\begin{aligned} \langle y_j^{(\alpha)} \cdot y_k^{(\alpha)} \rangle_\alpha &= 0, \qquad j \neq k, \qquad j, k = 1, 2, \cdots, m, \\ \langle y_j^{(\alpha)} \cdot z_i^{(\alpha)} \rangle_\alpha &= 0, \qquad j = 1, 2, \cdots, m, \qquad i = 1, 2, \cdots, n, \\ \langle z_i^{(\alpha)} \cdot z_k^{(\alpha)} \rangle_\alpha &= 0, \qquad i \neq k, \qquad i, k = 1, 2, \cdots, n, \end{aligned} \tag{A8.5.6}$$

where the superscript (α) indicates the "data values" of the random variables in the same sense as $x_1^{(\alpha)}$ is the data value of the random variable x_i. These two conditions can also be regarded as the conditions of normalization and orthogonality, if α is considered to designate the components of a vector.

The interesting part of the problem is that the search for the common factors y can be based on the observed correlations among the x's instead of the values of the x's themselves. From (A8.5.3), (A8.5.5), and (A8.5.6) we obtain

$$r_{ij} = \langle x_i^{(\alpha)} x_j^{(\alpha)} \rangle_\alpha = \sum_{k=1}^{m} a_{ik} a_{jk} + \delta_{ij} b_i^2. \tag{A8.5.7}$$

In particular, for $i = j$, we have

$$r_{ii} = \langle x_i^{(\alpha)2} \rangle_\alpha = a_i^2 + b_i^2 = 1, \tag{A8.5.8}$$

with

$$a_i^2 = \sum_{j=1}^{m} a_{ij}^2 , \tag{A8.5.9}$$

where a_i^2 is called "communality" of x_i and measures the degree to which the variable x_i can be attributed to the common factors. The total communality $\sum_{i=1}^{m} a_i^2$ is, so to speak, the degree of factorization of the entire problem.

The arbitrariness mentioned above manifests itself in three ways: in the value of m, in the values of the a_i's, and in the choice of the new variables y's for a given set of values of m and the a's. It is intuitively easy to expect that for a given n and r_{ij} a sufficiently large number of m will always make the "factorization" (A8.5.3) possible. There exists an algebraic theorem that gives the number N of independent conditions imposed on the "adjusted" correlation

$$\sigma_{ij} = r_{ij} - \delta_{ij} b_i = \sum_{k=1}^{m} a_{ik} a_{jk} \tag{A8.5.10}$$

as a function of n and m so that factorization (A8.5.3) is possible:

$$N = \binom{n - m + 1}{2}.$$

If N is equal to or less than n (the number of a's), factorization is possible for a given arbitrary correlation matrix r_{ij}, provided that we adjust the communalities a_i suitably. The third form of arbitrariness resides in the fact that if a set of y_j's is a solution of (A8.5.3) with (A8.5.6) we can apply any orthogonal transformation to this set of variables to obtain another set of y's satisfying all the requirements.

The so-called principal-factor solution is the oldest in the sense that the idea was proposed by Karl Pearson in 1910 (see [P-1, H-14, A-5]), long before the problem of factor analysis was formulated, and is the newest in the sense that it became practicable only after the advent of electronic computers. The major portion of what we learn as an integral part of factor analysis is something that happened during the time in between and has lost usefulness considerably by now.

Consider a rotation among the y's introduced in (A8.5.3). The result would be a linear transformation among the a_{ij} in such a way that the communalities of the x_i, $\sum_{j=1}^{m} a_{ij}^2 = a_i^2$, remain unchanged. The term a_{ij}^2 can be considered as the contribution to the communality of variable x_i from the common factor y_j. The sum $\sum_{i=1}^{m} a_{ij}^2$ is the contribution of the factor y_j to the total communality. If we aim at discovering those important common factors that are greatly responsible for the behavior of the observed variables, it is natural to set up the problem so as first to find the y_j that maximizes $\sum_{i=1}^m a_{ij}^2$, then to find the y_j that has the next largest value of $\sum_{i=1}^m a_{ij}^2$, and so on. It can be shown that this strategy amounts to finding eigenvectors of σ_{ij} in descending order of the eigenvalues. Since the total communality $\sum_{i=1}^n \sum_{j=1}^m a_{ij}^2$ is the trace of the matrix σ_{ij}, we have to stop the process when the sum of the eigenvalues becomes just the trace of the matrix. The remaining eigenvalues will be degenerate eigenvalues zero.

Psychologists are not usually satisfied with these optimum axes on the ground that they do not represent intuitively meaningful quantities. They apply an orthogonal or even oblique transformation within the space subtended by the y's until the axes satisfy a certain set of criteria. Some often-used criteria that originally were adopted for intuitive reasons can be formulated analytically, yet nobody can claim their "necessity" (such criteria are sometimes called principles of simple structure).

The Karhunen-Loève expansion was originally conceived in the framework of square-integrable functions in a continuous domain. But if we once expand such functions in terms of a complete set of orthogonal functions, each function can be expressed by a vector with an enumerable number of components. If we use the Fourier expansion and ignore high frequencies, the number of components becomes finite. In other words, if we take a finite number of sampling points, as we are bound to do in handling continuous functions on a computer, each curve can be represented by a vector with a finite number of components. The finiteness of the number of components makes the comparison with factor analysis immediately apparent.

If we make a correspondence between the vector representation of objects in the SELFIC method and the observed value set of variables in factor analysis, we note that a close analogy emerges between the two methods; in fact, the essential part of both methods resides in diagonalization of their corresponding matrices G and σ. If we ignore the z's in (A8.5.3), σ becomes r, which has essentially the same definition as G. [Compare (8.39) with (A8.5.7).] The difference, as we explain presently, seems to suggest a possible improvement in the methodology of factor analysis, although how much practical gain can be achieved is not easy to predict. There are other minor differences that can be eliminated easily. For example, in the SELFIC method we have $\sum_{i=1}^n x_i^2 = 1$, whereas in factor analysis we have $\langle \sum_{i=1}^n x_i^2 \rangle = n$. This is a matter of normalization. In practical applications of the SELFIC method it is useful for the purpose of pattern recognition to subtract the average vector so that $\sum_\alpha x_i^{(\alpha)} w^{(\alpha)} = 0$, which corresponds to (A8.5.1).

The basic methodological weakness of factor analysis is to be found in the artificial distinction between the specific variables z_i and the common factors y_j. Each factor y_j contributes to variable x_i to the extent quantitatively expressible by $\tau_{ij} = a_{ij}^2 / \sum_{i=1}^n a_{ij}^2$. Some factors y_j may have a widespread contribution, that is, a large value of $S_j = -\sum_{i=1}^{n} \tau_{ij} \log \tau_{ij}$, and some may have a sharply concentrated contribution, that is, a small value of S_j. The specific variable z_i is just an idealized case of the entropy $S_j = 0$. By rotation among the y_j-axes, the entropy value of each y-axis varies and in some cases can approach zero; such a y is not essentially different from a z. Hence there is no compelling reason to separate the z's from the y's.

An alternative way of expressing the same point is to consider the matrix σ_{ij}, which is an $n \times n$ symmetrical matrix. In principle, therefore, it could determine n eigenvectors. The reason why the method described above stops at m eigenvectors is that the diagonal elements are reduced so as to give room for the b's (i.e., for the z's), but these extra dimensions preserved for the z's must correspond to the $n - m$ dimensional subspace corresponding to the degenerate eigenvalue of zero in this space. Hence there is no justification for again apportioning a full n-dimensional subspace to the z's.†

This suggests that it would be very logical to assume that the space subtended by the y's and the z's is just n-dimensional, as in the case of the SELFIC expansion, and not $(n + m)$-dimensional, as often assumed in factor analysis. Then we do not need to separate the y's and z's. As a consequence the distinction between the matrix r_{ij} and σ_{ij} also disappears. An important byproduct of this point of view is that one need no longer use guesswork (as often done in factor analysis) in determining the appropriate values for the communalities a_j^2 (which in turn determine the m). We simply diagonalize the observed estimate of the r_{ij}. The distinction between the y's and z's then becomes a matter of degree; more precisely, it depends on the extent to which y_j extends its influence over many x's. The entropy S_j is a good measure of this degree, and may be called the "reach" of a common variable. A common variable of small reach is a specific variable.

This modification eliminates the main cause of abundant arbitrariness in factor analysis; more precisely, it eliminates the first two of the three kinds of arbitrariness mentioned above. The third form, namely, an arbitrary rotation among the discovered y's, could be retained in our modified version of factor analysis and be applied to a limited number of y's that correspond to large eigenvalues and whose "reaches" S_j are sufficiently large. The distinction between acceptable factors and unacceptable factors according to some ad hoc criteria, however, is in my opinion somewhat wanton, and is comparable to saying that the variable "linear polarization" is acceptable but the variable "elliptical polarization" is inacceptable in optics. (This example is not a metaphor; it stems from the same mathematical structure.) The factor with the largest eigenvalue, for instance, has a special distinction of its own and need not be watered down by mixing with some less conspicuous factors to become "acceptable."

† The only justifiable role left for the z's is to represent a noise peculiar to each variable x.

In conclusion, we submit that the "unique" approach of the SELFIC method can be applied to any factor analysis problem, and will definitely streamline the procedure getting rid of many unimportant complications peculiar to this tradition-laden discipline. This, of course, is not meant to preclude useful mathematical devices contrived especially for individual problems.

A8.6. THE SO-CALLED PERCEPTRON CONVERGENCE THEOREM

This appendix continues the discussion of Section 8.5. Let us introduce an $(N+1)$-dimensional vector $\{q_1, q_2, \cdots, q_N, q_{N+1}\}$ corresponding to each N-dimensional paradigm vector $\{y_1, y_2, \cdots, y_N\}$ by the definition

$$q_i = y_i, \qquad i = 1, 2, \cdots, N; \qquad q_{N+1} = 1 \quad \text{if} \quad y \in \text{I}$$

and

$$q_i = -y_i, \qquad i = 1, 2, \cdots, N; \qquad q_{N+1} = -1 \quad \text{if} \quad y \in \text{II}. \qquad \text{(A8.6.1)}$$

Then the condition that the paradigms of Class I and Class II are placed on the positive and negative sides, respectively, of the plane $\xi(y) = 0$ can be expressed by a single inequality,

$$F(q) = \sum_{i=1}^{N+1} c_i q_i > 0, \qquad \text{(A8.6.2)}$$

where the vector q is a paradigm vector of either class. If we introduce a dead zone of thickness $\varepsilon/(\sum_{i=1}^{N} c_i^2)^{1/2}$, the condition becomes

$$F(q) \equiv \sum_{i=1}^{N+1} c_i q_i > \varepsilon \qquad \text{(A8.6.3)}$$

for paradigms of both classes. The correction procedure [(8.148) and (8.149)] can also be written simply

$$c_i^{(\sigma+1)} = c_i^{(\sigma)} + \lambda^{(\sigma+1)} q_i^{(\sigma+1)}, \qquad i = 1, 2, \cdots, N+1, \qquad \text{(A8.6.4)}$$

where

$$\sum_{i=1}^{N+1} c_i^{(\sigma)} q_i^{(\sigma+1)} \leq \varepsilon. \qquad \text{(A8.6.5)}$$

The superscript (σ) has the following meaning. We start the training with $\sigma = 0$; that is, the initial coefficients are $c_i^{(0)}$, $i = 1, 2, \cdots, N+1$. Not every paradigm requires revision of the c's. We give the labels $\sigma = 1, 2, 3, \cdots$ only to the paradigms that require revision of the c's. The new coefficients revised by the σth paradigm are $c_i^{(\sigma)}$, $i = 1, 2, \cdots, N$. We attach a label σ to each λ, because λ can vary from one revision to the next. We assume at this stage only that

$$0 < l \leq \lambda^{(\sigma)} \leq L, \qquad \text{(A8.6.6)}$$

with some positive constants l and L. Those paradigms that do not require revision of the c_i do not appear in (A8.6.4) and (A8.6.5). It should be noted that between the σth and $(\sigma + 1)$st revisions there cannot be infinitely many paradigms that do not require revision, because, if there are, all paradigm types must appear in this infinite sequence requiring no revision, and hence there will be no $(\sigma + 1)$st revision.

A proof of the convergence of this successive revision is now given according to the method of Novikoff [N-2]. The main assumptions are (a) that there exists only a finite number of possible paradigm points in the space, (b) that if a paradigm point appears once in the training sequence the same paradigm point reappears after M paradigm showings for any integer M, (c) all paradigm points are in a finite region of the space, and (d) the paradigm points are linearly separable. An obvious lemma is that if the vector c is such that (A8.6.2) is satisfied for all paradigm types q, there exists a positive number ε such that (A8.6.3) is true for all paradigm types q. This is an obvious consequence of the assumption (a) that there is only a finite number of paradigm types. Another equally obvious, but useful, lemma is that if there is a solution for the c's satisfying (A8.6.3) with a certain value of ε and for all paradigm types, there is always a solution for the c's satisfying (A8.6.3) with another value of ε. This is because (A8.6.3) is invariant for multiplication of each side by a positive real number. This allows the use of an arbitrary threshold ε.

The convergence theorem states that there is a finite integer σ^* such that there will be no modification of the type (A8.6.3) and (A8.6.4) beyond σ^*, and that the $c_i^{(\sigma^*)}$ are a solution of (A8.6.3). The proof assumes the negation of the conclusion and shows that this assumption leads to a contradiction. On the one hand, we have, from (A8.6.4)–(A8.6.6),

$$\begin{aligned}\sum_{i=1}^{N+1} |c_i^{(\sigma+1)}|^2 &= \sum_{i=1}^{N+1} |c_i^{(\sigma)}|^2 + (\lambda^{(\sigma+1)})^2 \sum_{i=1}^{N+1} |q_i^{(\sigma+1)}|^2 + 2\lambda^{(\sigma+1)} \sum_{i=1}^{N+1} c_i^{(\sigma)} q_i^{(\sigma+1)} \\ &\leq \sum_{i=1}^{N+1} |c_i^{(\sigma)}|^2 + L^2 \sum_{i=1}^{N+1} |q_i^{(\sigma+1)}|^2 + 2L\varepsilon. \qquad \text{(A8.6.7)}\end{aligned}$$

Since Assumption (c) entails that $\sum_{i=1}^{N+1} |q_i^{(\sigma+1)}|^2$ is finite, the sum of the second and third terms of the right side is smaller than some positive constant A; that is, of squared length of the vector $\{c_i\}$ does not increase by a difference of more than A at each step. Hence, taking A also larger than $\sum_{i=1}^{N+1} |c_i^{(0)}|^2$, we can conclude that

$$\sum_{i=1}^{N+1} |c_i^{(\sigma+1)}|^2 \leq (\sigma + 2)A. \qquad \text{(A8.6.8)}$$

On the other hand, from the assumption (d) of the existence of a hyperplane of separation, we conclude the existence of values of coefficients c_i^+

such that

$$\sum_{i=1}^{N+1} c_i^+ q_i > \varepsilon \tag{A8.6.9}$$

for all possible paradigm points q. (Note: the second lemma is used here.) With the help of (A8.6.4), (A8.6.6), and (A8.6.9) we obtain, labeling by ρ those paradigms that call for revision,

$$\sum_{i=1}^{N+1} c_i^+ c_i^{(\sigma+1)} = \sum_{i=1}^{N+1} c_i^+ c_i^{(0)} + \sum_{\rho=1}^{\sigma+1} \lambda^{(\rho)} \sum_{i=1}^{N+1} (c_i^+ q_i^{(\rho)})$$

$$\geq \sum_{i=1}^{N+1} c_i^+ c_i^{(0)} + l(\sigma + 1)\varepsilon. \tag{A8.6.10}$$

Hence with a suitable positive number B we can write

$$\sum_{i=1}^{N+1} c_i^+ c_i^{(\sigma+1)} \geq (\sigma + 2)B. \tag{A8.6.11}$$

Since the scalar product of two vectors is never larger than the product of their lengths, we have

$$\left[\sum_{i=1}^{N+1} |c_i^{(\sigma+1)}|^2\right]^{1/2} \left[\sum_{i=1}^{N+1} |c_i^+|^2\right]^{1/2} \geq \sum_{i=1}^{N+1} c_i^+ c_i^{(\sigma+1)}.$$

Noting that the c_i^+ are constants, we conclude that

$$\sum_{i=1}^{N+1} |c_i^{(\sigma+1)}|^2 \geq (\sigma + 2)^2 B^*. \tag{A8.6.12}$$

with some positive number B^*. If the revision process goes on indefinitely, for a sufficiently large σ (A8.6.8) and (A8.6.12) will contradict each other. This shows that σ can never surpass a certain upper bound.

A direct corollary of this theorem, in view of Assumption (b), is that the final vector $c^{(\sigma*)}$ is a solution of (A8.6.9). For if (A8.6.9) is violated for any paradigm point, that point will appear again according to Assumption (b) and will cause this $c^{(\sigma*)}$ to change in view of (A8.6.4) and (A8.6.5). This completes the proof of the theorem.

The reader will probably agree that good intellectual work usually evokes a sense of optimality and uniqueness, which are consequences of a full exploitation of all possible modifications, variations, and other opportunities. From this viewpoint we have to say that the perceptron convergence theorem indeed leaves much to be desired. First, the parameter $\lambda^{(\sigma+1)}$ can take any arbitrary value between bounds. Second, the information provided by some of the paradigms may not be exploited at all in determination of the final plane. Third, the final plane may be placed in any "unbalanced" position between the two groups of points insofar as (A8.6.3) is satisfied. The following is just a suggestion to exploit the arbitrariness of the λ's to bring the plane to a "better-balanced" position, where the estimation of balance will take all the available paradigms into consideration. This will not make the theory unique or optimal, but it may represent a slight improvement.

A hyperplane is determined by the direction of its normal and its distance from the coordinate origin. The proposed algorithm is intended to adjust the latter at each stage of adaptive correction so that the plane bisects the space at a more desirable position relative to the two groups of given paradigm points. Let μ be the number of the paradigms of Class I, $q^{(\alpha)} \in \mathrm{I}$, and ν be the number of the paradigms of Class II, $q^{(\beta)} \in \mathrm{II}$, which have been revealed at a particular stage of the adaptive process.† If the same point has reappeared more than once, it must be counted the same number of times in μ or in ν. Let D_{I} be the distance of the center of mass of the Class I points from the plane. In other words, D_{I} is $(1/\mu)$ times the total sum of the distances of points of Class I from the plane. We define D_{II} similarly for Class II. The "desirable" position, as far as the "weight" balance is concerned, is defined by

$$\frac{D_{\mathrm{I}}}{D_{\mathrm{II}}} = \varphi(\mu/\nu), \qquad \text{(A8.6.13)}$$

where φ is an arbitrary function of the argument μ/ν, which the user can select according to his sense of balance. Perhaps it may be required that $\varphi(\mu/\nu) = 1$ for $\mu/\nu = 1$ for the sake of symmetry. More generally, we may require $\varphi(r) = 1/\varphi(1/r)$ if Class I and Class II are to be treated on an equal footing. If the values of μ and ν are considered as having no particular importance in deciding $D_{\mathrm{I}}/D_{\mathrm{II}}$, φ should be equated with a constant. If we are allowed to assume, in the spirit of linear separability, that the μ paradigm points of Class I and ν paradigm points of Class II form more or less separate volumes, it probably is natural to assume a monotonically increasing function for φ, because the larger the number of points, the larger the volume they will occupy. If we roughly assume an even density distribution of points of both classes in the N-dimensional space, it may be reasonable to assume

$$\varphi(r) = r^{(1/N)}, \qquad \text{(A8.6.14)}$$

for the volume of an N-dimensional sphere is proportional to the Nth power of its radius.

† The limiting values of μ and ν will be proportional to the actual numbers of paradigm types of classes I and II in the set of paradigms, if Condition (b) above is sharpened so that the showing is not only "recurrent" but also "ergodic"; that is, each type comes up with an equal frequency in the long run. It is assumed that the ratio of members of class I and class II remains the same not only during the training period but also during the application period.

It is obvious that indiscriminate enforcement of the rule (A8.6.13) in some cases could push that plane to a position where it no longer separates the two groups. Hence we have to add a proviso that (A8.6.13) will be enforced only insofar as convergence is ensured toward a plane that still separates the groups. If this cannot be guaranteed, we shall require only that the difference between $D_{\mathrm{I}}/D_{\mathrm{II}}$ and $\varphi(\mu/\nu)$ be made as small as possible.

Let us write

$$F^{(\sigma)}(q) \equiv \sum_{i=1}^{N+1} c_i^{(\sigma)} q_i = 0 \tag{A8.6.15}$$

for the plane at the σth stage of adaptation, and write $\mathrm{I}^{(\sigma)}$ $\mathrm{II}^{(\sigma)}$, respectively, for the collection of the μ paradigms of Class I and the collection of the ν paradigms of Class II that have so far been revealed at the σth stage. The collections of paradigms $\mathrm{I}^{(\sigma)}$ and $\mathrm{II}^{(\sigma)}$ may contain more than one representative of the same object if the object has appeared more than once. They also include paradigms that have caused no correction, according to (A8.6.4) and (A8.6.5). Then the distances of the two centers of mass from the plane at the σth stage will be given by

$$\begin{aligned} D_{\mathrm{I}}^{(\sigma)} &= \frac{\dfrac{1}{\mu}\left|\sum\limits_{\alpha\in\mathrm{I}^{(\sigma)}} F^{(\sigma)}(q^{(\alpha)})\right|}{\left(\sum\limits_{i=1}^{N} c_i^{(\sigma)2}\right)^{1/2}}, \\ D_{\mathrm{II}}^{(\sigma)} &= \frac{\dfrac{1}{\nu}\left|\sum\limits_{\beta\in\mathrm{II}^{(\sigma)}} F^{(\sigma)}(q^{(\beta)})\right|}{\left(\sum\limits_{i=1}^{N} c_i^{(\sigma)2}\right)^{1/2}}. \end{aligned} \tag{A8.6.16}$$

The condition that the relation (A8.6.13) be upheld at the σth stage is

$$\frac{\sum\limits_{\alpha\in\mathrm{I}^{(\sigma)}} F^{(\sigma)}(q^{(\alpha)})}{\sum\limits_{\beta\in\mathrm{II}^{(\sigma)}} F^{(\sigma)}(q^{(\beta)})} = \frac{\mu^{(\sigma)}}{\nu^{(\sigma)}}\,\varphi\left(\frac{\mu^{(\sigma)}}{\nu^{(\sigma)}}\right). \tag{A8.6.17}$$

We do not need to take the absolute value on the left side of (A8.6.17), provided that the center of mass of $\mathrm{I}^{(\sigma)}$ and the center of mass of $\mathrm{II}^{(\sigma)}$ lie on the opposite sides of the plane, because then we shall have

$$\left[\sum_{\alpha\in\mathrm{I}^{(\sigma)}} F^{(\sigma)}(q^{(\alpha)})\right]\cdot\left[\sum_{\beta\in\mathrm{II}^{(\sigma)}} F^{(\sigma)}(q^{(\beta)})\right] > 0. \tag{A8.6.18}$$

We added the superscript (σ) on μ and ν in (A8.6.17) to make clear that they are the numbers of the members of $\mathrm{I}^{(\sigma)}$ and $\mathrm{II}^{(\sigma)}$ at the σth stage.

The condition (A8.6.17) is desirable but cannot be expected to be satisfied by the usual adaptation procedure. The proposal here is to take advantage of the arbitrariness of λ at each corrective step to realize (A8.6.17) if possible. Suppose that we want to realize (A8.6.17) at the $(\sigma + 1)$st stage, which is defined by (A8.6.4). Substituting the latter in (A8.6.17) for $\sigma + 1$ with the help of (A8.6.15), we obtain

$$\lambda^{(\sigma+1)} = -\frac{\sum_{i=1}^{N+1} c_i^{(\sigma)}\left[\sum_{\alpha\in\mathrm{I}^{(\sigma+1)}} q_i^{(\alpha)} - \kappa^{(\sigma+1)} \sum_{\beta\in\mathrm{II}^{(\sigma+1)}} q_i^{(\beta)}\right]}{\sum_{i=1}^{N+1} q_i^{(\sigma+1)}\left[\sum_{\alpha\in\mathrm{I}^{(\sigma+1)}} q_i^{(\alpha)} - \kappa^{(\sigma+1)} \sum_{\beta\in\mathrm{II}^{(\sigma+1)}} q_i^{(\beta)}\right]}, \qquad \text{(A8.6.19)}$$

where

$$\kappa^{(\sigma+1)} = \frac{\mu^{(\sigma+1)}}{\nu^{(\sigma+1)}}\, \varphi\left(\frac{\mu^{(\sigma+1)}}{\nu^{(\sigma+1)}}\right). \qquad \text{(A8.6.20)}$$

If in addition to the values of $c_i^{(\sigma)}$ $(i = 1, 2, \cdots, N + 1)$, the machine keeps the updated values of $\sum_{\alpha\in\mathrm{I}^{(\sigma+1)}} q_i^{(\alpha)}$, $\sum_{\beta\in\mathrm{II}^{(\sigma+1)}} q_i^{(\beta)}$ $(i = 1, 2, \cdots, N + 1)$, $\mu^{(\sigma+1)}$, and $\nu^{(\sigma+1)}$, the right-hand side of (A8.6.19) can be calculated by using these quantities obtainable at the σth stage and the data about the newcomer, $q^{(\sigma+1)}$. With this new $\lambda^{(\sigma+1)}$ we can perform the correction according to (A8.6.4). If $\lambda^{(\sigma+1)}$ is outside the range (l, L), we should replace $\lambda^{(\sigma+1)}$ by l or L, whichever is closer to the calculated $\lambda^{(\sigma+1)}$. Since we need not store the entire collection of the past paradigms, but only three updated vectors instead of the usual one sums, we may still say that this is an adaptive method. It would be a good idea to check whether (A8.6.18) is satisfied, although for a sufficiently large σ it will always be satisfied. This algorithm was programmed successfully by my student, T. Kaminuma [K-2*a*].

The above method achieves "weight balance" in the sense that the distance of the hyperplane from the center of mass of class I and that from the center of mass of class II are so balanced that the rule (A8.6.13) is more or less satisfied. However, the normal of the hyperplane thus obtained lies often in a very unreasonable direction. Roughly speaking, it would be desirable to have the normal of the plane parallel to the straight line connecting the two centers of mass. Let the hyperplane obtained by the weight balance be expressed as $\sum_{i=1}^{N} \alpha_i x_i - p = 0$, with $\sum_{i=1}^{N} \alpha_i^2 = 1$. If we fix the intersection of this plane with the straight line connecting the centers of mass and rotate the plane until its normal coincides with the direction of the straight line, we obtain a "desirable" plane $\sum_{i=1}^{N} \alpha_i^* x_i - p^* = 0$, where

$$\alpha_i^* = \frac{\gamma_i}{\sqrt{\sum_{i=1}^{N} \gamma_i^2}} \qquad \text{(A8.6.20)}$$

$$p^* = \frac{1}{\sqrt{\sum_{i=1}^{N} \gamma_i^2}}\left[\frac{\left(p - \sum_{j=1}^{N} \alpha_j \gamma_{\mathrm{I}j}\right)}{\sum_{j=1}^{N} \alpha_j \gamma_j} \sum_{i=1}^{N} \gamma_i^2 + \sum_{i=1}^{N} \gamma_i \gamma_{\mathrm{I}i}\right] \qquad \text{(A8.6.21)}$$

with

$$\gamma_i = \gamma_{IIi} - \gamma_{Ii} \tag{A8.6.22}$$

In (A8.6.22) the two vectors $\{\gamma_{Ii}\}$ and $\{\gamma_{IIi}\}$ are the updated positions of the centers of mass of class I and class II which should include all the paradigms that have so far appeared:

$$\gamma_{Ii} = \sum_{\alpha \in I} q_i^{(\alpha)}/\mu \quad \text{and} \quad \gamma_{IIi} = -\sum_{\beta \in II} q_i^{(\beta)}/\nu \tag{A8.6.23}$$

A reasonable algorithm is not to jump from (α_i, p) to (α_i^*, p^*) at once, but to change (α_i, p) to $(\alpha_i \pm \delta\alpha_i^*, p \pm \delta p^*)$ with some positive constant $\delta < 1$, everytime a new arrival does not require a correction of the type (A8.6.4). (The upper sign if $\sum_i \alpha_i \alpha_i^* > 0$ and lower sign if $\sum_i \alpha_i \alpha_i^* < 0$.) This will permit the directional balance to reach a compromise with classificatory correction (A8.6.4) and weight balance, provided a suitable prescription is introduced to stop the directional correction at a certain stage. This algorithm has also been successfully carried out by Kaminuma.

A9.1. SIMULTANEOUS DIAGONALIZATION OF COMMUTING PROJECTION OPERATORS

Before the demonstration let us recall that any projection operator can be brought to a diagonal form by a suitable orthogonal transformation and that in its diagonal form the diagonal elements are either 0 or 1. This was done by using the coordinate system of which m coordinates subtend the m-dimensional subspace $\mathfrak{M}$, to which the projection operator corresponds.

Suppose that we are given $\mathcal{P}_1 = \mathcal{P}[\mathfrak{M}_1]$ and $\mathcal{P}_2 = \mathcal{P}[\mathfrak{M}_2]$, which commute with each other:

$$\mathcal{P}_1\mathcal{P}_2 - \mathcal{P}_2\mathcal{P}_1 = 0. \tag{A9.1.1}$$

Let us first, by a suitable transformation, bring $\mathcal{P}_1$ to a diagonal form $\mathcal{P}_1'$ and relabel the coordinates in such a way that all the 1's on the diagonal come to the upper left corner:

$$\mathcal{P}_I' = \left(\begin{array}{c|c} I & \mathbf{O} \\ \hline \mathbf{O} & \mathbf{O} \end{array}\right), \tag{A9.1.2}$$

where I is a diagonal matrix with only 1's on the diagonal. This means that the rows above the horizontal division line and the columns to the left of the vertical division line refer to the coordinates corresponding to $\mathcal{P}_1$. Let us assume that $\mathcal{P}_2'$ in this coordinate system has the form

$$\mathcal{P}_2' = \left(\begin{array}{c|c} A & B \\ \hline C & D \end{array}\right), \tag{A9.1.3}$$

using the same division lines of rows and columns as in (A9.1.2). If we write $\mathcal{P}_1'\mathcal{P}_2' - \mathcal{P}_2'\mathcal{P}_1' = 0$, which follows from (A9.1.1) on account of (9.91), we obtain

$$\left(\begin{array}{c|c} A & B \\ \hline \mathbf{O} & \mathbf{O} \end{array}\right) - \left(\begin{array}{c|c} A & \mathbf{O} \\ \hline C & \mathbf{O} \end{array}\right) = 0, \tag{A9.1.4}$$

which means that $B = \mathbf{O}$ and $C = \mathbf{O}$. Hence

$$\mathcal{P}_2' = \left(\begin{array}{c|c} A & \mathbf{O} \\ \hline \mathbf{O} & D \end{array}\right), \tag{A9.1.5}$$

where A is a matrix within the subspace $\mathfrak{M}_1$ and D is a matrix within a subspace $\mathfrak{M}_1^*$ orthogonal to $\mathfrak{M}_1$. To bring $\mathcal{P}_2$ to a real diagonal form, it suffices to apply a transformation within $\mathfrak{M}_1$ on $\mathcal{P}_2'$ so that A becomes diagonal and to apply another transformation within $\mathfrak{M}_1^*$ so that D becomes diagonal. Of course, in this final diagonal form there are only 0's and 1's on the diagonal. These last two transformations obviously do not change the form of $\mathcal{P}_1'$ (A9.1.2). Thus in the final coordinate systems both $\mathcal{P}_1'$ and $\mathcal{P}_2'$ are diagonal and have 0 and 1 on the diagonal. Unless $\mathcal{P}_1 = \mathcal{P}_2$, the positions where $\mathcal{P}_1'$ and $\mathcal{P}_2'$ have 1's do not entirely coincide. $\mathfrak{M}_1$ is a subspace defined by the coordinates for which $\mathcal{P}_1'$ has 1's, and $\mathfrak{M}_2$ is a subspace defined by the coordinates for which the final $\mathcal{P}_2'$ has 1's. Since the coordinates used are the same and are orthogonal, however, $\mathfrak{M}_1$ and $\mathfrak{M}_2$ belong to the same family of orthogonal subspaces.

A9.2. NONCOMMUTING PROJECTION OPERATORS

In order to derive (9.96) in the general case and also to have some insight into the situation in which $A_1 \sim A_2$ is not the case, we start the discussion anew, assuming that $\mathfrak{M}_1 = \mathfrak{M}(A_1)$ and $\mathfrak{M}_2 = \mathfrak{M}(A_2)$ do not belong to a family of orthogonal subspaces. We assume that $\mathfrak{M}_1$ is l-dimensional and is defined by l linearly independent vectors $\boldsymbol{\xi}^{(1)}, \boldsymbol{\xi}^{(2)}, \cdots, \boldsymbol{\xi}^{(l)}$, while $\mathfrak{M}_2$ is m-dimensional and is defined by m linearly independent vectors $\boldsymbol{\eta}^{(1)}, \boldsymbol{\eta}^{(2)}, \cdots, \boldsymbol{\eta}^{(m)}$. This does not mean that the $l + m$ vectors $\boldsymbol{\xi}^{(i)}$ and $\boldsymbol{\eta}^{(i)}$ are linearly independent. In general, there can be k (≥ 0) relations of the type

$$\sum_{i=1}^{l} a_i \boldsymbol{\xi}^{(i)} = \sum_{i=1}^{m} b_i \boldsymbol{\eta}^{(i)}. \tag{A9.2.1}$$

When we say there are k relations, we mean k *independent* relations, in the sense that none of the k relations can be derived from the rest of the k relations. To distinguish each of the k relations we attach a superscript (j) on the a's and b's ($j = 1, 2, \cdots, k$). Let us thus write

$$\boldsymbol{\zeta}^{(j)} \equiv \sum_{i=1}^{l} a_i^{(j)} \boldsymbol{\xi}^{(i)} = \sum_{i=1}^{m} b_i^{(j)} \boldsymbol{\eta}^{(i)}, \qquad j = 1, 2, \cdots, k. \tag{A9.2.2}$$

The fact that there are k independent relations of the form (A9.2.1) can now be translated as the linear independence of k vectors $\boldsymbol{\zeta}^{(j)}$. This means that the set of vectors $\mathbf{x}$ of the type

$$\mathbf{x} = \sum_{j=1}^{k} c_j \boldsymbol{\zeta}^{(j)} \tag{A9.2.3}$$

forms a subspace of k dimensions. The $\mathbf{x}$ of (A9.2.3) belongs, of course, to $\mathfrak{M}_1$ as well as $\mathfrak{M}_2$. Conversely, we now prove that any vector, say, $\mathbf{y}$, that belongs to both $\mathfrak{M}_1$ and $\mathfrak{M}_2$ can be written in the form of (A9.2.3) and therefore belongs to the space defined by the $\boldsymbol{\zeta}$'s. If such a $\mathbf{y}$ cannot be written in the form of (A9.2.3), it can be written as

$$\mathbf{y} = \sum_{j=1}^{k} c_j \boldsymbol{\zeta}^{(j)} + \mathbf{z}, \tag{A9.2.4}$$

where $\mathbf{z} = \mathbf{y} - \sum_{j=1}^{k} c_j \boldsymbol{\zeta}^{(j)}$ is nonzero and cannot be expressed as a linear combination of the $\boldsymbol{\zeta}$'s. Since $\mathbf{y} \in \mathfrak{M}_1$, $\mathbf{y} \in \mathfrak{M}_2$, $\boldsymbol{\zeta}^{(j)} \in \mathfrak{M}_1$, and $\boldsymbol{\zeta}^{(j)} \in \mathfrak{M}_2$, however, we have to conclude that $\mathbf{z} \in \mathfrak{M}_1$ and $\mathbf{z} \in \mathfrak{M}_2$; that is, $\mathbf{z}$ can be expressed in terms of the $\boldsymbol{\xi}$'s only as well as in terms of the $\boldsymbol{\eta}$'s only. Then the linear combination expressing $\mathbf{z}$ in terms of the $\boldsymbol{\xi}$'s and that in terms of the $\boldsymbol{\eta}$'s must have the form of (A9.2.2) and yet cannot be derived from the k quantities in that equation, since $\mathbf{z}$ is supposed to be not expressible in terms of the $\boldsymbol{\zeta}$'s. This would entail that there are more than k independent relations of the type (A9.2.2), and contradicts the assumption that k is the total number of the independent relations of the type (A9.2.2). This completes the proof that the subspace defined by the k $\boldsymbol{\zeta}$'s is identical with the set of all vectors, each of which is contained in both $\mathfrak{M}_1$ and $\mathfrak{M}_2$; that is, that this subspace is $\mathfrak{M}(A_1 \cap A_2)$, which for simplicity is denoted $\mathfrak{M}_{1\cap 2}$.

Let us now come back to $\mathfrak{M}_1$, in which one can take l independent vectors. One can choose k among them as $\boldsymbol{\zeta}^{(i)}$, $i = 1, 2, \cdots, k$. The remaining $l - k$ independent vectors can be expressed as a linear combination of the original $\boldsymbol{\xi}$'s. Let us again use the same symbols $\boldsymbol{\xi}^{(i)}$ ($i = 1, 1, 2, \cdots k, \cdots, l$) to express these l independent vectors thus chosen, so that the first k of them coincide with the $\boldsymbol{\zeta}$'s:

$$\boldsymbol{\xi}^{(i)} = \boldsymbol{\zeta}^{(i)}, \qquad i = 1, 2, \cdots, k. \tag{A9.2.5}$$

The other ξ's cannot be expressed in terms of $\boldsymbol{\zeta}$'s and hence are not in $\mathfrak{M}_{1\cap 2}$. Similarly, we choose m independent vectors in $\mathfrak{M}_2$ and denote them by the same symbols, $\boldsymbol{\eta}^{(i)}$ ($i = 1, 2, \cdots, k, \cdots, m$) in such a way that the first k

of them coincide with the $\boldsymbol{\zeta}$'s:

$$\boldsymbol{\eta}^{(i)} = \boldsymbol{\zeta}^{(i)}, \qquad i = 1, 2, \cdots, k. \tag{A9.2.6}$$

Thus we have in all $l + m - k$ vectors at hand, k of which, namely, $\boldsymbol{\zeta}^{(i)}$ $(j = 1, 2, \cdots, k)$, are in $\mathfrak{M}_{1\cap 2}$; $(l - k)$ of which, namely, $\boldsymbol{\xi}^{(i)}$ $(i = k + 1, k + 2, \cdots, l)$, are in $\mathfrak{M}_1$ but not in $\mathfrak{M}_{1\cap 2}$; and $(m - k)$ of which, namely, $\boldsymbol{\eta}^{(i)}$ $(i = k + 1, k + 2, \cdots, m)$, are in $\mathfrak{M}_2$ but not in $\mathfrak{M}_{1\cap 2}$. If we denote these $1 + m - k$ vectors by $\boldsymbol{\omega}^{(i)}$ $(i = 1, 2, \cdots, 1 + m - k)$, these $\boldsymbol{\omega}$'s are linearly independent. If they were not linearly independent, there should be at least one relation of the type

$$\sum_{i=1}^{1+m-k} a_i \boldsymbol{\omega}^{(i)} = \sum_{i=1}^{l} b_i \boldsymbol{\xi}^{(i)} + \sum_{i=k+1}^{m} c_i \boldsymbol{\eta}^{(i)} = 0, \tag{A9.2.7}$$

where not all a's are zero. Since the $\boldsymbol{\xi}$'s are linearly independent, $\sum_{i=1}^{l} b_i \boldsymbol{\xi}$ cannot be zero except when all the b's are zero. In this last case (A9.2.7) would entail that all c's are also zero, since the $\boldsymbol{\eta}$'s are independent. This would contradict the assumption that not all a's are zero. Hence we have to put

$$\sum_{i=1}^{l} b_i \boldsymbol{\xi}^{(i)} \neq 0.$$

This condition yields, in view of (A9.2.7),

$$\sum_{i=1}^{k} b_i \boldsymbol{\zeta}^{(i)} + \sum_{i=k+1}^{l} b_i \boldsymbol{\xi}^{(i)} = - \sum_{i=k+1}^{m} c_i \boldsymbol{\eta}^{(i)} \neq 0. \tag{A9.2.8}$$

If in this relation all the coefficients of $\boldsymbol{\zeta}^{(i)}$ are zero, we have

$$\sum_{i=k+1}^{l} b_i \boldsymbol{\xi}^{(i)} = - \sum_{i=k+1}^{m} \delta_i \boldsymbol{\eta}^{(i)} \neq 0. \tag{A9.2.9}$$

Because $(\boldsymbol{\zeta}^{(1)}, \boldsymbol{\zeta}^{(2)}, \cdots, \boldsymbol{\zeta}^{(k)}, \boldsymbol{\xi}^{(k+1)}, \cdots, \boldsymbol{\xi}^{(l)})$ are linearly independent, the left side of (A9.2.9) cannot be expressed only in terms of the $\boldsymbol{\zeta}$'s. This means that the quantity (A9.2.9) cannot be derived from (A9.2.2), implying that there are more than k relations of the type (A9.2.1). This contradicts the premise of our discussion. If, on the other hand, some coefficients of the $\boldsymbol{\zeta}^{(i)}$ in (A9.2.8) are nonzero, $\sum_{i=1}^{k} b_i \boldsymbol{\zeta}^{(i)} \neq 0$ because of the linear independence of the $\boldsymbol{\zeta}$'s. If this is the case,

$$\sum_{i=1}^{k} b_i \boldsymbol{\zeta}^{(i)} = - \sum_{i=k+1}^{l} b_i \boldsymbol{\xi}^{(i)} - \sum_{i=k+1}^{m} k_i \boldsymbol{\eta}^{(i)} \neq 0.$$

This last equation states that a nonzero vector in $\mathfrak{M}_{1\cap 2}$ is expressed in terms of the vectors that are not in $\mathfrak{M}_{1\cap 2}$, since the $\boldsymbol{\xi}^{(i)}$ $(i = k + 1, \cdots, l)$ and the $\boldsymbol{\eta}^{(i)}$ $(i = k + 1, \cdots, m)$ are supposed to be outside $\mathfrak{M}_{1\cap 2}$. This is absurd.

Hence all $1 + m - k$ vectors $\boldsymbol{\omega}^{(i)}$ $(i = 1, 2, \cdots, l + m - k)$ are linearly independent. Consequently the set of vectors $\mathbf{x}$

$$\mathbf{x} = \sum_{i=1}^{1+m-k} a_i \boldsymbol{\omega}^{(i)} \tag{A9.2.10}$$

form a subspace of $1 + m - k$ dimensions. This space must be identical with $\mathfrak{M}(A_1 \cup A_2) \equiv \mathfrak{M}_{1\cup 2}$. In fact, any linear combination of two vectors, one in $\mathfrak{M}_1$ and the other in $\mathfrak{M}_2$, has the form (A9.2.10), and any vector of the form (A9.2.10) can be written as a linear combination of one vector in $\mathfrak{M}_1$ and another vector in $\mathfrak{M}_2$. A consequence of this discussion is that if the numbers of dimensions of $\mathfrak{M}_1$, $\mathfrak{M}_2$, and $\mathfrak{M}_{1\cap 2}$ are l, m, and k, respectively, the number of dimensions of $\mathfrak{M}_{1\cup 2}$ is $l + m - k$. This is exactly the relation (9.96), since trace $\mathcal{T}[\mathfrak{M}_1] = l$, trace $\mathcal{T}[\mathfrak{M}_2] = m$, trace $\mathcal{T}[\mathfrak{M}_{1\cap 2}] = k$, and trace $\mathcal{T}[\mathfrak{M}_{1\cup 2}] = l + m - k$. We return to this topic presently.

We can of course take the $\boldsymbol{\zeta}$'s in $\mathfrak{M}_{1\cap 2}$ so that they are all mutually orthogonal. We can also make all the $\boldsymbol{\xi}$'s, including the $\boldsymbol{\zeta}$'s, mutually orthogonal. Similarly, we can make all the $\boldsymbol{\eta}$'s, including the $\boldsymbol{\zeta}$'s, mutually orthogonal. This however does not imply that all the$\boldsymbol{\omega}$'s, that is, both the $\boldsymbol{\xi}$'s and $\boldsymbol{\eta}$'s, can be mutually orthogonal. Suppose that we have made all the $\boldsymbol{\xi}$'s mutually orthogonal and all the $\boldsymbol{\eta}$'s mutually orthogonal, and write

$$\begin{aligned}
\mathcal{T}[\mathfrak{M}_1] &= \sum_{i=1}^{k} \mathcal{T}[\boldsymbol{\zeta}^{(i)}] + \sum_{i=k+1}^{l} \mathcal{T}[\boldsymbol{\xi}^{(i)}], \\
\mathcal{T}[\mathfrak{M}_2] &= \sum_{i=1}^{k} \mathcal{T}[\boldsymbol{\zeta}^{(i)}] + \sum_{i=k+1}^{m} \mathcal{T}[\boldsymbol{\eta}^{(i)}], \\
\mathcal{T}[\mathfrak{M}_{1\cap 2}] &= \sum_{i=1}^{k} \mathcal{T}[\boldsymbol{\zeta}^{(i)}].
\end{aligned} \tag{A9.2.11}$$

The commutativity of $\mathcal{T}[\mathfrak{M}_1]$ and $\mathcal{T}[\mathfrak{M}_2]$ reduces to the commutativity of $\mathfrak{K}_1 = \sum_{i=k+1}^{l} \mathcal{T}[\xi^{(i)}]$ and $\mathfrak{K}_2 = \sum_{j=k+1}^{m} \mathcal{T}[\boldsymbol{\eta}^{(i)}]$. The result of application of $\mathfrak{K}_1\mathfrak{K}_2$ on any arbitrary vector is a vector in the space defined by $\boldsymbol{\xi}^{(i)}$, with $i = k + 1, \cdots, l$, and the result of application of $\mathfrak{K}_2\mathfrak{K}_1$ on the same vector is a vector in the space defined by $\boldsymbol{\eta}^{(j)}$, with $j = k + 1, \cdots, m$. In order that these two resultant vectors may be the same, they must be zero, for if they are not, the vector to which these two are equal should be in $\mathfrak{M}_{1\cap 2}$ because it is in both $\mathfrak{M}_1$ and $\mathfrak{M}_2$. This is impossible, however, because it can be expressed in terms of $\xi^{(i)}$ with $i > k$ only. Therefore commutativity requires that $\mathfrak{K}_2\mathfrak{K}_1$ as well as $\mathfrak{K}_1\mathfrak{K}_2$ be identically zero; in other words the space defined by $\boldsymbol{\eta}^{(i)}$ $(j > k)$ and the space defined by $\boldsymbol{\xi}^{(i)}$ $(i > k)$ must be orthogonal. This result has already been obtained; if $\mathcal{T}[\mathfrak{M}_1]$ and $\mathcal{T}[\mathfrak{M}_2]$ commute then $\mathfrak{M}_1$ and $\mathfrak{M}_2$ belong to the same family of orthogonal subspaces. We should note further that we have, from (A9.2.11),

$$\mathfrak{T}[\mathfrak{M}_1]\mathfrak{T}[\mathfrak{M}_2] = \mathfrak{T}[\mathfrak{M}_{1\cap 2}] + \mathfrak{R}_1\mathfrak{R}_2. \tag{A9.2.12}$$

When $\mathfrak{T}[\mathfrak{M}_1]$ and $\mathfrak{T}[\mathfrak{M}_2]$ commute, $\mathfrak{R}_1\mathfrak{R}_2 = 0$; hence $\mathfrak{T}[\mathfrak{M}_1]\mathfrak{T}[\mathfrak{M}_2] = \mathfrak{T}[\mathfrak{M}_{1\cap 2}]$, which we obtained in (9.89).

It is easy to see that the infinite alternative product of $\mathfrak{T}[\mathfrak{M}_1]$ and $\mathfrak{T}[\mathfrak{M}_2]$ is equivalent to $\mathfrak{T}[\mathfrak{M}_{1\cap 2}]$:

$$\mathfrak{T}[\mathfrak{M}_{1\cap 2}] = \cdots \mathfrak{T}[\mathfrak{M}_1] \cdot \mathfrak{T}[\mathfrak{M}_2] \cdot \mathfrak{T}[\mathfrak{M}_1] \cdot \mathfrak{T}[\mathfrak{M}_2] \cdots. \tag{A9.2.13}$$

The reason is that by an application of $\mathfrak{T}[\mathfrak{M}_1] = \mathfrak{T}[\mathfrak{M}_{1\cap 2}] + \mathfrak{R}_1$ any vector $\mathbf{x}$ becomes $\mathfrak{T}[\mathfrak{M}_{1\cap 2}]\mathbf{x}$ plus a vector $\mathfrak{R}_1\mathbf{x}$ in the subspace defined by $\{\boldsymbol{\xi}^{(i)}\}$, $i = k + 1, \cdots, l$. By a subsequent application of $\mathfrak{T}[\mathfrak{M}_2] = \mathfrak{T}[\mathfrak{M}_{1\cap 2}] + \mathfrak{R}_2$ the former component remains unchanged while the latter component becomes $\mathfrak{R}_2\mathfrak{R}_1\mathbf{x}$, which is smaller in magnitude than a definite fraction of the magnitude of $\mathfrak{R}_1\mathbf{x}$. By an infinite reiteration of this process the first component remains $\mathfrak{T}[\mathfrak{M}_{1\cap 2}]\mathbf{x}$ and the second component vanishes. Since $\mathfrak{R}_1\mathfrak{T}[\mathfrak{M}_{1\cap 2}] = \mathfrak{R}_2\mathfrak{T}[\mathfrak{M}_{1\cap 2}] = 0$, (A9.2.13) is equivalent to

$$0 = \cdots \mathfrak{R}_1 \cdot \mathfrak{R}_2 \cdot \mathfrak{R}_1 \cdot \mathfrak{R}_2 \cdots. \tag{A9.2.14}$$

Compare this result with the result obtained at the end of Section 9.3.

A9.3. PLANCK'S CONSTANT AND THE COMMUTATION RELATION

Before discussing the connection between the model theory in Chapter 9 and genuine physical theory, let us repeat some of our previous results, rewriting them in Dirac's notation [D-8], which is now a standard notation in the literature. In fact, our notation in Chapter 8 is more or less in agreement with Dirac's.

Let us take two coordinate systems $\{\boldsymbol{\psi}_i\}$ and $\{\boldsymbol{\varphi}_a\}$ and express one of the $\boldsymbol{\varphi}$'s in terms of the $\boldsymbol{\psi}$'s. Then we have $\boldsymbol{\varphi}_a = \sum_i \boldsymbol{\psi}_i U_{ia}$, as in (9.49) or (9.159). The element U_{ia} of the unitary matrix U is nothing but the ith component in the $\boldsymbol{\psi}$-representation of the vector $\boldsymbol{\varphi}_a$. (By a "representation" is meant all mathematical expressions using some fixed coordinate system.) In other words, $U_{ia} = (\boldsymbol{\psi}_i, \boldsymbol{\varphi}_a)$, to use the notation of (9.100). In Dirac's notation U_{ia} is written as $\langle i \mid a\rangle$. The unitarity of U, that is, $U\bar{U} = \bar{U}U = 1$, means $\sum_a U_{ia}\bar{U}_{aj} = \delta_{ij}$ and $\sum_i \bar{U}_{ai}U_{ib} = \delta_{ab}$. In Dirac's notation $\sum_a \langle i \mid a\rangle\langle a \mid j\rangle = \delta_{ij}$ and $\sum_i \langle a \mid i\rangle\langle i \mid b\rangle = \delta_{ab}$. The quantity $\langle a \mid j\rangle$ is by definition the element of the inverse unitary transformation U^{-1}, but on account of the unitarity $U^{-1} = \bar{U}$, we have $\langle a \mid j\rangle = \langle j \mid a\rangle^*$. This is nothing but the second equality in (9.100). The matrix element A_{ij} of any operator in our previous notation is written $\langle i| A |j\rangle$ in Dirac's notation. Thus, for instance, $(\mathcal{T}[\boldsymbol{\varphi}_a])_{ij} = (\boldsymbol{\varphi}_a)_i(\boldsymbol{\varphi}_a)_j^* = U_{ia}U_{aj}$ is now written $\langle i| \mathcal{T}\ [\boldsymbol{\varphi}_a] \,|j\rangle = \langle i \mid a\rangle\langle a \mid i\rangle$. (Note: no summation is made over a.) Thus, if a physical quantity Q is expressed as $Q = \sum_a q_a \mathcal{T}[\boldsymbol{\psi}_a]$, our former notation gave $(Q)_{ij} = \sum_a U_{ia}q_a\bar{U}_{aj}$. In the new notation $\langle i| Q |j\rangle = \sum_a \langle i \mid a\rangle q_a\langle a \mid j\rangle$. When there is no degeneracy, each eigenvalue q_a is sufficient to specify a single quantum state $\boldsymbol{\psi}_a$. Hence we can use q_a itself as well as the label a. In this case $\langle i| Q |j\rangle = \sum_a \langle i \mid a\rangle a\langle a \mid j\rangle$. If two coordinate systems $\{\boldsymbol{\psi}_i\}$ and $\{\boldsymbol{\varphi}_a\}$ coincide, then, of course, we have $U_{ia} = \delta_{ia}$ or $\langle i \mid a\rangle = \delta_{ia}$. Similarly, in this case $\langle i| \mathcal{T}[\psi_k] \,|j\rangle = \langle i \mid k\rangle\langle k \mid j\rangle = \delta_{ik}\,\delta_{ij}$. Under the same assumption, if $Q = \sum_k q_k\mathcal{T}[\boldsymbol{\psi}_k]$, $\langle i| Q |j\rangle = \sum_k \delta_{ik}q_k\delta_{kj} = q_i\delta_{ij}$, or $i\delta_{ij}$ in the nondegenerate case.

The next step is to generalize our formalism so that it can be applied to the case in which the indices, such as i and a, are continuous variables. The following consideration is logically unfounded but inductively plausible. First, it is to be expected that the summations with regard to the indices should be replaced by integrations (see Section 8.2). Next, we have to generalize the Kronecker symbol δ_{ij} to the continuous case. We note that the Kronecker symbol δ_{ij} is actually a function only of the difference $i - j$, such that $\delta(i - j) = 0$ if $i - j \neq 0$ and $\sum_i^{\text{all}} \delta(i - j) = 1$. Hence a natural generalization of $\delta(i - j)$ to the continuous case would be given by a function $\delta(\xi)$, which has the properties $\delta(\xi) = 0$ for $\xi \neq 0$ and $\int_{-\infty}^{\infty}\delta(\xi)\, d\xi = 1$. Such a function is called Dirac's δ-function. The familiar formula $\sum_i f_i\delta_{ij} = f_j$ in terms of the Kronecker symbol is now replaced by a similar formula, $\int f(\xi)\delta(\xi - \eta)\, d\xi = f(\eta)$, which follows from the definition of the δ-function, provided that $f(\xi)$ is continuous at $\xi = \eta$. Thus we assume that all the formulas regarding the unitary transformations introduced above will be retained only by replacing the summation by an integration and the Kronecker symbol by Dirac's δ-function.

Later we use the formula

$$\int_{-\infty}^{\infty} e^{ia\xi}\, d\xi = \int_{-\infty}^{\infty} \cos a\xi\, d\xi = 2\pi\, \delta(a), \tag{A9.3.1}$$

which is a symbolic way of expressing the fact (proven by Dirichlet's integral) that

$$\lim_{x\to\infty} \int_{-\infty}^{\infty} f(a)\, da \int_{-x}^{x} \cos a\xi\, d\xi = \lim_{x\to\infty} 2\int_{-\infty}^{\infty} f(a)\frac{\sin ax}{a}\, da = 2\pi f(0),$$

whenever $f(a)$ is continuous at $a = 0$.

Now let us take two specific quantities, x and p, where x is the position along the x-axis of a particle, say, an electron, and p is its linear momentum in the same direction. These may be represented respectively (assuming nondegeneracy) as

$$\langle x'| \, x \,|x''\rangle = x'\delta(x' - x'') \tag{A9.3.2}$$

and

$$\langle p' | p | p'' \rangle = p'\delta(p' - p''). \tag{A9.3.3}$$

The most important element in the theory is the unitary transformations connecting various physical variables. In the present case we are interested in $\langle x' \mid p' \rangle$, which is as stated above the analogue of what was denoted by U_{ia}, which is the ith component in the $\boldsymbol{\psi}$-representation of the vector $\boldsymbol{\varphi}$. Similarly, $\langle x' \mid p' \rangle$ means the component in the x-representation of the quantum state that is the eigenstate of p corresponding to the eigenvalue p'.

The easiest way to obtain the unitary transformation $\langle x' \mid p' \rangle$ may be to invoke the Einstein-de Broglie relation

$$E = \hbar\omega \quad \text{and} \quad p = \frac{\hbar}{\lambda\!\!\!^{-}}. \tag{A9.3.4}$$

This relation implies that there are two aspects to any physical entity, the particle aspect and the wave aspect, which are so related to each other that the energy E and momentum p in the former aspect and the frequency (times 2π) ω and the wavelength (divided by 2π) $\lambda\!\!\!^{-}$ in the latter aspect are related by the equation (A9.3.4), where $\hbar$ is Planck's constant. This suggests that there is a wave whose nature is unknown but which can be expressed mathematically

$$\exp\left(i\frac{p}{\hbar}x - i\frac{E}{\hbar}t\right). \tag{A9.3.5}$$

We may interpret this, from our present point of view, as expressing a quantum state corresponding to a well-defined momentum and a well-defined energy in the representation of x and t. In other words, the factor $\exp[i(p/\hbar)x]$ must be proportional to $\langle x' \mid p' \rangle$:

$$\langle x' \mid p' \rangle = A \exp\left(i\frac{p'}{\hbar}x'\right), \tag{A9.3.6}$$

where the constant A can be determined by the condition

$$\int \langle x' \mid p' \rangle \langle p' \mid x'' \rangle \, dp' = \delta(x' - x'').$$

With the help of (A9.3.1) we obtain

$$\int AA^* \exp\left[i\frac{p'}{\hbar}(x' - x'')\right] dp' = AA^* 2\pi\, \delta(x' - x''). \tag{A9.3.7}$$

Hence $A = 1/\sqrt{2\pi}$. This allows us to find the matrix elements in the

x-representation of the physical quantity p by the use of the formula mentioned above.

$$\begin{aligned}\langle x'|\, p\, |x''\rangle &= \int \langle x' \mid p'\rangle p' \langle p' \mid x''\rangle\, dp' \\ &= \frac{1}{2\pi\hbar}\int p' \exp\left[i\,\frac{p'}{\hbar}(x' - x'')\right] dp' \\ &= \frac{1}{2\pi\hbar}\int \frac{\hbar}{i}\frac{\partial}{\partial x'} \exp\left[i\,\frac{p'}{\hbar}(x' - x'')\right] dp' \\ &= \frac{\hbar}{i}\frac{\partial}{\partial x'}\,\delta(x' - x''). \end{aligned} \tag{A9.3.8}$$

This is nothing but the well-known relation $p = (\hbar/i)(\partial/\partial x)$ in Dirac's notation. The relation $p = (\hbar/i)(\partial/\partial x)$ could, in an unrigorous way, have been obtained directly from (A9.3.5) by differentiating it with respect to x. Equation (A9.3.8) allows us to evaluate the degree of "noncommutativity" between the two quantities x and p by expressing $xp - px$ in the x-representation:

$$\begin{aligned}\langle x'|\, xp\, |x''\rangle &= \int \langle x'|\, x\, |x'''\rangle\langle x'''|\, p\, |x''\rangle\, dx''' \\ &= \frac{\hbar}{i}\int x'\,\delta(x' - x''')\,\frac{\partial}{\partial x'''}\,\delta(x''' - x'')\, dx''' \\ &= \frac{\hbar}{i}\,x'\,\frac{\partial}{\partial x'}\,\delta(x' - x''). \end{aligned} \tag{A9.3.9}$$

Similarly,

$$\begin{aligned}\langle x'|\, px\, |x''\rangle &= \frac{\hbar}{i}\int \frac{\partial}{\partial x'}\,\delta(x' - x''')x'''\,\delta(x''' - x'')\, dx''' \\ &= \frac{\hbar}{i}\frac{\partial}{\partial x'}\,\delta(x' - x'')x'' = \frac{\hbar}{i}\frac{\partial}{\partial x'}\,\delta(x' - x'')x' \\ &= \frac{\hbar}{i}\left(1 + x'\,\frac{\partial}{\partial x'}\right)\delta(x' - x''). \end{aligned} \tag{A9.3.10}$$

Hence we obtain

$$\langle x'|\, px - xp\, |x''\rangle = \frac{\hbar}{i}\,\delta(x' - x'') \tag{A9.3.11}$$

or, in the operator form,

$$px - xp = \frac{\hbar}{i}, \tag{A9.3.12}$$

which is the basic commutation relation of quantum mechanics. As we anticipated in S-73, Planck's constant $\hbar$ appears as a measure of non-commutativity (hence of non-Booleanity). It is known that every time we take a pair of "canonically conjugate" quantities, we have a commutation relation of the type (A9.3.12).

Furthermore, (A9.3.5) suggests that if we differentiate a quantum state (having a well-defined energy) with respect to t, we obtain a factor $-i(E/\hbar)$. Comparing this with our former result (S-73), we can conclude that when the quantum state has a well-defined energy we have

$$H\psi = -i\frac{E}{\hbar}\psi = \frac{\partial}{\partial t}\psi. \tag{A9.3.13}$$

This shows that $\hbar H$ must be an operator corresponding to the physical quantity "energy." The H-function used in Section 9.4 is $\hbar$ times the usual Hamiltonian. As a consequence the dynamic law in terms of a commutation relation [(9.144), for instance] also becomes proportional to Planck's constant.

A9.4. PROOF OF THE GENERALIZED GIBBS THEOREM

The purpose of this appendix is to provide a proof for Theorem 9.11. We use two lemmas mentioned earlier in the book, repeating them here for convenience of reference.

Lemma A9.4.1 If

$$p_i \geq 0, \qquad \sum_i p_i = 1, \qquad q_i \geq 0, \qquad \sum_i q_i = 1, \qquad i = 1, 2, \cdots,$$

we have

$$-\sum_i p_i \log q_i \geq -\sum_i p_i \log p_i, \tag{A9.4.1}$$

where the equality holds if, and only if, $p_i = q_i$ for all i.

Lemma A9.4.2 Let $\|A_{ji}\|$ be a matrix such that $A_{ji} \geq 0$, $\sum_j A_{ji} = 1$, and $\sum_i A_{ji} = 1$. If the probabilities p_i and q_i are related by the relation $q_j = \sum_i A_{ji} p_i$, we have

$$-\sum_i q_i \log q_i \geq -\sum_i p_i \log p_i. \tag{A9.4.2}$$

The equality in (A9.4.2) holds if, and only if, those p_i's that belong to the same "terminally connected family" have the same value.

Lemma A9.4.1 is the original Gibbs theorem and was introduced as Theorem 1.1. Lemma A9.4.2 is the prototype H-theorem, which was introduced as Theorem 5.3. It was also mentioned as Theorem 8.2 supplemented by Lemma A8.2.1. The definition of the term "terminally connected family" is given in Section 5.2 and Appendix 8.2. The proof of Theorem 9.11 is as follows.

Let Z_1 and Z_2 be written as $Z_1 = \sum_i u_i \mathfrak{P}[\varphi_i]$ and $Z_2 = \sum_j v_j \mathfrak{P}[\psi_j]$, and define A_{ji} by

$$A_{ji} = |(\psi_j, \varphi_i)|^2 = |T_{ji}|^2 = \langle j \,|\, \mathfrak{P}[\varphi_i] \,|\, j \rangle, \tag{A9.4.3}$$

where (ψ_j, φ_i) is of course the scalar product in the sense of (9.100) and can also be considered as the unitary transformation $T_{ji} = \langle j \,|\, i \rangle$ between $\{\varphi_i\}$ and $\{\psi_j\}$ (see, for instance, the first part of Appendix 9.3). The unitarity means that $(T_{ji})^* = (T^{-1})_{ij}$. Hence $A_{ji} = T_{ji}(T_{ji})^* = T_{ji}(T^{-1})_{ij}$ and $\sum_i A_{ji} = \sum_j A_{ji} = 1$, satisfying the double stochasticity conditions imposed on A_{ji} in Lemma A9.4.2.

If we take the diagonal sum in the ψ_j-representation, the left side of (9.189) becomes

$$-\text{trace}\,(Z_1 \log Z_2) = -\sum_j w_j \log v_j, \tag{A9.4.4}$$

with

$$w_j = \langle j |\, Z_1 \,| j \rangle = \sum_i A_{ji} u_i \left(\sum_j w_j = 1 \right). \tag{A9.4.5}$$

On account of Lemma A9.4.1 we have

$$-\sum_j w_j \log v_j \geq -\sum_j w_j \log w_j \tag{A9.4.6}$$

and on account of Lemma A9.4.2 we have

$$-\sum_j w_j \log w_j \geq -\sum_i u_i \log u_i = -\text{trace}\,(Z_1 \log Z_1). \tag{A9.4.7}$$

Combining (A9.4.4), (A9.4.6), and (A9.4.7), we get the first part of Theorem 9.11.

The equality in (9.189) will hold if, and only if, both equalities in (A9.4.6) and (A9.4.7) hold simultaneously. Let us demonstrate that the simultaneous equalities in (A9.4.6) and (A9.4.7) imply $Z_1 = Z_2$, because the converse is obvious. Let $\mathfrak{M}_\mu$ be an invariant subspace, that is, a subspace that is transformed into itself by the unitary transformation T_{ji} (A9.4.3), and let us assume that $\mathfrak{M}_\mu$ contains no invariant subspace except the origin and $\mathfrak{M}_\mu$ itself. Those indices i of $\{\varphi_i\}$ that correspond to an invariant subspace $\mathfrak{M}_\mu$ form a terminally connected family. [Recall the similar situation in A8.2.] Hence, according to Lemma A9.4.2, the equality in (A9.4.7) holds if, and only if, the u_i's are equal to a constant, say, p_μ, for the vectors φ_i belonging to each $\mathfrak{M}_\mu$. If we write $Z_3 = \sum_j w_j \mathfrak{P}[\psi_j]$, we shall find under the circumstances assumed, that those w_j [connected to the u_i by (A9.4.5)] corresponding to the ψ_j in $\mathfrak{M}_\mu$ are also all equal to p_μ. Hence we have $Z_1 = Z_3 = \sum_\mu p_\mu P[\mathfrak{M}_\mu]$. The projection operator $\mathfrak{P}[\mathfrak{M}_\mu]$ can be considered either as the sum of the projection operators $\mathfrak{P}[\varphi_j]$ corresponding to the φ_j belonging to $\mathfrak{M}_\mu$ or as the sum of the projection operators $\mathfrak{P}[\psi_j]$ corresponding to the ψ_j belonging to $\mathfrak{M}_\mu$ [see (9.76)]. Next, the assumed equality in (A9.4.6) implies, in view of Lemma A9.4.1, that $v_j = w_j$ for all j; that is, $Z_2 = Z_3$. Hence, if equality holds in both (A9.4.6) and (A9.4.7), we have $Z_1 = Z_3 = Z_2$.

A9.5. ENTROPIES IN OPTICAL CHANNELS

In 1956 Gamo [G-3, G-4] and Gabor [G-2] introduced independently what is sometimes called a "matrix theory" of optical images, which has since proved to be a useful tool in the analysis of image formation, particularly when the illumination is neither perfectly coherent nor completely incoherent but only partially coherent, which is actually the case in many practical applications. Their "matrix," however, is merely a discrete representation of a more classical mathematical entity known in optics under the name of "mutual intensity" (see, for instance, the text by Born and Wolf [B-7]). In the terminology of the present book, this matrix may be considered either as the autocorrelation function of Chapter 8 or as the density matrix (or state matrix) of Chapter 9. This appendix shows briefly how the entropy functions defined in Chapter 9 on the basis of non-Boolean information theory can be used in problems of optical channels with the help of the matrix theory of images.

Propagation of light through an optical system can, of course, be described by the electromagnetic field occupying that region of space and time. Let the σ-component ($\sigma = 1, 2, 3$) of the electric field at point (x, y, z) at time t be denoted by

$$E(x, y, z, t, \sigma), \tag{A9.5.1}$$

where the indices $\sigma = 1, 2, 3$ correspond respectively to the x-, y-, and z-directions. We assume E to be a complex quantity, so that not only the amplitude but the phase of the field can be expressed by a single quantity. Then each E can be considered a ψ-function (in the sense of Chapter 9) in the representation of x, y, z, and σ. The magnetic field can be calculated from (A9.5.1). Then the state matrix Z will be expressed as

$$\langle x, y, z, t, \sigma|\, Z \,|x', y', z', t', \sigma'\rangle = \frac{E(x, y, z, t, \sigma)\, E^*(x', y', z', t', \sigma')}{I_0}. \tag{A9.5.2}$$

The total intensity I_0 is inserted to secure the normalization

$$\text{trace}\, Z = 1, \tag{A9.5.3}$$

that is,

$$I_0 = \frac{1}{T}\int_0^T dt \int_{-\infty}^{\infty} dx \int_{-\infty}^{\infty} dy \int_{-\infty}^{\infty} dz \sum_{\sigma} |E(x, y, z, t, \sigma)|^2, \tag{A9.5.4}$$

where the average $(1/T)\int_0^T dt$ is redundant if the integrand is constant in time. In this case the total intensity I_0 is proportional to the sum of the energy $\hbar\omega$ of each photon involved. If the light is monochromatic, I_0 is directly proportional to the number of photons.

In order to discuss systematically the relationship between a many-photon state matrix and a one-photon state matrix, we have to extend our discussion of fermion ensembles in Section 9.5 to the case of boson ensembles, or use a second-quantized theory. However, the physical interaction and statistical dependence (due to quantum effect) between photons can be practically ignored; furthermore, the optical quantities in which we are usually interested are proportional to the intensity. For this reason we can use Z as defined in (A9.5.2) and define a physical quantity Q in such a way that trace $(Q \cdot Z)$ will give the expected value of this quantity "per average photon" and trace $(Q \cdot X)$ will give the expected value of this quantity in the actual field, where X is defined by

$$X = I_0 Z, \qquad \text{trace } X = I_0. \tag{A9.5.5}$$

Hereafter we assume the field to be monochromatic, but this does not represent any restriction because most of the useful results either refer to a single frequency anyway or are additive with respect to different frequencies.

Let us assume that these light waves are generated by several point sources, which will be labeled $\xi = 1, 2, 3, \cdots$. Then we may decompose E as

$$E = \sum_{\xi} E(x, y, z, t, \sigma; \xi), \tag{A9.5.6}$$

where each term represents a wave originating from a point source. We may then put each component in the form

$$E(x, y, z, t, \sigma; \xi) = A(\xi)F(x, y, z, t, \sigma; \xi), \tag{A9.5.7}$$

where the dependence of F on t must be

$$F(x, y, z, t, \sigma; \xi) \propto \exp\,(-i\omega t + i\theta_{\xi}(t)). \tag{A9.5.8}$$

The frequency ω is a constant common to all the point sources, and the phase $\theta_{\xi}(t)$, which depends on ξ, remains constant during a certain length of time (period of coherence) but changes discontinuously in a random fashion from time to time. The validity of (A9.5.8) is not impaired by the possibility that the light may reach the point (x, y, z) from the source ξ by more than one path. The difference in optical path lengths, which is assumed to be constant in time, can be incorporated in the factor of F other than the time-dependent factor explicitly given on the right side of (A9.5.8). We assume the normalization

$$\frac{1}{T}\int_0^T dt \int_{-\infty}^{\infty} dx \int_{-\infty}^{\infty} dy \int_{-\infty}^{\infty} dz \sum_{\sigma} |F(x, y, z, t, \sigma; \xi)|^2 = 1 \tag{A9.5.9}$$

so that $A(\xi)$ in (A9.5.7) becomes a measure of the strength of the source ξ. The integration with respect to t from 0 to T and the subsequent division by T has no actual effect, for $|F|^2$ does not depend on time. If we take two different sources, ξ and η, we have

$$\frac{1}{T}\int_0^T dt \int_{-\infty}^{\infty} dx \int_{-\infty}^{\infty} dy \int_{-\infty}^{\infty} dz \sum_{\sigma} F(x, y, z, t, \sigma; \xi)\, F^*(x, y, z, t, \sigma; \eta) = 0 \tag{A9.5.10}$$

because of the factor $\exp i\theta_\xi(t)$ in (A9.5.8) if T is large compared with the period of coherence of individual sources.

The X matrix can be written

$$X = \sum_{\xi\eta} A(\xi)\, A^*(\eta)\, F(\xi)\, F^*(\eta), \tag{A9.5.11}$$

where $F(\xi)$ is defined in (A9.5.7). The $F(\xi)$ are orthogonal and normalized in the sense of (A9.5.9) and (A9.5.10). They may not be "complete" if they refer only to the actual sources, but the $F(\xi)$ can be considered as part of an orthogonal-normalized complete system of four-dimensional functions. The time average $\underline{X}$ in the sense of $(1/T)\int_0^T dt$ of X becomes

$$\underline{X} = \frac{1}{T}\int_0^T (t|\, X\, |t)\, dt = \sum_{\xi} |A(\xi)|^2\, \mathcal{P}[F(\xi)], \tag{A9.5.12}$$

for all the nondiagonal elements drop off. $\mathcal{P}[F(\xi)]$ is the projection operator corresponding to the wave function $F(\xi)$. Because of (A9.5.9) and (A9.5.10) we have

$$\mathcal{P}[F(\xi)]\mathcal{P}[F(\eta)] = \delta(\xi, \eta)\mathcal{P}[F(\xi)] \tag{A9.5.13}$$

and

$$\text{trace}\, \mathcal{P}[F(\xi)] = 1, \tag{A9.5.14}$$

where the multiplication in (A9.5.13) and trace in (A9.5.14) are to be understood in a four-dimensional sense; that is, the summation or integration must be taken with respect to x, y, z, t, and σ. The integration with respect to t is to be understood in the sense of $(1/T)\int_0^L dt$. Thus the $|A(\xi)|^2$ are the eigenvalues of the matrix $\underline{X}$. The state matrix $\underline{Z}$ corresponding to $\underline{X}$ is then

$$\underline{Z} = \sum_{\xi} w(\xi)\, \mathcal{P}[F(\xi)], \tag{A9.5.15}$$

with

$$w(\xi) = \frac{|A(\xi)|^2}{\sum_{\xi} |A(\xi)|^2}\,. \tag{A9.5.16}$$

Although $\underline{X}$ and $\underline{Z}$ do not depend on time, the individual $F(\xi)$'s do.

The microscopic entropy defined by Z is

$$S(\underline{Z}) = -\text{trace}\,(\underline{Z} \log \underline{Z}) = -\sum_{\xi} w(\xi) \log w(\xi), \tag{A9.5.17}$$

where the trace is to be taken with respect to the five variables x, y, z, t, and σ. This quantity takes its minimum value 0 if there is only one source, that is, if the light field is completely coherent. If we fix the number N of sources, $S(\underline{Z})$ takes its maximum value $\log N$ if all the N sources have the same strength $|A(\xi)|$. This corresponds to the maximum incoherence. The idea of using the entropy $S(\underline{Z})$ as a measure of incoherence is one of the interesting proposals made by Gamo [G-3], [G-4].

Let us now consider an optical system (lens system) whose axis coincides with the z-axis, and discuss the electromagnetic field on an xy-plane for a fixed value of z. We shall also change the normalization, so that it refers only to the x- and y-coordinates. At the same time we drop the variable σ for simplicity, by considering only one polarization or assuming that the light is unpolarized. Thus the time average (A9.5.12) at a given z becomes

$$\langle x, y| \underline{X} |x', y'\rangle = \sum_{\xi} |A(\xi)|^2 F(x, y; \xi) F^*(x', y'; \xi). \qquad \text{(A9.5.18)}$$

The normalization is now defined by

$$\int dx \int dy\, |F(x, y, \xi)|^2 = 1 \qquad \text{(A9.5.19)}$$

and the cross-product

$$\int dx \int dy\, F(x, y, \xi) F^*(x, y, \eta), \qquad \xi \neq \eta \qquad \text{(A9.5.20)}$$

is not guaranteed to be zero because we do not integrate with respect to t in (A9.5.20). Hence $|A(\xi)|^2$ cannot be claimed to be eigenvalues of the two-dimensional $\underline{X}$. We further simplify the problem by fixing $y = 0$ and taking only the x-values as a variable, and hereafter consider the matrix

$$\langle x| \underline{X} |x'\rangle = \sum_{\xi} |A(\xi)|^2 F(x; \xi) F^*(x'; \xi), \qquad \text{(A9.5.21)}$$

with

$$\int dx\, |F(x; \xi)|^2 = 1. \qquad \text{(A9.5.22)}$$

The matrix $\underline{X}$ does not depend on time, but the individual $F(x; \xi)$ may contain time as a parameter. The matrix $\underline{X}$ is obviously Hermitian. Its diagonal element $\langle x| \underline{X} |x\rangle$ of course means the intensity of light at x. Hence $\underline{X}$ will be referred to as the intensity matrix. Although $|A(\xi)|^2$ may not be the eigenvalues of $\langle x| \underline{X} |x'\rangle$, it can easily be shown that the eigenvalues of $\underline{X}$ are all non-negative and add up to unity.

Since the frequencies of light in practical applications have an upper limit, the wave numbers (along the x-axis) also have an upper limit. Exploiting this fact, we can represent the matrix $\underline{X}$ in terms of a finite number of so-called

Shannon "sampling functions." Gamo's matrix is nothing but this representation of $\underline{X}$. The theorem underlying Shannon's expansion was discovered and rediscovered by many mathematicians before him, including Whittaker [W-34] and Someya [S-8].

In order to apply this theory to an optical channel, we have to define at least the following three procedures. First, if the light expressible as an $\underline{X}$ of the type (A9.5.21) falls on an object at position $z = z_0$, we have to know how to determine the average intensity matrix, say, $\underline{X}'$, immediately after the object. Second, we have to know how the intensity matrix $\underline{X}'$ at the position of the object $z = z_0$ is transformed to the intensity matrix $\underline{X}''$ at the position of the image, say, $z = z$, by passing through a lens system. Third, we have to know how to express the result of observation made on $\underline{X}''$ at $z = z$.

As regards the first process, namely, determination of the intensity matrix immediately after the object from the state immediately before the object, we have to note that the effect of the object on the intensity matrix must be expressible in terms of a transformation by a matrix that is diagonal in the x-representation, since at each position x the wave amplitude must be multiplied by a certain complex number, say,

$$g(x) = |g(x)| \exp [i\varphi(x)], \tag{A9.5.23}$$

where $|g(x)| \leq 1$ and $\varphi =$ real. Of course the first condition, $|g(x)| \leq 1$, means that the intensity of the light cannot increase by passing through the object, and the second condition, $\varphi =$ real, provides for the possible change in phase. Hence $\underline{X}$ will be transformed as

$$\underline{X} \rightarrow \underline{X}' = G\underline{X}\bar{G}, \tag{A9.5.24}$$

with

$$\langle x \,|G|\, x' \rangle = g(x)\, \delta(x - x') \tag{A9.5.25}$$

and

$$\langle x \,|\bar{G}|\, x' \rangle = g^*(x)\, \delta(x - x'). \tag{A9.5.26}$$

The matrix G itself is not Hermitian, but because of the transformation rule (A9.5.24) $\underline{X}'$ will be Hermitian if $\underline{X}$ is Hermitian. Furthermore, we can show that the eigenvalues of $\underline{X}'$ are non-negative, implying that $\underline{X}'$ satisfies the conditions of an intensity matrix.

The next question concerns the relationship between the intensity matrix $\underline{X}'$ immediately after the object and the intensity matrix $\underline{X}''$ at the image plane. Gamo has shown that as long as the lens system is nonabsorbing and nonreflecting, these two matrices $\underline{X}'$ and $\underline{X}''$ are connected by a unitary transformation:

$$\underline{X}'' = U\underline{X}'U^{-1}, \qquad \bar{U} = U^{-1}. \tag{A9.5.27}$$

Although a rigorous proof of this law requires a detailed discussion, the result is not unexpected. An interesting fact is that this result is true even in the presence of aberrations. It also holds when z_0 and z are two arbitrary positions, not necessarily identifiable as the object plane and image plane. Caution must be taken however, that the light bundle is kept within the apertures of the lenses so that no absorption takes place.

The third question concerns the observation made at the object plane. From our general consideration of Chapter 9, we can expect that if an observable quantity Q corresponds to a Hermitian operator, say, $Q = \sum_i q_i \mathfrak{I}[\varphi_i(x)]$, the expectation value of Q in a state matrix $\underline{Z}''$ will be given by

$$\langle Q \rangle = \text{trace}\,(Q \cdot \underline{Z}''). \tag{A9.5.28}$$

But many of the quantities in which we are interested in optics are proportional to the intensity (number of photons), and so it is more convenient to define $\langle Q \rangle$ as

$$\langle Q \rangle = \text{trace}\,(Q \cdot \underline{X}''). \tag{A9.5.29}$$

In fact, Gamo showed in detail that most of the observational results in optics can be expressed in the form of (A9.5.28) or (A9.5.29) with a Hermitian Q. His consideration includes observations made by blackening of photographic plates. When we are interested in the sharpness of the image, however, we should consider (A9.5.28) rather than (A9.5.29).

Let us consider an observational quantity Q, which is expressible as

$$Q = \sum q_\mu \mathfrak{I}[\mathfrak{M}_\mu], \tag{A9.5.30}$$

where the number of dimensions of the subspace $\mathfrak{M}_\mu$ is $D_\mu = \text{trace}\,\mathfrak{I}[\mathfrak{M}_\mu]$ and is not necessarily 1. We assume that the q's are nondegenerate in the sense that $q_\mu \neq q_\nu$ if $\mu \neq \nu$. The probability of obtaining the result q_μ in the light bundle $\underline{X}$ (which could be $\underline{X}'$ or $\underline{X}''$ in our earlier discussion) is

$$\begin{aligned} \Pr\{\mu \mid \underline{Z}\} &= \frac{\text{trace}\,(\mathfrak{I}[\mathfrak{M}_\mu] \cdot \underline{X})}{\text{trace}\,\underline{X}} \\ &= \text{trace}\,(\mathfrak{I}[\mathfrak{M}_\mu] \cdot \underline{Z}), \end{aligned} \tag{A9.5.31}$$

where $\underline{Z} = \underline{X}/\text{trace}\,\underline{X}$ is the normalized intensity matrix.

It is interesting to introduce as a measure of "diffuseness" of the results of the observation of Q on $\underline{Z}$ the entropy

$$S(Q, \underline{Z}) = -\sum_\mu \Pr\{\mu \mid \underline{Z}\} \log \Pr\{\mu \mid \underline{Z}\} \tag{A9.5.32}$$

or, more usefully, the relative entropy

$$S(Q, \underline{Z}) = -\sum_\mu \Pr\{\mu \mid \underline{Z}\} \log \frac{\Pr\{\mu \mid \underline{Z}\}}{D_\mu}. \tag{A9.5.33}$$

If $\underline{Z}$ is expressible as

$$\underline{Z} = \sum_\nu w_\nu \mathfrak{I}[\mathfrak{N}_\nu]/D_\nu, \tag{A9.5.34}$$

the probability Pr $\{\mu \mid \underline{Z}\}$ becomes

$$p(\mu) = \Pr\{\mu \mid \underline{Z}\} = \sum_{\nu} w_{\nu} \operatorname{trace}(\mathfrak{T}[\mathfrak{M}_{\mu}] \cdot \mathfrak{T}[\mathfrak{N}_{\nu}])/D_{\nu} \qquad \text{(A9.5.35)}$$

It is then easily seen that the diffuseness as expressed by $S(Q, \underline{Z})$ of (A9.5.33) takes its minimum value when the set of subspaces $\{\mathfrak{M}_{\mu}\}$ (of the observation) and the set of subspaces $\{\mathfrak{N}_{\mu}\}$ of the state coincide. Indeed, Pr $\{\mu \mid \underline{Z}\}$ then becomes simply

$$p^{(0)}(\mu) = w_{\mu}. \qquad \text{(A9.5.36)}$$

The transition matrix $A_{\mu\nu}$ that connects $p(\mu)$ and $p_0(\nu)$ by

$$p(\mu) = \sum_{\nu} A_{\mu\nu} p_0(\nu) \qquad \text{(A9.5.37)}$$

is then given by

$$A_{\mu\nu} = \frac{\operatorname{trace}(\mathfrak{T}[\mathfrak{M}_{\mu}] \cdot \mathfrak{T}[\mathfrak{N}_{\nu}])}{D_{\nu}}, \qquad \text{(A9.5.38)}$$

satisfying

$$\sum_{\mu} A_{\mu\nu} = 1 \quad \text{and} \quad \sum_{\nu} A_{\mu\nu} D_{\nu} = D_{\mu}. \qquad \text{(A9.5.39)}$$

This allows us to use the H-theorem [see (5.119), (5.129), 5.131)] and conclude that $S(Q, \underline{Z})$ calculated with $p^{(0)}(\mu)$ is the minimum. This minimum value is

$$S(Q^{(0)}, \underline{Z}) = -\sum_{\mu} w_{\mu} \log (w_{\mu}/D_{\mu}), \qquad \text{(A9.5.40)}$$

where $Q^{(0)}$ means the observation of the type (A9.5.30) whose subspaces agree with those of $\underline{Z}$.

Let us now consider the problem of aberration from this viewpoint of "diffuseness." Let us assume that the state $\underline{Z}'$ right after the object plane is given by the same expression as $\underline{Z}$ of (A9.5.34). The "best" observation that can be made at that position will be characterized by the minimum entropy given in (A9.5.40). Let V be the "ideal" unitary transformation, which transforms the state $\underline{Z}'$ at the object plane into an aberration-free state $\underline{Z}_I'' = V\underline{Z}'V^{-1}$ at the image plane. The freedom from aberration would imply that the same entropy is obtainable at the image plane as at the object plane. This may be interpreted as meaning that for a given Q_0 at the object plane there exists an observation $Q = V^{-1}Q_0V$ at the image plane, so that

$$S(Q_0, \underline{Z}') = S(VQ_0V^{-1}, V\underline{Z}'V^{-1}) = S(Q, \underline{Z}_I'') \qquad \text{(A9.5.41)}$$

in view of (A9.5.35). The observation $Q = VQ_0V^{-1}$ at the image plane is equivalent to the observation Q_0 at the object plane. Then the entropy of observation Q in the actual (not necessarily aberration-free) image produced by the unitary transformation U will be given by

$$S(Q, \underline{Z}'') = S(VQ_0V^{-1}, U\underline{Z}'U^{-1}) = S(WQ_0W^{-1}, \underline{Z}'), \quad \text{(A9.5.42)}$$

with

$$W = V^{-1}U. \quad \text{(A9.5.43)}$$

By the same argument as before, we can easily show that

$$S(WQ_0W^{-1}, \underline{Z}') \geq S(Q_0, \underline{Z}'). \quad \text{(A9.5.44)}$$

This increase may be considered as due to the aberration, and

$$A = \frac{S(WQ_0W^{-1}, \underline{Z}') - S(Q_0, \underline{Z}')}{S(Q_0, \underline{Z}')} \quad \text{(A9.5.45)}$$

may be used as a measure of aberration for the given $\underline{Z}'$. The use of entropy in optical channels is not limited to the problem of aberration, but we end the appendix here.

References

[A-1] Abbott, J. H., dissertation, University of Illinois, 1959.

[A-2] Abramson, N., *Information Theory and Coding*, McGraw-Hill, New York, 1963.

[A-3] Adler, R. L., *Proc. Am. Math. Soc.*, **12**, 924 (1961).

[A-4] Aizerman, M. A., E. M. Braverman, and L. I. Rozonoer, *Theoretical Foundations of the Method of Potential Functions in the Problem of Instructing Automata to Partition Input Situations into Classes*, Moscow, 1963; *The Probabilistic Problem Concerning the Instruction of Automata to Recognize Classes and the Method of Potential Functions*, Moscow, 1964.

[A-5] Anderson, T. W., *Introduction to Multivariate Statistical Analysis*, Wiley, New York, 1965.

[A-6] Araki, H., and M. M. Yanase, *Phys. Rev.*, **120**, 662 (1960).

[A-7] Atkinson, R. C., G. H. Bower, and E. J. Crothers, *An Introduction to Mathematical Learning Theory*, Wiley, New York, 1965.

[A-8] Ayer, A. J., *Language, Truth and Logic*, Victor Gollancz, London, New York, 1958.

[A-9] Ayer, A. J., *Philosophical Essays*, Macmillan, New York, 1954.

[B-1] Besson, L., *Compt. Rend.*, **178**, 1743 (1924).

[B-2] Birkhoff, G., and J. von Neumann, *Ann. Math., 2nd Ser.*, **37**, 823 (1936).

[B-3] Birkhoff, G., *Lattice Theory*, Am. Math. Soc., New York, 1948.

[B-3a] Blackwell, D., *Transactions of the First Prague Conference on Information Theory, etc. 1957*, Czechoslovakia Academy of Science, Prague.

[B-4] Bohr, N., *Nature*, **137**, 344 (1936).

[B-5] Boltzmann, L., *Vorlesungen über Gastheorie*, Leipzig, 1896–1898.

[B-6] Borel, E., *Compt. Rend.*, **197**, 1257, 1369 (1933).

[B-7] Born, M., and E. Wolf, *Principle of Optics*, Pergamon, New York, 1959.

[B-8] Brillouin, L., *Am. Scientist*, **37**, 554 (1949).

[B-9] Brillouin, L., *Am. Scientist*, **38**, 594 (1950).

[B-10] Brillouin, L., *J. Appl. Phys.*, **22**, 334 (1951).

[B-11] Brillouin, L., *J. Appl. Phys.*, **24**, 1151 (1953).

[B-12] Brillouin, L., *J. Appl. Phys.*, **25**, 887 (1954).

[B-13] Brillouin, L., *Science and Information Theory*, Academic, New York, 1956.

[B-14] Brillouin, L., *Vie, Matière et Observation*, Albin-Michel, Paris, 1959.

[B-15] Bush, R. R., and W. K. Estes, *Studies in Mathematical Learning Theory*, Stanford University Press, Palo Alto, 1959.
[B-16] Bush, R. R., and F. Mosteller, *Stochastic Model for Learning*, Wiley, New York, 1955.
[C-1] Carnap, R., *Meaning and Necessity*, University of Chicago Press, Chicago, 1956.
[C-2] Carnap, R., *Logical Foundations of Probability*, University of Chicago Press, Chicago, 1950.
[C-3] Carnap, R., and Y. Bar-Hillel, in E. C. Cherry (ed.), *Information Theory, Third London Symposium*, Academic, New York, 1956, p. 503.
[C-4] Cherry, E. C., *On Human Communication*, Wiley, New York, 1957.
[C-5] Chung, K. L., *Markov Chains with Stationary Transition Probabilities*, Springer, Berlin, 1960.
[C-6] Church, A., *J. Symbol. Logic*, **14,** 52 (1949).
[C-7] Cooley, J., *J. Phil.*, **56,** 297 (1959).
[D-1] Dauer, F., "Application of Clustering Technique to Grammatical Substructures." IBM Internal Publication, 1963.
[D-2] Davis, M., *Computability and Unsolvability*, McGraw-Hill, New York, 1958.
[D-3] Dedekind, R., *Werke II*, No. XXX, Braunchweig, 1930.
[D-4] de Finetti, B., *Ann. Inst. Henri Poincaré*, **7,** 1 (1937).
[D-5] de Finetti, B., in J. J. Good (ed.), *The Scientist Speculates*, Basic Books, New York, 1962, p. 357.
[D-6] Demers, P., *Can. J. Res.*, **22,** 27 (1944).
[D-7] Demers, P., *Can. J. Res.*, **23,** 47 (1945).
[D-8] Dirac, P. A. M., *The Principles of Quantum Mechanics*, 4th ed., Clarendon, Oxford, 1958.
[D-9] Doob, J. L., *Stochastic Processes*, Wiley, New York, 1953.
[D-10] Duhem, P., *The Aim and Structure of Physical Theory*, Part 2, Princeton University Press, Princeton, N.J., 1954.
[D-11] Dyson, F., *Math. Gaz.*, **30,** 231 (1946).
[E-1] Elias, P., in E. M. Grabbe (ed.), *Handbook of Automation, Computation and Control*, Vol. 1, Wiley, New York, 1958, Chapter 16.
[E-2] Estes, W. K., *Psychol. Rev.*, **57,** 94 (1950).
[E-3] Ehrenfest, P. and T., *Encyklopaedie der mathematischen Wissenschaften*, Vol. IV-4, Teubner, Leipzig, 1907–1911.
[F-1] Fano, R. M., *The Transmission of Information*, M.I.T. Research Laboratory of Electronics Technical Report No. 65, 1949.
[F-2] Fano, R. M., *Transmission of Information*, M.I.T. Press and Wiley, New York, 1961.
[F-3] Feinstein, A., *Foundations of Information Theory*, McGraw-Hill, New York, 1958.
[F-4] Feller, W., *An Introduction to Probability Theory and Its Applications*, Vol. 1, Wiley, New York, 1950.
[F-5] Finch, H. A., *Phil. Phenomenol. Res.*, **18,** 368 (1958).
[F-6] Fisher, R. A., *The Design of Experiments*, Oliver and Boyd, Edinburgh and London, 1935.
[F-7] Fitch, F. B., *Symbolic Logic—An Introduction*, Ronald, New York, 1952.
[F-8] Forester, J., in *Transactions of the 6th Conference on Cybernetics*, the Josiah Macy Jr. Foundation, 1949.
[F-9] Fréchet, M., *Recherches théoriques modernes sur le calcul des probabilités*, Vol. 2, Gauthiers-Villars, Paris, 1938.
[F-10] Frobenius, G., *Sitzungsber. Akad. Wiss. Berlin*, 456 (1912).

[F-11] Fujiwara, S., and Y. Nakata, *Geophysical Magazine*, Vol. 3, 1930, p. 27 (published by Central Meteorological Observatory, Tokyo).
[G-1] Gabor, D., *J. IEE*, **93,** 429 (1946).
[G-2] Gabor, D., *Proc. Symp. Astronomical Optics and Related Subjects, Amsterdam, 1956; Proc. Third London Symp. Information Theory, London, 1956.*
[G-3] Gamo, H., *J. Appl. Phys. Japan*, **25,** 431 (1956).
[G-4] Gamo, H., in E. Wolf (ed.), *Progress in Optics*, Vol. III, Interscience, New York, 1964, p. 189.
[G-5] Garner, W. R., *Uncertainty and Structure as Psychological Concepts*, Wiley, New York, 1962.
[G-6] Garner, W. R., and W. J. McGill, *Psychometrika*, **21,** 219 (1956).
[G-7] Gelernter, H. L., *Proc. First UNESCO Int. Conf. Information Processing, Paris, 1959.*
[G-8] Gibbs, J. W., *Elementary Principles in Statistical Mechanics*, Yale University Press, New Haven, Conn., 1902.
[G-9] Goldmen, S., *Information Theory*, Prentice-Hall, Englewood, N.J., 1953.
[G-10] Goodman, N., *Fact, Fiction and Forecast*, Harvard University Press, Cambridge, Mass., 1955.
[G-11] Goodman, N., *Phil. Phenomenol. Res.*, **19,** 429 (1959).
[G-12] Grünbaum, A., *Phil. Sci.*, **27,** 75 (1960).
[G-13] Grünbaum, A., in M. W. Wartofsky et al. (eds.), *Boston Studies in the Philosophy of Science*, D. Reidel Publishing Co., Dordrecht, The Netherlands, 1963.
[H-1] Halmos, P. R., *Measure Theory*, Van Nostrand, Princeton, N.J., 1950.
[H-2] Halmos, P. R., *Lectures on Ergodic Theory*, Mathematical Society of Japan, Tokyo, 1956.
[H-3] Harman, H. H., *Modern Factor Analysis*, University of Chicago Press, Chicago, 1960.
[H-4] Harris, T. E., *Pacific J. Math.*, **5,** 707 (1955).
[H-5] Hartley, R. V. L., *Bell System Tech. J.*, **7,** 535 (1928).
[H-6] Hartmanis, J., *Information and Control*, **2,** 199 (1959).
[H-7] Hempel, C. G., *J. Symbol. Logic*, **8,** 122 (1943).
[H-8] Hempel, C. G., *Mind* (*New Ser.*), **54,** 1, 97 (1945).
[H-9] Hempel, C. G., *Synthese*, **12,** 439 (1960).
[H-10] Hempel, C. G., and P. Oppenheim, *Phil. Sci.*, **12,** 98 (1945); **15,** 135 (1948).
[H-11] Hertz, H. R., *The Principles of Mechanics*, Dover, New York, 1956.
[H-12] Hopf, E., *Ergodentheorie* (Ergeb. Math. 5, No. 2.), J. Springer, Berlin, 1937.
[H-13] Hosiasson-Linderbaum, J., *J. Symbolic Logic*, **5,** 133 (1940).
[H-14] Hotelling, H., *J. Educ. Psychol.*, **24,** 417, 498 (1933).
[H-15] Huffman, D. A., *Proc. IRE*, **40,** 1098 (1952).
[H-16] Hume, D., *A Treatise on Human Nature, Being an Attempt to Introduce the Experimental Method of Reasoning into Moral Subjects*, London, 1739.
[H-17] Hume, D., *An Enquiry Concerning Human Understanding*, London, 1748.
[H-18] Husimi, K., *Proc. Phys.-Math. Soc. Japan*, **19,** 766 (1937).
[K-1] Kakutani, S., *Proc. Int. Congr. Mathematicians*, **2,** 128 (1950).
[K-2] Kamentsky, L. A., and C. N. Liu, *IBM J. Res. Develop.*, **7,** 2 (1963).
[K-2a] Kaminuma, T., T. Takekawa, and S. Watanabe, "Reduction of Clustering to Pattern Recognition," to appear in the *Journal of Pattern Recognition*.
[K-3] Kelly, T. L., *Harvard Study in Education Series*, No. 26, Harvard University Press, Cambridge, Mass., 1935, p. 146.
[K-4] Kemeny, J. G., and P. Oppenheim, *Phil. Sci.*, **19,** 307 (1952).
[K-5] Kemeny, J. G., *J. Phil.*, **52,** 722 (1955).

[K-6] Kemeny, J. G., and J. L. Snell, *Finite Markov Chains*, Van Nostrand, Princeton, N.J., 1960.
[K-7] Keynes, J. M., *A Treatise on Probability*, Macmillan, London, 1921.
[K-8] Khinchin, A. I., *Mathematical Foundation of Statistical Mechanics*, Dover, New York, 1949.
[K-9] Khinchin, A. I., *Mathematical Foundations of Information Theory*, Dover, New York, 1957.
[K-10] Kleene, S. C., *Introduction to Metamathematics*, Van Nostrand, Princeton, N.J., 1952.
[K-11] Kneale, W., *Probability and Induction*, Oxford University Press, Oxford, 1963.
[K-12] Kolmogorov, A. N., *Mat. Sborn.* (*New Ser.*), **1**, 607 (1936).
[K-13] Kolmogorov, A. N., *Foundations of the Theory of Probability*, Dover, New York, 1956.
[K-14] Kraft, L. K., M.S. thesis, Massachusetts Institute of Technology, 1949.
[K-15] Kuhn, H. W., *Proc. Symp. Appl. Math.*, **10**, 141 (1960).
[K-16] Kullbach, S., *Information Theory and Statistics*, Wiley, New York, 1959.
[K-17] Kullbach, S., M. Kupperman, and H. H. Ku, *Technometrics*, **4**, 573 (1962).
[L-1] Lamb, W. E., to appear in *Physics Today*, April or May, 1969.
[L-2] Lambert, P. F., Dissertation, Yale University, 1967.
[L-3] Landau, L. D., and E. M. Lifschitz, *Statistical Physics*, Pergamon, London, 1958.
[L-4] Lettvin, J. Y., H. R. Maturana, W. S. McCulloch, and W. H. Pitts, *Proc. IRE*, **47**, 1940 (1959).
[L-4a] Lipp, H. M., in H. L. Oestreicher and D. R. Moore (eds.), *Cybernetic Problems in BIONICS*, *Bionics Symposium 1966*, Gordon and Breach, New York, 1968, p. 769.
[L-5] Loève, M., *Probability Theory*, Van Nostrand, Princeton, N.J., 1963.
[L-6] Luce, R. D., *Individual Choice Behavior*, *A Theoretical Analysis*, Wiley, New York, 1959.
[M-1] Mach, E., *The Science of Mechanics*, 1883. Open Court Publishing Co., Chicago, 1902.
[M-2] MacKay, D., in H. von Foerster (ed.), *Proceedings of the Eighth Conference on Cybernetics*, Josiah Macy Jr. Foundation, New York, 1951, p. 222.
[M-3] Maker, P., *Duke Math. J.*, **6**, 27 (1940).
[M-4] Mandelbrot, B., *IRE Trans. Information Theory*, **IT-2**, 190 (1956).
[M-5] Maxwell, J. C., *The Scientific Papers*, University Press, Cambridge, England, 1890.
[M-6] McGill, W. J., *Psychometrika*, **19**, 97 (1954).
[M-7] Mitra, S. K., Dissertation, University of North Carolina, 1955.
[M-8] Moore, E. F., and C. E. Shannon, *J. Franklin Inst.*, **262**, 191, 281 (1956).
[M-9] Muroga, S., *J. Phys. Soc. Japan*, **8**, 484 (1953).
[N-1] Nilssen, N. J., *Learning Machines*, McGraw-Hill, New York, 1965.
[N-2] Novikoff, A. B. J., Stanford Research Institute Technical Report 116, 117, Project ESU-3605 (647), January 1963.
[N-3] Nyquist, H., *Bell System Tech. J.*, **3**, 324 (1924).
[O-1] Onicescu, O., *Act. Sci. Ind.*, **737**, Part 4, Section III, 29 (1935).
[O-2] Onicescu, O., *Nombres et Systèmes Aléatoires*, Eyrolles, Paris, 1964.
[P-1] Pearson, K., *Phil. Mag.*, **6**, 559 (1901).
[P-2] Peirce, C. S., *Monist*, **7**, 19 (1896); *Collected Papers*, Vol. 3, Harvard University Press, Cambridge, Mass., 1960, p. 279.
[P-3] Peirce, C. S., *Collected Papers*, Vol. 4, Harvard University Press, Cambridge, Mass, 1960.

[P-4] Perez, A., in *Transactions of the Second Prague Conference on Information Theory, Statistical Decision Functions and Random Processes, 1959*, published by Czechoslovak Academy of Science, 1960, Academic, New York, 1961, p. 413.

[P-5] Peterson, W. W., and M. O. Rabin, *IBM J. Res. Develop.*, **2**, 163 (1959).

[P-6] Pierce, J. R., *Symbols, Signals and Noise, the Nature and Process of Communication*, Harper, New York, 1961.

[P-7] Poincaré, H., *Science and Method*, Nelson and Sons, London, 1914.

[P-8] Popper, K. R., *The Logic of Scientific Discovery*, Hutchinson, London, 1959.

[Q-1] Quine, W. V. D., *Methods of Logic*, Holt, New York, 1950.

[Q-2] Quine, W. V. D., *From a Logical Point of View*, 2nd ed., Harvard University Press, Cambridge, Mass., 1961.

[R-1] Ramsey, F. P., *The Foundations of Mathematics*, Routledge and K. Paul, London, 1931.

[R-2] Reichenbach, H., *The Theory of Probability*, University of California Press, Berkeley, 1949.

[R-3] Reza, F. M., *An Introduction to Information Theory*, McGraw-Hill, New York, 1961.

[R-4] Roby, T. B., ESD-TDR-64-238, Decision Sciences Laboratory, Hanscom Field, Bedford, Mass., 1965.

[R-5] Rogers, D. J., and T. T. Tanimoto, *Science*, **132**, 1115 (1960).

[R-6] Rosenfeld, L., *Rendiconti della Scuola Internationale di Fisica*, Corso XIV, Varenna, 1960 (Zanichelli, Bologna, 1962), p. 1.

[R-7] Rothstein, J., *J. Appl. Phys.*, **23**, 1281 (1952).

[R-8] Rothstein, J., *Communication, Organization and Science*, Falcon's Wing Press, Indian Hills, Colorado, 1958.

[R-9] Ruyer, R., *Cybernetique et l'origine de l'information*, Flammarion, Paris, 1954.

[S-1] Savage, L. J., *Foundations of Statistics*, Wiley, New York, 1954.

[S-2] Scheffler, I., *Science*, **127**, 177 (1958).

[S-3] Scheffler, I., *Conditions of Knowledge*, an Introduction to Epistemology and Education, Scott, Foresman, Chicago, 1966.

[S-4] Schutzenberger, M. P., *Contribution aux applications statistiques de la théorie de l'information*, Vol. 3, Institut de Statistique de l'Universite de Paris, Paris, 1954.

[S-5] Shannon, C. E., *Bell System Tech. J.*, **30**, 50 (1951).

[S-6] Shannon, C. E., and W. Weaver, *Bell System Tech. J.*, **27**, 379, 623 (1948); *The Mathematical Theory of Communication*, University of Illinois Press, Urbana, 1949.

[S-7] Shuford, E. H., A. Albert, and H. E. Massengill, *Psychometrica*, **31**, 125 (1966).

[S-8] Someya, I., *Propagation of Wave Forms*, Kenshusha, Tokyo, 1949 (in Japanese).

[S-9] Steinbuch, K., *Kybernetik*, **2**, 148 (1965). See also [L-4a].

[S-10] Stückelberg, E. C. O., *Helv. Phys. Acta*, **25**, 577 (1952).

[S-11] Stumpers, F. L., *IRE Trans. Information Theory*, **PGIT-2** (1953); **IT-1** (1955); **IT-3** (1957); **IT-6** (1960).

[S-12] Szilard, L., *Z. Phys.*, **53**, 840 (1929).

[T-1] Takekawa, T., unpublished observation. See also [K-2a].

[T-2] Tanimoto, T. T., and D. J. Rogers, *Science*, **132**, 115 (1960).

[T-3] Tati, T., and S. Tomonaga, *Progr. Theoret. Phys.*, **3**, 391 (1948).

[T-4] Thurston, L. L., *Multiple Factor Analysis*, University of Chicago Press, Chicago, 1947.

[T-5] Toda, M., *Measurement of Subjective Probability Distribution* ESD-TDR-63-407, Decision Sciences Laboratory, Hanscom Field, Bedford, Mass., 1963.

[V-1] von Mises, R., *Probability, Statistics and Truth*, W. Hodge, London, 1939.

[V-2] von Neumann, J., *Z. Phys.*, **57,** 30 (1929).
[V-3] von Neumann, J., *Mathematische Grundlagen der Quantenmechanik*, Julius Springer, Berlin, 1932.
[V-4] von Neumann, J., in C. E. Shannon and J. McCarthy (eds.), *Automata Studies*, Princeton University Press, Princeton, N. J., 1956, p. 43.
[V-5] von Neumann, J., *Continuous Geometry*, Princeton University Press, Princeton, N.J., 1960.
[W-1] Watanabe, S., *Le deuxième théorème de la thermodynamique et la mécanique ondulataire*, Hermann et Cie, Paris, 1935; thesis, University of Paris.
[W-2] Watanabe, S., *Z. Phys.*, **113,** 482 (1939).
[W-3] Watanabe, S., *Kagaku*, **14,** 82, 122, 169 (1944).
[W-4] Watanabe, S., *Phys. Rev.*, **84,** 1008 (1951).
[W-5] Watanabe, S., in (no editor), *Louis de Broglie, Physician et penseur*, Albin Michel, Paris, 1952, p. 385.
[W-6] Watanabe, S., *IRE Trans. Information Theory*, **PGIT-4,** 84 (1954).
[W-7] Watanabe, S., *Rev. Mod. Phys.*, **27,** 26 (1955).
[W-8] Watanabe, S., *Rev. Mod. Phys.*, **27,** 40 (1955).
[W-9] Watanabe, S., *Rev. Mod. Phys.*, **27,** 179 (1955).
[W-10] Watanabe, S., in I. Prigogine (ed.), *Transport Process in Statistical Mechanics*, Interscience, New York, 1958, p. 285.
[W-11] Watanabe, S., *Nuovo Cimento*, Ser. X, **13** (suppl.), 576 (1959).
[W-12] Watanabe, S., *IBM J. Res. Develop.*, **4,** 66 (1960).
[W-13] Watanabe, S., *IBM J. Res. Develop.*, **4,** 208 (1960).
[W-14] Watanabe, S., *Information and Control*, **4,** 224 (1961).
[W-15] Watanabe, S., *Information and Control*, **4,** 291 (1961).
[W-16] Watanabe, S., *Synthese*, **13,** 261 (1961).
[W-17] Watanabe, S., Lecture, AAAS Annual Meeting, 1961, unpublished.
[W-18] Watanabe, S., in S. Dockx and P. Bernays (eds.), *Information and Prediction in Science*, Academic, New York, 1965, p. 39.
[W-19] Watanabe, S., *Progr. Theort. Phys. Supp.* (Extra No.), 135 (1965).
[W-20] Watanabe, S., *Proc. Int. Colloq. Foundations of Mathematics, Mathematical Machines, etc., Tihany, Hungary, 1962*, Hungarian Academy of Science, 1965.
[W-21] Watanabe, S., *Transactions of the Fourth Prague Conf. on Information Theory, Statistical Decision Function 1965*, published by the Czechoslovak Academy of Sciences, Prague, 1967.
[W-22] Watanabe, S., in J. T. Fraser (ed.), *The Voices of Time*, Braziller, New York, 1966.
[W-23] Watanabe, S., *Progr. Theoret. Phys. Suppl.*, Nos. 37 & 38 (1966), p. 350.
[W-24] Watanabe, S., in H. L. Oestreicher and D. R. Moore (eds.), *Cybernetic Problems in BIONICS, Bionics Symposium 1966*, Gordon and Breach, London, 1968.
[W-25] Watanabe, S., et alia in J. Tou (ed.), *Computer and Information Science*, Vol. II, Academic, New York, 1966, p. 91.
[W-26] Watanabe, S., *Proc. Int. Colloq. Information Theory, Debrecen, 1967*, Hungarian Academy of Science, in press.
[W-27] Watanabe, S., *IRE Trans. Information Theory*, PGIT Trans. **IT-8,** p. 246, 1962.
[W-28] Watanabe, S., *Verhandl. Schweiz. Naturforsch. Ges., Sitten*, 1963, p. 41.
[W-29] Watanabe, S., in S. Dockx (ed.), *Civilisation Technique et Humanisme*, Brussels, Office international de librairie, 1968, p. 19.
[W-30] Watanabe, S., in S. Watanabe (ed.), *Methodologies of Pattern Recognition*, Academic, New York, 1969.
[W-31] Watanabe, S., *Proc. Hawaiian Conf. on Philosophical Problems in Psychology 1968*, in press.

[W-31a] Watanabe, S., to appear in *Information and Control.*

[W-32] Watanabe, S., *Proc. IEEE*, in press.

[W-32a] Watanabe S. (in collaboration with T. Kaminuma and T. Takekawa) to appear in *the Proceedings of the International Symposium on Information Theory, San Remo, 1967*, published by IEEE, PGIT.

[W-32b] Watanabe, S., in *Proceedings of the IFIP Congress 1968*, Edinburgh.

[W-33] Watanabe, S., and C. Abraham, *Information and Control*, **3**, 248 (1960).

[W-34] Whittaker, E. T., *Proc. Roy. Soc.* (*Edinburgh*), **35**, 181 (1915).

[W-35] Wick, G. C., A. S. Wightman, and E. P. Wigner, *Phys. Rev.*, **88**, 101 (1952).

[W-36] Wiener, N., *Cybernetics*, 2nd ed., MIT Press and Wiley, New York, 1961.

[W-37] Wigner, E. P., *Z. Phys.*, **131**, 101 (1952).

[W-38] Wilks, S. S., *Ann. Math. Statist.*, **6**, 190 (1935).

[W-39] Winograd, S., and J. D. Cowan, *Reliable Computation in the Presence of Noise*, MIT Press and Wiley, New York, 1963.

[W-40] Wolfowitz, J., *Coding Theorems of Information Theory*, Springer, Berlin, 1961.

[W-41] Wong, E., Dissertation, Princeton University, 1959.

[W-42] Wong, E., in S. Watanabe (ed.), *Methodologies of Pattern Recognition*, Academic, New York, 1969.

[Y-1] Yaglom, I. M., and A. M. Yaglom, *Probability and Information*, Dover, New York, 1969.

[Y-2] Yanase, M. M., *Phys. Rev.*, **123**, 666 (1961).

[Y-3] Yoshida, K., and S. Kakutani, *Proc. Imp. Acad. Tokyo*, **15**, 165 (1939).

[Z-1] Zimmerman, S., *American Math. Monthly*, **66**, 690–693 (1959).

Index

MP 10/7/70